AF704379

OXFORD LECTURE SERIES IN MATHEMATICS AND ITS APPLICATIONS

Series Editor

JOHN BALL

Emeritus Editor

DOMINIC WELSH

OXFORD LECTURE SERIES IN MATHEMATICS AND ITS APPLICATIONS

For a full list of titles please visit

http://ukcatalogue.oup.com/nav/p/category/academic/series/mathematics/olsma.do

10. Pierre-Louis Lions: *Mathematical Topics in Fluid Mechanics, Vol. 2: Compressible Models*
11. W.T. Tutte: *Graph Theory As I Have Known It*
12. Andrea Braides and Anneliese Defranceschi: *Homogenization of Multiple Integrals*
13. Thierry Cazenave and Alain Haraux: *An Introduction to Semilinear Evolution Equations*
14. Jean-Yves Chemin: *Perfect Incompressible Fluids*
15. Giuseppe Buttazzo, Mariano Giaquinta, and Stefan Hildebrandt: *One-Dimensional Variational Problems: An Introduction*
16. Alexander I. Bobenko and Ruedi Seiler: *Discrete Integrable Geometry and Physics*
17. Doina Cioranescu and Patrizia Donato: *An Introduction to Homogenization*
18. E.J. Janse van Rensburg: *The Statistical Mechanics of Interacting Walks, Polygons, Animals and Vesicles*
19. Sergei B. Kuksin: *Analysis of Hamiltonian PDEs*
20. Alberto Bressan: *Hyperbolic Systems of Conservation Laws: The One-Dimensional Cauchy Problem*
21. Benoit Perthame: *Kinetic Formulation of Conservation Laws*
22. Andrea Braides: *Gamma-Convergence for Beginners*
23. Robert Leese and Stephen Hurley: *Methods and Algorithms for Radio Channel Assignment*
24. Charles Semple and Mike Steel: *Phylogenetics*
25. Luigi Ambrosio and Paolo Tilli: *Topics on Analysis in Metric Spaces*
26. Eduard Feireisl: *Dynamics of Viscous Compressible Fluids*
27. Antonín Novotný and Ivan Straškraba: *Introduction to the Mathematical Theory of Compressible Flow*
28. Pavol Hell and Jaroslav Nešetřil: *Graphs and Homomorphisms*
29. Pavel Etingof and Frederic Latour: *The Dynamical Yang-Baxter Equation, Representation Theory, and Quantum Integrable Systems*
30. Jorge L. Ramírez Alfonsín: *The Diophantine Frobenius Problem*
31. Rolf Niedermeier: *Invitation to Fixed-Parameter Algorithms*
32. Jean-Yves Chemin, Benoit Desjardins, Isabelle Gallagher, and Emmanuel Grenier: *Mathematical Geophysics: An Introduction to Rotating Fluids and the Navier-Stokes Equations*
33. Juan Luis Vázquez: *Smoothing and Decay Estimates for Nonlinear Diffusion Equations*
34. Geoffrey Grimmett and Colin McDiarmid: *Combinatorics, Complexity, and Chance*
35. Alessio Corti: *Flips for 3-folds and 4-folds*
36. Andreas Kirsch and Natalia Grinberg: *The Factorization Method for Inverse Problems*
37. Marius Ghergu and Vicentiu Radulescu: *Singular Elliptic Problems: Bifurcation & Asymptotic Analysis*
38. András Frank: *Connections in Combinatorial Optimization*
39. E.J. Janse van Rensburg: *The Statistical Mechanics of Interacting Walks, Polygons, Animals and Vesicles, 2nd Edition*

The Statistical Mechanics of Interacting Walks, Polygons, Animals and Vesicles

Second Edition

E.J. JANSE VAN RENSBURG

Professor of Mathematics, York University, Toronto

OXFORD
UNIVERSITY PRESS

OXFORD
UNIVERSITY PRESS

Great Clarendon Street, Oxford, OX2 6DP,
United Kingdom

Oxford University Press is a department of the University of Oxford.
It furthers the University's objective of excellence in research, scholarship,
and education by publishing worldwide. Oxford is a registered trade mark of
Oxford University Press in the UK and in certain other countries

© E.J. Janse van Rensburg 2015

The moral rights of the author have been asserted

First Edition published in 2000
Second Edition published in 2015

Impression: 1

All rights reserved. No part of this publication may be reproduced, stored in a retrieval system, or transmitted, in any form or by any means, without the prior permission in writing of Oxford University Press, or as expressly permitted by law, by licence or under terms agreed with the appropriate reprographics rights organization. Enquiries concerning reproduction outside the scope of the above should be sent to the Rights Department, Oxford University Press, at the address above

You must not circulate this work in any other form
and you must impose this same condition on any acquirer

Published in the United States of America by Oxford University Press
198 Madison Avenue, New York, NY 10016, United States of America

British Library Cataloguing in Publication Data
Data available

Library of Congress Control Number: 2014955314

ISBN 978–0–19–966657–7

Printed and bound by
CPI Group (UK) Ltd, Croydon, CR0 4YY

Links to third party websites are provided by Oxford in good faith and for information only. Oxford disclaims any responsibility for the materials contained in any third party website referenced in this work.

Dedicated to the women in my life: Katherine, my mother Lenie, my Daughters, Sister and Grandmothers, who have given much and taken little in return.

PREFACE

The central topic of this book is the modelling of polymer entropy on lattices. This is a classical field of lattice statistical mechanics models which include the self-avoiding walk, lattice trees and animals, and lattice surfaces. The analysis of lattice models of polymer entropy is fundamentally a combinatorial problem, namely, the counting of connected structures in a lattice. Many lattice models remain intractable and can only be analysed using sophisticated mathematical approaches.

Lattice models of the self-avoiding walk, or clusters such as lattice trees and animals, vesicles and surfaces, may be directed or undirected, interacting or free, confined in sub-lattices or interacting with a boundary. In each case the basic questions are similar in nature, namely, how many walks, paths or clusters are there of given size, what are the free energies, what are the scaling properties and can anything be said about the phase diagrams?

Determining the free energies and phase diagrams of lattice models poses in most cases challenging mathematical problems. A wide collection of approaches, including methods from mathematical physics, probability theory, combinatorics and the theory of phase transitions, has been used in some way or another in the very large scientific literature devoted to these models. Numerical approaches have become equally sophisticated and are for the most part based on exact enumeration or Monte Carlo methods.

The free energy of a lattice model defines its phase diagram, which may include several phases separated by critical lines and points. The phases are frequently representative of phase behaviour seen in polymeric systems, and universal thermodynamic scaling near critical points remains the subject of much research. Proving the existence of phase boundaries and critical points and calculating critical exponents pose significant difficulties in many models. It is both the mathematical challenges and the significant progress made which underlie the continuing popularity of lattice models of polymeric systems. Much has been shown, and even more remains to be discovered and seems within reach.

The first edition of this book was out of date within a few years of its publication in the year 2000, due to the fast pace of new results. This, and the omission of a chapter on Monte Carlo methods, made an update of the first edition a priority, and I hope that the new edition will improve on the first.

An undertaking of this size draws on the resources of many, and I have benefited and learned much over two decades in collaborations with Stuart Whittington, Enzo Orlandini, Carla Tesi, Andrew Rechnitzer, Claus Ernst, Yuanan Diao and Neal Madras. I am also grateful to Emmanuel Bradlow for his support more than thirty years ago, and to Ron Horgan for introducing me to the ideas and models in polymer entropy.

CONTENTS

DETAILED CONTENTS

1

LATTICE MODELS OF LINEAR AND RING POLYMERS

A linear polymer is a large molecule consisting of a backbone of atoms or groups of atoms (monomers) which are joined in a sequence by covalent bonds. Parts of a polymer joined by a single covalent bond may rotate freely relative to one another, since single covalent bonds permit free rotations. These rotational degrees of freedom contribute to the configurational entropy of the polymer – quantifying this configurational entropy is the fundamental problem in polymer entropy [117, 202–204].

A typical polymer is illustrated in figure 1.1. If all the monomers are chemically identical, then the polymer is a *homopolymer*, and, if they are of at least two different flavours, then it is a *heteropolymer*.

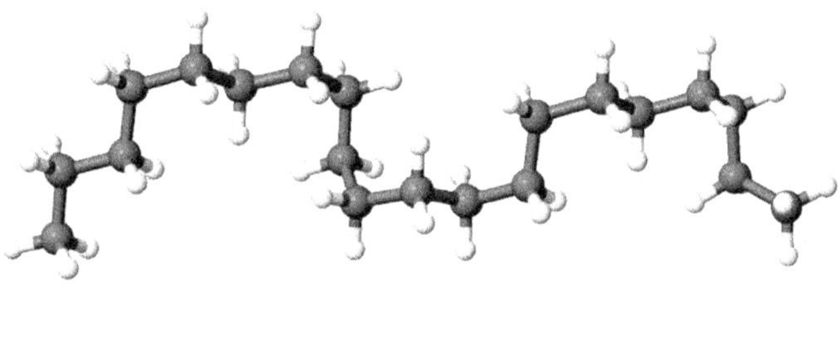

FIG. 1.1. An alkane with a backbone of twenty carbon atoms bound in a linear chain [600].

A *branched polymer* is formed when polymeric side-chains are attached to a polymer *backbone*. Branched polymers are found in different forms, including *trees* and *animals*, as well as *brushes*, *combs*, *stars*, *dendrimers*, and so on.

A popular model of linear polymer entropy is a random walk (see for example reference [176]). A random walk is a good model for the conformational degrees of freedom but fails to explain the asymptotic properties of a linear polymer in a good solvent because *excluded volume effects* are not included in the model.

Configurational entropy and excluded volume of a linear polymer can be modelled by a lattice *self-avoiding walk*. A self-avoiding walk is the union of a sequence of lattice edges or steps joining adjacent vertices along a path which

The Statistical Mechanics of Interacting Walks, Polygons, Animals and Vesicles, 2nd edition,
E.J. Janse van Rensburg. © E.J. Janse van Rensburg. Published in 2015 by Oxford University Press.

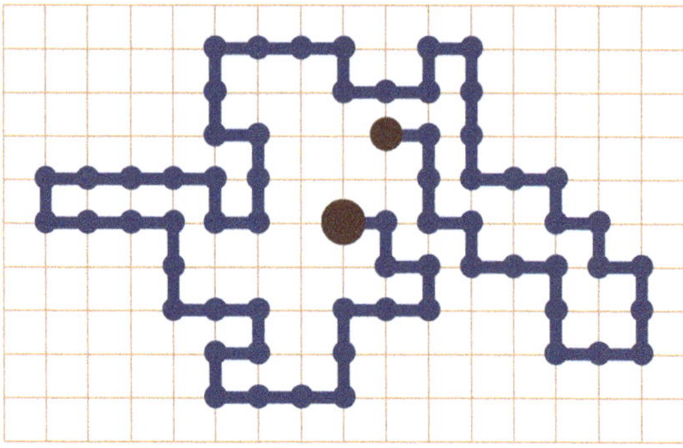

FIG. 1.2. A self-avoiding walk from the origin in $\mathbb{L}^2$. The walk is oriented away from its endpoint at the origin.

avoids itself. The number of such paths is the microcanonical partition function in the model and is a quantitative and relative measure of linear polymer entropy.

The self-avoiding walk model is obtained by endowing the set of self-avoiding walks of fixed given length with the uniform measure. This is a classical model about which a great deal is known and even more remains unknown [399].

1.1 The self-avoiding walk

Let $\mathbb{R}^d$ be d-dimensional real vector space and denote its standard basis by $\langle \vec{e}_1, \vec{e}_2, \ldots, \vec{e}_d \rangle$. The *hypercubic lattice* $\mathbb{L}^d$ is the graph $\mathbb{R}^d$ with vertex set $\mathbb{Z}^d$ and edge set formed by all unit length line segments between vertices in $\mathbb{Z}^d$. That is,

$$\mathbb{L}^d = \big\{ \langle \vec{v} \sim \vec{u} \rangle \mid \text{for } \vec{u}, \vec{v} \in \mathbb{L}^d \text{ with } \|\vec{u} - \vec{v}\|_2 = 1 \big\}. \tag{1.1}$$

If $\langle \vec{u} \sim \vec{v} \rangle \in \mathbb{L}^d$, then $\vec{u}, \vec{v} \in \mathbb{L}^d$ and $\vec{u}$ and $\vec{v}$ are *adjacent* while $\vec{u}$ (or $\vec{v}$) is *incident* with $\langle \vec{u} \sim \vec{v} \rangle$. If $\vec{v} \in \mathbb{L}^d$ is a vertex, then its Cartesian coordinates are $\langle \vec{v}(1), \vec{v}(2), \ldots, \vec{v}(d) \rangle$.

A *self-avoiding walk* ω of length n steps is a sequence of $n+1$ distinct vertices $\langle \vec{v}_i \rangle_{i=0}^n = \langle \vec{v}_0, \vec{v}_1, \vec{v}_2, \ldots, \vec{v}_n \rangle$ such that $\langle \vec{v}_{i-1} \sim \vec{v}_i \rangle$ is an edge in $\mathbb{L}^d$ for $i = 1, 2, \ldots, n$. The i-th edge in the walk is $\langle \vec{v}_{i-1} \sim \vec{v}_i \rangle$. Normally, the zeroth vertex is placed at the origin: $\vec{v}_0 = \vec{0}$. This induces a natural orientation in each walk, away from $\vec{0}$ (see figure 1.2).

The number of self-avoiding walks from $\vec{0}$ of length n is denoted by c_n. It can be checked that $c_0 = 1$ (a single vertex), and in $\mathbb{L}^d$, $c_1 = 2d$, $c_2 = 2d(2d-1)$, and $c_3 = 2d(2d-1)^2$. If $d = 2$, then $c_4 = 100$, and $c_5 = 284$. Notice that $c_{n+1} \le (2d-1)c_n$ for $n \ge 1$, since there are at most $2d-1$ choices for the $(n+1)$-th step. This shows that $c_n \le 2d(2d-1)^{n-1}$. By counting walks which only step in positive directions, $c_n \ge d^n$. This shows that c_n grows exponentially in n.

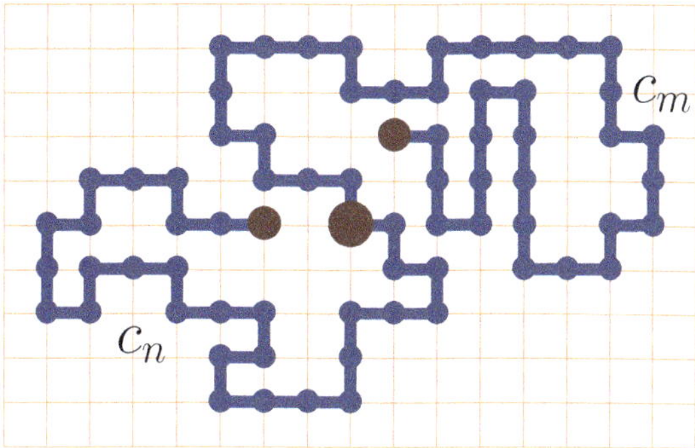

FIG. 1.3. The number of self-avoiding walks c_n is a *submultiplicative function.* To see this, place two walks in $\mathbb{L}^d$, the first of length n to $\vec{0}$, and the second of length m from $\vec{0}$. If the two walks avoid one another, then a (unique) walk of length $n+m$ is obtained. There are c_n choices for the first walk, and c_m choices for the second walk, and they are *concatenated* into at most c_{n+m} walks of length $n+m$. This shows that $c_n c_m \geq c_{n+m}$.

The function c_n is known for $n \leq 71$ [338] in the *square lattice* $\mathbb{L}^2$ (see reference [111]). In the *cubic lattice* $\mathbb{L}^3$, values for c_n are known for $n \leq 36$ [505] (see reference [108] as well).

Two walks can be concatenated as illustrated in figure 1.3. Place two walks, the first of length n to $\vec{0}$ and the second of length m from $\vec{0}$, in $\mathbb{L}^d$. If these two walks avoid one another, then a self-avoiding walk of length $n+m$ is obtained. There are c_n choices for the first walk, and c_m choices for the second walk. Since every walk of length $n+m$ is generated this way,

$$c_n c_m \geq c_{n+m}. \tag{1.2}$$

This shows that c_n is a submultiplicative function on $\mathbb{N}_0$. The following theorem is a corollary of theorem A.1 (in appendix A; see references [80, 260], and also references [255, 256, 258, 263] for additional results).

Theorem 1.1 *There exists a connective constant* κ_d *defined by*

$$\lim_{n\to\infty} \tfrac{1}{n} \log c_n = \inf_{n>0} \tfrac{1}{n} \log c_n = \kappa_d.$$

The growth constant μ_d *of the self-avoiding walk is defined by* $\kappa_d = \log \mu_d$ *and* $d \leq \mu_d \leq 2d-1$. *Consequently,* $c_n \geq \mu_d^n$ *and* $c_n = \mu_d^{n+o(n)}$. □

A lexicographic ordering of a finite set S of vertices in $\mathbb{L}^d$ by Cartesian coordinates gives a *bottom vertex* and a *top vertex* in S. A finite set E of edges in $\mathbb{L}^d$ can similarly be put in lexicographic order by ordering the coordinates of their midpoints. The *top edge* is the lexicographic top edge obtained in this way, and the *bottom edge* is the lexicographic bottom edge obtained in this way. Given a walk ω, it always has a bottom vertex $\vec{b}$ and a top vertex $\vec{t}$.

Lemma 1.2 $c_n \leq c_{n+2}$.

Proof Let $\vec{v}$ be the top vertex of a self-avoiding walk ω of length n.

If $\vec{v}$ is an endpoint of ω and $\vec{v} \neq \vec{0}$, then append the edges $\langle \vec{v} \sim (\vec{v}+\vec{e}_1)\rangle$ and $\langle (\vec{v}+\vec{e}_1) \sim (\vec{v}+2\vec{e}_1)\rangle$. This gives a walk ω' of length $n+2$.

If $\vec{v}$ is an endpoint of ω and $\vec{v} = \vec{0}$, then append $\langle \vec{0} \sim \vec{e}_1\rangle$ and $\langle \vec{e}_1 \sim (2\vec{e}_1)\rangle$ and translate the entire walk so that the origin is at $2\vec{e}_1$. This gives a walk with its top vertex as the endpoint at $\vec{0}$.

If $\vec{v}$ is not an endpoint, then there are two edges $\langle \vec{v} \sim \vec{u}\rangle$ and $\langle \vec{v} \sim \vec{w}\rangle$, incident with $\vec{v}$. It is necessarily the case that these two edges are orthogonal: $\langle \vec{v} \sim \vec{u}\rangle \perp \langle \vec{v} \sim \vec{w}\rangle$. One of these edges, say $\langle \vec{v} \sim \vec{u}\rangle$, is the top edge of ω, and it is necessarily orthogonal to the first standard basis vector $\vec{e}_1$. Delete $\langle \vec{v} \sim \vec{u}\rangle$ and replace it by three edges as follows:

$$\langle \vec{v} \sim \vec{u}\rangle \longrightarrow \langle \vec{v} \sim (\vec{v}+\vec{e}_1)\rangle, \langle (\vec{v}+\vec{e}_1) \sim (\vec{u}+\vec{e}_1)\rangle, \langle (\vec{u}+\vec{e}_1) \sim \vec{u}\rangle.$$

That is, replace the top edge by three edges: $| \to \sqsupset$. This gives a walk ω' of length $n+2$.

In each of the cases, ω' can be uniquely undone to recover ω. Since not every walk of length $n+2$ can be found in this way, $c_n \leq c_{n+2}$. □

It is more difficult to show that $c_n \leq c_{n+1}$ [441]. The limit

$$\lim_{n\to\infty} \frac{c_{n+2}}{c_n} = \mu_d^2 \tag{1.3}$$

exists [348] (see also reference [394] and theorem 7.17 in chapter 7). Showing that $\lim_{n\to\infty} \frac{c_{n+1}}{c_n} = \mu_d$ remains an open problem (and this limit is known not to exist in some lattices [256], but it can be shown to exist in the triangular lattice [348]). If this limit exists, then it must be equal to μ_d. There exists a constant $b > 0$ such that

$$\left| \frac{c_{n+2}}{c_n} - \mu_d^2 \right| \leq b n^{-1/3}. \tag{1.4}$$

This result is due to H Kesten [348, 349].

It is straightforward to see that $\mu_{d-1} \leq \mu_d$ and $\mu_d \geq d$ for all $d \geq 2$. If $d = 1$, then $c_n = 2$ for all $n \geq 1$, so $\mu_1 = 1$.

Theorem 1.3 $\mu_{d-1} < \mu_d$.

Proof Denote the number of self-avoiding walks of length n from $\vec{0}$ in $\mathbb{L}^{d-1}$ by $c_n^{(d-1)}$. Let ω be one such walk. Label the edges of ω by $1, 2, 3, \ldots, n$ from $\vec{0}$ and call edges with an even label *even*, and those with an odd label *odd*. Notice that all edges of ω are orthogonal to $\vec{e}_d$.

Any even edge of ω may be replaced by a $\sqcap$-conformation of three edges in the $\vec{e}_d$-th direction. Similarly, any odd edge of ω may be replaced by a $\sqcup$-conformation of three edges in the $-\vec{e}_d$-th direction, without creating self-intersections.

Choose $\lfloor \epsilon n \rfloor$ edges of ω, and replace the even edges by $\sqcap$-conformations and the odd edges by $\sqcup$-conformations, as above. Since there are $c_n^{(d-1)}$ choices for ω, and the result is a self-avoiding walk of length $n + 2\lfloor \epsilon n \rfloor$ in $\mathbb{L}^d$, this shows that

$$\binom{n}{\lfloor \epsilon n \rfloor} c_n^{(d-1)} \leq c_{n+2\lfloor \epsilon n \rfloor}^{(d)}.$$

Take the $\frac{1}{n}$-th power on both sides, and let $n \to \infty$ to obtain (see corollary D.2 in appendix D)

$$\left(\frac{1}{\epsilon^\epsilon (1-\epsilon)^{1-\epsilon}} \right) \mu_{d-1} \leq \mu_d^{1+2\epsilon} \quad \Rightarrow \quad \left(\frac{\mu^{-2\epsilon}}{\epsilon^\epsilon (1-\epsilon)^{1-\epsilon}} \right) \mu_{d-1} \leq \mu_d.$$

The factor in brackets on the right exceeds 1 for small $\epsilon > 0$. □

The best estimates of μ_2 and μ_3 were obtained from exact enumeration studies of self-avoiding walks. These are $\mu_2 = 2.63815856(3)$ [338] (see [111, 196, 244] for earlier estimates, and reference [246] for a result in the triangular lattice), and $\mu_3 = 4.6840401(50)$ [505] (see [108]). Monte Carlo simulations in reference [322] gave the estimate $\mu_3 = 4.68398(16)$, and in reference [271] the estimate $\mu_3 = 4.683907(22)$ is credited to AJ Guttmann. See table 1.1 for more results in the literature.

The value of μ_2 was estimated in the triangular lattice to high accuracy by I Jensen: $\mu_2 = 4.150797226(26)$ [339]. This improved the older estimate $\mu_2 = 4.1515$ in reference [406].

Remarkably, the value of μ_2 in the hexagonal lattice is known exactly: $\mu_2 = \sqrt{2+\sqrt{2}}$; this result was proven in reference [158] after it was conjectured in reference [436] using Coulomb gas techniques (see section 10.1).

The rate of convergence of $c_n^{1/n}$ to μ_d was studied by JM Hammersley [258, 263]. The expansion

$$\mu_d = 2d - 1 - \tfrac{1}{2d-1} - \tfrac{2}{(2d-1)^2} - \tfrac{11}{(2d-1)^3} - \tfrac{62}{(2d-1)^4} + O\Big(\tfrac{1}{(2d-1)^5}\Big) \qquad (1.5)$$

is due to ME Fisher and D Gaunt [194]. In high dimensions the expansion $\mu_d = 2d - 1 - \frac{1}{2d} - \frac{3}{(2d)^2} + O(\frac{1}{(2d)^3})$ becomes very accurate [271, 432, 513] (see also reference [271, 349]). In reference [108] this expansion was extended to

$$\begin{aligned}\mu_d = 2d - 1 &- \frac{1}{2d} - \frac{3}{(2d)^2} - \frac{16}{(2d)^3} - \frac{102}{(2d)^4} - \frac{729}{(2d)^5} - \frac{5533}{(2d)^6} - \frac{42229}{(2d)^7} \\ &- \frac{288761}{(2d)^8} - \frac{1026328}{(2d)^9} - \frac{21070667}{(2d)^{10}} - \frac{78028046816}{(2d)^{11}} + O\Big(\frac{1}{(2d)^{12}}\Big). \qquad (1.6)\end{aligned}$$

1.2 Lattice polygons

If a self-avoiding walk from $\vec{0}$ is conditioned to end at $\vec{0}$, then it is a *closed walk*. Such a walk is oriented in one of two directions and has a *root vertex* at $\vec{0}$. Lattice polygons are obtained by removing the root and orientation of closed walks.

TABLE 1.1. Numerical estimates of μ_d

Reference	$d = 2$	$d = 3$
Derrida (1981) [128]	2.63817(21)	−−
Enting *et al.* (1985) [178]	2.63816(10)	−−
Berretti *et al.* (1985) [37]	2.63820(34)	−−
Caracciolo *et al.* (1987) [94]	2.6375(29)	−−
Guttmann (1987) [245]	2.638155(11)	−−
Guttmann *et al.* (1988) [248]	2.6381585(10)	−−
Guttmann (1989) [246]	−−	4.68391(22)
MacDonald *et al.* (1992) [393]	−−	4.68404(9)
Meirovitch *et al.* (1993) [412]	2.63816(2)	−−
Grassberger (1993) [233]	−−	4.684007(22)
Caracciolo *et al.* (1990) [91]	2.63815(44)	−−
Nidras (1996) [434]	2.638164(14)	−−
Jensen *et al.* (1999) [340]	2.63815852927(1)	−−
Rechnitzer *et al.* (2002) [487]	2.63816(6)	−−
Jensen (2003) [337]	2.63815853029(3)	−−
Jensen (2004) [338]	2.63815856(3)	−−
Grassberger (2004) [236]	−−	4.6840386(11)
Clisby *et al.* (2007) [108]	−−	4.684043(12)
Janse van Rensburg *et al.* (2008) [322]	2.63805(12)	4.68398(16)
Beaton *et al.* (2011) [29]	2.6381585321(4)	−−
Schram *et al.* (2011) [505]	−−	4.6840401(50)
Clisby *et al.* (2012) [107]	2.63815853035(2)	−−
Clisby (2013) [106]	−−	4.684039931(27)

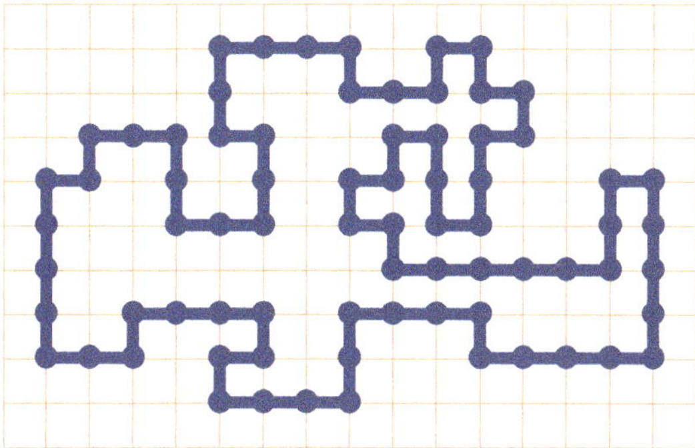

FIG. 1.4. A lattice polygon in the square lattice $\mathbb{L}^2$.

Let the two closed walks ω_1 and ω_2 be equivalent if there is a $\vec{v} \in \mathbb{Z}^d$ such that $\omega_2 = \omega_1 + \vec{v}$ or $\omega_2 = \widetilde{\omega_1} + \vec{v}$, where $\widetilde{\omega_1}$ is obtained by reversing the orientation of ω_1. The equivalence classes of closed walks are *lattice polygons* (see figure 1.4). The convention is that lattice polygons have neither a root nor an orientation.

The number of lattice polygons in $\mathbb{L}^d$ of length n steps is denoted p_n. Since the girth of $\mathbb{L}^d$ is 4, $p_n = 0$ if $n < 4$. $\mathbb{L}^d$ is a bipartite graph and, since polygons are cycles of $\mathbb{L}^d$, $p_{2n+1} = 0$ for $n \in \mathbb{N}$. Inspection shows that $p_4 = 1$, $p_6 = 2$ and $p_8 = 7$ in $\mathbb{L}^2$.

The function p_n is given for $n \leq 110$ in reference [337] (see [177, 178, 247, 340]). In the hexagonal lattice, p_n is known for $n \leq 82$ [179]. In $\mathbb{L}^3$ the lace expansion has been used to determine p_n for $n \leq 32$ in reference [108].

The number of closed walks of length n from the origin is $2np_n$. The number of *rooted polygons* of length n is np_n.

A lexicographic ordering of the vertices in a lattice polygon ω identifies its *top vertex* $\vec{t}_1$ and *bottom vertex* $\vec{b}_1$. The *top edge* $\langle \vec{t}_1 \sim \vec{t}_2 \rangle$ and *bottom edge* $\langle \vec{b}_1 \sim \vec{b}_2 \rangle$ of ω are found by a lexicographic ordering of edges by their midpoints. It is necessarily the case that $\langle \vec{t}_1 \sim \vec{t}_2 \rangle$ and $\langle \vec{b}_1 \sim \vec{b}_2 \rangle$ are unique and orthogonal to $\vec{e}_1$. Notice that the top edge is incident with the top vertex, and the bottom edge is incident with the bottom vertex.

Let ω_1 and ω_2 be two polygons in $\mathbb{L}^d$. Let the top vertex of ω_1 be $\vec{t}_1$ and let its top edge be $\langle \vec{t}_1 \sim \vec{t}_2 \rangle$. Let the bottom vertex of ω_2 be $\vec{s}_2$ and its bottom edge be $\langle \vec{s}_2 \sim \vec{s}_1 \rangle$. Then ω_1 and ω_2 can be concatenated into a single polygon if $\langle \vec{t}_1 \sim \vec{t}_2 \rangle$ is parallel to $\langle \vec{s}_2 \sim \vec{s}_1 \rangle$.

That is, if $\langle \vec{s}_2 \sim \vec{s}_1 \rangle \| \langle \vec{t}_1 \sim \vec{t}_2 \rangle$, then place ω_2 such that $\langle \vec{s}_2 \sim \vec{s}_1 \rangle = \langle \vec{t}_1 \sim \vec{t}_2 \rangle + \vec{e}_1$. Delete $\langle \vec{s}_2 \sim \vec{s}_1 \rangle$ and $\langle \vec{t}_1 \sim \vec{t}_2 \rangle$ and insert the two edges $\langle \vec{s}_2 \sim (\vec{s}_2 - \vec{e}_1) \rangle$ and $\langle \vec{t}_1 \sim (\vec{t}_1 + \vec{e}_1) \rangle$. This joins ω_1 and ω_2 into a single polygon. This is illustrated in figure 1.5 in $\mathbb{L}^2$.

If ω_1 has length n and ω_2 has length m, then the result is a polygon of length $n+m$. There are p_n choices for ω_1, and $\frac{1}{d-1} p_m$ choices for ω_2 since it must be chosen such that its bottom edge is parallel to the top edge of ω_1. This shows

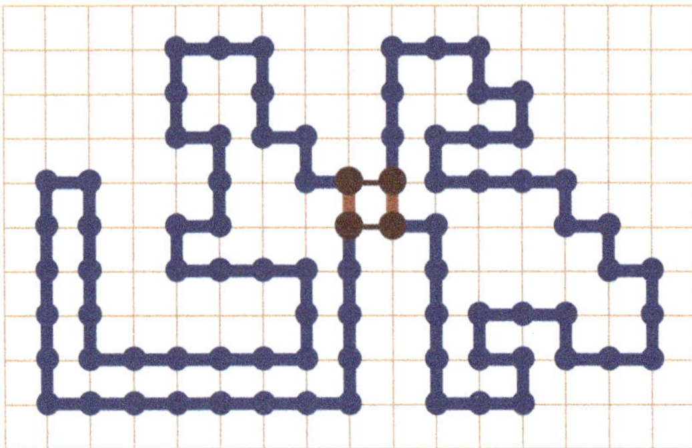

FIG. 1.5. Two polygons can be concatenated as shown above.

---

that, for even values of n and m,

$$p_n p_m \le (d-1) p_{n+m}. \tag{1.7}$$

That is, $-\log(\frac{1}{d-1} p_{2n})$ is a subadditive function on $\mathbb{N}$ for $n \ge 2$. It can be shown that $p_{2n} \ge 2^{n-2}$ if $n \ge 2$ but that $p_{2n-1} = 0$. This gives the following theorem (which is a corollary of theorem A.1; see reference [257]).

Theorem 1.4 *The connective constant κ_p for lattice polygons is defined by*

$$\lim_{n\to\infty} \tfrac{1}{2n} \log p_{2n} = \sup_{n>0} \tfrac{1}{2n} \log\Big(\tfrac{1}{d-1} p_{2n}\Big) = \kappa_p.$$

The growth constant μ_p of lattice polygons is given by $\kappa_d = \log \mu_p$. Consequently, $p_{2n} \le (d-1)\mu_p^{2n}$ and $p_{2n} = \mu_p^{2n+o(n)}$. □

The limit $\lim_{n\to\infty} \frac{1}{n} \log p_n$ does not exist since $p_n = 0$ if n is odd. The convention is to take this limit in $2\mathbb{N}$ (the even numbers) whenever it appears in a calculation. Therefore, write $\lim_{n\to\infty} \frac{1}{n} \log p_n = \log \mu_p$ for lattice polygons on bipartite lattices, with the understanding that the limit is taken in $2\mathbb{N}$. By deleting the bottom edge of polygons, $p_{2n} \le c_{2n-1}$. This shows that $\mu_p \le \mu_d$.

Enumeration of polygons for $n \le 110$ gives $\mu_p = 2.63815853031(3)$ in the square lattice [337] (see references [111, 248, 340]). It is known that $\mu_p = \mu_d$, so polygons and walks grow at the same exponential rate (in the even numbers [257, 263]; see section 7.1.4).

By replacing the top edge of polygons by three edges (that is: $|\to\sqsupset$), it follows that p_n is non-decreasing on even integers (see the proof of lemma 1.2).

Lemma 1.5 $p_{2n} \le p_{2n+2}$. □

A stronger result than theorem 1.4 is that the limit

$$\lim_{n\to\infty} \frac{p_{2n+2}}{p_{2n}} = \mu_p^2 \tag{1.8}$$

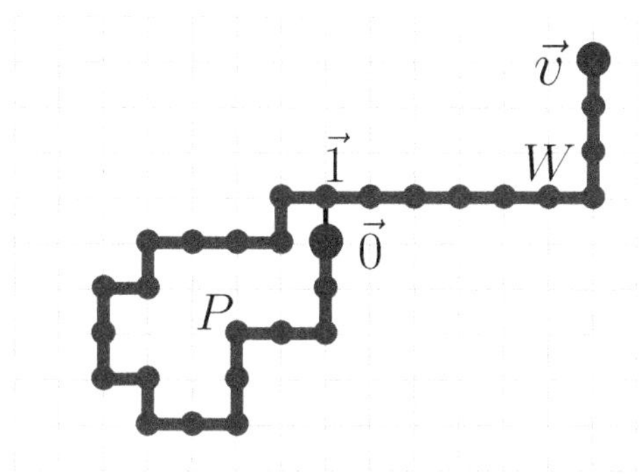

FIG. 1.6. A self-avoiding walk from $\vec{0}$ to the vertex $\vec{v}$ in the first quadrant. This walk was constructed by placing a polygon P such that its top edge is the edge joining the origin $\vec{0}$ to the vertex $\vec{1} = (0, 1, 0, \ldots)$. Next, the shortest path W from the origin to $\vec{v}$ through $\vec{1}$ was constructed. W and P have the edge $\langle \vec{0} \sim \vec{1} \rangle$ in common. If this edge is deleted, then a self-avoiding walk remains. This walk first traces out P from $\vec{0}$ to $\vec{1}$ and then continues along W to $\vec{v}$. Since the top edge of P must contain the top edge $\langle \vec{0} \sim \vec{1} \rangle$, there are $\frac{1}{d-1} p_n$ choices for P if it has length n. A similar construction can be used if $\vec{v}$ is not in the first quadrant.

---

exists [348, 349] (see reference [394] and theorem 7.17 in chapter 7). There exist constants b_1 and b_2 so that

$$b_1 n^{-1/2} \leq \left| \frac{p_{n+2}}{p_n} - \mu_d^2 \right| \leq b_2 n^{-1/3}. \tag{1.9}$$

This result is due to H Kesten [348].

1.3 Self-avoiding walks with fixed endpoints

The number of self-avoiding walks of length n from $\vec{0}$ to $\vec{v} \in \mathbb{L}^d$ is denoted by $c_n(\vec{v})$. Obviously, $c_n = \sum_{\vec{v} \neq 0} c_n(\vec{v})$, where the summation avoids $\vec{0}$.

To relate p_n and $c_n(\vec{v})$, let $\vec{v} = \langle v_1, v_2, \ldots, v_d \rangle$ and suppose without loss of generality that $\vec{v}$ is a vertex with $v_1 > 0$ and $v_2 \geq 0$ (that is, projected into the first two coordinates, $\vec{v}$ projects to a vertex in the first quadrant but not on the $\vec{e}_2$-axis).

A walk from $\vec{0}$ to $\vec{v}$ may be constructed using a polygon P of length n and a walk W from the $\vec{0}$ to $\vec{v}$ of length $\|\vec{v}\|_1$, as illustrated in figure 1.6. Since there are $\frac{1}{d-1} p_n$ choices for P if it has length n, this shows that

$$p_{n+2} \leq (d-1) c_{n+\|\vec{v}\|_1}(\vec{v}). \tag{1.10}$$

Take logarithms, divide by $n + \|\vec{v}\|_1$ and take the limit inferior of the right-hand side as $n \to \infty$ through even numbers. By theorem 1.4,

$$\mu_p \leq \liminf_{n\to\infty} \tfrac{1}{n} \log c_n(\vec{v}) \leq \limsup_{n\to\infty} \tfrac{1}{n} \log c_n(\vec{v}) \leq \mu_d. \tag{1.11}$$

Since $\mu_p = \mu_d$ [258, 263] (see theorem 7.7), this gives the following theorem.

Theorem 1.6 $\lim_{n\to\infty} \frac{1}{n} \log c_n(\vec{v}) = \log \mu_d$. □

Similar to p_n, $c_n(\vec{v})$ is non-decreasing on integers of the same parity as $\|\vec{v}\|_1$. The proof of this is similar to the cases for p_n and c_n. The following constructive proof is based on the proof in Madras and Slade [399].

Lemma 1.7 *If n has the same parity as $\|\vec{v}\|_1$, then there exists a constant C_0 (dependent on $\vec{v}$) such that for all $n \geq C_0$, $c_n(\vec{v}) \leq c_{n+2}(\vec{v})$.*

Proof Let $C \geq \|\vec{v}\|_\infty$ and $n \geq (2C+1)^d + 1$. Let Q be the cube with side-length $2C$ and centre at $\vec{0}$, and let $\omega = \langle \omega_0, \omega_1, \ldots, \omega_n \rangle$ be a self-avoiding walk from $\vec{0}$ to $\vec{v} = \omega_n$. Denote the j-th coordinate of ω_i by $\omega_i(j)$.

At least one point of ω must be outside Q, and, if $M = \max\{\|\omega_i\|_\infty \mid 0 \leq i \leq n\}$, then $M > C$, and both endpoints of ω are inside Q.

There exists a smallest j such that $M = \|\omega_j\|_\infty$. Then, ω_j is outside Q and is not an endpoint of ω. Moreover, there must be a coordinate of ω_j such that its absolute value is equal to M.

Without loss of generality, assume that $M = |\omega_j(1)|$ (the other cases are similarly treated). Then it follows that $M = |\omega_{j+1}(1)|$ by the geometry of $\mathbb{L}^d$. This shows that the edge $\langle \omega_j \sim \omega_{j+1} \rangle$ is outside Q, and $M = |\omega_j(1)| = |\omega_{j+1}(1)| = \|\omega_j\|_\infty$.

Assume without loss of generality that $M = \omega_j(1)$. The vertices $\ell_j = \omega_j + \vec{e}_1$ and $\ell_{j+1} = \omega_{j+1} + \vec{e}_1$ are both outside Q and disjoint with ω. Moreover, the sequence $L = \langle \omega_j, \ell_j, \ell_{j+1}, \omega_{j+1} \rangle$ is a self-avoiding walk of length three. Replacing the edge $\langle \omega_j \sim \omega_{j+1} \rangle$ by L increases the length of ω by two; in addition, L is also the only part of the new walk outside Q. Thus, the construction can be uniquely reversed by locating L on a polygon and deleting it. □

1.4 Scaling

Theorem 1.1 shows that $c_n = \mu_d^{n+o(n)}$. Assuming a power law correction to the exponential growth gives

$$c_n \simeq A n^{\gamma-1} \mu_d^n, \tag{1.12}$$

where γ is the *entropic exponent*. Since $c_n \geq \mu_d^n$ for all $n \in \mathbb{N}$, it must be the case that $\gamma \geq 1$.

In two dimensions the exact value of the entropic exponent is $\gamma = \frac{43}{32}$ [161]. In three dimensions the ϵ-expansion gives $\gamma = 1.1575(6)$ [376]. The mean field value is $\gamma = 1$. More estimates for γ are given in table 1.2.

The mean field value of γ is exact in four and higher dimensions and this is known rigorously if $d \geq 5$ [265, 266, 268]. The scaling relation (1.12) is modified by a logarithmic correction in four dimensions to $c_n \simeq A (\log n)^{1/4} \mu_d^n$ [78].

TABLE 1.2. Estimates of γ for self-avoiding walks

Reference	$d=2$	$d=3$
Guttmann (1987) [245]	1.34361(13)	1.162(2)
Guttmann (1989) [246]	--	1.161(2)
MacDonald *et al.* (1992) [393]	--	1.1585
Grassberger (1993) [233]	--	1.1608(3)
Nidras (1996) [434]	1.3414(26)	--
Caracciolo *et al.* (1998) [87]	--	1.1575(6)
Rechnitzer *et al.* (2002) [487]	1.345(2)	--
Jensen (2004) [338]	1.343745(15)	--
Clisby *et al.* (2007) [108]	--	1.1568(8)
Schram *et al.* (2011) [505]	--	1.15698(34)

The more complete and general form for equation (1.12) includes *analytic* and *confluent* corrections:

$$c_n \simeq A n^{\gamma-1} \mu_d^n \left(1 + \tfrac{a_1}{n} + \tfrac{a_2}{n^2} + \cdots + \tfrac{c_1}{n^{\Delta_1}} + \tfrac{c_2}{n^{\Delta_1+1}} + \cdots + \tfrac{d_1}{n^{\Delta_2}} + \cdots\right). \quad (1.13)$$

The terms $\frac{a_j}{n^j}$ are *analytic corrections* to scaling, and $\frac{c_1}{n^{\Delta_1}}$ is the first in a series of *confluent corrections* with confluent correction exponents $\Delta_1 < \Delta_2 < \cdots$ [161]. In two dimensions $\Delta_1 = \frac{3}{2}$ [90, 114, 338, 436, 438] (but see also the result in reference [500]), while in three dimensions it is thought that $\Delta_1 < 1$ (and in fact $\Delta_1 = 0.56(3)$ [380] or $\Delta_1 = 0.53(1)$ [105]; see also references [243, 380]).

Assigning the generating variable $t \in \mathbb{R}$ to the edges of walks gives the *susceptibility* of the self-avoiding walk:

$$\chi(t) = \sum_{n=0}^{\infty} c_n t^n. \quad (1.14)$$

This is also the generating function of the self-avoiding walk. The logarithm of the generating variable t may be considered a chemical potential; that is, $\log t = v$ is the chemical potential of edges in the grand canonical partition function (or generating function) of walks.

By theorem 1.1, $\chi(t)$ has radius of convergence $t_c = \frac{1}{\mu_d}$. The scaling relation in equation (1.12) suggests that

$$\chi(t) \sim (1 - \mu_d t)^{-\gamma}, \quad \text{as } t \nearrow \mu_d^{-1}. \quad (1.15)$$

Since c_n is submultiplicative, $\chi(t)$ is divergent at $t = t_c$.

1.4.1 *Scaling of polygons*

The scaling assumption for the number of self-avoiding walks from $\vec{0}$ to $\vec{v}$ is

$$c_n(\vec{v}) \simeq B n^{\alpha_s - 2} \mu_d^n, \tag{1.16}$$

where α_s is the *polygon entropic exponent.*

If $\vec{v} = \vec{0}$, then the scaling of *rooted polygons* or *closed walks* is

$$c_n(0) \simeq B n^{\alpha_s - 2} \mu_d^n. \tag{1.17}$$

The root contributes a factor of n to the number of polygons of length n. Thus, the number of polygons of length n scales as

$$p_n \simeq B_p n^{\alpha_s - 3} \mu_d^n. \tag{1.18}$$

It is implicitly understood that equation (1.18) is valid for even n. Since $p_n \leq (d-1)\mu_d^n$, it must be the case that $\alpha_s \leq 3$.

The exact value of the polygon entropic exponent in two dimensions is $\alpha_s = \frac{1}{2}$ (shown by conformal field theory and Coulomb gas techniques [165]). In three dimensions $\alpha_s = 0.237(2)$ [243, 380] (by the N-vector model and Monte Carlo simulations). The mean field value is $\alpha_s = 0$ (for $d \geq 4$) and this is rigorous if $d \geq 5$ [265, 266, 268]. In four dimensions there is a modification of the scaling by a logarithmic factor: $p_n \simeq B_p (\log n)^{1/4} n^{-3} \mu_d^n$. A more complete and general scaling form for p_n is similar to equation (1.13), involving analytic and confluent corrections to scaling.

1.4.2 *Metric scaling*

Let $\omega = \langle \omega_0, \omega_1, \ldots, \omega_n \rangle$ be a self-avoiding walk or a polygon in $\mathbb{L}^d$ with vertices ω_j. Denote the Cartesian coordinates of ω_j by $\omega_j = \langle \omega_j(1), \omega_j(2), \ldots, \omega_j(d) \rangle$. The *span* of ω in the $\vec{e}_1$ direction is given by

$$S_1(\omega) = \max_{0 \leq i \leq n} \left\{ \left| \omega_i(1) - \min_{0 \leq i \leq n} \omega_i(1) \right| \right\}. \tag{1.19}$$

The spans $S_k(\omega)$ in other directions $\vec{e}_k$ are similarly defined. The *total span* and *average span* of ω are defined by

$$S_T(\omega) = \sum_{i=1}^{d} S_i(\omega), \quad \text{and } S_a(\omega) = \tfrac{1}{d} S_T(\omega). \tag{1.20}$$

The *maximal span* is

$$S_m(\omega) = \max_k S_k(\omega). \tag{1.21}$$

The *radius of gyration* R_g of a walk ω in $\mathbb{L}^d$ is defined by

$$R_g^2(\omega) = \tfrac{1}{n^2} \sum_{i,j} \|\omega_i - \omega_j\|_2^2, \tag{1.22}$$

where $\|\cdot\|_2$ is the Euclidean norm.

TABLE 1.3. Estimates of ν for self-avoiding walks

Reference	$d = 2$	$d = 3$
Flory (1969) [204]	$\frac{3}{4}$	$\frac{3}{5}$
Derrida (1981) [128]	0.7503(2)	--
Privman *et al.* (1985) [481]	0.7500(15)	--
Berretti *et al.* (1985) [37]	0.759(11)	--
Rapaport (1985) [484]	--	0.592(2)
Caracciolo *et al.* (1987) [94]	0.750(11)	--
Guttmann (1987) [245]	0.750	0.592(2)
Guttmann *et al.* (1988) [248]	0.753(7)	--
Madras *et al.* (1988) [401]	0.7496(7)	0.592(3)
Guttmann (1989) [246]	--	0.592(3)
Caracciolo *et al.* (1990) [91]	0.7505(13)	--
MacDonald *et al.* (1992) [393]	--	0.5875(6)
Grassberger (1993) [233]	--	0.5850(15)
Li *et al.* (1995) [380]	0.74967(11)	0.5877(6)
Nidras (1996) [434]	0.7499(5)	--
Clisby *et al.* (2007) [108]	--	0.5876(5)
Clisby (2010) [105]	--	0.587597(7)
Schram *et al.* (2011) [505]	--	0.58772(17)

The *mean total span* $\langle S_T \rangle_n$ and *mean square radius of gyration* $\left\langle R_g^2 \right\rangle_n$ are obtained by taking the averages of $S_T(\omega)$ and $R_g^2(\omega)$ over all walks ω of length n (with uniform measure). That is, $\langle S_T \rangle_n = \langle S_T(\omega) \rangle_n$ and $\left\langle R_g^2 \right\rangle_n = \left\langle R_g^2(\omega) \right\rangle_n$, where $\langle \cdot \rangle_n$ is a simple average over all walks or polygons of length n.

It is obvious that $n^{1/d} \leq \langle S_T \rangle_n \leq n$, and $n^{2/d} \leq \left\langle R_g^2 \right\rangle_n \leq n^2$. Thus, it may be supposed that

$$\langle S_T \rangle_n \simeq C_S n^{\nu}, \text{ and } \left\langle R_g^2 \right\rangle_n \simeq C_R n^{2\nu}, \tag{1.23}$$

where $\nu \leq 1$. These scaling assumptions are valid in all dimensions $d \neq 4$. If $d \geq 4$, then ν takes its mean field value $\nu = \frac{1}{2}$. In four dimensions, the scaling is modified by a logarithmic term [78]; for example, in $d = 4$, $\left\langle R_g^2 \right\rangle_n \simeq C_R n \left(\log n\right)^{1/4}$.

In the above cases it is easily seen that $\nu \leq 1$. The lower bound $\nu \geq \frac{1}{d}$ is obvious for the mean square radius of gyration and the spans. Estimates of ν in $d = 2$ and $d = 3$ are listed in table 1.3.

The *mean end-to-end distance* between the endpoints of self-avoiding walks of length n is denoted by $\langle R_e \rangle_n$. Assuming that there is only one length scale in the model (as above), it should be the case that $\langle R_e \rangle_n \simeq C_e n^\nu$. In this case it is known that $\nu < 1$ [157], so the self-avoiding walk is sub-ballistic. It is not known, for $\langle R_e \rangle_n$, that $\nu \geq \frac{1}{d}$. The best lower bound $\nu \geq \frac{2}{3d}$ is due to N Madras [398].

The *mean square end-to-end distance* of walks of length n is denoted by $\langle R_e^2 \rangle_n$. If ω is a walk from $\vec{0}$, and $\vec{v}$ is a vertex in ω, then the mean square distance of $\vec{v}$ from $\vec{0}$, averaged over all choices for $\vec{v}$, is denoted by $\langle R_m^2 \rangle_n$. Several dimensionless ratios may be formed, including

$$\lim_{n\to\infty} \frac{\langle R_g^2 \rangle_n}{\langle R_e^2 \rangle_n} = \frac{R_g^2}{R_e^2}, \quad \text{and} \quad \lim_{n\to\infty} \frac{\langle R_m^2 \rangle_n}{\langle R_e^2 \rangle_n} = \frac{R_m^2}{R_e^2}. \tag{1.24}$$

These *limiting dimensionless amplitude ratios* are universal quantities. If $d = 2$, then conformal field theory predicts that [93, 96, 97]

$$\frac{123}{91}\frac{R_g^2}{R_e^2} - \frac{R_m^2}{R_e^2} + \frac{1}{4} = 0. \tag{1.25}$$

Estimates of the ratios were made in reference [380]. If $d = 2$, then

$$\frac{R_g^2}{R_e^2} = 0.140264(73), \quad \text{and} \quad \frac{R_m^2}{R_e^2} = 0.439605(38), \tag{1.26}$$

while, if $d = 3$,

$$\frac{R_g^2}{R_e^2} = 0.1599(2). \tag{1.27}$$

The results in $d = 2$ are consistent with the relation in equation (1.25). Similar tests were done by AJ Guttmann in references [245, 246].

Let $r = \|\vec{r}\|_2$ be the Euclidean distance from $\vec{0}$ to $\vec{r} \in \mathbb{L}^d$. The number of walks from $\vec{0}$ to $\vec{r}$ is $c_n(\vec{r})$ (see equation (1.16)), and this should be asymptotically isotropic (the same in all directions). The *end-to-end distribution function* is defined by $P_n(r) = c_n(\vec{r})/c_n$. Since the endpoints of walks of length n explore a volume of R_g^d,

$$P_n(r) \sim \frac{1}{R_g^d} F\left(\frac{r}{R_g}\right) \tag{1.28}$$

for some distribution F.

The short- and long-ranged behaviour of F is expected to be [190]

$$F(x) \sim x^g \text{ for small } x, \text{ and } F(x) \sim e^{-x^\delta} \text{ for large } x, \tag{1.29}$$

where $g = \frac{\gamma-1}{\nu}$ [132, 134], and $\delta = \frac{1}{1-\nu}$ [190].

The number of walks ending a distance $r = \|\vec{r}\|_2$ from $\vec{0}$ is $c_n(r) = \sum_{\|\vec{r}\|_2=r} c_n(\vec{r})$. These walks end on a spherical shell of radius r. Thus, $c_n(r) \sim r^{d-1} c_n(\vec{r})$, and, in terms of the end-to-end distribution function,

$$c_n(r) \sim \frac{r^{d-1}}{R_g^d} F\left(\frac{r}{R_g}\right) c_n. \tag{1.30}$$

1.4.3 *The Flory values*

An argument due to P Flory [204] gives surprisingly accurate estimates for the metric exponent ν of self-avoiding walks. Let $s_n(r)$ be the number of random walks of length n from $\vec{0}$ and ending at a distance r from the origin, and let $P_n(r)$ be the probability that a random walk, starting from the origin, has stepped a distance r away from the origin in n steps.

If s_n is the number of random walks from the origin of length n, then $P_n(r) = s_n(r)/s_n$, and the entropy of the random walk is proportional to $\log s_n(r)$. The distribution $P_n(r)$ can be approximated by the solution of the heat equation. If n is treated as a time variable, then

$$\frac{\partial}{\partial n} P_n(r) = D \triangle P_n(r), \tag{1.31}$$

which is a discrete heat equation with diffusion constant D and with the approximate solution

$$P_n(r) \approx \frac{1}{\sqrt{4\pi Dn}} e^{-r^2/4Dn}. \tag{1.32}$$

The total relative entropy can be estimated as $S_n(r) = \log P_n(r) = -A \frac{r^2}{4Dn} + C$, where A is a constant and where C is a term independent of r. This completes the first part of the argument.

The second part of a Flory argument is a 'mean field' estimate of the energy density. The density of monomers is proportional to $\frac{n}{r^d}$, so that each monomer is in a mean field of density $\frac{n}{r^d}$. Summing over the monomers gives an energy density proportional to $\frac{n^2}{r^d}$:

$$E_n(r) = B \frac{n^2}{r^d}. \tag{1.33}$$

The free energy of the walk is

$$F_n(r) = E_n(r) - T S_n(r) = B \frac{n^2}{r^d} + AT \frac{r^2}{4Dn} - CT. \tag{1.34}$$

Minimising $F_n(r)$ gives $r \sim n^{3/(d+2)}$. This shows that the metric exponent is

$$\nu = \frac{3}{d+2}. \tag{1.35}$$

This estimate is remarkably accurate: in $d = 2$, $\nu = \frac{3}{4}$ [436, 438], and, if $d = 4$, then $\nu = \frac{1}{2}$, its mean field value. In $d = 3$ the Flory value is $\frac{3}{5}$, while numerical simulations give $\nu = 0.5877(6)$ [233, 380]; and $\nu = 0.587597(7)$ [105] (see reference [16] as well). A more probabilistic approach to the Flory argument can be found in reference [399]. The Flory formula is also discussed in reference [117].

1.4.4 *Fisher's scaling relation*

Let t be the generating variable for edges in walks. The mean of the average span $\langle S_a \rangle_n$ (see equation (1.20)) over walks of all lengths is given by

$$S_t = \frac{\sum_n \langle S_a \rangle_n c_n t^n}{\sum_n c_n t^n} \sim |\log(\mu_d t)|^{-\nu} \sim (1 - \mu_d t)^{-\nu}. \tag{1.36}$$

S_t defines a length scale which diverges if t approaches the critical point: $S_t \to \infty$ as $t \to \frac{1}{\mu_d}$ from below.

A length scale can also be determined from the (thermodynamic) correlation functions. In particular, define the *two-point function*

$$C(\vec{v}; t) = \sum_{n=0}^{\infty} c_n(\vec{v}) t^n. \tag{1.37}$$

The two-point function of an $O(N)$-vector model is known to exhibit Ornstein-Zernike decay [193]. This behaviour can be expected to carry over into the $N = 0$ limit for the self-avoiding walk two-point function $C(\vec{v}; t)$ [101]:

$$C(\vec{v}; t) \sim \|\vec{v}\|_2^{-(d-1)/2} e^{-\|\vec{v}\|_2/\xi(t)}. \tag{1.38}$$

The rate of decay of $C(\vec{v}; t)$ defines a *correlation length* $\xi(t)$ by

$$\xi^{-1}(t) = \liminf_{\|\vec{v}\|_2 \to \infty} \left(\frac{-\log C(0, \vec{v}; t)}{\|\vec{v}\|_2} \right). \tag{1.39}$$

Comparison to equation (1.36) shows that

$$\xi(t) \sim (1 - \mu_d t)^{-\nu} \tag{1.40}$$

since both ξ and S_t are measures of the same length scale.

The largest contribution to the sum in equation (1.37) occurs when $\xi(t)$ is comparable in size to $\|\vec{v}\|_2$. That is, for fixed t, there is a scaling function g which decays exponentially fast such that

$$C(\vec{v}; t) \sim \frac{1}{\|\vec{v}\|_2^{d-2+\eta}} g\left(\frac{\|\vec{v}\|_2}{\xi(t)} \right). \tag{1.41}$$

This introduces the exponent η, also called the *anomalous dimension*.

The mean field value of η is zero, but in low dimensions it is positive. The exact value in two dimensions is $\eta = \frac{5}{24}$ [165], and, in three dimensions, $\eta = 0.031(4)$, a result obtained from the ϵ-expansion [87].

The susceptibility can be determined from the two-point function

$$\chi(t) \simeq \sum_{\vec{v}} \frac{1}{\|\vec{v}\|_2^{d-2+\eta}} g\left(\frac{\|\vec{v}\|_2}{\xi(t)} \right) \sim (\xi(t))^{2-\eta} \sim (1 - \mu_d t)^{-(2-\eta)\nu}. \tag{1.42}$$

In terms of the correlation length in equation (1.40), this is $\chi(t) \sim \xi(t)^{\gamma/\nu}$. Comparison with equation (1.15) gives

$$\gamma = (2 - \eta)\nu. \tag{1.43}$$

This is *Fisher's scaling relation* for the self-avoiding walk.

TABLE 1.4. Exact and estimated exponents for walks and polygons

d	$d=2$	$d=3$	$d \geq 4$ (mean field)
γ	$\frac{43}{32}$ [161]	1.15698(34) [233, 377]	1
α_s	$\frac{1}{2}$ [165]	0.237(2) [243, 377]	0
ν (Flory)	$\frac{3}{4}$ [204]	$\frac{3}{5}$ [204]	$\frac{1}{2}$
ν	$\frac{3}{4}$ [165]	0.587597(7) [105]	$\frac{1}{2}$
η	$\frac{5}{24}$ [165]	0.031(4) [87]	0
Δ_1	$\frac{3}{2}$ [90, 338]	0.470(25) [243, 377]	--
Δ_4	$\frac{155}{92}$	1.4603(12)	$\frac{3}{2}$

Fisher's relation is satisfied by the mean field values of the exponents, namely $\nu = \frac{1}{2}$, $\gamma = 1$ and $\eta = 0$. In two dimensions, the exact values $\nu = \frac{3}{4}$, $\gamma = \frac{43}{32}$ and $\eta = \frac{5}{24}$ of the exponents also satisfy the relation [161, 165, 436, 438]. The numerical estimates of the exponents in three dimensions are also consistent with the relation [87, 105, 505].

The ϵ-expansion for self-avoiding walk exponents was developed by P-G de Gennes [115] from the $N = 0$ limit in the $O(N)$-vector model. In $d = 4 - \epsilon$ dimensions,

$$\gamma = 1 + \tfrac{1}{8}\epsilon + \tfrac{13}{256}\epsilon^2 + O(\epsilon^3), \quad \text{and } \eta = \tfrac{1}{64}\epsilon^2\left(1 + \tfrac{17}{16}\epsilon\right) + O(\epsilon^4). \tag{1.44}$$

Assuming Fisher's scaling relation then gives the following expansion for the metric exponent:

$$\nu = \tfrac{1}{2}\left(1 + \tfrac{1}{8}\epsilon + \tfrac{15}{256}\epsilon^2\right) + O(\epsilon^3) \tag{1.45}$$

(see equation (2.119)).

1.4.5 *Hyperscaling*

Define $c_{n,m}(\vec{v})$ to be the number of pairs of self-avoiding walks, the first of length n from the origin, and the second of length m from $\vec{v}$, each intersecting the other at least once. Put $c_{n,m} = \sum_{\vec{v}} c_{n,m}(\vec{v})$. A scaling assumption for $c_{n,m}$ is

$$c_{n,m} \sim A_{n,m}\mu_d^{n+m} n^{2\Delta_4+\gamma-2} F\left(\tfrac{n}{m}\right), \tag{1.46}$$

for some universal scaling function F. For fixed ratios $\frac{n}{m}$, the dependence on n is the power law $n^{2\Delta_4+\gamma-2}$ (Δ_4 should not be confused with the confluent correction exponent in equation (1.13)).

The function $c_{n,n}$ can be estimated as follows: a walk of length n from $\vec{0}$ occupies a spherical volume V of size $O\left(n^{d\nu}\right)$. The second walk will typically intersect the first if it is started inside V. By equation (1.12),

$$c_{n,n} \sim c_n^2 n^{\nu d} \sim A^2 n^{2\gamma-2+\nu d} \mu_d^2. \tag{1.47}$$

Comparison to equation (1.46) then gives [380]

$$d\nu - 2\Delta_4 + \gamma = 0. \tag{1.48}$$

Estimates of Δ_4 are obtained by assuming hyperscaling and computing them from equation (1.48) (by computing ν and γ). In two dimensions the numerical estimate $2\Delta_4 - \gamma = 1.4999(2)$ was obtained [380]. This compares well with the estimate computed from hyperscaling in table 1.4.

The explicit dependence of equation (1.48) on d implies that the relation breaks down in high dimensions. Notice that, if $d\nu > 2$, then two walks in V will generally not intersect, so $c_{n,n} \approx c_n^2 n^2$. This gives the result $\Delta_4 = 1 + \frac{1}{2}\gamma$; by inserting the mean field value $\gamma = 1$, it follows that $\Delta_4 = \frac{3}{2}$. Substitution of this into equation (1.48) shows that the relation breaks down if $d\nu > 2$ (or if $d > 4$). For extended discussions of hyperscaling and the application of it to self-avoiding walks, see, for example, references [380, 399, 433].

The *interpenetration ratio* of the self-avoiding walk [117] is defined by

$$\Psi_n = 2\left(\frac{d}{12\pi}\right)^{d/2} \frac{c_{n,n}}{c_n^2 \left\langle R_g^2 \right\rangle_n^{d/2}}. \tag{1.49}$$

This is the *second virial coefficient* of the self-avoiding walk. The larger Ψ_n, the more difficult it is for two walks to stay in the same volume of space. Substitution of the scaling assumptions for c_n, $c_{n,n}$ and $\left\langle R_g^2 \right\rangle_n$ gives

$$\Psi_n \sim C n^{2\Delta_4 - \gamma - d\nu}. \tag{1.50}$$

If the hyperscaling in equation (1.48) is satisfied, then Ψ_n approaches a non-zero finite constant as $n \to \infty$. This limit is defined by

$$\Psi^* = \lim_{n\to\infty} \Psi_n. \tag{1.51}$$

Violations of the hyperscaling relation will result in Ψ_n either approaching zero or diverging to infinity. Determining Ψ^* is a sensitive test for hyperscaling in equation (1.48). In reference [380] extensive numerical simulations were done to estimate

$$\Psi^* = \begin{cases} 0.66296(43), & \text{in two dimensions;} \\ 0.2471(3), & \text{in three dimensions.} \end{cases} \tag{1.52}$$

1.4.6 *Josephson's hyperscaling relation*

The number of closed walks (rooted at $\vec{0}$) was assumed to have asymptotic behaviour $c_{2n}(\vec{0}) \sim B(2n)^{\alpha_s - 2}\mu_d^{2n}$ in equation (1.16). The number of closed walks can also be estimated putting pairs of walks together – this approach was developed in reference [399] and is as follows.

Let ω_1 and ω_2 be self-avoiding walks of length n, from $\vec{0}$ to $\vec{v}$. If these walks avoid one another (except at their endpoints), then they form a polygon of length $2n$ passing through the points $\vec{0}$ and $\vec{v}$. The number of such pairs of walks is

$$c_{2n}(\vec{0}) = \sum_{\vec{v}\in\mathbb{Z}^d} \sum_{\omega_1,\omega_2} I\Big(\omega_1 \cap \omega_2 = \{\vec{0},\vec{v}\}\Big), \tag{1.53}$$

where the indicator function $I(\cdot)$ is 1 if $\omega_1 \cap \omega_2 = \{\vec{0},\vec{v}\}$, and 0 otherwise.

Dominant contributions to $c_{2n}(\vec{0})$ should be from terms in the summation with $\|\vec{v}\|_2 = O(n^\nu)$, in which case there are $O(n^{d\nu})$ terms making a contribution.

The probability that two walks from $\vec{0}$ avoid one another is c_{2n}/c_n^2. Assuming that intersections between the walks occur close to $\vec{0}$ implies that the probability of two walks from $\vec{0}$, both ending site $\vec{v}$ and avoiding one another, is $(c_{2n}/c_n^2)^2$. This probability scales as $n^{2-2\gamma}$.

The probability that a walk ends at site $\vec{v}$ (with $\|\vec{v}\|_2$ of order n^ν) is inversely proportional to the volume of a sphere of radius n^ν. That is, $c_n(\vec{v}) \approx c_n\, n^{-d\nu} \sim A n^{\gamma-1-d\nu} \mu_d^n$.

Hence, if $\|\vec{v}\|_2$ is of order n^ν, then

$$c_{2n}(\vec{0}) \sim n^{d\nu}\, c_n^2(\vec{v}) \left(\tfrac{c_{2n}}{c_n^2}\right)^2 \sim n^{d\nu}\left(A n^{\gamma-1-d\nu}\mu_d^n\right)^2 n^{2-2\gamma} = A^2 n^{-d\nu}. \tag{1.54}$$

Since $c_{2n}(\vec{0})$ is the number of rooted polygons of length n, compare it to equation (1.17) to obtain Josephson's hyperscaling relation:

$$2 - \alpha_s = d\nu. \tag{1.55}$$

The exact value of α_s in two dimensions is $\frac{1}{2}$. Since $\nu = \frac{3}{4}$ exactly in two dimensions, these values satisfy Josephson's relation. In three dimensions the estimates $\alpha_s = 0.237(2)$ [243, 377] and $\nu = 0.587597(7)$ [105] are also consistent with the relation. The mean field values of the exponents similarly satisfy the relation, but above four dimensions the relation breaks down.

1.5 Walk and polygon generating functions

The self-avoiding walk generating function is given by

$$C(t) = \sum_{n=0}^{\infty} c_n t^n, \tag{1.56}$$

which is also the self-avoiding walk susceptibility $\chi(t)$ defined in equation (1.14). $C(t)$ is a formal power series in t with non-negative coefficients. Since $c_n = \mu_d^{n+o(n)}$ the dominant singularity of $C(t)$ is on the positive real axis at $t_c = \frac{1}{\mu_d}$ (see theorem 1.1). For $|t| < t_c$, the generating function is absolutely convergent, and it extends to a holomorphic function within its radius of convergence in the t-plane with dominant singularity at t_c on the positive real axis.

$C(t)$ is divergent at $t = t_c$; this is a consequence of theorem 1.1, and equation (1.15) suggests that

$$C(t) \sim (1 - \mu_d t)^{-\gamma}. \tag{1.57}$$

The self-avoiding walk entropic exponent γ is also called the *susceptibility exponent* (see equation (1.15)).

The polygon generating function is

$$P(t) = \sum_{n=0}^{\infty} p_n t^n, \tag{1.58}$$

where the summation is implicitly over even values of n (see equation (1.18)). Theorem 1.4 shows that $p_n = \mu_p^{n+o(n)}$ for even n, so the dominant singularity is on the positive real axis at $t_p = \frac{1}{\mu_p}$. Notice that $\mu_d = \mu_p$ (see theorem 7.7). The function $P(t)$ is absolutely convergent if $|t| < t_p$, and is a holomorphic function in the t-plane within its radius of convergence (which is given by the dominant singularity at t_p on the positive real axis).

Substituting equation (1.18) into equation (1.58) gives

$$P(t) \sim (1 - \mu_p t)^{2-\alpha_s}. \tag{1.59}$$

Generally, $\alpha_s \leq 2$, so $P(t)$ should be finite when $t = t_p$.

The correlation function $C(0, \vec{v})$ (see equation (1.37)) is the generating function of $c_n(\vec{v})$. If $\vec{v} = \vec{0}$, then the generating function $R(t)$ of rooted polygons is obtained:

$$R(t) = \sum_{n=0}^{\infty} n p_n t^n \sim (1 - \mu_p t)^{1-\alpha_s}. \tag{1.60}$$

The *specific heat* of the self-avoiding walk is the derivative of $R(t)$ to $\log t$, given by

$$\mathcal{C}_t = \sum_{n=0}^{\infty} n^2 p_n t^n \sim (1 - \mu_p t)^{-\alpha_s}, \tag{1.61}$$

and the polygon entropic exponent α_s is also called the *specific heat exponent.*

Equations (1.57) and (1.59) do not imply the postulated asymptotic expressions for c_n and p_n in equations (1.12) and (1.18). This would be the case if additional information were available, giving rise to a Tauberian theorem [274].

The behaviour of $C(t)$ and $P(t)$ is similar to the critical behaviour of grand partition functions in thermodynamic systems. The generating variable t may be related to a temperature T by $t = e^{\beta}$ (where $\beta = \frac{1}{k_B T}$ and k_B is the Boltzmann constant). The regime $t < t_c$ may be interpreted as a *phase of finite polygons or walks*, while $t > t_c$ is a *phase of infinite polygons or walks.* These phases are often also called the *subcritical* and *supercritical* phases separated by the critical point t_c about which thermodynamic scaling described by the exponents γ and α_s is seen.

Scaling in lattice walks and polygons has been studied using different approaches, including numerical approaches [37, 87, 94, 323, 337, 338, 380, 434, 435, 489], transfer matrix methods [601], Coulomb gas techniques [436–439], conformal field theory [95–97, 142, 165] and renormalisation group studies of $O(N)$-vector models [128, 243, 373–375, 377]. The existence of some critical exponents in five and higher dimensions was shown by the lace expansion [265, 266, 268, 270]. The mean field exponents are exact above the upper critical dimension $d_c = 4$ [269]. Experimental values for critical exponents have been obtained in three dimensions in several studies [136, 417, 473, 516].

1.6 Tadpoles, figure eights, dumbbells and thetas

Tadpoles, figure eights, dumbbells and thetas are examples of *uniform networks* in $\mathbb{L}^d$ (see figure 1.7). A tadpole is a lattice polygon head to which a self-avoiding walk tail is attached. The tadpole is rooted (at $\vec{0}$) in its single vertex of degree 1. The size of a tadpole is its total number of edges, and the head and tail may have different sizes. (If the head and tail have the same size, then the tadpole is *uniform*).

A tadpole is oriented along its tail from its root. The head of the tadpole is a polygon attached to the tail in vertex of degree 3. The head may be oriented in two ways, and the convention is that these are counted together as representing one conformation.

The number of tadpoles with a tail of length r and a head of size $n-r$ is $T_{r,n-r}$. For example, $T_{1,4} = 8$ in $\mathbb{L}^2$. Notice that $T_{0,n} = np_n$. The total number of (non-uniform) tadpoles of size n is

$$T_n = \sum_{r=0}^{n-4} T_{r,n-r}. \tag{1.62}$$

A relation between c_n, p_n and $T_{r,n-r}$ is obtained as follows. Let ω be a self-avoiding walk of length n from $\vec{0}$ in $\mathbb{L}^d$ and put $\sigma = 2d-1$. Then there are σ unoccupied edges incident with the last vertex of ω. Extending ω by one step along one of these edges gives two possibilities: (1) either the attempt is successful, or (2) a tadpole or a rooted lattice polygon is obtained. Since every self-avoiding walk of length $n+1$ can be obtained in this way,

$$c_{n+1} = \sigma c_n - 2(n+1)p_{n+1} - 2\sum_{r=1}^{n-3} T_{r,n-r+1}. \tag{1.63}$$

The factors of 2 account for the two orientations available for each of the polygons and tadpoles, respectively.

A *figure eight* is a network composed of two polygons sharing a single vertex (see figure 1.7(b)). That is, if ρ_1 and ρ_2 are two polygons of lengths n and m, respectively, and $\rho_1 \cap \rho_2 = \{\vec{v}\}$, then $\rho_1 \cup \rho_2$ is a figure eight of size $n+m$, and the two heads ρ_1 and ρ_2 are joined at $\vec{v}$. The convention is that $\vec{v}$ is placed at

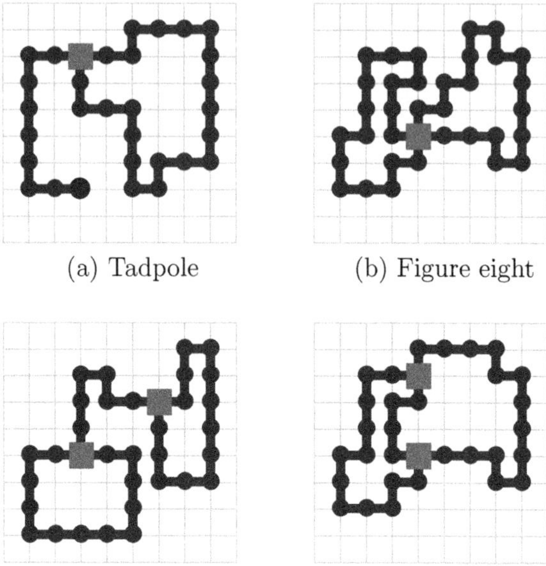

(a) Tadpole (b) Figure eight

(c) Dumbbell (d) Theta

FIG. 1.7. Examples of networks in the square lattice. Each network consists of *branches* (self-avoiding walks) joined at nodes. The connectivity of the network is given by an abstract graph. For example, a tadpole is obtained if the endpoint of a line segment is placed on a point in a cycle graph – the cycle is the *head* of the tadpole, and the line is its *tail*. A figure eight is composed of two heads joined at a node of degree 4. A dumbbell has two heads joined into a connected graph by a branch with an endpoint in each head. A theta is composed of three branches with joint endpoints in nodes of degree 3. If the branches of a network all have the same length, then the network is *uniform*.

$\vec{0}$, and the heads are not oriented. The number of figure eights with one head of size n and the other head of size m is $f_{n,m}$.

Dumbbells and *thetas* (see figures 1.7(c) and (d)) are similarly defined. A dumbbell is composed of two polygons, ρ_1 and ρ_2, (these are the heads) joined by a self-avoiding walk ω. That is, $\rho_1 \cap \rho_2 = \emptyset$, $\rho_1 \cap \omega = \vec{v}_1$ and $\rho_2 \cap \omega = \vec{v}_2$. Dumbbells are counted up to equivalence under translations in $\mathbb{L}^d$. The number of dumbbells with one head of size n, the other head of size m, and the walk joining them of length k is denoted $d_{k,n,m}$. Put $d_n = \sum_{k,m} d_{k,n-k-m,m}$.

A theta is composed of three self-avoiding walks or *branches* ω_1, ω_2 and ω_3 such that $\omega_i \cap \omega_j = \{\vec{v}_1, \vec{v}_2\}$ for each $i \neq j$. That is, the walks have two common endpoints and otherwise avoid one another. The endpoints have degree equal to 3. Thetas are counted up to equivalence under translation in $\mathbb{L}^d$, similar to dumbbells. The number of thetas with branches of sizes $\{k, n, m\}$ is $\theta_{k,n,m}$. Put $\theta_n = \sum_{k,m} \theta_{k,n-k-m,m}$.

Let τ be a tadpole with a tail ω and a head ρ. Then ω is a self-avoiding walk, and ρ is a lattice polygon such that $\omega \cap \rho = \{\vec{v}\}$ is the junction between ω and ρ. Denote the free end of the tail of τ by $\vec{u}$.

The tail ω may be extended by appending an edge $\langle \vec{u} \sim \vec{w} \rangle$ on $\vec{u}$ in one of σ available directions. If $\langle \vec{u} \sim \vec{w} \rangle \cap \tau = \{\vec{u}\}$, then $\langle \vec{u} \sim \vec{w} \rangle \cup \tau$ is a tadpole.

The other possibilities are that $\langle \vec{u} \sim \vec{w} \rangle \cap \tau = \{\vec{u}, \vec{w}\}$. If $\vec{w} = \vec{v}$, then $\langle \vec{u} \sim \vec{w} \rangle \cup \tau$ is a figure eight. If $\vec{w} \in \omega$, then a dumbbell was formed, and, if $\vec{w} \in \rho$, then a theta was obtained.

Notice that the orientation of τ induces an orientation for figure eights (or dumbbells or thetas) if $\langle \vec{u} \sim \vec{w} \rangle$ is appended on $\vec{u}$ as above. There are four possible orientations of the figure eights (each head has two orientations), as well as a choice for which head to traverse first. This gives a factor of 8. Similarly, there is a factor of 8 for dumbbells and a factor of 12 for thetas. Accounting for these orientations, and noting that there are $T_{r,n-r}$ choices for τ above, the result is

$$2T_{r,n-r} = 2\sigma T_{r-1,n-r} - 8 f_{r,n-r} - 8 \sum_{j=1}^{r-4} d_{r-j,j,n-r} - 12 \sum_{j=1}^{n-r-1} \theta_{r,j,n-r-j}. \quad (1.64)$$

If $r = 1$, then

$$2T_{1,n-1} = 2(n-1)(\sigma-1)p_{n-1} - 12 \sum_{j=1}^{n-1} \theta_{1,j,n-1-j}. \quad (1.65)$$

Summing $T_{r,n-r}$ over r using equations (1.64) and (1.65) gives

$$\sum_{r=1}^{n-4} T_{r,n-r} = \sigma \sum_{r=1}^{n-5} T_{r,n-r-1} + (n-1)(\sigma-1)p_{n-1} - 4 f_n - 4 d_n - 6\theta_n. \quad (1.66)$$

Determine $c_n - \sigma c_{n-1}$ from equation (1.63):

$$\begin{aligned} &c_n - 2\sigma c_{n-1} + \sigma^2 c_{n-2} \\ &\quad = 2\sigma(n-1)p_{n-1} - 2np_n - 2\sum_{r=1}^{n-4} T_{r,n-r} + 2\sigma \sum_{r=1}^{n-5} T_{r,n-r-1}. \end{aligned} \quad (1.67)$$

Replace the summations over $T_{r,n-r}$ in the above by using equation (1.63). This gives *Sykes' counting theorem* [542]:

$$c_n - 2\sigma c_{n-1} + \sigma^2 c_{n-2} = 2(n-1)p_{n-1} - 2np_n + 8 f_n + 8 d_n + 12\theta_n. \quad (1.68)$$

Define the figure eight, dumbbell and theta generating functions by $F(t)$, $D(t)$ and $\theta(t)$, respectively. Then, by multiplying the above by t^n and summing over n, the following relation between generating functions is obtained [252]:

$$(1-\sigma t)^2 C(t) = 2t(t-1)P'(t) + 8F(t) + 8D(t) + 12\theta(t). \quad (1.69)$$

1.6.1 Growth constants

The growth constants μ_d and μ_p of walks and polygons were defined in theorems 1.1 and 1.4. By theorem 7.7:

$$\lim_{n\to\infty} \tfrac{1}{n} \log c_n = \lim_{n\to\infty} \tfrac{1}{2n} \log p_{2n} = \log \mu_d, \tag{1.70}$$

so that $\mu_p = \mu_d$.

An *fl-walk* ω is a self-avoiding walk with end-vertices which are also its bottom and top vertices (lexicographic first and last vertices). The number of fl-walks of length n from $\vec{0}$ is denoted by c_n^{fl}. Clearly, $c_n^{fl} \leq c_n$, while two fl-walks can be concatenated into a single fl-walk (place the bottom vertex of the second on the top vertex of the first). Since each walk has two orientations, this shows that

$$c_n^{fl} c_m^{fl} \leq 2 c_{n+m}^{fl}. \tag{1.71}$$

By cutting a polygon in its bottom and top vertices, two fl-walks are obtained:

$$p_n \leq \sum_{m=1}^{n-1} c_m^{fl} c_{n-m}^{fl} \leq 2n c_n^{fl} \leq 2n c_n. \tag{1.72}$$

Take logarithms, divide by n and let $n \to \infty$. This shows (by equation (1.70)) that

$$\lim_{n\to\infty} \tfrac{1}{n} \log c_n = \lim_{n\to\infty} \tfrac{1}{n} \log c_n^{fl} = \lim_{n\to\infty} \tfrac{1}{2n} \log p_{2n} = \log \mu_d. \tag{1.73}$$

A theta may be created by capping the top edge of a polygon by three edges in a $\sqsupset$-conformation. This shows that $p_n \leq \theta_{n-1,1,3} \leq \theta_{n+3}$. A theta may be turned into a self-avoiding walk by placing one of its vertices of degree 3 at the origin and deleting one edge at each vertex of degree 3 in the appropriate way. One theta can give rise to twelve different walks (two choices for placing a vertex of degree at the origin, three choices for deleting an edge at the origin, and then two choices for deleting an edge at the other vertex of degree 3). This shows that $\theta_n \leq \frac{1}{12} c_{n-2}$. Hence, $p_n \leq \theta_{n+3} \leq c_{n+1}$. Taking logarithms, dividing by n and letting $n \to \infty$,

$$\lim_{n\to\infty} \tfrac{1}{n} \log \theta_n = \log \mu_d. \tag{1.74}$$

A tadpole may be formed by placing the bottom vertex of an fl-walk on the top vertex of a polygon. This shows that $c_n^{fl} p_n \leq T_{n,n} \leq T_{2n}$, and $c_{n+1}^{fl} p_n \leq T_{n+1,n} \leq T_{2n+1}$, where parity effects in T_n are taken into account. By deleting an edge incident on the vertex of degree 3 in the head of a tadpole, a walk is recovered. There are two choices for deleting the edges, and each deletion gives two choices for a root in the walk. Hence, $2T_n \leq 2c_{n-1}$.

A figure eight may be created by concatenating two polygons by placing the top vertex of the first on the bottom vertex of the second. This shows that $p_n^2 \leq f_{2n}$. On the other hand, a figure eight can be turned into a walk by removing two edges from its vertex of degree 4 in the proper way (there are four ways of doing this, and each way gives rise to two walks by choosing an endpoint as root). Hence, $4f_n \leq 2c_{n-2}$.

Dumbbells can be created by concatenating fl-walks and polygons to see that $p_n c_n^{fl} p_n \leq d_{n,n,n} \leq d_{3n}$, or $p_{n+1} c_n^{fl} p_n \leq d_{n+1,n,n} \leq d_{3n+1}$, and also $p_{n+2} c_n^{fl} p_n \leq d_{n+1,n,n} \leq d_{3n+2}$. On the other hand, by deleting two edges in the obvious way, $4d_n \leq 2c_{n-2}$. Together, these arguments show that

$$\lim_{n\to\infty} \tfrac{1}{n} T_n = \lim_{n\to\infty} \tfrac{1}{n} f_n = \lim_{n\to\infty} \tfrac{1}{n} d_n = \lim_{n\to\infty} \tfrac{1}{n} \theta_n = \lim_{n\to\infty} \tfrac{1}{n} c_n = \log \mu_d. \tag{1.75}$$

The above results suggest the following scaling for each of the generating functions of these quantities, similar to equation (1.57):

$$g(t) \sim (1-\mu_d t)^{-\gamma_g}, \text{ for } g \in \{T, f, d, \theta\}, \tag{1.76}$$

for tadpoles (T), figure eights (f), dumbbells (d) and thetas (θ). The walk generating function $C(t)$ is given by equation (1.57), and the polygon generating function by equation (1.59). For the entropic exponent γ of walks, equation (1.69) implies that

$$\gamma = \max\{\alpha_s - 1, \gamma_F, \gamma_D, \gamma_\theta\}. \tag{1.77}$$

That is, at least one of $\alpha_s - 1$, γ_F, γ_D and γ_θ is equal to γ.

1.7 Knotted lattice polygons

Let S be the (topological) circle, and suppose that k is a placement of S in $\mathbb{L}^3$. Then k is a lattice polygon, and its placement induces an injection $k : S \to \mathbb{R}^3$ representing S. The injection k is a *knot* – and the image of k in $\mathbb{R}^3$ will also be called a knot and denoted by k. Since knots in $\mathbb{L}^3$ are polygons, they may be oriented by giving an orientation to the image of k (which is a finite 1-complex in $\mathbb{R}^3$ since it is piecewise linear). Finite piecewise linear embeddings of S in $\mathbb{L}^3$ are realisations of tame knots since they can be extended to an embedding of a solid torus in $\mathbb{R}^3$.

Two embeddings k_1 and k_2 of S are equivalent if there exists an orientation preserving homeomorphism $H : (\mathbb{R}^3, k_1) \to (\mathbb{R}^3, k_2)$. In other words, k_1 and k_2 are equivalent if there is an orientation preserving isotopy taking k_1 to k_2.

An equivalence class of embeddings is a *knot type* denoted by K, and k is a representative of K. A member $k \in K$ is also called a knot.

The collection of all knot types K is denoted $\mathcal{K}(S)$. Amongst the knot types in $\mathcal{K}$ is the *unknot*, whose realisations are homeomorphic to the (planar) geometric circle in $\mathbb{R}^3$. Since S has placements in $\mathbb{L}^3$ which are planar, the equivalence class of planar embeddings also defines the unknot.

In the case of ring polymers, the above defines equivalence classes of knotted ring polymers. Determining the entropy of knotted ring polymers is an instance of the polymer entropy problem.

Polymer entropy of knotted ring polymers is an old problem [118, 174, 175]. The Frisch-Wasserman-Delbruck conjecture [126, 212] proposed that sufficiently

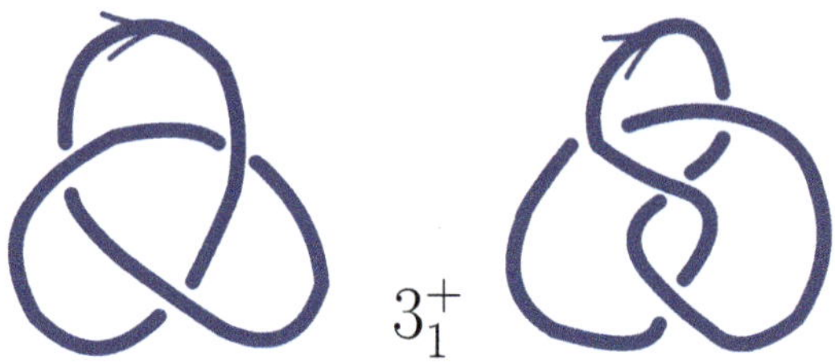

3_1^+

FIG. 1.8. A regular projection of a right-handed trefoil knot is illustrated on the left. On the right is a different regular projection of this knot. The right-handed trefoil knot is denoted by 3_1^+ in Alexander-Briggs notation.

---

long ring polymers will be knotted with probability approaching 1 as the length of the polymers becomes asymptotically long. This was proven for randomly sampled Gaussian polygons in $\mathbb{R}^3$ [343], for equilateral polygons in $\mathbb{R}^3$ [145] and for some continuum models in reference [146].

The Frisch-Wasserman-Delbruck conjecture in $\mathbb{L}^3$ states that lattice polygons (endowed with the uniform measure) will be knotted with probability 1 in the scaling limit. This conjecture was proven in references [466, 539]. There is a considerable body of numerical work done on the Frish-Wasserman-Delbruck conjecture [296, 328, 360, 420, 421, 566]. Experimental studies of knotted ring polymers were done in references [498, 499, 512].

Lattice models of the polymer entropy problem in knotted ring polymers were studied using numerical simulations in references [296, 328, 449, 452, 485], but surprisingly little is known beyond some basic results [301, 456].

1.7.1 *Lattice knots*

Knots (including lattice knots) can be studied by examining their projections into $\mathbb{R}^2$, which are called *knot diagrams.* In figure 1.8 two examples of a knot diagram are shown.

A connected segment (or curve) in a knot k is an *arc* in 3-space, with orientation induced by the orientation of the knot. A *projection* of a tame knot into a geometric plane is the *projected image* of the knot. Short arcs in a tame knot project to finite length curves in the plane.

The projection of a tame knot k in $\mathbb{R}^3$ into a plane with normal unit vector $\hat{\xi}$ may in general have singular points (or *multiple points*, where two or more points in the knot project to the same point). A *double point* in a projection is a multiple point which is the image of exactly two points in the knot.

A double point is *regular* if it is the crossing point in the projection where the projected image of the first arc passes transversely over the image of the second arc. The set of vectors $\hat{\xi}$ where there are singular points which are not double points, or where double points are not regular, is a set of zero measure [82]. Projections of knots where all singular points are regular double points are said to be *regular knot projections.*

FIG. 1.9. Left-handed (−) and right-handed (+) crossings in a regular knot diagram.

A *regular projection* of a knot is a *knot projection* if the curve is oriented and the orientation is projected into the regular projection of the curve, and if, at each double point in the projection, the over-passing arc is indicated (see figure 1.9). This is normally done by removing a small arc from the under-passing arc at the double point in the projection. The double points in a regular projection of k are collectively called the *crossings* in the projection.

Realisations of lattice knots k of type K project to sets of regular knot projections, and amongst these are realisations of k and projected classes of diagrams with a minimal number of crossings (double points). This is a *minimal knot diagram* and its number of crossings is the *minimal crossing number* $C(k)$ of the knot.

The crossings in a knot diagram are signed. This may be detected by using a right-hand rule. Take the over-passing arc at the crossing in the right hand with the thumb pointing in the direction of the arrow orienting the knot. If fingers curl underneath the over-passing arc in the same direction as the under-passing arc, then the crossing is *positive* or *right-handed.* Otherwise, it is *negative* or *left-handed.* Notice that the sign of a crossing is independent of the orientation of the knot.

Knot types are labelled using Alexander-Briggs notation [79]. The unknot, which is the only knot with minimal crossing number equal to zero, is denoted by $0_1 \equiv \emptyset$ in this notation (since it can be realised as a geometric circle with zero crossings in the plane).

The *trefoil knot* is the simplest non-trivial knot type. It has a minimal crossing number equal to 3 (see figure 1.8). Since it is the first knot on three crossings, it is denoted by 3_1 (the first knot on three crossings).

Generally, Alexander-Briggs notation assigns the m-th knot on n crossings as n_m. For example, the figure eight knot is the first (and only) knot on four crossings, and it is denoted by 4_1.

A realisation of a trefoil knot is not homeomorphic to its mirror image if orientation is preserved. The knot is said to be *chiral.* A *right-handed trefoil* has minimal knot projection with three positive crossings (see figure 1.8) – it is denoted by 3_1^+. Its mirror image is the *left-handed trefoil* denoted by 3_1^-. In

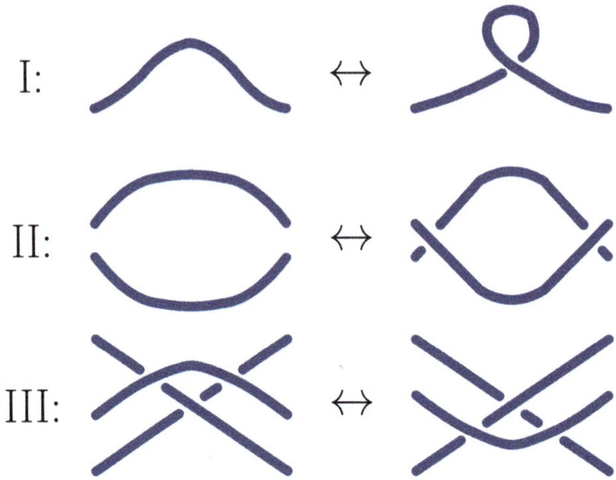

FIG. 1.10. Reidemeister moves. Two knots have the same knot type if and only if their regular knot projections are equivalent under a finite sequence of Reidemeister moves and an orientation preserving homeomorphism of $\mathbb{R}^2$ [492].

contrast the figure eight knot 4_1 is homeomorphic to its mirror image and is said to be *amphicheiral* or *achiral.*

In figure 1.8 two regular projections of the right-handed trefoil knot (the knot 3_1^+) are illustrated. The orientation of the projected knot is indicated by using an arrow. Since lattice knots are piecewise embeddings of the circle in $\mathbb{R}^3$, all lattice knots are tame and have regular projections.

Knot diagrams are manipulated by three Reidemeister moves (see figure 1.10) which involve one, two or three arcs in the projection. Reidemeister's theorem states that two knots have the same knot type if and only if their knot diagrams are equivalent under application of Reidemeister moves and an orientation preserving homeomorphism of $\mathbb{R}^2$ [492].

The *crossing number of a knot type* is the minimal number of crossings in any of the knot projections representing it. The crossing number of the unknot is zero. The trefoil knot 3_1 represented in figure 1.8 has crossing number 3. Knot types are listed in knot tables with increasing crossing numbers [495] in Alexander-Briggs notation.

The *connected sum* of the two knot types K_1 and K_2 is obtained by taking a representative knot of each type, separated by an infinite plane P and with each knot having a short arc in P. These arcs are identified and then deleted. This gives a *compound knot* with type $K_1 \# K_2$ and with *components* of knot types K_1 and K_2. *Prime knot types* cannot be represented as the connected sum of two non-trivial knot types.

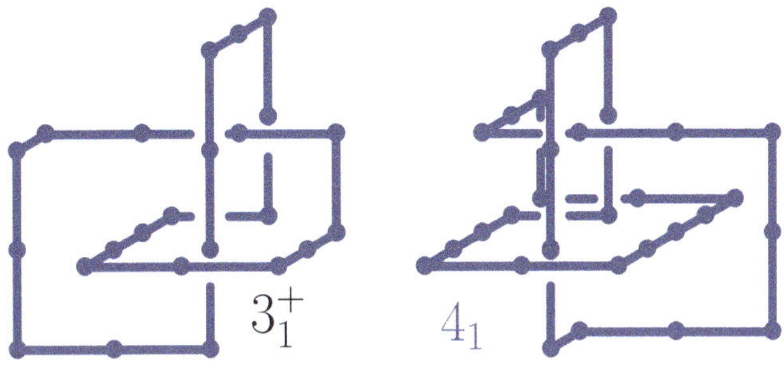

FIG. 1.11. Two projected lattice knots. On the left is shown a right-handed trefoil (3_1^+) of length 24, and on the right is a figure eight knot (4_1) of length 30. These are minimal length representations of knot types 3_1^+ and 4_1 in the cubic lattice – lattice knots of type 3_1^+ and of length less than 24 cannot be realised in $\mathbb{L}^3$, and, similarly, lattice knots of type 4_1 and length less than 30 can not be realised in $\mathbb{L}^3$ [143, 503].

The *square knot* is the connected sum $3_1^+\#3_1^-$ (and it has prime components 3_1^+ and 3_1^-). The *granny knot* is a chiral knot of knot type either $3_1^+\#3_1^+$ or $3_1^-\#3_1^-$.

Compound knot types can be decomposed recursively into a connected sum of prime knots. This decomposition is unique and is a prime decomposition theorem for knots. Define N_K to be the number of prime knots in the decomposition of a knot type K into prime knots types. If K is a prime knot type, then $N_K = 1$. If K is not a prime knot type, then it is a compound knot type, and $N_K > 1$. For example, if $K = 3_1^+\#3_1^-$, then $N_K = 2$.

Let $p_n(K)$ be the number of lattice polygons (equivalent under translation in $\mathbb{L}^3$) of knot type K. Then $p_n(K) \le p_n$. Concatenating a polygon of length n and knot type K, and a polygon of length m and knot type L (see figure 1.5 and equation (1.7)) gives the supermultiplicative inequality [539]

$$p_n(K)p_m(L) \le 2p_{n+m}(K\#L), \tag{1.78}$$

where $K\#L$ is the connected sum of the knot types K and L.

In the case that $K = L = 0_1 = \emptyset$ (the unknot), then

$$p_n(\emptyset)p_m(\emptyset) \le 2p_{n+m}(\emptyset). \tag{1.79}$$

Since $p_n(\emptyset) \le p_n$, it follows by theorems A.1 and 1.4 that the growth constant $\mu_\emptyset$ of unknotted polygons exists so that $p_n(\emptyset) = \mu_\emptyset^{n+o(n)}$ for all even values $n \in 2\mathbb{N}$.

Theorem 1.8 *The limit*

$$\lim_{n\to\infty} \tfrac{1}{n} \log p_n(\emptyset) = \log \mu_\emptyset$$

exists and $\mu_\emptyset \leq \mu_3$. *In addition,* $p_n(\emptyset) \leq 2\mu_\emptyset^n$. □

It is known that $\mu_\emptyset < \mu_3$ (see corollary (7.32)), a result based on Kesten's pattern theorem [348] for self-avoiding polygons [466, 539] (see appendix C).

1.7.2 *Lattice knots and knot probabilities*

The probability of an unknotted polygon in the set of lattice polygons of length n is $\Pr_n(\emptyset) = p_n(\emptyset)/p_n$. By theorem 1.8,

$$\Pr_n(\emptyset) = \left(\frac{\mu_\emptyset}{\mu_3}\right)^{n+o(n)}. \qquad (1.80)$$

Since $\mu_\emptyset < \mu_3$, this shows that unknotted polygons are exponentially rare in the set of lattice polygons. Numerical simulations [328, 585] give the estimate

$$\log \mu_3 - \log \mu_\emptyset = (4.15 \pm 0.32) \times 10^{-6}. \qquad (1.81)$$

In the face-centred cubic lattice, $\log \mu_3 - \log \mu_\emptyset = (7.6 \pm 0.9) \times 10^{-6}$ [296, 332].

The growth constants for polygons of knot type K are given by

$$\log \upsilon_K = \liminf_{n\to\infty} \tfrac{1}{n} \log p_n(K) \leq \limsup_{n\to\infty} \tfrac{1}{n} \log p_n(K) = \log \mu_K. \qquad (1.82)$$

It is not known whether these limits exist (that is, $\upsilon_K = \mu_K$), except for the unknot (when $K = \emptyset$). By equation (1.78), $\mu_\emptyset \leq \upsilon_K$, and $\mu_K \leq \mu_{K\#L}$ [524].

Let K^N denote a compound knot type composed of N copies of K. Choose K in equation (1.78) to be $K^{\lfloor \epsilon n \rfloor}$ and choose L to be $K^{\lfloor \epsilon m \rfloor}$, for $\epsilon > 0$ and small. Then

$$p_n\left(K^{\lfloor \epsilon n \rfloor}\right) p_m\left(K^{\lfloor \epsilon m \rfloor}\right) \leq 2 p_{n+m}\left(K^{\lfloor \epsilon n \rfloor + \lfloor \epsilon m \rfloor}\right). \qquad (1.83)$$

Then the limit

$$\log \mathcal{K}(\epsilon) = \lim_{n\to\infty} \tfrac{1}{n} \log p_n\left(K^{\lfloor \epsilon n \rfloor}\right) \qquad (1.84)$$

exists (see section 3.3.1 and theorem 3.9).

$\mathcal{K}(\epsilon) \leq \mu_3$, and Kesten's pattern theorem (see appendix C, theorem C.5) implies that $\mathcal{K}(\epsilon) < \mu_3$. On the other hand, $p_n(K^{\lfloor \epsilon n \rfloor}) = p_n(\emptyset)$ if $\epsilon < \frac{1}{n}$. This shows that $\mathcal{K}(0) = \mu_\emptyset$.

Observe that $p_n(K^{\lfloor \epsilon n \rfloor}) = p_n(K)$ if $\lfloor \epsilon n \rfloor = 1$. Hence, there exists a sequence $\langle \epsilon_n \rangle$ such that $\epsilon_n \to 0$ and $\limsup_{n\to\infty} \frac{1}{n} \log p_n(K^{\lfloor \epsilon_n n \rfloor}) = \log \mu_K \geq \log \mu_\emptyset$. This implies that $\lim_{\epsilon\to 0^+} \mathcal{K}(\epsilon) = \mathcal{K}(0^+) \geq \log \mu_\emptyset$.

If $\mathcal{K}(0^+) = \mathcal{K}(0)$, then it follows that $\lim_{n\to\infty} \frac{1}{n} \log p_n(K) = \log \mu_K = \log \mu_\emptyset$ exists. This is an open problem.

1.7.3 *The scaling of lattice knots*

A reasonable assumption for the scaling of $p_n(K)$ suggested by equation (1.18) is

$$p_n(K) \simeq A_K n^{\alpha_K - 3} \mu_K^n. \qquad (1.85)$$

This defines α_K, the *entropic exponent of lattice knots of knot type* K. Numerical estimates of the growth constant μ_K are given in table 1.5. For the unknot, $\alpha_\emptyset \approx \alpha_s = 0.237(4)$ [301].

TABLE 1.5. Amplitudes of knotted polygons [301]

K	μ_K	ν_K	B_K	N_K
0_1	4.6852	0.588(8)	0.103(28)	−−
3_1	4.6832	0.599(8)	0.1032(16)	0.86(14)
4_1	4.6833	0.603(10)	0.0967(22)	0.83(14)
6_2	4.6844	0.586(10)	0.0842(12)	0.93(7)
$3_1^+\#3_1^+$	4.6800	0.604(20)	0.0889(42)	−−
$3_1^+\#4_1$	4.6841	0.596(12)	0.089(12)	−−

The mean square radius of gyration of lattice knots has asymptotic behaviour

$$\langle R_n^2 \rangle_K \simeq B_K n^{2\nu_K}, \tag{1.86}$$

where B_K is an amplitude, and ν_K is the metric exponent.

A scaling argument relating α_K to the polygon entropic exponent α_s (see equation (1.18)) is based on figure 1.12.

Suppose that ω is a lattice knot of prime knot type K. Let S be a geometric sphere which intersects ω in exactly two points. S cuts ω into two arcs, each of which can be closed by a curve on S into an embedded circle. One of these arcs closed by a curve on S is a knot of type K. Let its length be m_K and suppose that ω has length n edges. Notice that there is always an intersection of S and K such that K is cut into two arcs, one of which closes in a knot of type K.

Define M_K to be the infimum of m_K over all possible intersections of geometric spheres S which cut ω in exactly two points into two arcs, one of which closes into a knot of type K. Define n_K to be the expected value of M_K taken uniformly over all lattice knots of length n and of knot type K. Intuitively, n_K is the expected length of the shortest 'knotted arc' in the polygon.

Heuristically, there are two possibilities. Either $\frac{n_K}{n} \to C_K$ for some constant $C_K > 0$ (so that $n_K = \theta(n)$), or, if not, then $n_K = o(n)$.

Consider the case $n_K = o(n)$ first. In this case the shortest knotted arc becomes a root on an unknotted closed curve in the scaling limit. This shows that $p_n(K) \sim n p_n(\emptyset)$ and so it is expected that $\mu_K = \mu_\emptyset$ and $\alpha_K = \alpha_\emptyset + 1$ in equation (1.85). In addition, $\langle R_n^2 \rangle_K \simeq B_K (n - n_K)^{2\nu_\emptyset} \simeq B_\emptyset n^{2\nu_K}$. This shows that $\nu_K = \nu_\emptyset$, and $B_\emptyset = B_K$, because $\frac{n_K}{n} \to C_K = 0$. That is, the 'shortest knotted arc' containing the knot K shrinks to a point on an unknotted curve in the scaling limit, and the effects of K are seen in corrections to scaling.

If $n_K = \theta(n)$, then the average length of the 'shortest knotted arc' of ω containing K grows proportional to n. The expected diameter $d_n(K)$ of the smallest sphere containing the shortest knotted arc grows at least as fast as $\theta(n^{1/3})$ and, on average, at the rate $\theta(n_K^\nu)$. If the root mean square radius of gyration $\langle R_n^2 \rangle_K^{1/2}$

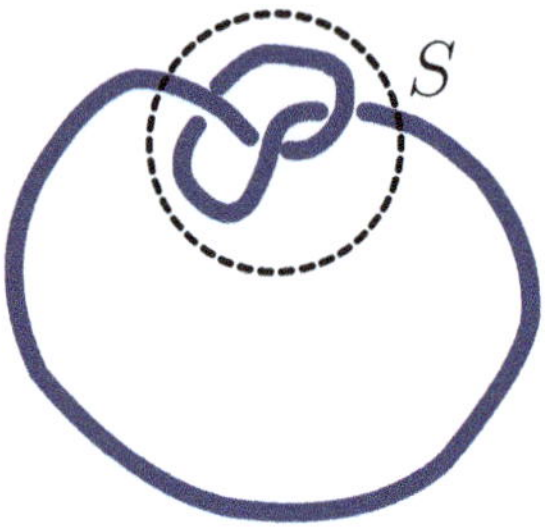

FIG. 1.12. A schematic diagram of a knotted polygon with a knotted ball-pair consisting of a 3-ball S containing a knotted arc (both endpoints of the arc are in the boundary ∂S of S). The boundary ∂S separates the knotted polygon into two arcs, one of which will be knotted if it is closed by an arc on ∂S.

---

of ω grows faster than $d_n(K)$, then in the scaling limit the shortest knotted arc becomes a root on an unknotted closed curve of asymptotic length $(1-C_K)n$. In this case $\nu_\emptyset = \nu_K$, and $(1-C_K)^{2\nu} B_\emptyset = B_K$. This shows that $B_K < B_\emptyset$. Since the scaling limit is a rooted unknotted closed curve (the knotted arc can occur at $O(n)$ locations on a closed curve), it follows that $\alpha_K = \alpha_\emptyset + 1$.

Finally, there is the possibility that $\langle R_n^2 \rangle_K^{1/2}$ does not grow faster than $d_n(K)$. In this case it is not possible to derive a relationship between the critical exponents or amplitudes of lattice knots and the unknotted lattice knot.

If the last possibility is ignored, then the above indicates that

$$\nu_K = \nu_\emptyset, \quad \text{and } \alpha_K = \alpha_\emptyset + 1 \tag{1.87}$$

while $A_K \leq A_\emptyset$, and $B_K \leq B_\emptyset$.

This argument generalises for compounded knot types K. If K has N_K prime components, then it should be the case that

$$\nu_K = \nu_\emptyset \ \text{ and } \alpha_K = \alpha_\emptyset + N_\mathrm{K} \tag{1.88}$$

while $A_K \leq A_\emptyset$, and $B_K \leq B_\emptyset$.

Numerical evidence that $\nu_K = \nu_\emptyset$, and $B_K = B_\emptyset$, was presented in reference [331]. Numerical evidence that $A_K = A_\emptyset$, and $\alpha_K = \alpha_\emptyset + N_\mathrm{K}$, can be found in references [125, 451, 452]; see references [301, 313, 322, 485] for additional results, and in particular reference [18] for high quality numerical data.

Definitions of a prime knot contained in a knotted arc in a lattice polygon are given in reference [405], where numerical data suggest that the length of this arc grows on average as $n^{0.75}$ in a polygon of length n. Numerical estimates for μ_K, ν_K, B_K and N_K are given in table 1.5 [301].

Monte Carlo simulations of lattice knot statistics have also considered the issue of detecting knot types efficiently [124, 328, 420, 422, 449, 485, 566] and computing the incidence of knot types as a function of length or other parameters [450, 545]. Knotted arcs in self-avoiding walks were addressed in references [327]

TABLE 1.6. Estimates of $M_{K/L}$

K/L	$M_{K/L}$	K/L	$M_{K/L}$
$3_1/4_1$	27.0 ± 2.2	$4_1/5_1$	16.2 ± 1.5
$3_1/5_1$	399 ± 44	$4_1/5_2$	9.0 ± 1.2
$3_1/5_2$	216 ± 24	$5_2/5_1$	2.4 ± 0.3

and [405], while random linking in confined geometries was studied in reference [444]; see reference [456] for a review.

Theoretical work on the effect of knots on the physical properties of polymers was done by Edwards [174, 175] and de Gennes [118]. Experimental work on the occurrence of knots as a function of the degree of polymerisation can be found in reference [512] (see references [498, 499, 551, 573] for more experimental results).

1.7.4 *Knot frequency*

The data in table 1.5 as well as the scaling assumption in equation (1.85) and the arguments leading to equations (1.87) and (1.88) suggest that, if K is a prime knot type, then

$$\lim_{n\to\infty} \frac{p_n(K)}{n p_n(\emptyset)} = 1. \tag{1.89}$$

Existence of this limit (taken through even integers) is an open question, but it does propose that the *amplitude ratio* of prime knot types to the unknot is always equal to 1.

The amplitude ratios of knotted polygons of types K and L are defined by

$$\lim_{n\to\infty} \frac{n^{-N_K} p_n(K)}{n^{-N_L} p_n(L)} = \frac{A_K}{A_L} = M_{K/L}\,. \tag{1.90}$$

This ratio measures the frequency of knot types K and L relative to one another in the asymptotic regime.

Numerical estimates of $M_{K/L}$ are given in table 1.6 for prime knot types to five crossings [325]. The value for $M_{3_1/4_1}$ suggests that trefoil components are approximately twenty-seven times more likely to occur than figure eight components in asymptotically long polygons. Lower estimates for these ratios were obtained in reference [15], where it was estimated that $M_{3_1/4_1} = 22.5(4)$, $M_{3_1/5_1} = 232(9)$, and $M_{3_1/5_2} = 150(5)$.

1.7.5 *Minimal length of lattice knots*

There is a minimal value of n, denoted n_K, such that $p_n(K) = 0$ for all $n < n_K$ and $p_n(K) > 0$ if $n = n_K$. The number n_K is the *minimal length* of lattice knots of type K [318].

Since the girth of $\mathbb{L}^3$ is 4, and polygons of length 4 are unknotted, it follows that $n_\emptyset = 4$. In addition, $p_4(\emptyset) = 3$, since there are three distinct

TABLE 1.7. Minimal length lattice knots

K	n_K	Number	Symmetry
$\emptyset$	4	3	3^1
3_1^+	24	1664	$4^1 8^2 12^7 24^{65}$
4_1	30	3648	24^{152}
5_1^+	34	3336	$12^6 24^{136}$
5_2^+	36	57456	$12^4 24^{2392}$
6_1^+	40	3072	$12^2 24^{127}$
6_2^+	40	16416	24^{648}
6_3	40	3552	24^{148}
$3_1^+\#3_1^+$	40	15312	$12^2 24^{637}$
$3_1^+\#3_1^-$	40	109440	24^{4560}

polygons of length 4. Determining n_K for non-trivial knot types is in general quite difficult. It is known that $n_{3_1} = 24$ [143], $p_{24}(3_1^+) = 1664$ [503], and $p_{24}(3_1) = p_{24}(3_1^+) + p_{24}(3_1^-) = 3328$. It is also known that $n_{4_1} = 30$, and $n_{5_1} = 34$ [503].

Each collection of minimal length lattice knots of type K can be partitioned into symmetry classes (equivalent under rotations in $\mathbb{L}^3$). There are three minimal length unknotted polygons of length 4 in the same symmetry class, since they can be rotated into one another. This is denoted by 3^1 (one class of three members). The trefoil knot has symmetry classes $4^1 8^2 12^7 24^{65}$ (one class of four members, two classes of eight, seven classes of twelve, and sixty-five classes each with twenty-four members) [503].

Numerical studies using Monte Carlo algorithms have been used to sieve and count minimal length lattice knots [326, 503]; estimates are given in table 1.7.

The minimal length n_K of knot types may be used as a measure of knot complexity [318]. This function is subadditive over the connected sum of knot types.

Lemma 1.9 *Let K and L be two knot types. Then $n_{K\#L} \leq n_K + n_L$.*

Proof Let ω_1 and ω_2 be polygon representatives of lengths n_K and n_L of the two knot types K and L, respectively. Concatenate ω_1 and ω_2 (see figure 1.5) to obtain a polygon of length $n_K + n_L$ and knot type $K\#L$. □

The existence of positive lower bounds on n_K is given by the following lemma.

Lemma 1.10 *Let K and L be two knot types. There exist constants $\alpha_K \geq 0$ and $\alpha_L \geq 0$ such that*

$$n_{K\#L} \geq \alpha_K + \alpha_L, \quad \textit{and}\, n_{K^N} \geq N\alpha_K,$$

where K^N *is the connected sum of* N *knots of type* K*. If* $K \neq \emptyset$*, then* $\alpha_K > 0$.

Proof The bridge number of a projection of a knot K is found by considering a regular projection of the knot and counting the number of over-passes. The *bridge index* of K, denoted $b(K)$, is the least bridge number taken over all possible regular projections of K [507].

It is known that $b(K)-1$ is additive over the connected sum of tame knot types: if K and L are tame knots, then $b(K)+b(L) = b(K\#L)+1$ [495]. Put $\alpha_K = 4(b(K)-1)$. Then α_K is additive over the connected sum of knot types. Moreover, if $K \neq \emptyset$, then $b(K) \geq 2$ so $\alpha_K > 0$.

For each tame knot type K, $2\pi b(K) = c(K)$, where $c(K)$ is the infimum of the total curvature of every realisation of K [495].

Let $C(K)$ be the curvature of a given realisation of K. Then $c(K) \leq C(K)$, and $C(K) \geq 2\pi b(K)$. If a lattice knot is considered, then contributions to the curvature are in units of $\frac{\pi}{2}$, since excluded angles between successive edges only have sizes 0 or $\frac{\pi}{2}$.

A lattice knot of curvature $C(K)$ has at least $\frac{2}{\pi}C(K)$ edges, one for each contribution of $\frac{\pi}{2}$ to $C(K)$. Thus,

$$n_K \geq \tfrac{2}{\pi}C(K) \geq \tfrac{2}{\pi}c(K) = 4b(K) > \alpha_K.$$

From the additivity of α_K, it follows that

$$n_{K\#L} \geq \alpha_{K\#L} = \alpha_K + \alpha_L.$$

The second inequality is similarly obtained. □

A corollary of the last lemma is the following theorem.

Theorem 1.11 *Let* K *be an arbitrary knot type. Then the minimal lattice edge index* κ_K *of* K *is defined by*

$$\lim_{N\to\infty} \tfrac{1}{N} n_{K^N} = \inf_{N\geq 1} \tfrac{1}{N} n_{K^N} = \kappa_K.$$

It is also the case that $n_{K^N} \geq N\kappa_K$ *and, if* $K \neq \emptyset$*, then* $\kappa_K > 0$.

Proof By lemma 1.9, $n_{K^N\#K^M} \leq n_{K^N} + n_{K^M}$. By lemma 1.10, n_{K^N} grows at least linearly with N if K is not the unknot. By theorem A.1 (in appendix A) the result follows. □

It is known that $\kappa_{3_1} \leq 17$ [318]. There are no estimates of κ_K available when K is a non-trivial knot type.

1.7.6 *Flory theory for lattice knots*

Flory theory for lattice knots minimises an approximate free energy to estimate the metric exponent [241]. The topological constraint (the knot type) is accommodated by confining the lattice polygon to a tube T of diameter r tied into a knot. For example, in the case of the unknot, the tube is a geometric torus (of genus 1).

Assume that T has length L and diameter D. The ratio $p = \frac{L}{D}$ is the *extent* of the tube, and the longer the tube, or the smaller D, the larger its extent. More complex knot types will have a larger extent [144].

The average diameter of a knotted polygon is denoted by r. The volume of T is proportional to $LD^2 = cr^3$, for some constant c. This shows that

$$L \simeq c^{1/3} r p^{2/3}, \text{ and } D \simeq c^{1/3} r p^{-1/3}. \tag{1.91}$$

The Flory arguments leading to equation (1.34) give the free energy

$$F_n(r) = B\frac{n^2}{r^3} + AT\frac{r^2}{4D'n} - CT + f_n(r), \tag{1.92}$$

where $\{A, B, C\}$ are constants, T is the temperature, D' is a diffusion constant, and where $f_n(r)$ is a contribution to $F_n(r)$ due to the fact that the lattice walk is stretched along the tube and confined to it.

The stretching of the walk along its confining tube gives an elastic energy contribution to the free energy. A Hooke term $\frac{L^2}{R^2}$ should be expected, where $R \propto \sqrt{n}$ is the Gaussian drift of the walk in the tube; the walk is assumed to break up into independent balls of size R arranged along the tube [117] (these are called Pincus balls [465]).

A second contribution to $f_n(r)$ is noted since the walk is squeezed by the walls of the tube, and this limits the size of the independent balls. This is similarly an elastic energy proportional to $\frac{R^2}{D^2}$.

Putting these terms together with equation (1.92) gives

$$F_n(r) = B\frac{n^2}{r^3} + AT\frac{r^2}{4D'n} - CT + \frac{k_1 L^2}{R^2} + \frac{k_2 R^2}{D^2}. \tag{1.93}$$

Assuming that each ball is a Gaussian chain of size $R \simeq b\sqrt{n}$ and using equation (1.91) gives

$$F_n(r) \simeq B\frac{n^2}{r^3} + AT\frac{r^2}{4D'n} - CT + \frac{k_1 c^{2/3} r^2 p^{4/3}}{b^2 n} + \frac{k_2 b^2 n p^{2/3}}{c^{2/3} r^2}. \tag{1.94}$$

Since $r = O(n^\nu)$, and $\nu \geq \frac{1}{2}$, the last term may be ignored for large values of n. Minimising this to r and solving for r gives

$$r \simeq C\left(\frac{(c_0 + c_1 p^{4/3})^{4/5} n^{3/5}}{c_0 + c_1 p^{4/3}}\right), \tag{1.95}$$

where c_i and C are constants. This recovers the Flory value of $\nu = \frac{3}{5}$ for knotted polygons, but the amplitude is modified.

For small values of n, the knot is swollen so that p is large; in this case, the above reduces to

$$r \simeq C_1 n^{3/5} p^{-4/15} \tag{1.96}$$

for a constant C_1 and where p decreases with increasing n. For large n, the argument in figure 1.12 suggests the knot is contained in a small ball within the tube in the scaling limit. This implies that $p \to$ a constant in the scaling limit. In other words, absorbing the constant into the coefficient (and rescaling p), then expanding about $p = 1$, the above crosses over into

$$r \simeq C_2 n^{3/5} (1 - O(p-1)) \tag{1.97}$$

for a constant C_2 as p approaches 1, for n asymptotic large.

For minimal length lattice knots (see section 1.7.5), $n = p = n_K$, and equation (1.96) simplifies to

$$r \simeq C_1 n_K^{1/3}. \tag{1.98}$$

This is not an unexpected result, since minimal length lattice knots tend to minimise surface area and assume compact conformations of volume n, and diameter proportional to $n_K^{1/3}$.

2

LATTICE MODELS OF BRANCHED POLYMERS

Lattice animals are connected subgraphs of $\mathbb{L}^d$ and are models of branched polymer entropy.

An animal A is *weakly embedded* in $\mathbb{L}^d$ if it is a single vertex, or is connected and is the union of a set of edges in $\mathbb{L}^d$ (see figure 2.1). Weakly embedded animals are called *bond-animals* or *edge-animals*. An edge in A is a *cut-edge* if deleting it leaves A connected. That is, $\langle \vec{v} \sim \vec{w} \rangle \in A$ is a cut-edge of an animal A if $A \setminus \{\langle \vec{v} \sim \vec{w} \rangle\}$ has more components than A.

If every edge in A is a cut-edge, then A is *acyclic* and is a *tree* (see figure 2.2). Edges in A which are not cut-edges are *cycle edges*. A set of cycle edges in an animal is independent if deleting them leaves a connected animal. The *cyclomatic index* of A is the largest number of independent cycle edges in A. A tree has cyclomatic index equal to 0 (see figure 2.3).

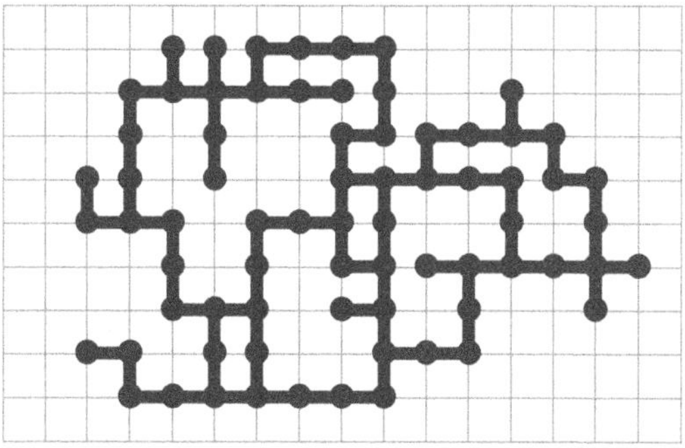

FIG. 2.1. A lattice animal in $\mathbb{L}^2$. Lattice animals are connected subgraphs of a lattice (for example, connected subgraphs of $\mathbb{L}^d$). The cyclomatic index of a lattice animal is the number of independent cycles in its underlying graph. The cyclomatic index of this animal is 6.

The degree of a vertex $\vec{v} \in A$ is denoted by $\delta(\vec{v})$ and is the total number of edges in A incident with $\vec{v}$. That is, it is the number of edges of the form $\langle \vec{v} \sim \vec{u} \rangle \in A$. Notice that $1 \leq \delta(\vec{v}) \leq 2d$. If $\delta(\vec{v}) = 1$, then $\vec{v}$ is an *end-vertex* and the edge $\langle \vec{v} \sim \vec{u} \rangle \in A$ incident with $\vec{v}$ is a *leaf*.

An animal A is *strongly embedded* if, for any pair of vertices $\vec{v}, \vec{w} \in A$ such that $\langle \vec{v} \sim \vec{w} \rangle \in \mathbb{L}^d$, it is necessarily the case that $\langle \vec{v} \sim \vec{w} \rangle \in A$. Such animals are

The Statistical Mechanics of Interacting Walks, Polygons, Animals and Vesicles, 2nd edition, E.J. Janse van Rensburg. © E.J. Janse van Rensburg. Published in 2015 by Oxford University Press.

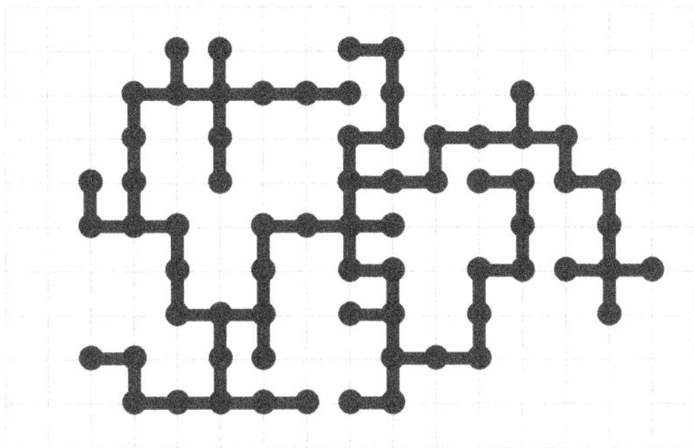

FIG. 2.2. A lattice tree is an acyclic lattice animal. The underlying graph of a lattice tree is an abstract tree with cyclomatic index equal to 0.

---

connected section graphs in $\mathbb{L}^d$ and are called *site-animals*. An acyclic site-animal is a *site-tree*.

A *node* in an animal A is a vertex $\vec{v}$ with $\delta(\vec{v}) \neq 2$. In general, an animal A is composed of nodes joined by self-avoiding walks into a connected branched structure. The nodes are *branch points*, and the self-avoiding walks between nodes are *branches*. More precisely, if $\vec{v}$, and $\vec{w}$, are two nodes in A, then there is a path $\omega = \langle \vec{u}_0, \vec{u}_1, \ldots, \vec{u}_n \rangle$ such that $\langle \vec{u}_{i-1} \sim \vec{u}_i \rangle \in A$ for $i = 1, 2, \ldots, n$, where $\vec{v} = \vec{u}_0$, and $\vec{w} = \vec{u}_n$. If ω is a self-avoiding walk in A such that the degrees of $\vec{u}_j$ are $\delta(\vec{u}_j) = 2$ for $j = 1, 2, \ldots, n-1$, but $\delta(\vec{u}_0) \neq 2$ and $\delta(\vec{u}_n) \neq 2$, then ω is a *branch* of A.

The branches of A join the nodes into a structure which has the connectivity of an abstract graph $\mathcal{G}$ (where two nodes are adjacent if they are the endpoints of a branch). The animal A is said to have the *connectivity* or *topology* of $\mathcal{G}$ and is a *network*. The network is *monodispersed* or *uniform* if all its branches have the same length, or otherwise is *polydispersed*. Networks may be weakly or strongly embedded in $\mathbb{L}^d$.

Nodes in a network are either end-vertices of degree $\delta = 1$ or are nodes of degree $\delta \in \{3, 4, \ldots, 2d\}$ (where several branches meet). The *size* of a network is its total number of edges n. The strict definition of uniformity in networks of size n may be relaxed to networks where branches are almost the same length (within a constant difference) [502] or classes of networks of all branches of length $O(n)$. Such networks have the same scaling properties as uniform networks of the same topology.

In polydisperse networks the branches fluctuate in size from almost zero to a maximum which is $O(n)$. If there are m branches, then this increases the entropy by a factor $O(n^{m-1})$.

2.1 Lattice animals and lattice trees

Bond- or edge-animals are usually counted by *size* (number of edges) but can also be counted by *order* (number of vertices). This gives two ensembles of bond-animals. Similarly, site-animals are usually counted by order but may instead be counted by size. This gives four different ensembles for bond- and site-animals, which do still belong to the same universality class. There are no simple relationships between these ensembles of bond- and site-animals counted by size or order.

Bond-animals (or just *animals* (see figure 2.1)) are counted up to equivalence under translation in the lattice. Let a_n be the number of animals of size n edges. Then $a_0 = 1$, $a_1 = 2$, $a_2 = 6$, $a_3 = 22$, and $a_4 = 88$ in $\mathbb{L}^2$. Similarly, let t_n be the number of bond-trees of size n edges so that $t_n = a_n$ if $n \leq 3$, and $t_4 = 87$ in $\mathbb{L}^2$. Then $c_n \leq 2t_n \leq 2a_n$ so a_n and t_n increase at least exponentially with n. The problem of counting lattice animals is related to cell growth problems [353, 354].

The *bottom vertex* $\vec{b}(A)$ and *top vertex* $\vec{t}(A)$ of an animal A are obtained by a lexicographic ordering of the vertices in ω.

Lemma 2.1 *There exists a finite constant $K > 0$ such that $a_n \leq K^n$ for $n \in \mathbb{N}$. That is, a_n grows exponentially with n (since $p_n \leq a_n$).*

Proof Let A be an animal of size n. Label the bottom vertex of A as 1.

Label the rest of the vertices in the animal recursively by starting from the vertex with label 1. Suppose that the current vertex being examined has label ℓ (initially $\ell = 1$) and suppose the last label assigned is i. Consider the unlabelled vertices adjacent to the current vertex ℓ and label them in lexicographic order as $i+1, i+2, \ldots$. Then, increment $\ell \to \ell+1$ and continue until all vertices are assigned a label. This gives a unique labelling to each animal A.

The above labelling of A is a (unique) coding of A by $2dn$ binary digits b_{ij} ($1 \leq i \leq n$, and $1 \leq j \leq 2d$). The digits are assigned as follows. The digit b_{ij} corresponds to the vertex with label i. If there is an edge in the j-th direction ($1 < j \leq 2d$) to a vertex adjacent to i and with label larger than i, then $b_{ij} = 1$. Otherwise, $b_{ij} = 0$. Since there are n edges, this shows that exactly n of the $2dn$ digits b_{ij} are equal to 1; the rest are 0.

The animal A can be recovered from the b_{ij}. Label a vertex 1 and consider b_{1j}. If $b_{1j} = 1$, then add an edge in the j-th direction. Label the new vertices created in this way from the lexicographic least by $2, 3, \ldots$. Proceed recursively by considering next the vertex with label 2 and the digits b_{2j}.

The number of binary sequences b_{ij} is an upper bound on the number of animals. Since the number of binary sequences of length $2dn$ with exactly n digits equal to 1 is $\binom{2dn}{n}$, it follows that $a_n \leq \binom{2dn}{n} \leq K^n$ (see corollary D.2 in appendix D). □

Two animals, A_1 and A_2, can be concatenated into a single animal. This is done as follows: let $\vec{t}(A_1)$ be the lexicographic top vertex of A_1, and $\vec{b}(A_2)$ be the lexicographic bottom vertex of A_2. Place A_2 such that $\vec{b}(A_2) = \vec{t}(A_1) + \vec{e}_1$. Inserting a single edge $\langle \vec{t}(A_1) \sim \vec{b}(A_2) \rangle$ joins the animals. If A_1 has size n, then it can be chosen in a_n ways, and, similarly, if A_2 has size m, then it can be chosen in a_m ways. The concatenated animal has size $n+m+1$, and it can have at most a_{n+m+1} conformations. This shows that

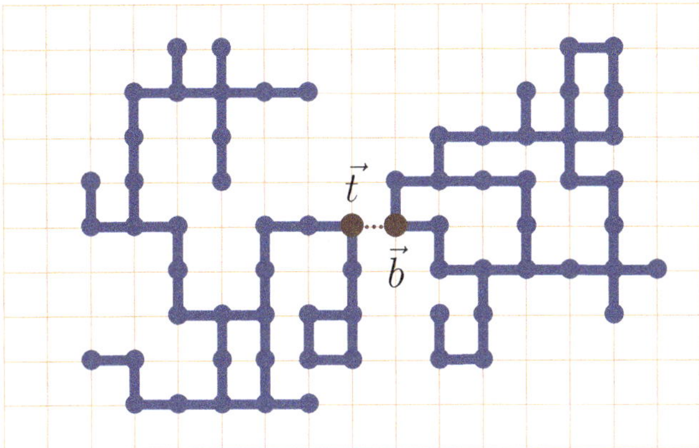

FIG. 2.3. Two lattice animals can be concatenated by placing the top vertex $\vec{t}$ of the first next to the bottom vertex $\vec{b}$ of the second. The animals are joined into a single animal by inserting the edge $\langle \vec{t} \sim \vec{b} \rangle$. If the first animal has size n, then it can be chosen in a_n ways and, if the second animal has size m, then it can be chosen in a_m ways.

$$a_n a_m \leq a_{n+m+1}. \tag{2.1}$$

That is, a_{n-1} is a supermultiplicative function on $\mathbb{N}$. A similar argument shows that

$$t_n t_m \leq t_{n+m+1} \tag{2.2}$$

for the number of bond-trees in $\mathbb{L}^d$. Together with lemma 2.1, a corollary of theorem A.1 in appendix A is the existence of growth constants for trees and animals (see references [354, 357]).

Theorem 2.2 *There exist growth constants $\tau_d > 0$ and $\lambda_d > 0$ in $\mathbb{L}^d$ defined by*

$$\lim_{n\to\infty} \tfrac{1}{n} \log t_n = \sup_{n>0} \tfrac{1}{n} \log t_{n-1} = \log \tau_d$$

and

$$\lim_{n\to\infty} \tfrac{1}{n} \log a_n = \sup_{n>0} \tfrac{1}{n} \log a_{n-1} = \log \lambda_d.$$

Thus, $t_{n-1} \leq \tau_d^n$ and $a_{n-1} \leq \lambda_d^n$. Moreover, $\tau_d \leq \lambda_d < \infty$ by lemma 2.1. □

Theorem 2.3 $\tau_{d-1} \leq \tau_d - 2 + \frac{4}{\tau_d}$, *and* $\lambda_{d-1} \leq \lambda_d - 2 + \frac{4}{\lambda_d}$.

Proof Let $t_n^{(d)}$ the number of trees in $\mathbb{L}^d$ and let $\mathbb{L}^{d-1}$ be the $(d-1)$-dimensional sub-lattice of $\mathbb{L}^d$ orthogonal to $\vec{e}_d$ and containing $\vec{0}$.

Let T be a tree in $\mathbb{L}^{d-1}$. Choose $\lfloor \epsilon n \rfloor$ vertices $\vec{v} \in T$ and add the edge $\langle \vec{v} \sim (\vec{v}+\vec{e}_d)\rangle$ or $\langle \vec{v} \sim (\vec{v}-\vec{e}_d)\rangle$ at each. Since there are $t_n^{(d-1)}$ choices for T, and the outcome is a tree in $\mathbb{L}^d$, this shows that

$$\binom{n+1}{\lfloor \epsilon n \rfloor} 2^{\lfloor \epsilon n \rfloor} t_n^{(d-1)} \leq t_{n+\lfloor \epsilon n \rfloor}^{(d)}.$$

Take the $\frac{1}{n}$-power of this and take $n \to \infty$. This gives

$$\left(\frac{2^\epsilon \tau_d^{-\epsilon}}{\epsilon^\epsilon (1-\epsilon)^{1-\epsilon}} \right) \tau_{d-1} \leq \tau_d.$$

The factor on the left is equal to $1 + \frac{2}{\tau_d} > 1$ when $\epsilon = \frac{2}{2+\tau_d}$. This gives $\tau_{d-1} \leq \frac{\tau_d}{1+2\tau_d^{-1}} \leq \tau_d - 2 + \frac{4}{\tau_d}$ as claimed. This completes the proof for trees. A similar proof works for λ_d. □

It is known that $\tau_d < \lambda_d$ [220, 402] (see section 11.1.2).

Let A be an animal and suppose that $\langle \vec{v} \sim \vec{w} \rangle \notin A$. Then $\langle \vec{v} \sim \vec{w} \rangle$ is a *perimeter edge* of A if either $\vec{v} \in A$ or $\vec{w} \in A$ (or both). If both $\vec{v} \in A$ and $\vec{w} \in A$, then $\langle \vec{v} \sim \vec{w} \rangle$ is a *contact* in A. If exactly one of $\vec{v} \in A$ or $\vec{w} \in A$, then $\langle \vec{v} \sim \vec{w} \rangle$ is a *solvent contact.*

The *perimeter* of a lattice animal is the set of all its perimeter edges. The perimeter is also called the *hull* of the animal. If an animal has s solvent contacts and k contacts, then its perimeter has size $\rho = s + k$.

If an animal has size n edges, and order v vertices, cyclomatic index equal to c with s solvent contacts and k contacts, then the numbers $\{v, n, c, s, k\}$ are not independent. By Euler's relation and hand shaking,

$$v - n + c = 1, \quad \text{and } 2n + 2k + s = 2dv. \tag{2.3}$$

If $a_n(c)$ is the number of bond-animals of size n edges and c independent cycles, then $a_n(0) = t_n$.

The growth constants of trees and animals have been approximated using a $\frac{1}{d}$-expansion. Gaunt and Rushkin [221] gave the expansion

$$\log \lambda_d = \log \delta + 1 - \tfrac{1}{2\delta} - \tfrac{13}{6\delta^2} - O(\delta^{-3}) \tag{2.4}$$

for the growth constant of bond-animals (where $\delta = 2d-1$).

Longer expansions were developed in references [219, 461]:

$$\begin{aligned} \log \tau_d &= \log \delta + 1 - \frac{1}{2\delta} - \frac{8}{3\delta^2} - \frac{85}{12\delta^3} - \frac{931}{20\delta^4} - \frac{2777}{10\delta^5} + O(\delta^{-6}), \quad \text{and} \\ \log \lambda_d &= \log \delta + 1 - \frac{0}{\delta} - \frac{55}{24\delta^2} - \frac{53}{24\delta^3} - \frac{39693}{960\delta^4} - \frac{44303}{240\delta^5} + O(\delta^{-6}) \end{aligned} \tag{2.5}$$

for the growth constants of bond-trees and bond-animals, and

$$\log \Lambda_0 = \log \delta + 1 - \frac{5}{2\delta} - \frac{13}{6\delta^2} - \frac{191}{12\delta^3} - \frac{139}{5\delta^4} - \frac{1585}{3\delta^5} + O(\delta^{-6}), \quad \text{and}$$
$$\log \Lambda = \log \delta + 1 - \frac{2}{\delta} - \frac{79}{24\delta^2} - \frac{317}{24\delta^3} - \frac{18321}{320\delta^4} - \frac{123307}{240\delta^5} + O(\delta^{-6}) \qquad (2.6)$$

for the growth constants Λ_0 of site-trees, and Λ of site-animals. Numerical estimates of the growth constants are given in table 2.1.

2.1.1 *Submultiplicativity of t_n*

The function t_n is supermultiplicative (see equation (2.2)). It also satisfies a generalised submultiplicative inequality [294]. To show this, define a *subtree* S of a tree T as subgraph of T which is joined to $T\backslash S$ at a single vertex. Thus $T\backslash S$ is itself also a tree. The next lemma shows that it is possible to pick 'large' subtrees from any tree.

Lemma 2.4 *Let T be a tree of size n edges and let $m \leq n$ be a positive integer. Then there is a subtree S in T of size k such that $\lceil \frac{m}{2d} \rceil \leq k \leq m$.*

Proof If $m = n$, then T is itself the subtree.

Assume that $m < n$. Choose a vertex p_1 in T. There are $2d$ subtrees incident with p_1 (some of which could be empty). Let these subtrees be $\langle U_1, U_2, \ldots, U_{2d} \rangle$ and suppose that U_i has u_i edges. Obviously, $\sum_{i=1}^{2d} u_i = n$.

If $u_i < \frac{m}{2d}$ for all i, then $\sum_{i=1}^{2d} u_i < m < n$. This is a contradiction; so, by the pigeonhole principle, there is at least one subtree U_i of size $u_i \geq \frac{m}{2d}$. Since u_i is an integer, this shows that $u_i \geq \lceil \frac{m}{2d} \rceil$. If $u_i \leq m$ as well, then the proof is completed, so suppose that $u_i > m$. Without loss of generality, let $i = 1$ (relabel the subtrees, if this is necessary).

To finish the proof, it will be shown that a subtree B of size b can be found such that $u_1 > b \geq \lceil \frac{m}{2d} \rceil$. Recursively finding such a subtree B will eventually produce a subtree of size less than or equal to m (and greater than or equal to $\lceil \frac{m}{2d} \rceil$).

Let p_2 be any vertex in U_1 but distinct from p_1. Let the subtrees incident with p_2 be $\langle V_1, V_2, \ldots, V_{2d} \rangle$, where V_{2d} is the subtree which contains p_1. Let the size of subtree V_i be v_i. Each subtree V_i is a subtree in U_1, except for V_{2d}. Thus, $\sum_{i=1}^{2d-1} v_i < u_1$.

The proof proceeds via a case analysis.

Case 1: Suppose that $\sum_{i=1}^{2d-1} v_i \geq m$. Then there is an i such that $v_i \geq \frac{m}{2d-1}$, and thus $v_i \geq \lceil \frac{m}{2d} \rceil$. Let this subtree be B of size $b = v_i$. Then $u_1 > b \geq \lceil \frac{m}{2d} \rceil$.

Case 2: Suppose that $\sum_{i=1}^{2d-1} v_i < m$. There are two subcases which must be considered.

Subcase 2a: Suppose that $\sum_{i=1}^{2d-1} v_i \geq \lceil \frac{m}{2d} \rceil$. Then the union of all the subtrees $B = V_1 \cup V_2 \cup \cdots \cup V_{2d-1}$ is a branch in the desired size range.

Subcase 2b: Suppose that $\sum_{i=1}^{2d-1} v_i < \lceil \frac{m}{2d} \rceil$. Let $B' = V_1 \cup V_2 \cup \cdots \cup V_{2d-1}$. Add edges to B' on the path from p_2 to p_1. The first edge to be added to B' is incident with p_2. Let the newly augmented subtree be B_1 and let it be joined to the rest of the tree in the vertex p_2^1. Incident with p_2^1, there are $2d - 2$ subtrees of

TABLE 2.1. Estimates of growth constants for trees and animals

Reference	d=2	d=3	d=4
bond-trees:	τ_2	τ_3	τ_4
Guttmann *et al.* [249]	5.208(4)	10.629(6)	--
Gaunt *et al.* [222]	5.14	10.53	16.2
De'Bell *et al.* [123]	5.1427(6)	10.547(3)	--
Gaunt *et al.* [217]	5.140(2)	--	--
Caracciolo *et al.* [89]	5.1434(13)	10.5439	--
Janse van Rensburg *et al.* [320]	5.1439(25)	--	--
Hsu *et al.* [283]	5.14276(2)	10.54646(10)	--
bond-animals:	λ_2	λ_3	λ_4
Hsu *et al.* [283]	5.201789(4)	10.61539(6)	--
site-trees:	Λ_0	Λ_0	Λ_0
Gaunt *et al.* [222]	3.795	7.85	12.7
Jensen [335]	3.795254(8)	--	--
site-animals:	Λ	Λ	Λ
Guttmann *et al.* [249]	4.065(5)	8.33(2)	--
Guttmann *et al.* [251]	4.062591(9)	--	--
Jensen *et al.* [335, 341]	4.062570(78)	--	--
Jensen [336]	4.0625696(5)	--	--
Hsu *et al.* [284]	4.06257(1)	8.34664(1)	13.30125(4)
Hsu *et al.* [283]	4.062568(13)	8.346641(17)	13.30125(8)

T (none of which contains p_1, and some of which may be empty). Augment B_1 by adding these branches to it one at a time. There are three possible outcomes.

(i) If all the subtrees are added to B_1, then it is still smaller than $\lceil \frac{m}{2d} \rceil$. In this case rename the subtree B' and (recursively) repeat subcase 2b. Since the tree is finite, this can occur only a finite number of times.

(ii) A subtree may be added which increases the size of B' from below $\lceil \frac{m}{2d} \rceil$ to above m. In this case the added subtree has size $b \geq m - \lceil \frac{m}{2d} \rceil > \lceil \frac{m}{2d} \rceil$. Thus, since it is a subtree in U_1, it must be that $u_1 > b \geq \lceil \frac{m}{2d} \rceil$.

(iii) The only other possible outcome is that B' will be in the desired size-range once some subtrees have been added. □

Corollary 2.5 *Let $m \leq n$ and let T be a tree of size n edges. Then k subtrees $\langle B_1, B_2, \ldots, B_k \rangle$ (where B_i has b_i edges) can be pruned from T such that*

1. $\sum_{i=1}^{k} b_i = m$,
2. $m - \sum_{i=1}^{l} b_i \geq b_{l+1} \geq \lceil \frac{1}{2d}(m - \sum_{i=1}^{l} b_i) \rceil$, *and*
3. $k \leq \frac{\log m}{\log(2d/(2d-1))}$.

Proof By lemma 2.4, it is possible to remove one edge from any tree. This proves the first claim. The second claim is an immediate consequence of lemma 2.4. It remains to prove the third claim.

Recursively prune k subtrees from T such that a total of m edges are removed. The worst possible situation occurs if the least number of edges are pruned from the tree when each subtree is pruned; this maximises k. That is, if p edges must be removed, then only $\lceil \frac{p}{2d} \rceil$ are cut, by lemma 2.4.

Suppose that, at the j-th step, p_j edges must be cut but only a subtree of size $\lceil \frac{p_j}{2d} \rceil$ is removed. Then, at the $(j+1)$-th step, at most $p_j - \lceil \frac{p_j}{2d} \rceil = \lfloor \frac{(2d-1)p_j}{2d} \rfloor$ edges remain to be pruned.

Iterating k times shows that at most $\lfloor \frac{2d-1}{2d} \lfloor \frac{2d-1}{2d} \lfloor \cdots \lfloor \frac{(2d-1)m}{2d} \rfloor \cdots \rfloor \rfloor \rfloor$ edges remain to be removed. A total of m edges will be removed if this expression is equal to 1 after $k-1$ iterations, and 0 after k iterations. Thus, after $k-1$ iterations, $\lfloor \frac{2d-1}{2d} \lfloor \frac{2d-1}{2d} \lfloor \cdots \lfloor \frac{(2d-1)m}{2d} \rfloor \cdots \rfloor \rfloor \rfloor = 1$. Hence, it is sufficient that k is large enough that $m\left(\frac{2d-1}{2d}\right)^k \leq 1$. Thus, $k \leq \frac{\log m}{\log(2d/(2d-1))}$ is sufficient. □

Theorem 2.6 *The function t_n satisfies*

$$t_{n+m} \leq [(n+m)m^2]^{\alpha \log m} t_n t_m \text{ and } t_n t_m \geq (n+m)^{-3\alpha \log(n+m)} t_{n+m},$$

where $\alpha = \frac{\log 8}{\log(2d/(2d-1))}$.

Proof Let T be a tree of size $n+m$.

Apply corollary 2.5 to cut at most $\alpha \log m$ subtrees from T containing exactly m edges. Any of these subtrees can be put back in T in at most $(n+m+1)(m+1)$ ways (select a vertex in each and identify them).

Concatenate the subtrees as they are pruned into a new tree containing m edges. The total number of ways this tree can be cut into the original subtrees

TABLE 2.2. Estimates of θ for branched polymers

Reference	2	3	4
Gaunt *et al.* [222]	1.00(2)	1.55(5)	1.9(1)
Caracciolo *et al.* [89]	1.001(24)	1.501(12)	--
Janse van Rensburg *et al.* [320]	1.014(22)	--	--
Hsu *et al.* [284]	--	--	1.835(6)

is at most $(m+1)^{\alpha \log m}$, since the tree must be cut in at most $\alpha \log m$ places, and, for each cut, there is a selection of at most $m+1$ vertices to choose for a cut.

Thus, $t_{n+m} \leq [(n+m+1)(m+1)^2]^{\alpha \log m} t_n t_m$. Replace $(n+m+1)$ by $2(n+m)$ and $(m+1)$ by $2m$ to obtain the claimed inequality. □

Take the logarithm of the submultiplicative inequality in theorem 2.6. Then $\log t_n$ satisfies the hypothesis of a_n in theorem A.4 (appendix A) where the function g_n is given by $\log g_n = 3\alpha \log^2 n$. Notice that

$$\sum_{m=2n}^{\infty} \frac{\log^2 m}{m(m+1)} \leq \int_{2n}^{\infty} \frac{\log^2(x-1)}{x(x-1)}\, dx \leq \int_{n}^{\infty} \left(\frac{\log x}{x}\right)^2 dx. \tag{2.7}$$

Compute the last integral and use the bound in theorem A.4 to obtain

$$\tfrac{1}{n} \log t_n \geq \log \tau_d - \tfrac{9\alpha}{n} \log^2 n - \tfrac{24\alpha}{n} \log n - \tfrac{24\alpha}{n}. \tag{2.8}$$

This gives the following bound on t_n.

Theorem 2.7 *The number of trees t_n in the hypercubic lattice is bounded by*

$$e^{-24\alpha} n^{-24\alpha} e^{-9\alpha \log^2 n} \tau_d^n \leq t_n \leq \tau_d^n,$$

where $\alpha = \frac{\log 8}{\log(2d/(2d-1))}$. □

2.1.2 *Scaling of lattice animals and lattice trees*

Theorem 2.2 shows that $t_n = \tau_d^{n+o(n)}$, and $a_n = \lambda_d^{n+o(n)}$. There is ample numerical evidence that both

$$t_n \simeq A_t n^{-\theta} \tau_d^n (1 + c_t n^{-\Delta_1}), \text{ and } a_n \simeq A_a n^{-\theta} \lambda_d^n (1 + c_a n^{-\Delta_1}), \tag{2.9}$$

where θ is the *branched polymer entropic exponent* [81, 191, 287, 365, 460] and Δ_1 is the dominant *confluent correction exponent* (see equation (1.13) for walks). (Since it is thought that $\Delta_1 < 1$, it gives dominant correction to scaling for trees and animals.) It was estimated that, in $d = 2$, $\Delta_1 = 0.9(1)$ [283] (but in

TABLE 2.3. Estimates of ν for branched polymers

Reference	2	3	4
Stauffer [532]	0.660(7)	--	--
Peters *et al.* [464]	0.65	--	--
Redner [490]	0.57(6)	0.45(6)	--
Family [185]	0.649(9)	0.51	--
Caracciolo *et al.* [89]	0.640(4)	0.495(9)	--
Janse van Rensburg *et al.* [307]	0.637(12)	0.4960(52)	0.420(17)
Janse van Rensburg *et al.* [308]	0.6442(17)	0.5016(13)	--
You *et al.* [602]	0.642(10)	0.498(10)	0.415(11)
Duarte [156]	0.656(15)	--	--
Vujic [570]	0.6513(3)	--	--
You *et al.* [603]	0.645(27)	--	--
You *et al.* [604]	0.645(30)	--	--
Janse van Rensburg *et al.* [320]	0.6437(35)	--	--
Hsu *et al.* [284]	0.6412(5)	--	0.4163(30)

reference [602] $\Delta_1 = 0.65(20)$, while in reference [1] $\Delta_1 = 0.85(10)$). If $d = 3$, then it was estimated that $\Delta_1 = 0.75(8)$ (reference [283]), but in reference [602] $\Delta_1 = 0.54(12)$, and in $d = 4$ $\Delta_1 = 0.46(11)$.

The asymptotic formulae for t_n and a_n can be made rigorous in high dimensions [264, 267]. The upper critical dimension of branched polymers is $d_c = 8$ [45, 214, 264, 389, 390]. The mean field value for the entropic exponent is $\theta = \frac{5}{2}$ [45, 61, 267]. The ϵ-expansion for θ is [460]

$$\theta = \frac{5}{2} - \frac{1}{12}\epsilon - \frac{79}{3888}\epsilon^2 + \left(\frac{\zeta(3)}{81} - \frac{10445}{1259712}\right)\epsilon^3 + O(\epsilon^4), \tag{2.10}$$

where $\epsilon = 8 - d$. Numerical estimates of θ are listed in table 2.2.

The metric exponent ν of lattice trees or animals is defined in terms of the scaling of metric quantities (similar to equations (1.21), (1.20) or (1.22) for self-avoiding walks). The mean field value of ν is $\nu = \frac{1}{4}$ [45, 267, 287, 389, 390]. Estimates of ν are listed in table 2.3.

By theorem 2.2, $\theta \geq 0$, and a result in reference [396] improves this to $\theta \geq \nu$ if $d = 2$, and $\theta \geq \frac{d-1}{d}$ in d dimensions.

Dimensional reduction gives a remarkable connection between the scaling of branched polymers in d dimensions and critical behaviour in an Ising model in an imaginary magnetic field in $d-2$ dimensions. In particular, there are relations between the exponent σ (associated with the Yang-Lee edge singularity) in the Ising model and the branch polymer exponents θ and ν [460]. These relations are $\nu = \frac{\sigma+1}{d-2}$, and $\theta = \sigma + 2$. This shows that

$$\theta = (d-2)\nu + 1. \tag{2.11}$$

Since the Ising model can be solved exactly in low dimensions, 'exact values' for θ can be obtained in two dimensions when $\sigma = -1$ (so that $\theta = 1$), and for θ and ν in three dimensions when $\sigma = -\frac{1}{2}$ ($\theta = \frac{3}{2}$, and $\nu = \frac{1}{2}$). In four dimensions $\sigma = -\frac{1}{6}$ ($\theta = \frac{11}{6}$ and $\nu = \frac{5}{12}$).

There are numerous numerical studies of lattice trees [227, 283, 307, 308, 320, 490, 508]. Renormalisation group studies can be found in references [129, 131].

2.1.3 *Flory theory for branched polymers*

A Flory argument for the metric exponent of branched polymers can be found in reference [288]. Let $t_n(\{\vec{0},\vec{r}\})$ be the number of trees of size n containing $\vec{0}$ and the lattice site $\vec{r}$. Assume that the probability of a tree to extend a distance $r = |\vec{r}|$ is asymptotically Gaussian with covariance $\sigma^2 \sim n^{2\nu}$. Then

$$\frac{t_n(\{\vec{0},\vec{r}\})}{t_n(\{\vec{0}\})} \sim e^{\frac{-r^2}{a\sqrt{n}}}, \tag{2.12}$$

where the mean field value $\nu = 1/4$ was used. Hence, the relative configurational entropy is given approximately by $S \approx \log(t_n(\{\vec{0},\vec{r}\})/t_n(\{\vec{0}\})) = C_0 - \frac{r^2}{a\sqrt{n}}$.

On the other hand, the density of monomers in a lattice tree is proportional to $\frac{n}{r^d}$, and each monomer finds itself in a background field of monomers of density $\frac{n}{r^d}$. Thus, the energy is proportional to $\frac{n^2}{r^d}$. Taken together, this gives the free energy

$$\mathcal{T} = \frac{bn^2}{r^d} + \frac{r^2}{a\sqrt{n}} + \mathrm{C}. \tag{2.13}$$

Take the derivative to r to find the minimum in $\mathcal{T}$. This is achieved when $r^{d+2} \propto n^{\frac{5}{2}}$. In other words, the Flory value of the metric exponent ν is

$$\nu = \frac{5}{2(d+2)}. \tag{2.14}$$

In two dimensions this shows that $\nu = \frac{5}{8}$; this value is slightly less than the numerical results in table 2.3, so the Flory value in two dimensions is too small. The Flory value is accurate in three and four dimensions, and it also gives the correct mean field value when $d = 8$.

By equation (2.11), the Flory value of θ is obtained:

$$\theta = \frac{7d-6}{2(d+2)}. \tag{2.15}$$

TABLE 2.4. Estimates of ρ for branched polymers

Reference	2	3	4
Janse van Rensburg *et al.* [307]	0.737(13)	0.6540(32)	0.6106(28)
Janse van Rensburg *et al.* [308]	0.74257(85)	0.6587(16)	--
You *et al.* [602]	0.738(10)	0.653(6)	0.607(6)
You *et al.* [603]	0.737(27)	--	--
You *et al.* [604]	0.737(10)	--	--
Janse van Rensburg *et al.* [320]	0.74000(62)	--	--

2.1.4 *The mean longest path in lattice trees*

Consider a lattice tree T of size n edges and let $P_n(T)$ be the length of the longest self-avoiding path in T. Then $P_n(T)$ can be determined as follows. Let $\vec{v}$ be a vertex in T and find the longest (self-avoiding) walk in T from $\vec{v}$. Next, determine the longest path $P_n(T)$ in T from the vertex $\vec{w}$. Then $P_n(T)$ has endpoints (say) $\vec{w}$ and $\vec{u}$ and it is also the longest self-avoiding walk in T [137, 138]. Endow the set of trees of size n with the uniform measure and denote the average of $P_n(T)$ over all trees of size n by $\langle P_n \rangle$.

The scaling of $\langle P_n \rangle$ is a measure of the way in which a tree differs from a self-avoiding walk and it is the expected number of edges in the longest path. Assume that

$$\langle P_n \rangle \simeq A_p n^\rho. \tag{2.16}$$

The expected distance between the endpoints of the longest path is $O(n^\nu)$. Since the longest path is a self-avoiding walk, it is expected that $\nu \leq \rho \leq 1$. Since self-avoiding walks and trees do not belong in the same universality class, it should be the case that $\rho < 1$. Estimates of ρ are shown in table 2.4.

Let T be a tree of n edges; then, select an edge uniformly in T and delete it. This divides the tree into two subtrees, the smaller of which will be called a *branch*. Let $b_n(T)$ be the average size of a branch in T (computed by deleting each edge once and computing the average size of the resulting branch). Then the expected value of $b_n(T)$ (uniformly over all trees) should scale as

$$\langle b_n \rangle \simeq A_b n^\epsilon \tag{2.17}$$

with n, where $0 \leq \epsilon \leq 1$.

The following heuristic argument shows that $\epsilon = \rho$ [307]. Assume that a tree T has the structure of a longest path with $\Theta(n^\rho)$ edges and with $\Theta(n^\rho)$ smaller side-branches sprouting from the vertices along the longest path. The total number of edges in the smaller branches is $n - \Theta(n^\rho)$, and so each smaller side-branch has an average size of $\Theta(n^{1-\rho})$.

TABLE 2.5. Branched polymer exponents

d	2	3	4	≥ 8
θ	1 [460]	$\frac{3}{2}$ [460]	$\frac{11}{6}$ [460]	$\frac{5}{2}$ [266]
ν	0.64115(5) [335]	$\frac{1}{2}$ [460]	$\frac{5}{12}$ [460]	$\frac{1}{4}$ [45]
ρ	0.7400(1) [320]	0.654(4) [307]	0.610(3) [307]	0.52(6) [307]
Δ_1	0.9(1) [284]	0.75(8) [284]	0.57(8) [284]	0.47(7) [284]

Deleting a uniformly selected edge on the longest path gives two subtrees, each with $\Theta(n)$ edges. If instead an edge is deleted in a side-branch off the longest path, then the size of the branch picked is $\Theta(n^\sigma)$, where σ satisfies $0 \leq \sigma \leq 1-\rho$.

Thus, computing the average of b_n gives

$$\langle b_n \rangle \sim \tfrac{1}{n} \left(\Theta(n^\rho)\Theta(n) + \Theta(n-n^\rho)\Theta(n^\sigma) \right) \sim \Theta(n^\rho) + \Theta(n^\sigma). \tag{2.18}$$

Comparison to equation (2.17) shows that $\epsilon = \max\{\rho, \sigma\}$, where $\sigma \leq 1-\rho$.

On the other hand, recall that each side-branch off the longest path has an average size of $\Theta(n^{1-\rho})$. Deleting an edge uniformly in a side-branch gives two components, and the component which does not touch the longest path in the tree has size $\Theta(n^\sigma)$, by the definition of σ.

Assume that the statistics of the side-branch is the same as that of a tree. If an edge is deleted in a tree of size n with one labelled *leaf*, there is a resulting subtree or component disjoint with the labelled leaf. This component has size $\Theta(n^{\sigma/(1-\rho)})$ (to see this, replace $n^{1-\rho}$ in the last paragraph by n).

The probability that the labelled edge is in the smaller branch is given by $\frac{1}{n}\langle b_n \rangle = \Theta(n^{\epsilon-1})$. Thus, the expected size of the branch which does not contain the labelled leaf is

$$\Theta(n^{\sigma/(1-\rho)}) = (n - \Theta(n^\epsilon))\Theta(n^{\epsilon-1}) + \Theta(n^\epsilon)(1-\Theta(n^{\epsilon-1})) = \Theta(n^\epsilon). \tag{2.19}$$

This shows that $\sigma = \epsilon(1-\rho) < \epsilon$. Combining this with $\epsilon = \max\{\rho, \sigma\}$ gives $\rho = \epsilon$.

The exponent ρ was estimated in reference [320] to have value $\rho = 0.74000(62)$ in two dimensions. Estimates of the values of branched polymer exponents are displayed in table 2.5.

2.1.5 *Positive trees*

The half-lattice $\mathbb{L}_+^d$ is a subset of $\mathbb{L}^d$ defined by

$$\mathbb{L}_+^d = \{ \langle \vec{v} \sim \vec{w} \rangle \mid \vec{v}, \vec{w} \in \mathbb{L}^d, \, \vec{v}(d) \geq 0, \text{ and } \vec{w}(d) \geq 0 \}, \tag{2.20}$$

where $\vec{v}(d)$ is the d-th Cartesian coordinate of the vertex $\vec{v} \in \mathbb{L}^d$. The boundary of $\mathbb{L}_+^d$ is the $(d-1)$-dimensional sub-lattice normal to $\vec{e}_d$ containing $\vec{0}$ and it is denoted by $\partial \mathbb{L}_+^d$.

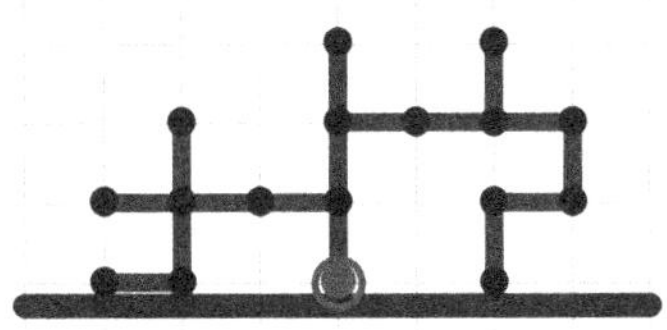

FIG. 2.4. A positive tree in $\mathbb{L}_+^d$ rooted at $\vec{0} \in \partial\mathbb{L}_+^d$. Vertices of the tree in $\partial\mathbb{L}_+^d$ are *visits*.

A *positive tree* is a lattice tree in $\mathbb{L}_+^d$ with its root at $\vec{0} \in \mathbb{L}_+^d$ (see figure 2.4).

Denote the number of positive trees of size n edges by t_n^+, and the number of rooted trees in $\mathbb{L}^d$ by t_n. Then $\mathrm{t}_n^+ \leq \mathrm{t}_n$, and $\mathrm{t}_n = (n+1)t_n$.

The number of positive trees is expected to have the asymptotic behaviour

$$\mathrm{t}_n^+ \simeq A_t n^{-1-\theta_1} \lambda_d^n, \tag{2.21}$$

similar to equation (2.9). The exponent θ_1 is a *surface entropic exponent* for branch polymers. A scaling argument in reference [123] shows that

$$\theta_1 = \theta - 1, \tag{2.22}$$

so the exact value of θ_1 follows from the Flory values in equation (2.15):

$$\theta_1 = \frac{5(d-2)}{2(d+2)}. \tag{2.23}$$

The equality in equation (2.22) may be seen heuristically as follows. Recall that the number of lattice trees of size n (up to equivalence under translations in $\mathbb{L}^d$) is denoted by t_n. Select trees of size n which can be placed in $\mathbb{L}_+^d$ so that they have exactly one vertex in $\partial\mathbb{L}_+^d$. By deleting the one vertex in $\partial\mathbb{L}_+^d$, a positive tree of size $n-1$ is obtained, which can be rooted in the vertex incident with the deleted edge. This gives a positive tree in the half-lattice $\mathbb{L}_+^d + \vec{e}_d$ with root in $\partial\mathbb{L}_+^d + \vec{e}_d$ (translating $\partial\mathbb{L}_+^d$ by $\vec{e}_d$). This shows that $t_n \geq \mathrm{t}_{n-1}^+$.

On the other hand, any rooted positive tree of size n may be translated one step away from $\partial\mathbb{L}_+^d$ to form a tree. This shows that $t_n \leq \mathrm{t}_n^+$.

Thus, $t_{n+1} \geq \mathrm{t}_n^+ \geq t_n$ and, since $t_n \sim n^{-\theta}\lambda_d^n$ (see equation (2.9)), it follows that $\theta = 1 + \theta_1$.

2.1.6 *Tree and animal generating functions*

The generating function of lattice trees is defined by

$$T(t) = \sum_{n=0}^{\infty} t_n t^n. \tag{2.24}$$

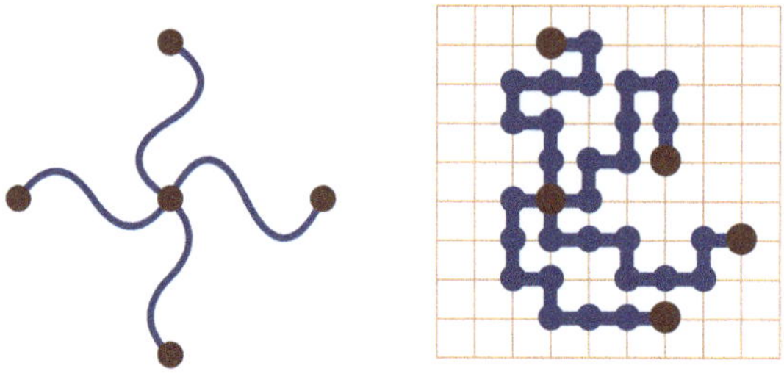

FIG. 2.5. A representation of a uniform star polymer. On the left the abstract star graph represents the connectivity of the star polymer on the right: four linear polymers joined in a central vertex of degree 4 to form the star. On the right is a lattice network model of the star polymer.

By equation (2.9), it should be the case that

$$T(t) \sim (1 - \tau_d t)^{\theta - 1}. \tag{2.25}$$

The animal generating function is similarly defined and has asymptotic behaviour:

$$A(t) = \sum_{n=0}^{\infty} a_n t^n \sim (1 - \lambda_d t)^{\theta - 1}. \tag{2.26}$$

The critical value of t in the tree and animal generating functions is $t_c = \tau_d^{-1}$ for trees, and $t_c = \lambda_d^{-1}$ for animals. Observe that both $T(t_c)$ and $A(t_c)$ are finite for $d > 2$, since $\theta > 1$ if $d > 2$, but that $\theta = 1$ if $d = 2$.

2.2 Stars, combs, brushes and uniform networks

A uniform *star polymer* has a central monomer (or vertex) which is the junction where a number of linear polymers are joined into a single macromolecule. The linear polymers are the *branches* or *arms* of the star. The star is uniform if the branches have the same lengths. A star graph may be used to represent the connectivity of a star. In figure 2.5 a star graph with four arms is illustrated.

Networks may have the connectivity of *combs* and *brushes*. These are similarly represented in terms of graphs. A comb has a linear polymer backbone, from which is hung a number of branches. This is illustrated in figure 2.6(b). A brush is similarly defined, except that the vertices along the backbone of the graph may have degree larger than 3; see figures 2.6(c) and (d).

The branches in each of the branched polymers represented in figures 2.5 and 2.6 are assumed to be self-avoiding walks of (almost) equal length and combined total length n. This assumption gives classes of *monodispersed or uniform branched polymers of specified topology.*

A graphical representation of a rooted ring polymer is given in figure 2.7(a). This is the first example in a class of uniform networks which may have cycles.

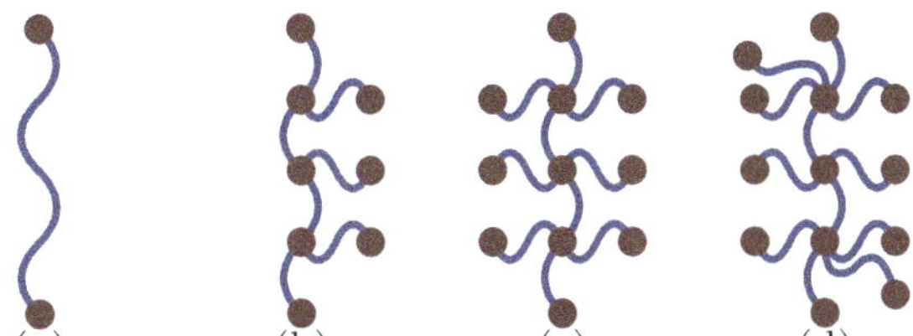

FIG. 2.6. Walks, combs and brushes. (a) An abstract representation of a self-avoiding walk. (b) A representation of a comb. This comb has seven branches, and each branch is a self-avoiding walk. If all the branches have the same length, then the comb is uniform. (c) A brush with internal vertices of degree 4 and with ten branches. (d) A more general representation of a brush.

Other examples are *thetas* (figure 2.7(b)), *tadpoles* (figure 2.7(c)) and *watermelons* (figure 2.7(d)).

2.2.1 *Scaling of uniform branched polymers*

Let $\mathcal{G}$ be an abstract connected graph (and possibly a multigraph) representing the connectivity of a (monodispersed) uniform branched polymer. Two uniform animals α_1 and α_2 in $\mathbb{L}^d$ of specified topology $\mathcal{G}$ are equivalent if $\alpha_1 = \alpha_2 + \vec{v}$ for some vector $\vec{v}$.

Let $c_n(\mathcal{G})$ be the number of distinct uniform animals of specified topology $\mathcal{G}$ with each branch of length n and total size $n|\mathcal{G}|$ (where $|\mathcal{G}|$ is the size or number of branches in the graph $\mathcal{G}$). The standard scaling assumption for $c_n(\mathcal{G})$ is [519, 526, 587]

$$c_n(\mathcal{G}) \simeq A_{\mathcal{G}}\, n^{\gamma(\mathcal{G})-1} \mu_d^n, \tag{2.27}$$

where $A_{\mathcal{G}}$ is an amplitude, μ_d is the self-avoiding walk growth constant (see theorem 1.1), and $\gamma(\mathcal{G})$ is the *entropic exponent* of the uniform network.

If the network is a self-avoiding walk, then it may either be represented as in figure 2.6(a) with two endpoints of degree 1 (denoted by $\mathcal{G} \equiv 1$), or it may be considered a 2-star graph (with two branches and a central node of degree 2; see figure 2.5), denoted by $\mathcal{G} \equiv 2$. In these cases, either

$$c_n(1) \simeq A_1 n^{\gamma(1)-1} \mu_d^n, \quad \text{or} \quad c_n(2) \simeq A_2 n^{\gamma(2)-1} \mu_d^n. \tag{2.28}$$

It is customary to assign the exponent σ_1 to each endpoint (node of degree 1) in figure 2.6(a), and σ_2 to vertices of degree 2. In other words,

$$\gamma(1) - 1 = 2\sigma_1, \quad \text{and}\ \gamma(2) - 1 = \sigma_2 + 2\sigma_1. \tag{2.29}$$

Since $\gamma(2) = \gamma(1)$, this shows that $\sigma_2 = 0$, and $\gamma(1) = 2\sigma_1 + 1$.

An f-star (see figure 2.5) is represented by putting $\mathcal{G} \equiv f$. In this case there is one node of degree f, and f nodes of degree 1. This shows that

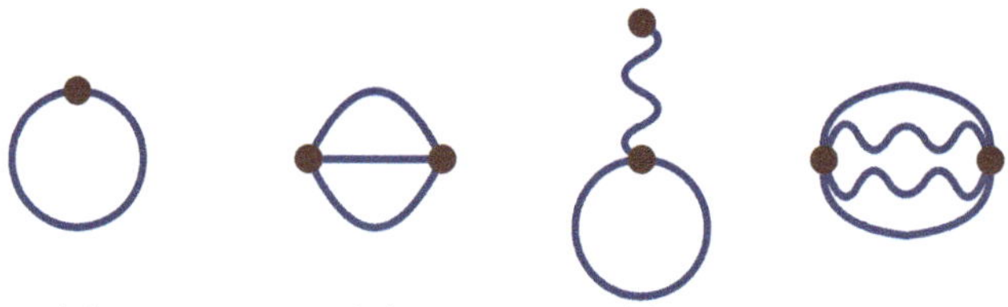

FIG. 2.7. Uniform networks. (a) A graphical representation of a rooted polygon. (b) A theta. (c) A tadpole. (d) A watermelon with four branches. If a watermelon has three branches, then it is a theta.

$$\gamma(f) - 1 = \sigma_f + f\sigma_1. \tag{2.30}$$

In the above there is an implicit assumption that σ_f is independent of the connectivity of the uniform branched polymer; the presence of other nodes of degree (say) ϵ does not affect the value of σ_f. Since the distance between nodes is $O(n^\nu)$, this seems reasonable for large values of n. Numerical testing of equation (2.30) was done in reference [282].

The entropic exponents for more general networks can be expressed in terms of the σ_f exponents of f-stars. For example, if $\mathcal{H}$ is a comb (see figure 2.6(b)) with two nodes of degree 3 and four nodes of degree 1, then $\mathcal{H}$ may be created from two 3-stars by identifying a node of degree 1 in the first star with a node of degree 1 in the second star. Since we assume that

$$c_n(\mathcal{H}) \simeq A_\mathcal{H} n^{\gamma(\mathcal{H})-1} \mu_d^n = A_\mathcal{H} n^{2\sigma_3 + 4\sigma_1} \mu_d^n, \tag{2.31}$$

the result is that

$$\gamma(\mathcal{H}) - 1 = 2\sigma_3 + 4\sigma_1 = 2\left(\gamma(3) - 1 - 3\sigma_1\right) + 4\sigma_1 = 2\gamma(3) - \gamma(1). \tag{2.32}$$

In other words, to obtain $\gamma(\mathcal{H})$, all that is needed are the entropic exponents of 3-stars and self-avoiding walks.

If all the σ_f for f-stars are known, then the value of the entropic exponents of *acyclic* networks can be determined by 'gluing' together uniform f-stars. In particular, if $\mathcal{H}$ is a tree graph, then the entropic exponent of the uniform polymer network corresponding to $\mathcal{H}$ is given by

$$\gamma(\mathcal{H}) - 1 = \sum_{f \geq 1} m_f \sigma_f, \tag{2.33}$$

where m_f is the number of nodes in $\mathcal{H}$ of degree f (and where $\sigma_2 = 0$).

More generally, $\mathcal{G}$ has cyclomatic index $c(\mathcal{G}) > 0$ and so a different approach is needed. Cycles are formed in a polymer network by joining two nodes of degree 1 into a loop. Normally, a free endpoint of a walk explores a volume of size $O(R^d)$, where R is the end-to-end length of the walk. Since $R \sim n^\nu$, the reduction in

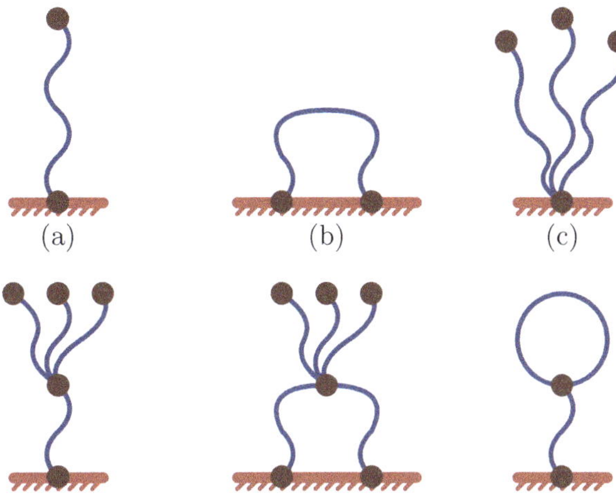

FIG. 2.8. Uniform networks attached to a hard wall. (a) A graphical representation of the self-avoiding walk with one endpoint attached to the wall; this is a *positive walk*. (b) A loop (a self-avoiding walk with both endpoints attached to the wall). (c) A 3-star with its node of degree 3 attached to the wall. (d) A 4-star with one end-vertex (node of degree 1) attached to the wall. (e) A 5-star polymer with two end-vertices attached to the wall. (f) A tadpole with its end-vertex attached to the wall.

---

entropy by fixing the endpoint with another to form a loop is proportional to $n^{-d\nu}$. In other words, for each cycle in $\mathcal{G}$, a term $d\nu$ should be subtracted from the entropic exponent. This argument modifies equation (2.33) to

$$\gamma(\mathcal{G}) - 1 = \sum_{f \geq 1} m_f \sigma_f - c(\mathcal{G}) d\nu. \tag{2.34}$$

If $\mathcal{G}$ is itself just a loop (a rooted ring polymer), then comparison to equation (1.16) gives $2 - \alpha_s = d\nu$. This recovers the hyperscaling relation in equation (1.55).

2.2.2 *Uniform networks near a hard wall*

Let $\mathbb{L}^d_+$ be the positive half-lattice with boundary $\partial\mathbb{L}^d_+$ (see equation (2.20)). Suppose that a (monodispersed) network in $\mathbb{L}^d_+$ of specified topology $\mathcal{G}$ is attached to $\partial\mathbb{L}^d_+$ (which is a *hard wall*; see figure 2.8). Associate with each node $\vec{v}_f \in \partial\mathbb{L}^d_+$, of degree f in $\mathcal{G}$, an exponent σ'_f.

As with the arguments leading to equation (2.29), it may be shown that $\sigma'_2 = 0$, and all the σ'_f can be obtained by examining f-stars with their central node attached to $\partial\mathbb{L}^d_+$.

A self-avoiding walk from $\vec{0} \in \partial\mathbb{L}^d_+$ in $\mathbb{L}^d_+$ is a *positive walk* (see figure 2.8(a)). A suitable scaling assumption for positive walks is

$$c_n^+(1) \simeq A_1 n^{\gamma_1 - 1} \mu_d^n, \tag{2.35}$$

where $c_n^+(1) \equiv c_n^+$ is the number of positive walks of length n.

Similar to the arguments in section 2.2.1, the walk has one end-vertex (a node of degree 1) in the hard wall and one free end-vertex, so

$$\gamma_1 - 1 = \sigma_1 + \sigma_1'. \tag{2.36}$$

An f-star in $\mathbb{L}^d_+$ with a central node at $\vec{0} \in \partial\mathbb{L}^d_+$ (see figure 2.8(c)) has the entropic exponent γ_f given by

$$\gamma_f - 1 = f\sigma_1 + \sigma_f', \tag{2.37}$$

since there are f nodes of degree 1 in bulk, and one node of degree f attached to $\partial\mathbb{L}^d_+$.

A *surface loop* (or *surface bridge*; sometimes just called a *loop*) is formed when both endpoints of a walk are in $\partial\mathbb{L}^d_+$ (see figure 2.8(b)). Attaching a vertex to $\partial\mathbb{L}^d_+$ reduces the volume it can explore from $O(n^{d\nu})$ to $O(n^{(d-1)\nu})$ and reduces its entropy by a factor $O(n^\nu)$. This shows that ν should be subtracted from the entropic exponent for each surface loop in the network. The number of surface loops of length n is assumed to be given by

$$c_n(11) \simeq A_1 n^{\gamma_{11} - 1} \mu_d^n, \tag{2.38}$$

and, since there are one surface loop and two endpoints in the wall,

$$\gamma_{11} - 1 = 2\sigma_1' - \nu. \tag{2.39}$$

Comparison of these results to equations (2.29) and (2.36) gives Barber's scaling relation [22]:

$$\gamma + \nu = 2\gamma_1 - \gamma_{11}. \tag{2.40}$$

In $d = 2$ the numerical simulations in reference [23] gave $\gamma_1 = 0.945(5)$ (and $\gamma_1 = 0.9551(3)$ [412]), and $\gamma_{11} = -0.19(3)$. This shows that $2\gamma_1 - \gamma_{11} = 2.08(4)$, while the exact value is $\gamma + \nu = \frac{67}{32} = 2.09375\ldots$. Notice that the exact values in $d = 2$ are $\gamma_1 = \frac{61}{64}$ [95], and $\gamma_{11} = -\frac{3}{16}$ [168]. An ϵ-expansion by J Reeve and AJ Guttmann gave $\gamma_{11} \neq \nu - 1$ [491] (see reference [77]).

In $d = 3$ the numerical simulations in reference [278] gave $\gamma_1 = 0.679(2)$, and $\gamma_{11} = -0.383(5)$, following on the earlier estimates that $\gamma_1 = 0.687(5)$ and $\gamma_{11} = -0.38(2)$ [387, 416]. Combining the numerical estimates in reference [278] gives $2\gamma_1 - \gamma_{11} = 1.741(9)$. The estimates in table 2.6 give $\gamma + \nu = 1.74458(35)$. The mean field values of the surface exponents for self-avoiding walks are $\gamma_1 =$

TABLE 2.6. Surface exponents for the self-avoiding walk

d	2	3	Mean Field
γ	$\frac{43}{32}$ [161]	1.15698(34) [233, 377]	1
ν	$\frac{3}{4}$ [138]	0.587597(7) [105]	$\frac{1}{2}$
γ_1	$\frac{61}{64}$ [95]	0.697(2) [278]	$\frac{1}{2}$
γ_{11}	$-\frac{3}{16}$ [168]	$-0.383(5)$ [278]	$-\frac{1}{2}$

$\frac{1}{2}$, and $\gamma_{11} = -\frac{1}{2}$ [39, 40]. These results are consistent with Barber's scaling relation.

Expressions for the entropic exponents of general networks $\mathcal{G}$ near a hard wall may be computed by 'gluing' together nodes to form cycles or putting them in the hard wall to form loops. For each cycle formed in this way, a term $d\nu$ is subtracted from the entropic exponent, and, for each surface loop, a term ν is subtracted from the entropic exponent. If m'_f is the number of nodes of degree f attached to the wall, and m_f is the number of free nodes of degree f, then

$$\gamma_{\mathcal{G}} - 1 = \sum_f m_f \sigma_f + \sum_f m'_f \sigma'_f - c(\mathcal{G})\, d\nu - \ell(\mathcal{G})\,\nu, \tag{2.41}$$

where the cyclomatic index of $\mathcal{G}$ is $c(\mathcal{G})$ and there are $\ell(\mathcal{G})$ surface loops.

Suppose that N_i end-vertices (nodes of degree 1) of a network $\mathcal{G}$ are attached to a hard wall. For two different values of N_i, the entropic exponents are $\gamma_{\mathcal{G}}(N_1)$ and $\gamma_{\mathcal{G}}(N_2)$. From equation (2.41), it follows that

$$\gamma_{\mathcal{G}}(N_1) - \gamma_{\mathcal{G}}(N_2) = (N_2 - N_1)(\sigma_1 - \sigma'_1) + (N_2 - N_1)\nu. \tag{2.42}$$

The right-hand side is independent of $\mathcal{G}$ and only dependent on the difference in the number of vertices of attached end-vertices.

Choose $N_1 = N_2 - 1$ in equation (2.42) and denote $\overline{\gamma}_N = \gamma_{\mathcal{G}}(N)$. Then it follows that $\overline{\gamma}_{N-1} - \overline{\gamma}_N = (\sigma_1 - \sigma'_1) + \nu$. Inductively, $\overline{\gamma}_0 - \overline{\gamma}_N = N(\overline{\gamma}_0 - \overline{\gamma}_1)$; see references [121, 122, 161, 168].

2.3 Conformal invariance

Conformal invariance is a powerful tool for computing exact values of scaling exponents of networks in two dimensions [95].

By theorem 1.1, the self-avoiding walk generating function $C(t)$ (see equation (1.56)) is divergent at the critical point. This is generally also true for correlation functions of the self-avoiding walk, so the local structure of the lattice becomes irrelevant and the model becomes lattice (and scale) invariant. The Hamiltonian $\mathcal{H}$ of the model has a fixed point value $\mathcal{H}^*$ (a function of field operators) which determines the properties of the correlation functions at the critical point. At the fixed point the (real space) field operators ϕ_j and correlation functions transform in a simple way given by a scaling relation and scaling exponents. That is, if $\vec{r} = w\vec{r}\,'$, then the correlation functions rescale as

$$\langle \phi_1(\vec{r}_1)\phi_2(\vec{r}_2)\cdots\rangle = \prod_{i\geq 1} w^{-x_i} \langle \phi_1(\vec{r}_1^{\,\prime})\phi_2(\vec{r}_2^{\,\prime})\cdots\rangle . \tag{2.43}$$

The rescaling is controlled by the scaling exponents in the factor w^{-x_i}.

The scaling in equation (2.43) is uniform, since $\vec{r} = w\vec{r}\,'$, and the fields rescale *globally*. The factor w^{-x_i} and the scaling exponent x_i are associated with the rescaling of each field $\phi_i(x_i)$. If the model is isotropic and homogeneous, then the correlation function is translationally and rotationally invariant and it rescales as above.

A more general rescaling $w = w(\vec{r})$ is non-uniform. In this event the fields ϕ_i scale *locally* in equation (2.43). This local scaling should be preserved even if $w(\vec{r})$ is a rescaling which is a combination of a dilation and rotation (but which preserves angles); in two dimensions this is the case when $w(\vec{r})$ is a conformal transformation.

Conformal transformations are a very general set of transformations, as every analytic function w on $\mathbb{C}$ corresponds to a conformal transformation $z \to w(z)$ (except when $w'(z) = 0$).

Putting $\vec{r} = (r_1, r_2)$, and $x = r_1 + ir_2$ so that the fields ϕ_i are functions of the complex variable z, equation (2.43) becomes

$$\langle \phi_1(z_1,\overline{z}_1)\phi_2(z_2,\overline{z}_2)\cdots\rangle = \prod_{i\geq 1} |w'(z_i)|^{-x_i} \langle \phi_1(w_1,\overline{w}_1)\phi_2(w_2,\overline{w}_2)\cdots\rangle . \tag{2.44}$$

Here, $|w'(z)|$ is the Jacobian of the conformal transformation, $w_j = w(z_j)$, and, as in equation (2.43), the rescaling induces a local rescaling of the fields controlled by a scaling exponent x_i.

Let $\vec{x} = (x_1, x_2)$, let $z_1 = x_1 + ix_2$, let $\vec{y} = (y_1, y_2)$ and let $z_2 = y_1 + iy_2$ in the complex plane and define the self-avoiding walk two-point function

$$C(x,y) \equiv C(z_1,\overline{z}_1,z_2,\overline{z}_2) = \langle \phi(z_1,\overline{z}_1)\phi(z_2,\overline{z}_2)\rangle$$

in $\mathbb{C}$ (see equation (1.37)). Then the correlation function $C(z_1,\overline{z}_1,z_2,\overline{z}_2)$ transforms as

$$C(z_1,\overline{z}_1,z_2,\overline{z}_2) = |w'(z_1)|^{b/2}\, |w'(z_2)|^{b/2}\, C(w_1,\overline{w}_1;w_2,\overline{w}_2), \tag{2.45}$$

where the exponent b is introduced to describe the rescaling of the fields.

2.3.1 *Exact results in two dimensions*

Let $C_w(\vec{r},\vec{r}\,')$ be the two-point correlation function of a *polydispersed network* $\mathcal{G}$ with connectivity as in the watermelon graph, with f branches (self-avoiding

TABLE 2.7. Exact values of network exponents in $d = 2$

$x_1 = \frac{5}{48}$	$\sigma_1 = \frac{11}{64}$	$\nu = \frac{3}{4}$	$\eta = \frac{5}{24}$
$x_2 = \frac{2}{3}$	$\sigma_2 = 0$	$\gamma = \frac{43}{32}$	$\alpha_s = \frac{5}{2}$
$x_3 = \frac{77}{48}$	$\sigma_3 = -\frac{29}{64}$	$\gamma(3) = \frac{17}{16}$	$\gamma_{W,3} = -\frac{29}{32}$
$x_4 = \frac{35}{12}$	$\sigma_4 = -\frac{19}{16}$	$\gamma(4) = \frac{1}{2}$	$\gamma_{W,4} = -\frac{23}{8}$

walks between the nodes; see figure 2.7(d)) and with one node at $\vec{r}$ and the other at $\vec{r}\,'$.

By equation (2.45), $C_w(\vec{r}, \vec{r}\,')$ decays as

$$C_w(\vec{r}, \vec{r}\,') \sim |\vec{r} - \vec{r}\,'|^{-2x_f}, \tag{2.46}$$

where x_f is a scaling exponent associated with the vertices in $\mathcal{G}$. If $f = 1$, then this is the self-avoiding walk correlation function, and, from equation (1.41), $2x_1 = \eta$.

The *susceptibility* of the model is defined by

$$\chi = \int C_w(\vec{r}, \vec{r}\,')\, d\vec{r}\,'. \tag{2.47}$$

Assuming that the integral is dominated by contributions where $|\vec{r} - \vec{r}\,'| = O(\xi)$ (where ξ is the correlation length), it follows that $\chi \sim \xi^{d-2x_f}$. By equations (1.40) and (1.15), $\chi \sim \xi^{\gamma/\nu}$, so the entropic exponent of the watermelon network $\mathcal{G}$ is given by

$$\gamma_{W,f} = \nu(d - 2x_f). \tag{2.48}$$

On the other hand, by equation (2.41),

$$\gamma_{W,f} = 1 + 2\sigma_f - (f-1)d\nu + (f-1), \tag{2.49}$$

where the term $(f-1)$ is added to account for the polydispersed nature of the network. This gives the f-star exponent σ_f by

$$2\sigma_f = (d\nu - 1)f - 2\nu x_f. \tag{2.50}$$

It remains to determine the x_f. These are given exactly (but not rigorously) by conformal dimensions $h_{r,s}$ associated with operators $\phi_{r,s}$ in a conformal field theory with central charge $C = 0$ [152, 500]. The identification of these operators with those in a Virasoro algebra allows the x_f to be given by the zeros of the Kac determinant [344]:

$$h_{r,s} = \tfrac{1}{24}\left((3p - 2q)^2 - 1\right). \tag{2.51}$$

In this particular case $x_f = 2h_{p+\frac{1}{2},\frac{3}{2}}$ if $f = 2p+1$, and $x_f = 2h_{p+2,3}$ if $f = 2p$ [168]. Simplification gives

$$x_f = \tfrac{1}{48}\left(9f^2 - 4\right). \tag{2.52}$$

For example, since $2x_1 = \eta$, this gives the exact value $\eta = \frac{5}{24}$ for the anomalous dimension of the self-avoiding walk in $d = 2$. Som exact values of x_f in two dimensions are displayed in table 2.7.

Since $\sigma_2 = 0$, it follows from equation (2.50) that $\sigma_2 = 2\nu - 1 - x_2\nu = 0$. This gives $\nu = \frac{3}{4}$ because $x_2 = \frac{2}{3}$, and, by Fisher's scaling relation (equation (1.43)), $\gamma = \frac{43}{32}$ for the self-avoiding walk.

With ν determined, the exponents σ_f can be determined exactly from equation (2.50), and, by exploiting equation (2.34), also the entropic exponents of any uniform network. For example, $x_1 = \frac{5}{48}$, so $\sigma_1 = \frac{11}{64}$.

The entropic exponent α_s for polygons can be determined by putting $f = 2$ in equation (2.48). This gives $\gamma_{W,2} = \frac{1}{2}$. A polydispersed watermelon has two branch points and, if $f = 2$, then this is a polygon with two roots. The entropic exponent for such polygons is $\alpha_s - 1$, and this is equal to $\gamma_{W,2} - 1$. Thus, it follows that $\alpha_s = \frac{1}{2}$.

For a uniform star polygon of f arms, equation (2.34) gives the formula

$$\gamma_f = 1 + f\sigma_1 + \sigma_f = \tfrac{1}{64}(68 + 9f(3 - f)). \tag{2.53}$$

This has been tested numerically in references [218, 234, 586, 594]. More generally, the entropic exponent of a network $\mathcal{G}$ is given by equation (2.34). Putting $f = 1$ gives the self-avoiding entropic exponent $\gamma = \frac{43}{32}$ (this is also the result for $f = 2$).

The entropic exponent of uniform watermelons of f branches follows from equation (2.48):

$$\gamma_{W,f} = 1 + 2\sigma_f - \tfrac{1}{2}(f - 1) = \tfrac{3}{4}(2 - 2x_f) = \tfrac{3}{2} - \tfrac{1}{32}(9f^2 - 4). \tag{2.54}$$

Putting $f = 1$ gives $\gamma_{W,1} = 1 + 2\sigma_1 = \frac{3}{2} - \frac{5}{32} = \frac{43}{32}$, as expected.

2.3.2 *Conformal invariance near a hard wall*

Consider the correlation function $C(z_1, \overline{z}_1; z_2, \overline{z}_2)$ in equation (2.45) near a hard wall (where $z_i = x_i + iy_i$). The conformal map

$$\frac{z'}{|z'|^2} = \frac{z}{|z|^2} + a \tag{2.55}$$

with $a = (\epsilon, 0)$ maps hyperspheres into hyperspheres and is translationally invariant. The Jacobian of the transformation $z \to w(z) = z'$ is

$$|w'(z)| = 1 + 4\epsilon x + O(\epsilon^2), \tag{2.56}$$

while to first order $z' = x' + iy'$ is given by

$$x' - x = \epsilon(x^2 - y^2), \text{ and } y' - y = 2\epsilon xy. \tag{2.57}$$

To first order in ϵ, this shows that equation (2.45) becomes

$$C(x_1 - x_2, y_1, y_2) = (1 + 4\epsilon x_1)^{b/2}(1 + 4\epsilon x_2)^{b/2} C(x_1' - x_2', y_1', y_2') + O(\epsilon^2). \quad (2.58)$$

Put $u = x_1 - x_2$ and expand this to $O(\epsilon)$. Note in particular that $x_1' - x_2' = u'$ with $u' = u + \epsilon(x_1^2 - x_2^2 - y_1^2 + y_2^2)$, as well as $y_1' = y_1 + \epsilon(x_1 + x_2 + u)y_1$ and $y_2' = y_2 + \epsilon(x_1 + x_2 - u)y_2$. The result is that

$$\begin{aligned}\left(u(x_1 + x_2) - y_1^2 + y_2^2\right) \tfrac{\partial}{\partial u} C + y_1 \left(x_1 + x_2 + u\right) \tfrac{\partial}{\partial y_1} C \\ + y_2 \left(x_1 + x_2 - u\right) \tfrac{\partial}{\partial y_2} C + 2b \left(x_1 + x_2\right) C = 0.\end{aligned} \quad (2.59)$$

However, $C(x_1 - x_2, y_1, y_2)$ is independent of $x_1 + x_2$, so the above separates into two equations:

$$u \tfrac{\partial}{\partial u} C + y_1 \tfrac{\partial}{\partial y_1} C + y_2 \tfrac{\partial}{\partial y_2} C + 2bC = 0; \quad \text{and} \quad (2.60)$$

$$\left(y_2^2 - y_1^2\right) \tfrac{\partial}{\partial u} C + u y_1 \tfrac{\partial}{\partial y_1} C - u y_2 \tfrac{\partial}{\partial y_2} C = 0. \quad (2.61)$$

The first of the differential equations above shows that

$$C(u, y_1, y_2) = u^{-2b} \Psi\left(\tfrac{y_1}{u}, \tfrac{y_2}{u}\right) \quad (2.62)$$

for some scaling function Ψ. Comparison to equation (1.41) shows consistency of this result with the general scaling form assumed in these models.

Inserting the above solution into the second of the differential equations and defining $\zeta_i = \frac{1}{u} y_i$ gives

$$\zeta_1 \left(1 + \zeta_1^2 + \zeta_2^2\right) \tfrac{\partial}{\partial \zeta_1} \Psi - \zeta_2 \left(1 - \zeta_1^2 + \zeta_2^2\right) \tfrac{\partial}{\partial \zeta_2} \Psi = \tfrac{2b}{u^2} \Psi \quad (2.63)$$

for Ψ. The right-hand side can be ignored if $\epsilon > 0$ is small since $u = O\left(\frac{1}{\epsilon}\right)$ is large. The resulting partial differential equation has the general solution

$$\Psi = \widetilde{\Psi}\left(\tfrac{1+\zeta_1^2+\zeta_2^2}{\zeta_1 \zeta_2}\right). \quad (2.64)$$

That is, the function Ψ in equation (2.62) is a function of one variable. This gives the scaling form

$$C(u, y_1, y_2) = u^{-2b} \widetilde{\Psi}\left(\tfrac{u^2+y_1^2+y_2^2}{y_1 y_2}\right) \quad (2.65)$$

for the two-point function.

It remains to determine the asymptotic form of $\widetilde{\Psi}$ and to identify the exponent b. Equation (1.41) suggests that $C(u, y_1, y_2) \sim u^{-\eta_\parallel}$ if $y_1 = O(1)$, and $y_2 = O(1)$, and for u large. Thus, $C(u, y_1, y_2) \sim u^{-2b} \widetilde{\Psi}(u^2) = u^{-2b} u^{\eta - \eta_\parallel}$ and this implies that $-2b + \eta = 0$ when $b = \frac{1}{2}\eta$.

Since $\widetilde{\Psi}$ is a function of u^2, this shows that

$$\widetilde{\Psi}(t) \sim t^{(\eta-\eta_{\|})/2}. \tag{2.66}$$

On the other hand, if $y_1 = O(1)$, and $y_2 = R\cos\theta$, and $u = R\sin\theta$ with R large, then $C(R,\theta) \sim R^{-\eta_\perp}$. Substitution in equation (2.65) and then taking R large shows that

$$\widetilde{\Psi}(R) \sim R^{(\eta-\eta_\perp)}. \tag{2.67}$$

Comparing equations (2.66) and (2.67) gives the scaling relation [39]

$$2\eta_\perp = \eta + \eta_{\|}. \tag{2.68}$$

The angular dependence of $\widetilde{\Psi}$ is $\widetilde{\Psi}(R,\theta) \sim (\cos\theta)^{-\eta+\eta_\perp}$ by equation (2.66).

2.3.3 *Exact exponents near a hard wall*

If the two nodes of a watermelon network are close to a hard wall, then an argument similar to that leading to equation (2.48) gives the relation

$$\gamma'_{W,f} = \nu(d - 2x'_f) \tag{2.69}$$

between the entropic exponent $\gamma_{W,f}$, the metric exponent ν and the exponent x'_f (which rescales fields near the wall). A comparison to equation (2.41) gives

$$2\sigma'_f = (d\nu - 1)f - 2\nu x'_f \tag{2.70}$$

for the f-star exponents σ'_f with a central node in the wall. The exponent x'_f relates the decay of the correlation function as above:

$$C'_w(r,r') \sim |r - r'|^{-2x'_f} \tag{2.71}$$

and it is the case that $2x'_1 = \eta_{\|}$ (see equation (2.46)).

Identifying the exponent x'_f with $h_{r,s}$ by choosing $r = f+1$ and $s = 1$ in equation (2.51) gives

$$x'_f = h_{f+1,1} = \tfrac{1}{8}f(3f+2). \tag{2.72}$$

This produces the spectrum of exponents for networks close to a wall [168]. In particular, if $f = 1$, then $x'_1 = \frac{5}{8}$, and $\sigma'_1 = -\frac{7}{32}$. This shows that $\eta_{\|} = 2h_{2,1} = \frac{5}{4}$ and it follows that $\eta_\perp = \frac{35}{48}$ from equation (2.68). The value of γ_1 follows from equation (2.36).

It can be shown that $x'_2 = 2$, and $\sigma'_2 = -1$, from which it follows that $\gamma_{11} = -\frac{3}{16}$ by equation (2.39). Other exponents are given in table 2.8. The values of γ_1 and γ_{11} satisfy the Barber scaling relation in equation (2.40).

More generally, the entropic exponent of a network $\mathcal{G}$ near a wall is given by equation (2.41). If $\mathcal{G}$ is an f-star with its vertex attached to the hard wall, then the entropic exponent is given by equation (2.37). This simplifies to

$$\gamma_f = 1 + \tfrac{3}{64}f(5-6f). \tag{2.73}$$

If $f = 1$, then $\gamma_1 = \frac{61}{64}$ is the entropic exponent of a half-space walk. Putting $f = 2$ gives the exponent for a walk with its middle vertex attached to the wall, namely, $\gamma_2 = \frac{11}{32}$. If $f = 3$, then $\gamma_3 = -\frac{53}{64}$, while $\gamma_4 = -\frac{41}{16}$ for $f = 4$.

TABLE 2.8. Exact values of surface network exponents in $d = 2$

$x'_1 = \frac{5}{8}$	$\sigma'_1 = -\frac{7}{32}$	$\eta_\parallel = \frac{5}{4}$	$\eta_\perp = \frac{35}{48}$
$x'_2 = 2$	$\sigma'_2 = -1$	$\gamma_1 = \frac{61}{64}$	$\gamma_{11} = -\frac{3}{16}$
$x'_3 = \frac{33}{8}$	$\sigma'_3 = -\frac{75}{32}$	$\gamma_2 = \frac{11}{32}$	$\gamma'_{W,3} = -\frac{75}{16}$
$x'_4 = 7$	$\sigma'_4 = -\frac{17}{4}$	$\gamma_3 = -\frac{53}{64}$	$\gamma'_{W,4} = -9$

A uniform watermelon with one node near a wall can similarly be shown to have the entropic exponent

$$\gamma_f^W = 1 + (d\nu - 1)f - \nu x_f - \nu x'_f - (f-1)d\nu = \tfrac{1}{16}(41 - 19f - \tfrac{27}{4}f^2). \tag{2.74}$$

For example, if $f = 1$, then the exponent $\gamma_1 = \gamma_1^W = \frac{61}{64}$ is recovered and, if $f = 2$, then $\gamma_2^W = -\frac{3}{2}$ is the exponent of a polygon rooted in the wall.

The scaling of a self-avoiding walk near an excluded line was considered in reference [555]. This is a special case of a network in a wedge. More generally, a conformal transformation to a network in a wedge of angle θ with a vertex of degree f in the vertex of the wedge [168] gives the γ-exponent of the network as

$$\gamma_\mathcal{G}(\theta) = \gamma_\mathcal{G} - \nu\left(\tfrac{\pi}{\theta} - 1\right)x'_f. \tag{2.75}$$

For example, if $\mathcal{G}$ is an f-star with its vertex of degree f in the vertex of the wedge, then this equation reduces to

$$\gamma_{wedge,f} = 1 + \tfrac{27}{64}f - \tfrac{3\pi}{32\theta}f(3f+2). \tag{2.76}$$

Similarly, if a uniform watermelon is placed with one node near the vertex of the wedge, then its entropic exponent is

$$\gamma_{wedge,f}^W = \tfrac{1}{16}(41 - 16f - \tfrac{9}{4}f^2) - \tfrac{3\pi}{32\theta}f(3f+2). \tag{2.77}$$

Putting $f = 1$ and $\theta = \pi$ in the last two cases recovers the exponent $\gamma_1 = \frac{61}{64}$. Putting $f = 2$ in the last equation gives the entropic exponent of polygons rooted in the vertex of a wedge, namely $\gamma_{wedge,2}^W = -\frac{3\pi}{2\theta}$.

2.4 The Edwards model

The Edwards model for a polymer chain [173] is constructed by assigning a probability measure to paths from $\vec{0}$ in $\mathbb{R}^d$. Let $\vec{r}(t)$ be the position vector of a polymer chain in $\mathbb{R}^d$, parameterised by $t \in [0, \ell]$ such that $\vec{r}(0) = \vec{0}$. Each conformation of the polymer is assigned a measure $P\{\vec{r}\} = e^{-S(\vec{r})}$, where the *action* is given by

$$S(\vec{r}) = \tfrac{1}{2}\int_0^\ell \left[\tfrac{d}{dt}\vec{r}(t)\right]^2 dt + \tfrac{1}{2}b\int_0^\ell dt \int_0^\ell dt'\, \delta(\vec{r}(t) - \vec{r}(t'))\,. \tag{2.78}$$

The parameter b is the *excluded volume* parameter, and δ is the Dirac δ-function.

The partition function is given by

$$Z_b = \int \mathcal{D}\{\vec{r}\} P\{\vec{r}\}, \tag{2.79}$$

where the integral is over all continuous paths from $\vec{0}$. This integral is ill defined but a more precise definition may be given by integrating over all Brownian paths from the origin with the Wiener measure μ. If the paths are continuous functions $\vec{r}(t)$, then the probability measure ν on the paths is defined in terms of μ by

$$\frac{d\nu}{d\mu} = \frac{1}{Z_b} e^{-S_1}, \tag{2.80}$$

where

$$S_1 = \tfrac{1}{2} b \int_0^\ell dt \int_0^\ell dt' \, \delta(\vec{r}(t) - \vec{r}(t')) \tag{2.81}$$

is the *interaction term* in $S(\vec{r})$, and the normalising constant Z_b is

$$Z_b = \int e^{-S_1} \, d\mu. \tag{2.82}$$

The existence of the probability measure ν was proven by Westwater in references [576–578, 578]. Additional results on the model (including a central limit theorem for the distribution of the endpoint of the path in one dimension) were proven in reference [553].

SF Edwards developed a self-consistent field approach for this model [174] (see also reference [579] for a lattice self-avoiding walk derivation).

The general structure of Z_b in equation (2.79) is similar to that of the path integral in quantum field theory [188]. It may be analysed using renormalisation group and dimensional reduction methods.

Rescale $\vec{r} \to \sqrt{\ell}\,\vec{\rho}$, and $t \to x\ell$. This reduces $S(\vec{r})$ in equation (2.78) to

$$S(\vec{\rho}) = \tfrac{1}{2} \int_0^1 \left[\tfrac{d}{dx} \vec{\rho}(x)\right]^2 dx + \tfrac{1}{2} z \int_0^1 dx \int_0^1 dx' \, (2\pi)^{d/2} \, \delta(\vec{\rho}(x) - \vec{\rho}(x')) \tag{2.83}$$

where the *coupling strength* of the interacting term is

$$z = (2\pi)^{-d/2} \, b \ell^{2-d/2}. \tag{2.84}$$

This gives the upper critical dimension of the model: if $d > 4$, then $z \to 0$ as $\ell \to \infty$, and the coupling is irrelevant. If $d < 4$, on the other hand, then $z \to \infty$ as $\ell \to \infty$; this shows that the coupling is relevant in the physics of the model as the continuum limit is taken (when $\ell \to \infty$). The case $d = 4$ is marginal.

A *free path* with partition function Z_0 is obtained if $b = 0$. This gives suitable normalisation $\mathcal{Z}_b = \frac{Z_b}{Z_0}$ for Z_b. An ϵ-expansion (where $\epsilon = 4 - d$) of $\mathcal{Z}_b$ in the coupling z would be

$$\mathcal{Z}_b = 1 + a_1(\epsilon) z + a_2(\epsilon) z^2 + a_3(\epsilon) z^3 + \cdots, \tag{2.85}$$

where the $a_i(\epsilon)$ are singular in ϵ. These coefficients are usually expanded in $\frac{1}{\epsilon}$ and then approximately computed in perturbation theory.

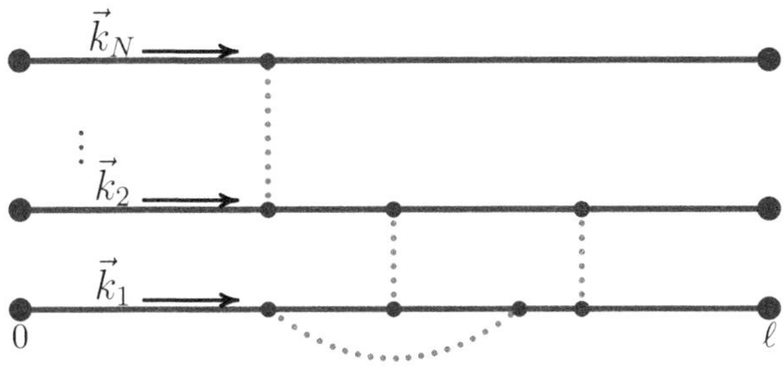

FIG. 2.9. A diagrammatic representation of N polymer lines, each carrying momentum $\vec{k}_i$. Interactions between the lines are indicated by dotted arcs and lines.

A similar approach may be taken to the metric scaling of the walk. In the case of free paths (when $z = 0$ in the above) the (normalised) mean square displacement between $\vec{0}$ and the other endpoint of the walk has scaling $R_0^2 = \ell$ (since $R_0^2 \sim \ell^{2\nu}$, this gives $\nu = \frac{1}{2}$).

Once the interaction is switched on, the walk *swells* or expands and this is seen in the metric scaling by the appearance of a *swelling factor* $\mathfrak{X}_0$. Thus,

$$R_b^2 = \mathfrak{X}_0 \ell. \tag{2.86}$$

The swelling factor may be expanded similarly in an ϵ-expansion as

$$\mathfrak{X}_0 = \frac{R_b^2}{R_0^2} = 1 + r_1(\epsilon) z + r_2(\epsilon) z^2 + \cdots . \tag{2.87}$$

As for the coefficient $a_i(\epsilon)$ in equation (2.85), the $r_i(\epsilon)$ may be determined approximately by an expansion in perturbation theory.

2.4.1 *Perturbation expansion of the Edwards model*

The goal of using perturbation theory is to calculate the coefficients in equations (2.85) and (2.87). The basic approach is to expand the interaction term in the action $S(\vec{r})$ and then to compute the terms one by one by integrating the resulting Gaussian integrals.

Proceed by noting that

$$\delta(\vec{r}) = \frac{1}{(2\pi)^d} \int d\vec{q} e^{i\vec{q}\cdot\vec{r}} \tag{2.88}$$

in equation (2.83) and that the exponential in equation (2.82) can be expanded in powers of z. The resulting integrals are Gaussian, and integration over the Wiener measure amounts to the calculation of Green's functions for the Brownian chain. In Fourier space,

$$\frac{\int \mathcal{D}\{\vec{r}\} e^{-S_0(\vec{r})} e^{i\vec{q}\cdot(\vec{r}(t)-\vec{r}(s))}}{\int \mathcal{D}\{\vec{r}\} e^{-S_0(\vec{r})}} = e^{-\frac{1}{2}q^2|s-t|}, \tag{2.89}$$

where $S_0(\vec{r}) = \frac{1}{2}\int_0^\ell \left[\frac{d}{dt}\vec{r}(t)\right]^2 dt$, and $q^2 = \vec{q}\cdot\vec{q}$.

Expanding the interaction term $e^{-S_1(\vec{r})}$ in powers b produces a number of terms at $O(b^n)$. These terms can be conveniently classified by using Feynman-style rules and diagrams.

It is more convenient to consider a model incorporating N chains as illustrated in figure 2.9. The action in equation (2.78) becomes

$$S_N(\vec{\rho}) = \tfrac{1}{2}\sum_{i=1}^{N}\int_0^{\ell_i}\left[\tfrac{d}{dx}\vec{r}_i(x)\right]^2 dx + \frac{b}{2}\sum_{i,j=1}^{N}\int_0^{\ell_i} dx \int_0^{\ell_j} dx'\,\delta(\vec{r}_i(x)-\vec{r}_j(x')) \quad (2.90)$$

for N chains of lengths $\langle \ell_1, \ell_2, \ldots, \ell_N\rangle$. If the chains have different lengths, the model is said to be *polydispersed*; otherwise, if $\ell_i = \ell_j$, it is *monodispersed.* In what follows the monodispersed model will be examined. The approach to a polydispersed system is similar and, if the chains have the same length ℓ up to $o(\ell)$ terms, results will be the same as for the monodispersed model.

An example of a term in the perturbation expansion is illustrated in figure 2.9. Polymer backbone lines are in solid and each carries a momentum vector $\vec{k}_i$. Interactions between the polymer backbones are denoted by dotted lines. A factor $(-b)$ is associated with each interaction, so the diagram in figure 2.9 has a factor $(-b)^4$; it is a *fourth order* contribution.

Examination of the terms in the perturbation series gives a set of rules whereby the terms can be recovered from a set of diagrams. In the case of the Edwards model the rules are as follows:

- A diagram consists of N polymer *backbones* or *propagators* carrying momenta $\vec{k}_i$. A symmetry factor for the diagram is determined by assuming the polymer chains are discernible. *Interactions* between points (called *vertices*) on the backbones are dotted lines joining pairs of vertices.
- Each polymer backbone is a line from left to right parameterised by $t_i \in [0,\ell]$. The point $t_i = 0$ corresponds to $\vec{r}_i = \vec{0}$, and $\vec{r}_i(\ell)$ is the other endpoint of the polymer. Momentum $\vec{k}_i$ enters the line at $t_i = 0$ and exits at $t_i = \ell$.
- A factor $(-b)$ is inserted for each interaction within the polymer or polymer backbones. If the total weight is $O(b^n)$, then the diagram is n-th order.
- Enumerate the number of independent loops in the diagram and associate with each loop a momentum $\vec{q}_j$.
- Conserve momentum at each vertex and determine the total momentum flow in each branch of the diagram.
- The Green's function in equation (2.89) propagates total momentum $\sum \vec{k}_m$ along the polymer backbone. Along each section of the backbone between consecutive interaction vertices at t_i and t_{i+1}, propagate the total momentum $\sum \vec{k}_m$ by the factor $e^{-\frac{1}{2}\left(\sum \vec{k}_m\right)^2 (t_{i+1}-t_i)}$.
- Integrate over the locations of vertices t_i in the diagram along the backbone while maintaining the topological integrity of the diagram.

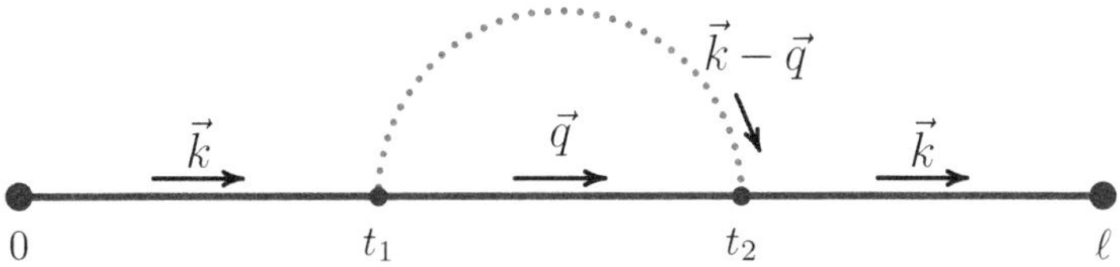

FIG. 2.10. The $O(b)$ term in the perturbation expansion of the reduced Edwards partition function. This diagram has one loop corresponding to an interaction between points $\vec{r}(t_1) = \vec{r}(s_2)$ along the polymer backbone. The loop carries momentum $\vec{q}$. The contribution of this diagram is $z\left(1+\frac{2}{\epsilon}\right)$.

- Integrate over all internal momenta $\vec{q}_j$ with measure $\frac{1}{(2\pi)^d} d^d\vec{q}_j$.
- Integrate the external momentum flow $\vec{k}_i$ through the diagram with the measure $(2\pi)^d\delta(\vec{k}_i)\times\frac{1}{(2\pi)^d} d^d\vec{k}_j$.

Integrating the momenta in a diagram is most usefully done by noting that the Gaussian integral in d dimensions has value

$$\int \frac{d^d\vec{k}}{(2\pi)^d}\, e^{-\frac{1}{2}k^2 t - i\vec{k}\cdot\vec{x}} = \frac{1}{(2\pi t)^{d/2}} e^{-\frac{1}{2}x^2/t}, \tag{2.91}$$

where $x^2 = \vec{x}\cdot\vec{x}$, and $t > 0$.

The zeroth order contribution is equal to 1.

Consider, as an example, the first order contribution in a model of a single chain illustrated in figure 2.10. There is a factor $(-b)$ for the interaction, and the loop carries total momentum $\vec{k}-\vec{q}$. Add a factor $e^{-\frac{1}{2}k^2(t_1-0)}$ for the backbone from 0 to t_1, where $k^2 = \vec{k}\cdot\vec{k}$, and then $e^{-\frac{1}{2}q^2(t_2-t_1)}$ for the backbone from t_1 to t_2, and, finally, $e^{-\frac{1}{2}k^2(\ell-t_2)}$ for the part from t_2 to ℓ. Integrating over $\{t_1, t_2\}$ and the momenta gives the contribution

$$-b\int \frac{d\vec{k}}{(2\pi)^d}(2\pi)^d\delta(\vec{k})\int\frac{d\vec{q}}{(2\pi)^d}\int_0^\ell dt_1\int_{t_1}^\ell dt_2\, e^{-\frac{1}{2}k^2 t_1} e^{-\frac{1}{2}q^2(t_2-t_1)} e^{-\frac{1}{2}k^2(\ell-t_2)}.$$

This simplifies to

$$I_1 = -b\int \frac{d\vec{k}}{(2\pi)^d}(2\pi)^d\delta(\vec{k})\int\frac{d\vec{q}}{(2\pi)^d}\int_0^\ell dt_1\int_{t_1}^\ell dt_2\, e^{-\frac{1}{2}(q^2-k^2)(t_2-t_1)} e^{-\frac{1}{2}k^2\ell}.$$

Substituting $t_2 - t_1 = s$, and $t_1 = t$, and partially integrating over t_1 and t_2 gives

$$I_1 = -b\int \frac{d\vec{k}}{(2\pi)^d}(2\pi)^d\delta(\vec{k})\int\frac{d\vec{q}}{(2\pi)^d}\int_0^\ell dt\,(\ell-t)e^{-\frac{1}{2}(q^2-k^2)t} e^{-\frac{1}{2}k^2\ell}. \tag{2.92}$$

Next, integrate over momentum $\vec{q}$ by using equation (2.91) (integration over external momentum $\vec{k}$ is simplified by the delta function $\delta(\vec{k})$, which has the effect of substituting $\vec{k}=\vec{0}$). The result is

$$I_1 = -b\int_0^\ell (\ell - t)(2\pi t)^{-d/2}\,dt = -z\int_0^1 (1-x)x^{-d/2}\,dx. \tag{2.93}$$

The integral (2.93) is singular because of ultraviolet divergences at short length scales. This is regularised by the introduction of an ultraviolet cut-off in a process of dimensional regularisation. Notice that, for $d < 2$, the integral (and partition function) is finite and may be evaluated explicitly. Dimensional regularisation is a process whereby the result for $d < 2$ (free of ultraviolet divergences) is continued analytically to $d \geq 2$ [135].

In the case of the integral I_1, an ultraviolet cut-off s_0 is introduced to get

$$I_1^r(s_0) = -b\int_{s_0}^\ell (\ell - t)(2\pi t)^{-d/2}\,dt. \tag{2.94}$$

This integral is divergent as $s_0 \to 0$. The effects of the singularity is explored by regularising in an ϵ-expansion near $d = 4$. Put $d = 4 - \epsilon$ and evaluate the integral. This gives

$$I_1^r(s_0) = \frac{-4(2\pi)^{-d/2}b\ell^{2-d/2}}{\epsilon(\epsilon-2)} + C(s_0), \tag{2.95}$$

where $C(s_0)$ is a function of the cut-off.

Substituting this into the perturbation expansion and recovering the partition function gives a general result [135] of the form

$$\mathcal{Z}_b \simeq e^{\ell C_0(s_0)}\,\mathcal{Z}_r \tag{2.96}$$

for equation (2.85). In other words, the dependence on the cut-off s_0 factors from the partition function and leaves the regularised partition function $\mathcal{Z}_r$ which is the physical object of interest. The factor $e^{\ell C_0(s_0)}$ is a bulk entropy term where $C_0(s_0)$ sets the 'length scale' of monomers in the model ($e^{C_0(s_0)}$ may be considered analogous to the growth constant in the model, that is, analogous to μ_d in equation (1.12)).

The dimensionally regularised partition function $\mathcal{Z}_r$ has scaling

$$\mathcal{Z}_r \sim \ell^{\gamma-1} = \ell^{2\sigma_1}, \quad \text{or by equation (2.84)}, \quad \mathcal{Z}_r \sim z^{2(\gamma-1)/\epsilon} = z^{4\sigma_1/\epsilon}, \tag{2.97}$$

where γ is the usual entropic exponent of self-avoiding walks, and σ_1 is the exponent corresponding to the endpoint of a walk (see equation (2.29)). Thus, regularisation leaves only the power law behaviour in $\mathcal{Z}_r$.

To determine the expansion for the swelling factor, consider the integration over the 'Brownian measure' in equation (2.79). For an external momentum $\vec{p}$,

$$\mathcal{R}(\vec{p}) = \tfrac{1}{\mathcal{Z}_b}\int \mathcal{D}\{\vec{r}\}P\{\vec{r}\}e^{i\vec{p}\cdot(\vec{r}(0)-\vec{r}(\ell))}. \tag{2.98}$$

The mean square end-to-end distance of the walk is given by

$$R_b^2 = -2\ell \left(\frac{d}{dp^2} \mathcal{R}(\vec{p}) \right) \Big|_{\vec{p}=0}. \tag{2.99}$$

In a perturbative approach, the above is implemented by replacing the normalisation $\mathcal{Z}_b$ with $\mathcal{Z}_0$ and by expanding $P\{\vec{r}\}$ in the parameter b.

Applying the rules for the diagram in figure 2.10 shows that the first order correction to $\mathcal{R}(\vec{p})$ in momentum space is

$$-b \int \frac{d\vec{k}}{(2\pi)^d} (2\pi)^d \delta(\vec{k}) \int \frac{d\vec{q}}{(2\pi)^d} \int_0^\ell dt\, (\ell - t) e^{-\frac{1}{2}(q^2-k^2)t} e^{-\frac{1}{2}k^2\ell} e^{\frac{1}{2}p^2 t},$$

where the factor $e^{\frac{1}{2}p^2(t_1-t_2)}$ was inserted for the swelling between the points t_1 and t_2. Taking the derivative to p^2 and multiplying by -2 gives, after putting $p^2 = 0$,

$$J_1 = b \int \frac{d\vec{k}}{(2\pi)^d} (2\pi)^d \delta(\vec{k}) \int \frac{d\vec{q}}{(2\pi)^d} \int_0^\ell dt\, t(\ell - t) e^{-\frac{1}{2}(q^2-k^2)t} e^{-\frac{1}{2}k^2\ell}. \tag{2.100}$$

Integrating this as above with a cut-off s_0 gives

$$J_1 = \frac{4(2\pi)^{-d/2} b \ell^{2-d/2}}{\epsilon(\epsilon+2)} + C_s(s_0) \tag{2.101}$$

as the first order contribution to the swelling factor $\mathfrak{X}_0 = 1 + J_1 + O(b^2)$.

Since $\mathfrak{X}_0$ is a dimensionless expansion factor in a metric quantity, it follows by equation (2.87) that its scaling is

$$\mathfrak{X}_0 \sim \ell^{2\nu-1} \text{ or, by equation (2.84), } \mathfrak{X}_0 \sim z^{2(2\nu-1)/\epsilon}, \tag{2.102}$$

where ν is the metric exponent of the walk.

It remains to extract the exponents γ and ν from the results in equations (2.95) and (2.101).

2.4.2 *Direct renormalisation in the Edwards model*

Since the calculation is only at the one-loop level, substitute $z = (2\pi)^{-d/2} b \ell^{2-d/2}$ and expand equation (2.95): $I_1^r = 2z\frac{1}{\epsilon} + z + O(\epsilon)$. This gives for the regularised partition function an expansion similar to equation (2.85) at the one-loop level:

$$\mathcal{Z}_r = 1 + z\left(2\frac{1}{\epsilon} + 1\right) + O(z^2). \tag{2.103}$$

Similarly, equation (2.101) yields

$$\mathfrak{X}_0 = 1 + z\left(2\frac{1}{\epsilon} - 1\right) + O(z^2) \tag{2.104}$$

for the swelling factor.

Define the scaling functions for $\mathcal{Z}_r$ and $\mathfrak{X}_0$ by

$$\sigma_1(z,\epsilon) = \tfrac{1}{2}\tfrac{\partial}{\partial \log \ell} \log \mathcal{Z}_r, \quad \text{and } \nu(z,\epsilon) = \tfrac{1}{2}\tfrac{\partial}{\partial \log \ell} \log \mathfrak{X}_0 \tag{2.105}$$

from equations (2.97) and (2.102), respectively. Noting that $z = (2\pi)^{-d/2} b \ell^{2-d/2}$, the above may be written as derivatives to z:

$$\sigma_1(z,\epsilon) = \tfrac{1}{2}\epsilon z \tfrac{\partial}{\partial z} \log \mathcal{Z}_r, \quad \text{and } \nu(z,\epsilon) = \tfrac{1}{2}\epsilon z \tfrac{\partial}{\partial z} \log \mathfrak{X}_0. \tag{2.106}$$

The scaling functions are Laurent series in ϵ and divergent series of z. In renormalising the theory, scaling functions should become regular in ϵ and the (renormalised) interaction parameter (that is, have double Taylor series expansion in ϵ and a renormalised interaction z_R). This observation gives a renormalisation principle for the Edwards model (see references [133] and [166]).

Minimal renormalisation principle: it is sufficient to introduce a minimally renormalised parameter z_R of the form

$$z_R = z + \sum_{n=2}^{\infty} \left(\sum_{j=1}^{n} \frac{a_j(n)}{\epsilon^j} \right) z^n \tag{2.107}$$

such that any scaling function becomes free of singularities in ϵ to all orders in z_R. The implementation is to replace z by z_R and to determine the a_j by removing singularities in ϵ order by order in z in the scaling function.

The simplest approach to renormalisation is to compute the fixed point of the (dimensionally reduced) second virial coefficient g. This virial coefficient is defined by expanding the osmotic pressure Π of a dilute solution of polymer chains of concentration C:

$$\beta\Pi = C + \tfrac{1}{2} g C^2 \left(\tfrac{2\pi}{d} \langle R^2 \rangle\right)^{d/2} + \cdots, \tag{2.108}$$

where $\beta = 1/k_B T$ (T is temperature, and k_B is Boltzmann's constant). The factor $\langle R^2 \rangle$ is the mean square radius of gyration.

The virial coefficient may be computed in perturbation theory by considering interactions between two polymer backbones. In particular, the partition function in this case is

$$\mathcal{Z}_b(\ell,\ell) = -g\, \mathcal{Z}_b^2 \left(\tfrac{2\pi}{d} \langle R^2 \rangle\right)^{d/2}. \tag{2.109}$$

This partition function can be determined to the one-loop level by computing the diagrams in figure 2.11. This gives

$$\mathcal{Z}_b(\ell,\ell) = -z(2\pi\ell)^{d/2}[1 + O(\tfrac{z^2}{\epsilon^2})]. \tag{2.110}$$

The first three diagrams are *I-reducible*; that is, they can be cut into two parts by cutting an interaction line between the two polymer backbones. Alternatively, these diagrams can be obtained by joining two polymer backbones with

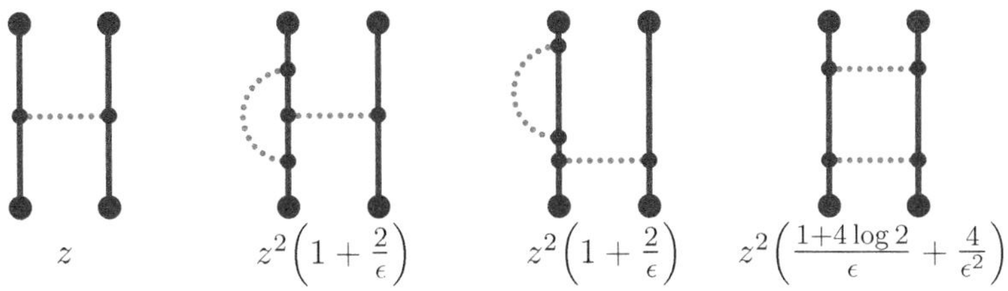

FIG. 2.11. Terms in the perturbation expansion to order $O(b^2)$ (or to one loop) for the partition function of two chains. The first three diagrams from the left are I-reducible since they can be split into independent diagrams by cutting a single interaction line. The right-most diagram is not I-reducible.

---

an interaction line. This process proceeds by first inserting a vertex in each polymer backbone and then joining them by an interaction line from one vertex to the other. This is illustrated in figure 2.12.

The effect of inserting a vertex in a polymer backbone is to split the propagator in half by inserting a new vertex. That is, $e^{\frac{1}{2}p^2 s} \to e^{\frac{1}{2}p^2 s_1} e^{\frac{1}{2}p^2 s_1}$ (where $s_1 + s_2 = s$). A factor of $\sqrt{-b}$ is also associated with the new vertex. This reduces to $\sqrt{-b}e^{\frac{1}{2}p^2 s}$, and hence the effect of the vertex is only the multiplicative factor $\sqrt{-b}$.

In other words, I-reducible diagrams are computed by cutting them into irreducible parts and then computing the parts independently. Finally, a factor of $(-b)$ is inserted for each interaction line which is cut.

Comparison of equations (2.109) and (2.110) gives

$$g \mathcal{Z}_b^2 \mathfrak{X}_0^{d/2} = z[1 + O(\tfrac{z^2}{\epsilon^2})]. \tag{2.111}$$

Inserting the results in equation (2.103) and (2.104) and solving for g gives

$$g = z - \tfrac{8}{\epsilon} z^2. \tag{2.112}$$

Inverting the series to obtain $z(g)$ gives $z = g + \frac{8}{\epsilon} g^2$.

The Wilson function [595] associated with g is

$$\begin{aligned} W(g, \epsilon) &= \tfrac{1}{2} \tfrac{\partial}{\partial \log \ell} g = \tfrac{1}{2} z\epsilon \tfrac{\partial}{\partial z} g \\ &= \tfrac{1}{2} z\epsilon \left(1 - \tfrac{16}{\epsilon} z\right) = \tfrac{1}{2} g\epsilon \left(1 - \tfrac{8}{\epsilon} g + O(g^2)\right). \end{aligned} \tag{2.113}$$

The fixed point g^* is determined by $W(g^*, \epsilon) = 0$. This gives the fixed point $g^* = \frac{\epsilon}{8}$ with the result that, to leading order in ϵ, the renormalised value of z is $z_R = g^* = \frac{\epsilon}{8}$ as $z \to \infty$.

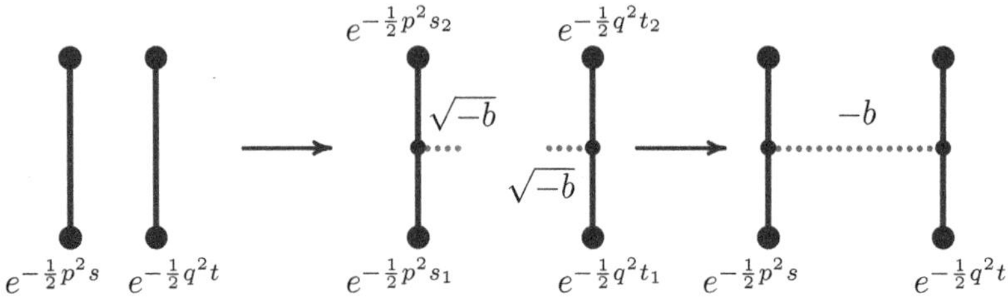

FIG. 2.12. An I-reducible diagram can be created by inserting a vertex in each of two polymer backbones and then joining them with an interaction. Since zero momentum is transferred between the backbones, the interaction vertex splits the backbone in two parts, so $e^{\frac{1}{2}p^2 s} \to e^{\frac{1}{2}p^2 s_1} e^{\frac{1}{2}p^2 s_1}$, plus a factor of $\sqrt{-b}$ for the backbone on the left. Since $s_1 + s_2 = s$, this becomes $\sqrt{-b}\, e^{\frac{1}{2}p^2 s}$. A similar argument shows that the contribution from the other backbone is $\sqrt{-b}\, e^{\frac{1}{2}q^2 s}$. Putting this together shows that the only effect of inserting the new interaction between the two polymer backbones is an overall factor of $(-b)$. Otherwise, the contribution is the product of the contributions of the two (independent) propagators.

Determining the scaling functions in equation (2.106) and then substituting $z = z_R$ gives the first order ϵ-expansion for the exponents σ_1 and ν. The expansion for the exponent γ follows from equation (2.97):

$$\nu = \tfrac{1}{2} + \tfrac{1}{16}\epsilon, \quad \text{and } \gamma = 1 + \tfrac{1}{8}\epsilon. \tag{2.114}$$

The ϵ-expansion can be extended to higher orders by computing the contribution of diagrams to higher orders in b. These diagrams are illustrated in figure 2.13.

A diagram is *reducible* if it can be cut in a propagator between two vertices into two independent diagrams. If a diagram is not reducible, then it is *irreducible* (the middle and right-most diagrams in figure 2.13 are irreducible).

An example of a reducible diagram is the left-most diagram in figure 2.13. The contribution of a reducible diagram is obtained by computing the contributions of each constituent irreducible part and then multiplying them together.

Computing the second virial coefficient to $O(z^3)$ (see reference [135]) requires the calculation of seven irreducible and I-irreducible diagrams with three interaction lines (see for example figure 2.11). The result is

$$g = z + z^2 \left(2 + 4\log 2 - \tfrac{8}{\epsilon}\right) + z^3 \left(\tfrac{64}{\epsilon^2} - (15 + 64\log 2)\tfrac{1}{\epsilon}\right). \tag{2.115}$$

The partition function and swelling factor to order $O(z^2)$ are obtained by examining the contribution of diagrams in figure 2.13 in addition to the lower order contribution in figure 2.10. This gives

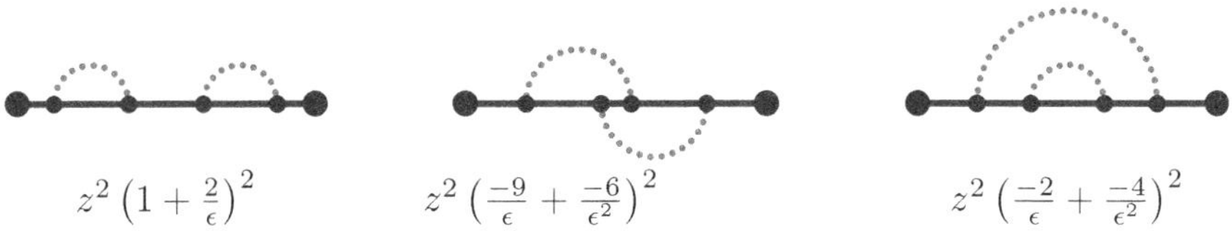

FIG. 2.13. Second order diagrams in the perturbation expansion for a single chain. The left-most diagram is reducible since it can be cut into two independent polymer backbones between the two interactions. The middle and right-most diagrams are irreducible.

$$\mathcal{Z}_r = 1 + z\left(\frac{2}{\epsilon}+1\right) - z^2\left(\frac{7}{\epsilon}+\frac{6}{\epsilon^2}\right), \quad \text{and} \tag{2.116}$$

$$\mathfrak{X}_0 = 1 + z\left(\frac{2}{\epsilon}-1\right) + z^2\left(\frac{11}{2\epsilon}-\frac{6}{\epsilon^2}\right). \tag{2.117}$$

The renormalised fixed point for g to $O(\epsilon^2)$ can be determined from equation (2.115). This gives

$$g^* = \tfrac{1}{8}\epsilon + \tfrac{1}{16}\left(\tfrac{25}{16} + \log 2\right)\epsilon^2 + \cdots . \tag{2.118}$$

The resulting values for the critical exponents are

$$\nu = \tfrac{1}{2}\left(1 + \tfrac{1}{8}\epsilon + \tfrac{15}{256}\epsilon^2 + \cdots\right), \quad \text{and } \gamma = 1 + \tfrac{1}{8}\epsilon + \tfrac{13}{256}\epsilon^2 + \cdots . \tag{2.119}$$

These results are consistent with the field theoretic calculations in reference [373] and can be compared with the expansions given in equations (1.44) and (1.45).

2.4.3 *The ϵ-expansion and uniform stars*

The scaling of the regularised partition function of a single polymer backbone in equation (2.97) suggests that σ_1 is associated with endpoints or end-vertices of the polymer. That is, since the polymer has two endpoints, consider that

$$\widehat{\mathcal{Z}}_r(1) \sim \sqrt{\mathcal{Z}_r} \sim \ell^{\sigma_1} \tag{2.120}$$

is a reduced and regularised partition function associated with one end-vertex in the polymer. In a general sense, the exponent σ_1 is a scaling function or operator (see equation (2.105)) associated with an endpoint of a linear polymer.

Generalising this to an f-star (see figure 2.5) suggests the scaling

$$\mathcal{Z}_r(f) \sim \ell^{\sigma_f + f\sigma_1} \tag{2.121}$$

for the regularised partition function of a monodispersed network. The exponent σ_f is a scaling function corresponding to the central node of the star, which is a branching point of degree f. This is consistent with equation (2.30). As with the argument preceding and following equation (2.29), it is expected that $\sigma_2 = 0$.

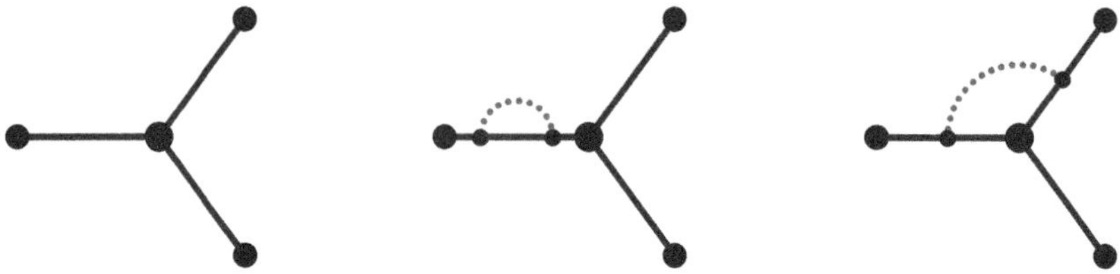

FIG. 2.14. Diagrams in the perturbation expansion for a star polymer to first order in z. In the middle diagram there is an interaction in one branch of the star, and in the right-most diagram, an interaction between two branches.

For more general monodispersed networks, the scaling of the regularised partition function may be determined from equation (2.34):

$$\mathcal{Z}_r(\mathcal{G}) \sim \ell^{\sum m_i \sigma_i - c(\mathcal{G}) d\nu}, \tag{2.122}$$

where m_i is the number of vertices in the network of degree i, and $c(\mathcal{G})$ is the cyclomatic index of the network.

To determine the scaling of the regularised partition function of a network, it is only necessary to determine the scaling functions σ_i associated with nodes of degree i in the network. These are star polymer scaling functions and, in the context of the Edwards model, they can be determined in an ϵ-expansion.

In order to determine the scaling of star networks, reduce the partition function by factoring the end-vertices in the arms of the star. This gives a reduced (and regularised) partition function $\widehat{\mathcal{Z}}_r(f)$ for the core (central node) of the star, defined by

$$\widehat{\mathcal{Z}}_r(f) = \mathcal{Z}_r(f) \times [\mathcal{Z}_r(1)]^{-f/2} \sim \ell^{\sigma_f} \tag{2.123}$$

in view of equations (2.120) and (2.121).

The rules for calculating a perturbative diagram of a network are the same as set out in section 2.4.1, with the minor modification that a network of fixed (graphical) topology and f branches is built from f polymer backbone lines. Interactions can be added in any branch or between any two branches.

In the case of a uniform star polymer, the terms to first order in z are illustrated in figure 2.14. Application of the rules of the perturbation expansion shows that the contribution of the middle diagram in figure 2.14 is

$$J_1 = -zf \int_0^1 (1-x)x^{-d/2}\, dx = \frac{-4zf}{(4-d)(2-d)} = zf\Big(1 + \tfrac{2}{\epsilon} + O(\epsilon)\Big). \tag{2.124}$$

Observe the appearance of the symmetry factor f, since there are f branches and the interaction can occur on anyone of these.

The contribution of the right-most diagram can be determined in the same way. It is given by

$$J_2 = -\tfrac{z}{2} f(f-1) \int_0^1 \int_0^1 (x+y)^{-d/2}\, dx dy$$
$$= \frac{-2zf(f-1)\left(2^{2-d/2}-2\right)}{(4-d)(2-d)} = zf(f-1)\left(\tfrac{1}{2}(\log 2 - 1) - \tfrac{1}{\epsilon} + O(\epsilon)\right). \qquad (2.125)$$

Noting that the regularised partition function for the core contribution of an f-star is $\widehat{\mathcal{Z}}_r(f) = 1 + J_1 + J_2$ to first order in ϵ, it follows that

$$\widehat{\mathcal{Z}}_r(f) = 1 + z\left(\tfrac{1}{\epsilon} f(2-f) + \tfrac{1}{2} f + \tfrac{1}{2} f(f-1)(\log 2 - 1)\right) + O(z^2). \qquad (2.126)$$

To determine the exponent σ_f in equation (2.123), define the scaling function (see equation (2.105)) as

$$\sigma_f(z,\epsilon) = \tfrac{\partial}{\partial \log \ell} \log \widehat{\mathcal{Z}}_r(z,\epsilon) = \tfrac{1}{2} z\epsilon \tfrac{\partial}{\partial z} \log \widehat{\mathcal{Z}}_r(z,\epsilon). \qquad (2.127)$$

Substitution and writing this in terms of the second virial coefficient via equation (2.112) gives

$$\sigma_f(g,\epsilon) = \tfrac{1}{2} g f(2-f) + O(g^2). \qquad (2.128)$$

At the fixed point $g = \frac{\epsilon}{8}$ the leading order result for the exponent σ_f is found:

$$\sigma_f = \tfrac{1}{16} \epsilon f(2-f) + O(\epsilon^2). \qquad (2.129)$$

Observe, to $O(\epsilon^2)$, that $\sigma_2 = 0$, and if $d = 4$, then $\sigma_f = 0$. If $d = 3$, then $\sigma_1 = \frac{1}{16}$ and this shows that $\gamma = \frac{9}{8}$ (by equation (2.29)), while $\sigma_3 = -\frac{3}{16}$, and $\gamma(3) = 1$ to first order in ϵ (by equation (2.30)).

3

INTERACTING LATTICE CLUSTERS

3.1 The free energy of lattice clusters

A *lattice cluster* is a connected subgraph of $\mathbb{L}^d$ composed of edges (and vertices), or, in the case of lattice surfaces, composed of *plaquettes*.

A set of lattice clusters is an *interacting model* if an *energy* is assigned to each cluster and if the model is endowed with a Boltzmann measure which gives rise to a partition function (and thus a probability measure on the space of clusters). These models include interacting self-avoiding walks, polygons, trees or animals – see the examples of self-avoiding walk models in figure 3.1.

The fundamental quantity in an interacting model of clusters is the *microcanonical partition function* $q_n(m)$; this is the number of conformations (or states) of size n and energy m.

The (canonical) partition function is given by

$$Z_n(z) = \sum_{m=0}^{M_n} q_n(m) z^m. \tag{3.1}$$

Note that $Z_n(z)$ is a function of a Boltzmann factor or *activity* $z = e^\beta$, where $\beta = \frac{1}{k_B T}$ is inverse temperature in suitable (lattice) units (and k_B is the Boltzmann constant). The factor z is also the *generating variable* of the energy and is *conjugate* to the energy m. The maximum energy over a set of clusters of size n is assumed to be a finite number M_n (without loss of generality, the minimum value of the energy on a set of clusters of size n is assumed to be 0). That is, assume that $q_n(0) > 0$ and that $q_n(M_n) > 0$.

In most lattice models the maximum in energy is bounded linearly in the *size* n of the clusters (so that $M_n = O(n)$). These models are *linear in energy*, so $\epsilon_m = \limsup_{n\to\infty} \frac{1}{n} M_n < \infty$.

Models with $M_n = \Theta(n^2)$ (for example, two-dimensional polygons with energy given by enclosed area) may also be considered; in this case the model is said to have a *quadratic energy*.

The *intensive limiting free energy* $\mathcal{F}(z)$ is defined as a limit:

$$\mathcal{F}(z) = \lim_{n\to\infty} \tfrac{1}{n} \log Z_n(z). \tag{3.2}$$

This is the *thermodynamic limit* of the model. The *energy* and *specific heat* are thermodynamic quantities which are derivatives of the free energy. These are defined by

The Statistical Mechanics of Interacting Walks, Polygons, Animals and Vesicles, 2nd edition, E.J. Janse van Rensburg. © E.J. Janse van Rensburg. Published in 2015 by Oxford University Press.

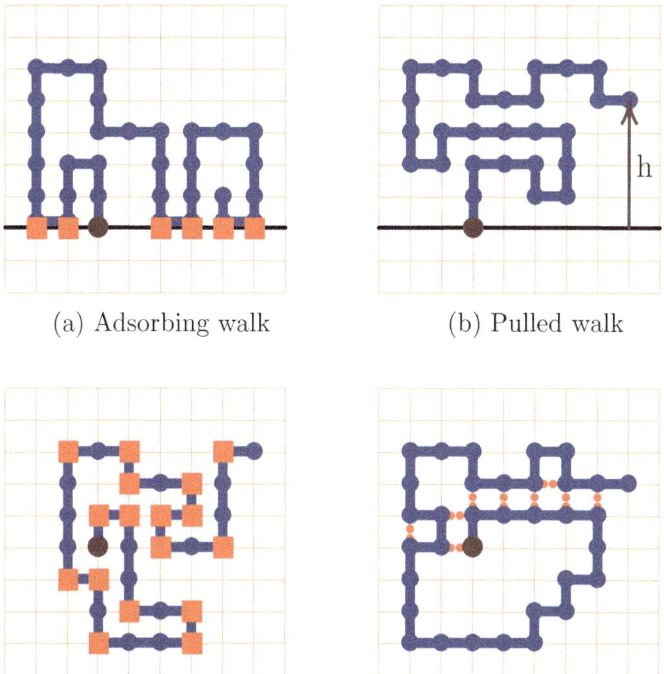

(a) Adsorbing walk (b) Pulled walk

(c) Stiff walk (d) Collapsing walk

FIG. 3.1. Examples of models of an interacting walk. (a) A positive walk from $\vec{0}$ adsorbing in the boundary $\partial\mathbb{L}_+^d$ of the half-lattice $\mathbb{L}_+^d$ (see equation (2.20)). Note that $\partial\mathbb{L}_+^d$ is also the *adsorbing wall or boundary*, or, if $d = 2$, the *adsorbing line*. Vertices in the walk which are also in $\partial\mathbb{L}_+^d$ are *visits* but the vertex at $\vec{0}$ is by convention not a visit. This walk has five visits and so its energy is $v = 5$. (b) A positive walk from $\vec{0}$ in $\mathbb{L}_+^d$ *pulled* by its endpoint (at height h above $\partial\mathbb{L}_+^d$). The height h of the endpoint of this walk is $h = 4$. (c) A *stiff walk* with a *bending energy*. If there is an energy gain for each pair of adjacent edges which are perpendicular, then the walk is *stiff*. This walk has $m = 17$ right angles. (d) Pairs of vertices in the walk which are nearest neighbour in the lattice but not nearest neighbour along the walk are *contacts*. This is a model of *collapsing walks*.

$$\mathcal{E}(z) = z\tfrac{d}{dz}\mathcal{F}(z), \quad \text{and } \mathcal{C}(z) = z\tfrac{d}{dz}\mathcal{E}(z). \tag{3.3}$$

The existence of the limiting free energy is a consequence of the supermultiplicativity of the partition function. That is, in many models it is known that the partition function satisfies a supermultiplicative inequality of the kind

$$Z_n(z)\,Z_m(z) \le Z_{n+m}(z), \tag{3.4}$$

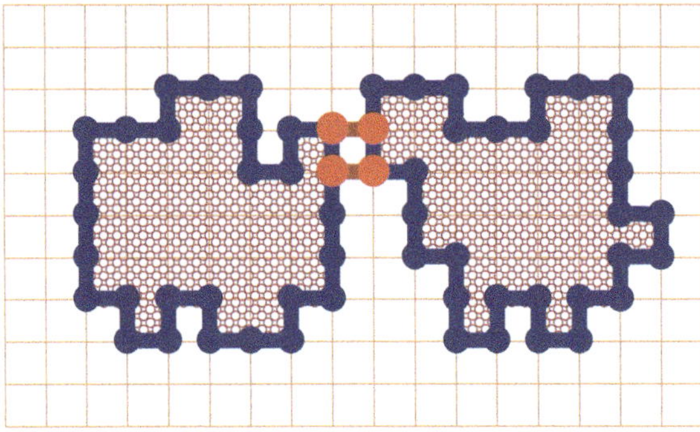

FIG. 3.2. Lattice clusters are concatenated by first finding the *top* and *bottom* edges or vertices by a lexicographic labelling. The clusters are placed such that the bottom of the second is next to the top of the first, as shown. The concatenation is completed by a local construction using edges and vertices to connect the two clusters into one.

---

in which case the model is said to be *supermultiplicative.* In some cases the partition function satisfies more general supermultiplicative relations, in which case the model will also be supermultiplicative.

Supermultiplicative relations of the type in equation (3.4) frequently arise when *lattice clusters* are concatenated, as illustrated schematically in figure 3.2. If this concatenation can be successfully completed and if the partition function is bounded by an exponential in z, then the *intensive limiting free energy* of the model exists. This is a consequence of theorem A.1 in appendix A.

Theorem 3.1 *Suppose that the partition function $Z_n(z)$ of an interacting model satisfies a supermultiplicative relation $Z_n(z)Z_m(z) \leq Z_{n+m}(z)$, where $Z_n(z) = \sum_{m=0}^{M_n} q_n(m) z^m$. If there exists a $\lambda < \infty$ such that $Z_n(1) = q_n \leq \lambda^n$, then $Z_n(z) \leq \lambda^n \max\{1, z^n\}$, and the limiting free energy*

$$\mathcal{F}(z) = \lim_{n\to\infty} \tfrac{1}{n} \log Z_n(z) = \sup_{n>0} \tfrac{1}{n} \log Z_n(z)$$

exists and is finite. $\mathcal{F}(z)$ is the intensive limiting free energy of the model. □

Suppose that two clusters, the first of size n_1 and energy k, and the second of size n_2 and energy $m-k$, are concatenated as illustrated in figure 3.2. There are $q_{n_1}(k)$ choices for the first cluster, and $q_{n_2}(m-k)$ choices for the second cluster. If the resulting cluster has size n_1+n_2 and energy m, then

$$\sum_{k=0}^{m} q_{n_1}(k) q_{n_2}(m-k) \leq q_{n_1+n_2}(m). \tag{3.5}$$

Multiplying this by z^m and summing over m gives the relation

$$\left(\sum_{m_1\geq 0} q_{n_1}(m_1)\,z^{m_1}\right)\left(\sum_{m_2\geq 0} q_{n_2}(m_2)\,z^{m_2}\right) \leq \sum_{m\geq 0} q_{n_1+n_2}(m)\,z^m, \tag{3.6}$$

which is the supermultiplicative relation seen in equation (3.4) for the partition function.

More general relations for the partition function may be of the forms such that $a_n = -\log Z_n(z)$ satisfies the relations in theorems A.2, A.3 or A.4 (see appendix A). This guarantees the existence of a limiting free energy.

Theorem 3.2 *Suppose that $Z_n(z)$ is the partition function of an interacting model and that, for any (fixed) value of z, the function $a_n = -\log Z_n(z)$ satisfies the relation and conditions in theorems A.2, A.3 or A.4. Then the limit*

$$\mathcal{F}(z) = \lim_{n\to\infty} \tfrac{1}{n}\log Z_n(z)$$

exists. The limit $\mathcal{F}(z)$ is finite if $Z_n(1) \leq \lambda^n$ for some finite $\lambda > 0$, and it is the intensive limiting free energy of the model. □

3.2 Free energies and generating functions

The *intensive finite size free energy* of an interacting model is given by

$$F_n(z) = \tfrac{1}{n}\log Z_n(z). \tag{3.7}$$

The free energy is a density (measured per edge or per unit size). The *extensive free energy* is defined by $F_n^e(z) = \log Z_n(z)$. The *limiting free energy* is

$$\mathcal{F}(z) = \lim_{n\to\infty} \tfrac{1}{n}\log Z_n(z) \tag{3.8}$$

and this limit exists in supermultiplicative models by virtue of theorems 3.1 or 3.2, or by other means. The limiting free energy may be infinite.

In models with a linear energy the limiting free energy $\mathcal{F}(z)$ may be asymptotic to $\log z$:

$$\mathcal{F}(z) \simeq \mathcal{S} + \mathcal{M}\log z. \tag{3.9}$$

The number $\mathcal{S}$ is the *limiting entropy* of the model; this definition is very natural if $\log z$ is interpreted as an inverse temperature, and $\mathcal{M}$ is interpreted as an energy density. Observe that $\mathcal{F}(z) \geq 0$ if $z > 0$ (since it was assumed that $q_n(0) \geq 1$), and $\mathcal{F}(z) \geq \limsup_{n\to\infty} \frac{1}{n}\log z^{M_n} = \epsilon_m \log z$ since $q_n(M_n) \geq 1$.

By the definition of the partition function in models of interacting clusters (see equation (3.1)), it follows that $\log Z_n(z)$ is a convex function of $\log z$. This is seen by using the Cauchy-Schwartz inequality.

$$\begin{aligned} Z_n(z_1)Z_n(z_2) &= \sum_{m_1} q_n(m_1)\,z_1^{m_1} \sum_{m_2} q_n(m_2)\,z_2^{m_2} \\ &\geq \left(\sum_m q_n(m)\,(\sqrt{z_1 z_2})^m\right)^2 = \left(Z_n(\sqrt{z_1 z_2})\right)^2. \end{aligned} \tag{3.10}$$

Hence, $\log Z_n(z_1) + \log Z_n(z_2) \geq 2\log Z_n(\sqrt{z_1 z_2})$. Divide by 2 to see that $\log Z_n(z)$ is convex in $\log z$. This gives the following theorem.

Theorem 3.3 *Suppose that $Z_n(z) = \sum_m q_n(m) z^m$ is the partition model of an interacting model. Then $\log Z_n(z)$ is a convex function of $\log z$. If the limiting free energy $\mathcal{F}(z) = \lim_{n\to\infty} F_n(z)$ exists, then it is convex in $\log z$ and is the limit of a sequence of convex functions.*

Proof Since $Z_n(z)$ is a polynomial in z and analytic in $(0,\infty)$, $F_n(z)$ is continuous, analytic and a convex function of $\log z$ on $(0,\infty)$. If the limit in equation (3.8) exists, then it is a limit of a sequence of convex functions. By lemma B.8 (appendix B) $\mathcal{F}(z)$ is a convex function of $\log z$. □

As a result of the above theorem, $\mathcal{F}(z)$ is measurable and differentiable almost everywhere. Hence, the energy $\mathcal{E}(z)$ and specific heat $\mathcal{C}(z)$ defined in equation (3.3) exist for almost every $z > 0$ (see theorem B.5 and corollary B.6, appendix B).

The first derivative of $\mathcal{F}(z)$ to $\log z$ is non-decreasing by theorem B.3 (appendix B). Moreover, if $\mathcal{F}(z)$ is non-analytic, then it is non-analytic on a countable set of points in $(0,\infty)$ (by theorem B.7, appendix B). By theorem B.9 in appendix B it follows that the sequence of first derivatives of $F_n(z)$ converges almost everywhere to the first derivative of $\mathcal{F}(z)$:

$$\lim_{n\to\infty} z\tfrac{d}{dz} F_n(z) = z\tfrac{d}{dz}\mathcal{F}(z) = \mathcal{E}(z), \text{ for ae } z. \tag{3.11}$$

Convexity of the limiting free energy $\mathcal{F}(z)$ is a general property in models with partition functions given by equation (3.1). $\mathcal{F}(z)$ has left- and right-derivatives everywhere. Define the left- and right-derivatives

$$\mathcal{E}_-(z) = \tfrac{d^-}{dz}\mathcal{F}(z), \text{ and } \mathcal{E}_+(z) = \tfrac{d^+}{dz}\mathcal{F}(z). \tag{3.12}$$

At each point z where $\mathcal{E}(z)$ is continuous, $\mathcal{E}_-(z) = \mathcal{E}_+(z) = \mathcal{E}(z)$; that is, $\mathcal{E}_-(z) = \mathcal{E}_+(z)$ almost everywhere (except perhaps at a finite or countable number of isolated points where $\mathcal{F}(z)$ is non-analytic). In addition, $\mathcal{E}_-(z)$ is lower semicontinuous, and $\mathcal{E}_+(z)$ is upper semicontinuous.

Phase transitions occur at non-analytic points z_c of $\mathcal{F}(z)$. If $\mathcal{E}_-(z_c) = \mathcal{E}_+(z_c) = \mathcal{E}(z_c)$, then the phase transition at $z = z_c$ is *continuous* (so-called because $\mathcal{E}(z)$ is continuous at $z = z_c$).

If, on the other hand, $\mathcal{E}_-(z_c) < \mathcal{E}_+(z_c)$, then there is a *first order transition* at $z = z_c$. At these points $\mathcal{E}(z_c)$ is not defined.

Define a hysteresis by defining the energy of the model for increasing z by $\mathcal{E}_-(z)$, and for decreasing z by $\mathcal{E}_+(z)$. At first order transitions at a critical point $z = z_c$ it is seen that $\mathcal{E}_-(z_c) < \mathcal{E}_+(z_c)$. This is consistent with *latent heat* in the model: decreasing z through z_c gives a jump discontinuity of magnitude $H = \mathcal{E}_+(z_c) - \mathcal{E}_-(z_c)$ in the energy density, equal to the *specific latent heat*.

Theorem 3.3 generalises to models with more than one activity. For example, consider the partition function $Z_n(x, y, \ldots, z)$ of a model dependent on $(x, y, \ldots, z)$. In this case theorem 3.3 may be stated as follows.

Theorem 3.4 *Suppose that*

$$Z_n(x,y,\ldots,z) = \sum_{\{a,b,\ldots,c\}} q_n(a,b,\ldots,c)x^a y^b \ldots z^c$$

is the partition function of an interacting model. If the limiting free energy

$$\mathcal{F}(x,y,\ldots,z) = \lim_{n\to\infty} \tfrac{1}{n} \log Z_n(x,y,\ldots,z)$$

exists, then $\mathcal{F}(x,y,\ldots,z)$ is the limit of a sequence of convex multivariate functions and is itself a convex multivariate function of $\{\log x, \log y, \ldots, \log z\}$.

Proof The proof exploits Holder's inequality. Notice that for $\lambda \in (0,1)$,

$$\left(\sum_{\{a_1\ldots c_1\}} q_n(a_1,\ldots,c_1){x_1}^{a_1}\cdots {z_1}^{c_1}\right)^{\lambda}\left(\sum_{\{a_2\ldots c_2\}} q_n(a_2,\ldots,c_2){x_2}^{a_2}\cdots {z_2}^{c_2})\right)^{1-\lambda}$$
$$\geq \sum_{\{a\ldots c\}} q_n(a,\ldots,c)\left({x_1}^{\lambda}{x_2}^{1-\lambda}\right)^a\left({y_1}^{\lambda}{y_2}^{1-\lambda}\right)^b\cdots\left({z_1}^{\lambda}{z_2}^{1-\lambda}\right)^c.$$

Take logarithms and divide by n to see that $\frac{1}{n}\log Z_n(x,y,\ldots,z)$ is a convex function of $(\log x, \log y, \ldots, \log z)$. Next, let $n\to\infty$. This shows that

$$\mathcal{F}\big({x_1}^{\lambda}{x_2}^{1-\lambda}, {y_1}^{\lambda}{y_2}^{1-\lambda}, \ldots, {z_1}^{\lambda}{z_2}^{1-\lambda}\big)$$
$$\leq \lambda\mathcal{F}(x_1,y_1,\ldots,z_1) + (1-\lambda)\mathcal{F}(x_2,y_2,\ldots,z_2).$$

This completes the proof. □

3.2.1 *Generating functions and critical points*

The *generating function* $G(z,t)$ (also called the *grand partition function*) is given by

$$G(z,t) = \sum_{n\geq 0} Z_n(z)t^n = \sum_{n\geq 0}\sum_{m} q_n(m)z^m t^n. \tag{3.13}$$

The *thermodynamic potential* of the model is $R(z,t) = \log G(z,t)$.

The radius of convergence $t_c(z)$ of $G(z,t)$ defines a *critical curve* in the *singularity diagram* (or *phase diagram*) of the model. The critical curve divides the zt-plane into a *finite* and an *infinite* phase. The finite phase is the *subcritical* phase, and the infinite phase is the *supercritical* phase.

Derivatives of $R(z,t)$ are thermodynamic quantities in the model and they are generally singular along the critical curve.

The generating variable t is conjugate to n (it is more precise to say that $\log t$ is conjugate to n), and $\log t = \mu_{ch}$ may be considered a *chemical potential.*

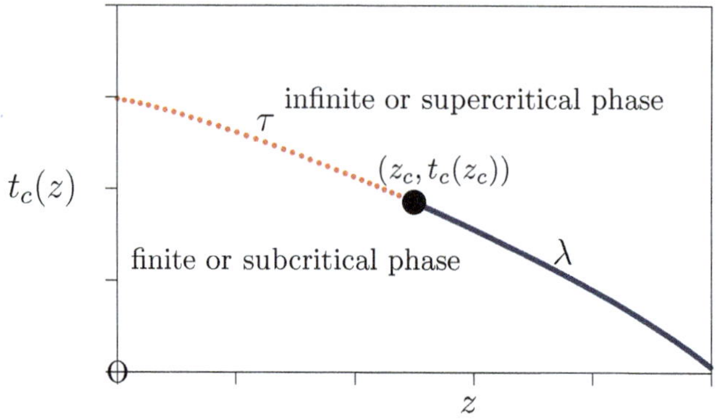

FIG. 3.3. The (grand canonical) phase diagram is a plot of the radius of convergence $t_c(z)$ of the generating function against the activity z. It divides the zt-plane into a finite phase and an infinite phase. A critical point is located at a non-analyticity at $z = z_c$ in $t_c(z)$. It separates the curve into two parts, each corresponding to a different type of singularity in $G(z, t)$ and to a different phase in the model. The model undergoes a *phase transition* at the *critical point* $z = z_c$.

If the limiting free energy exists, then it may be computed from the radius of convergence of $G(z, t)$ by using equation (3.13):

$$\mathcal{F}(z) = -\lim_{n\to\infty} \tfrac{1}{n} \log Z_n(z) = -\log t_c(z). \tag{3.14}$$

In some models the free energy is not known to exist, and the limit above may be replaced by a lim sup.

Non-analyticities in $\mathcal{F}(z)$ correspond to non-analyticities in $t_c(z)$. These are called *critical points* or (sometimes) *multicritical points*, and they separate phases on either side of the critical curve $t_c(z)$. A generic critical curve with a single non-analyticity at z_c is illustrated in figure 3.3. The critical point divides the critical curve into two parts, each corresponding to a different type of singularity in the generating function. The regimes on either side of z_c are *phases*, and a *phase transition* occurs at the *critical point* $z = z_c$.

3.2.2 *Free energies of interacting models in wedges*

Consider a supermultiplicative model of interacting clusters with partition function $Z_n(z)$ satisfying equation (3.4) and suppose that $Z_n(z) \leq Z_{n+1}(z)$. Clusters in this model may be confined to subsets of $\mathbb{L}^d$. For example, in $\mathbb{L}^2$, clusters may be confined to a *wedge* [261] or to a *slit* (a *slab* in $\mathbb{L}^d$).

Define a generalised wedge W in $\mathbb{L}^d$ by

$$W = \{\vec{x} \in \mathbb{Z}^d \,|\, \vec{x}(j) \geq 0 \text{ for } 1 \leq j \leq d-1, \text{ and } 0 \leq \vec{x}(d) \leq f(\vec{x}(1), \cdots, \vec{x}(d-1))\},$$

where $\vec{x}(j)$ is the j-th Cartesian component of $\vec{x}$, and f is a non-negative function with the property that

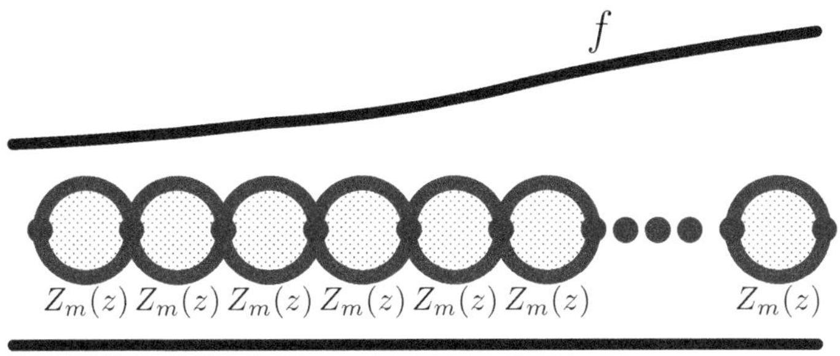

FIG. 3.4. Clusters in a wedge may be concatenated in linear strings. By first taking the number of clusters in the string to infinity, and then the size of each constituent cluster to infinity, a lower bound on the limiting free energy of clusters in wedges is shown to be the limiting free energy of any constituent cluster. In this illustration the clusters are parallel to the bottom hyperplane of a generalised wedge f where $\lim_{\|\vec{v}\|_\infty \to \infty} f(\vec{v}) = \infty$, and $f(\vec{v}) \geq n$ for all $\|\vec{v}\|_\infty \geq 0$, ($n$ is large enough to fit a walk or polygon of size n under the graph of f in $\mathbb{R}^d$).

$$\lim_{\|\vec{v}\|_\infty \to \infty} f(\vec{v}) = \infty, \tag{3.15}$$

where $\vec{v} \in \mathbb{Z}^{d-1}$, and $\|\cdot\|_\infty$ is the $(d-1)$-dimensional L_∞-norm. By equation (3.15), a cluster of arbitrary finite size n can be placed in W.

Fix the origin $\vec{0}$ in the wedge such that all clusters of size m can be rooted at $\vec{0}$. Let the partition function of this model be $Z_m^W(z)$. This shows that $Z_m^W(z) = Z_m(z)$ for *this* particular value of m. With increasing $n \geq m$, the confinement of the wedge will give $Z_n^W(z) \leq Z_n(z)$. Assume that, in the wedge, $Z_n^W(z) \leq Z_{n+1}^W(z)$ for all $n \geq m$.

Suppose that clusters of size m can be concatenated in a linear string of arbitrary length in one of the $d-1$ directions parallel to the bottom hyperplane, as illustrated in figure 3.4. That is, it is possible to place a first cluster of size m to intersect $\vec{0}$ and then to concatenate a string of N clusters along a direction which runs off to infinity in W. By equation (3.4), this shows that

$$(Z_m(z))^N \leq Z_{mN}^W(z). \tag{3.16}$$

Let $n = mN + r \geq mN$ for $r \in \{0, 1, \ldots, n-1\}$. Since $Z_{mN}^W(z) \leq Z_{mN}(z)$, and $Z_n(z) \leq Z_{n+1}(z)$,

$$(Z_m(z))^N \leq Z_{mN}^W(z) \leq Z_{mN+r}(z). \tag{3.17}$$

Take logarithms, divide by $n = mN + r$ and let $N \to \infty$ while choosing r such that the limit inferior of $\frac{1}{n} \log Z_n^W(z)$ is obtained. This shows that

$$\tfrac{1}{m} \log Z_m(z) \leq \liminf_{n\to\infty} \tfrac{1}{n} \log Z_n^W(z) \leq \limsup_{n\to\infty} \tfrac{1}{n} \log Z_n^W(z) \leq \mathcal{F}(z). \tag{3.18}$$

Take $m \to \infty$ (and use the result in theorem 3.1) to obtain the following theorem.

Theorem 3.5 *Let $Z_n(z)$ be the partition function of a model of interacting clusters and suppose that $Z_n(z)Z_m(z) \leq Z_{n+m}(z)$. Let W be a generalised wedge in $\mathbb{Z}^d$ defined by*

$$W = \{\vec{x} \in \mathbb{Z}^d \mid \vec{x}(j) \geq 0 \text{ for } 1 \leq j \leq d-1 \text{ and } 0 \leq \vec{x}(d) \leq f(\vec{x}(1), \ldots, \vec{x}(d-1))\},$$

where f is a non-negative function such that, if $\vec{v} \in \mathbb{Z}^{d-1}$, then

$$\lim_{\|\vec{v}\|_\infty \to \infty} f(\vec{v}) = \infty.$$

Suppose that the partition function of clusters confined to W is $Z_n^W(z) \leq Z_n(z)$. Then the free energy of clusters rooted in W is

$$\lim_{n\to\infty} \tfrac{1}{n} \log Z_n^W(z) = \lim_{n\to\infty} \tfrac{1}{n} \log Z_n(z) = \mathcal{F}(z),$$

where $\mathcal{F}(z)$ is the limiting free energy of the unconfined model. □

3.3 The microcanonical density function

The Legendre transform of $\mathcal{F}(z)$ is given by

$$\log \mathcal{P}(\epsilon) = \inf_{z\in\mathbb{R}^+} \{\mathcal{F}(z) - \epsilon \log z\}, \tag{3.19}$$

where $\mathbb{R}^+$ represents positive real numbers; $\mathcal{P}(\epsilon)$ is the *(microcanonical) density function* of the model.

By putting $f = \mathcal{F}$ and putting $x = \log z$ in equation (B.12) (appendix B), it follows that $\log \mathcal{P}(\epsilon)$ is concave in ϵ by lemma B.10 and by lemma B.11 that

$$\mathcal{F}(z) = \sup_{\epsilon\in E}\{\log \mathcal{P}(\epsilon) + \epsilon \log z\}, \tag{3.20}$$

where the supremum is taken over the essential domain E of $\mathcal{P}(\epsilon)$.

The function $\mathcal{P}(\epsilon)$ is related to the microcanonical partition function $q_n(m)$ in theorem 3.6.

Theorem 3.6 *Suppose that*

$$\mathcal{F}(z) = \lim_{n\to\infty} \tfrac{1}{n} \log \sum_{m=0}^{M_n} q_n(m)\, z^m$$

exists and is finite. Then there is a sequence of integers $\langle \delta_m \rangle$ and an $\epsilon \geq 0$ in the essential domain E of $\mathcal{P}$ such that

$$\lim_{n\to\infty} \tfrac{1}{n}\delta_n = \epsilon, \text{ and } \log \mathcal{P}(\epsilon) = \lim_{n\to\infty} \tfrac{1}{n} \log q_n(\delta_n).$$

The number ϵ is a function of z.

Proof Assume that $\mathcal{F}(z)$ is non-negative and finite for all $z \in \mathbb{R}^+$.

Suppose that $z > 0$. Let δ_n be the smallest value of m which maximises $q_n(m) z^m$. Then $0 \leq \delta_n \leq M_n$ for each n, and the partition function $Z_n(z)$ (see equation (3.1)) is bounded by

$$q_n(\delta_n) z^{\delta_n} \leq Z_n(z) \leq (M_n + 1)\, q_n(\delta_n) z^{\delta_n}. \tag{3.21}$$

Take logarithms, divide by n and let $n \to \infty$. This shows that

$$\mathcal{F}(z) = \lim_{n\to\infty} \tfrac{1}{n} \log\big(q_n(\delta_n) z^{\delta_n}\big) \tag{3.22}$$

exists. Choose an $\epsilon \geq 0$ such that

$$0 \leq \liminf_{n\to\infty} \tfrac{1}{n}\delta_n \leq \epsilon \leq \limsup_{n\to\infty} \tfrac{1}{n}\delta_n \leq \limsup_{n\to\infty} \tfrac{1}{n} M_n = \epsilon_m.$$

Multiply equation (3.21) by $z^{-\lfloor \epsilon n \rfloor}$. Take logarithms, divide by n and take $n \to \infty$ to see that

$$\lim_{n\to\infty} \tfrac{1}{n} \log\Big(q_n(\delta_n) z^{\delta_n - \lfloor \epsilon n \rfloor}\Big) = \mathcal{F}(z) - \epsilon \log z > -\infty$$

since $\epsilon \leq \epsilon_m$. Taking the infimum for $z > 0$ gives

$$-\infty < \log \mathcal{P}(\epsilon) = \inf_{z \in \mathbb{R}^+} \Big\{ \lim_{n\to\infty} \tfrac{1}{n} \log\Big(q_n(\delta_n) z^{\delta_n - \lfloor \epsilon n \rfloor}\Big)\Big\} < \infty$$

because $\mathcal{F}(z)$ is finite, convex and non-negative.

However, the limit in equation (3.22) is finite. Assuming that $\limsup_{n\to\infty} \frac{1}{n}\delta_n > \epsilon$ shows that $\log \mathcal{P}(\epsilon) = +\infty$ if $z > 1$, and $\log \mathcal{P}(\epsilon) = -\infty$ if $z < 1$; this is a contradiction.

Similarly, if $\liminf_{n\to\infty} \frac{1}{n}\delta_n < \epsilon$, then $\log \mathcal{P}(\epsilon) = -\infty$ if $z > 1$, and $\log \mathcal{P}(\epsilon) = \infty$ if $z < 1$. This is also a contradiction.

Hence, if $\log \mathcal{P}(\epsilon)$ is finite, then $\epsilon = \liminf_{n\to\infty} \frac{1}{n}\delta_n = \limsup_{n\to\infty} \frac{1}{n}\delta_n$.

The case that $z = 0$ is approached in a similar way. □

Define $\epsilon_m = \limsup_{n\to\infty} \frac{1}{n} M_n$, where M_n is defined in equation (3.1).

Theorem 3.7 *Suppose that, for every $\epsilon \in (0, \epsilon_m)$, there is a sequence $\langle \delta_n \rangle$ such that $\lim_{n\to\infty} \frac{1}{n}\delta_n = \epsilon$ and there exists a density function such that $\log \mathcal{P}(\epsilon) = \lim_{n\to\infty} \frac{1}{n} \log q_n(\delta_n)$. Then the limiting free energy $\mathcal{F}(z) = \lim_{n\to\infty} \frac{1}{n} \log Z_n(z)$ exists and it is a convex function of $\log z$ for $z \in (0, \infty)$.*

Proof Let $\epsilon \in (0, \epsilon_m)$ and consider the sequence $\langle \delta_n \rangle$ such that $\lim_{n\to\infty} \frac{1}{n}\delta_n = \epsilon$. Then, for each $z \in (0, \infty)$,

$$Z_n(z) = \sum_{m=0}^{M_n} q_n(m) z^m \geq q_n(\delta_n) z^{\delta_n}.$$

Take logarithms, divide by n and take the limit inferior on the left-hand side as $n \to \infty$. This shows that

$$\liminf_{n\to\infty} \tfrac{1}{n} \log Z_n(z) \geq \log \mathcal{P}(\epsilon) + \epsilon \log z.$$

In particular,

$$\liminf_{n\to\infty} \tfrac{1}{n} \log Z_n(z) \geq \sup_{\epsilon\in(0,\epsilon_m)} \{\log \mathcal{P}(\epsilon) + \epsilon \log z\}.$$

On the other hand, let κ_n be that value of m which maximises $q_n(m)z^m$. Then $Z_n(z) \leq (M_n+1)p_n(\kappa_n)z^{\kappa_n}$. Take logarithms, divide by n and take the limit superior on the left-hand side as $n \to \infty$. The limit superior of the right-hand side cannot exceed $\mathcal{P}(\epsilon)z^\epsilon$ for all $\epsilon \in (0, \epsilon_m)$ since $\mathcal{P}(\epsilon)$ is a concave and continuous function of ϵ. Hence,

$$\limsup_{n\to\infty} \tfrac{1}{n} \log Z_n(z) \leq \sup_{\epsilon\in(0,\epsilon_m)} \{\log \mathcal{P}(\epsilon) + \epsilon \log z\}.$$

This proves that

$$\mathcal{F}(z) = \lim_{n\to\infty} \tfrac{1}{n} \log Z_n(z) = \sup_{\epsilon\in(0,\epsilon_m)} \{\log \mathcal{P}(\epsilon) + \epsilon \log z\}.$$

Convexity of $\mathcal{F}(z)$ follows from properties of the Legendre transform (see lemma B.10 in appendix B). This completes the proof. □

If $\mathcal{P}(\epsilon)$ is known to exist, then a free energy $\mathcal{F}(z)$ can be found by equation (3.20) and by lemmas B.10 and B.11 (appendix B).

If $\log \mathcal{P}(\epsilon)$ is strictly concave, then the supremum in $\sup_{\epsilon\in E}\{\log \mathcal{P}(\epsilon) + \epsilon \log z\}$ is found either as $\epsilon \to 0^+$, or $\epsilon \to \epsilon_M$, or at a unique $\epsilon_* \in (\epsilon_m, \epsilon_M)$. This is not necessarily the case if $\log \mathcal{P}(\epsilon)$ is not strictly concave.

As $z \to \infty$, the supremum in $\sup_{\epsilon\in(0,\epsilon_m)}\{\log \mathcal{P}(\epsilon) + \epsilon \log z\}$ is realised at non-decreasing values of ϵ. In the asymptotic regime, from equation (3.20),

$$\mathcal{F}(z) \gtrsim \log \mathcal{P}(\epsilon_m^-) + \epsilon_m^- \log z \tag{3.23}$$

as $z \to \infty$. This shows that the limiting entropy $\mathcal{S}$ of the model is bounded from below by the left-limit of $\log \mathcal{P}(\epsilon)$ as $\epsilon \to \epsilon_m^-$.

3.3.1 *The density function as a limit*

Theorem 3.6 suggests that a good definition for $\mathcal{P}(\epsilon)$ should be

$$\mathcal{P}(\epsilon) = \lim_{n\to\infty} (q_n(\lfloor \epsilon n \rfloor))^{1/n}. \tag{3.24}$$

In this section a justification is given in supermultiplicative models which are linear in energy (see theorem 3.9).

Let $q_n(m)$ be the microcanonical partition function and put $q_n = \sum_m q(m)$. A suitable starting point is the generalisation of equation (1.7) to

$$q_{n_1}(m_1)q_{n_2}(m_2) \leq q_{n_1+n_2}(m_1+m_2). \tag{3.25}$$

Models with this type of supermultiplicativity have an *additive* energy.

Some generality will be lost by further assumptions but most supermultiplicative models with an additive energy can be reformulated to be consistent with the assumptions. For example, the parity of n or m may be important, and so on.

Thus, assume that

- there is a $K > 0$ such that $0 \leq q_n(m) < K^n$ for all m;
- $q_n(m) > 0$ for $0 \leq m \leq A_n$, where A_n is an non-decreasing function of n;
- and $\liminf_{n\to\infty} \frac{1}{n} A_n > 0$, and $\limsup_{n\to\infty} \frac{1}{n} A_n < \infty$.

Define

$$\epsilon_m = \limsup_{n\to\infty} \tfrac{1}{n} A_n. \tag{3.26}$$

Note than the second assumption may be relaxed by not requiring $q_n(0) > 0$ but that no generality is lost by imposing this assumption.

Summing equation (3.25) over m_1 and m_2 gives

$$q_{n_1}q_{n_2} \leq (1 + A_{n_1+n_2})q_{n_1+n_2}. \tag{3.27}$$

Since $q_n \leq A_n K^n$ by the assumptions above, by theorem A.4, appendix A,

$$\lim_{n\to\infty} \tfrac{1}{n} \log q_n = \log \mu < \infty. \tag{3.28}$$

Lemma 3.8 *Given $\epsilon \in (0, \epsilon_m)$, there exists a z_{n_1,n_2}, dependent on ϵ and with $|z_{n_1,n_2}| \leq 1$, such that*

$$q_{n_1}(\lfloor \epsilon n_1 \rfloor)\, q_{n_2}(\lfloor \epsilon n_2 \rfloor + z_{n_1,n_2}) \leq q_{n_1+n_2}(\lfloor \epsilon(n_1+n_2)\rfloor)$$

Proof Either $\lfloor \epsilon n_1 \rfloor + \lfloor \epsilon n_2 \rfloor = \lfloor \epsilon(n_1+n_2)\rfloor$, or $\lfloor \epsilon n_1 \rfloor + \lfloor \epsilon n_2 \rfloor = \lfloor \epsilon(n_1+n_2)\rfloor - 1$, for fixed $\epsilon > 0$. Define $z_{n_1,n_2} = 0$ in the first case, and $z_{n_1,n_2} = 1$ in the second case. Then $\lfloor \epsilon n_1 \rfloor + \lfloor \epsilon n_2 \rfloor - z_{n_1,n_2} = \lfloor \epsilon(n_1+n_2)\rfloor$. In equation (3.25) put $m_1 = \lfloor \epsilon n_1 \rfloor$ and, $m_2 = \lfloor \epsilon n_2 \rfloor + z_{n_1,n_2}$. □

Theorem 3.9 *If $\epsilon \in (0, \epsilon_m)$, then*

$$\mathcal{P}(\epsilon) = \lim_{n\to\infty} \left(q_n(\lfloor \epsilon n \rfloor)\right)^{1/n}.$$

In addition, $\log \mathcal{P}(\epsilon)$ is a concave function of ϵ; moreover, for each n and for $\epsilon \in (0, \epsilon_m)$, there is an $\eta_n \in \{0,1\}$ such that $q_n(\lfloor \epsilon n \rfloor + \eta_n) \leq [\mathcal{P}(\epsilon)]^n$. The function $\log \mathcal{P}(\epsilon)$ is concave on $(0, \epsilon_m)$ and, if $\log \mathcal{P}(\epsilon)$ is defined on the closed interval $[0, \epsilon_m]$, then it is concave on this interval.

Proof Choose $n_1+n_2=n$ where $n_1=n-k$, and $n_2=k$ in lemma 3.8. This gives

$$q_{n-k}(\lfloor \epsilon(n-k)\rfloor)q_k(\lfloor \epsilon k\rfloor + z_{n-k,k}) \leq q_n(\lfloor \epsilon n\rfloor). \tag{3.29}$$

Let $N,m\in\mathbb{N}$ and assume that N is large. Put $n=Nm+r$, with r chosen such that $N_0\leq r<N_0+m$, for some fixed large N_0. Assume that m is large, so $n\gg N_0$, and $n\gg N$.

If $k=r$ (and $n=Nm+r$), then equation (3.29) becomes

$$q_{Nm}(\lfloor \epsilon Nm\rfloor)q_r(\lfloor \epsilon r\rfloor + z_{Nm,r}) \leq q_{Nm+r}(\lfloor \epsilon(Nm+r)\rfloor). \tag{3.30}$$

Increase N_0 until $\lfloor \epsilon r\rfloor \geq |z_{Nm,r}|$. Since $\epsilon>0$, and $|z_{Nm,r}|\leq 1$, this is always possible.

Apply equation (3.30) recursively to $q_{Nm}(\lfloor \epsilon Nm\rfloor)$ with $k=m$, and with $n=Nm$, $(N-1)m$, $\ldots$. The result is

$$\begin{aligned}
q_n(\lfloor \epsilon n\rfloor) &= q_{Nm+r}(\lfloor \epsilon(Nm+r)\rfloor)\\
&\geq q_{Nm}(\lfloor \epsilon Nm\rfloor)q_r(\lfloor \epsilon r\rfloor + z_{Nm,r})\\
&\geq q_{(N-1)m}(\lfloor \epsilon(N-1)m\rfloor)q_m(\lfloor \epsilon m\rfloor + z_{(N-1)m,m})q_r(\lfloor \epsilon r\rfloor + z_{Nm,r})\\
&\geq \cdots\\
&\geq \left[\prod_{j=1}^{N} q_m(\lfloor \epsilon m\rfloor + z_{m(N-j),m})\right] q_r(\lfloor \epsilon r\rfloor + z_{Nm,r}),
\end{aligned}$$

where $z_{0,m}=0$ (this defines $z_{0,m}$).

Choose η_m to be that value of $\ell\in\{-1,0,1\}$ which minimises $q_m(\lfloor \epsilon m\rfloor+\ell)$. Then $|\eta_m|\leq 1$, and the $z_{m(N-j),m}$ in the above may be replaced by η_m. This gives

$$q_{Nm+r}(\lfloor \epsilon(Nm+r)\rfloor) \geq [q_m(\lfloor \epsilon m\rfloor + \eta_m)]^N q_r(\lfloor \epsilon r\rfloor + z_{Nm,r}).$$

Take logarithms of this and divide by $n=Nm+r$. Fix the value of m and take the limit inferior as $n\to\infty$ of the left-hand side. Then $N\to\infty$ while $N_0\leq r<N_0+m$. This gives

$$\liminf_{n\to\infty} \tfrac{1}{n}\log q_n(\lfloor \epsilon n\rfloor) \geq \tfrac{1}{m}\log q_m(\lfloor \epsilon m\rfloor + \eta_m). \tag{3.31}$$

Take the limit superior as $m\to\infty$ of the right-hand side to find that

$$\liminf_{n\to\infty} \tfrac{1}{n}\log q_n(\lfloor \epsilon n\rfloor) \geq \limsup_{m\to\infty} \tfrac{1}{m}\log q_m(\lfloor \epsilon m\rfloor + \eta_m). \tag{3.32}$$

Next, put $n_1+n_2=m$, where $n_1=m-k$, and $n_2=k$, with $m_1+m_2=\lfloor \epsilon m\rfloor+\eta_m$ and $m_1=\lfloor \epsilon(m-k)\rfloor$ in equation (3.25). Then

$$q_{m-k}(\lfloor \epsilon(m-k)\rfloor)q_k(\lfloor \epsilon m\rfloor - \lfloor \epsilon(m-k)\rfloor + \eta_m) \leq q_m(\lfloor \epsilon m\rfloor + \eta_m).$$

Define $\delta_{m,k}$ by $\lfloor \epsilon k\rfloor + \delta_{m,k} = \lfloor \epsilon m\rfloor - \lfloor \epsilon(m-k)\rfloor$. Then $|\delta_{m,k}|\leq 1$. Hence,

$$q_{m-k}(\lfloor \epsilon(m-k) \rfloor) q_k(\lfloor \epsilon k \rfloor + \delta_{m,k} + \eta_m) \leq q_m(\lfloor \epsilon m \rfloor + \eta_m).$$

Choose k large enough that $\lfloor \epsilon k \rfloor > 2 \geq |\delta_{m,k}| + |\eta_m|$. Since $\epsilon > 0$, this is always possible. Fix k, take logarithms, divide the above by m and then take the limit superior on the left-hand side as $m \to \infty$. This shows that

$$\limsup_{m\to\infty} \tfrac{1}{m} \log q_m(\lfloor \epsilon m \rfloor) \leq \limsup_{m\to\infty} \tfrac{1}{m} \log q_m(\lfloor \epsilon m \rfloor + \eta_m). \tag{3.33}$$

Comparison of equation (3.33) with equation (3.32) proves the existence of the limit. It follows from equation (3.31) that $\log q_m(\lfloor \epsilon m \rfloor + \eta_m) \leq m \log \mathcal{P}(\epsilon)$ for some $\eta_m \in \{0, 1\}$.

Finally, to see that $\log \mathcal{P}(\epsilon)$ is concave, put $k = \lfloor \delta n \rfloor$ for $\delta \in (0,1)$ in equation (3.29), take logarithms, divide by n and take $n \to \infty$. Proving log-concavity at the endpoints of the interval $[0, \epsilon_m]$ is similarly done by considering $q_n(0)$ and $q_n(\lfloor \epsilon_m n \rfloor)$. This completes the proof. □

If $\mathcal{P}(\epsilon)$ exists as a limit in theorem 3.9, then the limiting free energy exists.

Corollary 3.10 *If $\mathcal{P}(\epsilon) = \lim_{n\to\infty} (q_n(\lfloor \epsilon n \rfloor))^{1/n}$ exists for every $\epsilon \in (0, \epsilon_m)$, then the free energy $\mathcal{F}(z) = \lim_{n\to\infty} \frac{1}{n} \log Z_n(z)$ exists for all finite $z > 0$.*

Proof There is a most popular value of m, denoted m^*, in equation (3.1) such that

$$q_n(\lfloor \epsilon n \rfloor) z^{\lfloor \epsilon n \rfloor} \leq Z_n(z) \leq (M_n + 1) q_n(m^*) z^{m^*}.$$

By taking logarithms, dividing by n and taking the limit inferior as $n \to \infty$, this shows that

$$\mathcal{P}(\epsilon) + \epsilon \log z \leq \liminf_{n\to\infty} \tfrac{1}{n} \log Z_n(z). \tag{3.34}$$

It remains to consider the upper bounds on $\limsup_{n\to\infty} \frac{1}{n} \log Z_n(z)$. Let $\mathcal{K} = \{n_k\}$ be an infinite set of integers such that

$$\limsup_{n\to\infty} \tfrac{1}{n} \log Z_n(z) = \lim_{k\to\infty} \tfrac{1}{n_k} \log Z_{n_k}(z) \leq \limsup_{k\to\infty} \tfrac{1}{n_k} \log(q_{n_k}(m^*) z^{m^*}).$$

It follows that, for infinitely many values of $n_k \in \mathcal{K}$,

$$Z_{n_k}(z) \leq (M_{n_k} + 1) q_{n_k}(m^*) z^{m^*}. \tag{3.35}$$

Since $m^* \in \{0, 1, \ldots, M_n\}$, and $M_n = O(n)$, there exists an $\epsilon^* \in [0, \epsilon_m]$ such that

$$\limsup_{k\to\infty} \tfrac{1}{n_k} m^* = \epsilon^*.$$

Hence, $m^* = \lfloor \epsilon^* n_k \rfloor + o(n_k)$ for infinitely many $n_k \in \mathcal{K}$ in an infinite set $\mathcal{N} \subseteq \mathcal{K}$. By equation (3.25), since $\lfloor \epsilon^*(n+\ell) \rfloor - m^* \in \{0, 1, \ldots, M_n\}$ for infinitely many $\ell > 0$,

$$q_n(m^*) q_\ell(\lfloor \epsilon^*(n+\ell) \rfloor - m^*) \leq q_{n+\ell}(\lfloor \epsilon^*(n+\ell) \rfloor).$$

This shows that $\limsup_{n\to\infty}(q_n(m^*))^{1/n} \leq \mathcal{P}(\epsilon^*)$. Use this result on the right hand side of equation (3.35) (by putting $n = n_k$ and taking $k \to \infty$). This gives

$$\limsup_{n\to\infty} \tfrac{1}{n} \log Z_n(z) = \lim_{k\to\infty} \tfrac{1}{n_k} \log Z_{n_k}(z) \leq \mathcal{P}(\epsilon^*) + \epsilon^* \log z. \tag{3.36}$$

Thus, by equations (3.34) and (3.36),

$$\mathcal{P}(\epsilon) + \epsilon \log z \leq \liminf_{n\to\infty} \tfrac{1}{n} \log Z_n(z) \leq \limsup_{n\to\infty} \tfrac{1}{n} \log Z_n(z) \leq \mathcal{P}(\epsilon^*) + \epsilon^* \log z.$$

Take the supremum on the left-hand side to complete the proof. □

3.3.1.1 *Essential domains:* The assumptions, lemma 3.8 and theorem 3.9 show that the essential domain of $\mathcal{P}(\epsilon)$ is $E \supseteq (0, \epsilon_m)$. There is no loss of generality if it is assumed that $E = (0, \epsilon_m)$.

In more general models, define

$$A_n = \min\{m \mid q_n(m) > 0\}, \text{ and } B_n = \max\{m \mid q_n(m) > 0\},$$

and suppose that $\epsilon_0 = \liminf_{n\to\infty} \frac{1}{n} A_n$, and $\epsilon_m = \limsup_{n\to\infty} \frac{1}{n} B_n$. Then the essential domain of $\mathcal{P}(\epsilon)$ is $E \supseteq (\epsilon_0, \epsilon_m)$. The closure of E is $\overline{E} = [\epsilon_0, \epsilon_m]$ with $0 \leq \epsilon_0 \leq \epsilon_m < \infty$. It follows that $\epsilon_0 = \inf E$, and $\epsilon_m = \sup E$. Since $\log \mathcal{P}(\epsilon)$ is concave in ϵ over E, it is the case that

$$\limsup_{n\to\infty} \tfrac{1}{n} \log q_n(A_n) \leq \lim_{\epsilon\to\epsilon_0^+} \log \mathcal{P}(\epsilon), \text{ and } \limsup_{n\to\infty} \tfrac{1}{n} \log q_n(B_n) \leq \lim_{\epsilon\to\epsilon_m^-} \log \mathcal{P}(\epsilon).$$

If these are equalities, and the left-hand sides are limits, then $\mathcal{P}(\epsilon)$ is continuous on the closure of E. If one or both of these are strict inequalities, then there are phase transitions at $z = 0$ or $z = \infty$ (if $z = e^{1/k_B T}$, then these are phase transitions in the model at infinite or zero temperature).

Generally, define the supremum of $\mathcal{P}$ by $\log \mu = \sup_\epsilon \log \mathcal{P}(\epsilon)$. By theorem 3.6, there exists a sequence $\langle \sigma_n \rangle$ such that $\log \mu = \lim_{n\to\infty} \frac{1}{n} \log q_n(\sigma_n)$, with $\frac{1}{n}\sigma_n \in E$ and hence $\lim_{n\to\infty} \frac{1}{n}\sigma_n = \epsilon_* \in \overline{E}$. In most models it will be the case that $\epsilon_* \in E$.

The general appearance of $\log \mathcal{P}(\epsilon)$ is illustrated in figure 3.5.

3.3.1.2 *Non-linear models:* Suppose that $\sum_m q_n(m) = q_n \leq K^n$ for some $K > 1$, and $q_n(m) > 0$ for $0 \leq m \leq \vartheta_n$, where $n = o(\vartheta_n)$. This model is not linear in its energy. The partition function $Z_n(z) = \sum_m q_n(m) z^m$ is bounded as follows:

$$\max\left\{1, z^{\vartheta_n}\right\} \leq Z_n(z) \leq \max\left\{q_n, q_n z^{\vartheta_n}\right\}. \tag{3.37}$$

Take logarithms, divide by ϑ_n and take $n \to \infty$ to see that

$$\mathcal{F}(z) = \lim_{n\to\infty} \tfrac{1}{\vartheta_n} \log Z_n(z) = \begin{cases} 0, & \text{if } z \leq 1; \\ \log z, & \text{if } z > 1. \end{cases} \tag{3.38}$$

Thus, $\mathcal{F}$ requires a different normalisation (compare this for example to equation (3.8)). Observe that there is a phase transition at $z = 1$ and that it is first order (because the first derivative of $\mathcal{F}(z)$ is discontinuous at $z = 1$).

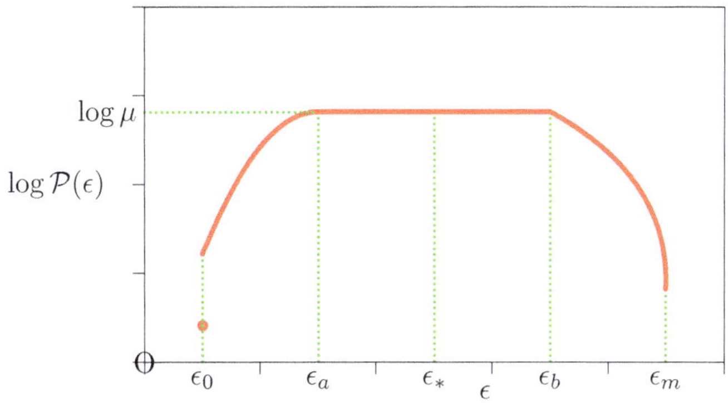

FIG. 3.5. The density function is log-concave in ϵ. It has a supremum μ in the closure of its essential domain E. In this example it realises its supremum on the interval $[\epsilon_a, \epsilon_b]$, has a jump discontinuity at ϵ_0, is non-analytic with continuous derivative at ϵ_a, is linear on (ϵ_a, ϵ_b), has a discontinuous first derivative at ϵ_b and is left-continuous with infinite left-derivative at ϵ_m.

In this model the density function can be defined by using equation (3.19):

$$\mathcal{P}(\epsilon) = 1, \quad \text{for } \epsilon \in [0, 1]. \tag{3.39}$$

The density of the energy (normalised by ϑ_n) in the limit as $n \to \infty$ is ϵ, and the definition of $\mathcal{P}(\epsilon)$ as a limit is

$$\log \mathcal{P}(\epsilon) = \lim_{n\to\infty} \tfrac{1}{\vartheta_n} \log q_n(\lfloor \epsilon\vartheta_n \rfloor) = 0. \tag{3.40}$$

This limit necessarily exists, because $1 \leq q_n \leq K^n$, and $n = o(\vartheta_n)$.

3.3.1.3 *Regular models:* Suppose $q_n(m)$ satisfies a generalised supermultiplicative inequality of the form

$$q_{n_1}(m_1) q_{n_2}(m_2) \leq \sum_{i=-\ell}^{\ell} q_{n_1+n_2}(m_1 + m_2 + i), \tag{3.41}$$

where ℓ may be a constant or a monotonic increasing function of $n_1 + n_2$.

A supermultiplicative model satisfying equation (3.41) with $q_n(m) < K^n$ for some constant K is said to be *regular* if $\ell = o(n_1 + n_2)$, and if there exists a constant $k \geq 0$ such that

$$q_n(m) \leq q_{n+k}(m), \quad \text{and } q_n(m) \leq q_{n+k}(m+1), \tag{3.42}$$

where both $q_{n+k}(m) > 0$ and $q_{n+k}(m+1) > 0$.

If a model is regular, then equation (3.41) becomes

$$q_{n_1}(m_1)q_{n_2}(m_2) \leq (2\ell+1)q_{n_1+n_2+2\ell k}(m_1+m_2+\ell). \tag{3.43}$$

It follows from theorem A.2 (appendix A) that the limit

$$\log \mathcal{P}(\epsilon) = \lim_{n\to\infty} \tfrac{1}{n} \log q_n(\lfloor \epsilon n \rfloor) \tag{3.44}$$

exists. By corollary 3.10 and equation (3.20), the model has a convex free energy.

3.3.1.4 *Joint density functions:* Let $q_n(m_1, m_2)$ be the number of clusters of size n with energy m_1 with respect to one property, and energy m_2 with respect to another.

Define the partition function $Z_n(z, m_2)$ in the usual way:

$$Z_n(z, m_2) = \sum_{m_1} q_n(m_1, m_2) z^{m_1}. \tag{3.45}$$

In a supermultiplicative model of clusters the partition function may satisfy the inequality

$$Z_{n_1}(z, m_1) Z_{n_2}(z, m_2) \leq Z_{n_1+n_2}(z, m_1+m_2). \tag{3.46}$$

This relation is similar to equation (3.25). Putting $m_1 = \lfloor \delta n_1 \rfloor$, and $m_2 = \lfloor \delta n_2 \rfloor$, it follows from theorem 3.9, that

$$\mathcal{F}_\delta(z) = \log \mathcal{P}_z(\delta) = \lim_{n\to\infty} \tfrac{1}{n} \log\left(Z_n(z, \lfloor \delta n \rfloor)\right) \tag{3.47}$$

exists and may be considered either the free energy at z or the microcanonical density function conjugate to m_2 at density δ defined in the essential domain E of $\mathcal{P}_z(\delta)$.

In these circumstances a *joint density function* may be defined by

$$\log \mathcal{P}(\epsilon, \delta) = \inf_{z>0} \left\{\mathcal{F}_\delta(z) - z \log \epsilon\right\}. \tag{3.48}$$

Alternatively, a two-parameter free energy with y conjugate to m_2 may be defined in the canonical way:

$$\mathcal{F}(z, y) = \sup_{\delta\in E} \left\{\log P_z(\delta) + \delta \log y\right\}. \tag{3.49}$$

3.3.2 *Properties of the density function*

In view of section 3.3.1.1 assume that the essential domain of $\mathcal{P}(\epsilon)$ is E with closure $\overline{E} = [\epsilon_0, \epsilon_m]$. Normally, $\epsilon_0 = 0$; but assume in what follows that $0 \leq \epsilon_0 < \epsilon_m < \infty$.

The density function is continuous at interior points of E because it is concave. It may be non-analytic at points of E; these points would signal phase transitions in the underlying thermodynamic model. Additional information about phase behaviour may be extracted by examining $\mathcal{P}(\epsilon)$ at boundary points ϵ_0 and ϵ_m of E.

Generally, the limiting free energy may be determined by solving for the supremum in equation (3.20).

3.3.2.1 *The function* $\log \mathcal{P}(\epsilon)$ *is not continuously differentiable:* The free energy of the model is given by equation (3.20). For small values of $z > 0$, the supremum may be realised when $\epsilon = \epsilon_0$ so that $\mathcal{F}(z) = \log \mathcal{P}(\epsilon_0^+) + \epsilon_0 \log z$ (for z sufficiently small), in which case $\epsilon^*(z) = \epsilon_0$, a constant function.

Similarly, for large values of $z > 0$, the supremum may be realised with $\epsilon = \epsilon_m$, and so $\epsilon^*(z) = \epsilon_m$, a constant function for these values of z.

Otherwise, the supremum is realised at points in the set (ϵ_0, ϵ_m).

Since $\mathcal{F}(z)$ is defined for every $z \geq 0$, there exists an $\epsilon^*(z)$ for every $z > 0$ which realises the supremum

$$\mathcal{F}(z) = \sup_{\epsilon \in (\epsilon_0, \epsilon_m)} \{\log \mathcal{P}(\epsilon) + \epsilon \log z\}. \tag{3.50}$$

By our assumption, $\epsilon^*(z) \in (\epsilon_0, \epsilon_m)$.

The function $\log \mathcal{P}(\epsilon) + \epsilon \log z$ is a concave function of ϵ on (ϵ_0, ϵ_m) and so it has left- and right-derivatives everywhere in (ϵ_0, ϵ_m). Let $\epsilon_-^s(z)$ be the solution of

$$\tfrac{d^-}{d\epsilon} (\log \mathcal{P}(\epsilon) + \epsilon \log z) = 0. \tag{3.51}$$

Similarly, let $\epsilon_+^s(z)$ be the solution of

$$\tfrac{d^+}{d\epsilon} (\log \mathcal{P}(\epsilon) + \epsilon \log z) = 0. \tag{3.52}$$

Suppose the domain of $\epsilon_-(z)$ is A_-, and the domain of $\epsilon_+(z)$ is A_+.

Since $\log \mathcal{P}(\epsilon)$ is differentiable almost everywhere in (ϵ_0, ϵ_m), it must be the case that $A_- \triangle A_+ = (A_- \setminus A_+) \cup (A_+ \setminus A_-)$ is a zero measure set. Thus, $\epsilon_-(z) = \epsilon_+(z)$ for all $z \in A_- \cap A_+$ and in fact for almost all $z \in A_- \cup A_+$. Hence,

$$\begin{aligned} \mathcal{F}(z) &= \log \mathcal{P}(\epsilon_-(z)) + \epsilon_-(z) \log z, \quad \text{for } z \in A_-, \text{ and} \\ \mathcal{F}(z) &= \log \mathcal{P}(\epsilon_+(z)) + \epsilon_+(z) \log z, \quad \text{for } z \in A_+. \end{aligned}$$

The energy density is the derivative of $\mathcal{F}(z)$ to $\log z$. Define $\mathcal{E}_-(z) = z \frac{d^-}{dz} \mathcal{F}(z)$ for $z \in A_-$, and $\mathcal{E}_+(z) = z \frac{d^+}{dz} \mathcal{F}(z)$ for $z \in A_+$. Then

$$\begin{aligned} \mathcal{E}_-(z) &= z \tfrac{d^-}{dz} \mathcal{F}(z) \\ &= \left[z \tfrac{d^-}{dz} \epsilon_-(z) \right] \tfrac{d^-}{d\epsilon} \log \mathcal{P}(\epsilon) \Big|_{\epsilon = \epsilon_-(z)} + z \log z \tfrac{d^-}{dz} \epsilon_-(z) + \epsilon_-(z) \\ &= -z \log z \tfrac{d^-}{dz} \epsilon_-(z) + z \log z \tfrac{d^-}{dz} \epsilon_-(z) + \epsilon_-(z) \\ &= \epsilon_-(z). \end{aligned} \tag{3.53}$$

This implies in particular that $\mathcal{E}_-(z) = \epsilon_-(z) = \lim_{n \to \infty} \frac{1}{n} \langle m \rangle_n$, similar to equation (3.57) for all $z \in A_-$.

Note that $\mathcal{E}_-(z)$ is a lower semicontinuous function which may have discontinuities at isolated critical points where the model undergoes a first order phase transition (see equation (3.12) and the arguments thereafter). The specific heat is given by $\mathcal{C}_-(z) = z\frac{d^-}{dz}\mathcal{E}_-(z)$ whenever $\mathcal{E}_-(z)$ is differentiable. Since $\mathcal{F}(z)$ is convex, it follows that $\mathcal{E}_-(z)$ is monotone and thus differentiable almost everywhere.

The upper semicontinuous energy $\mathcal{E}_+(z)$ may be defined on A_+ by $\mathcal{E}_+(z) = \epsilon_+(z) = \lim_{n\to\infty} \frac{1}{n} \langle m \rangle_n$ for all $z \in A_+$. This gives rise to the specific heat $\mathcal{C}_+(z)$ on A_+.

Whenever $\mathcal{F}(z)$ is differentiable, $\mathcal{E}_-(z) = \mathcal{E}_+(z) = \lim_{n\to\infty} \frac{1}{n} \langle m \rangle_n$ in concordance with equation (3.57), in which case $\mathcal{C}_-(z) = \mathcal{C}_+(z)$ whenever $\mathcal{E}_-(z)$ and $\mathcal{E}_+(z)$ are differentiable. The energy of the model is $\mathcal{E}(z) = \mathcal{E}_-(z)$ for ae z, and the specific heat is $\mathcal{C}(z) = \mathcal{C}_-(z)$ for ae z.

If $\mathcal{F}(z)$ is not differentiable at a critical point z_c, then $\mathcal{E}_-(z_c) \leq \mathcal{E}_+(z_c)$. This is a critical point in the model, which is first order if $\mathcal{E}_-(z_c) < \mathcal{E}_+(z_c)$. The *latent heat* in the model is $H(z_c) = \mathcal{E}_+(z_c) - \mathcal{E}_-(z_c)$ and this is strictly positive at first order transitions. In the case that $\mathcal{E}_-(z_c) = \mathcal{E}_+(z_c)$ the model undergoes a *continuous phase transition*.

3.3.2.2 *The function* $\log \mathcal{P}(\epsilon)$ *is continuously differentiable:* If $\mathcal{P}(\epsilon)$ is continuously differentiable, then the above simplifies. $\mathcal{F}(z)$ can be determined by solving for ϵ in

$$\tfrac{d}{d\epsilon}\left(\log \mathcal{P}(\epsilon) + \epsilon \log z\right) = 0, \quad \text{which simplifies to } \tfrac{d}{d\epsilon} \log \mathcal{P}(\epsilon) = -\log z. \tag{3.54}$$

If $\log \mathcal{P}(\epsilon)$ is a strictly concave function of ϵ, then $\frac{d}{d\epsilon} \log \mathcal{P}(\epsilon)$ is monotonic decreasing and continuous by the assumptions. Hence, if there is a solution $\epsilon = \epsilon^*(z)$, then it is continuous and unique.

Substituting $\epsilon = \epsilon^*(z)$ in equation (3.20) gives $\mathcal{F}(z)$:

$$\mathcal{F}(z) = \log \mathcal{P}(\epsilon^*(z)) + \epsilon^*(z) \log z. \tag{3.55}$$

The *energy density* of the model is $\mathcal{E}(z) = z\frac{d}{dz}\mathcal{F}(z)$. By arguing as in equation (3.53), it follows that, for almost every $z \in \mathbb{R}^+$, $\mathcal{E}(z) = \epsilon^*(z)$. That is, the function $\epsilon^*(z)$ is the energy density of the model, given explicitly by

$$\epsilon^*(z) = \lim_{n\to\infty} z\tfrac{d}{dz}\left(\tfrac{1}{n} \log Z_n(z)\right) = z\tfrac{d}{dz}\mathcal{F}(z) = \mathcal{E}(z) \ \text{ for ae } z \in \mathbb{R}^+, \tag{3.56}$$

in terms of the partition function (see equation (3.1) and theorem B.9 in appendix B), since the derivative and limit of a sequence of convex functions with a convex limit may be interchanged.

The function $\mathcal{E}(z)$ is the expectation of $\frac{m}{n}$ with respect to the Boltzmann distribution in equation (3.1) (which is a probability measure) in the limit $n \to \infty$. Explicitly, the above shows that for almost all $z \in \mathbb{R}^+$,

$$\mathcal{E}(z) = \epsilon^*(z) = \lim_{n\to\infty} \frac{\sum_m m q_n(m) z^m}{n \sum_m q_n(m) z^m} = \lim_{n\to\infty} \tfrac{1}{n}\langle m \rangle_n = z\tfrac{d}{dz}\mathcal{F}(z). \tag{3.57}$$

Since $\mathcal{F}(z)$ is convex in $\log z$, $\mathcal{E}(z)$ is monotone non-decreasing in $\log z$. Hence, $\mathcal{E}(z)$ is differentiable almost everywhere (and in fact, is differentiable whenever $\mathcal{F}(z)$ is analytic). This shows that the second order derivative

$$\mathcal{C}(z) = z\tfrac{d}{dz}\mathcal{E}(z) \tag{3.58}$$

of $\mathcal{F}(z)$ exists for ae $z > 0$. The derivative $\mathcal{C}(z)$ is the variance of $\mathcal{E}(z)$ and is the *specific heat* of the model.

3.3.2.3 *The case* $\log \mathcal{P}(\epsilon) = \kappa\epsilon + \delta$*:* Suppose that $\log \mathcal{P}(\epsilon)$ is linear in $[\epsilon_1, \epsilon_2]$; say, $\log \mathcal{P}(\epsilon) = \kappa\epsilon + \delta$ for $\epsilon \in [\epsilon_1, \epsilon_2]$, where $\epsilon_0 \leq \epsilon_1 < \epsilon_2 \leq \epsilon_m$. Since $\log \mathcal{P}(\epsilon)$ is concave, $\log \mathcal{P}(\epsilon) \leq \kappa\epsilon + \delta$ on (ϵ_0, ϵ_m). Hence, the supremum in equation (3.20) is realised at $\epsilon \leq \epsilon_1$ if $\log z < -\kappa$, and at $\epsilon \geq \epsilon_2$ if $\log z > -\kappa$. This, in particular, shows that

$$\mathcal{E}(z) \begin{cases} \leq \epsilon_1, & \text{if } \log z < -\kappa; \\ \geq \epsilon_2, & \text{if } \log z > -\kappa. \end{cases} \tag{3.59}$$

Thus, $\mathcal{E}(z)$ has a jump discontinuity at $\log z = -\kappa$, which is a jump discontinuity in the first derivative of $\mathcal{F}(z)$. This is a first order transition.

3.3.2.4 *The case* $\mathcal{P}(\epsilon_0) < \mathcal{P}(\epsilon_0^+)$*, or* $\mathcal{P}(\epsilon_m^-) > \mathcal{P}(\epsilon_m)$*:* There is a jump discontinuity in the density function at the left endpoint of its essential domain E. Then

$$\mathcal{F}(0) = \log \mathcal{P}(\epsilon_0) < \log \mathcal{P}(\epsilon_0^+) \leq \mathcal{F}(z), \quad \text{for all } z > 0. \tag{3.60}$$

This shows a jump discontinuity in the free energy itself, which may be interpreted as a zeroth order transition at zero temperature (for another example of a claimed zeroth order transition, see reference [85]). As a practical matter, it is not possible to reach $\mathcal{F}(0)$ by lowering $z \to 0^+$; the system will approach the limiting free energy $\mathcal{F}(0^+) = \log \mathcal{P}(\epsilon_0^+)$. The jump discontinuity at ϵ_0 may be ignored by *defining* $\log \mathcal{P}(\epsilon_0) = \mathcal{P}(\epsilon_0^+)$. A similar observation can be made if $\mathcal{P}(\epsilon_m^-) > \mathcal{P}(\epsilon_m)$, except that the transition occurs at $z = \infty$.

3.3.2.5 *A finite right-derivative at* ϵ_0*, or a finite left-derivative at* ϵ_m*:* Assume that $\mathcal{P}(\epsilon)$ is right-continuous at $\epsilon = \epsilon_0$. Denote the right-derivative of $\mathcal{P}(\epsilon)$ by $D^+\mathcal{P}(\epsilon) = \frac{d^+}{d\epsilon}\mathcal{P}(\epsilon)$ and put $D^+\mathcal{P}(\epsilon_0) = \lim_{\epsilon \to \epsilon_0^+} D^+\mathcal{P}(\epsilon)$.

Define the function $\mathcal{Q}(\epsilon) = \log \mathcal{P}(\epsilon) + \epsilon \log z$. Suppose that $D^+\mathcal{P}(\epsilon_0)$ is finite at ϵ_0: that is, $-\infty < D^+(\epsilon_0) < \infty$. Then $\mathcal{Q}(\epsilon)$ is concave, and

$$D^+\mathcal{Q}(\epsilon_0) = D^+ \log \mathcal{P}(\epsilon_0) + \log z. \tag{3.61}$$

Define

$$\log z_c = -D^+ \log \mathcal{P}(\epsilon_0). \tag{3.62}$$

If $z < z_c$, then $D^+\mathcal{Q}(\epsilon_0) \leq 0$ and, since $\mathcal{Q}(\epsilon)$ is concave, it is monotonic and non-increasing. Hence, its (global maximum) is realised at $\epsilon = \epsilon_0$. By equation (3.20), the free energy is given by

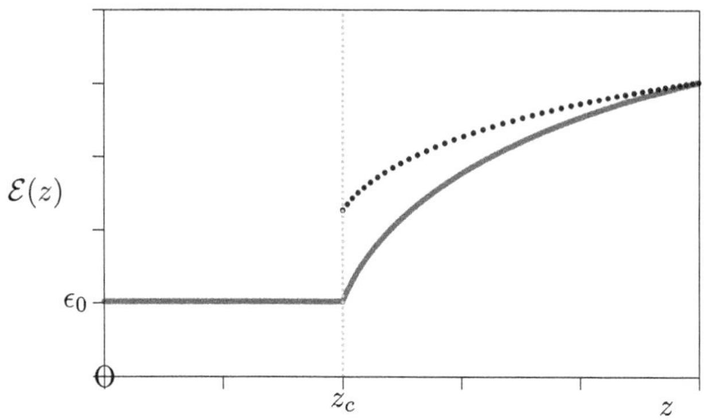

FIG. 3.6. The energy (density) $\mathcal{E}(z)$ of model with finite right-derivative at ϵ_0. There is a critical point at $z_c = -D^+ \log \mathcal{P}(\epsilon_0)$; $\mathcal{E}(z) = \epsilon_0$ for all $z < z_c$ and either is continuous with a positive right-derivative at $z = z_c$ (solid curve; this corresponds to a continuous transition in the model) or it has a jump discontinuity at $z = z_c$ (broken curve; this corresponds to a first order transition in the model). The height in the jump is the latent heat. Models in which $\mathcal{E}(z)$ has a profile of being constant either on $z < z_c$ or on $z > z_c$ are *asymmetric models.*

$$\mathcal{F}(z) = \mathcal{Q}(\epsilon_0) = \log \mathcal{P}(\epsilon_0) + \epsilon_0 \log z, \quad \text{for } z < z_c. \tag{3.63}$$

On the other hand, if $z > z_c$, then $D^+\mathcal{Q}(\epsilon_0) > 0$, and the global maximum is realised at $\epsilon^* > \epsilon_0$: this, in particular, shows that

$$\begin{aligned}\mathcal{F}(z) &= \mathcal{Q}(\epsilon^*) = \log \mathcal{P}(\epsilon^*) + \epsilon^* \log z \\ &> \mathcal{Q}(\epsilon_0) = \log \mathcal{P}(\epsilon_0) + \epsilon_0 \log z, \quad \text{for } z > z_c.\end{aligned} \tag{3.64}$$

This shows that $\mathcal{F}(z)$ is non-analytic at z_c. Hence, there is a phase transition at $z = z_c$ in the model.

If z_c defined in equation (3.62) is non-zero and finite, then the global maximum of $\mathcal{Q}(\epsilon)$ is realised at ϵ_0 for all $z < z_c$. In this phase, the free energy is determined at the limiting contribution of $q_n(\lfloor \epsilon_0 n \rfloor)$ in the partition function (see equations (3.1) and (3.8)). For $z > z_c$, the free energy is determined by the limiting contributions of $q_n(\lfloor \epsilon^* \rfloor)$, as seen in equation (3.64). This defines the function $\epsilon^*(z)$ by $\epsilon^*(z) = \epsilon_0$ if $z < z_c$, and $\epsilon^*(z) = \epsilon^*$ for $z > z_c$.

Note that $\mathcal{E}(z) = \epsilon^*(z)$ is the limiting energy density of the model, as seen in equation (3.56). It is monotonic non-decreasing; in the circumstances here it is constant and equal to ϵ_0 when $z < z_c$, and then either has a jump discontinuity or a positive right-derivative at $z = z_c$. The general shape of $\mathcal{E}(z)$ is illustrated in figure 3.6.

Similarly, assume that $\mathcal{P}(\epsilon)$ is left-continuous at $\epsilon = \epsilon_m$. Denote the left-derivative of $\log \mathcal{P}(\epsilon)$ by $D^- \log \mathcal{P}(\epsilon) = \frac{d^-}{d\epsilon} \log \mathcal{P}(\epsilon)$, and put $D^- \log \mathcal{P}(\epsilon_m) = \lim_{\epsilon \to \epsilon_m^-} (D^- \log \mathcal{P}(\epsilon))$. If z_d is defined by

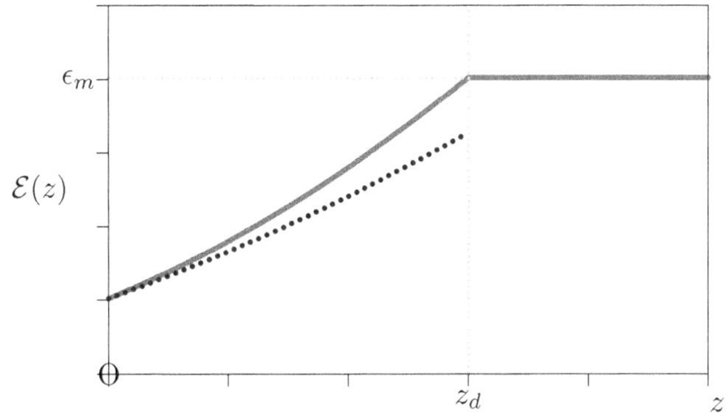

FIG. 3.7. The energy (density) $\mathcal{E}(z)$ of a model with a finite left-derivative at ϵ_m. There is a critical point at the negative left-derivative of the density function $\mathcal{P}(\epsilon)$: $z_d = -D^-\mathcal{P}(\epsilon_m)$. It follows that $\mathcal{E}(z) = \epsilon_m$ for all $z > z_d$, and it is either continuous with a positive left-derivative at $z = z_d$ (solid curve; this corresponds to a continuous transition in the model) or it has a jump discontinuity at $z = z_d$ (broken curve; this corresponds to a first order transition in the model). Models in which $\mathcal{E}(z)$ has a profile of being constant either on $z < z_c$ or on $z > z_c$ are *asymmetric models.*

$$\log z_d = -D^- \log \mathcal{P}(\epsilon_m) \tag{3.65}$$

and $0 < z_d < \infty$, then a set of arguments similar to the above show that there is a phase transition at $z = z_d$ in the model. The typical shapes of the energy density $\mathcal{E}(z)$ are illustrated in figure 3.7; $\mathcal{E}(z)$ is increasing for $z < z_d$, and $\mathcal{E}(z) = \epsilon_m$ for $z > z_d$. The transition may either be continuous or first order.

In some models the right- and left-derivatives at ϵ_0 and ϵ_m may be infinite. This rules out transitions at finite values of z but there may still be transitions at 0 or at infinite z.

3.3.2.6 *A jump discontinuity in the derivative of $\mathcal{P}(\epsilon)$:* Consider the case where the first derivative of $\mathcal{P}(\epsilon)$ has a jump discontinuity, say at ϵ_b (see figure 3.5). That is, assume that

$$D^-\mathcal{P}(\epsilon_b) = \tfrac{d^-}{d\epsilon}\mathcal{P}(\epsilon)\Big|_{\epsilon=\epsilon_b} > \tfrac{d^+}{d\epsilon}\mathcal{P}(\epsilon)\Big|_{\epsilon=\epsilon_b} = D^+\mathcal{P}(\epsilon_b). \tag{3.66}$$

Put $\mathcal{Q}(\epsilon) = \log \mathcal{P}(\epsilon) + \epsilon \log z$, then $\mathcal{Q}(\epsilon)$ is a concave function of ϵ on (ϵ_0, ϵ_m) for every $z \in (0, \infty)$. Then

$$D^-\mathcal{Q}(\epsilon)\Big|_{\epsilon=\epsilon_b} = \frac{D^-\mathcal{P}(\epsilon_b)}{\mathcal{P}(\epsilon_b)} + \log z, \quad \text{and } D^+\mathcal{Q}(\epsilon)\Big|_{\epsilon=\epsilon_b} = \frac{D^+\mathcal{P}(\epsilon_b)}{\mathcal{P}(\epsilon_b)} + \log z.$$

Interpolate linearly between these derivatives; define the function

$$\zeta(\lambda) = -\tfrac{1}{\mathcal{P}(\epsilon_b)}\left(\lambda D^-\mathcal{P}(\epsilon_b) + (1-\lambda) D^+\mathcal{P}(\epsilon_b)\right). \tag{3.67}$$

Putting $z = \zeta(\lambda)$ and computing $D^-\mathcal{Q}(\epsilon)|_{\epsilon=\epsilon_b}$ and $D^+\mathcal{Q}(\epsilon)|_{\epsilon=\epsilon_b}$ then shows that

$$D^-\mathcal{Q}(\epsilon)|_{\epsilon=\epsilon_b} = \tfrac{1-\lambda}{\mathcal{P}(\epsilon_b)} \left(D^-\mathcal{P}(\epsilon_b) - D^+\mathcal{P}(\epsilon_b)\right) > 0, \text{ and} \tag{3.68}$$

$$D^+\mathcal{Q}(\epsilon)|_{\epsilon=\epsilon_b} = \tfrac{\lambda}{\mathcal{P}(\epsilon_b)} \left(D^+\mathcal{P}(\epsilon_b) - D^-\mathcal{P}(\epsilon_b)\right) < 0. \tag{3.69}$$

Since $\mathcal{Q}(\epsilon)$ is continuous, the positive left-derivative and negative right-derivative imply that it has a maximum at ϵ_b for every value of $z = \zeta(\lambda)$ and for $\lambda \in (0, 1)$.

In other words, for $z \in (\zeta(0), \zeta(1))$, $\mathcal{F}(z) = \log \mathcal{P}(\epsilon_b) + \epsilon_b \log z$. This shows that $\mathcal{F}(z)$ is linear in $\log z$ on the interval $I = (\zeta(0), \zeta(1))$. Extend I, if it is necessary so that it is the largest open interval on which $\mathcal{F}(z)$ is linear in $\log z$. If I is a finite or semi-infinite interval, then $\mathcal{F}(z)$ is non-analytic at its endpoints, signalling phase transitions. Generally, the interval I will be semi-infinite.

3.4 Integrated density functions

In some models the supermultiplicative inequality for $q_n(m)$ in equation (3.25) may not be valid. A weaker inequality is

$$q_{n_1}(m_1) q_{n_2}(m_2) \le \sum_{i=-s}^{s} q_{n_1+n_2}(m_1 + m_2 + i), \tag{3.70}$$

where s is a constant. Define, as before, $q_n = \sum_m q_n(m)$.

A few more basic assumptions are made about $q_n(m)$, namely,

- there is a $K > 0$ such that $0 \le q_n(m) < K^n$ for each m;
- $q_n(m) > 0$ for $0 \le m \le A_n$, where A_n is an non-decreasing function of n;
- and $\liminf_{n\to\infty} \frac{1}{n} A_n > 0$, and $\limsup_{n\to\infty} \frac{1}{n} A_n < \infty$. That is, assume that there exist a finite c and an N_0 such that $0 < A_n \le cn$ for all $n \ge N_0$.

Define

$$\epsilon_m = \limsup_{n\to\infty} \tfrac{1}{n} A_n. \tag{3.71}$$

By summing equation (3.70) over m_1 and m_2,

$$q_{n_1} q_{n_2} \le (2s + 1)(A_{n_1+n_2} + 1) q_{n_1+n_2}. \tag{3.72}$$

This shows that the growth constant μ exists and is finite:

$$\lim_{n\to\infty} \tfrac{1}{n} \log q_n = \log \mu, \tag{3.73}$$

by theorem A.4 in appendix A. Proceed by defining

$$q_n(\le m) = \sum_{i\le m} q_n(i), \text{ and } q_n(\ge m) = \sum_{i\ge m} q_m(i). \tag{3.74}$$

A suitable definition for integrated density functions would be

$$\mathcal{P}(\le\epsilon) = \lim_{n\to\infty} \left(q_n(\le \lfloor \epsilon n\rfloor)\right)^{1/n}, \text{ and } \mathcal{P}(\ge\epsilon) = \lim_{n\to\infty} \left(q_n(\ge \lfloor \epsilon n\rfloor)\right)^{1/n}. \tag{3.75}$$

This is justified in corollary 3.14.

Lemma 3.11 *Let s be defined as in equation (3.70). Then*

$$q_{n_1}(\leq m_1 - s)q_{n_2}(\leq m_2 - s) \leq f_c(n_1,n_2)q_{n_1+n_2}(\leq m_1 + m_2 - s), \quad \text{and}$$
$$q_{n_1}(\geq m_1 + s)q_{n_2}(\geq m_2 + s) \leq g_c(n_1,n_2)q_{n_1+n_2}(\geq m_1 + m_2 + s)$$

for functions $f_c(n_1,n_2)$ and $g_c(n_1,n_2)$ which are linear in s, n_1 and n_2.

Proof Observe that $q_n(\leq m) \leq q_n(\leq m+1)$, and that $q_n(m) = 0$ if $m > cn$. This shows that summing equation (3.70) over m_1 and m_2 gives

$$q_{n_1}(\leq m_1)q_{n_2}(\leq m_2) \leq c(n_1+n_2+1)(2s+1)q_{n_1+n_2}(\leq m_1+m_2+s).$$

Replace m_1 by $m_1 - s$ and replace m_2 by $m_2 - s$ to get the first inequality with $f_c(n_1,n_2) = c(n_1+n_2+1)(2s+1)$ linear in each of s, n_1 and n_2 as claimed.

The proof of the second inequality follows a similar set of arguments. □

A corollary of lemma 3.11 is that $\mathcal{P}(\leq\epsilon)$ and $\mathcal{P}(\geq\epsilon)$ exist.

Theorem 3.12 *Let $\epsilon \in (0,\epsilon_m)$. The integrated density functions are defined by*

$$\log\mathcal{P}(\leq\epsilon) = \lim_{n\to\infty}\tfrac{1}{n}\log q_n(\leq\lfloor\epsilon n\rfloor - s), \text{ and } \log\mathcal{P}(\geq\epsilon) = \lim_{n\to\infty}\tfrac{1}{n}\log q_n(\geq\lceil\epsilon n\rceil + s).$$

The function s is defined in equation (3.70).

Proof In the first inequality in lemma 3.11 choose $m_1 = \lfloor\epsilon n_1\rfloor$, and $m_2 = \lfloor\epsilon n_2\rfloor$. Since $\lfloor a\rfloor + \lfloor b\rfloor = \lfloor a+b\rfloor - 1$ or $\lfloor a\rfloor + \lfloor b\rfloor = \lfloor a+b\rfloor$, the fact that $q_n(\leq m) \leq q_n(\leq m+1)$ gives $q_{n_1+n_2}(\leq\lfloor\epsilon n_1\rfloor + \lfloor\epsilon n_2\rfloor - s) \leq 2q_{n_1+n_2}(\leq\lfloor\epsilon(n_1+n_2)\rfloor)$.

In other words,

$$q_{n_1}(\leq\lfloor\epsilon n_1\rfloor - s)q_{n_2}(\leq\lfloor\epsilon n_2\rfloor - s) \leq 2f_c(n_1,n_2)q_{n_1+n_2}(\leq\lfloor\epsilon(n_1+n_2)\rfloor - s).$$

By theorem A.4 in appendix A, the limit $\lim_{n\to\infty}\frac{1}{n}\log q_n(\leq\lfloor\epsilon n\rfloor - s)$ exists.

A similar argument shows that the second limit exists. □

Put $n = n_1 = n_2$, let $m_1 = \lfloor\epsilon n\rfloor$, and let $m_2 = \lfloor\delta n\rfloor$ in lemma 3.11. Then

$$q_n(\leq\lfloor\epsilon n\rfloor - s)q_n(\leq\lfloor\delta n\rfloor - s) \leq 2f_c(n,n)q_{2n}(\leq\lfloor(\epsilon+\delta)n\rfloor - s), \tag{3.76}$$

similar to the argument in the proof of theorem 3.12. Take logarithms, divide by n and let $n \to \infty$ to see that $\mathcal{P}(\leq\epsilon)$ is log-concave. A similar argument shows that $\mathcal{P}(\geq\epsilon)$ is log-concave.

3.4.1 *Existence of integrated density functions*

The log-concavity of the integrated density functions, together with their definition in theorem 3.12, implies that these functions are measurable, continuous and differentiable almost everywhere. As before, assume that the integrated density function is defined on the interval $(0,\epsilon_m)$ (that is, $\epsilon_0 = 0$).

Theorem 3.13 *Suppose that* $\langle \delta_n \rangle$ *is a sequence in* $\mathbb{N}_0$ *such that* $0 \leq \delta_n \leq A_n$ *for all* $n \geq N_0 \in \mathbb{N}$. *If the limit* $\lim_{n\to\infty} \frac{1}{n}\delta_n = \delta \in (0, \epsilon_m)$ *exists, then*

$$\log \mathcal{P}(\leq\delta) = \lim_{n\to\infty} \tfrac{1}{n} \log q_n(\leq\delta_n), \quad \textit{and } \log \mathcal{P}(\geq\delta) = \lim_{n\to\infty} \tfrac{1}{n} \log q_n(\geq\delta_n).$$

Proof Since $\delta < \epsilon_m$, there is a β such that $\delta < \beta < \epsilon_m$. Let $s > 0$ and choose n large enough that $\lfloor \beta n \rfloor - s \geq \delta_n$ for all n larger than (say) N_1. Then $q_n(\leq\lfloor \beta n\rfloor - s) \geq q_n(\leq\delta_n)$, which gives $\log \mathcal{P}(\leq\beta) \geq \limsup_{n\to\infty} \frac{1}{n} \log q_n(\leq\delta_n)$. Take $\beta \to \delta^+$. Since $\mathcal{P}(\leq\epsilon)$ is log-concave, its right-limit exists:

$$\log \mathcal{P}(\leq\delta^+) \geq \limsup_{n\to\infty} \tfrac{1}{n} \log q_n(\leq\delta_n). \tag{3.77}$$

On the other hand, since $\delta > 0$, there is a β' such that $0 < \beta' < \delta$. Increase n until $\lfloor \beta' n\rfloor \leq \delta_n$ for all n larger than (say) N_2. Then $q_n(\leq\lfloor \beta' n\rfloor - s) \leq q_n(\leq\lfloor \beta' n\rfloor) \leq q_n(\leq\delta_n)$. It follows that $\log \mathcal{P}(\leq\beta') \leq \liminf_{n\to\infty} \frac{1}{n} \log q_n(\leq\delta_n)$. Take $\beta' \to \delta^-$. Since $\mathcal{P}(\leq\epsilon)$ is log-concave, its left-limit exists. Hence,

$$\log \mathcal{P}(\leq\delta^-) \leq \liminf_{n\to\infty} \tfrac{1}{n} \log q_n(\leq\delta_n).$$

Compare this to equation (3.77) above and use continuity of $\mathcal{P}(\leq\delta)$ and the squeeze theorem for limits to conclude that $\lim_{n\to\infty} \frac{1}{n} \log q_n(\leq\delta_n)$ exists.

This establishes the existence of the first claimed limit in the statement of the theorem. The existence of the second limit in the statement of the theorem is proven using a similar argument. □

The corollary of theorem 3.13 is the justification of the definition of the integrated density functions in equation (3.75).

Corollary 3.14 *The integrated density functions are defined on* $(0, \epsilon_m)$ *by*

$$\mathcal{P}(\leq\epsilon) = \lim_{n\to\infty} \left(q_n(\leq\lfloor\epsilon n\rfloor)\right)^{1/n}, \quad \textit{and } \mathcal{P}(\geq\epsilon) = \lim_{n\to\infty} \left(q_n(\geq\lfloor\epsilon n\rfloor)\right)^{1/n}.$$

This proves equation (3.75). □

Assume without loss of generality that the integrated density functions are defined on $(0, \epsilon_m)$ (that is, $\epsilon_0 = 0$). Since $q_n(\leq\lfloor\epsilon n\rfloor) + q_n(\geq\lfloor\epsilon n\rfloor + 1) = q_n$, an immediate consequence of corollary 3.14 is the following.

Corollary 3.15 *Either* $\mathcal{P}(\leq\epsilon)$, $\mathcal{P}(\geq\epsilon)$ *or both, are equal to* μ *(see equation (3.73)). That is,* $\mu = \max\{\mathcal{P}(\leq\epsilon), \mathcal{P}(\geq\epsilon)\}$ *for every* $\epsilon \in (0, \epsilon_m)$. □

In other words, there exists an interval $(\epsilon_1, \epsilon_2) \subseteq (0, \epsilon_m)$ defined by

$$\epsilon_1 = \inf\{\epsilon \mid \mathcal{P}(\leq\epsilon) = \mu\}, \quad \text{and } \epsilon_2 = \sup\{\epsilon \mid \mathcal{P}(\geq\epsilon) = \mu\}, \tag{3.78}$$

such that $\mathcal{P}(\leq\epsilon)$ is strictly increasing in (ϵ_0, ϵ_1) and equal to μ in $[\epsilon_1, \epsilon_m)$, whereas $\mathcal{P}(\geq\epsilon)$ is equal to μ in $(\epsilon_0, \epsilon_2]$ and is strictly decreasing in (ϵ_2, ϵ_M).

3.4.2 *Existence of the density function*

The density function may be defined in terms of integrated density functions. Let $\{\epsilon_0, \epsilon_1, \epsilon_2, \epsilon_m\}$ be as defined in section 3.3.2 and equation (3.78) and suppose that $\epsilon_0 = 0$ without loss of generality.

Theorem 3.16 *Suppose that $\epsilon \in (0, \epsilon_m)$. Then there exists a sequence of integers $\langle \delta_n \rangle$ such that*

$$\log \mathcal{P}(\epsilon) = \lim_{n\to\infty} \tfrac{1}{n} \log q_n(\delta_n) = \min \{\log \mathcal{P}(\leq\epsilon), \log \mathcal{P}(\geq\epsilon)\}.$$

Moreover, $\lim_{n\to\infty} \frac{1}{n}\delta_n = \epsilon$.

Proof Suppose that $\epsilon \in (0, \epsilon_1]$ and choose an integer $m = \delta_n \in [0, \lfloor \epsilon n \rfloor]$ which maximises $q_n(m)$ over this set. Then

$$q_n(\leq\lfloor\epsilon n\rfloor) \leq (\lfloor\epsilon n\rfloor + 1)\, q_n(\delta_n) \leq (\lfloor\epsilon n\rfloor + 1)\, q_n(\leq\delta_n) \leq (\lfloor\epsilon n\rfloor + 1)\, q_n(\leq\lfloor\epsilon n\rfloor).$$

This shows that the limit $\lim_{n\to\infty} \frac{1}{n} \log q_n(\delta_n) = \log \mathcal{P}(\leq\epsilon)$ exists for all choices of $\epsilon \in (0, \epsilon_1]$.

It remains to show that $\lim_{n\to\infty} \frac{1}{n}\delta_n = \epsilon$. Observe that $\delta_n \leq \lfloor \epsilon n \rfloor$ for $n \in \mathbb{N}$. Suppose that $\liminf_{n\to\infty} \frac{1}{n}\delta_n = \kappa < \epsilon$. Then there is a sequence $\langle n_i \rangle$ in $\mathbb{N}$ such that $\lim_{i\to\infty} \frac{1}{n_i}\delta_{n_i} = \kappa$. Since $\kappa < \epsilon \leq \epsilon_1$, it must be the case that $\mathcal{P}(\leq\kappa) < \mathcal{P}(\leq\epsilon)$.

It will be shown that the assumption that $\liminf_{n\to\infty} \frac{1}{n}\delta_n = \kappa < \epsilon$ gives a contradiction. For each $i \in \mathbb{N}$,

$$q_{n_i}(\leq\lfloor\epsilon n_i\rfloor) \leq (\lfloor\epsilon n_i\rfloor + 1)\, q_{n_i}(\delta_{n_i}).$$

Take logarithms, divide by n_i and take $i \to \infty$. By theorem 3.13 and by corollary 3.14,

$$\mathcal{P}(\leq\epsilon) = \lim_{i\to\infty} \tfrac{1}{n_i} \log q_{n_i}(\leq\lfloor\epsilon n_i\rfloor) \leq \lim_{i\to\infty} \tfrac{1}{n_i} \log q_{n_i}(\delta_{n_i}) = \mathcal{P}(\kappa).$$

This is a contradiction. Hence, $\kappa = \epsilon$, and $\mathcal{P}(\leq\epsilon) = \mathcal{P}(\epsilon)$, for all $\epsilon \in (0, \epsilon_1]$. A similar proof works for the case that $\epsilon \in (\epsilon_2, \epsilon_m)$.

It remains to consider the case that $\epsilon \in (\epsilon_1, \epsilon_2)$. If $\epsilon_1 = \epsilon_2$, then the proof is done, so suppose that $\epsilon_1 < \epsilon_2$.

Claim: If $\epsilon = \frac{1}{2}(\epsilon_1 + \epsilon_2)$, then there exists a sequence $\langle \delta_n \rangle$ such that

$$\log \mathcal{P}(\epsilon) = \lim_{n\to\infty} \tfrac{1}{n} \log q_n(\delta_n).$$

Moreover, $\lim_{n\to\infty} \frac{1}{n}\delta_n = \epsilon$.

PROOF OF CLAIM: Choose the sequence $\langle x_n \rangle$ such that $\lim_{n\to\infty} \frac{1}{n} x_n = \epsilon_1$ and choose the sequence $\langle y_n \rangle$ such that $\lim_{n\to\infty} \frac{1}{n} y_n = \epsilon_2$ in equation (3.70). Then

$$q_n(x_n) q_n(y_n) \leq \sum_{i=-s}^{s} q_{2n}(x_n + y_n + i). \tag{3.79}$$

By the above, $\langle x_n \rangle$ and $\langle y_n \rangle$ can be chosen such that

$$\lim_{n\to\infty} \tfrac{1}{n} \log q_n(x_n) = \log \mathcal{P}(\epsilon_1), \quad \text{and} \quad \lim_{n\to\infty} \tfrac{1}{n} \log q_n(y_n) = \log \mathcal{P}(\epsilon_2).$$

Choose i to maximise the right-hand side of equation (3.79), say, $i = \sigma$ (a function of (n, x_n, y_n) and $|\sigma| \leq s$). The above becomes

$$q_n(x_n) q_n(y_n) \leq (2s+1) q_{2n}(x_n + y_n + \sigma).$$

Take logarithms, divide by n and take the limit inferior on the right-hand side as $n \to \infty$. By the choices of x_n and y_n,

$$\log \mathcal{P}(\epsilon_1) + \log \mathcal{P}(\epsilon_2) \leq 2 \liminf_{n\to\infty} \tfrac{1}{2n} \log q_{2n}(x_n + y_n + \sigma) \leq 2 \log \mu. \tag{3.80}$$

By putting $n_2 = n+1$ in equation (3.70) instead, the set of arguments above gives

$$\log \mathcal{P}(\epsilon_1) + \log \mathcal{P}(\epsilon_2) \leq 2 \liminf_{n\to\infty} \tfrac{1}{2n+1} \log q_{2n+1}(x_n + y_{n+1} + \sigma') \leq 2 \log \mu. \tag{3.81}$$

Construct the sequence $\langle \delta_n \rangle$ by $\delta_{2n} = x_n + y_n + \sigma$, and $\delta_{2n+1} = x_n + y_{n+1} + \sigma'$. Then $\lim_{n\to\infty} \frac{1}{n} \delta_n = \frac{\epsilon_1 + \epsilon_2}{2} \in (\epsilon_1, \epsilon_2)$.

Since $\mathcal{P}(\epsilon_1) = \mathcal{P}(\epsilon_2) = \mu$ in equations (3.80) and (3.81), the result is that

$$\liminf_{n\to\infty} \tfrac{1}{n} \log q_n(\delta_n) = \log \mu$$

at the midpoint of (ϵ_1, ϵ_2). Since $\limsup_{n\to\infty} \frac{1}{n} \log q_n(\delta_n) \leq \log \mu$, this establishes existence of the sequence $\langle \delta_n \rangle$ for $\epsilon = \frac{\epsilon_1 + \epsilon_2}{2}$. △

The proof of the claim bisected the interval (ϵ_1, ϵ_2). By recursively bisecting (ϵ_1, ϵ_2), there is a dense subset $D \subset (\epsilon_1, \epsilon_2)$ such that, if $\epsilon \in D$, then there exists a sequence $\langle \delta_n \rangle$ such that $\log \mathcal{P}(\epsilon) = \lim_{n\to\infty} \frac{1}{n} \log q_n(\delta_n) = \log \mu$, and $\lim_{n\to\infty} \frac{1}{n} \delta_n = \epsilon$. Since the real line is complete, this extends to all $\epsilon \in (\epsilon_1, \epsilon_2)$. Hence, $\mathcal{P}(\epsilon) = \mu$ for all $\epsilon \in (\epsilon_1, \epsilon_2)$. □

Since $\lim_{n\to\infty} \frac{1}{n} \delta_n = \epsilon$ in theorem 3.16, the sequence $\langle \delta_n \rangle$ can be chosen such that $\delta_n = \lfloor \epsilon n \rfloor + o(n)$.

These results are taken together in theorem 3.17.

Theorem 3.17 *Suppose that $q_n(m)$ satisfies the supermultiplicative inequality in equation (3.70) and the assumptions of section 3.4. Then there is a sequence $\langle \delta_n \rangle$ such that*

$$\log \mathcal{P}(\epsilon) = \lim_{n\to\infty} \tfrac{1}{n} \log q_n(\delta_n) = \min \{\log \mathcal{P}(\leq \epsilon), \log \mathcal{P}(\geq \epsilon)\},$$

where $\delta_n = \lfloor \epsilon n \rfloor + o(n)$. Moreover, $\sup \log \mathcal{P}(\epsilon) = \log \mu$, *and* $\log \mathcal{P}(\epsilon)$ *is a concave function of* ϵ *on its essential domain.* □

3.5 Combinatorial examples

3.5.1 *Chromatic polynomials of complete and path graphs*

A proper n-colouring of a graph G is an assignment of n colours to the vertices of G such that adjacent vertices have different colours. The *chromatic polynomial* of G is the number of distinct proper n-colourings of G.

The chromatic polynomial of the *unlabelled* complete graph of order k, K_k, is the binomial coefficient $\binom{n}{k}$ (choose k from n colours to assign to the vertices in one way). The partition function of this is

$$Z_n(z) = \sum_{k=0}^{n} \binom{n}{k} z^k = (1+z)^n. \tag{3.82}$$

The generating function can be computed directly:

$$G(z,t) = \sum_{n=0}^{\infty} Z_n(z) t^n = \frac{1}{1-(1+z)t}. \tag{3.83}$$

The function $G(z,t)$ has dominant singularity on the positive real axis at $t_c = \frac{1}{1+z}$, and the free energy, energy and specific heat of the model are

$$\mathcal{F}(z) = \log(1+z),\ \mathcal{E}(z) = \frac{z}{1+z}, \quad \text{and } \mathcal{C}(z) = \frac{z}{(1+z)^2}, \text{ respectively.} \tag{3.84}$$

The free energy $\mathcal{F}(z)$ is a convex function of $\log z$. The density function is obtained through equation (3.19); the infimum is realised at $z = \frac{\epsilon}{1-\epsilon}$ so that

$$\mathcal{P}(\epsilon) = \frac{1}{\epsilon^\epsilon (1-\epsilon)^{1-\epsilon}}. \tag{3.85}$$

It follows that $\log \mathcal{P}(\epsilon)$ is concave on its essential domain $E = [0,1]$.

In the case of (unlabelled) path graphs, consider the situation that there are n colours from which k colours are chosen. The chromatic polynomial is given by $\frac{1}{2}\binom{n}{k} k(k-1)^{n-1}$. Multiplying by $z^k t^n$, summing n from k to ∞ and then summing k from 0 to ∞ gives the generating function

$$G(z,t) = \tfrac{1}{2} \sum_{k=0}^{\infty} \frac{k(k-1)^{k-1}(zt)^k}{(1-(k+1)t)^{k+1}}. \tag{3.86}$$

The generating function is singular for $t = \frac{1}{k+1}$ for all $k \in \mathbb{N}_0$. These singularities accumulate at $t=0$, which is an essential singularity in the t-plane. Hence, the radius of convergence is $t_c = 0$, and the free energy is $\mathcal{F}(z) = \infty$.

3.5.2 *Stiff random walk in $\mathbb{L}^d$*

The number of random walks of length n from $\vec{0}$ with k corners (where adjacent steps are perpendicular) in the d-dimensional hypercubic lattice $\mathbb{L}^d$ is $r_n(k) = \binom{n-1}{k}(d-1)^k 2^{n-1}$.

By concatenating two walks,

$$r_{n_1}(k_1)r_{n_2}(k_2) \leq r_{n_1+n_2}(k_1+k_2) + r_{n_1+n_2}(k_1+k_2+1), \tag{3.87}$$

since one extra corner may be created at the point of concatenation. This should be compared with equation (3.70). By corollary 3.14, the integrated density functions can be determined:

$$\mathcal{P}(\leq\epsilon) = \begin{cases} \frac{2(d-1)^\epsilon}{\epsilon^\epsilon(1-\epsilon)^{1-\epsilon}}, & \text{if } \epsilon \leq \frac{d-1}{d}; \\ 2d, & \text{if } \epsilon > \frac{d-1}{d}, \end{cases} \tag{3.88}$$

and

$$\mathcal{P}(\geq\epsilon) = \begin{cases} 2d, & \text{if } \epsilon \leq \frac{d-1}{d}; \\ \frac{2(d-1)^\epsilon}{\epsilon^\epsilon(1-\epsilon)^{1-\epsilon}}, & \text{if } \epsilon \geq \frac{d-1}{d}. \end{cases} \tag{3.89}$$

By theorem 3.17, it follows that

$$\mathcal{P}(\epsilon) = \frac{2(d-1)^\epsilon}{\epsilon^\epsilon(1-\epsilon)^{1-\epsilon}} \tag{3.90}$$

in this model. The essential domain is $E = [0, 1]$.

The free energy can be determined from equation (3.20). Simplification gives $\mathcal{F}(z) = \log 2 + \log(1+(d-1)z)$. The energy and specific heat may be determined as well: $\mathcal{E}(z) = \frac{(d-1)z}{1+(d-1)z}$, and $\mathcal{C}(z) = \frac{(d-1)z}{[1+(d-1)z]^2}$.

3.5.3 *Lobb numbers*

Lobb numbers were introduced in reference [361] and are given by

$$L_{n,m} = \frac{2m+1}{n+m+1}\binom{2n}{n+m}. \tag{3.91}$$

The generating function is $G(z,t) = \sum_{n=0}^{\infty}\sum_{m=0}^{n} L_{n,m}\, z^m t^n$ and this simplifies to

$$G(z,t) = \frac{2\left(1-3t+(1-t)\sqrt{1-4t}\right)}{\left(1-\sqrt{1-4t}\right)\left(1-(z+3)t+(1-(z+1)t)\sqrt{1-4t}\right)}. \tag{3.92}$$

The radius of convergence of $G(z,t)$ in the t-plane is determined by branch points at $t = \frac{1}{4}$ as well as by a simple pole when the denominator vanishes. This gives $t_c = \min\left\{\frac{1}{4}, \frac{z}{(z+1)^2}\right\}$. The free energy is

$$\mathcal{F}(z) = \begin{cases} \log 4, & \text{if } z \leq 1; \\ 2\log(z+1) - \log z, & \text{if } z > 1. \end{cases} \tag{3.93}$$

There is a critical point $z_c = 1$. The density function can be computed from $\mathcal{F}(z)$ and is given by

$$\mathcal{P}(\epsilon) = \frac{4}{(1+\epsilon)^{1+\epsilon}(1-\epsilon)^{1-\epsilon}} \tag{3.94}$$

with essential domain $E = [0, 1]$.

The right-derivative $\mathcal{P}(\epsilon)$ at $\epsilon = 0$ is $\log z_c = -D^+\mathcal{P}(0) = 0$ and this gives a critical value $z_c = 1$ (see equation (3.62)). This shows that $\mathcal{F}(z) = \log 4$ for $z \in (0, 1]$.

3.5.4 *Lah numbers*

Lah numbers [372] are given by

$$L(n,m) = \frac{n!}{m!}\binom{n-1}{m-1}, \tag{3.95}$$

and were introduced in reference [388]. Summing $\sum_{m=0}^{n} L(n,m)\, z^m$ gives the partition function

$$Z_n(z) = n!e^{-z/2}\mathrm{M}(-n, \tfrac{1}{2}, z), \tag{3.96}$$

where $\mathrm{M}(\kappa, \mu, z)$ is the Whittaker M-function of order κ [514]. This gives the generating function as an infinite series. The free energy can be determined from an asymptotic expression for $\mathrm{M}(-n, \frac{1}{2}, z)$ [388]: for $n > 0$, and $z > 0$,

$$\mathrm{M}(-n, \tfrac{1}{2}, z) = \frac{z^{1/4}}{2\sqrt{\pi n^{3/2}}}\, e^{2\sqrt{nz}}(1 + o(1)). \tag{3.97}$$

Substitution into equation (3.96) shows that $\log Z_n(z) = \Theta(n \log n)$ and, if the model is normalised with $n!$, then $\log(Z_n(z)/n!)) = \Theta(\sqrt{n})$.

This model is not linear in energy, and the free energy is computed in the first case by dividing by $n \log n$ and then taking $n \to \infty$. This gives $\mathcal{F}(z) = 1$. In the second case divide by $\sqrt{n}$ and then take $n \to \infty$ to obtain the alternative model $\mathcal{F}(z) = 2\sqrt{z}$, which is convex in $\log z$. In this case $\log \mathcal{P}(\epsilon) = 2\epsilon(1 - \log \epsilon)$, and the energy density is $\mathcal{E}(z) = \sqrt{z}$, while $\mathcal{C}(z) = \frac{1}{2}\sqrt{z}$. The function $\log \mathcal{P}(\epsilon)$ is concave in ϵ, and its essential domain is $E = [0, \infty)$.

3.5.5 *Dyck paths and Narayana numbers*

Define the half-lattice

$$\mathbb{L}^2_+ = \{\langle \vec{v} \sim \vec{w}\rangle \mid \vec{v}, \vec{w} \in \mathbb{L}^2,\ \vec{v}(2) \geq 0, \text{ and } \vec{w}(2) \geq 0\} \tag{3.98}$$

in $\mathbb{L}^2$, where $\vec{v}(2)$ is the second Cartesian coordinate of the vertex $\vec{v}$. The boundary of $\mathbb{L}^2_+$ is denoted by $\partial\mathbb{L}^2_+$ and is a *hard wall*. A Dyck path is a fully directed path in $\mathbb{L}^2_+$ from $\vec{0}$ and is constrained to end in $\partial\mathbb{L}^2_+$. The path gives north-east and south-east steps of length $\sqrt{2}$ (that is, it steps along vectors $\{(1,1), (1,-1)\}$; this is also the *step-set* of the path (see figure 3.8)).

The number of Dyck paths is given by Catalan numbers; the number of Dyck paths of length $2n$ is denoted by d_{2n}, and $d_{2n} = C_n = \frac{1}{n+1}\binom{2n}{n}$.

The generating function of Dyck paths is the Catalan generating function

$$G(t) = \frac{1 - \sqrt{1 - 4t^2}}{2t^2} = \frac{2}{1 + \sqrt{1 - 4t^2}}. \tag{3.99}$$

The function $G(t)$ is singular at $t = \frac{1}{2}$; this shows that the number of Dyck paths of length n grows exponentially at the rate $2^{n+o(n)}$.

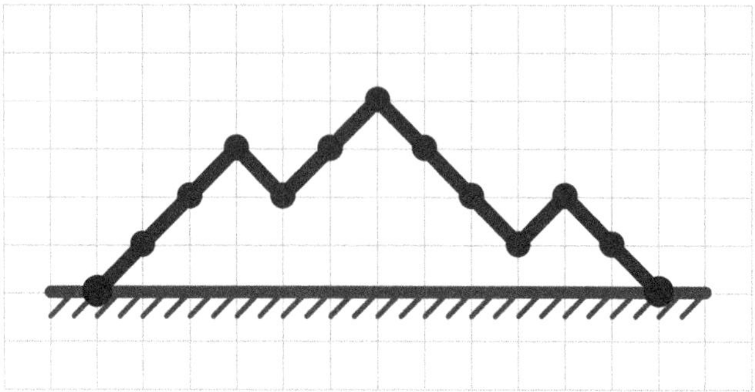

FIG. 3.8. A Dyck path. The path steps from the origin in $\mathbb{L}^2_+$ along north-east or south-east steps and it may return any number of times to $\partial\mathbb{L}^2_+$. It is constrained to end in a vertex in $\partial\mathbb{L}^2_+$. The number of Dyck paths of length $2n$ steps is given by Catalan numbers. A Dyck path has subpaths which are *rises*. Along a rise the path steps only north-east. A *peak* is a vertex in the path where a rise turns over and starts to descend. In this example the path has three peaks. The number of Dyck paths of length $2n$ and with k peaks is given by the Narayana number.

Dyck paths may be counted with respect to *peaks*. A vertex in a Dyck path is a peak if it is preceded by a north-east step and succeeded by south-east step. The number of Dyck paths of length $2n$ and with k peaks is given by the Narayana number [431]

$$N_{n,k} = \frac{1}{n}\binom{n}{k}\binom{n}{k-1}. \tag{3.100}$$

The partition function of Dyck paths with peaks weighted by z is given by

$$Z_{2n}(z) = \sum_{k=0}^{n} N_{n,k} z^k = N_n(z), \tag{3.101}$$

where $N_n(z)$ is the Narayana polynomial of degree n [44]. The generating function is a series over hypergeometric functions:

$$G(z,t) = \sum_{n=0}^{\infty} {}_2F_1\left([n, n+1], [2]; t^2\right) (t^2 z)^n, \tag{3.102}$$

where t is conjugate to the number of steps in the paths.

The free energy may be obtained by identifying the dominant terms in the summation of equation (3.101). In this respect, corollary D.2 in appendix D is useful. Note that $N_{n_1,k_1} N_{n_1,k_2} \leq N_{n_1+n_2,k_1+k_2}$ by concatenating two paths. This model has an additive energy and satisfies the assumptions of section 3.3.1. By theorem 3.9,

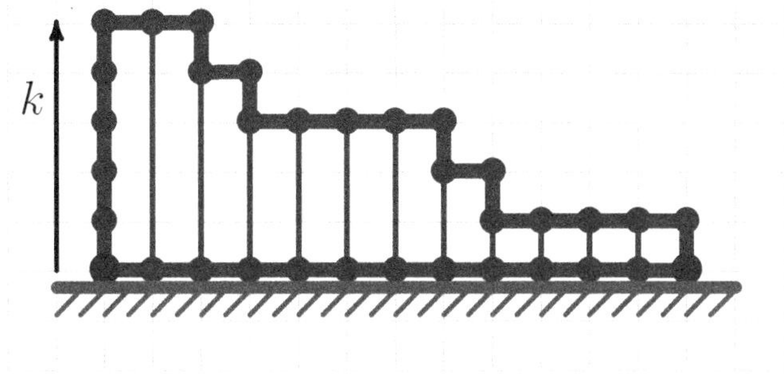

FIG. 3.9. A partition polygon represented as a Ferrer's diagram. Each column is a part in the partition, and the largest part is left-most.

$$\mathcal{P}(\epsilon) = \lim_{n\to\infty} \left(N_{n,\lfloor \epsilon n \rfloor}\right)^{1/2n} = \frac{1}{\epsilon^\epsilon (1-\epsilon)^{1-\epsilon}}. \tag{3.103}$$

The essential domain is $E = [0,1]$.

Observe that $\mathcal{P}(\epsilon)$ is defined for $\epsilon \in [0,1]$ in this model but that the power $\frac{1}{2n}$ is taken in the above. This shows that ϵ is conjugate to $\sqrt{z}$ in equation (3.101) with the result that

$$\mathcal{F}(z) = \sup_{0<\epsilon<1} \left\{\log \mathcal{P}(\epsilon) + \epsilon \log \sqrt{z}\right\} = \log(1+\sqrt{z}). \tag{3.104}$$

The free energy $\mathcal{F}(z)$ is analytic, and the energy and specific heat of the model is

$$\mathcal{E}(z) = \frac{\sqrt{z}}{2(1+\sqrt{z})}, \quad \text{and } \mathcal{C}(z) = \frac{\sqrt{z}}{4(1+\sqrt{z})^2}. \tag{3.105}$$

Notice that $\mathcal{E}(1) = \frac{1}{4}$, implying that $\frac{1}{4}$ of vertices are peaks in the limit that the length of the path is infinite and $z = 1$.

3.5.6 *Partition polygons*

A partition of $n \in \mathbb{N}$ with its largest part of size k is illustrated in figure 3.9 as a *Ferrers diagram*, which will be called a *partition polygon*. The perimeter of the partition polygon (including the edges along its base) is a lattice polygon enclosing an area of size n unit squares in $\mathbb{L}^2_+$ with the property that it intersects horizontal and vertical lines almost always in zero or two points. Such lattice polygons are *two-way convex* or *fully directed*. The *height* of the partition polygon is the height k of its largest part.

Let $p_n(k)$ be the number of partition polygons of height k and area n. (Then $p_n(k)$ is also the number of partitions of n with largest part k.) It is known that [531]

$$\sum_{n=0}^{\infty} p_n(k) q^n = \frac{q^k}{(q)_k}, \tag{3.106}$$

where the q-Pochhammer symbol is defined in equation (D.42) in appendix D. The generating function can be determined from the above. By using a partition identity (see for example [8]),

$$G(y,q) = \sum_{k=0}^{\infty} \frac{y^k q^k}{(q)_k} = \frac{1}{(yq,q)_\infty}, \tag{3.107}$$

where y is conjugate to height, and q is conjugate to area (see equation (D.43) in appendix D).

The generating function $G(y,q)$ has simple poles along curves $yq^n = 1$ in the yq-plane for $n \in \mathbb{N}$. These curves accumulate on the line $q = 1$, which is a line of essential singularities in $G(y,q)$. The radius of convergence is given by $q_c = 1$ if $y \leq 1$, and $q_c = \frac{1}{y}$ if $y > 1$. Hence, the free energy of this model is

$$\mathcal{F}(y) = -\log q_c(y) = \begin{cases} 0, & \text{if } y \leq 1; \\ \log y, & \text{if } y > 1. \end{cases} \tag{3.108}$$

The density function can be determined from $\mathcal{F}(y)$ by using equation (3.19). This gives $\mathcal{P}(\epsilon) = 1$ for $\epsilon \in E = [0,1]$. The energy is $\mathcal{E}(z) = 0$ if $z < 1$, and $\mathcal{E}(z) = 1$ if $z > 1$. The specific heat is singular when $z = 1$, but 0 otherwise.

3.5.7 *Queens on a chessboard*

This is a celebrated problem concerned with the disposition (the placing of pieces so that they do not attack one another) of k queens on an $n \times n$ chessboard. If the disposition is ignored, then there are $\binom{n^2}{k}$ ways of placing k queens on the board. The partition function is $(1+z)^{n^2}$, and the generating function is given by

$$G(z,t) = \sum_{n=0}^{\infty} ((1+z)t)^{n^2} = \tfrac{1}{2}\left(1 + \phi((1+z)t)\right) \tag{3.109}$$

by equations (D.66) and (D.69). In particular, equation (D.69) shows that

$$G(z,t) = \tfrac{1}{2}\left(1 + ((1+z)^2t^2;(1+z)^2t^2)_\infty(-(1+z)t;(1+z)^2t^2)^2_\infty\right). \tag{3.110}$$

There is an essential singularity in the generating function when $(1+z)t = 1$, and the free energy of the model is $\mathcal{F}(z) = \log(1+z)$. This model has the same free energy as computed in equation (3.84) for the chromatic polynomials of unlabelled complete graphs.

3.5.8 *Walks unzipping from an adsorbing surface*

Let $q_n(k)$ be the number of self-avoiding walks from $\vec{0}$ in $\mathbb{L}^d_+$ with the *adsorbing boundary* $\partial\mathbb{L}^d_+$ (see equation (3.98)). The first k steps are in $\partial\mathbb{L}^d_+$, and the last $n-k$ steps are in $\mathbb{L}^d_+$ (and are disjoint with $\partial\mathbb{L}^d_+$). This is a model of an adsorbed grafted polymer undergoing desorption by unzipping from one endpoint. The walk in this model with $d = 2$ is illustrated in figure 3.10.

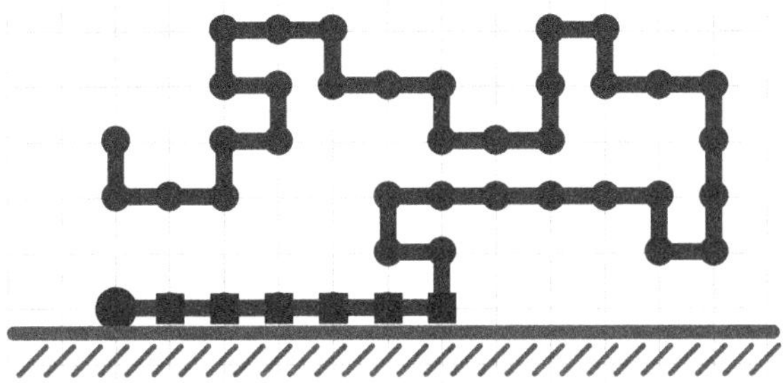

FIG. 3.10. An unzipping self-avoiding walk. The walk gives its first k steps in the hyperplane $Z = 0$ and, upon stepping out of the plane into the positive half-lattice $\mathbb{L}_+^d$, is constrained not to visit the hyperplane $Z = 0$ again.

A *positive walk* $\omega = \langle \omega_0, \omega_1, \ldots, \omega_n \rangle$ of length n is a walk from the origin in $\mathbb{L}_+^d$ (where $\omega_j(k)$ is the k-th vertex of ω_j). Then $\omega_0(d) \leq \omega_i(d)$ for $0 \leq i \leq n$. Let $c_n^{(d-1)}$ be the number of self-avoiding walks from $\vec{0}$ in $\partial\mathbb{L}_+^d$ and let c_n^h be the number of half-space walks of length n from $\vec{0}$ in $\mathbb{L}_+^d$ (these walks step away from $\partial\mathbb{L}_+^d$ and never return to it).

By deleting the edge where the walk steps out of $\partial\mathbb{L}_+^d$, it follows that

$$q_n(k) = c_k^{(d-1)} c_{n-k-1}^h. \tag{3.111}$$

In order to analyse equation (3.111), consider first the function c_n^h. Clearly, $c_n^h \leq c_n$ (where c_n is the number of self-avoiding walks from the origin; see section 1.1). By theorem 1.1, $\limsup_{n\to\infty} \frac{1}{n} \log c_n^h \leq \log \mu_d$.

On the other hand, every self-avoiding walk from the origin can be translated in the $\vec{e}_d$ direction until it has at least one vertex in $\partial\mathbb{L}_+^d$, but is disjoint with $\mathbb{Z}_-^{d-1} = \{\vec{v} \in \mathbb{Z}^d \mid \vec{v}(d) < 0\}$. This walk may be uniquely rooted at its first vertex in $\partial\mathbb{L}_+^d$, which cuts it into two self-avoiding walks from the root into the half-lattice $\mathbb{L}_+^d$. This shows that

$$c_n \leq (n+1) \sum_{\ell=0}^{n} c_{\ell+1}^h c_{n-\ell+1}^h. \tag{3.112}$$

There is a value of ℓ which maximises the summand on the right-hand side, say ℓ^*. This gives $c_n \leq (n+1)^2 c_{\ell^*+1}^h c_{n-\ell^*+1}^h$. Take logarithms and divide by n. This shows that

$$\tfrac{1}{n} \log c_n \leq \tfrac{2}{n} \log(n+1) + \tfrac{\ell^*+1}{n} \left(\tfrac{1}{\ell^*+1} \log c_{\ell^*+1}^h \right) + \tfrac{n-\ell^*+1}{n} \left(\tfrac{1}{n-\ell^*+1} \log c_{n-\ell^*+1}^h \right).$$

Take the limit as $n \to \infty$ on the left-hand side. Then either ℓ^*, or $n - \ell^*$, or both increase to infinity. This shows that $\log \mu_d \leq \limsup_{n\to\infty} \frac{1}{n} \log c_n^h$ and thus that

$$\limsup_{n\to\infty} \tfrac{1}{n} \log c_n^h = \log \mu_d. \tag{3.113}$$

The existance of $\lim_{n\to\infty} \frac{1}{n} \log c_n^h = \log \mu_d$ will be shown in section 7.1.1 (corollary 7.3).

By equation (3.111), the density function of this model is

$$\log \mathcal{P}(\epsilon) = \lim_{n\to\infty} \tfrac{1}{n} \log q_n(\lfloor \epsilon n \rfloor) = \epsilon \log \mu_{d-1} + (1-\epsilon) \log \mu_d. \tag{3.114}$$

Hence, $\log \mathcal{P}(\epsilon)$ is linear in ϵ. By equation (3.59), this model has a first order unzipping transition at $z_c = \frac{\mu_d}{\mu_{d-1}}$. The limiting free energy can be obtained from the Legendre transform:

$$\mathcal{F}(z) = \begin{cases} \log \mu_d, & \text{if } z < z_c; \\ \log \mu_{d-1} + \log z, & \text{if } z \geq z_c. \end{cases} \tag{3.115}$$

The energy of the model is $\mathcal{E}(z) = 0$ if $z < z_c$, and $\mathcal{E}(z) = 1$ if $z > z_c$; this is a discontinuity characteristic of first order phase transitions.

4

SCALING, CRITICALITY AND TRICRITICALITY

Thermodynamic systems are dependent on temperature and exhibit phase transitions at critical values of the temperature. Scaling laws describe the behaviour of thermodynamic quantities (such as magnetisation or specific heat) close to critical values of the temperature. The thermodynamic quantities may have singular dependence on the temperature T at the critical point T_c of general form $|T - T_c|^x$, where x is a *scaling exponent*. The numerical value of x characterises the scaling, and exhibits *universality*; that is, many systems, although different in microscopic details, have the same macroscopic scaling behaviour close to T_c.

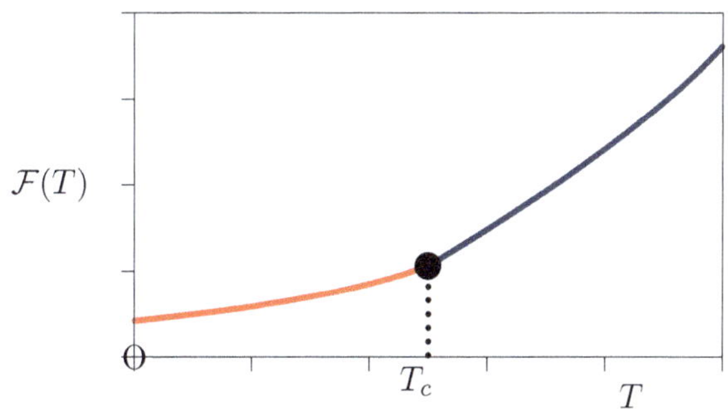

FIG. 4.1. The limiting free energy of an interacting model. A non-analyticity at $T = T_c$ signals a phase transition in the model at the *critical point* T_c.

Phase transitions occur at non-analyticities in the limiting free energy $\mathcal{F}(T)$ of an interacting model of clusters (see equation (3.2)). In figure 4.1 a schematic drawing of $\mathcal{F}(T)$ is displayed, with a non-analyticity at T_c.

The limiting free energy $\mathcal{F}(T)$ is a convex function which decomposes into an analytic background $\mathcal{F}_b(T)$ and a singular part $\mathcal{F}_{sing}(T)$ such that $\mathcal{F}(T) = \mathcal{F}_{sing}(T) + \mathcal{F}_b(T)$. The singular part $\mathcal{F}_{sing}(T)$ has the critical behaviour

$$\mathcal{F}_{sing}(T) \sim |T - T_c|^{2-\alpha}, \tag{4.1}$$

where α is the *specific heat exponent*. Ignoring the analytic term, the above is often simply stated as $\mathcal{F}(T) \sim |T - T_c|^{2-\alpha}$. The difference $\tau = T - T_c$ is a *scaling field* and it vanishes at the critical point $T = T_c$. The analytic background term may be ignored, so

The Statistical Mechanics of Interacting Walks, Polygons, Animals and Vesicles, 2nd edition, E.J. Janse van Rensburg. © E.J. Janse van Rensburg. Published in 2015 by Oxford University Press.

$$\mathcal{F}(T) \sim |\tau|^{2-\alpha} \tag{4.2}$$

means that the singular part of $\mathcal{F}(T)$ is proportional to $|\tau|^{2-\alpha}$ close to the critical point $\tau = 0$.

The scaling of the energy and specific heat of the model can be determined from equations (4.2), (3.57) and (3.58):

$$\mathcal{E}(T) \sim |\tau|^{1-\alpha}, \text{ and } \mathcal{C}(T) \sim |\tau|^{-\alpha}. \tag{4.3}$$

If $\alpha = 1$, then the energy $\mathcal{E}(T)$ approaches a finite constant at $\tau = 0$, while $\mathcal{C}(T)$ is divergent. This is a *first order transition* [7, 289]. *Continuous transitions* generally have $\alpha < 1$, while $\alpha > 1$ is ruled out by the convexity of the free energy.

4.1 Tricritical scaling

The theory of tricritical points provides a set of principles whereby the scaling of the generating function and other thermodynamic quantities can be analysed. This was reviewed extensively in reference [371] and further developed for lattice models of walks and polygons in references [71, 72].

Consider a model of lattice clusters with generating function

$$G(z,t) = \sum_{n,m} q_n(m)\, z^m t^n, \tag{4.4}$$

where $q_n(m)$ is the microcanonical partition function (the number of lattice clusters of size n and energy m). The partition function of the model is $Z_n(z) = \sum_{m=0}^{\infty} q_n(m)\, z^m$ (see equation (3.13)).

The function $G(z,t)$ is a power series in t with non-negative coefficients for $z > 0$, so its dominant singularity (radius of convergence) is an isolated point located at the point $t_c(z)$ on the positive real axis in the t-plane. The limiting free energy of the model is $\mathcal{F}(z) = -\log t_c(z)$ (see equation (3.14)). A plot of $t_c(z)$ against z is similar to figure 3.3.

The function $t_c(z)$ is the *critical curve* in the phase diagram. Derivatives of $\log G(z,t)$ are the thermodynamic functions of the model. These are singular along the critical curve. Non-analyticities in the free energy $\mathcal{F}(z)$ (and in $t_c(z)$) are *critical points* in the model.

Critical points divide $t_c(z)$ into parts, each corresponding to a different type of singularity in $G(z,t)$. In some models, $t_c(z)$ is a curve of simpler singularities (branch points or poles) which transitions to a curve of more complicated singularities (for example, essential singularities) at a critical point z_c. In this case z_c is a *multicritical point.*

The critical point z_c on the critical curve $t_c(z)$ in the phase diagram is analysed by setting up scaling axes or fields through the critical point. These *scaling fields* fix a coordinate system for describing the scaling behaviour of thermodynamic quantities close to the critical point and curve. The scaling along different

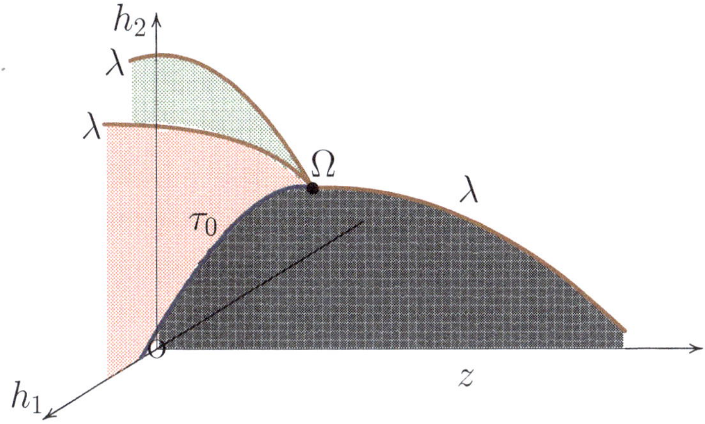

FIG. 4.2. The geometry of a tricritical phase diagram. The horizontal axis labelled by z is a temperature axis, while the two axes normal to it are other fields coupling to clusters. The τ_0-curve of triple points ends in the tricritical point Ω, where it branches into three λ-isotherms of continuous transitions. Each λ-isotherm and the τ_0-curve bound a critical surface of first order transitions.

---

fields is compared by using *crossover exponents*, and this ties the entire description into a self-consistent theory of multicritical or tricritical points. Tricritical points are found in the phase diagrams of the spin - 1 Ising model [574] and the anti-ferromagnetic Ising model [504].

The general nature of a tricritical point is exposed when more scaling fields (say h_1 and h_2) are introduced in the model (h_1 and h_2 may, for example, be chemical potentials or other scaling fields which couple to the clusters). In this case a higher dimensional phase diagram appears (such as in figure 4.2). The (z, h_2)-plane in this figure is similar to the curve in figure 3.3. The critical curve $t_c(z)$ is exposed as part of a broader structure, and it is composed of two parts: a critical curve τ_0, which is a curve of triple points where three phases meet in the phase diagram, and an isotherm λ, which is the boundary of a sheet of first order phase transitions; at the tricritical point Ω, the triple points branch into three isotherms bounding sheets of critical points.

If $h_1 = 0$ in figure 4.2, then the phase diagram becomes similar to the illustration in figure 4.3 (this may be compared to figure 3.3). In this case the critical curve consists of the τ_0-curve and the λ-isotherm (τ_0 is the locus of triple points which ends in the tricritical point at Ω). Along τ_0 the generating function has more complex singularities (for example essential singularities) but along the λ-isotherm it has simpler non-analyticities, such as branch points or poles.

The region below the critical curve is marked *finite clusters*, and that above the curve is marked *infinite clusters*, because the generating function is domi-

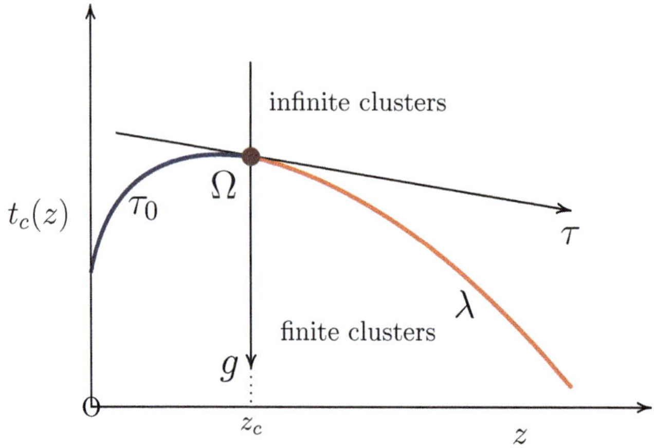

FIG. 4.3. The critical curve $t_c(z)$ consists of a τ_0-curve meeting a λ-isotherm at the tricritical point Ω. The *scaling fields* (τ, g) are set up with an origin at the tricritical point as shown: the τ-axis is tangent to the λ-isotherm at Ω, and the g-axis is transverse to the λ-isotherm at Ω.

---

nated in each of these regions by finite and infinite clusters, respectively. The phase of infinite clusters is not unphysical (in the way that the infinite cluster in percolation is not unphysical) but requires different analytical techniques than those presented here.

4.1.1 *Shift-exponents for supermultiplicative models*

The generating function $G(z,t)$ of a model of lattice clusters may exhibit critical behaviour that is similar to that observed for the polygon generating function (see equation (1.59)). The phase diagram may be similar to the illustration in figure 4.3 and may include a tricritical point whose influence is felt in a *tricritical scaling region.*

Since the limiting free energy $\mathcal{F}(z) = -\log t_c(z)$ is a convex function of $\log z$, the critical curve $t_c(z)$ in figure 4.3 is differentiable almost everywhere. If it is differentiable in a neighbourhood of the tricritical point Ω, then the tangent line to $t_c(z)$ exists in an open interval containing Ω. The τ-scaling axis is set up along the tangent of the τ_0-curve at Ω, as illustrated in figure 4.3. In the case that $t_c(z)$ is not differentiable at Ω, then the τ-axis has a gradient equal to the left-derivative of $t_c(z)$ at Ω (on the τ_0-side of Ω in the critical curve).

Denote the τ_0-curve by $t_\tau(z) = t_c(z)$ if $z < z_c$, and the λ-isotherm by $t_\lambda(z) = t_c(z)$ for $z > z_c$. That is, $t_\tau(z)$ is the critical curve along τ_0, and $t_\lambda(z)$ is the critical curve along λ. Assuming that the critical curve is differentiable at Ω, $t_\tau(z)$ and $t_\lambda(z)$ may be expanded in z about $z = z_c$ to obtain

$$\begin{aligned} t_\tau(z) &= t_\lambda(z_c) - a_\lambda(z - z_c) - b_\tau\,|z_c - z|^{\psi_\tau}, \quad \text{and} \\ t_\lambda(z) &= t_\lambda(z_c) - a_\lambda(z - z_c) - b_\lambda\,|z - z_c|^{\psi_\lambda}. \end{aligned} \tag{4.5}$$

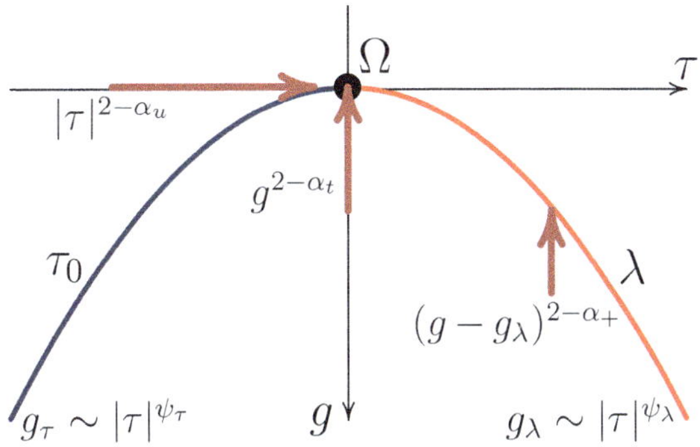

FIG. 4.4. Tricritical scaling of the generating function.

The exponents ψ_λ and ψ_τ are *shift-exponents* describing the shape of the critical curve on either sides of Ω. The shift-exponents are true tricritical exponents, because their values are dependent on the nature of the singularity at Ω. If the τ_0-line is a straight line, then the model is called *asymmetric*, and $\psi_\tau = 1$.

The expansions in equation (4.5) provide a natural coordinate system for the critical curve. Denote these coordinates by (τ, g). By equation (4.5), choose coordinates $\tau = z - z_c$, and choose

$$g = t_\lambda(z_c) - t_\tau(z) - a_\lambda \tau \quad \text{along the } \tau_0\text{-curve, and} \tag{4.6}$$

$$g = t_\lambda(z_c) - t_\lambda(z) - a_\lambda \tau \quad \text{along the } \lambda\text{-isotherm,} \tag{4.7}$$

as illustrated in figure 4.3. The τ-coordinate runs along the tangent to the critical curve at Ω, and the g-coordinate runs transverse to the tangent line through Ω. The τ_0-curve is found for $\tau < 0$, and the λ-isotherm for $\tau > 0$. The tricritical point is located at the origin in the $g\tau$-plane.

Close to Ω the critical curve $t_c(z)$ is given approximately by

$$g_\tau(t) \simeq b_\tau |\tau|^{\psi_\tau}, \quad \text{if } \tau < 0, \text{ and } g_\lambda(t) \simeq b_\lambda \tau^{\psi_\lambda}, \quad \text{if } \tau > 0 \tag{4.8}$$

in the gt-plane (see figure 4.4).

4.1.2 *Tricritical scaling of the generating function*

The singularity along the λ-isotherm in the generating function $G(z,t)$ of an interacting lattice cluster is of a simpler nature. In analogy with equation (1.59), it may be assumed to be

$$G_s(\tau, g) \simeq A_+(\tau)(g - g_\lambda(\tau))^{2-\alpha_+}, \tag{4.9}$$

where G_s denotes the *singular part of the generating function*. The exponent α_+ describes the nature of the singularity along the λ-isotherm.

In models of lattice clusters it is expected that $\alpha_+ \leq 2$, as seen for polygons. The relation of α_+ to a metric exponent ν_+ is given by Josephson's hyperscaling relation (see equation (1.55)),

$$2 - \alpha_+ = d\nu_+, \tag{4.10}$$

in low dimensions (see the arguments made in section 1.4.6). The exponent $\frac{1}{\nu_+}$ is a fractal dimension of the cluster near the λ-isotherm.

Along the τ_0-curve, the transition to the infinite phase may be a first order transition; this may show itself as a jump discontinuity in the first derivative of the thermodynamic potential $\log G(\tau, g)$ as this behaviour is consistent with an essential singularity in $G(\tau, g)$ [7, 289].

In models of lattice clusters where the singular parts of the thermodynamic potential are finite along the τ_0-curve, the restriction to the τ_0-curve exhibits singular behaviour when the tricritical point is approached. This is described by

$$G_s(\tau, g) \simeq S_- |\tau|^{2-\alpha_u} \text{ along the } \tau_0\text{-line as } \tau \to 0^-, \tag{4.11}$$

where S_- contains the g-dependence of $G_s(\tau, g)$. The exponent α_u is the third tricritical exponent (the shift-exponents being the first two).

The scaling of the generating function when the tricritical point is approached transversely along the g-axis is given by

$$G_s(\tau, g) \simeq A_\tau g^{2-\alpha_t} \text{ along the } g\text{-axis as } g \to 0^+, \tag{4.12}$$

where A_τ contains the τ-dependence of $G_s(\tau, g)$, and α_t is a tricritical exponent.

4.1.3 *Crossover scaling*

The scaling of the singular part of the generating function $G_s(\tau, g)$ in equations (4.11) and (4.12) must be reconciled. This is done by assuming *crossover scaling* behaviour between the t-axis and the g-axis, and controlled by the *crossover exponent* ϕ. This exponent is introduced to find a self-consistent description of scaling near the tricritical point.

The scaling of $G_s(\tau, g)$ along curves $g^{-\phi}|\tau| = C$ (which are transverse to the critical curve for constants $C > 0$) should be the power law $G_s \sim g^{2-\alpha_t}$. Taking $g \to 0^+$ requires that $|\tau| \to 0$ along these curves, which suggests the introduction of a scaling function $f(x)$ such that

$$G(\tau, g) \simeq A_\tau g^{2-\alpha_t} f(g^{-\phi}|\tau|), \tag{4.13}$$

where f is dependent on the sign of τ and where $f(0) = 1$ in order to recover equation (4.12) when $\tau = 0$. This assumption implies that the shape of the critical curve close to Ω is invariant under a rescaling of coordinates by $g \to xg$ and $\tau \to x^\phi \tau$, for some $x > 0$.

Notice that, as $g \to 0^+$, $g^{-\phi}|\tau| \to \infty$. Thus, assuming that $f(x) \simeq x^u$ as $x \to \infty$ and comparing equation (4.13) with equation (4.11) gives $u = 2 - \alpha_u$.

Comparison with equation (4.12) gives $(2-\alpha_t)-\phi(2-\alpha_u)=0$, which gives the tricritical scaling relation relating ϕ with α_t and α_u:

$$\phi = \frac{2-\alpha_t}{2-\alpha_u}. \tag{4.14}$$

The scaling of the generating function in equation (4.13) is good close to Ω but deteriorates with distance from Ω as the influence of the tricritical point lessens. The qualitatively defined region where $G(\tau,g)$ has scaling given approximately by equation (4.13) is the *tricritical scaling region.*

Equation (4.13) may be cast as

$$\begin{aligned} G_s(\tau,g) &\simeq A_\tau |\tau|^{(2-\alpha_t)/\phi} (g|\tau|^{-1/\phi})^{2-\alpha_t} f(g^{-\phi}|\tau|) \\ &= A_\tau |\tau|^{2-\alpha_u} F(g^{-\phi}|\tau|), \end{aligned} \tag{4.15}$$

where $F(x)=x^{-(2-\alpha_u)}f(x)$, by using equation (4.14). Define $\overline{F}(x)=F(x^{-1/\phi})$. Then this becomes $G_s(\tau,g)\simeq A_\tau|\tau|^{2-\alpha_u}\overline{F}(g|\tau|^{-1/\phi})$. Taking derivates gives

$$\tfrac{\partial^n}{\partial g^n} G_s(\tau,g) = A_\tau |\tau|^{(2-\alpha_u-n\Delta)} \overline{F}^{(n)}(g|\tau|^{-1/\phi}), \tag{4.16}$$

where $\Delta=\frac{1}{\phi}$ is the *gap exponent.*

The singularity in $G(\tau,g)$ along the λ-isotherm in figure 4.4 is given by equation (4.9). The scaling function f in equation (4.13) cannot grow unbounded on approach to the λ-isotherm because that would invalidate the scaling in equation (4.9). Thus, $g^{-\phi}|\tau|$ is approximately equal to a constant on the approach to the λ-isotherm. This shows that the λ-isotherm has the shape $g_\lambda(\tau)\simeq b_\lambda|\tau|^{1/\phi}$ in the τg-plane. Comparing this to equation (4.8) shows that

$$\psi_\lambda = \tfrac{1}{\phi} = \Delta. \tag{4.17}$$

A similar argument may be made along the τ_0-curve (replace λ by τ_0 in the above). In some cases the conclusion is that $\psi_\lambda=\psi_\tau=1/\phi$. These are the *symmetric* models.

Many models exhibit a different geometry of singular curves near the tricritical point (than that shown in figure 4.4). If the τ_0-curve and λ-isotherm meet at an angle, then the τ-axis is chosen as the tangent to the τ_0-curve at Ω, and the g-axis is set up transverse to this as before.

The τ_0-curve and λ-isotherm may also meet with different shift-exponents: $\psi_\lambda\neq\psi_\tau$; these models are *asymmetric*, and more complex multicritical scaling crossover behaviour is seen near the multicritical point.

In some simpler models the τ_0-curve is a curve of branch points and meets a λ-isotherm in the tricritical point. The analysis of these models proceeds as above, with the additional assumption that (similar to equation 4.9),

$$G_s(\tau,g) \simeq A_-(\tau)(g-g_\tau(\tau))^{2-\alpha_-} \text{ along the } \tau_0\text{-curve.} \tag{4.18}$$

Provided that $G(\tau,g)$ is finite along the τ_0-curve, the exponent α_- is expected to be less than or equal to 2 and is related to a fractal dimension $\frac{1}{\nu_-}$ analogous to equation (4.10), namely, $2-\alpha_-=d\nu_-$.

4.1.4 *Tricritical scaling for submultiplicative models*

The self-avoiding walk has the generating function $C(t)$ given by equation (1.57). The function $C(t)$ is divergent at its critical point $t_c = \frac{1}{\mu_d}$. This divergence is characterised by the entropic exponent γ of self-avoiding walks (such that $C(t) \sim (1-\mu_d t)^{-\gamma}$ (see equation (1.57))). Notice that $\gamma \geq 1$ (by theorem 1.1 and the submultiplicativity of c_n).

Similar to this, there are models where the singular part of the generating function $G(\tau, z)$ is divergent along the λ-isotherm in figure 4.4. An appropriate scaling assumption for the singular part of the generating function is

$$G_s(\tau, g) \simeq A_+(\tau)(g - g_\lambda(t))^{-\gamma_+} \text{ along the } \lambda\text{-isotherm}, \tag{4.19}$$

where $\gamma_+ > 0$. Since the singular part of $G(t, g)$ is divergent in these models, it is also dominating the analytic background term close to the critical curve. As in equation (4.10), a metric exponent for clusters along the λ-isotherm is given by

$$\gamma_+ = d\nu_+. \tag{4.20}$$

Some models may have a simpler singularity along the τ_0-curve, in which case

$$G_s(\tau, g) \simeq A_-(\tau)(g - g_\tau(\tau))^{-\gamma_-} \text{ along the } \tau_0\text{-curve}, \tag{4.21}$$

and the metric exponent ν_- may be defined by $\gamma_- = d\nu_-$.

Tricritical scaling for these models is obtained by modifying the assumptions in equations (4.11) and (4.12) to

$$G_s(\tau, g) \simeq S_-|\tau|^{-\gamma_u} \text{ along the } \tau_0\text{-line } (\tau \to 0^-), \text{ and} \tag{4.22}$$
$$G_s(\tau, g) \simeq A_\tau g^{-\gamma_t} \text{ along the } g\text{-axis } (g > 0). \tag{4.23}$$

The introduction of the crossover exponent ϕ gives the tricritical scaling form

$$G_s(\tau, g) \simeq A_\tau g^{-\gamma_t} f(g^{-\phi}|\tau|), \tag{4.24}$$

where $f(0) = 1$ in order to recover equation (4.23) when $\tau = 0$. These assumptions give the relation

$$\phi = \frac{\gamma_t}{\gamma_u}, \tag{4.25}$$

and equation (4.24) becomes

$$G_s(\tau, g) \simeq A_\tau |\tau|^{-\gamma_u} F(g^{-\phi}|\tau|). \tag{4.26}$$

The gap exponent is defined as before by $\psi_\lambda = \frac{1}{\phi} = \Delta$.

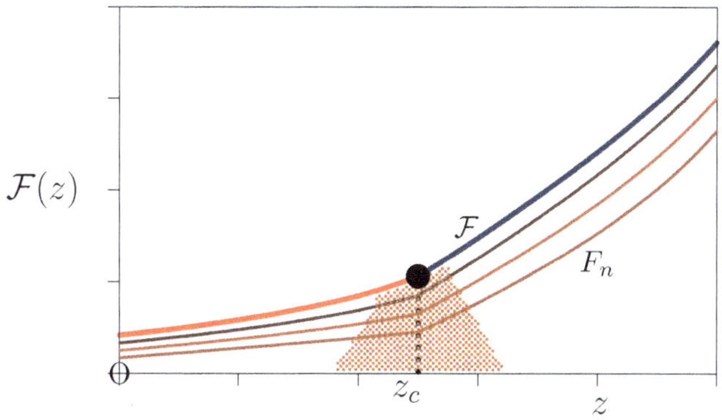

FIG. 4.5. Finite size scaling of the finite size intensive free energy F_n as it approaches the limiting free energy $\mathcal{F}$. The presence of the critical point (marked with a $\bullet$) affects the scaling of the F_n.

4.2 Finite size scaling

Let $Z_n(z)$ be the partition function of a model of interacting lattice clusters and define the finite size free energy $F_n(z) = \frac{1}{n}\log Z_n(z)$. Assume that $F_n \to \mathcal{F}$ as $n \to \infty$, where $\mathcal{F}$ is the limiting free energy of the model [21, 476].

The scaling of F_n as a function of n is the *finite size scaling* of the model. Finite size scaling is developed by choosing $g = \frac{1}{n}$ as a scaling field. Taking the limit $n \to \infty$ is the same as taking $g \to 0^+$. A second scaling field τ is a function of the activity z; normally, $\tau = z - z_c$, where z_c is the critical point in the model.

The combination $g^{-\phi}|\tau|$ in equation (4.26) becomes $n^{\phi_c}|\tau|$, where ϕ_c is the *finite size crossover exponent.*

4.2.1 *Finite size scaling of the free energy*

The basic assumption underlying finite size scaling is that increases ('scaling') in n is compensated by rescaling of the z-axis in figure 4.5 in such a way that the singular part of the free energy is invariant under this scaling (that is, the finite size free energies F_n are scaled images of the limiting free energy $\mathcal{F}$).

Assume that F_n is a function of the combined variable $n^{\phi_c}(z - z_c) = n^{\phi_c}\tau$. This gives the finite size scaling relation

$$F_n(z) \sim \tfrac{1}{n} f(n^{\phi_c}\tau) \tag{4.27}$$

for the finite size free energy, where $\tau = z - z_c$, and f is the *finite size scaling function.* Putting $\overline{f}(x) = x^{-1/\phi_c} f(x)$ gives the relation

$$F_n(z) \sim \tau^{1/\phi_c}\left(n^{-1}\tau^{-1/\phi_c} f(n^{\phi_c}\tau)\right) = \tau^{1/\phi_c}\overline{f}(n^{\phi_c}\tau). \tag{4.28}$$

Taking $n \to \infty$ gives the limiting free energy $\mathcal{F}(z) \sim \tau^{1/\phi_c}$, and this should be compared with equation (4.2) to obtain the hyperscaling relation

$$2 - \alpha = \frac{1}{\phi_c}, \tag{4.29}$$

where α is the specific heat exponent. By taking the derivatives of $F_n(z)$ in equation (4.27), the finite size scaling of the energy and specific heat is obtained:

$$E_n(z) \sim n^{\phi_c - 1} f'(n^{\phi_c}\tau), \text{ and } C_n(z) \sim n^{2\phi_c - 1} f''(n^{\phi_c}\tau). \tag{4.30}$$

The specific heat has a peak when $\tau = 0$ (at the critical point) of height

$$H_n = C_n(z_c) \sim n^{2\phi_c - 1} f''(0) \sim n^{\alpha\phi_c}. \tag{4.31}$$

Notice that $H_n \sim$ constant if $\phi_c = \frac{1}{2}$ and $\alpha = 0$.

4.2.2 *Finite size scaling of the partition function*

The convergence of F_n to $\mathcal{F}$ is controlled by a rescaling of the z-axis about z_c at a rate given by the finite size crossover exponent ϕ_c. Since $F_n(z) = \log Z_n(z)$ (where $Z_n(z)$ is the partition function), the most suitable assumption for the scaling of $Z_n(z)$ is

$$Z_n(z) \simeq h(n^{\phi_c}\tau) \tag{4.32}$$

for some scaling function h.

The generating function of $Z_n(z)$ is $G(z,t)$ (which has different singular behaviour along the λ-isotherm, at the tricritical point Ω and along the τ_0-curve). Thus, the asymptotic behaviour of $Z_n(z)$ as $n \to \infty$ should also be different for $z > z_c$, for $z = z_c$ and for $z < z_c$.

From equations (4.9) and (4.12), it can be inferred that the partition function should have the asymptotic behaviour given by

$$Z_n(z) \simeq B_+ n^{\alpha_+ - 3} [t_\lambda(z)]^{-n} \text{ along the } \lambda\text{-isotherm (where } z > z_c), \tag{4.33}$$

where $t_\lambda(z)$ is the critical value of t in $G(z,t)$ along the λ-isotherm. The choice of the exponent $(\alpha_+ - 3)$ makes this scaling assumption compatible with equation (4.9). In particular, along the λ-isotherm

$$G(\tau, g) = \sum_{n\geq 0} Z_n(z) t^n \sim \sum_{n\geq 0} n^{\alpha_+ - 3} \left(\tfrac{t}{t_\lambda(z)}\right)^n \sim (t_\lambda(z) - t)^{2-\alpha_+}. \tag{4.34}$$

By equation (4.8), $g = b_\lambda \tau^{\psi_\lambda}$, and $t_\lambda(z) = t_\lambda(z_c) - a_\lambda\tau - b_\lambda t^{\psi_\lambda}$ by equation (4.7). Hence, if $t \lesssim t_\lambda(z)$, then $t_\lambda(z) - t = t_\lambda(z_c) - a_\lambda\tau - t_\lambda(z) - b_\lambda t^{\psi_\lambda} \approx 0$ by equation (4.5). This may be written as $t_\lambda(z_c) - a_\lambda\tau - t_\lambda(z) \approx b_\lambda t^{\psi_\lambda}$. By equation (4.7), $g = t_\lambda(z_c) - a_\lambda\tau - t_\lambda(z)$, so this becomes $g - g_\lambda(t)$. Hence, the above becomes $G(\tau, g) \sim (g - g_\lambda(t))^{2-\alpha_+}$, consistent with equation (4.9).

At the tricritical point, it may similarly be expected that

$$Z_n(z_c) \simeq B_t n^{\alpha_t - 3} (t_c(z_c))^{-n}. \tag{4.35}$$

Notice the appearance of *bulk entropy* in the description given by $\log t_c(z_c)$ (this is analogous to the connective constant for self-avoiding walks or polygons, for example). The scaling of the partition function is given by the power law factor $n^{\alpha_t - 3}$.

The rescaling of the t-axis describes the crossover between these different asymptotic regimes along the λ-isotherm and at the tricritical point. Since $Z_n(z)$ is a function of only the scaled variable $n^{\phi_c}\tau$, a natural assumption for the partition function which takes the above into account will be

$$Z_n(z) \simeq B_\lambda h_\lambda(n^{\phi_c}\tau) n^{\alpha_t - 3} \mu_+^{-[n^{\phi_c}\tau]^{1/\phi_c}}, \tag{4.36}$$

where $h_\lambda(x)$ is a scaling function to be determined.

The asymptotic shape of $h_\lambda(x)$ can be guessed by examining the behaviour of Z_n along the λ-isotherm and comparing it with the behaviour at z_c. In particular, $h_\lambda(x)$ is approximately equal to a constant if x is small, and $h_\lambda(x) \approx x^{(\alpha_+ - \alpha_t)/\phi_c}$ as $x \to \infty$.

Along the τ_0-curve, arguments similar to the above can be made to find the scaling assumption

$$Z_n(z) \sim B_\tau h_\tau(n^{\phi_c}\tau) n^{\alpha_t - 3} \mu_-^{-[n^{\phi_c}\tau]^{1/\phi_c}}, \tag{4.37}$$

where $h_\tau(x) \sim |x|^{(\alpha_- - \alpha_t)/\phi_c}$ as $x \to -\infty$. In the case of asymmetric models, $\mu_- = 1$, but it may take other values in symmetric models.

The above scaling forms for Z_n are incomplete. In asymmetric models of clusters where there is a phase transition to a phase of compact clusters of low entropy, there is a significant contribution to the free energy from the entropy of the surface conformations of clusters. The surface area of the collapsed cluster is expected to grow proportional to $n^{(d-1)/d}$, and a factor of the form $\mu_s^{n^\sigma}$ with $\sigma = 1 - \frac{1}{d}$ must be included in the compact phase to account for this contribution [70]. The factor μ_s is an effective free energy per unit area, while the critical exponent σ is defined along the τ_0-curve. Generally, the exponent σ may not be equal to $1 - \frac{1}{d}$, if the collapsed cluster is not fully compact but has a 'surface' with a fractal dimension.

Take the above together, absorb the factors in μ_+ and μ_- in the scaling functions h_λ and h_τ and define the finite size scaling form for the partition function by (compare this to equation (1.18)):

$$Z_n(z) \simeq n^{\alpha_t - 3} \hat{h}(n^{\phi_c}\tau) \tag{4.38}$$

where

$$\hat{h}(x) \simeq \begin{cases} B_\lambda x^{(\alpha_+ - \alpha_t)/\phi_c} \mu_+^{x^{1/\phi_c}}, & \text{if } x \to \infty; \\ \text{a constant}, & \text{if } x = 0; \\ B_\tau |x|^{(\alpha_- - \alpha_t)/\phi_c} \mu_s^{|x|^{\sigma/\phi_c}} \mu_-^{|x|^{1/\phi_c}}, & \text{if } x \to -\infty. \end{cases} \tag{4.39}$$

The finite size scaling of the free energy may be obtained from this result by taking logarithms and dividing by n. For asymptotic values of n, this gives

$$F_n(z) \simeq \tfrac{1}{n} f_d(n^{\phi_c}\tau), \tag{4.40}$$

where f_d is the extensive free energy, and F_n is the intensive free energy. The limiting free energy is $\lim_{n\to\infty} F_n(z) = \mathcal{F}(z)$ and it is the *thermodynamic* limit of the model. The extensive free energy has the following scaling properties:

$$f_d(x) \approx \begin{cases} x^{1/\phi_c}, & \text{as } x \to \infty; \\ |x|^{\sigma/\phi_c}, & \text{if } x \to -\infty, \text{ and } \mu_- = 1; \\ |x|^{1/\phi_c}, & \text{if } x \to -\infty, \text{ and } \mu_- > 1. \end{cases} \tag{4.41}$$

Observe that asymmetric models with $\mu_- = 1$ are listed separately; in this case the surface contribution free energy in the low entropy phase is seen in the appearance of the exponent σ [459].

In symmetric models, the bulk contributions to the entropy will dominate the surface terms. This is, for example, seen in percolation. Define the partition function of percolation clusters of size n at density p that an edge or bond is open. Let $q = 1-p$. Then the probability $P_n(p)$ that the open cluster at the origin has size n is given by

$$P_n(p) = \sum_{c,k} A_n(c,k) p^n q^{s+k}, \tag{4.42}$$

where $A_n(c,k)$ is the number of lattice animals containing $\vec{0}$ of size n with c cycles and k nearest-neighbour contacts (and s contacts between occupied and unoccupied sites). The total perimeter of a cluster is $s+k$, and direct computation shows that $\langle s+k\rangle = \frac{q}{p}n + q\frac{d}{dq}\log P_n(p)$.

Let p_c be the critical percolation probability. The usual finite size scaling assumption is $P_n(p) = \hat{f}((p_c - p)n^{\sigma_P})$, where σ_P is the percolation crossover exponent. If the expected perimeter length is computed, then

$$\langle s+k\rangle \approx \tfrac{q}{p}n + qn^{\sigma_P}\hat{f}'(0) \text{ as } p \to p_c^-. \tag{4.43}$$

The term $\frac{q}{p}n$ is a bulk contribution to the expected perimeter length and it dominates the slower growing contribution $qn^{\sigma_P}\hat{f}'(0)$ with increasing n.

4.2.3 *Scaling of the limiting free energy*

Along the entire critical curve the finite size free energy F_n approaches the limiting free energy $\mathcal{F}$ as $n \to \infty$. By equations (4.40) and (4.41), to leading order,

$$\mathcal{F}(z) \simeq C_\tau \tau^{1/\phi_c} \text{ as } \tau \to 0^+ \text{ along the } \lambda\text{-isotherm.} \tag{4.44}$$

This reaffirms the result in equation (4.28). The phase transition is signalled by a non-analyticity in $t_c(z)$ at $z = z_c$ (when $\tau = 0$), and this is seen in $\mathcal{F}$.

If the first derivative of $\mathcal{F}$ is not continuous at z_c, then the transition is classified as *first order*, and if a higher derivative is not continuous, then it is classified as a *continuous phase transition.*

Comparing equation (4.44) to equation (4.2) gives a relation between the specific heat exponent and the finite size crossover exponent in equation (4.29).

On the other hand, observe that $\log t_c(z) = -\mathcal{F}(z)$. Along the λ-isotherm it follows by equation (4.5) that $t_c(z) = t_\lambda(z) = t_\lambda(z_c) - a_\lambda \tau - b_\lambda |\tau|^{\psi_\lambda}$. Assuming that $\psi_\lambda < 1$ gives

$$\log t_c(z) = \log t_c(z_c) + \log\left(1 - a'_\lambda \tau - b'_\lambda |\tau|^{\psi_\lambda}\right) \simeq \log t_c(z_c) - a'_\lambda \tau - b'_\lambda |\tau|^{\psi_\lambda}.$$

Comparing this to equation (4.44) shows the relation

$$\psi_\lambda = \frac{1}{\phi_c} \tag{4.45}$$

between the shift-exponent and the finite size crossover exponent. Comparison to equations (4.2) and (4.17) gives

$$2 - \alpha = \psi_\lambda = \frac{1}{\phi} = \frac{1}{\phi_c}. \tag{4.46}$$

Thus, $\phi = \phi_c$, and this connects the scaling of the generating function to the scaling of the partition function. The relation $2 - \alpha = \frac{1}{\phi}$ is a hyperscaling relation and it relates a thermodynamic exponent α to the shift-exponent ψ_λ (and to the shape of the critical curve close to the tricritical point). This establishes a direct link between the combinatorial properties of the model, and its thermodynamic behaviour.

4.3 Homogeneity of the generating function

The correlation length in a system undergoing a continuous phase transition sets a length scale which is typically large compared to microscopic details. Thus, if the length scale is very large, then a model should be invariant under a spatial dilation followed by a rescaling of the length scale to compensate for the dilation.

Fix the scaling field t and let the correlation length $\xi(g)$ be a function of g. The length $\xi(g)$ is divergent along the λ-isotherm and at Ω. This is described by a power law relation

$$\xi(g) \simeq (g - g_\lambda(t))^{-\nu_+}. \tag{4.47}$$

The exponent ν_+ is the same as in equation (4.10) (since there is only one length scale). At the tricritical point, $\tau = g_\lambda(0) = 0$, and the above becomes

$$\xi(g) \simeq g^{-\nu_t}, \tag{4.48}$$

where ν_t is the *tricritical metric exponent.*

Rescale lengths by rescaling all vectors $\vec{r}$ by $\vec{r} \to \frac{1}{s}\vec{r}$. Then the scaling fields τ and g must be rescaled to compensate for the change in length scale. This is achieved by introducing two scaling exponents y_g and y_t:

$$g \to s^{y_g} g, \quad \text{and } \tau \to s^{y_t} \tau. \tag{4.49}$$

Invert the first to $s = g^{-1/y_g}$ to induce scale changes s by using using the scaling field g.

In the vicinity of Ω, any vector $\vec{r}$ is measured in terms of $\xi(g)$, and a change in the correlation length can be compensated for by a corresponding change in s. In particular, since $\xi(g) \simeq g^{-\nu_t}$ and since $s \propto \xi(g)$, their exponents must be the same. Thus,

$$y_g = \frac{1}{\nu_t}. \tag{4.50}$$

Scale invariance of the generating function is implemented by the scaling assumption for the singular part of the generating function under a dilation of space by s in d dimensions:

$$G_s(\tau, g) \simeq s^{-d} G_s(s^{y_t}\tau, s^{y_g} g). \tag{4.51}$$

The generating function is said to transform *homogeneously* under dilation of space.

The scaling of $G_s(\tau, g)$ in supermultiplicative models along the critical curve is given by equations (4.9), (4.11) and (4.12). The consistency of these with equation (4.51) shows that, if $s > 1$, then both y_g and y_t are positive exponents. Scaling of τ and g by powers of s drives the generating function away from the tricritical point, and its singular part (which vanishes on the critical curve) becomes larger.

Eliminate s from equation (4.51) by putting $s = cg^{-1/y_g}$. This gives

$$G_s(\tau, g) \simeq g^{d\nu_t} G_s(g^{-y_t/y_g}\tau, c), \tag{4.52}$$

where c is a constant and where τ is kept fixed. Comparison to equation (4.13) (see equation (4.50)) gives

$$2 - \alpha_t = d\nu_t, \quad \text{and } \phi = \frac{y_t}{y_g} = y_t \nu_t. \tag{4.53}$$

By equation (4.14), the relation

$$2 - \alpha_u = \frac{d}{y_t} \tag{4.54}$$

is obtained. These relations should not be confused with the hyperscaling law in equation (4.46). Replacing ϕ in equation (4.46) with the result in equation (4.53) gives $2 - \alpha = \frac{1}{y_t \nu_t}$. The exponent $\frac{1}{\nu_t}$ may be interpreted as the 'fractal dimension' of the cluster at the tricritical point.

The gap exponent is given by the ratio $\Delta = \frac{y_g}{y_t}$, and, from equation (4.40), the derivatives of $G(\tau, g)$ to g will obey power law scaling relations with exponents separated by Δ.

In submultiplicative models the appropriate assumption for homogeneity is suggested by equation (4.24). Assume that the generating function scales as

$$G(\tau, g) \simeq s^d G(s^{y_t}\tau, s^{y_g} g). \tag{4.55}$$

Since the singular part of the generating function is divergent along $g_\lambda(t)$, it dominates the analytic background, which may be ignored.

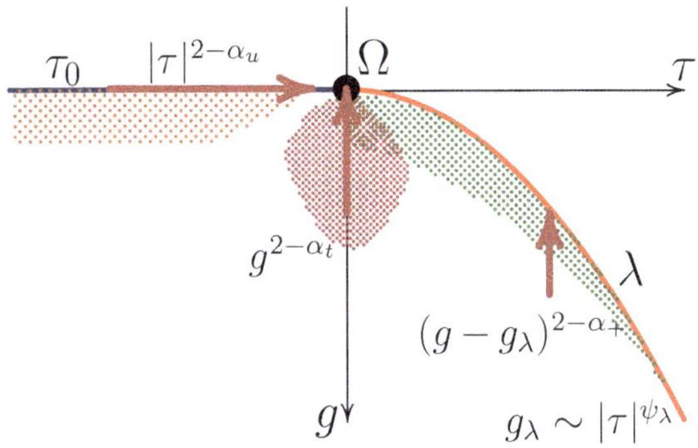

FIG. 4.6. The tricritical scaling phase diagram. The ϵ-asymptotic regions are represented by the shaded areas. If the ϵ-asymptotic regions of nearby points overlap, then the scaling is said to be asymptotically complete.

Eliminating s by putting $s = cg^{-1/y_g}$ in equation (4.55) and invoking equations (4.24), (4.25) and (4.26) gives

$$G_s(\tau, g) \simeq g^{-d\nu_t} G_s(g^{-y_t/y_g}\tau, c), \tag{4.56}$$

which gives the relations

$$\phi = \frac{y_t}{y_g} = y_t\nu_t, \quad \gamma_t = d\nu_t, \quad \text{and } \gamma_u = \frac{d}{y_t}. \tag{4.57}$$

4.4 Uniform asymptotics and the finite size scaling function

4.4.1 *Asymmetric tricriticality and ϵ-asymptotics*

Uniform asymptotics for the generating function can be developed to give a more formal description of tricritical scaling (see references [70, 71]). As above, let $G(z,t) = \sum_{n\geq 0} Z_n(z)t^n$ be the generating function of a model of an interacting clusters with radius of convergence $t_c(z)$ (in the t-plane) and partition function $Z_n(z)$. Denote the tricritical point at $z = z_c$ by Ω and suppose the τ_0-curve is given by $t_\tau(z) = t_c(z)$ with $z < z_c$, and the λ-isotherm is given by $t_\lambda(z) = t_c(z)$ with $z > z_c$.

Define the function

$$\omega_c(z) = \begin{cases} t_c(z), & \text{if } z < z_c; \\ t_c(z_c), & \text{otherwise.} \end{cases} \tag{4.58}$$

Define the modified generating function

$$G_A(z,t) = \sum_{n\geq 0} Z_n(z)(\omega_c(z)t)^n. \tag{4.59}$$

Then the radius of convergence of $G_A(z,t)$ is

$$t_c(z) = \begin{cases} 1, & \text{if } z < z_c; \\ \dfrac{t_c(z)}{t_c(z_c)}, & \text{otherwise.} \end{cases} \tag{4.60}$$

The critical curve $t_c(z)$ of $G_A(z,t)$ is plotted in figure 4.6 and corresponds to the tricritical phase diagram of an asymmetric model: the τ_0-curve is a constant function of z for $z < z_c$.

Assume that $t_c(z)$ is differentiable at $t = t_c$. Then the scaling fields (τ, g) can be chosen by putting $\tau = z - z_c$ and $g = 1 - t$.

The *scaling regime* in the vicinity of Ω can be defined precisely by introducing ϵ-asymptotic regions in the phase diagram.

Definition 4.1 (ϵ-Asymptotic regions) *Suppose that $u(x) \sim p(x)$ as $x \to x_0^-$. Then the ϵ-asymptotic region of $u(x)$ with respect to $p(x)$, $\Delta_p^u(\epsilon)$, is defined as follows: let $\epsilon > 0$ and let $x_1 = \min\{x \mid |\frac{u(y)}{p(y)} - 1| < \epsilon,\ \forall y \in (x, x_0)\}$. Then $\Delta_p^u(\epsilon) = (x_1, x_0)$, and the width of the ϵ-asymptotic interval is $\delta_p^u(\epsilon) = x_0 - x_1$.* □

The asymptotic region $\Delta_p^u(\epsilon)$ is that interval close to x_0 where $p(x)$ is a good approximation to $u(x)$. There are ϵ-asymptotic regions for scaling on approach to each point on the critical curve $t_c(z)$. The ϵ-asymptotic regions for approach to the tricritical point may or may not overlap with the regions for points nearby on the critical curve. This may be problematic, because in that instance the scaling hypothesis may be inadequate.

Therefore, assume that the scaling hypotheses are adequate and that the ϵ-asymptotic regions of nearby points overlap along the critical curve $t_c(z)$ in the vicinity of Ω. This assumption is called *asymptotic completeness.*

4.4.2 *Extended tricriticality*

A set of necessary conditions for asymptotic completeness of tricritical scaling extends the definition of tricriticality to *extended tricriticality.*

Consider the interval $I = (z_0, z_1)$, chosen such that the tricritical point $z_c \in I$. Assume that z_0 and z_1 are far enough from z_c that scaling close to the τ_0-curve and λ-isotherm is valid.

Assume that the analytic background term in the generating function $G_A(z,t)$ is zero or absent, and the symbol $\sim$ will be defined to mean precisely that if $f \sim g$, then the singular parts of f and g are asymptotic.

Definition 4.2 (Tricriticality) *Let $G_A(z,t)$ be the generating function defined in equation (4.59) and let its radius of convergence be $t_c(z)$ (see equation (4.60)).*

Let $0 < z_0 < z_c < z_1$ and define the interval $I_- = (z_0, z_c)$, define the interval $I_+ = (z_c, z_1)$ and define the interval $I = (z_0, z_1)$. Define the λ-isotherm and τ_0-curve as before and introduce the scaling fields (τ, g). Let the λ-isotherm be given by $g = g_\lambda(\tau)$, and let the τ_0-curve be given by $g = g_\tau(\tau) = 1$ in these coordinates.

Tricritical scaling is said to occur at the point $(1, z_c)$ if

(1) *$G_A(z,t)$ has a radius of convergence $t_c(z) > 0$, with $t_c(z) = 1$ if $z \leq z_c$, and $t_c(z) < 1$ and analytic if $z \in I_+$;*

(2) *there exist positive numbers D and ϕ such that*

$$g_\lambda(\tau) \sim D t^{1/\phi} \quad \text{as } \tau \to 0^+, \tag{4.61}$$

defines the crossover exponent as a shift-exponent which describes the shape of the λ-isotherm as it approaches the critical point (see equation (4.8));

(3) *there exists a number $2-\alpha_+$ and a function $A_+(\tau)$, analytic in I_+, such that*

$$G_A(z,t) \sim A_+(\tau)(g - g_\lambda(\tau))^{2-\alpha_+} \quad \text{as } g \to g_\lambda(\tau)^+ \tag{4.62}$$

(compare this to equation (4.9));

(4) *there exist numbers A_τ and $2-\alpha_t$ such that, if $z = z_c$, then*

$$G_A(z_c,t) \sim A_\tau g^{2-\alpha_t}, \quad \text{as } g \to 0^+ \tag{4.63}$$

(compare this to equation (4.12));

(5) *assume that $G_A(z,t)$ is finite on the τ_0-line; in fact, assume more by requiring the existence of an analytic function $A_-(\tau)$ in I_- such that*

$$G_A(z,t) \sim A_-(\tau) \quad \text{on the } \tau_0\text{-line}; \tag{4.64}$$

moreover, there exist numbers S_- and $2-\alpha_u$ (compare this to equation (4.11)) such that

$$A_-(\tau) \sim S_-|\tau|^{2-\alpha_u} \tag{4.65}$$

and where $2-\alpha_u = \frac{1}{\phi}(2-\alpha_t)$;

(6) *there exists a number S_+ such that*

$$A_+(\tau) \sim S_+ \tau^{[(\alpha_+-2)/\phi]+(2-\alpha_u)} \quad \text{as } \tau \to 0^+; \tag{4.66}$$

(7) *there exists a tricritical scaling function $f(x)$ such that*

$$G_A(z,t) \sim A_\tau g^{2-\alpha_t} f(A_s g^{-\phi}\tau); \tag{4.67}$$

this tricritical scaling assumption is the same as equation (4.13); the scaling function $f(x)$ must be analytic in an interval $(-\infty, x_0)$ where $x_0 = A_s D^{-\phi}$; moreover, there are the following properties for consistency against items (1) through (6):
in the first place,

$$f(0) = 1; \tag{4.68}$$

second, the following asymptotic formula must hold for $f(x)$:

$$f(x) \sim G_-(-x)^{2-\alpha_u} \quad \text{as } x \to -\infty \tag{4.69}$$

(see the arguments following equation (4.13)), and

$$f(x) \sim G_+(x_0 - x)^{2-\alpha_+} \quad as\ x \to x_0^-; \tag{4.70}$$

the constants G_+ *and* G_- *should be given by*

$$G_+ = S_+ D^{\alpha_t - \alpha_+} (x_0 \phi)^{\alpha_+ - 2} A_\tau^{-1}, \quad and\ G_- = S_- D^{\alpha_t - 2} x_0^{\alpha_u - 2} A_\tau^{-1}$$

to be compatible with the scaling assumptions in equations (4.62) and (4.65); and

(8) *assume that neither equation (4.62) nor equation (4.65) need be uniform and observe that equation (4.67) cannot be uniform either.* □

The first goal is to develop a scaling assumption which gives uniform asymptotics for $G_A(z,t)$ in the interval I defined in definition 4.2. The definition above is still inadequate; in particular, equation (4.67) must be adapted to account for the scaling in equation (4.65) and especially in equation (4.62). This can be achieved by slightly changing the assumption in equation (4.67) to *extended tricritical scaling.* The extended tricritical scaling assumption will give the asymptotic behaviour of the generating function as the critical line is approached for any fixed $z \in I$.

Theorem 4.3 (Extended tricritical scaling) *Let* $G_A(z,t)$ *exhibit tricritical scaling as in definition 4.2 and let the scaling fields be* (τ, g). *Suppose that* $z \in I$ *is fixed but arbitrary. Then there exist real-valued functions* $d(\tau)$ *and* $h(\tau)$ *such that*

$$G_A(z,t) \sim A_\tau d(\tau) g^{2-\alpha_t} f(A_s g^{-\phi} h(\tau)),$$

where

$$d(\tau) \sim 1, \quad and\ h(\tau) \sim t \quad as\ \tau \to 0.$$

Moreover, $h(\tau)$ *is monotonic and has an inverse in* I.

Proof Define

$$h(\tau) = \begin{cases} D^{-\phi} (g_\lambda(\tau))^\phi, & \text{if } z \in I_+; \\ \tau, & \text{if } z \in I_- \cup \{z_c\}. \end{cases}$$

The function $g_\lambda(\tau)$ is analytic in I_+; thus, $h(\tau)$ is analytic in I_- and in I_+. By equation (4.61), $g_\lambda(\tau) \sim D\tau^{1/\phi}$; this shows that $h(\tau) \sim \tau$ as $\tau \to 0$. If $g \to [g_\lambda(\tau)]^+$, then the argument of f in equation (4.67) approaches $x_0 = A_s D^{-\phi}$, and, by equation (4.70), the correct behaviour is found along the λ-isotherm.

Since $g_\lambda(\tau)$ is monotonic in I_+, $h(\tau)$ is as well; therefore, $h(\tau)$ is invertible in I. The function $d(\tau)$ can also be explicitly defined:

$$d(\tau) = \begin{cases} A_+(\tau) (x_0 \phi)^{\alpha_+ - 2} A_\tau^{-1} G_+^{-1} (g_\lambda(\tau))^{\alpha_t - \alpha_+}, & \text{if } z \in I_+; \\ 1, & \text{if } z = z_c; \\ A_-(\tau) A_s^{\alpha_u - 2} |\tau|^{\alpha_u - 2} A_\tau^{-1} G_-^{-1}, & \text{if } z \in I_-. \end{cases}$$

Use equations (4.61), (4.65) and (4.66) to show that the assumed asymptotic form has the required behaviour along the τ_0-line and in approach to the tricritical point. This completes the proof. □

In theorem 2.3 the approach to the tricritical point was with $\tau = 0$ as $g \to 0^+$, and the scaling is given by equation (4.64). It will also be useful to be able to approach the tricritical point along other trajectories; in particular, consider the family of curves indexed by $q \in (-\infty, x_0]$:

$$A_s g^{-\phi} h(\tau) = q \ \text{ as } g \to 0^+. \tag{4.71}$$

If q approaches $x_0 = A_s D^{-\phi}$, then the approach to the tricritical point is along the λ-isotherm. If q approaches $-\infty$, then the approach will be along the τ_0-curve.

Since $h(\tau)$ is invertible in I, solve for τ in equation (4.71) and consider the curve $(g, \frac{1}{h}(qg^\phi/A_s))$ parameterised by g and indexed by q. If g is sufficiently small, then $\frac{1}{h}(qg^\phi/A_s) \in I$, and, as $g \to 0^+$, the tricritical point is approached. By theorem 4.3 and equation (4.67),

$$\begin{aligned} G_A(g, \tfrac{1}{h}(qg^\phi/A_s)) &\sim A_\tau d(\tfrac{1}{h}(qg^\phi/A_s)) g^{2-\alpha_t} f(A_s g^{-\phi} h(\tfrac{1}{h}(qg^\phi/A_s))) \\ &\sim A_\tau g^{2-\alpha_t} f(q) \end{aligned} \tag{4.72}$$

as $g \to 0^+$. Thus, for any $q \in (-\infty, x_0)$, the correct scaling behaviour is found on approaching the tricritical point along any of these curves. This gives the following lemma.

Lemma 4.4 *Let $q \in (-\infty, x_0]$ be fixed and consider the curve $A_s g^{-\phi} h(\tau) = q$ as $g \to 0^+$. Then*

$$G_A(g, \tfrac{1}{h}(qg^\phi/A_s)) \sim A_\tau d(\tfrac{1}{h}(qg^\phi/A_s)) g^{2-\alpha_t} f(q) \ \text{ as } g \to 0^+. \qquad \square$$

4.4.3 *Uniform asymptotics for the generating function*

The requirement that ϵ-asymptotic regions overlap (see figure 4.6) is necessary for the description of tricritical scaling using tricritical exponents and a tricritical scaling function. If this fails (that is, if there are wedges incident with the tricritical point which are not covered by ϵ-asymptotic regions of either the tricritical point, the λ-isotherm or the τ_0-curve), then in the model there may be additional singularities and other critical behaviours, which are not modelled by the tricritical scaling assumption. To avoid this possibility, assume that tricritical scaling is *asymptotically complete*; that is, the ϵ-asymptotic regions overlap around the tricritical point.

To make this more precise, consider the level curve defined by equation (4.71) and suppose that $q \in (0, x_0)$. Fix a point $z \in I_+$ and consider the separation between the corresponding g-coordinates for points on the curve in equation (4.71), and the critical curve. This separation is given by

$$\left(\tfrac{1}{q} A_s h(\tau)\right)^{1/\phi} - \left(\tfrac{1}{x_0} A_s h(\tau)\right)^{1/\phi}. \tag{4.73}$$

Asymptotic completeness requires that the width of the ϵ-asymptotic region of the λ-isotherm at this value of z be wider than this separation for some fixed value of q and for all $z \in I_+$. In other words, the wedge formed by the λ-isotherm and the level curve must be completely contained in the ϵ-asymptotic region of the λ-isotherm for some finite and fixed value of q and for all $z \in I_+$.

The same requirement is made on the ϵ-asymptotic region of the τ_0-line.

Definition 4.5 (Asymptotic completeness) *Let $G_A(\tau, g)$ be a generating function which exhibits tricritical scaling as set out in definition (4.2) and which exhibits extended tricritical scaling as set out in theorem (4.3) and lemma (4.4).*

Let the ϵ-asymptotic regions of the τ_0-line be $\Delta^-(\epsilon)$, and of the λ-isotherm be $\Delta^+(\epsilon)$, with corresponding widths $\delta^-(\epsilon, t)$ and $\delta^+(\epsilon, t)$, respectively.

If there exists a $q_0 \in (0, x_0)$ and a $q_1 \in (-\infty, 0)$, both dependent on ϵ, such that, for all $z \in I_+$,

$$\delta^+(\epsilon, \tau) \geq \left(\tfrac{1}{q_0} A_s h(\tau)\right)^{1/\phi} - \left(\tfrac{1}{x_0} A_s h(\tau)\right)^{1/\phi},$$

and, for all $z \in I_-$,

$$\delta^-(\epsilon, \tau) \geq \left(\tfrac{1}{q_1} A_s h(\tau)\right)^{1/\phi},$$

then $G_A(t, g)$ satisfies a condition of asymptotic completeness. □

It is possible to prove that, if $G_A(\tau, g)$ satisfies the hypothesis in definition 4.2 and if it is asymptotically complete, then there exists a uniform tricritical scaling relation for $G_A(\tau, g)$. It will be helpful to define the function

$$H(\tau, g) = A_\tau d(\tau) g^{2-\alpha_t} f(A_s g^{-\phi} h(\tau)), \tag{4.74}$$

and this will be used in the next theorem.

Theorem 4.6 *Let $G_A(\tau, g)$ be a generating function which exhibits tricritical scaling as in definition 4.2 and is also asymptotically complete (definition 4.5).*

Then $G_A(\tau, g)$ has extended tricritical scaling uniformly for all $z \in I$. In particular,

$$G_A(\tau, g) \sim A_\tau d(\tau) g^{2-\alpha_t} f(A_s g^{-\phi} h(\tau)),$$

uniformly as g approaches the critical line for all $z \in I$ and with $d(\tau)$ and $h(\tau)$ as defined in theorem 4.3.

Proof Define the ϵ-asymptotic regions of $H(\tau, g)$ by $\Delta_H^+(\epsilon)$ and $\Delta_H^-(\epsilon)$ on I_+ and I_- respectively.

By asymptotic completeness, $\Delta_H^+(\epsilon)$ contains the region between the curve $A_s g^{2-\alpha_t} h(\tau) = p_0$, the critical curve $g = g_\lambda(\tau)$ and the curve $g - g_\lambda(\tau) = k$, for some non-zero constant k, where p_0 is close enough to x_0; say, for p_0 in the interval (Q_0, x_0).

Since $f(x) \sim G_+(x_0 - x)^{2-\alpha_+} = f_+(x)$ (equation (4.70)), there is an asymptotic region with $|f(x) - f_+(x)| \leq \epsilon' |f_+(x)| = \epsilon' f_+(x)$. Thus,

$$\begin{aligned} H(\tau, g) &= A_\tau d(\tau) g^{2-\alpha_t} \left(f(A_s g^{-\phi} h(\tau)) - f_+(A_s g^{-\phi} h(\tau)) \right) \\ &\qquad\qquad + A_\tau d(\tau) g^{2-\alpha_t} f_+(A_s g^{-\phi} h(\tau)) \\ &< A_\tau d(\tau) g^{2-\alpha_t} \epsilon' f_+(A_s g^{-\phi} h(\tau)) + A_\tau d(\tau) g^{2-\alpha_t} f_+(A_s g^{-\phi} h(\tau)) \\ &= (1+\epsilon') A_+ (g - g_\lambda(\tau))^{2-\alpha_+} + (1+\epsilon') O((g - g_\lambda(\tau))^{3-\alpha_+}), \qquad (4.75) \end{aligned}$$

where $A_\tau d(\tau) g^{2-\alpha_t} f_+(A_s g^{-\phi} h(\tau)) = A_+ (g - g_\lambda(\tau))^{2-\alpha_+} + O((g - g_\lambda(\tau))^{3-\alpha_+})$ as $g \to (g_\lambda(\tau))^+$ was used (see equation (4.70)). Observe that

$$\begin{aligned} &A_\tau d(\tau) g^{2-\alpha_t} f_+(A_s g^{-\phi} h(\tau)) \qquad (4.76) \\ &\quad = A_+ (g/g_\lambda(\tau))^{2-\alpha_t} (g_\lambda(\tau))^{2-\alpha_+} \phi^{\alpha_+ - 2} (1 - (g_\lambda(\tau)/g)^\phi)^{2-\alpha_+}. \end{aligned}$$

Expand the exponents:

$$1 - \left(\tfrac{g}{g_\lambda(\tau)} \right)^\phi \approx \phi \log \left(\tfrac{g}{g_\lambda(\tau)} \right), \quad \text{and} \qquad (4.77)$$

$$\left(\tfrac{g}{g_\lambda(\tau)} \right)^{2-\alpha_t} \approx 1 + (2 - \alpha_t) \log \left(\tfrac{g}{g_\lambda(\tau)} \right). \qquad (4.78)$$

Substitute these in the previous expressions to obtain

$$\begin{aligned} &A_+ (g - g_\lambda(\tau))^{2-\alpha_+} \left(1 + (2 - \alpha_t) \log \left(\tfrac{g}{g_\lambda(\tau)} \right) \right) \\ &\qquad = A_+ (g - g_\lambda(\tau))^{2-\alpha_+} + O((g - g_\lambda(\tau))^{3-\alpha_+}). \qquad (4.79) \end{aligned}$$

By rearranging terms in equation (4.75),

$$\frac{H(\tau, g)}{A_+ (g - g_\lambda(\tau))^{2-\alpha_+}} - 1 < \epsilon' + (1+\epsilon') O(g - g_\lambda(\tau)). \qquad (4.80)$$

Choose $\epsilon' = \frac{\epsilon - O(g - g_\lambda(\tau))}{1 + O(g - g_\lambda(\tau))}$, and g sufficiently close to $g_\lambda(\tau)$ so that $\epsilon' > 0$. Then $H(\tau, g) \sim A_+ (g - g_\lambda(\tau))^{2-\alpha_+}$ as g approaches the λ-isotherm. This is valid if p_0 is chosen in the interval (Q_0, x_0) and such that $f(x)$ and $f_+(x)$ are ϵ'-asymptotic.

In particular, suppose the ϵ-asymptotic regions of $f(x)$ and $f_+(x)$ have width $\delta_f^{f_+}(\epsilon)$; then the level curve determined by p_0 must be in the region determined by the interval $(g_\lambda(\tau) + \delta_f^{f_+}(\epsilon), g_\lambda(\tau))$, for all $z \in I_+$. But this means that p_0 must be chosen in the interval $(A_s (g_\lambda(\tau) + \delta_f^{f_+}(\epsilon))^{-\phi} h(\tau), x_0)$. The intersection of this interval with (Q_0, x_0) gives possible values for p_0. Let q_2 be one such value.

Next, it must be verified that the region between the level curve defined by q_2 and $g_\lambda(\tau)$ is contained in some ϵ-asymptotic region of $G_A(\tau, g)$ and $H(\tau, g)$. By lemma 4.4, there is a q_0 so that, if q_3 is the maximum of q_2 and

q_0, then the region between the level curve determined by q_3 and $g_\lambda(\tau)$ is in the (say) 3ϵ-asymptotic region of $G_A(\tau, g)$ and $H(t, g)$ (this is $\Delta_G^H(3\epsilon)$), and $H(\tau, g) \sim A_+(g - g_\lambda(\tau))^{2-\alpha_+}$ as g approaches the λ-isotherm. Thus, it follows that $G_A(\tau, g) \sim H(\tau, g) \sim A_+(g - g_\lambda(\tau))^{2-\alpha_+}$ as g approaches the λ-isotherm within the region determined by the level curve indexed by q_3 and $g_\lambda(\tau)$, for all $z \in I_+$.

Finally, the asymptotics are uniform. To see this, first notice that, by lemma 4.4, there exists a width $\delta_G^H(\epsilon)$ for each $q \in (0, x_0)$ such that $|\frac{G(\tau,g)}{H(\tau,g)} - 1| < \epsilon$ if $g - g_\lambda(\tau) < \delta_G^H(\epsilon)$, where $\delta_G^H(\epsilon)$ may be dependent on $z \in I_+$. Second, $\delta_G^H(\epsilon) > 0$ for all $q \in (0, x_0)$, by asymptotic completeness, and $\delta_G^H(\epsilon)$ is strictly positive for all $q \in (0, q_3]$. Third, this ϵ-asymptotic region is different from the region defined by the ϵ-asymptotic region of $f(x)$ and $f_+(x)$; however, these are both contained in $\Delta_G^H(3\epsilon)$. Thus, they have a common neighbourhood.

Therefore, choose some $\delta'(\epsilon)$ independent of z so that the region defined by $|g - g_\lambda(\tau)| < \delta'(\epsilon)$ is contained in $\delta_G^H(3\epsilon)$ for all $z \in I_+$. In other words, there is a $\delta'(\epsilon) > 0$ independent of z so that $G_A(t, g) \sim H(t, g)$ uniformly in I_+.

The proof follows similar arguments for $z \in I_-$ along the τ_0-line. □

Theorem 4.6 is due to R Brak and AL Owzcarek [70], and its significance is that, with the assumption of asymptotic completeness, and the properties of tricritical scaling as set out in definition 4.2, it is possible to find extending functions which will give a uniform asymptotic approximation of the generating function.

4.4.4 *The finite size scaling function*

Recall that $G_A(z, t)$ is the generating function of $Z_n(z) \times (\omega_c(z))^n$ (see equation (4.59)). Without loss of generality, suppose that $\omega_c(z) = 1$. The finite size scaling of $Z_n(z)$ must be reconciled with the tricritical scaling of $G_A(z, t)$; this requires describing the connection between the finite size crossover exponent ϕ_c in equation (4.38), and the crossover exponent ϕ in equation (4.13) (see equation (4.46)).

The partition function can be extracted from $G_A(z, t)$ by the contour integral

$$Z_n(z) = \frac{1}{2\pi i} \oint_C \frac{G_A(z,t)}{t^{n+1}}\, dt, \tag{4.81}$$

where C is a contour which circles $\vec{0}$ in the t-plane. In general, not enough is known about $G_A(\tau, g)$ to make the execution of this integral possible.

Suppose that $z \in I$ and suppose that $G_A(\tau, g)$ satisfies the conditions in definition 4.2, and the uniform asymptotics in theorem 4.6 ($G_A(\tau, g) \sim H(\tau, g)$). Can $G_A(\tau, g)$ be replaced by $H(\tau, g)$ in equation (4.81) to find an asymptotic formula for $Z_n(z)$? This is possible if another assumption about the singularity structure of $G_A(\tau, g)$ and $H(\tau, g)$ is made. This assumption is a Darboux condition, which will allow the use of Darboux's theorem (see theorem D.18 in appendix D, and reference [443]) to find the asymptotics of $Z_n(z)$ from $H(\tau, g)$.

Since both $G_A(\tau, g)$ and $H(\tau, g)$ are analytic in the annulus $0 < |t| < t_c(z)$, it is only necessary to assume that the difference of derivatives has a finite number of singularities (at t_i) such that

$$G_A^{(m)}(\tau, g) - H^{(m)}(\tau, g) = O((-t_i)^{\sigma_j - 1}), \tag{4.82}$$

where the σ_j are positive exponents. With this assumption, Darboux's theorem states that

$$Z_n(z) = \frac{1}{2\pi i} \oint_C \tfrac{H(\tau,g)}{t^{n+1}}\, dt + o(t_c(z)^{-n} n^{-m}), \tag{4.83}$$

where the contour circles $\vec{0}$ in the t-plane and excludes all singularities of $H(\tau, g)$.

If $H(\tau, g)$ is substituted from equation (4.74), then

$$Z_n(z) = \frac{A_\tau d(\tau)}{2\pi i} \oint_C \frac{(1-t)^{2-\alpha_t}}{t^{n+1}} f(A_s(1-t)^{-\phi} h(\tau))\, dt + o(t_c(z)^{-n} n^{-m}), \tag{4.84}$$

where $g = 1 - t$ was used. Change variables $1 - t = \frac{q}{n}$. The contour now circles the point $q = n$; denote this by C_n. Then the above becomes

$$Z_n(z) = -\frac{A_\tau d(\tau) n^{\alpha_t - 3}}{2\pi i} \oint_{C_n} q^{2-\alpha_t} (1 - \tfrac{q}{n})^{1-n} f(A_s n^\phi h(\tau) q^{-\phi})\, dq \\ + o(t_c(z)^{-n} n^{-m}). \tag{4.85}$$

Put $n^\phi h(\tau) = \zeta$; then $z = z_c + h^{-1}(zn^\phi)$ (since $\tau = z - z_c$). Increasing n in equation (4.85) while ζ is fixed implies that z approaches z_c. Thus, $t_c(z)^{-n} = (t_c(z_c + h^{-1}(\zeta n^\phi)))^{-n}$ in equation (4.85) and, as $n \to \infty$, this approaches a constant. Therefore, the correction term can be ignored in the asymptotic limit and only the evaluation of the integral in equation (4.85) should be of concern.

Equations (4.69) and (4.70) show that the scaling function $f(x)$ has radius of convergence $x_0 = A_s D^{-\phi}$ and that it has a Taylor expansion about $x = 0$. Since $\zeta = n^\phi h(\tau)$, it follows that $A_s n^\phi h(\tau) q^{-\phi} = A_s \zeta q^{-\phi}$, and

$$f(A_s \zeta q^{-\phi}) = \sum_{m=0}^{\infty} \tfrac{1}{m!} f^{(m)}(0) (A_s \zeta q^{-\phi})^m \tag{4.86}$$

if $\zeta < \frac{q^\phi}{A_s}$.

The contour C_n in equation (4.85) can be chosen such that the series is uniformly convergent in a compact domain of C_n. Parameterise C_n by $n + Re^{i\theta}$ and choose r_0 such that $0 < R < r_0 < n$. Then

$$\left| \tfrac{1}{m!} f^{(m)}(0) (A_s \zeta q^{-\phi})^m \right| \leq \left| \tfrac{1}{m!} f^{(m)}(0) (A_s \zeta (n - r_0)^{-\phi})^m \right|, \tag{4.87}$$

and the series

$$\sum_{m=0}^{\infty} \tfrac{1}{m!} \left| f^{(m)}(0) (A_s \zeta (n - r_0)^{-\phi})^m \right| \tag{4.88}$$

is convergent if $|\zeta| < |x_0 (n - r)^\phi A_s^{-1}|$ (where $x_0 = A_s D^{-\phi}$).

Pick ζ appropriately, substitute equation (4.86) into equation (4.85), interchange the series and the integral and evaluate the integral (it is a β-function). The result is

$$Z_n(z) = -A_\tau d(\tau) \sum_{m=0}^{\infty} \frac{\Gamma(m\phi + \alpha_t - 4 + n)}{\Gamma(m\phi + \alpha_t - 2)\Gamma(n-1)} f^{(m)}(0)(A_s\zeta)^m n^{-m\phi}. \quad (4.89)$$

By equation (4.40), the finite size scaling function can be computed from

$$\hat{h}(\zeta) = \lim_{n\to\infty} n^{3-\alpha_t} Z_n(z). \quad (4.90)$$

Since

$$\lim_{n\to\infty} \frac{n^{3-\alpha_t - m\phi}\Gamma(m\phi + \alpha_t - 4 + n)}{\Gamma(n-1)} = 1, \quad (4.91)$$

the series in equation (4.89) is absolutely convergent if $|\zeta| < |x_0(n-r)^\phi / A_s|$, where $x_0 = A_s D^{-\phi}$ (see equation (4.88)). In this case the sum and the limit can be interchanged in equation (4.91). Evaluating the limit shows that

$$\hat{h}(\zeta) = -A_\tau d(\tau) \sum_{m=0}^{\infty} \frac{f^{(m)}(0)(A_s\zeta)^m}{\Gamma(m\phi + \alpha_t - 2)} \quad (4.92)$$

Notice that $\hat{h}(\zeta)$ is an entire function since it is a convergent power series, which also converges absolutely.

Since the radius of convergence of the scaling function $f(x)$ in equation (4.86) is $x_0 = \limsup_{m\to\infty} |f^{(m)}(0)|^{-1/m}$, the radius of convergence of equation (4.92) is determined by

$$\limsup_{m\to\infty} \left| \frac{f^{(m)}(0)(A_s\zeta)^m}{\Gamma(m\phi + \alpha_t - 2)} \right|^{-1/m} = 0, \quad (4.93)$$

provided that $\phi > 0$ (this is guaranteed by the second point in definition 4.2). In other words, the introduction of a finite size scaling crossover exponent equal to the crossover exponent in equation (4.46) is justified, as is the finite size scaling assumption in equation (4.37).

5

DIRECTED LATTICE PATHS

Directed lattice models of polymers (see reference [479]) are related to classical models of directed paths in enumerative combinatorics.

Define a square lattice with north-east and south-east edges by

$$\mathbb{J}^2 = \{\langle \vec{u} \sim \vec{v} \rangle \bigm| \vec{u}, \vec{v} \in \mathbb{Z}^2, \text{ and } \|\vec{u} - \vec{v}\|_2 = \sqrt{2}\}, \tag{5.1}$$

and the half-lattice

$$\mathbb{J}^2_+ = \{\langle \vec{u} \sim \vec{v} \rangle \in \mathbb{J}^2 \bigm| \vec{u}(2) \geq 0, \text{ and } \vec{v}(2) \geq 0\}, \tag{5.2}$$

where $\vec{v}(2)$ is the second Cartesian coordinate of the vertex $\vec{v} \in \mathbb{Z}^2$. The boundary of $\mathbb{J}^2_+$ is $\partial \mathbb{J}^2_+ = \{\vec{v} \in \mathbb{J}^2_+ \mid \vec{v}(2) = 0\}$ and is also called a *hard wall*.

The simplest directed model of a polymer is the (fully) directed path from $\vec{0}$ in $\mathbb{J}^2$ giving north-east and south-east steps of length $\sqrt{2}$. There are 2^n directed paths of length n (steps). The model is more interesting if the path is confined to the half-lattice $\mathbb{J}^2_+$. A directed path in $\mathbb{J}^2_+$ from $\vec{0}$ and ending in a vertex $\vec{v} \in \partial \mathbb{J}^2_+$ is a *Dyck path* (see figure 5.1).

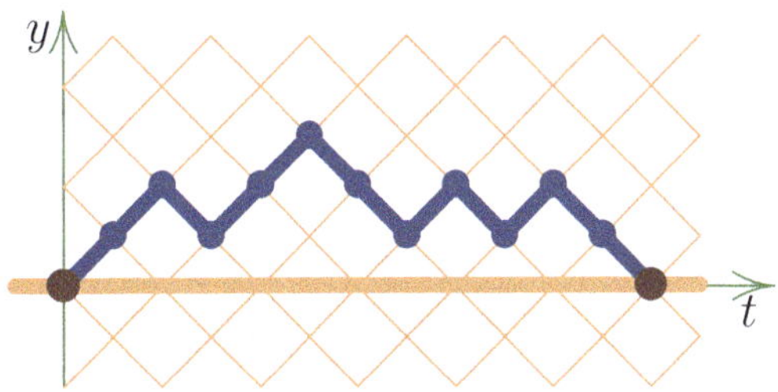

FIG. 5.1. A Dyck path in $\mathbb{J}^2_+$. The path gives north-east and south-east steps of length $\sqrt{2}$ in the first quadrant and ends in the horizontal axis, which is a *hard wall.*

5.1 Dyck paths

The generating function of Dyck paths can be determined by a decomposition or factorisation of Dyck paths as shown in figure 5.2. Each Dyck path either has length 0 (is a single vertex at $\vec{0}$) *or* has a first return to $\partial \mathbb{J}^2_+$, where it renews (as

The Statistical Mechanics of Interacting Walks, Polygons, Animals and Vesicles, 2nd edition, E.J. Janse van Rensburg. © E.J. Janse van Rensburg. Published in 2015 by Oxford University Press.

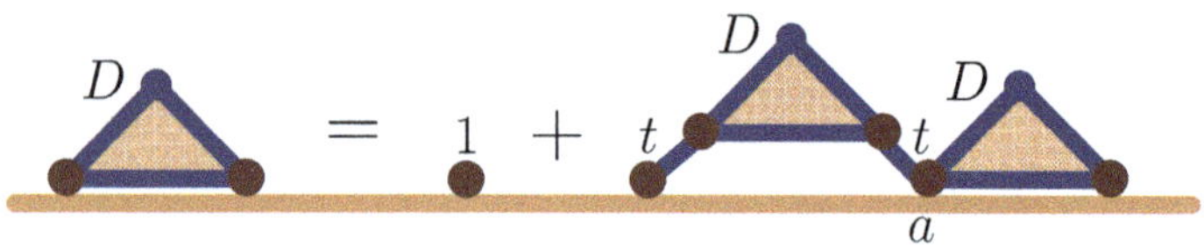

FIG. 5.2. Factorisation of the Dyck path generating function.

a Dyck path). The section of the path from $\vec{0}$ to its first return is an *excursion*. If t is the length generating variable, then this decomposition shows that the generating function $D(t)$ satisfies the functional recurrence

$$D(t) = 1 + t^2\, D^2(t), \tag{5.3}$$

since the generating function of the first excursion is $t^2 D(t)$.

Factorisation methods such as the above generalise to other models of lattice paths. The method is implemented by finding a first 'narrow point' where the path renews. Cutting the path in this point gives a renewal equation for the generating function. This is also called the *wasp-waist method* [58, 69, 231, 469, 488].

The recurrence in equation (5.3) is quadratic in $D(t)$ and has roots

$$D^{\pm}(t) = \frac{2}{1 \pm \sqrt{1-4t^2}}. \tag{5.4}$$

Series expansion of $D^{\pm}(t)$ shows that $D^{+}(t) \equiv D(t)$ is a power series with coefficients in $\mathbb{N}_0$ counting paths. This is the *physical root* of equation (5.3).

The function $D(t)$ is also the Catalan generating function. This shows that the number of Dyck paths of length $2n$ is given by

$$d_{2n} = \frac{1}{n+1}\binom{2n}{n}. \tag{5.5}$$

Catalan numbers are given by $C_n = d_{2n}$ (see references [187, 531]). By putting $a = 0$ in theorem D.1 in appendix D, it follows that

$$d_{2n} \sim \frac{4^n}{\sqrt{\pi\, n^3}}. \tag{5.6}$$

A directed path giving north-east and south-east steps from $\vec{0}$ in $\mathbb{J}^2_+$ is a *ballot path* with the generating function $D_B(t)$. A ballot path is a *tail* if it steps into $\mathbb{J}^2_+$ from $\vec{0}$ and never returns to $\partial\mathbb{J}^2_+$. The generating function of a tail is denoted by $D_T(t)$. When the first north-east step of a tail is removed, then a ballot path remains. That is, $D_T(t) = t\,D_B(t)$.

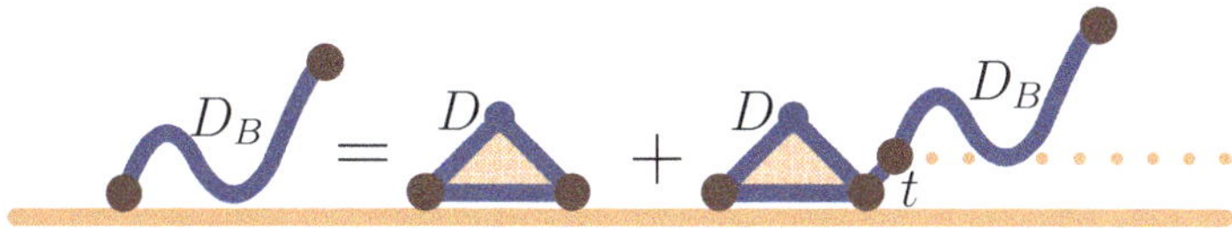

FIG. 5.3. The factorisation of ballot paths.

Ballot paths are factorised by noting that each ballot path is a Dyck path or a Dyck path followed by a tail (see figure 5.3). That is,

$$D_B(t) = D(t) + D(t)\,D_T(t) = D(t) + t\,D(t)\,D_B(t). \tag{5.7}$$

Since $D(t)$ is known, this can be solved to give

$$D_B(t) = \frac{1}{2t}\left(\sqrt{\frac{1+2t}{1-2t}} - 1\right), \text{ and } D_T(t) = \frac{1}{2}\left(\sqrt{\frac{1+2t}{1-2t}} - 1\right). \tag{5.8}$$

The number of ballot paths from the origin of length n and ending in a vertex at height h above $\partial\mathbb{J}^2_+$ is

$$d_n(h) = \frac{2h+2}{n+h+2}\binom{n}{\frac{1}{2}(n+h)}, \tag{5.9}$$

where n and h must have the same parity.

5.1.1 *Algebraic language formulation of Dyck paths*

An *algebraic language* is defined as follows: let X be any finite non-empty set (this is the *alphabet*). Define X^* to be the free monoid generated by the concatenation of the elements in X. Elements of X^* are strings of letters in X, each called a *word*, and the empty word e is the identity. A *language* is a subset of X^*.

Let N and X be finite disjoint sets. A *production* on (N, X) is an ordered pair of elements $(\alpha, \beta) \in N \times (N \cup X)^*$, (also denoted $(\alpha \to \beta)$). For example, if $s \in N$, and $x \in X$, then $(s \to xsx)$ is a production on (N, X).

An *algebraic grammar* can be defined by using productions on (N, X). This is a 4-tuple $\mathcal{G} = (N, X, P, s)$, where N and X are finite disjoint sets; P is a set of productions on (N, X), and $s \in N$. Starting at s, define the set of words in $(N \cup X)^*$ recursively by applying the productions in P.

The subset of words in X^* generated by $\mathcal{G}$ is the *algebraic language* generated by $\mathcal{G}$, denoted $L(\mathcal{G})$.

If $\mathcal{G} = (N, X, P, s)$ is an algebraic grammar, then N is called the *non-terminal set*, X is the *terminal set*, P denotes the *production rules*, and s represents the *start element*. If $w \in L(\mathcal{G})$ is a word, then $|w|_x$ is the number of occurrences of the letter x in w.

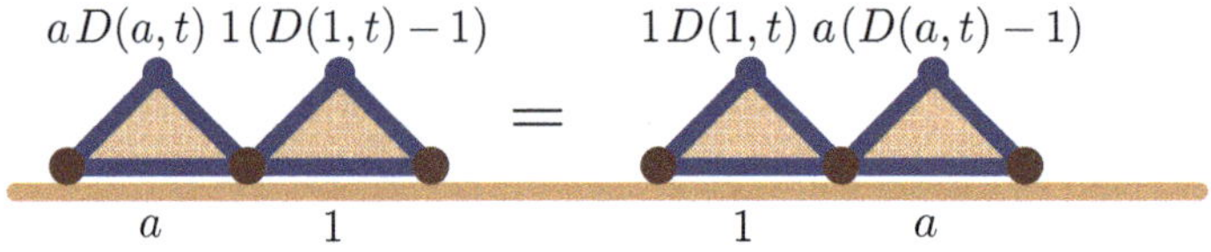

FIG. 5.4. The exchange relation for Dyck paths.

The *Dyck language* is obtained as follows [127]. Let $N = \{\mathfrak{d}\}$, let $X = \{\mathfrak{D}, \mathfrak{C}\}$, let $P = \{(\mathfrak{d} \to \mathfrak{D}\,\mathfrak{d}\,\mathfrak{C}\,\mathfrak{d}), (\mathfrak{d} \to e)\}$ and let $s = \mathfrak{d}$. Then the algebraic language is

$$L(\mathcal{G}) = \{e, \mathfrak{D}\mathfrak{C}, \mathfrak{D}\mathfrak{D}\mathfrak{C}\mathfrak{C}, \mathfrak{D}\mathfrak{C}\mathfrak{D}\mathfrak{C}, \ldots\}. \tag{5.10}$$

The language $L(\mathcal{G})$ is a non-commutative algebra over integers, with concatenation as the product.

The Dyck language has the following two properties: (1) If u is a prefix (left-factor) of a word w, then $|u|_{\mathfrak{D}} \geq |u|_{\mathfrak{C}}$, and (2) for any word w, $|w|_{\mathfrak{D}} = |w|_{\mathfrak{C}}$. These correspond to steps in a Dyck path, where $\mathfrak{D}$ is a north-east step, and $\mathfrak{C}$ is a south-east step.

Each Dyck word is generated by the production $(w \to \mathfrak{D}w\mathfrak{C}w)$, and the empty word e is the start element. This shows that Dyck words are recursively generated by

$$w = e + \mathfrak{D}w\mathfrak{C}w. \tag{5.11}$$

Sending $\mathfrak{D} \to t$, and $\mathfrak{C} \to t$, and defining $D(t) = \sum_{n\geq 0} w_n\, t^n$, where w_n is the number of words of length n, gives the recurrence $D(t) = 1 + t^2 D^2(t)$, which was also obtained in equation (5.16). For additional results, see, for example, references [127, 391, 605].

5.1.2 *Adsorbing Dyck paths*

Adsorbing Dyck paths is a standard model for directed polymer adsorption [64, 99, 121, 205, 478, 479, 559, 582]. The model is defined as follows. A *visit* in a Dyck path is a return of the path to $\partial\mathbb{J}^2_+$. The generating variable of visits is a (see figure 5.2). The generating function of adsorbing Dyck paths can be obtained by the factorisation in figure 5.2. Each Dyck path either has length 0 *or* it has a first return to $\partial\mathbb{J}^2_+$, where it creates a visit and then renews itself. This shows that the generating function is

$$D(a,t) = 1 + a t^2 D(1,t) D(a,t), \tag{5.12}$$

where t is the length or edge generating variable.

Putting $a = 1$ gives $D(1,t) \equiv D^+(t)$ (see equation (5.4)). Then solve for $D(a,t)$ in equation (5.12):

$$D(a,t) = \frac{2}{2 - a\left(1 - \sqrt{1-4t^2}\right)} = 1 + a\,t^2 + (a + a^2)\,t^4 + \cdots. \tag{5.13}$$

$D(a,t)$ satisfies the *exchange relation* [488]

$$a\,D(a,t)\,(D(1,t)-1) = 1\,D(1,t)\,(D(a,t)-a), \tag{5.14}$$

which betrays a symmetry under the exchange $1 \leftrightarrow a$. A proof-by-picture is given in figure 5.4.

The radius of convergence of $D(a,t)$ in the t-plane is seen from equation (5.13):

$$t_c(a) = \begin{cases} \frac{1}{2}, & \text{if } a \leq 2; \\ \frac{1}{a}\sqrt{a-1}, & \text{if } a > 2. \end{cases} \tag{5.15}$$

The free energy of the model is

$$\mathcal{D}(a) = -\log t_c(a) = \begin{cases} \log 2, & \text{if } a \leq 2; \\ \log(a) - \frac{1}{2}\log(a-1), & \text{if } a > 2. \end{cases} \tag{5.16}$$

$\mathcal{D}(a)$ is non-analytic at $a = 2$. This gives the critical point $a_c = 2$ in the model.

There are singularities in $D(a,t)$ at $t = \pm\frac{1}{2}$ and when $t = \frac{1}{a}\sqrt{a-1}$. If $a < 2$, then the dominant singularity are branch points at $t = \pm\frac{1}{2}$ and, if $a > 2$, then there is a simple pole at $t = \frac{1}{a}\sqrt{a-1}$. The partition function of this model is $D_n(a) = \sum_v d_n(v)\,a^v$ (where $d_n(v)$ is the number of Dyck paths with v visits). The asymptotic behaviour of the partition function is (see section 5.1.6):

$$D_n(a) \simeq \begin{cases} \dfrac{2^{n+3}}{\sqrt{2\pi n^3}\log^2(a-1)}, & \text{if } a < 2; \\ \dfrac{2^{n+1}}{\sqrt{2\pi n}}, & \text{if } a = 2; \\ \dfrac{a-2}{a-1}\left(\dfrac{a^2}{a-1}\right)^{n/2}, & \text{if } a > 2. \end{cases} \tag{5.17}$$

The derivative of the free energy $\mathcal{D}(a)$ to $\log a$ is the *density of visits* given by

$$\mathcal{E}(a) = \begin{cases} 0, & \text{if } a \leq 2; \\ \dfrac{a-2}{2(a-1)}, & \text{if } a > 2. \end{cases} \tag{5.18}$$

This result shows that $\mathcal{E}(a)$ is continuous and that it is strictly positive for $a > 2$. The model undergoes a continuous transition at $a_c = 2$ from a phase of desorbed paths (with zero density of visits) to a phase of adsorbed paths (with a positive density of visits).

The microcanonical density function of visits in this model can be determined exactly:

$$\mathcal{P}_D(\epsilon) = \sqrt{1-2\epsilon}\left(\frac{2-2\epsilon}{1-2\epsilon}\right)^{1-\epsilon}, \quad \text{for } \epsilon \in \left(0, \tfrac{1}{2}\right). \tag{5.19}$$

Naturally, $\mathcal{D}(a) = \sup_\epsilon \{\mathcal{P}_D(\epsilon) + \epsilon \log a\}$, and it may be checked that this supremum is realised when $\epsilon = \mathcal{E}(a)$ so that $\mathcal{D}(a) = \mathcal{P}_D \circ \mathcal{E}(a) + \mathcal{E}(a)\,\log a$. The finite

TABLE 5.1. Adsorbing Dyck path tricritical exponents

ϕ	α	γ_t	γ_u	y_t	ν_t	γ_+	ν_+	γ_-
$\frac{1}{2}$	0	$\frac{1}{2}$	1	2	$\frac{1}{4}$	1	$\frac{1}{2}$	$-\frac{1}{2}$

right-derivative of $\mathcal{P}_D(\epsilon)$ at $\epsilon = 0^+$ is consistent with a continuous transition of the model (see section 3.3.2.5).

Scaling axes (τ, g) may be set up for the generating function. The appropriate geometry is similar to figure 4.4, with the origin at the critical point $(a_c, t_c) = (2, \frac{1}{2})$, so the τ-axis is tangent to the critical curve $t_c(a)$ at the critical point, and the g-axis is transverse to the critical curve at the critical point. That is, $\tau = 2 - a$, and $g = 1 - 4t^2$.

The τ_0-curve and λ-isotherm in figure 4.4 can be read from equation (5.15). The τ_0-curve is given by $t_c(a) = \frac{1}{2}$ if $a < 2$, while the λ-isotherm is given by $\frac{1}{a}\sqrt{a-1}$ for $a > 2$. Along the λ-isotherm the generating function is divergent as $t \nearrow t_c(a)$ (or $g \to 0^+$); it has a simple pole in the t-plane. This shows that $\gamma_+ = 1$ in equation (4.19), from which it follows that $\nu_+ = \frac{1}{2}$ by equation (4.20).

Since

$$D(a,t) = \frac{2 - a(1 + \sqrt{1-4t^2})}{2(1 - a + a^2t^2)}, \tag{5.20}$$

the singular part of $D(a,t)$ has a square root singularity as $g \to 0^+$ (or as $t \to \frac{1}{2}$ on approaching the τ_0-curve). This shows that $\gamma_- = -\frac{1}{2}$ in equation (4.21).

Along the λ-isotherm (for $a > 2$) the critical curve has the shape given by

$$g_\lambda(\tau) = \tfrac{1}{2} - \tfrac{1}{a}\sqrt{a-1} = \frac{(\sqrt{\tau+1}-1)^2}{2(\tau+2)} \approx \frac{\tau^2}{8(\tau+2)}, \quad \text{if } \tau < 0 \text{ is small.} \tag{5.21}$$

Comparison to equation (4.8) shows that the shift-exponent is $\psi_\lambda = 2$. Since $g = 0$ for $a < 2$, it is similarly concluded that $\psi_\tau = 0$. This shows that the crossover exponent is $\phi = \frac{1}{2}$ by equation (4.17), so the specific heat exponent is $\alpha = 0$ by equation (4.46).

Putting $a = 2$ (so that $\tau = 0$) gives $D(2,t) = \frac{2}{\sqrt{1-4t^2}} \sim g^{-1/2}$ if $t < \frac{1}{2}$, and so $g > 0$. This shows that $\gamma_t = \frac{1}{2}$ in equation (4.23).

Putting $t = \frac{1}{2}$ (so that $g = 0$) gives $D(a, \frac{1}{2}) = \frac{2}{2-a} \sim \tau^{-1}$ if $a < 2$, and so $\tau > 0$. By equation (4.22), $\gamma_u = 1$. These results verify equation (4.25).

Finally, by equation (4.57), $y_t = \frac{2}{\gamma_u} = 2$, and $\nu_t = \frac{\gamma_t}{2} = \frac{1}{4}$. Notice that equation (5.20) becomes

$$D(a,t) = \frac{2(\tau - a\sqrt{g})}{\tau^2 - a^2 g} = \frac{2}{\tau + a\sqrt{g}} = \frac{2\tau^{-1}}{1 + a\tau^{-1}\sqrt{g}} = \tau^{-1} f_D(g^{-1/2}\tau) \tag{5.22}$$

for a scaling function $f_D(x) = \frac{2x}{x+a}$, as expected, giving the combination $g^{-\phi}\tau$ with $\phi = \frac{1}{2}$. Observe that $\phi = y_t\nu_t$. The tricritical exponents for adsorbing Dyck paths are listed in table 5.1.

5.1.3 *A constant term formulation for adsorbing Dyck paths*

The partition function of adsorbing Dyck paths can be obtained by solving a difference equation with a suitable boundary condition using a *constant term formulation* [62–65, 73]. This is best exploited by noting that there is a bijection between Dyck paths and random walks on $\mathbb{N}_0$; visits are returns to 0.

Let $v_n(j; j_0)$ be the number of random walks on $\mathbb{N}_0$ of length n from $j_0 \geq 0$ to $j \geq 0$. Let a be the generating variable of returns or visits to 0, with the convention that the first vertex is not a visit if $j_0 = 0$. Let $v_n(j; j_0) \equiv v_n(j)$ so that $v_0(j_0) = 1$. Then $v_n(j)$ satisfies the difference equation

$$\begin{aligned} v_n(j) &= v_{n-1}(j-1) + v_{n-1}(j+1), && \forall j \geq 1 \text{ and } \forall n \geq 1; \\ v_n(0) &= a\, v_{n-1}(1), && \forall n \geq 1. \end{aligned} \tag{5.23}$$

These can be solved by a *constant term formula* [64].

If $k \in (-\pi, \pi]$, then

$$v_n(j) = (e^{ik} + e^{-ik})^n \left(A_1 e^{ijk} + A_2 e^{-ijk}\right) \tag{5.24}$$

is a solution of equation (5.23). This is the *Bethe ansatz*, and the coefficients A_1 and A_2 must be determined from boundary conditions, while k must be integrated in $(\pi, \pi]$.

Substitution of (5.24) into (5.23) shows that

$$\frac{A_1}{A_2} = -\frac{(e^{ik} + e^{-ik}) - a\, e^{-ik}}{(e^{ik} + e^{-ik}) - a\, e^{ik}} = \mathcal{S}(k). \tag{5.25}$$

$\mathcal{S}(k)$ is the *scattering function* of the walk off the origin.

Substitution gives a solution of equation (5.23):

$$v_n(j) = A_2(k; j_0)\, (e^{ik} + e^{-ik})^n \left(e^{-ijk} + \mathcal{S}(k) e^{ijk}\right). \tag{5.26}$$

The function $A_2(k; j_0)$ is an arbitrary function of k which must be fixed by the initial condition, namely, that the walk starts at j_0.

A reversal of the walk occurs when $k \to -k$ in equation (5.26). Replacing k by $-k$, and n by m, gives the partition function of paths of length m stepping backwards from height j to height j_0 in m steps:

$$w_m(j_0) = A_2(-k; j)\, (e^{-ik} + e^{ik})^m \left(e^{ij_0k} + \mathcal{S}(-k) e^{-ij_0k}\right). \tag{5.27}$$

Comparing equations (5.26) and (5.27), putting $m = n$ and noticing that $\mathcal{S}(-k) = (\mathcal{S}(k))^{-1}$ give

$$A_2(k;j_0) = \left(e^{ij_0k} + \mathcal{S}(-k)e^{-ij_0k}\right). \tag{5.28}$$

Substitute $A_2(k;j_0)$ in equation (5.26) and integrate $k \in (-\pi,\pi]$ to obtain an integral expression for $v_n(j)$:

$$v_n(j) = C_0 \int_{-\pi}^{\pi} \left(e^{ik} + e^{-ik}\right)^n \left(e^{i(j-j_0)k} + \mathcal{S}(k)e^{i(j+j_0)k}\right)\, dk, \tag{5.29}$$

where symmetry of the integral was used and a factor of 2 was adsorbed into C_0.

The substitution $\zeta = e^{ik}$ can be made to turn equation (5.29) into a contour integral with the contour C around the unit circle in the ζ-plane. Put $\overline{\zeta} = \frac{1}{\zeta}$; then

$$v_n(j) = C_0 \oint_C (\zeta + \overline{\zeta})^n \left(\zeta^{j-j_0} - \left(\tfrac{\zeta+\overline{\zeta}-a\overline{\zeta}}{\zeta+\overline{\zeta}-a\zeta}\right)\zeta^{j+j_0}\right) \frac{d\zeta}{i\zeta}. \tag{5.30}$$

Evaluation of this integral gives an expression for $v_n(j)$. Denote the integrand in the last integral above by S. By the residue theorem, the integral selects the constant term in a Laurent expansion of $i\zeta S$, up to a constant factor (a factor of $2\pi i$).

Define the operator $\mathrm{CT}[\cdot]$ to select the constant term in the Luarent series of its argument. Then the solution of the difference equation (5.23) is given by

$$v_n(j) = C_0\, \mathrm{CT}\left[(\zeta + \overline{\zeta})^n \left(\zeta^{j-j_0} - \left(\tfrac{\zeta+\overline{\zeta}-a\overline{\zeta}}{\zeta+\overline{\zeta}-a\zeta}\right)\zeta^{j+j_0}\right)\right] \tag{5.31}$$

where C_0 is a constant to be determined.

If $j = j_0 = 0$, then $v_n(0)$ is the partition function of adsorbing Dyck paths; the constant C_0 can be fixed by considering this case. Since only even length paths are encountered, replace n by $2n$ and expand $(\zeta+\overline{\zeta})^{2n}$ to find for $j = j_0 = 0$ that

$$v_{2n}(0;0) = C_0 \sum_{\ell=0}^{n} \frac{2\ell+1}{n+\ell+1}\binom{2n}{n+\ell}(a-1)^\ell. \tag{5.32}$$

This is the partition function of Dyck paths of length $2n$ where a is the generating variable of visits. Taking the $\ell = 0$ term gives $v_0(0,0) = C_0 + \cdots$. Since $v_0(0,0) = 1 + \cdots$ this fixes $C_0 = 1$.

The partition function for arbitrary j and j_0 can be determined by expanding the factors in equation (5.31) and determining the constant term:

$$\begin{aligned} v_n(j;j_0) &= \binom{n}{\frac{n+j-j_0}{2}} - \binom{n}{\frac{n+j+j_0}{2}} \\ &+ a \sum_{\ell=0}^{(n-j-j_0)/2} \left(\binom{n}{\frac{n+j+j_0}{2}+\ell} - \binom{n}{\frac{n+j+j_0}{2}+\ell+1}\right)(a-1)^\ell. \end{aligned} \tag{5.33}$$

The first two binomial terms count paths starting at height j_0 and ending at height j while avoiding the origin 0. The summation counts paths similarly, but

TABLE 5.2. Adsorbing ballot path tricritical exponents

ϕ	α	γ_t	γ_u	y_t	ν_t	γ_+	ν_+	γ_-
$\frac{1}{2}$	0	1	2	1	$\frac{1}{2}$	1	$\frac{1}{2}$	$\frac{1}{2}$

these have at least one return (or visit) to 0. In the case of Dyck paths, these correspond to paths avoiding the adsorbing boundary $\partial\mathbb{J}^2_+$, and paths with at least one return to $\partial\mathbb{J}^2_+$, respectively.

The generating function of paths with at least one visit may be recovered by introducing the generating variable t conjugate to length. The above simplifies to

$$C(t,a;j_0,j) = \frac{(2t)^{j+j_0}}{(1+\sqrt{1-4t^2})^{j+j_0}} \left(\frac{2a}{2-a(1-\sqrt{1-4t^2})}\right). \tag{5.34}$$

Putting $j_0 = j = 0$ and dividing by a gives equation (5.13).

Similarly, putting $j_0 = 0$, dividing by a and summing over j gives adsorbing ballot paths from $\vec{0}$ in $\mathbb{J}^2_+$ with endpoints at arbitrary heights. This generating function is

$$D_B(a,t) = \frac{(1-2t+\sqrt{1-4t^2})}{(1-2t)(2-a(1-\sqrt{1-4t^2}))}. \tag{5.35}$$

Put $a = 1$ to find $D_B(t)$ in equation (5.8). Similarly, putting $a = 0$ gives the generating function of tails $D_T(t)$ in equation (5.8).

The generating function $D_B(a,t)$ may be written in the form

$$D_B(a,t) = \frac{4t}{(\sqrt{1-4t^2}+2t-1)(2-a(1-\sqrt{1-4t^2}))}. \tag{5.36}$$

The radius of convergence is either $t_c = \frac{1}{2}$ if $a \leq 2$ or $t_c = \frac{1}{a}\sqrt{a-1}$ if $a > 2$ (see equation (5.15)). The critical point is located at $(a_c, t_c) = (2, \frac{1}{2})$, and scaling axes may be set up at this point. The scaling is similar to that of Dyck paths, and the scaling fields are again $\tau = 2 - a$, and $g = 1 - 4t^2$.

Along the λ-isotherm the generating function has a simple pole so that $\gamma_+ = 1$ in equation (4.19), and $\nu_+ = \frac{1}{2}$ by equation (4.20). On approach to the τ_0-curve the generating function has a square root singularity since

$$D_B(a,t) = \left(1 + \sqrt{\frac{1+2t}{1-2t}}\right) \frac{(2-a-a\sqrt{1-4t^2})}{4(1-a+a^2t^2)}. \tag{5.37}$$

This shows that $\gamma_- = \frac{1}{2}$ in equation (4.21).

Since the critical curve is the same as for adsorbing Dyck paths (see equation (5.15)), $\psi_\lambda = 2$, and $\psi_\tau = 0$, so the crossover exponent is $\phi = \frac{1}{2}$, and $\alpha = 0$, by equations (4.17) and (4.46). Moreover, from equation (5.35),

$$D_B(a,t) = \frac{1}{\tau^2} \frac{4t\,(1+2t)}{(1+2t+\sqrt{g})\,\tau^{-1}\sqrt{g}(1-a\tau^{-1}\sqrt{g})} \sim \tau^{-2} F(g^{-1/2}\tau) \tag{5.38}$$

for some scaling function $F(x)$. This shows that $\gamma_u = 2$, and $\gamma_t = 1$, by equations (4.26) and (4.23). Hence, $y_t = \frac{2}{\gamma_u} = 1$, and $\nu_t = \frac{\gamma_t}{2} = \frac{1}{2}$; therefore $\phi = y_t \nu_t$ by equation (4.57). Exponents for ballot paths are listed in table (5.2).

5.1.4 *The Temperley method and ballot paths*

The Temperley method [544] can be used to find recurrences by constructing directed clusters a slice (or a column) at a time. In the case of ballot paths of even length the method is implemented as follows.

Let $C_e(a, y, t)$ be the generating function of adsorbing ballot paths of even length, with generating variable t conjugate to length n, with a conjugate to the number of visits v and with y conjugate to h, the height of the end-vertex of the path. Putting $y = e^{f/T}$, where T is temperature, introduces a *vertical pulling force* f in the model. For large f (or small T), the path is pulled from the adsorbing surface into a *ballistic phase*. In other words, in this ensemble the paths can be adsorbed, desorbed or ballistic (and thus desorbed). This is a model of *pulled adsorbed ballot paths*.

Note that $C_e(1, 1, t)$ is the generating function of even length ballot paths (compare this to $D_B(t)$ in equation (5.8)). Also, $C_e(a, 0, t) = D(a, t)$, the generating function of adsorbing Dyck paths given in equation (5.13).

The Temperley method is implemented by extending a ballot path by two steps to find a recurrence for $C_e(a, y, t)$, as shown in figure 5.5. Let $s \equiv y^2$ and append two north-east steps to paths in $C_e(a, y, t)$; then a factor of $t^2 s$ should multiply the generating function.

To account for a step in the north-east direction followed by a step in the south-east direction, or for a step in the south-east direction followed by a step in the north-east direction, the factor $(t^2 + t^2)$ should multiply the generating function.

Finally, two steps may be appended in the south-east direction. This multiplies the generating function by the factor $t^2 \overline{s}$, where $\overline{s} = \frac{1}{s}$.

Taking these cases together shows that $(t^2 s + 2t^2 + t^2 \overline{s})$ generates two additional steps.

The function $C_e(a, y, t)$ generates an empty path (a single vertex at $\vec{0}$) or generates a path which was obtained by adding two steps to paths already in $C_e(a, y, t)$. This gives the recurrence

$$C_e(a, y, t) = 1 + t^2 (s + 2 + \overline{s})\, C_e(a, y, t) + [\text{corrections for new visits}].$$

It remains to account for new visits and to subtract paths stepping outside $\mathbb{J}^2_+$.

Next, subtract and add back (with weight a) those paths which step first north-east and then south-east from height 0. This gives the term $t^2 (a - 1)\, C_e(a, 0, t)$.

A new visit may also be generated when two south-east edges are attached to a path which ends in a vertex at height 2; the number of such paths is $\frac{\partial}{\partial s} C_e \mid_{s=0}$. Thus, the term $t^2 (a-1) \frac{\partial}{\partial s} C_e \mid_{s=0}$ subtracts out paths stepping into the adsorbing line and then adds them back with weight a.

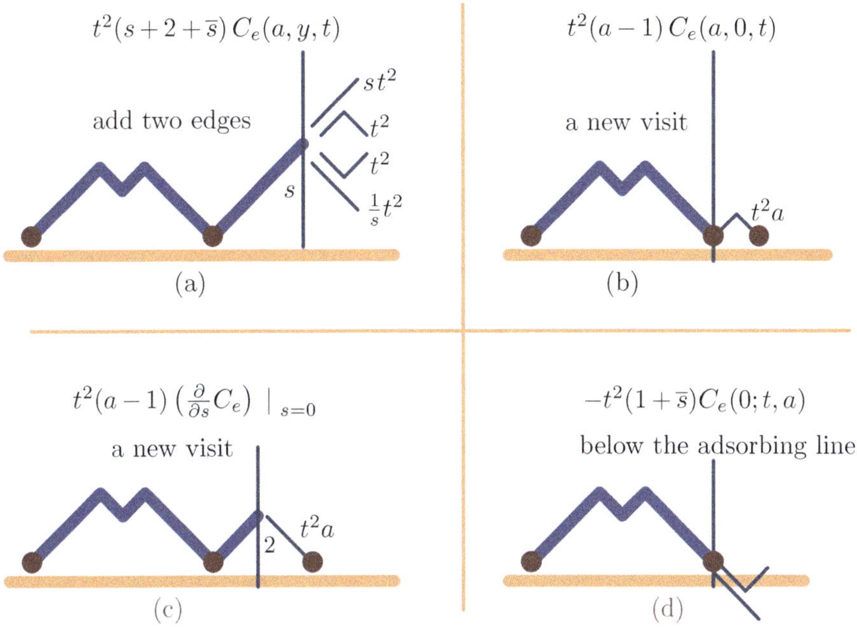

FIG. 5.5. The generating function $C_e(a,y,t)$ of even length adsorbing ballot paths can be determined by using the Temperley method to add two more steps to paths as shown above. (a) The path is extended by adding two more steps in four ways; this is generated by $t^2(s+2+\overline{s})\,C_e(s;t,a)$. New visits are created in cases (b) and (c). Since $C_e(a,0,y)$ are directed paths with the final vertex at height 0, the case in (b) creates a new visit; this corresponds to the term $t^2(a-1)\,C_e(0;t,a)$ (subtract out paths and add them back with the right weight). Similarly for case (c), subtract and add back paths with weight a if they end at height 0. In case (d), paths stepping below the adsorbing line are subtracted out.

Finally, subtract paths stepping below the adsorbing line by adding the term $-t^2(1+\overline{s})\,C_e(a,0,t)$; see figure 5.5 for a full explanation.

This gives the mixed differential and functional recursion

$$C_e(a,y,t) = 1 + t^2(s+2+\tfrac{1}{s})\,C_e(a,y,t) + t^2(a-2-\tfrac{1}{s})\,C_e(a,0,t) + t^2(a-1)\,\tfrac{\partial}{\partial s}C_e\,\Big|_{s=0}. \tag{5.39}$$

It remains to solve this recurrence for $C_e(a,y,t)$. Observe that, if $C_e(a,0,t)$ and $\frac{\partial}{\partial s}C_e\,\Big|_{s=0}$ are known, then $C_e(a,y,t)$ is also known.

In addition, $\lim_{s\to 0^+}\frac{1}{s}\,(C_e(a,y,t)-C_e(a,0,t)) = \frac{\partial}{\partial s}C_e\,|_{s=0}$ (where $s \equiv y^2$). Thus, taking $s \to 0^+$ in equation (5.39) gives

$$C_e(a,0,t) = 1 + at^2\, C_e(a,0,t) + at^2 \tfrac{\partial}{\partial s} C_e \,\Big|_{\,s=0}. \qquad (5.40)$$

Solve for $\frac{\partial}{\partial s} C_e \,\Big|_{\,s=0}$ in terms of $C_e(a,0,t)$ and substitute this into equation (5.39):

$$a\left(1 - t^2(s+2+\tfrac{1}{s})\right) C_e(a,y,t) = 1 + \left(a - 1 - at^2(1+\tfrac{1}{s})\right) C_e(a,0,t). \qquad (5.41)$$

The generating function $C_e(a,0,t)$ is of even length paths with endpoints at height 0; these are adsorbing Dyck paths, so $C_e(a,0,t) = D(a,t)$ in equation (5.13). Hence, $C_e(a,y,t)$ can be determined from equation (5.41).

If $D(a,t)$ is not known, then equation (5.41) can be solved by the *kernel method* [19, 56, 186, 423, 424]. This is implemented as follows. The coefficient of $C_e(a,y,t)$ on the left-hand side of equation (5.41) is the *kernel*, given by

$$K(s) = a\left(1 - t^2(s+2+\overline{s})\right), \qquad (5.42)$$

where s is given the status of being the *catalytic variable*. If $s_\pm$ is a root of $K(s)$, then the left-hand side of equation (5.41) is 0, and $C_e(a,0,t)$ may be found. The roots of $K(s)$ are

$$s_\pm = \frac{1 \pm \sqrt{1-4t^2}}{2t^2} - 1. \qquad (5.43)$$

Putting $s = s_-$, or putting $s = s_+$, *kills* the kernel $K(s)$, in which case it follows from equation (5.41) that

$$C_e(a,0,t) = \frac{s_\pm}{at^2(1+s_\pm) + s_\pm(1-a)}. \qquad (5.44)$$

Substituting in $s = s_-$ and simplifying gives a generating function which is a power series with positive integer coefficients. The substitution $s = s_+$ gives an 'unphysical' result (the series expansion of $C_e(a,0,t)$ will have negative or non-integer coefficients). Thus, the correct choice is $s = s_-$, and this recovers the generating function $D(a,t)$ in equation (5.13).

The full solution for $C_e(a,y,t)$ is obtained from equation (5.41) by replacing s by y^2:

$$C_e(a,y,t) = \frac{2t^2 - y^2\left(1 - 2t^2 + \sqrt{1-4t^2}\right)}{\left(t^2(y^2+1)^2 - y^2\right)\left(2 - a(1-\sqrt{1-4t^2})\right)}. \qquad (5.45)$$

The generating function $C_o(a,y,t)$ for paths of odd length can be determined from $C_e(a,y,t)$ by appending a single step. This gives the relation

$$C_o(a,y,t) = t\,(y+\overline{y})\; C_e(a,y,t) - t\overline{y}\, C_e(a,0,t), \qquad (5.46)$$

where $\overline{y} = \frac{1}{y}$. Adding $C_o(a,y,t)$ to $C_e(a,y,t)$ gives the full generating function:

$$C(a,y,t) = \frac{2\left(1 - 2t^2 + \sqrt{1-4t^2}\right)}{\left(1 - 2at^2 + \sqrt{1-4t^2}\right)\left(1 - 2yt + \sqrt{1-4t^2}\right)}. \qquad (5.47)$$

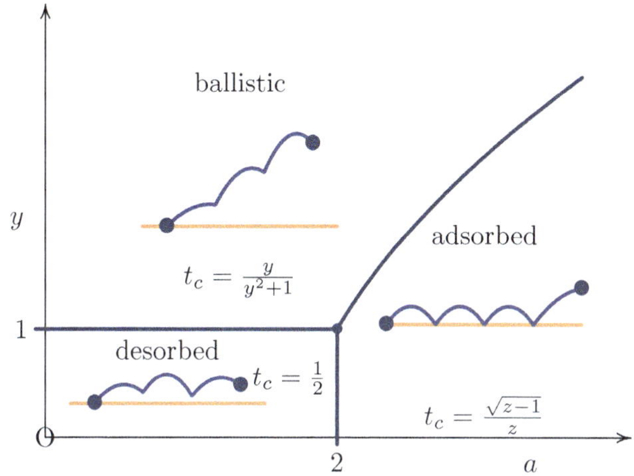

FIG. 5.6. The phase diagram of pulled adsorbing ballot paths. There are three phases: (1) a *desorbed* phase, when y and a are small; (2) a *ballistic* phase, if y is large; and (3) an *adsorbed* phase, if a is large. These three phases meet along three critical curves. The phases also meet at a critical point located at $y = 1$, and $a = 2$, in the phase diagram.

Putting $y = 1$ in this expression recovers $D_B(a, t)$ in equation (5.35), and putting $a = 1$ recovers $D_B(t)$ in equation (5.8). Putting $a = 1$ in equation (5.47) gives

$$C(1, y, t) = \frac{2}{1 - 2yt + \sqrt{1 - 4t^2}}, \tag{5.48}$$

the generating function of *pulled directed paths*.

Singularities in equation (5.47) give the radius of convergence of $C(a, y, t)$:

$$t_c(a, y) = \min\left\{\tfrac{1}{2}, \tfrac{y}{y^2+1}, \tfrac{\sqrt{a-1}}{a}\right\}. \tag{5.49}$$

This evaluates to

$$t_c(a, y) = \begin{cases} \frac{1}{2}, & \text{if } y \leq 1, \text{ and } a \leq 2; \\ \frac{y}{y^2+1}, & \text{if } y^2 \geq \max\{1, a - 1\}; \\ \frac{\sqrt{a-1}}{a}, & \text{if } a \geq 2, \text{ and } y^2 \leq a - 1. \end{cases} \tag{5.50}$$

This gives three phases in the model, and the phase diagram is illustrated in figure 5.6. The free energy is given by $\mathcal{D}(a, y) = -\log t_c(a, y)$.

If both y and a are small, then $t_c(a, y) = \frac{1}{2}$, and the path is in a *desorbed* phase (the free energy is not a function of either a or y). If $y^2 \geq \max\{1, a - 1\}$, then the path is in a *ballistic* phase, and the free energy is only a function of y. Finally, if $a > 2$, and $y^2 \leq a - 1$, then the path is in an *adsorbed* phase (and the free energy is only a function of a).

TABLE 5.3. Pulled ballot path tricritical exponents

ϕ	α	γ_t	γ_u	y_t	ν_t	γ_+	ν_+	γ_-
$\frac{1}{2}$	0	$\frac{1}{2}$	1	2	$\frac{1}{4}$	1	$\frac{1}{2}$	$-\frac{1}{2}$

5.1.5 *Scaling of the generating function of pulled ballot paths*

The scaling of the generating function of pulled paths, $C(1,y,t)$ in equation (5.48), is uncovered by choosing scaling axes $\tau = 1-y$, and $g = 1-4t^2$. The τ_0-axis is $t_c(y) = \frac{1}{2}$ for $y < 1$, and the λ-isotherm is $t_c(y) = \frac{y}{y^2+1}$ for $t > 1$.

Along the λ-isotherm the generating function $C(1,y,t)$ has a simple pole, so $\gamma_+ = 1$ in equation (4.19). This shows that $\nu_+ = \frac{1}{2}$ by equation (4.20).

On approach to the τ_0-curve the generating function has a square root singularity as $g \to 0^+$ (or $t \to \frac{1}{2}$). This shows that $\gamma_- = -\frac{1}{2}$ in equation (4.21).

Along the λ-isotherm (for $y > 1$) the critical curve has the shape given by

$$g_\lambda(\tau) = \tfrac{1}{2} - \tfrac{y}{y^2+1} = \tfrac{\tau^2}{2(1+(1+\tau)^2)} \approx \tfrac{\tau^2}{4(1+\tau)} \quad \text{if } \tau < 0 \text{ is small.}$$

Comparison to equation (4.8) shows that the shift-exponent is $\psi_\lambda = 2$. Since $g = 0$ for $y < 1$, it follows that $\psi_\tau = 0$. Thus, the crossover exponent is $\phi = \frac{1}{2}$ by equation (4.17), and the specific heat exponent is $\alpha = 0$ by equation (4.46).

Putting $y = 1$ (so that $\tau = 0$) gives $C(1,1,t) = D_B(t)$ (see equation (5.8)) if $t < \frac{1}{2}$ and so $g > 0$. This shows that $\gamma_t = \frac{1}{2}$ in equation (4.23).

Next consider the factor $(1 - 2yt + \sqrt{1-4t^2})$ in equation (5.48). Expanding to leading order in the scaling fields gives $2t\tau + \sqrt{g} + \cdots$. On approach to the critical point,

$$C(1,y,t) \sim \frac{1}{\tau(2t + \tau^{-1}\sqrt{g})} + O(1) \sim \tau^{-1} F(g^{-\frac{1}{2}}\tau) \tag{5.51}$$

for some scaling function F. By equation (4.26), the identification $\gamma_u = 1$ is made. By equation (4.57), $y_t = \frac{2}{\gamma_u} = 2$, and $\nu_t = \frac{\gamma_t}{2} = \frac{1}{4}$; therefore, $\phi = y_t \nu_t$ (see table 5.3).

Tricritical scaling at the critical adsorption point $a = 2$ is scaling at the *special (adsorption) point.* The generating function is

$$C(2,y,t) = \frac{1 + \sqrt{1-4t^2}}{\sqrt{1-4t^2}\,(1 - 2yt + \sqrt{1-4t^2})}. \tag{5.52}$$

The critical curve is given as above. There is a simple pole along the λ-isotherm, so $\gamma_+ = 1$, and $\nu_+ = \frac{1}{2}$. Along the τ_0-curve there is a square root divergence, so $\gamma_- = \frac{1}{2}$.

With the same scaling axes as before, the tricritical exponents can be extracted and are given in table 5.4. As above, $\psi_\lambda = 2$, and $\phi = \frac{1}{2}$. Putting $y = 1$

TABLE 5.4. Pulled ballot paths special point exponents

ϕ	α	γ_t	γ_u	y_t	ν_t	γ_+	ν_+	γ_-
$\frac{1}{2}$	0	1	2	1	$\frac{1}{2}$	1	$\frac{1}{2}$	$\frac{1}{2}$

gives $D_B(2,t)$ in equation (5.36) so that $C(2,1,t) \sim \frac{1}{g}$ with exponent $\gamma_t = 1$. Expanding in the scaling fields at the special point gives

$$C(2,y,t) \sim \frac{1}{\tau^2} \frac{1+\sqrt{g}}{\tau^{-1}\sqrt{g}\,(2t+\tau^{-1}\sqrt{g}+\cdots)} \sim \tau^{-2}\, G(g^{-\frac{1}{2}}\tau) \tag{5.53}$$

for some scaling function G. This shows that $\gamma_u = 2$. Hence, $y_t = \frac{2}{\gamma_u} = 1$, and $\nu_t = \frac{\gamma_t}{2} = \frac{1}{2}$. Then $\phi = y_t \nu_t = \frac{1}{2}$, as expected (see table 5.4).

5.1.6 *The partition function of adsorbing Dyck paths*

Put $j = j_0 = 0$ in equation (5.33) to find the Dyck path partition function:

$$D_n(a) = \sum_{m=0}^{\lfloor n/2 \rfloor} \frac{4m+2}{n+2(m+1)} \binom{n}{\lfloor n/2 \rfloor + m} (a-1)^m. \tag{5.54}$$

Put $j = 0$ and sum over $j_0 \geq 0$ instead to obtain the adsorbing ballot path partition function

$$T_n(a) = \sum_{m=0}^{\lfloor n/2 \rfloor} \binom{n}{\lceil n/2 \rceil + m} (a-1)^m. \tag{5.55}$$

Scaling of $D_n(a)$ and $T_n(a)$ follows from their asymptotic behaviour.

5.1.6.1 *Case 1: $1 < a < 2$.* The asymptotics of the partition functions in equations (5.54) and (5.55) are determined by approximating the sums by integrals and the binomial coefficients by gamma functions. For asymptotic values of n, these integrals can be estimated using a saddle point method. Numerical work show that the saddle points are located close to $m = 0$ and have a width proportional to $\sqrt{n}$.

In the summands above, substitute $n = \frac{1}{\epsilon^2}$, and $m = \frac{\alpha}{\epsilon}$ (so that $m = O(\sqrt{n})$). Expand the results in ϵ and simplify. In the case of $T_n(a)$, this results in

$$T_n(a) \simeq \frac{2^{3+1/\epsilon^2}\epsilon}{\sqrt{2\pi}} \int_0^\infty (a-1)^{\alpha/\epsilon}\, e^{-2\alpha^2}\, \alpha\, d\alpha. \tag{5.56}$$

Evaluation gives $T_n(a) \simeq 2^{n-1} e^{n \log^2(a-1)/8} \left(1 + \mathrm{erf}(\frac{1}{4}\sqrt{2n}\,|\log(a-1)|)\right)$ after the substitution $\epsilon^2 = \frac{1}{n}$. Expanding this asymptotically in n and keeping only the leading order term yields

$$T_n(a) \simeq \frac{2^{n+1}}{\sqrt{\pi n}\,|\log(a-1)|}. \tag{5.57}$$

The same procedure may be applied to $D_n(a)$. It follows that

$$D_n(a) \simeq \frac{2^{1+1/\epsilon^2}}{\sqrt{2\pi}} \int_0^\infty (a-1)^{\alpha/\epsilon}\, e^{-2\alpha^2}\, d\alpha, \tag{5.58}$$

which after evaluation, expanding asymptotically in n and keeping leading order terms gives

$$D_n(a) \simeq \frac{2^{n+3}}{\sqrt{2\pi n^3}\,\log^2(a-1)}. \tag{5.59}$$

5.1.6.2 *Case 2: $a = 2$.* The partition functions evaluate exactly in terms of factorials and gamma functions if $a = 2$:

$$T_n(2) = 2^{n-1}\left(\frac{\Gamma\left(\frac{n+1}{2}\right)}{\sqrt{\pi}\,\Gamma\left(\frac{n+2}{2}\right)} + 1\right), \text{ and } D_n(2) = \frac{n!}{(n/2)!(n/2)!}. \tag{5.60}$$

Asymptotic formulae follow by equations (D.13) or (D.14) in appendix D (for example, $D_n(2) \simeq \frac{2^{n+1}}{\sqrt{2\pi n}}$).

5.1.6.3 *Case 3: $a > 2$.* Asymptotic formulae are obtained by exploring the saddle point (located at δn in the asymptotic regime with $\delta \simeq \frac{1}{2a}(a-2)$) in the summands of $D_n(a)$ and $T_n(a)$). Numerical work shows the width of the saddle point to be proportional to $\sqrt{n}$. Use Stirling's approximation in the summands, put $m = \frac{1}{2a}(a-2)n + \alpha\sqrt{n}$, and $n = \frac{1}{\epsilon^2}$, expand in ϵ to $O(1)$ and simplify. This gives

$$T_n(a) \simeq \frac{\epsilon\, n}{\sqrt{2\pi n}} \left(\frac{a}{\sqrt{a-1}}\right)^{(\epsilon^2+1)/\epsilon^2} \int_{-\infty}^\infty e^{-\frac{\alpha^2 a^2}{2(a-1)}}\, d\alpha, \tag{5.61}$$

and

$$D_n(a) \simeq \frac{\epsilon\, n}{\sqrt{2\pi n}} \left(\frac{a-2}{a-1}\right) \left(\frac{a}{\sqrt{a-1}}\right)^{(\epsilon^2+1)/\epsilon^2} \int_{-\infty}^\infty e^{-\frac{\alpha^2 a^2}{2(a-1)}}\, d\alpha. \tag{5.62}$$

Integrating α over $(-\infty, \infty)$ gives the saddle point approximations

$$T_n(a) \simeq \left(\frac{a}{\sqrt{a-1}}\right)^n, \text{ and } D_n(a) \simeq \frac{a-2}{a-1}\left(\frac{a}{\sqrt{a-1}}\right)^n. \tag{5.63}$$

Better approximations are found by keeping more terms in the expansions above. Expanding to $O(\epsilon)$ gives the approximations

$$T_n(a) \simeq \frac{\epsilon\sqrt{n}}{\sqrt{2\pi}} \left(\frac{a}{\sqrt{a-1}}\right)^{(\epsilon^2+1)/\epsilon^2} \int_{-\infty}^\infty e^{-\frac{\alpha^2 a^2}{2(a-1)} - A_T\epsilon}\, d\alpha, \tag{5.64}$$

where

$$A_T = \frac{\left(\alpha^2 a^4 - (4\alpha^2-3)a^3 + (4\alpha^2+9)a^2 - 18\,a + 12\right)\alpha a}{6\,(a-1)^2(a-2)},$$

and

$$D_n(a) \simeq \frac{\epsilon\sqrt{n}}{\sqrt{2\pi}}\left(\frac{a-2}{a-1}\right)\left(\frac{a}{\sqrt{a-1}}\right)^{(\epsilon^2+1)/\epsilon^2}\int_{-\infty}^{\infty} e^{-\frac{\alpha^2 a^2}{2(a-1)} - A_D\epsilon}\,d\alpha, \tag{5.65}$$

where

$$A_D = \frac{\left(\alpha^2 a^3 - (2\alpha^2+3)a^2 + 9\,a - 6\right)\alpha a}{6\,(a-1)^2}.$$

Expand the exponential $e^{-A_T\epsilon} = 1 - A_T\epsilon + \frac{1}{2}A_T^2\epsilon^2 + O(\epsilon^3)$, and similarly expand $e^{-A_D\epsilon} = 1 - A_D\epsilon + \frac{1}{2}A_D^2\epsilon^2 + O(\epsilon^3)$ in the above and integrate to α. This gives improved asymptotic expressions for $T_n(a)$ and $D_n(a)$ if a is not too large. Simplification gives

$$T_n(a) \simeq \frac{f_1(a)}{12n\,(a-1)}\left(\frac{a}{\sqrt{a-1}}\right)^n, \tag{5.66}$$

where $f_1(a) = a^2 + 4(3n-1)a - 4(3n-1)$, and

$$D_n(a) \simeq \frac{f_2(a)}{12n\,(a-1)^2(a-2)}\left(\frac{a}{\sqrt{a-1}}\right)^n, \tag{5.67}$$

where $f_2(a) = a^4 + 4(3n-2)a^3 - 30(2n-1)a^2 + 16(2a-1)(3n-1)$ (see for example reference [317]; and equation (5.17) for large n).

5.1.7 *The partition function of adsorbing directed paths*

Define the partition function of adsorbing directed paths from equation (5.33) by

$$D_n(\alpha,\beta,a) = \binom{n}{\frac{n-\lfloor\alpha n\rfloor+\lfloor\beta n\rfloor}{2}} - \binom{n}{\frac{n+\lfloor\alpha n\rfloor+\lfloor\beta n\rfloor}{2}} \tag{5.68}$$

$$+\, a\sum_{\ell=0}^{(n-\lfloor\alpha n\rfloor-\lfloor\beta n\rfloor)/2}\left(\binom{n}{\frac{n+\lfloor\alpha n\rfloor+\lfloor\beta n\rfloor}{2}+\ell} - \binom{n}{\frac{n+\lfloor\alpha n\rfloor+\lfloor\beta n\rfloor}{2}+\ell+1}\right)(a-1)^\ell.$$

This is the partition function of directed paths of length n with endpoints at heights $\lfloor\alpha n\rfloor$ and $\lfloor\beta n\rfloor$ [303, 304].

The asymptotic behaviour of the first two terms can be determined using, for example, the expressions in equation (D.14) in appendix D. This leaves the summation, which may be approximated by using a saddle point method as

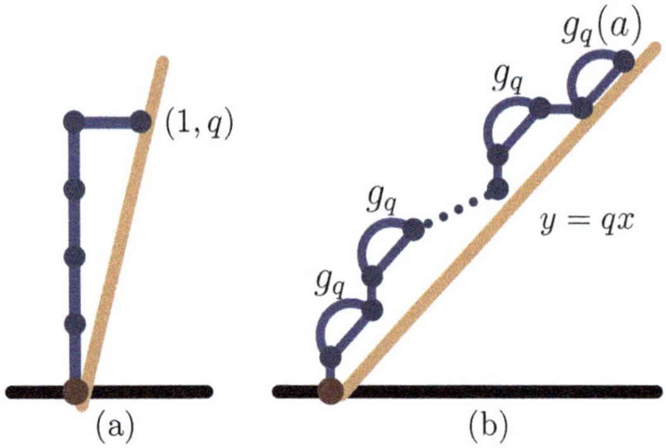

FIG. 5.7. Directed paths above the line $y = rx$ in $\mathbb{L}^2_+$. (a) The shortest path in the case that $r = q \in \mathbb{Q}$. (b) The factorisation of a directed path above the line $Y = qX$.

---

above. Putting $A = \alpha + \beta$ and simplifying the result finally gives the following asymptotic expression for the partition function:

$$D_n(\alpha,\beta,a) \simeq \binom{n}{\frac{n-\lfloor \alpha n\rfloor+\lfloor \beta n\rfloor}{2}} - \binom{n}{\frac{n+\lfloor \alpha n\rfloor+\lfloor \beta n\rfloor}{2}}$$
$$+ \begin{cases} \frac{2^n aA}{\sqrt{8\pi n}} \left[\log\left(\frac{(a-1)(1-A)}{1+A}\right)\right]^{-1} \left(\frac{(1-A)^{A-1}}{(1+A)^{A+1}}\right)^{n/2}, & \text{if } 1 < a < a_c; \\ \frac{2^n A}{2(1-A^2)} \left(\frac{(1-A)^{A-1}}{(1+A)^{A+1}}\right)^{n/2}, & \text{if } a = a_c; \\ \frac{a(a-2)}{a-1}\left(\frac{a^2}{(a-1)^{A+1}}\right)^{n/2}, & \text{if } a > a_c. \end{cases} \quad (5.69)$$

The adsorption critical point is $a_c = \frac{2}{1-\alpha-\beta}$.

5.2 Directed paths above the line $y = rx$

Let $\mathbb{L}^2_+$ be the two-dimensional half-lattice (see equation (2.20)) and let the Cartesian coordinates of $\vec{v} \in \mathbb{L}^2_+$ be $(\vec{v}(1), \vec{v}(2))$. Let $r > 0$ and define the set $S_r = \{\vec{v} \in \mathbb{L}^2_+ \mid \vec{v}(2) \geq r\vec{v}(1) \geq 0\}$. Define the wedge

$$W_r = \{\langle \vec{v} \sim \vec{w}\rangle \in \mathbb{L}^2_+ \mid \vec{v}, \vec{w} \in S_r\}. \quad (5.70)$$

Then W_r is the wedge in $\mathbb{L}^2_+$ formed by the y-axis and the line $y = rx$ in the Cartesian plane. The line $y = rx$ will be called the *adsorbing wall.*

Suppose that $r \in \mathbb{Q}$ and let ω be a fully directed path giving north or east steps from $\vec{0} \in W_r$, constrained to step only on vertices in W_r. Assume that ω has endpoint in the line $y = rx$. Denote the generating function of these paths from $\vec{0}$ in W_r to a vertex in the line $y = rx$ by $g_r(a,t)$ (where a is the generating variable of returns or visits to the line $y = rx$). Define $g_r(t) \equiv g_r(1,t)$.

If $r = q \in \mathbb{N}$, then the shortest non-empty path from $\vec{0} \in W_q$ to a vertex in $y = qx$ is illustrated in figure 5.7(a). This path starts from $\vec{0}$ and has the endpoint $(1, q)$. A functional recurrence can be found for the generating function $g_q(a,t)$ by using a factorisation method. That is, each such path is either empty, or it visits the lines $y = qx + j$ one last time for $j = 1, 2, 3, \ldots, q$ before it returns to the line $y = qx$ for the first time to renew as a path above $y = qx$. This is illustrated in figure 5.7(b) and it shows that the generating function satisfies

$$g_q(a,t) = 1 + a t^{q+1} (g_q(t))^q g_q(a,t). \tag{5.71}$$

If $q = 1$, then the generating function of adsorbing Dyck paths is recovered; see equation (5.12).

This approach generalises to the case where $r = \frac{q}{p} \in \mathbb{Q}$ (and where (q,p) are relative primes). Let $g_{q/p}(a,t)$ be the generating function.

Let $e_{q/p}$ be the number of paths of length $p+q$ from $\vec{0}$ to $(p,q) \in W_{q/p}$. Then the generating function satisfies the recurrence

$$g_{q/p}(a,t) = 1 + e_{q/p} a t^{p+q} \left(g_{q/p}(t)\right)^{q+p-1} g_{q/p}(a,t). \tag{5.72}$$

This shows that $g_{q/p}(a,t)$ is an algebraic function, as $g_{q/p}(t) = g_{q/p}(1,t)$ is the root of a polynomial of degree $q+p$.

If $a = 1$, then the recurrence reduces to

$$g_{q/p}(t) - 1 = e_{q/p} t^{p+q} \left(g_{q/p}(t)\right)^{q+p}. \tag{5.73}$$

Denote the radius of convergence of $g_{q/p}(t)$ by $t_{q/p}$ and put $g^* = g_{q/p}(t_{q/p})$.

The right-hand side of equation (5.73) is a curve if it is plotted against $g = g_{q/p}(t)$. In fact, the function $y = e_{q/p}\,(tg)^{p+q}$ is a convex curve if plotted against g. The tangent to this curve passing through the point $(t_{q/p}, g^*)$ is

$$(p+q)\, e_{q/p} t_{q/p}^{p+q} \,(g^*)^{p+q-1}\, g - (p+q-1)\, e_{q/p} t_{q/p}^{p+q}\, (g^*)^{p+q}. \tag{5.74}$$

Comparison to the left-hand side of equation (5.73) shows that this tangent is equal to $g - 1$ if both

$$(p+q)\, e_{q/p} t_{q/p}^{p+q} \,(g^*)^{p+q-1} = 1, \quad \text{and } (p+q-1)\, e_{q/p} t_{q/p}^{p+q}\, (g^*)^{p+q} = 1.$$

Solve these equations simultaneously for $t_{q/p}$ and g^* to obtain

$$g_{q/p}(t_{q/p}) = g^* = \frac{p+q}{p+q-1}, \quad \text{and } t_{q/p} = e_{q/p}^{-1/(p+q)}\ \frac{(p+q-1)^{1-1/(p+q)}}{p+q}.$$

Substituting these results into equation (5.72) shows that the critical adsorption point in this model is

$$a_{q/p} = p + q. \tag{5.75}$$

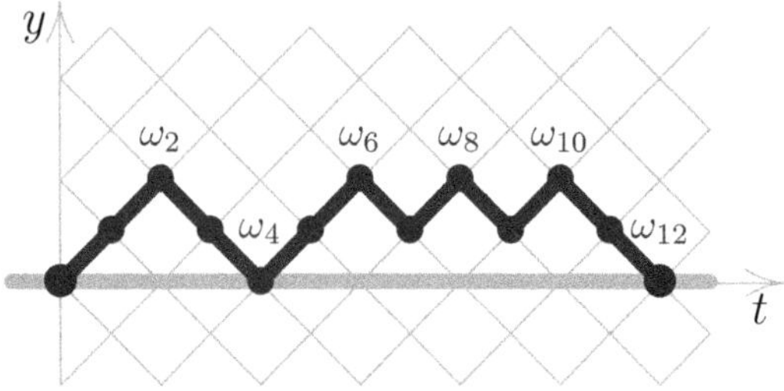

FIG. 5.8. A Dyck path model of an adsorbing linear copolymer. Even labelled vertices may adsorb and represent comonomers of different types. This is modelled by colouring the vertices of the Dyck path such that different comonomers are represented with different colours.

For example, if $p = q = 1$, then the model of adsorbing Dyck paths in section 5.1.2 is recovered. Taking $\frac{q}{p} \to r \in \mathbb{I}$ (the irrational numbers) gives the results [297]

$$g_r(t_r) = 1, \quad \text{and } t_r = \frac{r^{t/(1+r)}}{1+r}, \quad \text{and } a_r = \infty. \tag{5.76}$$

That is, the path cannot adsorb (since $a_{q/p} \to a_r = \infty$).

5.3 Dyck path models of adsorbing copolymers

A polymer with two or more different types of monomers along its backbone is a copolymer. If the distribution of comonomers along the polymer backbone is random, then the polymer is a *random copolymer*. If comonomers along the polymer chain are fixed in position, then the copolymer is *quenched*.

A Dyck path model (in $\mathbb{J}^2_+$) of an adsorbing linear copolymer is obtained by colouring the vertices of the Dyck path with colours $\chi \in \Omega$ (where Ω is a set of colours). The interaction of the path with the adsorbing line $\partial\mathbb{J}^2_+$ depends on the colouring, since visits of different colours are given weights in the partition function depending on their colour [286, 319, 407, 488, 528, 582].

This model is constructed as follows. Let $\omega = \langle \omega_0, \omega_1, \omega_2, \ldots, \omega_n \rangle$ be a Dyck path with *even vertices* ω_{2j}. Assign colour $\chi_j \in \Omega$ to vertex ω_{2j} for $j = 1, 2, \ldots$ (see figure 5.8; by convention, the vertex labelled 0 does not get a colour). This gives a colouring $\chi = \langle \chi_1, \chi_2, \ldots \rangle$ of ω.

The colouring χ is periodic if it is a periodic sequence. If χ is of the form $\langle \chi_1^{n_1}, \chi_2^{m_2} \rangle^*$ where the $*$ indicates that the sequence repeats an arbitrary number of times, then it is a *diblock colouring* of two colours: χ_1 repeats in a block n_1 times, followed by χ_2 in a block n_2 times, and then the entire sequence repeats periodically. This is also an *alternating block colouring*. More generally, the colouring $\left\langle \chi_1^{n_1}, \chi_2^{m_2}, \ldots, \chi_p^{m_p} \right\rangle^*$ is a *block colouring* of period $n_1 + n_2 + \cdots + n_p$. Block colourings (with a finite number of colours) are necessarily periodic. An *alternating colouring* is a periodic diblock colouring with two colours and period 2.

5.3.1 *Coloured Dyck paths in the annealed ensemble*

A random colouring of a Dyck path with two colours may be generated by independently assigning the even vertices one of two colours: 0 or 1. Colour 0 is assigned to a vertex with probability p, and colour 1 is assigned with probability $q = 1-p$. In this case the colours along the even vertices of the path are binomially and independently distributed. In the *annealed ensemble* the partition function is averaged over the set of all possible colourings.

Suppose that there are $d_n(v)$ Dyck paths of v visits and that $w \leq v$. Then the generating function in the annealed ensemble is

$$D_a(a,b,t;p) = \sum_{n=0}^{\infty} \sum_{v=0}^{n/2} \sum_{w=0}^{v} \binom{v}{w} d_n(v)\,(pa)^w (qb)^{v-w} t^n, \tag{5.77}$$

where vertices of type 0 are weighted by a, and vertices of type 1 are weighted by b. This evaluates to

$$D_a(a,b,t;p) = \sum_{n=0}^{\infty} \sum_{v=0}^{n/2} d_n(v)\,(pa+qb)^v t^n = \frac{2(pa+qb)}{2-(pa+qb)(1-\sqrt{1-4t^2})}$$

by equation (5.13). The radius of convergence is given by

$$t_c(z;p) = \begin{cases} \frac{1}{2}, & \text{if } pa+qb \leq 2; \\ \frac{\sqrt{pa+qb-1}}{pa+qb}, & \text{if } pa+qb > 2. \end{cases} \tag{5.78}$$

The critical point is $pa+qb=2$. This solves the model completely.

5.3.2 *Quenched models of Dyck path copolymer adsorption*

Suppose that $\Omega = \{0,1\}$ and that χ_i is selected independently and uniformly from Ω (in general, more than two colours may be used).

Assign colour χ_j to the vertices ω_{2j} and *fix* the colouring to obtain a *quenched* colouring of the path by $\chi = \langle \chi_1, \chi_2, \ldots \rangle$, where $\chi_j \in \Omega$.

Let $d_n(v, v_a \mid \chi)$ be the number of Dyck paths of length n coloured by χ and with v visits of which v_a visits are 0-coloured. Since the number of visits with colour 1 is $v - v_a$, the partition function is given by

$$D_n(a,b \mid \chi) = \sum_{v,v_a} d_n(v, v_a \mid \chi)\, a^{v_a}\, b^{v-v_a}, \tag{5.79}$$

where a is conjugate to visits with colour 0, and b is conjugate to visits with colour 1. In the limit that $n \to \infty$, the quenched limiting free energy is

$$\mathcal{D}_{qu}(a,b \mid \chi) = \lim_{n\to\infty} \tfrac{1}{n} \log D_n(a,b \mid \chi), \ \text{ if this limit exists.} \tag{5.80}$$

Note that $\mathcal{D}_{qu}(a,b \mid \chi)$ is a function on the space of binary integer sequences $\langle \chi_i \rangle$ or binary expansions of real numbers in $[0,1]$. If there are N colours, then it is a function on N-ary expansions of real numbers in $[0,1]$. Thus, $\mathcal{D}_{qu}(a,b\,|\,\chi)$ is a function on Lebesgue measure space $[0,1]$. This defines integration over sequences so that averages with respect to colouring can be computed (using the Lebesgue measure).

The existence of the average quenched free energy follows by concatenation of paths. Take two adsorbing Dyck paths of lengths n and m, the first coloured by ϕ, and the second coloured by ψ. By placing the first vertex of the second on the last vertex of the first (see figure 1.3), it follows that

$$\sum_{w,w_a} d_n(w,w_a\,|\,\phi)\,d_m(v-w,v_a-w_a\,|\,\psi) \leq d_{n+m}(v,v_a\,|\,\phi\psi), \tag{5.81}$$

where $\phi\psi$ is the concatenation of the colourings in the obvious way. Multiply this by $a^{v_a}\,b^{v-v_a}$ and sum over $\{v,v_a\}$. This gives the supermultiplicative inequality $D_n(a,b\,|\,\phi)\,D_m(a,b\,|\,\psi) \leq D_{n+m}(a,b\,|\,\phi\psi)$. Take logarithms and then the average over ϕ and ψ. Since the colours are identically distributed and independent, the result is that

$$\langle \log D_n(a,b\,|\,\chi)\rangle_\chi + \langle \log D_m(a,b\,|\,\chi)\rangle_\chi \leq \langle \log D_{n+m}(a,b\,|\,\chi)\rangle_\chi\,. \tag{5.82}$$

By theorem A.1 in appendix A, the quenched average limiting free energy exists:

$$\mathcal{D}_{qu}(a,b) = \lim_{n\to\infty} \frac{1}{n}\,\langle\langle \log D_n(a,b\,|\,\chi)\rangle_\chi\,. \tag{5.83}$$

5.3.3 *Quenched adsorbing alternating Dyck paths*

Assign the alternating sequence of colours $\alpha = \langle 0,1,0,1,\ldots\rangle$ to the even vertices of a Dyck path and quench this colouring. Let a be the generating variable of visits of colour 0, and b be the generating variable of visits of colour 1. By convention, the starting point of the path at $\vec{0}$ is not assigned a colour (and has weight 1).

The generating function of Dyck paths of length 0 mod 4, denoted C_4, can be determined from Catalan numbers (see equation (5.5)):

$$C_4 = \sum_{n=0}^{\infty} d_{4n}\,t^{4n} = \frac{\sqrt{2}}{\sqrt{1+\sqrt{1-16\,t^4}}}\,. \tag{5.84}$$

The generating function C_2 of Dyck paths of length 2 mod 4 may be determined as follows. Every path contributing to C_2 makes an excursion until its first return to the adsorbing line. Here the path renews either as a path of length 0 mod 4 (in which case the first excursion has length 2 mod 4 and generating function t^2C_4), or a path of length 2 mod 4 (in which case the first excursion has length 0 mod 4 and generating function t^2C_2). This gives the factorisation and result that

$$C_2 = t^2\left(C_2^2 + C_4^2\right) = \frac{8t^2}{(1+\sqrt{1-16\,t^4})\left(2+\sqrt{2-2\sqrt{1-16\,t^4}}\right)}. \tag{5.85}$$

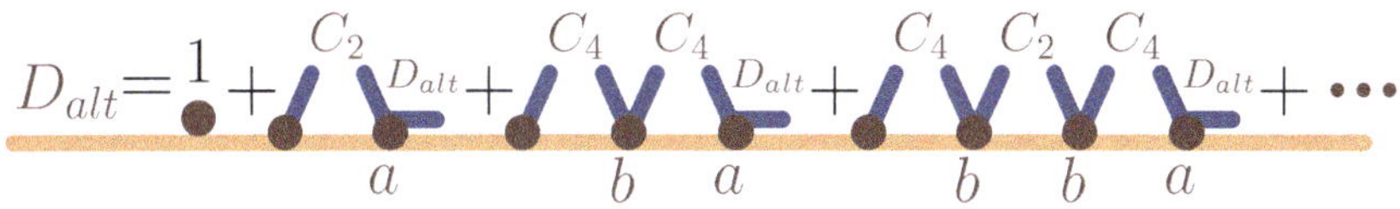

FIG. 5.9. The adsorbing quenched alternating Dyck path is either a single A-visit, or has an excursion which returns in an A-visit or a B-visit. If it returns in a B-visit, then it may continue making excursions which return in B-visits until a return to an A-visit is made. Once a return to an A-visit is made, the path renews.

To determine the generating function $D_{alt}(a,b,t)$ of adsorbing quenched alternating Dyck paths, the decomposition in figure 5.9 is used. This gives $D_{alt}(a,b,t)$ in terms of the generating functions C_2 and C_4:

$$D_{alt} = 1 + at^2C_2D_{alt} + \left(\frac{abt^4C_4^2}{1-bt^2C_2}\right)D_{alt} \tag{5.86}$$

with solution

$$D_{alt}(a,b,t) = \frac{1-bt^2C_2}{(1-at^2C_2)(1-bt^2C_2)-abt^4C_4^2}. \tag{5.87}$$

Substituting $bt^2 = s^2$ shows that $D_{alt}(a,b,t) = D_{alt}(\frac{a}{b},1,s)$. In other words, it is only necessary to consider the case $b=1$, where the radius of convergence is

$$t_c(a) = \begin{cases} \frac{1}{2}, & \text{if } a \leq 2+\sqrt{2}; \\ \frac{1}{2a}\left[\sqrt{\sqrt{8a^3-20a^2+16a}-4}\right], & \text{if } a \geq 2+\sqrt{2}. \end{cases} \tag{5.88}$$

Adsorption of the path occurs at the critical point $a_c = 2+\sqrt{2}$.

The generating functions of adsorbing quenched Dyck paths are in general complicated expressions. In some cases the exact (but non-rigorous) critical point may be found by using the asymptotic approximations in equation (5.69).

For example, let $\gamma = \langle 0^p, 1^q\rangle$ be the colouring of a Dyck path model of a diblock-copolymer of length n with $p = \lfloor \frac{n}{2} \rfloor$, and $q = \lceil \frac{n}{2} \rceil$. If visits coloured by 0 are generated by a, and visits coloured by 1 are generated by b, then the partition function of the model can be determined from equation (5.69). By examining the asymptotic rate of growth of the product $D_{\lfloor n/2 \rfloor}(0,\beta,a)\,D_{\lceil n/2 \rceil}(\beta,0,b)$ and taking the supremum over β, the limiting free energy of the model is

$$\mathcal{D}_B(a,b) = \begin{cases} \log 2, & \text{if } a \leq 2, \text{ and } b \leq 2; \\ \frac{1}{2}\log 2 + \frac{1}{2}\log a + \frac{1}{4}\log(a-1), & \text{if } a > 2, \text{ and } b \leq 2; \\ \frac{1}{2}\log 2 + \frac{1}{2}\log b + \frac{1}{4}\log(b-1), & \text{if } a \leq 2, \text{ and } b > 2; \\ \frac{1}{2}\log(ab) + \frac{1}{4}\log\left((a-1)(b-1)\right), & \text{if } a > 2, \text{ and } b > 2. \end{cases} \tag{5.89}$$

The free energies of other models can be similarly determined.

5.3.4 *The adsorption of a* $\{\{01^{p-1}\}^*0\}$ *coloured Dyck path*

Consider a family of periodic quenches $\chi_p = \left\{\{01^{p-1}\}^*0\right\}$ and colour Dyck paths with these as before, where 0-visits have activity a, and 1-visits have activity b.

Let $d_n(v, v_a)$ be the number of Dyck paths with v visits (of either colour) of which v_a visits are of colour 0 (and $v - v_a$ visits of colour 1). Introduce activities z conjugate to v, and y conjugate to v_a. The generating function of the model is

$$D(y,z,t \mid \chi_p) = \sum_{n=0}^{\infty} \sum_{v,v_a} d_n(v,v_a)\, y^{v_a} z^v\, t^n. \tag{5.90}$$

The ab-ensemble is obtained from this by substituting $z \to b$ and by substituting $y \to \frac{a}{b}$.

Introduce the generating function

$$U_p(z,t) = \tfrac{1}{p} \sum_{j=0}^{p-1} D(z, \beta^j t), \text{ where } \beta = e^{i\pi p}, \tag{5.91}$$

namely, the generating function of Dyck paths of length 0 mod $2p$ with z conjugate to visits, and t conjugate to edges. ($D(z,t)$ is the generating function of adsorbing Dyck paths in equation (5.13).)

Let $L(y,z,t \mid \chi_p)$ be the generating function of Dyck paths coloured by χ_p and of length 0 mod $2p$, with visits (of any colour) generated by z, and with 0-visits generated by y. Notice that $U_p(z,t) = L(1,z,t \mid \chi_p)$.

By following the arguments leading to equation (5.14) for Dyck paths of length 0 mod $2p$ and partially coloured by χ_p, the following *exchange relation* is obtained:

$$y\, L(y,z,t \mid \chi_p)\,(U_p(z,t) - z) = (L(y,z,t \mid \chi_p) - yz)\, U_p(z,t) \tag{5.92}$$

(see equation (5.14)). Solving for $L(y,z,t \mid \chi_p)$ gives

$$L(y,z,t \mid \chi_p) = \frac{yz\, U_p(z,t)}{yz + (1-y)\, U_p(z,t)}. \tag{5.93}$$

Each Dyck path counted by $D(y,z,t \mid \chi_p)$ may be factored by decomposing it at its last 0-visit into a Dyck path of length 0 mod $2p$ and coloured by χ_p, and then into a remainder of arbitrary length and with no A-visits. Thus,

$$D(y,z,t \mid \chi_p) = L(y,z,t \mid \chi_p)\, D(0,z,t \mid \chi_p). \tag{5.94}$$

Put $y = 1$ and solve

$$D(0,z,t \mid \chi_p) = \frac{D(1,z,t \mid \chi_p)}{L(1,z,t \mid \chi_p)} = \frac{D(z,t)}{U_p(z,t)}, \tag{5.95}$$

since $U_p(z,t) = L(1,z,t \mid \chi_p)$. Substitution of equations (5.94) and (5.95) in equation (5.93) gives the result

$$D(y,z,t \mid \chi_p) = \frac{yz\,D(z,t)}{yz + (1-y)\,U_p(z,t)}. \tag{5.96}$$

The function $D(a,1,t \mid \chi_p)$ generates paths with only 0-visits adsorbing. Taking $y \to \frac{1}{a}$ and taking $z \to a$ instead gives the generating function of adsorbing Dyck paths with the complementary colouring $\widetilde{\chi}_p = \{10^{p-1}\}^* 1$ and with adsorbing 0-visits.

Consider the model with generating function $D(a,1,t \mid \chi_p)$. Let $a_c(p)$ be the critical adsorption point (in this model with colouring χ_p). Clearly, $a_c(1) = 2$ by equation (5.15), and $a_c(2) = 2+\sqrt{2}$ by equation (5.88). Determining $a_c(p)$ for larger values of p quickly becomes difficult.

An asymptotic expression for $a_c(p)$ can be developed [488]. There are two sources of singularities in $D(a,1,t \mid \chi_p)$; namely, square root singularities in $D(1,t)$ and $U_p(1,t)$, and simple poles whenever the denominator vanishes; this occurs when

$$U_p(1,t) = \frac{a}{a-1}. \tag{5.97}$$

By equation (5.91) and by equation (5.13), $U_p(1,t)$ simplifies to

$$U_p(1,t) = -\frac{1}{p}\sum_{j=0}^{p-1} \frac{\sqrt{1-4\,t^2\beta^{2j}}}{2\,t^2\beta^{2j}}, \text{ where } \beta = e^{i\pi p}. \tag{5.98}$$

The singularity in $U_p(1,t)$ is a branch point on the real axis at $t = \frac{1}{2}$. This determines the radius of convergence for $D(a,1,t \mid \chi_p)$ at small values of a.

If a is large, then simple poles in the t-plane given by the solution of equation (5.97) determine the radius of convergence. These singularities coalesce with the branch point when $\frac{a}{a-1} = U_p(1,\frac{1}{2})$. That is, the critical adsorption activity is

$$a_c(p) = \frac{U_p(1,\frac{1}{2})}{U_p(1,\frac{1}{2})-1}. \tag{5.99}$$

By determining the asymptotics of $U_p(\frac{1}{2},1)$, the asymptotic behaviour of $a_c(p)$ can be determined.

Dyck paths are counted by Catalan numbers, and $U_p(1,t)$ is the generating function of Dyck paths of length 0 mod $2p$. Thus,

$$U_p(1,\tfrac{1}{2}) = 1 + \sum_{n\geq 1} \binom{2np}{np} \frac{4^{-np}}{np+1}. \tag{5.100}$$

Use the bound in equation (D.7) in appendix D on Catalan numbers and perform the summations over n. This gives (where $\zeta(x)$ is the usual ζ function)

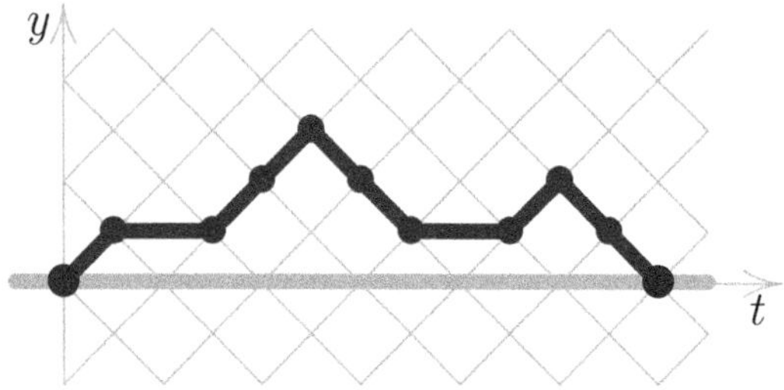

FIG. 5.10. A Motzkin path in $\mathbb{J}^2_+$ gives north-east, east or south-east steps as shown. The path steps on vertices $(x, y) \in \mathbb{J}^2_+$ which have even parity ($x+y$ is even) and it is constrained to end in a vertex in $\partial\mathbb{J}^2_+$.

$$U_p(1, \tfrac{1}{2}) \simeq 1 + \tfrac{1}{\sqrt{\pi}} p^{-3/2} \left(\zeta\left(\tfrac{3}{2}\right) - \tfrac{9}{8p} \zeta\left(\tfrac{5}{2}\right) + \tfrac{145}{128p^2} \zeta\left(\tfrac{7}{2}\right) + O\left(\tfrac{1}{p^3}\right) \right). \tag{5.101}$$

Substitution into equation (5.99) gives to the first decaying term

$$a_c(p) \simeq \frac{\sqrt{\pi}\, p^{3/2}}{\zeta\left(\frac{3}{2}\right)} + \frac{9\sqrt{\pi}\, \zeta\left(\frac{5}{2}\right) p^{1/2}}{8\, \zeta^2\left(\frac{3}{2}\right)} + 1 + O(p^{-1/2}). \tag{5.102}$$

See reference [488].

5.4 Motzkin paths

A Motzkin path [429] is a directed path from $\vec{0}$ in $\mathbb{J}^2_+$ (see equation (5.2)) with north-east, east or south-east steps and with last vertex in $\partial\mathbb{J}^2_+$ (figure 5.10). See for example references [69, 298, 302, 305, 563] for Motzkin path models of directed polymers.

A *visit* of a Motzkin path is vertex in the path which is also in $\partial\mathbb{J}^2_+$. Edges in a Motzkin path which are also in $\partial\mathbb{J}^2_+$ are *edge-visits*.

The generating function of Motzkin paths is denoted by $M(t)$, and if visits have generating variable a, then it is denoted $M(a, t)$. The function $M(a, t)$ can be determined by the factorisation of Motzkin paths in figure 5.11. This gives the recurrence

$$M(a, t) = 1 + at\, M(a, t) + at^2 M(1, t)\, M(a, t). \tag{5.103}$$

Putting $a = 1$ and solving for $M(t) \equiv M(1, t)$ gives $M(t)$:

$$M(t) = \frac{2}{1 - t + \sqrt{(1+t)(1-3t)}}. \tag{5.104}$$

An asymptotic approximation to m_n, the number of Motzkin paths of length n, can be determined from $M(t)$. This was done in reference [319]:

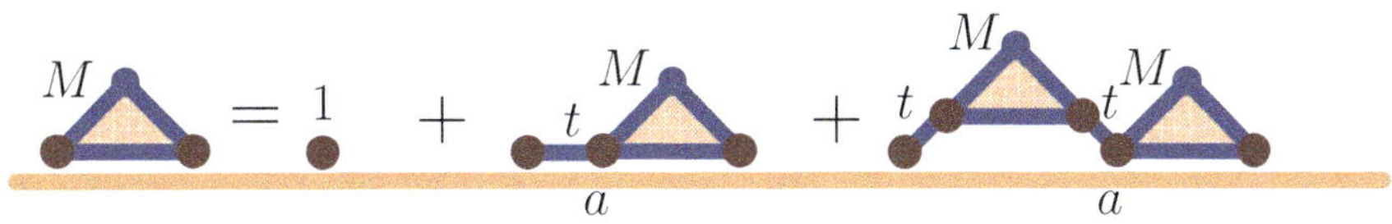

FIG. 5.11. The factorisation of Motzkin paths.

$$m_n = \frac{3^{n+3/2}}{2\sqrt{\pi n^3}}\Big(1 - \tfrac{39}{16n} + \tfrac{2665}{512n^2} + O(n^{-3})\Big). \tag{5.105}$$

Motzkin paths are related to the Motzkin algebraic language. In the terminology of section 5.1.1, let the non-terminal set of an algebraic language $L(\mathcal{M})$ be $N = \{\mathfrak{d}\}$ and let the terminal set be $X = \{\mathfrak{D}, \mathfrak{C}, \mathfrak{A}\}$. The production rules for 1-*coloured Motzkin words* are $P = \{(\mathfrak{d} \to \mathfrak{D}\,\mathfrak{d}\,\mathfrak{C}\,\mathfrak{d}), (\mathfrak{d} \to \mathfrak{A}\,\mathfrak{d}), (\mathfrak{d} \to e)\}$. The starting word is $s = \mathfrak{d}$. For any prefix u of a Motzkin word w, $|u|_\mathfrak{D} \geq |u|_\mathfrak{C}$, (since the path steps in $\mathbb{J}^2_+$).

If instead $X = \{\mathfrak{D}, \mathfrak{C}, \mathfrak{A}, \mathfrak{B}\}$, then the production rules for 2-*coloured Motzkin words* are $P = \{(\mathfrak{d} \to \mathfrak{D}\,\mathfrak{d}\,\mathfrak{C}\,\mathfrak{d}), (\mathfrak{d} \to \mathfrak{A}\,\mathfrak{d}), (\mathfrak{d} \to \mathfrak{B}\,\mathfrak{d}), (\mathfrak{d} \to e)\}$. For any prefix u of a 2-coloured Motzkin word w, $|u|_\mathfrak{D} \geq |u|_\mathfrak{C}$.

Consequently, words in Motzkin languages are generated by the recursions

$$\begin{aligned} w &= e + \mathfrak{A}w + \mathfrak{D}w\mathfrak{C}w, && \text{1-coloured;} \\ w &= e + \mathfrak{A}w + \mathfrak{B}w + \mathfrak{D}w\mathfrak{C}w, && \text{2-coloured.} \end{aligned}$$

Sending $\mathfrak{A} \to t$, sending $\mathfrak{B} \to t$, sending $\mathfrak{D} \to t$ and sending $\mathfrak{C} \to t$ gives the generating function $M(t)$ in equation (5.104) for 1-coloured Motzkin paths, and

$$M_2(t) = \frac{2}{1 - 2t + \sqrt{1-4t}} \tag{5.106}$$

for 2-coloured Motzkin paths. Notice that $M_2(t)$ is related to $D(t)$, the generating function of Dyck paths, by $D(t) = 1 + t^2\, M_2(t^2)$. This may be seen as follows: the map h defined by $h(\mathfrak{D}) = \mathfrak{D}\mathfrak{D}$, by $h(\mathfrak{C}) = \mathfrak{C}\mathfrak{C}$, by $h(\mathfrak{A}) = \mathfrak{U}\mathfrak{C}$ and by $h(\mathfrak{B}) = \mathfrak{D}\mathfrak{D}$ defines a bijection $f(w) = \mathfrak{D}\, h(w)\, \mathfrak{C}$ between 2-coloured Motzkin words and Dyck words.

5.4.1 *Adsorbing Motzkin paths*

By equations (5.103) and (5.104), the generating function of adsorbing Motzkin paths is

$$M(a,t) = \frac{2}{2 - a(1+t) + a\sqrt{(1+t)(1-3t)}}. \tag{5.107}$$

Observe that $M(a,t)$ satisfies the exchange relation [319]

$$a\,M(a,t)(M(1,t) - 1) = 1\,M(1,t)(M(a,t) - 1). \tag{5.108}$$

This can be demonstrated using a proof-by-picture, similar to the case for the exchange relation of Dyck paths in figure 5.4.

The radius of convergence of $M(a,t)$ gives the free energy

$$\mathcal{M}(a) = \begin{cases} \log 3, & \text{if } a \le \frac{3}{2}; \\ \log(2a) - \log\left(1 - a + \sqrt{(a+3)(a-1)}\right), & \text{if } a > \frac{3}{2}. \end{cases} \tag{5.109}$$

The free energy is non-analytic at $a_c = \frac{3}{2}$ (the adsorption critical point in the model).

The scaling of $\mathcal{M}(a)$ is similar to the scaling of the adsorbing Dyck path free energy $\mathcal{D}(a)$. The asymptotic behaviour of the partition function $M_n(a)$ can be extracted from $M(a,t)$. This is given by

$$M_n(a) \simeq \begin{cases} \dfrac{\sqrt{3}\,a\,3^{n+1}}{2\sqrt{\pi n^3}(3-2a)^2}, & \text{if } a < \frac{3}{2}; \\ \dfrac{2\sqrt{3}\,3^{n-1}}{\sqrt{\pi n}}, & \text{if } a = \frac{3}{2}; \\ \dfrac{C+a-3+\sqrt{2}\sqrt{C(a-3)+a^2-2a+3}}{8C(C-a+1)}\left(\dfrac{a}{C-a+1}\right)^n, & \text{if } a > \frac{3}{2}, \end{cases} \tag{5.110}$$

where $C = \sqrt{(a+3)(a-1)}$ (see equation (5.17) for a comparison to Dyck paths). Taking logarithms of the above, dividing by n and letting $n \to \infty$ recovers $\mathcal{M}(a)$ in equation (5.109).

The number of Motzkin paths of length n steps (denoted by m_n) is given by

$$m_n = \frac{1}{n+1}\binom{n+1}{1}_2 \tag{5.111}$$

in terms of (central) trinomial coefficients (see section D.2 and equations (D.20) and (D.21) in appendix D).

A Motzkin path has a *lifted endpoint* if its terminal vertex has height $h > 0$ steps above $\partial\mathbb{J}^2_+$. The number of Motzkin paths with their endpoints at height h is

$$m_n(h) = \frac{h+1}{n+1}\binom{n+1}{h+1}_2. \tag{5.112}$$

The generating function of Motzkin paths with a lifted endpoint at height h is

$$M_h(t) = \frac{2(2t)^h}{(1-t+\sqrt{(1+t)(1-3t)})^{h+1}}. \tag{5.113}$$

5.4.2 *Constant term formulation of adsorbing Motzkin paths*

A constant term method can be used to find the partition function of Motzkin paths with lifted endpoints (see figure 5.12). In this model there will be two classes of visits by vertices: the first class consists of those visits to the adsorbing

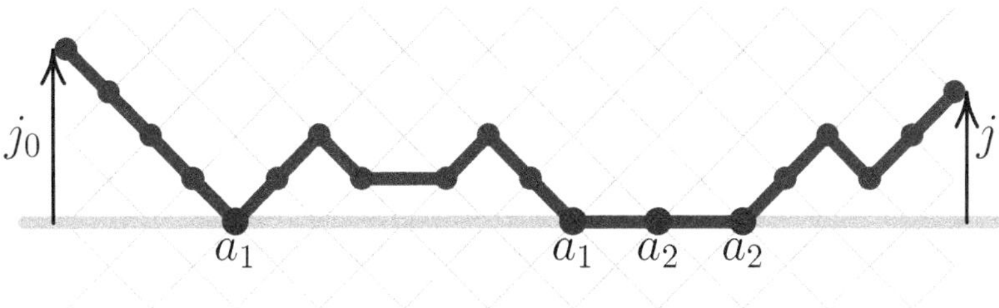

FIG. 5.12. An adsorbing Motzkin path with lifted endpoints. The path starts at a vertex at height j_0 above the adsorbing line and terminates at a vertex at height j. Visits of the path to the adsorbing line are put into two classes: the first class contains the collection of visits created by returns of the path to the adsorbing line (these are generated by a_1). The second class contains the collection of visits created when the path steps from a visit in the adsorbing line into a new visit in the adsorbing line (these are generated by a_2). In this example the path has two visits in the first class and two visits in the second class.

---

line where the path *returns* to $\mathbb{J}_+^2$, and the second class consists of those visits where the path steps from a visit (already *in* $\mathbb{J}_+^2$) to a new visit in $\mathbb{J}_+^2$ (these are edge-visits). Visits in the first class are generated by a_1, and visits in the second class by a_2.

This model is equivalent to an adsorbing Motzkin path with vertex-visits weighted by a and edge-visits weighted by b (replace a_1 by a, and a_2 by ab, where a generates vertex-visits, and b generates edge-visits).

Let $w_n(j)$ be the number of Motzkin paths of length n, starting in a vertex at height j_0 and with last vertex at height j. Then $w_n(j)$ satisfies the set of recurrences given by

$$\begin{aligned}
&w_n(j) = w_{n-1}(j-1) + w_{n-1}(j) + w_{n-1}(j+1), && \text{for } j \geq 1, \text{ and } n \geq 1;\\
&w_n(0) = a_2 w_{n-1}(0) + a_1 w_{n-1}(1), && \text{for } n \geq 1;\\
&w_0(j_0) = 1. && (5.114)
\end{aligned}$$

Notice the creation of visits in the middle recurrence. A path stepping from height 0 into a visit gets a factor of a_2, while a path stepping down from height 1 into a visit gets a factor a_1.

A trial solution (using the Bethe ansatz) of the recurrences above is

$$w_n(j) = \Lambda^n \left(A_1 e^{ijk} + A_2 e^{-ijk} \right). \tag{5.115}$$

It remains to determine A_1, A_2 and Λ as well as integrating $k \in (-\pi, \pi]$.

Substituting $w_n(j)$ into equation (5.114) shows that

$$\frac{A_1}{A_2} = -\frac{\Lambda - a_2 - a_1 e^{-ik}}{\Lambda - a_2 - a_1 e^{ik}} = \mathcal{S}(k), \tag{5.116}$$

where $\mathcal{S}(k)$ is a scattering function. This also fixes $\Lambda = e^{ik} + 1 + e^{-ik}$.

This gives a general solution for equation (5.114):

$$w_n(j) = A_2(k;j_0)\left(e^{ik}+1+e^{-ik}\right)^n \left(e^{-ijk} + \mathcal{S}(k)\, e^{ijk}\right). \tag{5.117}$$

The coefficient $A_2(k;j_0)$ remains to be determined.

The reversal of the path is obtained by letting $k \to -k$ in the above. If the path steps backwards to a terminal vertex at height j_0 in m steps, then

$$w_m(j_0) = A_2(-k;j)\left(e^{-ik}+1+e^{ik}\right)^m \left(e^{ij_0k} + \mathcal{S}(-k)\, e^{-ij_0k}\right). \tag{5.118}$$

Comparing equations (5.117) and (5.118) with $n = m$ gives the appropriate solution for $A_2(k;j_0)$:

$$A_2(k;j_0) = \left(e^{ij_0k} + \mathcal{S}(-k)\, e^{-ij_0k}\right). \tag{5.119}$$

Substitute $A_2(k;j_0)$ in equation (5.117) and notice that $\mathcal{S}(-k) = (\mathcal{S}(k))^{-1}$. Integrate $k \in (\pi, \pi]$ to find that the number of paths of length n from height j_0 to height j is given by

$$w_n(j) = C_0 \int_{-\pi}^{\pi} \Lambda^n \left(e^{i(j-j_0)k} + \mathcal{S}(k)\, e^{i(j+j_0)k}\right) dk, \tag{5.120}$$

where C_0 is a constant to be determined by the initial condition in equation (5.114), and where Λ is as defined before.

It remains to integrate over k. Substitute $\zeta = e^{ik}$. Then the integrand in equation (5.120) becomes

$$\Lambda^n \left(\zeta^{j-j_0} - \zeta^{j+j_0} - \left[\frac{(\zeta^2-1)a_1}{(1-a_1)\zeta^2 + (1-a_2)\zeta + 1}\right]\zeta^{j+j_0}\right). \tag{5.121}$$

This shows that

$$\begin{aligned} w_n(j) = C_0 \oint_C \Lambda^n \zeta^{j-j_0}\, \tfrac{d\zeta}{i\zeta} - C_0 \oint_C \Lambda^n \zeta^{j+j_0}\, \frac{d\zeta}{i\zeta} \\ - a_1 C_0 \oint_C \left[\frac{\Lambda^n(\zeta^2-1)\zeta^{j+j_0}}{(1-a_1)\zeta^2 + (1-a_2)\zeta + 1}\right] \frac{d\zeta}{i\zeta}, \end{aligned} \tag{5.122}$$

where a factor of 2 was absorbed into C_0 and where C is the circular contour of radius one anticlockwise rotation about the origin in the ζ-plane.

The first two integrals can be done by expanding Λ^n using trinomial coefficients (see equation (D.20) in appendix D). This shows that

$$C_0 \oint_C \Lambda^n \zeta^{j-j_0} \tfrac{d\zeta}{i\zeta} = 2\pi C_0 \binom{n}{j-j_0}_2, \quad \text{and } C_0 \oint_C \Lambda^n \zeta^{j+j_0} \frac{d\zeta}{i\zeta} = 2\pi C_0 \binom{n}{j+j_0}_2.$$

This leaves the last integral in equation (5.122). Substitute $\zeta = \overline{\xi} = \frac{1}{\xi}$ and simplify. Then the integral becomes

$$\oint_C \frac{\Lambda^n (\zeta^2-1)\zeta^{j+j_0}}{(1-a_1)\zeta^2+(1-a_2)\zeta+1}\,\frac{d\zeta}{i\zeta} = \oint_C \frac{\Lambda^n (1-\xi^2)\overline{\xi}^{j+j_0}}{\xi^2+(1-a_2)\xi+(1-a_1)}\,\frac{d\xi}{i\xi}.$$

Factor the denominator by putting $\xi^2+(1-a_2)\xi+(1-a_1) = (\xi-\lambda_+)(\xi-\lambda_-)$, where the roots are given by

$$\lambda_\pm = \tfrac{1}{2}\left(a_2 - 1 \pm \sqrt{(1-a_2)^2 - 4(1-a_1)}\right). \tag{5.123}$$

Thus, the third integral in equation (5.122) becomes

$$\oint_C \frac{\Lambda^n (1-\xi^2)\,\overline{\xi}^{j+j_0}}{\xi^2+(1-a_2)\xi+(1-a_1)}\,\frac{d\xi}{i\xi} = \sum_{\ell=-n}^{n} \binom{n}{\ell}_2 \oint_C \frac{(1-\xi^2)\,\overline{\xi}^{j+j_0-\ell}}{(\xi-\lambda_+)(\xi-\lambda_-)}\,\frac{d\xi}{i\xi}$$

once Λ^n is expanded using trinomial coefficients.

The integral can be done using the residue theorem. For large values of a_2, the contour may be expanded to capture the poles at $\lambda_\pm$. This gives

$$\begin{aligned}\oint_C \frac{(1-\xi^2)\,\overline{\xi}^{j+j_0-\ell}}{(\xi-\lambda_+)(\xi-\lambda_-)}\,\frac{d\xi}{i\xi} &= 2\pi\left(\frac{\lambda_+{}^{\ell-1-j-j_0} - \lambda_-{}^{\ell-1-j-j_0}}{\lambda_+-\lambda_-}\right)\mathrm{I}_{\ell-1-j-j_0} \\ &\quad - 2\pi\left(\frac{\lambda_+{}^{\ell+1-j-j_0} - \lambda_-{}^{\ell+1-j-j_0}}{\lambda_+-\lambda_-}\right)\mathrm{I}_{\ell+1-j-j_0},\end{aligned} \tag{5.124}$$

where I_N is an indicator function defined by $\mathrm{I}_N = 1$ if $N \geq 0$, and $\mathrm{I}_N = 0$ if $N < 0$.

Adsorbing the constant 2π into C_0, and simplifying the expressions give the following expression for the partition function of adsorbing Motzkin paths with lifted endpoints:

$$\begin{aligned}M_n(j_0,j,a) &= C_0\binom{n}{j-j_0}_2 - C_0\binom{n}{j+j_0}_2 \\ &\quad + \frac{a_1 C_0}{\lambda_+-\lambda_-}\sum_{\ell=j+j_0-1}^{n}\binom{n}{\ell}_2\left(\lambda_+{}^{\ell+1-j-j_0} - \lambda_-{}^{\ell+1-j-j_0}\right) \\ &\quad - \frac{a_1 C_0}{\lambda_+-\lambda_-}\sum_{\ell=j+j_0+1}^{n}\binom{n}{\ell}_2\left(\lambda_+{}^{\ell-1-j-j_0} - \lambda_-{}^{\ell-1-j-j_0}\right).\end{aligned} \tag{5.125}$$

Putting $n = 0$ and putting $j = j_0$ in the above shows that $M_0(j, j, a) \equiv w_0(j) = C_0 = 1$.

If the discriminant in equation (5.123) is negative, then the roots are a complex conjugate pair with real part which moves outside the unit disk if $a_2 > 3$. If the discriminant is positive, then the two roots are on the real axis.

If $j = j_0 = 0$, then $w_n(j)$ above grows asymptotically at an exponential rate determined by the largest root λ_+. By equation (D.28) and theorem D.7 in appendix D, this shows that the limiting free energy in the model is

$$\mathcal{M}(a_1, a_2) = \begin{cases} \log 3, & \text{if } a_1 + a_2 \leq 3; \\ \log \left(\dfrac{a_1 + a_2(a_1 - 2) + a_1\sqrt{(a_2-1)^2 + 4(a_1-1)}}{2(a_1 - 1)} \right), & \text{if } a_1 + a_2 > 3. \end{cases}$$

This shows that the critical curve in the model is $a_1 + a_2 = 3$.

If $a_1 = a$ and if $a_2 = ab$, then the model reduces to an adsorbing Motzkin path with adsorbing vertices weighted by a and adsorbing edges weighted by b. The critical curve is given by $a(1+b) = 3$. Putting $b = 1$ recovers the free energy in equation (5.16).

The free energy of Motzkin paths with lifted endpoints was determined in reference [302]. Let λ_1 be a root of

$$\lambda^2 + (1 - a_2)\lambda + (1 - a_1) = 0. \tag{5.126}$$

(The solution of this quadratic is exactly $\lambda_\pm$ in equation (5.123).) Then the free energy of adsorbing Motzkin paths with lifted endpoints is given by

$$\mathcal{M}_{\alpha\beta}(a_1, a_2) = \begin{cases} \log \left(\dfrac{2^{\alpha-\beta}\left(4 - 3(\alpha-\beta) + \sqrt{4 - 3(\alpha-\beta)^2}\right)}{(1-\alpha+\beta)^{1-\alpha+\beta}\left(\alpha - \beta + \sqrt{4 - 3(\alpha-\beta)^2}\right)^{1+\alpha-\beta}} \right), & \text{if } \lambda \in [0, \lambda_1]; \\ \log \left(\dfrac{2^{\alpha+\beta} a_1 \left(1 + a_2(a_1 - 2) + \sqrt{(a_2-1)^2 + 4(a_1-1)}\right)}{2(a_1-1)\left(a_1 a_2 - 1 + \sqrt{(a_2-1)^2 + 4(a_1-1)}\right)^{\alpha+\beta}} \right), & \text{if } \lambda \geq \lambda_1. \end{cases}$$

The critical curve is obtained by requiring that $\mathcal{M}_{\alpha\beta}(a_1, a_2)$ is continuous.

5.5 Partially directed paths

A self-avoiding path from $\vec{0}$ in $\mathbb{L}^2$, giving only east, north or south steps (and never steps to the west) such that a north step cannot immediately be followed by a south step, and a south step cannot immediately be followed by a north step, is a *partially directed path* (see figure 5.13).

Denote the number of partially directed paths of n steps from the origin by r_n and define the generating function of partially directed paths by

$$R(t) = \sum_{n=0}^{\infty} r_n \, t^n. \tag{5.127}$$

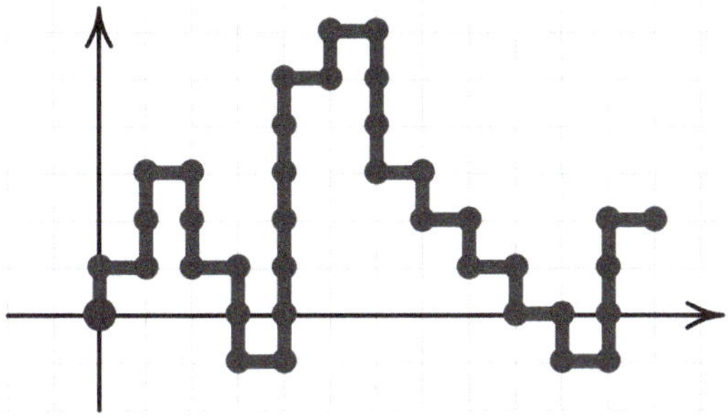

FIG. 5.13. A partially directed path in $\mathbb{L}^2$ from $\vec{0}$ gives steps in the east, north or south directions.

The function $R(t)$ can be determined using the Temperley method. Let $R_r(s,t)$ be the related generating function of partially directed paths which end in an east step or are empty paths (single vertices), and where t generates east steps, and s generates north or south steps. If ρ is any such path, then edges may be appended at its endpoint as illustrated in figure 5.14. Notice that $R(t) = \frac{1}{t}(R_r(t,t) - 1)$.

By figure 5.14, the generating function R_r satisfies the recurrence

$$R_r = 1 + t R_r + \left(\tfrac{st}{1-s}\right) R_r + \left(\tfrac{st}{1-s}\right) R_r \tag{5.128}$$

since $\frac{st}{1-s}$ generates at least one north or south step, followed by one east step. Solving this recurrence for R_r and then finding R gives

$$R(r) = \frac{1+t}{1-2t-t^2}. \tag{5.129}$$

By expanding this in powers of t,

$$r_n = \tfrac{1}{2}(1+\sqrt{2})^{n+1} + \tfrac{1}{2}(1-\sqrt{2})^{n+1} \tag{5.130}$$

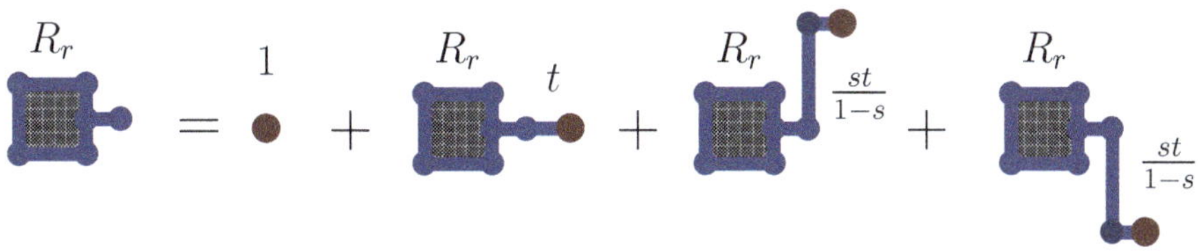

FIG. 5.14. The Temperley method for partially directed paths. Either a single east step is appended, a sequence of north steps followed by a single east step is appended or a sequence of south steps, followed by a single east step, is appended.

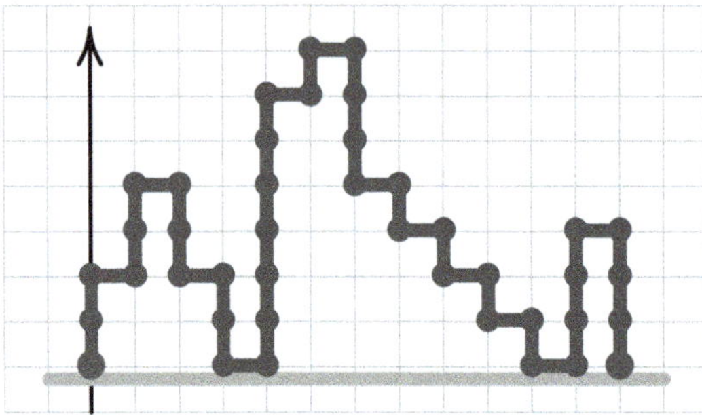

FIG. 5.15. A bargraph path in $\mathbb{L}^2_+$ is a partially directed path with endpoints in $\partial\mathbb{L}^2_+$.

for the number of partially directed paths of length n from the origin.

A partially directed path in the half-lattice $\mathbb{L}^2_+$ (see equation (2.20)) with endpoints in the adsorbing line $\partial\mathbb{L}^2_+$ is a *bargraph* path (see figure 5.15).

The factorisation in figure 5.16 gives a recurrence for the generating function $B \equiv B(s,t)$ of bargraph paths (where t generates east steps, and s generates north or south steps):

$$B = 1 + tB + s^2(B-1) + s^2t(B-1)B. \tag{5.131}$$

Solving for $B(s,t)$ gives two roots: one root is a Laurent series in t, and the second root is a power series in t which is the generating function

$$B(s,t) = \tfrac{1}{2s^2t}\Big((1-t)(1-s^2) - \sqrt{(1-s^2)((1-t)^2 - s^2(1+t)^2)}\Big) \tag{5.132}$$

of bargraph paths. The asymptotics for b_n, the number of bargraphs of length n, can be determined from equation (5.132). This was done in reference [319]:

$$b_n = \frac{\beta^{n+3/2}}{\sqrt{\pi n^3}}\left(1 - \tfrac{21\sqrt{2}+36}{16n} + \tfrac{5(252\sqrt{2}+349)}{256n^2} + O(n^{-3})\right) \tag{5.133}$$

where $\beta = 1+\sqrt{2}$.

The generating function $B_a(a,s,t)$ of adsorbing bargraphs, with visits to the adsorbing line weighted by a, can similarly be determined from figure 5.16.

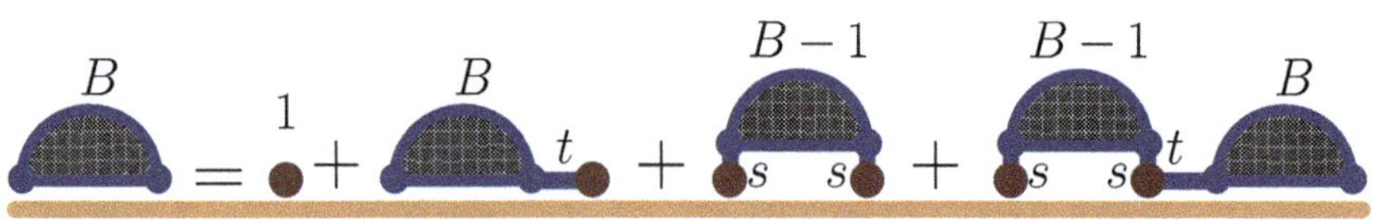

FIG. 5.16. The factorisation of the bargraph generating function.

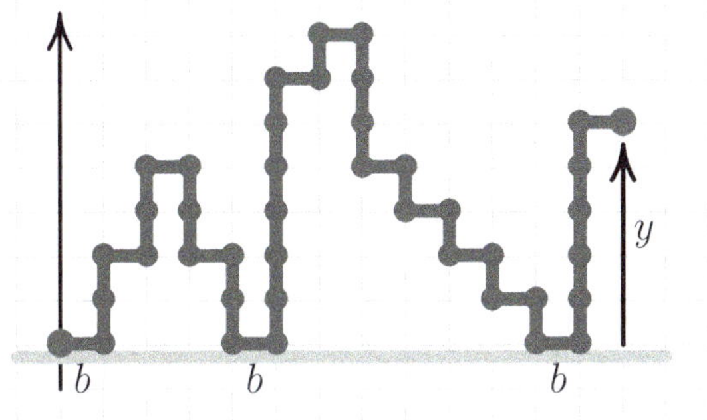

FIG. 5.17. A pulled and adsorbing partially directed path. The path is weighted by the number of edge-visits it makes to the adsorbing line and by the height of its endpoint. The generating variable b is conjugate to edge-visits, and the generating variable y is conjugate to the height of the endpoint.

Assuming again the convention that the first vertex (at $\vec{0}$) is not weighted, the generating function $B_a(a, s, t)$ satisfies the recurrence

$$B_a = 1 + taB_a + s^2a(B-1) + s^2ta(B-1)B_a \tag{5.134}$$

with B given by equation (5.132). This gives the solution

$$\begin{aligned} B_a(a, s, t) &= \frac{1 + s^2a\,(B-1)}{1 - ta\,(1 + s^2(B-1))} \qquad (5.135) \\ &= \frac{2t + a((1-t) + s^2(1+t)) - a\sqrt{(1-s^2)((1-t)^2 - s^2(1+t)^2)}}{t\,(2 - a(1-s^2)(1+t)) + a\sqrt{(1-s^2)((1-t)^2 - s^2(1+t)^2)}}. \end{aligned}$$

The function $B_a(a, s, t)$ has a square root singularity similar to the generating function of the adsorbing Dyck path $D(a, t)$ (see equation (5.20)) and the adsorbing Motzkin path $M(a, t)$ (see equation (5.107)). This shows that the scaling of this model will have similar properties to those models.

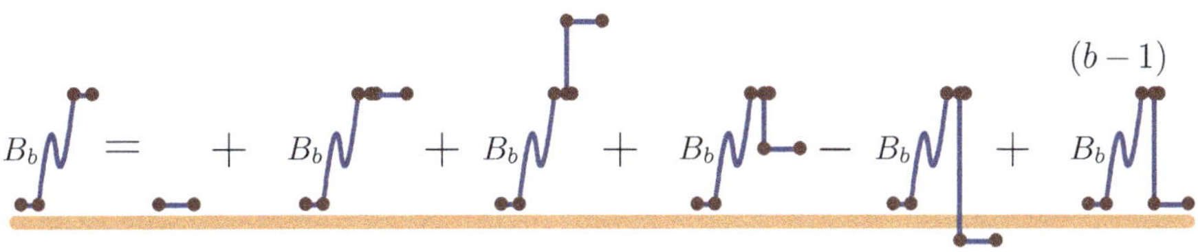

FIG. 5.18. The Temperley method for pulled adsorbing partially directed paths.

5.5.1 *Pulled and adsorbing partially directed paths*

Let $r_n(w,h)$ be the number of partially directed paths of length n starting in a horizontal edge in $\vec{0} \in \mathbb{L}^2_+$, having w edges in the adsorbing line (edge-visits) and ending in horizontal step in a terminal vertex of height h. Introduce the following generating variables: y conjugate to height; b conjugate to the number of edge-visits; s conjugate to vertical steps (north- or south-steps); and t conjugate to horizontal steps.

Denote the generating function of the model by $B_b(y) \equiv B_b(y,b,s,t)$. The model is illustrated in figure 5.17 and was analysed in reference [457].

Notice that $B_b(s)$ (obtained by $y \to s$) is the generating function of bargraph paths ending in a south step in the adsorbing line at height 0.

The generating function $B_b(y)$ can be obtained by finding a recurrence using the Temperley method. This is illustrated in figure 5.18. The recurrence is

$$B_b(y) = bt + \left(t + \frac{sty}{1-sy} + \frac{st\overline{y}}{1-s\overline{y}}\right) B_b(y) - \left(\frac{st\overline{y}}{1-s\overline{y}}\right) B_b(s) + (b-1)t\, B_b(s).$$

Notice that y is the *catalytic variable* and that $\overline{y} = \frac{1}{y}$. This recurrence can be solved using the kernel method [19, 56, 186, 423, 424]. The coefficient of $B_b(y)$ is the kernel, given by

$$K(y) = 1 - \frac{t(1-s^2)}{1-s(y+\overline{y})+s^2}. \tag{5.136}$$

The roots of the kernel are given by

$$y_p = \tfrac{1}{2s}\left(1-t+s^2+ts^2 \pm \sqrt{(1-s^2)((1-t)^2-s^2(1+t)^2)}\right) \tag{5.137}$$

and the same discriminant seen in equation (5.132) is seen here. The choice of negative sign before the radical shows that y_p enumerates a class of paths, so that this is the 'physical root'.

Substituting $y = y_p$ in the recurrence solves for $B_b(s)$:

$$\begin{aligned} B_b(s) &= \frac{b(y_p - s)}{s-(b-1)(y_p-s)} \qquad (5.138)\\ &= \frac{2bt(1-s^2)}{1+t-s^2(1+t)-2bt(1-s^2)+\sqrt{(1-s^2)((1-t)^2-s^2(1+t)^2)}}, \end{aligned}$$

from which a solution for $B_b(y)$ is obtained.

Define the discriminant $D = (1-s^2)((1-t)^2-s^2(1+t)^2)$; then

$$B_b(y) = \frac{bt(1-sy)\left((s^2-1)(s(1+t)-y(1-t))+(y-s)\sqrt{D}\right)}{(ys^2(1+t)-s(1+y^2)+y(1-t))\left((1-s^2)(1+t-2bt)+\sqrt{D}\right)}.$$

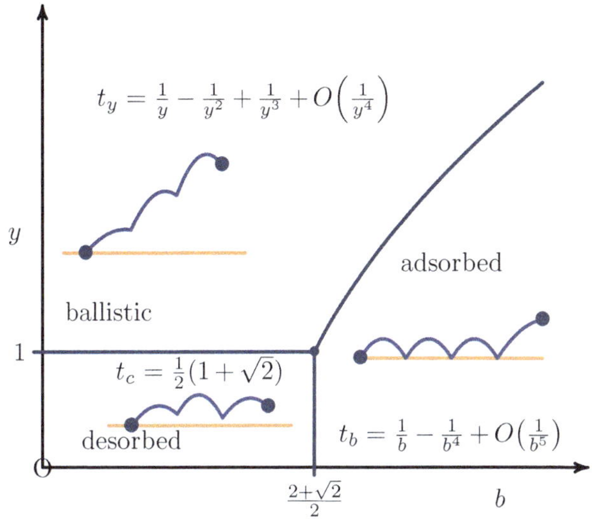

FIG. 5.19. The phase diagram of pulled adsorbing partially directed paths.

Putting $s = t$ gives the generating function of vertically pulled adsorbing partially directed paths:

$$B_c(b, y) = \frac{bt(1-yt)\left(t^4 + (1+y)t(t^2-t-1) + y + (y-t)\sqrt{D}\right)}{(yt^2(1+t) - (1+y+y^2)t + y)\left((1-(2b-1)t)(1-t^2) + \sqrt{D}\right)}$$

where $D = (1-t^4)(1-2t-t^2)$.

There are several sources of singularities in $B_c(b,y)$. The discriminant vanishes when $t = \pm 1$ or $t = \frac{1}{1\pm\sqrt{2}}$, and there are poles when the factors in the denominator vanishes. The dominant singularity is at $t_c = \frac{1}{1+\sqrt{2}}$ and the factor containing b vanishes when $b = 1 + \sqrt{2^{-1}}$, while the factor containing y vanishes when $y = 1$.

This shows that, for small y, there is an adsorption transition at $b = 1 + \sqrt{2^{-1}}$, and, for small b, a transition to a ballistic phase when $y = 1$.

Determining a transition to a ballistic phase for large values of b is more difficult. The factor containing b vanishes when $t = t_b$, and the factor containing y vanishes when $t = t_y$. For large b, $t_b < t_c$, and, for large y, $t_y < t_c$. Thus, the free energy in the model is determined by these singularities. Asymptotic expansions show that $t_b = \frac{1}{b} - \frac{1}{b^4} + O(b^{-5})$ and $t_y = \frac{1}{y} - \frac{1}{y^2} + \frac{1}{y^3} + O(y^{-4})$. These are equal to leading order when $y_c(b) = b - 1 + \frac{2}{b^2} + O(b^{-3})$. This gives the phase diagram in figure 5.19 for the model. The critical curve separating the ballistic and adsorbed phases is asymptotic to $y = b - 1$.

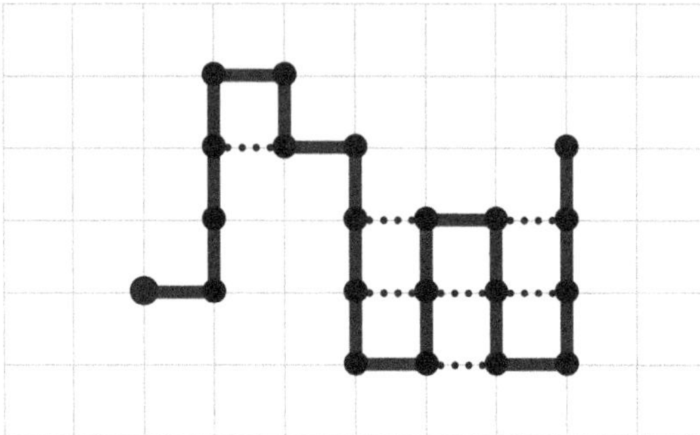

FIG. 5.20. A partially directed path with contacts in $\mathbb{L}^2$. The contacts are the dotted horizontal lattice edges.

5.5.2 *Collapsing partially directed paths*

An edge $\langle \vec{u} \sim \vec{v} \rangle$ is a *contact* in a walk or path ω if $\vec{u}, \vec{v} \in \omega$ but $\langle u \sim v \rangle \notin \omega$. A walk counted by length and contacts is a model of a *self-interacting* polymer in a good solvent which may undergo a *collapse transition* at a θ-point and precipitate out of solution [208, 210]. In figure 5.20 a partially directed path with contacts is illustrated (see references [67, 68]).

Let $r_n(m)$ be the number of partially directed paths with m nearest-neighbour contacts. The partition function of this model is given by $Z_n(x) = \sum_{m \geq 0} r_n(m)\, x^m$. This model has a thermodynamic limit (this follows from the concatenation of two paths in a manner similar to that seen in figure 3.2 and then using arguments similar to those in theorem 3.1). Denote the generating function of the model by

$$K(x,t) = \sum_{n \geq 0} Z_n(x)\, t^n \tag{5.139}$$

so that x is conjugate to contacts, and t is conjugate to edges.

Define the generating function $K_p(x,t)$ of paths with a first step in the east direction, followed by p edges in the north or south directions. Then $K \equiv K(x,t) = \sum_{p \geq 0} K_p(x,t)$.

The Temperley method can be used to find a recurrence for $K_p(x,t) \equiv K_p$, by *prepending* $\lrcorner$-shaped sequences of edges to paths in K_p (see figure 5.21). This gives

$$K_p = t^{p+1} \left[2 + \sum_{q=0}^{p} (1+x^q)\, K_q + (1+x^p) \sum_{q=p+1}^{\infty} K_q \right]. \tag{5.140}$$

By direct substitution, K_p satisfies the homogeneous second order recurrence

$$K_{p+2} - (1+x)t\, K_{p+1} - (1-x)x^{p+1}t^{p+3}\, K_{p+1} + xt^2\, K_p = 0. \tag{5.141}$$

Solutions of this recurrence will have the general product form $A(x,t)\, K_p$, where $A \equiv A(x,t)$ is independent of p. Thus, denote the solutions of equation (5.141)

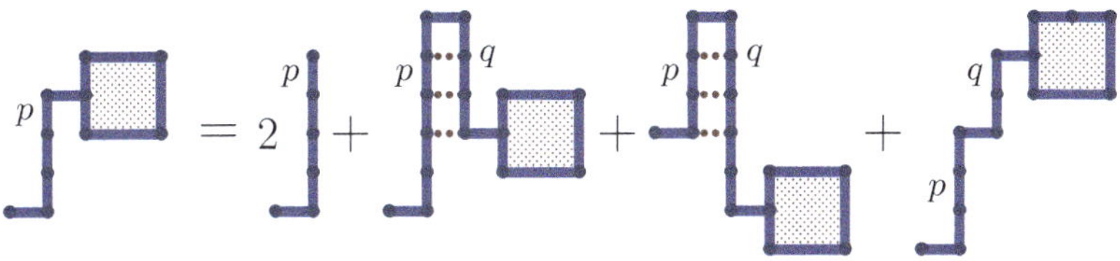

FIG. 5.21. The Temperley method gives the above classification for paths counted by the generating function $K_p(x,t)$. The source term is a single horizontal step followed by p vertical up or vertical down steps. This gives the factor of 2 just after the equal sign. The remaining three cases are obtained by arguing as follows. The first $p+1$ steps are followed by an east step, then by *either* q vertical steps in the opposite direction to the first p vertical steps (with either $p \geq q$ or with $p < q$), *or* by q steps in the same direction as the first p vertical steps.

---

by g_p, so that $2AK_0 = g_0$, and $AK_1 = g_1$; K_0 will have an extra factor of 2, since putting $p = 0$ in figure 5.21 shows the contributions to K_0 to be doubly counted. Since paths contributing to K_0 give two initial steps in the east direction,

$$K_0 = t + tK. \tag{5.142}$$

For K_1 in equation (5.140),

$$K_1 = (2 + t - xt)t^2 + (1 + t + x - xt)t^2 K = u + vK, \tag{5.143}$$

where the functions u and v are coefficients defined in the obvious way.

Given g_0 and g_1, solve simultaneously for α and K from equations (5.142) and (5.143):

$$K = -\frac{2t g_1 - u g_0}{2t g_1 - v g_0}. \tag{5.144}$$

It remains to determine g_0 and g_1. Notice that knowing g_0 and g_1 gives K to within an unknown multiplicative factor A, which must also be determined to find a full generating function.

Let $y = xt$ and let $K_p = \alpha\, g_p$; then equation (5.141) simplifies to

$$g_{p+2} - (2y + (t - y)(1 + y^{p+1}t))\, g_{p+1} + yt\, g_p = 0. \tag{5.145}$$

This type of recurrence has the general solution

$$g_p = \lambda^p \sum_{m=0}^{\infty} r_m(y)\, y^{mp} \tag{5.146}$$

with $r_0(y) = 1$. It remains to determine $r_m(y)$ and λ. Substituting and collecting terms gives

$$\sum_{m=0}^{\infty} (y^m\lambda - y)\,(y^m\lambda - t)\; r_m(y)\, y^{mp} + \lambda(y-t)\, ty \sum_{m=0}^{\infty} r_m(y)\, y^m y^{(m+1)p} = 0.$$

Examining the case $p = 0$ and the $m = 0$ term shows that $(\lambda - y)(\lambda - t) = 0$, so there are two possible solutions for λ: namely, $\lambda = y$, or $\lambda = t$. Replacing $m \to (m+1)$ in the first series above and then collecting the coefficients of $y^{(m+1)p}$ gives a recurrence for $r_m(y)$:

$$r_{m+1}(y) = \frac{\lambda(t-y)ty^{m+1}\, r_m(y)}{(\lambda y^{m+1} - y)\,(\lambda y^{m+1} - t)}. \tag{5.147}$$

Since $r_0(y) = 1$, this can be solved by iteration. This shows that

$$r_m(y) = \frac{\lambda^m (t-y)^m y^{\binom{m}{2}}}{(\lambda y t^{-1}; y)_m\, (\lambda; y)_m}, \tag{5.148}$$

where $(t;q)_n$ denotes the q-Pochhammer function defined in equation (D.41) in appendix D.

It remains to fix the choice for λ. Note that $K_1 = 2t^2 + O(t^3)$, which approaches 0 as $t \to 0^+$. If $\lambda = y$, then $g_1 = y\,(1 + O(t))$, which does not approach 0 as $t \to 0^+$ (with $xt = y$ fixed). Therefore, the choice that $\lambda = y$ can be ruled out. Choosing $\lambda = t$ gives the solutions in terms of q-deformed Bessel functions (see equation (D.48) in appendix D):

$$g_0 = \sum_{m=0}^{\infty} \frac{(1-x)^m t^{2m}\, (xt)^{\binom{m}{2}}}{(t;xt)_m (xt;xt)_m} = J(t, xt, (x-1)t^2); \tag{5.149}$$

$$g_1 = t \sum_{m=0}^{\infty} \frac{(1-x)^m t^{2m}\, (xt)^{\binom{m+1}{2}}}{(t;xt)_m (xt;xt)_m} = J(t, xt, x(x-1)t^3). \tag{5.150}$$

Notice that if $x = 0$, then $g_0 = 1$, and $g_1 = t$, and, by equation (5.144), $K(0,t) = -\frac{2t^2-u}{2t^2-v} = \frac{t}{1-t}$. This is singular when $t = 1$. Since K is a power series in x, it is non-decreasing with x, so the radius of convergence in the t-plane is $t_c(x) \le 1$ for $x \ge 0$.

Proceed by defining $h = \frac{g_0}{g_1}$. Using this in equation (5.144) gives

$$K = \frac{v - u}{v - \frac{2t}{h}}. \tag{5.151}$$

This has poles at zeros of the denominator: both g_0 and g_1 have poles when $t^{n+1}x^n = 1$ for $n \in \mathbb{N}_0$. Since $g_0 \ne g_1$, there are points between these poles where $vh = 2t$. In this case the denominator is 0, and there is a singularity in K. For fixed values of t, the zeros in the denominator accumulate on $xt = 1$; that is, the curve $xt = 1$ is a curve of essential singularities in K.

Other singular points can be extracted as follows. Define

$$h_0(x,t,s) = \sum_{m=0}^{\infty} \frac{(1-x)^m s^m t^{2m} (xt)^{\binom{m}{2}}}{(t;xt)_m (xt;xt)_m}. \tag{5.152}$$

Then $h_0(x,t,1) = g_0$, and $t h_0(x,t,xt) = g_1$ in equation (5.149).

Moreover, $h(x,t) = \frac{g_0}{g_1} = \frac{h_0(x,t,1)}{t\, h_0(x,t,xt)}$. By substitution, $h_0(x,t,s)$ satisfies

$$h_0(x,t,s) - \left(1 + \tfrac{1}{x} + st^2(1-x)\right) h_0(x,t,xts) + \tfrac{1}{s}\, h_0(x,t,(xt)^2 s) = 0. \tag{5.153}$$

Define $H(s) = \frac{h_0(x,t,s)}{h_0(x,t,xts)}$ and let $\alpha_p = 1 + \frac{1}{x} + st^2(1-x)(xt)^p$. Then $H(1) = h(x,t)$.

Division of equation (5.153) by $h_0(x,t,xts)$ shows that

$$H(s) = \alpha_0 \left(1 - \frac{1}{x\alpha_0 H(xts)}\right). \tag{5.154}$$

Along the critical curve $xt = 1$ this becomes a quadratic for $H(s)$. In fact, if $s = 1$, then $\alpha_0 = 1 + \frac{1}{x^2}$ and solving for $H(1)$ gives

$$H(1) = \tfrac{1}{2x^2}\left(1 + x^2 \pm \sqrt{(1-x)(1+x+3x^2-x^3)}\right). \tag{5.155}$$

Since $x H(1) = h(x,t)$ along the curve $xt = 1$ in equation (5.151), it follows that K has a square root singularity on the curve of essential singularities $xt = 1$. Approaching this critical point along $xt = 1$ is also an approach to the critical point along the τ_0-curve. By equation (4.12), $2 - \alpha_u = \frac{1}{2}$.

In these circumstances $v = \frac{x^2+1}{x^3}$ in equation (5.151). The denominator in equation (5.151) vanishes if $h = xH(1)$ along the curve $xt = 1$; this occurs when $vh = 2t$, and this shows

$$\frac{(x-1)\sqrt{(1-x)(1-x-3x^2-x^3)}}{x(x^2+1)} = 0. \tag{5.156}$$

This shows critical values on $xt = 1$ for either $x = 1$ or when $1 + x + 3x^2 - x^3 = 0$. It can be checked that the singularity at $x = 1$ cancels in equation (5.151), so the roots of the cubic must be considered. This gives the critical point at $x_c = 3.382975\ldots$.

If $x > x_c$, then the critical curve is $t_c(x) = \frac{1}{x}$; this is a curve of essential singularities. If $x < x_c$, then $t_c(0) = 1$, and $t_c(x)$ must meet the curve $xt = 1$ at the critical point $\left(x_c, \frac{1}{x_c}\right)$, since the free energy is convex and continuous. There is a non-analyticity at x_c, corresponding to the collapse transition in the model.

TABLE 5.5. Collapsing partially directed path tricritical exponents

ϕ	α	$2-\alpha_t$	$2-\alpha_u$	y_t	ν_t	$2-\alpha_-$
$\frac{2}{3}$	$\frac{1}{2}$	$\frac{1}{3}$	$\frac{1}{2}$	4	$\frac{1}{6}$	$\frac{1}{4}$

The function K has no singularities if $x > x_c$, and $0 \leq t < \frac{1}{x}$. This may be shown by writing equation (5.154) as a continued fraction:

$$H(s) = \alpha_0 - \cfrac{\alpha_0}{x\,\alpha_0\alpha_1 \left(1 - \cfrac{1}{x\,\alpha_1\alpha_2\left(1 - \frac{1}{x\,\alpha_2\alpha_3(1-\cdots)}\right)}\right)} = \alpha_0(1-C), \qquad (5.157)$$

where C is defined to simplify the expression. By Worpitzky's theorem [572], this converges if $\inf_{p\geq 0} |x\,\alpha_p\alpha_{p+1}| \geq 4$, in which case $\left|C - \frac{4}{3}\right| \leq \frac{2}{3}$. It follows that $-x\,\alpha_0 \leq h \leq \frac{x}{3}\alpha_0$. There is an singularity in K if the denominator in equation (5.151) vanishes. By using monotonicity and calculating the bounds on h explicitly at some values of t and x (where $x > x_c$, and $xt < 1$), it follows that there are no other singularities in K if $x > x_c$, and $xt < 1$.

Derivatives of g_0 and g_1 can be taken in order to find the derivative of K. Using a symbolic computations programme, this approach gives the gap exponent $\Delta = \frac{3}{2}$ (see equation (4.16)). Thus, the crossover exponent is $\phi = \frac{2}{3}$; therefore $2-\alpha_t = \frac{1}{3}$ by equation (4.14). By equation (4.14), it follows that $2-\alpha_u = \frac{1}{2}$, as argued below equation (5.155).

The rest of the tricritical exponents may be computed from these results. It follows that $\nu_t = \frac{1}{y_g} = \frac{1}{6}$ by equation (4.53) and so $y_t = 4$. The specific heat exponent is $\alpha = \frac{1}{2}$ by equation (4.29). The tricritical exponents for this model are listed in table 5.5 (see reference [459]). Scaling in the collapsed polymer phase was examined in reference [459] and it was found that $2-\alpha_- = \frac{1}{4}$ in equation (4.18), and $\sigma = \frac{1}{2}$ in equations (4.38) and (4.41).

5.6 Staircase polygons

A staircase polygon consists of two fully directed paths of the same length with common endpoints but which avoid one another otherwise (see figure 5.22). Staircase polygons are also known as *parallelogram polygons* [60, 64, 127] and have been studied at least since the 1950s (see for example [544]).

The perimeter generating function of staircase polygons are related to Catalan numbers. To see this, notice that the number of directed paths from $\vec{0}$ to the vertex $(r, n-r)$ in $\mathbb{L}^2$ is $\binom{n}{r}$. Thus, the total number of pairs of directed paths from $\vec{0}$ has the generating function

$$g(t) = \sum_{n=0}^{\infty}\sum_{r=0}^{n}\binom{n}{r}^2 t^{2n} = \sum_{n=0}^{\infty}\binom{2n}{n}t^{2n} = \frac{1}{\sqrt{1-4z^2}}. \qquad (5.158)$$

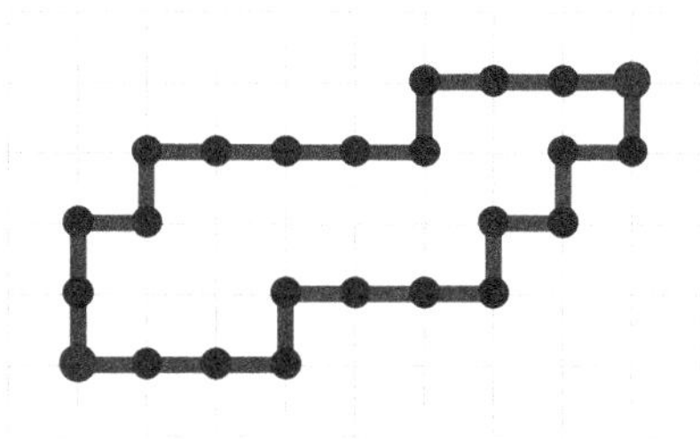

FIG. 5.22. A staircase polygon.

Each realisation of a pair of directed paths from $\vec{0}$ to $(r, n-r)$ generates a sequence of staircase polygons or double edges, as shown in figure 5.23. Each path has two ways of circling the perimeter of each staircase polygon in the chain, so the generating function $g(t)$ will count all conformations twice, except for the case when $n = 0$ or when $n = 1$.

Let $H_0(t)$ be the generating function of staircase polygons (including the degenerate cases of twice-counted double edges). Then

$$g(t) = 1 + H_0(t) + H_0^2(t) + \cdots = \frac{1}{1 - H_0(t)}. \tag{5.159}$$

The perimeter generating function of staircase polygons, $H(t)$, is related to $H_0(t)$ by $H(t) = \frac{1}{2}\left(H_0(t) - 2t^2\right)$ (where $2t^2$ is the contribution of double edges to $H_0(t)$). Solving for $H(t)$ gives

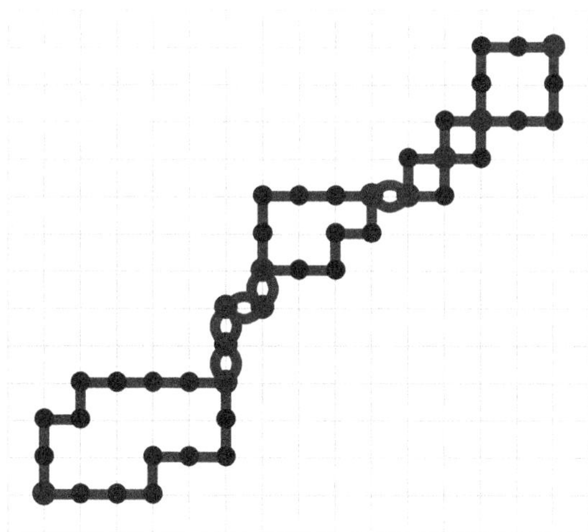

FIG. 5.23. Two directed paths with common endpoints form a chain of staircase polygons and double edges.

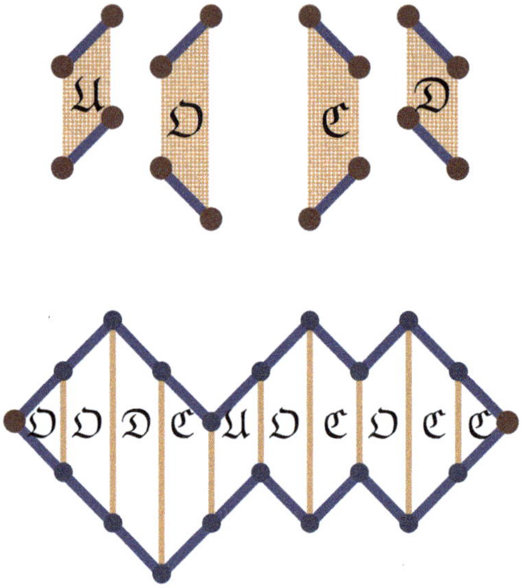

FIG. 5.24. Coding a staircase polygon in an algebraic language.

$$H(t) = \tfrac{1}{2}\left(1 - 2t^2 - \sqrt{1-4t^2}\right) = \frac{2t^4}{1 - 2t^2 + \sqrt{1-4t^2}}. \tag{5.160}$$

This is related to the Catalan generating function (see equation (5.4)).

Staircase polygons can be coded into an algebraic language, as shown in figure 5.24. The polygon is rotated by 45° so that it is oriented from a source vertex on the left to a sink vertex on the right. Two directed paths start in the source vertex and end in the sink vertex, each avoiding the other and taking only north-east or south-east steps. By joining vertices between the two Dyck paths by vertical lines as illustrated, the staircase polygon becomes decomposed into trapezoidal slabs (labelled by $\{\mathfrak{O},\mathfrak{U},\mathfrak{D},\mathfrak{C}\}$), illustrated in figure 5.24.

Across slabs labelled $\mathfrak{O}$ the paths move away from each other, and across slabs labelled $\mathfrak{C}$ they move closer to one another. Similarly, across slabs labelled $\mathfrak{U}$ both paths take a north-east step, and across slabs labelled $\mathfrak{D}$ both take a south-east step.

This coding maps each staircase polygon to a word in an algebraic language. This word necessarily starts in a $\mathfrak{O}$ and terminates in a $\mathfrak{C}$. These letters may be removed and it would still be possible to recover the polygon from the remainder; a staircase polygon of length $2k+2$ will be coded by a word of length $k-1$.

Suppose that w is a word coding a staircase polygon. Since the path is constrained to be closed, $|w|_{\mathfrak{O}} = |w|_{\mathfrak{C}}$. In addition, since the top Dyck path cannot intersect the bottom Dyck path, for any prefix u of w, $|u|_{\mathfrak{O}} \geq |u|_{\mathfrak{C}}$.

Comparison to the constraints on the words in a 2-coloured Motzkin language (see section 5.4) shows that words in the staircase polygon language comprise a subset of the words in the 2-coloured Motzkin language. On the other hand, if w is a 2-coloured Motzkin word of length $k-1$, then any prefix u of w satisfies $|u|_{\mathfrak{O}} \geq |u|_{\mathfrak{C}}$ and corresponds to a staircase polygon.

Thus, there is a bijection between staircase polygons of length $2k+2$ and 2-coloured Motzkin words of length $k-1$ so the generating function for staircase polygons can be obtained from the generating function of 2-coloured Motzkin paths $M_2(t)$ (see equation (5.106)); each letter in the language corresponds to two edges in the staircase polygon (so $t \to t^2$), and by prepending and appending two edges each to close the two directed paths into a single polygon. This shows that $H(t) = t^4\, M_2(t^2)$.

The coding of staircase polygons into an algebraic language also relates them to square lattice random walks. Code a lattice walk as follows: $\mathfrak{O}$ is a north step, $\mathfrak{C}$ is a south step, $\mathfrak{U}$ is an east step, and $\mathfrak{D}$ is a west step. Then 2-coloured Motzkin words are paths from $\vec{0}$ in the half-lattice $\mathbb{L}^2_+$ (see equation (2.20)), constrained to terminate in a vertex in the boundary $\partial\mathbb{L}^2_+$. The mapping of directed path models to random walks in subsets of the square lattice was shown in reference [187].

5.6.1 *The partition function of adsorbing staircase polygons*

An adsorbing staircase polygon in $\mathbb{J}^d_+$ is illustrated in figure 5.25. The left-most vertex in the polygon has the Cartesian coordinates $(-1,1)$, and $\vec{0}$ is necessarily a visit, weighted by a (the visit generating variable). These staircase polygons are said to be *grafted* to $\partial\mathbb{J}^d_+$.

The staircase polygon in figure 5.25 can be coded into words of a 2-coloured Motzkin language as explained in section 5.6 (see figure 5.24). If u is any prefix of a word w corresponding to a staircase polygon, then w and u have the following properties:

(1) $|w|_{\mathfrak{U}} \geq |w|_{\mathfrak{D}}$, and $|w|_{\mathfrak{O}} = |w|_{\mathfrak{C}}$;
(2) $|u|_{\mathfrak{U}} \geq |u|_{\mathfrak{D}}$, and $|u|_{\mathfrak{O}} \geq |u|_{\mathfrak{C}}$; and
(3) $|u|_{\mathfrak{C}} + |u|_{\mathfrak{U}} \geq |u|_{\mathfrak{O}} + |u|_{\mathfrak{D}}$.

In addition, 2-coloured Motzkin words may be represented by random walks in $\mathbb{L}^2_+$. Let $\mathfrak{O}$ be a north step, $\mathfrak{C}$ a south step, $\mathfrak{U}$ an east step and $\mathfrak{D}$ a west step. Then the constraints above give a random walk from $\vec{0} \in \mathbb{L}^2_+$ for each adsorbing polygon. The random walk has the following properties:

(1) the walk starts at $\vec{0} \in \mathbb{L}^2_+$;
(2) the walk steps on vertices in $\mathbb{L}^2_+$ *below or on* the *main diagonal*;
(3) each visit of the random walk to the main diagonal generates a visit of the bottom path of the staircase polygon to $\partial\mathbb{L}^d_+$; and
(4) the walk must terminate in $\partial\mathbb{L}^d_+$ to close the polygon.

Thus, grafted staircase polygons are in one-to-one correspondence with random walks from $\vec{0}$ in the wedge $\mathcal{W}$ below the main diagonal of the half-lattice $\mathbb{L}^d_+$ (see figure 5.26).

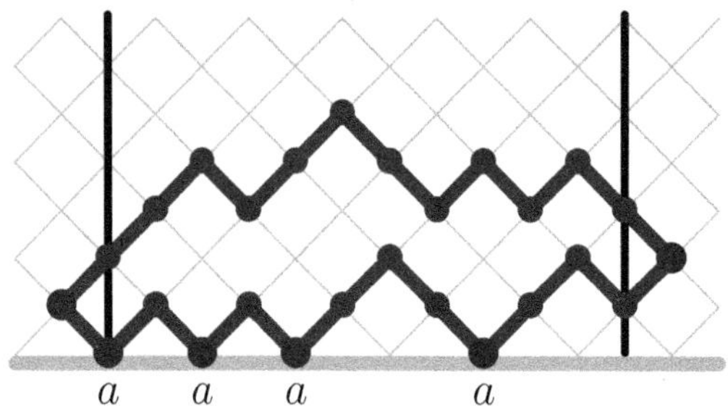

FIG. 5.25. An adsorbing staircase polygon in $\mathbb{J}_+^2$. The bottom path is rooted at the left-most visit.

The wedge $\mathcal{W}$ is the *principal wedge*, and the random walk adsorbs into the main diagonal which is part of the boundary of $\mathcal{W}$ (denoted by $\partial\mathcal{W}$, and since visits are weighted by a). Thus, adsorbing staircase polygons are related to a model of random walks in subgraphs of $\mathbb{L}^d$; this is a rich field with interesting problems (see for example references [410, 423, 424]).

Walks of length n from $\vec{0}$ in $\mathcal{W}$ and ending in the lattice site with coordinates (j, l) correspond to pairs of directed paths in $\mathbb{J}_+^d$ avoiding each other. The bottom directed path starts in $\vec{0}$ and ends in the vertex $(n, j-l)$, and the top path starts from $(0, 2)$ and ends in the vertex $(n, j+l+2)$. Visits of the random walks to the main diagonal corresponds to visits of the bottom path to $\mathbb{J}_+^d$, where it picks up a factor a.

If the walk ends in (n, j), then the paths have endpoints (n, j) and $(n, j+2)$; by prepending two edges, and appending two edges, a staircase polygon is recovered (see figure 5.27).

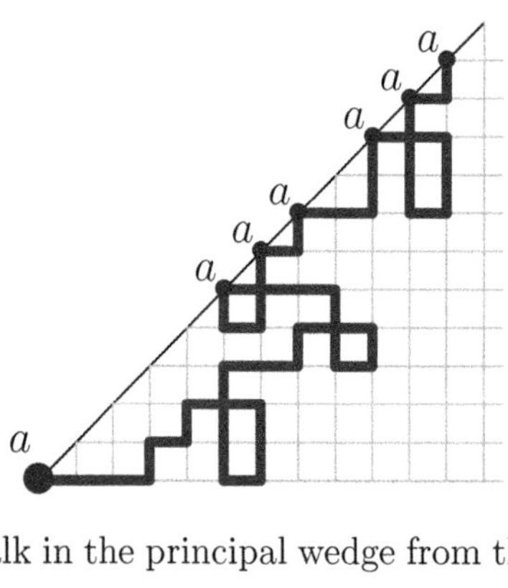

FIG. 5.26. A random walk in the principal wedge from the origin and adsorbing in the main diagonal.

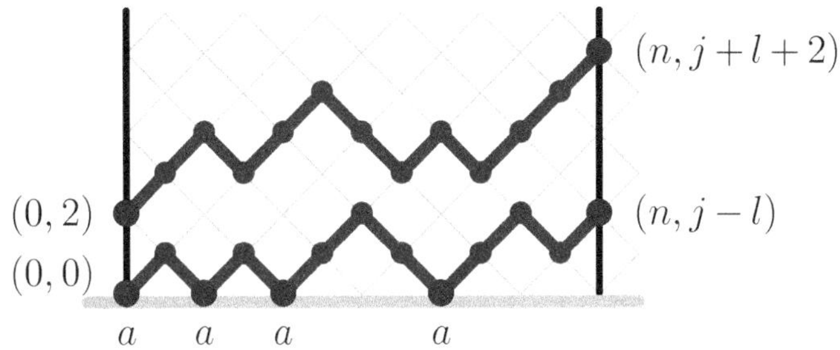

FIG. 5.27. A pair of directed paths in $\mathbb{J}^2_+$ with fixed endpoints. The paths avoid one another.

A random walk in $\mathcal{W}$ (which adsorbs on the main diagonal) can be formulated by a system of partial difference equations with suitable boundary conditions. Let $r_n(j,l)$ be the number of random walks of length n which arrive at site (j,l) after starting at site (j_0,l_0). Then,

$$\begin{aligned} r_n(j,l) &= r_{n-1}(j-1,l) + r_{n-1}(j+1,l) && (5.161) \\ &\qquad + r_{n-1}(j,l-1) + r_{n-1}(j,l+1), && \forall 1 \leq l \leq j; \\ r_n(j,0) &= r_{n-1}(j-1,0) + r_{n-1}(j+1,0) + r_{n-1}(j,1), && \forall j \geq 1; \\ r_n(j,j) &= a\left(r_{n-1}(j+1,j) + r_{n-1}(j,j-1)\right), && \forall j \geq 1. \end{aligned}$$

The initial condition is $r_0(0,0) = a$, since the walk of zero length has one visit at $\vec{0}$. Notice that $r_n(j,l)$ is also the number of pairs of Dyck paths avoiding each other in $\mathbb{J}^d_+$; one path from $\vec{0}$ to its end-vertex at height $j-l$, and the other path from the vertex $(0,2)$ to its end-vertex at height $j+l+2$. As noted above, staircase polygons are recovered when $l=0$, in which case the addition of four edges closes the two paths into a polygon. If $j=0$ as well, then the last vertex of the bottom path is also a visit, and the resulting staircase polygon is said to be grafted at both ends.

The system of equations (5.161) can be solved by a constant term method. Put $\zeta_m = e^{ik_m}$, and put $\overline{\zeta}_m = e^{-ik_m}$. Exhaust all possible combinations of ζ_1 and ζ_2 by assuming the following general solution $r_n(j,l)$ via a Bethe ansatz:

$$\begin{aligned} r_n(j,l) = \Lambda^n \Big(& A_1^{11}\zeta_1^j\zeta_2^l + A_1^{21}\overline{\zeta}_1^j\zeta_2^l + A_1^{12}\zeta_1^j\overline{\zeta}_2^l + A_1^{21}\overline{\zeta}_1^j\overline{\zeta}_2^l \\ & + A_2^{11}\zeta_2^j\zeta_1^l + A_2^{21}\overline{\zeta}_2^j\zeta_1^l + A_2^{12}\zeta_2^j\overline{\zeta}_1^l + A_2^{22}\overline{\zeta}_2^j\overline{\zeta}_2^l \Big). \end{aligned} \tag{5.162}$$

Substitution into equation (5.161) gives

$$\Lambda = \zeta_1 + \overline{\zeta}_1 + \zeta_2 + \overline{\zeta}_2. \tag{5.163}$$

Substitution into the second and third equations in (5.161) (these are boundary conditions) gives relations amongst the constants A_i^{jk}:

$$A_1^{12} = -\overline{\zeta}_2^2 A_1^{11};\ A_1^{21} = -\zeta_2^2 A_1^{22};\ A_2^{12} = -\overline{\zeta}_1^2 A_2^{11};\ \text{and } A_2^{21} = -\zeta_1^2 A_2^{22}. \tag{5.164}$$

Define the scattering function

$$\mathfrak{T}(\alpha,\beta) = -\frac{\alpha+\bar{\alpha}+\beta+\bar{\beta}-a(\alpha+\beta)}{\alpha+\bar{\alpha}+\beta+\bar{\beta}-a(\bar{\alpha}+\bar{\beta})}. \tag{5.165}$$

Then the following relations are found:

$$A_2^{11} = \mathfrak{T}_{1\bar{2}}\, A_1^{11};\ A_2^{12} = \mathfrak{T}_{\overline{12}}\, A_1^{21};\ A_2^{21} = \mathfrak{T}_{12}\, A_1^{12};\ \text{and } A_2^{22} = \mathfrak{T}_{\overline{1}2}\, A_1^{22}, \tag{5.166}$$

where the notation was simplified by $\mathfrak{T}_{12} = \mathfrak{T}(\zeta_1,\zeta_2)$, by $\mathfrak{T}_{1\bar{2}} = \mathfrak{T}(\zeta_1,\overline{\zeta}_2)$, and so on. The eight relations in equations (5.164) and (5.166) can be used to eliminate all but one of the A_i^{jk}. This gives

$$\begin{aligned} r_n(j,l) = A_1^{11}\Lambda^n\Big(\Big(\zeta_1^j\zeta_2^l+\zeta_2^j\zeta_1^l\mathfrak{T}_{1\bar{2}}\Big) - \overline{\zeta}_2^2\Big(\zeta_2\overline{\zeta}_2^l+\overline{\zeta}_2^j\zeta_1^l\mathfrak{T}_{12}\Big) \\ -\overline{\zeta}_1^2\mathfrak{T}_{1\bar{2}}\Big(\zeta_2^j\overline{\zeta}_1^l+\overline{\zeta}_1^j\zeta_2^l\mathfrak{T}_{12}\Big)+\overline{\zeta}_1^2\overline{\zeta}_2^2\mathfrak{T}_{12}\Big(\overline{\zeta}_2^j\overline{\zeta}_1^l+\overline{\zeta}_1^j\overline{\zeta}_2^l\mathfrak{T}_{1\bar{2}}\Big)\Big). \end{aligned} \tag{5.167}$$

Assuming that the random walk starts at the site (j_0,l_0), then the proper choice for A_1^{11} is the complex conjugate of $r_0(j_0,l_0)$, multiplied by a constant C_0. Expanding the product gives sixty-four terms. Integration over k_1 and k_2 allows the use of symmetries of the integrand which reduces the sixty-four terms to just eight. The final result is

$$\begin{aligned} r_n(j,l;j_0,l_0) = 8\,C_0\int_{-\pi}^{\pi}\int_{-\pi}^{\pi}\Lambda^n\,\zeta_1^j\zeta_2^l\Big(\overline{\zeta}_1^{j_0}\overline{\zeta}_2^{l_0}+\mathfrak{T}_{\bar{1}2}\overline{\zeta}_1^{l_0}\overline{\zeta}_2^{j_0}-\zeta_2^2\overline{\zeta}_1^{j_0}\zeta_2^{l_0} \\ -\zeta_2^2\mathfrak{T}_{\overline{12}}\overline{\zeta}_1^{l_0}\zeta_2^{j_0}-\zeta_1^2\mathfrak{T}_{\bar{1}2}\mathfrak{T}_{\overline{12}}\zeta_1^{j_0}\overline{\zeta}_2^{l_0}-\zeta_1^2\mathfrak{T}_{\bar{1}2}\zeta_1^{l_0}\overline{\zeta}_2^{j_0} \\ +\zeta_1^2\zeta_2^2\mathfrak{T}_{\bar{1}2}\mathfrak{T}_{\overline{12}}\zeta_1^{j_0}\zeta_2^{l_0}+\zeta_1^2\zeta_2^2\mathfrak{T}_{\overline{12}}\zeta_1^{l_0}\zeta_2^{j_0}\Big)\,dk_1dk_2. \end{aligned} \tag{5.168}$$

Putting $\zeta_j = e^{ix_j}$ turns the above into a contour integral and shows that the effect of the integrals is to select the constant term in the integrand.

The case that $j_0 = l_0 = 0$ corresponds to a random walk from the origin in $\mathcal{W}$, and this gives the following constant term formula for $r_n(j,l)$ (where $\mathrm{CT}[\cdot]$ selects the constant term):

$$\begin{aligned} r_n(j,l) =& 8\,C_0\,\mathrm{CT}\Big[\Lambda^n{\zeta_1}^j{\zeta_2}^{l+1}\Big(\overline{\zeta}_2(1+\mathfrak{T}_{\bar{1}2})(1+{\zeta_1}^2{\zeta_2}^2\mathfrak{T}_{\overline{12}}) \\ & -\zeta_2(1+\mathfrak{T}_{\overline{12}})(1+\zeta_1^2\overline{\zeta}_2^2\mathfrak{T}_{\bar{1}2})\Big)\Big]. \end{aligned} \tag{5.169}$$

This may be simplified by changing variables: $\zeta_1 = \eta_1\eta_2$, and $\zeta_2 = \eta_1\overline{\eta}_2$. Substitution and simplification gives $\Lambda = (\eta_1+\overline{\eta}_1)(\eta_2+\overline{\eta}_2)$; $\mathfrak{T}_{\bar{1}2} = \mathfrak{S}(\eta_2)$; and $\mathfrak{T}_{\overline{12}} = \mathfrak{S}(\eta_1)$, where

$$\mathfrak{S}(\eta) = -\frac{\Lambda - a\,\overline{\eta}}{\Lambda - a\,\eta}. \tag{5.170}$$

This gives the following expression for $r_n(j,l)$:

$$r_n(j,l) = 8\,C_0\,\mathrm{CT}\Big[\Lambda^n {\eta_1}^{j+l} {\eta_2}^{j-l}\Big((1+\mathfrak{S}(\eta_2))\,(1+{\eta_1}^4\mathfrak{S}(\eta_1)) - {\eta_1}^2\overline{\eta_2}^2\,(1+\mathfrak{S}(\eta_1))\,(1+{\eta_2}^4\mathfrak{S}(\eta_2))\Big)\Big]. \qquad (5.171)$$

Putting $n = 0$ and putting $j = l = 0$ gives $r_0(0,0) = 8C_0 a$, from which it is concluded that $C_0 = \frac{1}{8}$ since $r_0(0,0) = a$. Selecting constant terms in the above using a symbolic computations programme gives $r_0(0,0) = a$; $r_2(0,0) = a^2$; $r_4(0,0) = a^2 + 2a^3$; and $r_6(0,0) = 3a^2 + 6a^3 + 5a^4$ (see for example reference [64]).

If $a = 1$, the constant term can be extracted from equation (5.171). Put $U_n(j,l) = r_n(j,l) \mid_{a=1}$. Then $U_n(j,l)$ is given by

$$\begin{aligned} &U_n(j,l) && (5.172)\\ &= \mathrm{CT}\Big[(\eta_1+\overline{\eta}_1)^n\,(\eta_2+\overline{\eta}_2)^n\,\eta_1^{j+l}\eta_2^{j-l}\,(1-\eta_1^2)\,(1-\eta_2^2)\,(\eta_1^2-\eta_2^2)\,(\eta_1^2-\overline{\eta}_2^2)\Big] \\ &= \frac{(l+1)(j+2)(j-l+1)(j+l+3)}{(n+1)(n+2)(n+3)^2}\binom{n+3}{\frac{1}{2}(n+j-l)+2}\binom{n+3}{\frac{1}{2}(n+j+l)+3}. \end{aligned}$$

The function $U_n(j,l)$ is the number of random walks of length n starting from $\vec{0}$ in $\mathcal{W}$ and terminating in (j,l). In the context of staircase polygons, this is the number of pairs of directed paths in $\mathbb{J}^2_+$ (see figure 5.27) starting in $(0,0)$ and $(0,2)$, respectively, and ending in vertices with coordinates $(n, j-l)$ and $(n, j+l+2)$.

By expanding denominators in equation (5.171) and simplifying the resulting expression, it is found that

$$r_n(j,l) = a\sum_{m_1=0}^{n}\sum_{m_2=0}^{n+1} U_n(j+m_1+m_2, l+m_1-m_2)\,(a-1)^{m_1+m_2}. \qquad (5.173)$$

If $a = 1$, then this solution corresponds to random walks in $\mathcal{W}$ with a reflecting boundary condition. If $a = 0$, then the solution corresponds to adsorbing boundary conditions.

The function $r_n(j,0)$ is the partition function of staircase polygons if four edges are added to close the two paths into a staircase polygon.

Putting $l = 0$ and summing over j gives a model of adsorbing staircase polygons with final endpoints at any height. If n is even, then the summation must be over even values of j. The result is

$$s_{2n}(a) = \sum_{j=0}^{n} r_{2n}(2j,0) = \frac{a}{n+1}\binom{2n}{n}\sum_{k=0}^{n}\frac{k+1}{n+1}\binom{2n+2}{n-k}(a-1)^k. \qquad (5.174)$$

Multiplying this by t^{4n}, summing over n and inserting an extra factor of t^4 for the first two and last two edges give the generating function of grafted adsorbing staircase polygons of perimeter length $4n$ in $\mathbb{J}^2_+$. The result is

$$H^e(a,t) = at^4 \sum_{k=0}^{\infty} C_k \left[t^4(a-1)\right]^k \qquad (5.175)$$
$$\times {}_3F_2\left([k+1, k+\tfrac{3}{2}, k+\tfrac{1}{2}]; [2k+3, k+2]; 16t^4\right),$$

where C_k is Catalan's number, and ${}_3F_2(\cdot)$ is a generalised hypergeometric function (in standard notation).

The generating function for staircase polygons of perimeter length $4n+2$ can be found by summing over odd values of j in equation (5.173). The result is

$$H^o(a,t) = 2\,at^6 \sum_{k=0}^{\infty} \frac{2k+1}{k+2} C_k \left(t^4(a-1)\right)^k \qquad (5.176)$$
$$\times {}_3F_2([k+1, k+\tfrac{3}{2}, k+\tfrac{1}{2}]; [2k+3, k+2]; 16t^4).$$

The generalised hypergeometric functions in equations (5.175) and (5.176) are convergent on the closed unit disk in the complex plane for any k, with an essential singularity at $16t^4 = 1$, or $t = \pm\frac{1}{2}$. To find the radius of convergence $t_c(a)$ of these generating functions, examine equation (5.174). This gives

$$t_c(a) = \lim_{n\to\infty} \left(s_{2n}(a)\right)^{-1/2n} = \frac{\sqrt[4]{a-1}}{\sqrt{2a}}. \qquad (5.177)$$

The density function of visits can be computed from this using equation (3.19):

$$\mathcal{P}_S(\epsilon) = \sqrt[4]{6-16\epsilon}\left(\frac{2-4\epsilon}{1-4\epsilon}\right)^{1/2-\epsilon} \quad \text{with } \epsilon \in [0, \tfrac{1}{4}]. \qquad (5.178)$$

The right-derivative at $\epsilon = 0$ is $\frac{d^+}{d\epsilon}\mathcal{P}(\epsilon)\Big|_{\epsilon=0} = -\log 2$. By equation (3.62), this shows that the critical point is at $a_c = 2$. The radius of convergence is given by $t_c(a) = \frac{1}{2}$ if $a \leq 2$ and, by equation (5.177), if $a \geq 2$. The continuity of the first derivative of $t_c(a)$ at $a = 2$ shows that the adsorption transition is continuous. Expanding the critical curve about $a_c = 2$ gives the shift-exponent, which shows that the crossover exponent is $\phi = \frac{1}{2}$.

Putting $a = 2$ in equation (5.175) gives

$$H^e(2,t) = \tfrac{2}{\pi} \sum_{n=0}^{\infty} \frac{\Gamma(n+\frac{1}{2})\,\Gamma(n+\frac{3}{2})}{\Gamma(n+2)\,\Gamma(n+2)} \left(16t^4\right)^n. \qquad (5.179)$$

Since $\lim_{n\to\infty}\left[n^2\,\frac{\Gamma(n+1/2)\Gamma(n+3/2)}{\Gamma(n+2)\Gamma(n+2)}\right] = 1$, it follows that $H^e(2,t)$ is convergent for $0 \leq t \leq \frac{1}{2}$. Expanding $H^e(2,t)$ as $t \to t_c^-$ gives

$$H^e(2,t) \sim \tfrac{2}{\pi}\left(\tfrac{\pi^2}{6} - (1-16t^4)(1-\log(1-16t^4))\right).$$

The singular part of $H^e(2,t)$ suggests that the assignment $2-\alpha_t = 1$ may be made, from which it follows that $2-\alpha_u = \frac{1}{2}$, by equation (4.14).

5.6.2 *Asymptotics of adsorbing staircase polygons*

The partition function of adsorbing staircase polygons is given by equation (5.173). Putting $l = j$ gives a model of two directed paths in $\mathbb{J}^2_+$ where the bottom path ends in the vertex $(n, 0)$ (see figure 5.27). Each of the paths must have even length, so replace n by $2n$. Then the partition function of two Dyck paths as illustrated in figure 5.27 is obtained; each path is of length $2n$, with the bottom path terminating in $(2n, 0)$, and the top path terminating in $(2n, 2j+2)$. Explicitly, the expression in equation (5.173) simplifies to

$$s_n(a;j) = \sum_{m_1=0}^{n} \sum_{m_2=0}^{j} K(m_1, m_2, j, n) \times \binom{2n+3}{n+m_2+2}\binom{2n+3}{n+m_1+j+3}(a-1)^{m_1+m_2}, \tag{5.180}$$

where

$$K(m_1, m_2, j, n) = \frac{(2m_2+1)(2m_1+2j+3)(m_1+m_2+j+2)(m_1-m_2+j+1)}{(2n+1)(2n+2)(2n+3)^2}.$$

Consider the case that $a = 2$. For fixed values of j, $s_n(2;j)$ simplifies to

$$s_n(2;j) = \frac{2(2n+1)(j+1)^2 C_n}{(n+j+1)(n+j+2)} \binom{2n}{n+j}. \tag{5.181}$$

where $C_n = \frac{1}{n+1}\binom{2n}{n}$ is Catalan's number. If this is summed over $j \in [0, n]$, then

$$\sum_{j=0}^{n} s_n(2;j) = \frac{\Gamma(n+\frac{1}{2})16^n}{\sqrt{\pi}\,\Gamma(n+2)}. \tag{5.182}$$

Asymptotics can be developed for equation (5.181). Substitute $n = \frac{1}{\epsilon}$, and substitute $j = \frac{\delta}{\epsilon}$. Expand $\log s_n(2;j)$ asymptotically in ϵ. This gives leading order terms

$$\tfrac{1}{\epsilon}\log 16 + \log(4\epsilon\delta^2) - \left(\tfrac{1}{\epsilon} - \tfrac{\delta}{\epsilon} + \tfrac{1}{2}\right)\log(1-\delta) - \left(\tfrac{1}{\epsilon} + \tfrac{\delta}{\epsilon} + \tfrac{1}{2}\right)\log(1+\delta) + O(\epsilon).$$

Exponentiate this, replace $\epsilon = \frac{1}{n}$, replace $\delta = \frac{j}{n}$ and simplify. This shows that

$$s_n(2;j) \simeq \frac{4\,j^2\,16^n}{\pi n(n+j)^2\left(1-\frac{j}{n}\right)^{n-j+1/2}\left(1+\frac{j}{n}\right)^{n+j+1/2}}. \tag{5.183}$$

If $j = 0$, then closed staircase polygons are obtained. In that case

$$s_n(2;0) = \frac{4 \cdot 16^n}{\pi n^3}\left(1+o(1)\right). \tag{5.184}$$

5.6.2.1 *Asymptotics for $a < 2$:* Put $m_1 + m_2 = k$ and sum over m_1 in equation (5.180):

$$s_n(a,j) = \sum_{k=0}^{n} \frac{(j+k+2)}{(n+1)^2} \binom{2n}{n} \binom{2n+2}{n+j+k+3} (a-1)^k \tag{5.185}$$

$$- \sum_{k=0}^{n} \left[\frac{(j+k+2)(n-2j(k+1)-2k-1)}{2(n+1)^2(2n+1)} \right] \binom{2n+2}{n+k+2} \binom{2n+2}{n+j+2} (a-1)^k.$$

That is, there is a single summation (over k) left in each of the terms; this summation will be approximated by an integral later.

Use gamma functions to represent the binomial coefficients above. The two summands become

$$C_1 = \frac{8\,(j+k+2)\,\Gamma^2\!\left(n+\frac{3}{2}\right)(a-1)^k\,16^n}{\pi\,(n+1)(2n+1)\,\Gamma(n+j+k+4)\,\Gamma(n-j-k)}, \quad \text{and} \tag{5.186}$$

$$C_2 = \frac{2\,(j+k+2)(2j(k+1)-n+2k+1)\,\Gamma^2(2n+2)\,(a-1)^k}{(2n+1)\,\Gamma(n+j+3)\,\Gamma(n-j+1)\,\Gamma(n+k+3)\,\Gamma(n-k+1)}, \tag{5.187}$$

where C_1 is the summand of the first summation in equation (5.185), and C_2 is the summand of the second summation.

Substitute $n = \frac{1}{\epsilon}$, substitute $j = \frac{\delta}{\epsilon}$ and substitute $k = \frac{\alpha}{\epsilon}$. Approximate the gamma functions by using Stirling's approximation. Take logarithms and expand the resulting expressions in ϵ. The leading term in C_1 is

$$\log\left(\frac{16\,(1-\delta-\alpha)^{\delta+\alpha-1}(a-1)^{\alpha}}{(1+\delta+\alpha)^{\delta+\alpha+1}} \right). \tag{5.188}$$

Taking the derivative and solving for α gives

$$\alpha_m = \max\{\tfrac{a-2}{a} - \delta, 0\} \tag{5.189}$$

since α_m cannot be negative.

A similar approach to C_2 gives the leading term in an expansion in ϵ:

$$\log\left(\frac{16\,(1-\delta)^{\delta-1}(1-\alpha)^{\alpha-1}(a-1)^{\alpha}}{(1+\alpha)^{\alpha+1}(1+\delta)^{\delta+1}} \right). \tag{5.190}$$

Taking the derivative and solving for α gives

$$\alpha_M = \max\{\tfrac{a-2}{a}, 0\} \tag{5.191}$$

since α_M cannot be negative.

These results for α_m and α_M give the dominating terms in k for C_1 and C_2 as n becomes large. This is useful when approximating these summands.

Suppose that $a < 2$; in this case both $\alpha_m = 0$ and $\alpha_M = 0$, so the summands are dominated by terms with $k \ll j$, and j small. Thus, substitute $j = \frac{\delta}{\epsilon}$, and

$n = \frac{1}{\epsilon^2}$, and expand to $O(\epsilon)$. Summing over k, replacing $\epsilon = \frac{1}{\sqrt{n}}$ and replacing $\delta = \frac{j}{\sqrt{n}}$ show that to leading order C_1 and C_2 cancel.

This shows that higher order terms must be determined in this case. Expanding to $O(\epsilon^{12})$, summing over k, combining the contributions of C_1 and C_2 and then extracting the leading order terms give

$$\sum_k (C_1 + C_2) \simeq \frac{4a(j+1)(2j^2(2-a)+j(8-a)+6)e^{-j^2/n}\,16^n}{\pi n^5 (a-2)^4}. \tag{5.192}$$

The above can be simplified by taking only the fastest growing terms in each factor. This shows that

$$s_n(a;j) = \sum_k (C_1 + C_2) \simeq \begin{cases} \dfrac{8aj^3 e^{-j^2/n}\,16^n}{\pi n^5 (2-a)^3}, & \text{if } j = O(\sqrt{n}); \\ \dfrac{24\,a\,16^n}{\pi n^5 (2-a)^4}, & \text{if } j = 0. \end{cases} \tag{5.193}$$

5.6.2.2 *Asymptotics for $a > 2$:* Numerical experimentation on C_1 and C_2 in equations (5.186) and (5.187) shows that C_1 is dominated by terms with $k = O(n)$, and $j = O(\sqrt{n})$, while $j + k = O(n)$ with a spread of the peak proportional to $\sqrt{n}$. This is in particular confirmed by the result in equation (5.189), which indicates that the dominant values of j and k in equation (5.186) are at $k = \frac{n}{2}(a-2) - j$, and $j = O(\sqrt{n})$.

Put $n = \frac{1}{\epsilon^2}$, put $k = \frac{a-2}{a\epsilon^2} - \frac{\delta}{\epsilon} + \frac{\alpha}{\epsilon}$ and put $j = \frac{\delta}{\epsilon}$ in C_1 in equation (5.186). By integrating α, the summation over k in equation (5.185) will be approximated.

Take logarithms in equation (5.186) and expand in ϵ to $O(1)$. Exponentiating and integrating the resulting expression give the asymptotic expression

$$\sum_k C_1 = \frac{a(a-2)}{(a-1)^3} \left(\frac{a^2}{a-1}\right)^n \frac{4^n}{\sqrt{\pi n^3}(a-1)^j} \left(1 - \tfrac{9}{8n}(1+o(1))\right) \tag{5.194}$$

where the substitution $\delta = \frac{j}{\sqrt{n}}$ was made. Expanding to higher order in ϵ before integrating gives sub-leading corrections.

Some care is needed for approximating $\sum_k C_2$. Approximating to leading order gives incorrect results, and so higher order terms must be included.

The arguments preceding equation (5.191) show that, if $n = \frac{1}{\epsilon^2}$, then the appropriate choices for j and k are $k = \frac{a-2}{a\epsilon^2} + \frac{\alpha}{\epsilon}$, and $j = \frac{\delta}{\epsilon}$, respectively. Substitute this into equation (5.187), take the logarithm and expand to $O(\epsilon)$. Exponentiate the result and expand in ϵ to $O(\epsilon^2)$. Integrate $\alpha \in (-\infty, \infty)$. This gives

$$\sum_k C_2 = \frac{a-2}{(a-1)^2} \left(\frac{a^2}{a-1}\right)^n \frac{e^{-j^2/n} 4^n}{\sqrt{\pi n^3}} \times \\ \left(a(2j+1) - 4(j+1) - \tfrac{2}{n}(a-2)j^2 + o(\tfrac{1}{n})\right). \tag{5.195}$$

The approximation for $\sum_k C_1$ peaks sharply at $j = 0$, while that for $\sum_k C_2$ dominates the contribution from C_1 when $j > 0$; that is, if $j > 0$, then $\sum_k C_1 +$

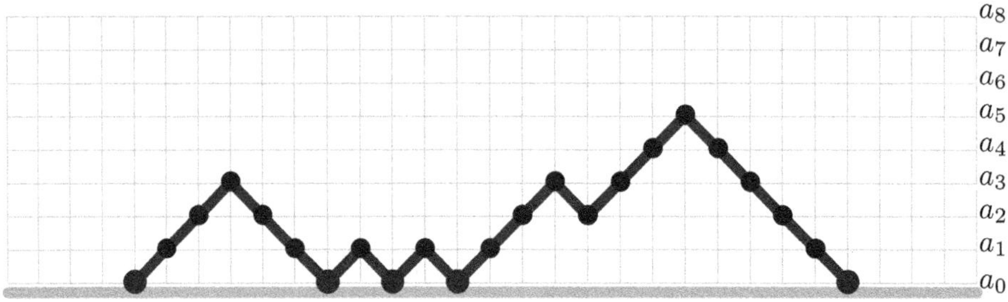

FIG. 5.28. A Dyck path in a layered environment. Each vertex of the path in a horizontal line a height h above the adsorbing line has weight a_h.

$\sum_k C_2$ gives the approximation. Taken together and keeping only the leading order terms

$$s_n(a;j) \simeq \begin{cases} \frac{a-2}{(a-1)^2}\left(\frac{a^2}{a-1}\right)^n \frac{4^n}{\sqrt{\pi n^3}} & \\ \quad \times\left(\frac{a}{(a-1)^{j+1}} + 2j(a-2)\,e^{-j^2/n}\right), & \text{if } j = O(\sqrt{n}); \\ \frac{(a-2)^3}{(a-1)^3}\left(\frac{a^2}{a-1}\right)^n \frac{4^n}{\sqrt{\pi n^3}}\left(1+o\left(\tfrac{1}{n}\right)\right), & \text{if } j = 0. \end{cases}$$

If $j = 0$, then equations (5.194) and (5.195) gives

$$s_n(a;0) = \frac{(a-2)^3}{(a-1)^3}\left(\frac{a^2}{a-1}\right)^n \frac{4^n}{\sqrt{\pi n^3}}\left(1+o\left(\tfrac{1}{n}\right)\right). \tag{5.196}$$

5.7 Dyck paths in a layered environment

A Dyck path in $\mathbb{J}^2_+$ is in a layered environment if vertices in the path are weighted by their height above $\partial\mathbb{J}^2_+$, as illustrated in figure 5.28. That is, if v is a vertex in a Dyck path at height h, then it is weighted by a_h (see references [6, 286]). Denote the generating function of this model by $D_0(a_0, a_1, a_2, \ldots)$.

A Dyck path can be displaced vertically by a distance h. Such a path will have end-vertices in $\partial\mathbb{J}^2_+ + h$, and the generating function $D_h(a_h, a_{h+1}, a_{h+2}, \ldots)$.

Simplify the notation by putting $D_h(a_h, a_{h+1}, a_{h+2}, \ldots) \equiv D_h(a_h)$ for $h = 0, 1, 2, \ldots$. The standard factorisation of Dyck paths in figure 5.2 gives

$$D_0(a_0) = 1 + a_0 a_1 t^2 D_0(a_0) D_1(a_1). \tag{5.197}$$

This can be (recursively) developed into the continued fraction

$$D_0(a_0) = \cfrac{1}{1 - \cfrac{a_0 a_1 t^2}{1 - \cfrac{a_1 a_2 t^2}{1 - \cfrac{a_2 a_3 t^2}{1 - \cdots}}}}. \tag{5.198}$$

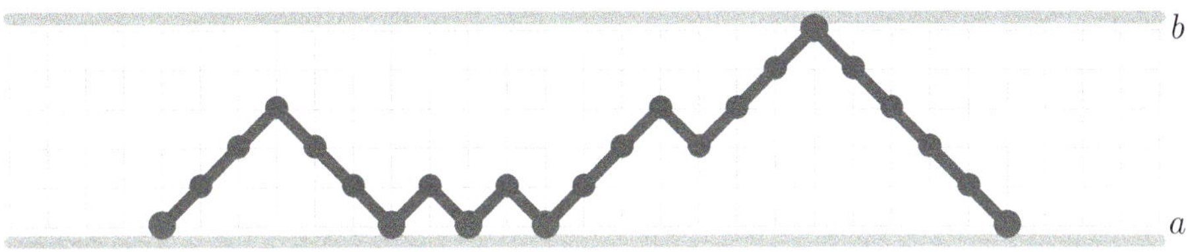

FIG. 5.29. A Dyck path confined to a slit.

The weights a_h are still undefined and may be chosen to realise several different models. The simplest choice is $a_h = 1$ for all $h \geq 0$, in which case the Dyck path generating function in equation (5.4) is recovered:

$$D^+(t) = \cfrac{1}{1 - \cfrac{t^2}{1 - \cfrac{t^2}{1 - \cfrac{t^2}{1 - \cdots}}}}. \tag{5.199}$$

5.7.1 *Adsorbing directed paths in a slit*

Define the *slit* in $\mathbb{J}^2_+$ by

$$\mathbb{S}_w = \left\{ \langle \vec{u} \sim \vec{v} \rangle \in \mathbb{J}^2 \mid 0 \leq \vec{u}(2) \leq w, \text{ and } 0 \leq \vec{v}(2) \leq w \right\}, \tag{5.200}$$

where $\vec{v}(j)$ is the j-th Cartesian coordinate of the vertex $\vec{v}$. Let the *bottom wall* of $\mathbb{S}_w$ be $B(\mathbb{S}_w) = \{\vec{v} \in \mathbb{S}_w \mid \vec{v}(2) = 0\}$ and define the *top wall* $T(\mathbb{S}_w) = \{\vec{v} \in \mathbb{S}_w \mid \vec{v}(2) = w\}$. By convention, $\vec{0} \in B(\mathbb{S}_w)$ and the boundary of $\mathbb{S}_w$ is $\partial\mathbb{S}_w = B(\mathbb{S}_w) \cup T(\mathbb{S}_w)$.

A model of directed paths in $\mathbb{S}_w$ is found by choosing $a_1 = a_2 = \cdots = a_{w-1} = 1$ and $a_{w+1} = 0$ in figure 5.28. Put $a_0 = a$ and put $a_w = b$ so that visits

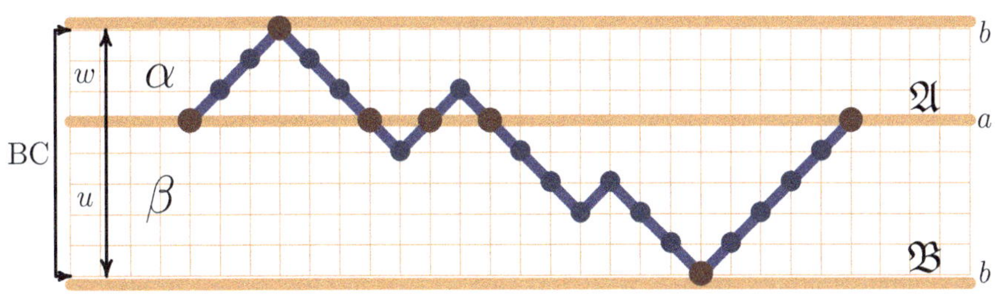

FIG. 5.30. A Dyck path confined to a slit with periodic boundary conditions. The edges in the α layer are weighted by α, and in the β layer by β. Visits to the defect line $\mathfrak{A}$ are weighted by a and visits to $\mathfrak{B}$ by b.

to $B(\mathbb{S}_w)$ are weighted by a, and visits to $T(\mathbb{S}_w)$ are weighted by b. This is a model of directed paths adsorbing in the walls of $\mathbb{S}_w$ [6, 74] (see figure 5.29).

The generating function is

$$D_w(a,b,t) = \cfrac{1}{1 - \cfrac{at^2}{1 - \cfrac{t^2}{1 - \cfrac{t^2}{1 - \cfrac{\cdots}{1-bt^2}}}}}, \tag{5.201}$$

and the endpoints of the path are in $B(\mathbb{S}_w)$.

The model in figure 5.29 can generalised to the model illustrated in figure 5.30. This is a directed path in a periodic layered environment of two layers labelled α and β. The path starts in $\vec{0}$ located in a *defect line* between the α and β layers and terminates in a vertex again in a defect line. Visits to one defect line are weighted by a, while visits to the *other* defect line are weighted by b.

Thus, distinguish between two defect lines in the model: the first defect line is at an interface where the β phase switches to α in the vertical direction; it is denoted by $\mathfrak{A}$. The other defect line is denoted by $\mathfrak{B}$. In the model, directed paths start in a vertex located in one of the two defect lines.

The α and β layers have widths w and u, respectively, and alternate in the vertical direction (or have a period boundary condition in this direction; see figure 5.30). The edges of the path in the α layer are weighted by α, and in the β layer by β.

In order to determine the generating function of the model illustrated in figure 5.30, introduce the following generating functions of directed paths in the periodic $\alpha\beta$-layered environment in figure 5.30:

$$\begin{aligned} g_0 &= \text{directed paths ending in } \mathfrak{A}; \\ g_w &= \text{directed paths ending in } \mathfrak{B}; \\ h_j &= \text{directed paths ending at height } j \text{ above } \mathfrak{A},\ j \in \{0,1,\ldots,w\}; \\ k_j &= \text{directed paths ending at height } -j \text{ below } \mathfrak{A},\ j \in \{0,1,\ldots,u\}. \end{aligned}$$

Notice that $h_0 = k_0 = g_0$.

The Temperley method (append a single step to each of the generating functions) gives the following set of linear recurrences:

$$\begin{aligned} g_0 &= a + a\alpha h_1 + a\beta k_1, & g_w &= b + b\alpha h_{w_a-1} + b\beta k_{w_b-1}, \\ h_1 &= \alpha g_0 + \alpha h_2, & k_1 &= \beta g_0 + \beta k_2, \\ h_2 &= \alpha h_1 + \alpha h_3, & k_2 &= \beta k_1 + \beta k_3, \\ &\cdots & &\cdots \\ h_{w-1} &= \alpha h_{w-2} + \alpha g_w, & k_{u-1} &= \beta k_{u-2} + \beta g_w. \end{aligned}$$

This set of recurrences can be solved using the Dyck path generating function. In particular, define the functions $\{p, q\}$ by

$$p = \alpha t(1+p^2), \quad \text{and } q = \beta t(1+q^2). \tag{5.202}$$

Then the roots of the above equations are given by equation (5.4) (but with t replaced by αt and βt respectively). The solution of the recurrences is given in theorem 5.1.

Theorem 5.1 *Let p and q be roots of the quadratics in equation (5.202). Define the functions*

$$c_0 = \frac{p^w(1-p^2)}{(1+p^2)(1-p^{2w})} + \frac{q^u(1-q^2)}{(1+q^2)(1-q^{2u})},$$

and

$$c_s = \frac{p^2-p^{2w}}{(1+p^2)(1-p^{2w})} + \frac{q^2-q^{2u}}{(1+q^2)(1-q^{2u})}.$$

Then the generating functions for paths in the periodic $\alpha\beta$-layered environment in figure 5.30 are given by

$$g_0 = \frac{a(1-b(c_s-c_0))}{1-(a+b)c_s+ab(c_s^2-c_0^2)}, \quad \textit{and } g_w = \frac{b(1-a(c_s-c_0))}{1-(a+b)c_s+ab(c_s^2-c_0^2)}.$$

The generating functions h_i and k_i are given by

$$h_i = \frac{a(1-b(c_s-c_0))(p^{w-i}-p^{i-w})+b(1-a(c_s-c_0))(p^i-p^{-i})}{(p^w-p^{-w})(1-(a+b)c_s+ab(c_s^2-c_0^2))}$$

and

$$k_i = \frac{a(1-b(c_s-c_0))(q^{u-i}-q^{i-u})+b(1-a(c_s-c_0))(q^i-q^{-i})}{(q^u-q^{-u})(1-(a+b)c_s+ab(c_s^2-c_0^2))}.$$

Proof This can be verified by substitution of the above into the recurrences followed by simplification. □

Notice that $c_s = c_s^{(1)} + c_s^{(2)}$, where

$$c_s^{(1)} = \frac{p^2-p^{2w}}{(1+p^2)(1-p^{2w})}, \quad \text{and } c_s^{(2)} = \frac{q^2-q^{2u}}{(1+q^2)(1-q^{2u})}. \tag{5.203}$$

The generating function $c_s^{(1)}$ generates directed paths from $\vec{0}$ into the α layer and which terminate in the first return to the defect line $\mathfrak{A}$ (without intersecting $\mathfrak{B}$). Similarly, $c_s^{(2)}$ generates directed paths from $\vec{0}$ into the β layer and which terminate in the first return to the defect line $\mathfrak{A}$ (without intersecting $\mathfrak{B}$).

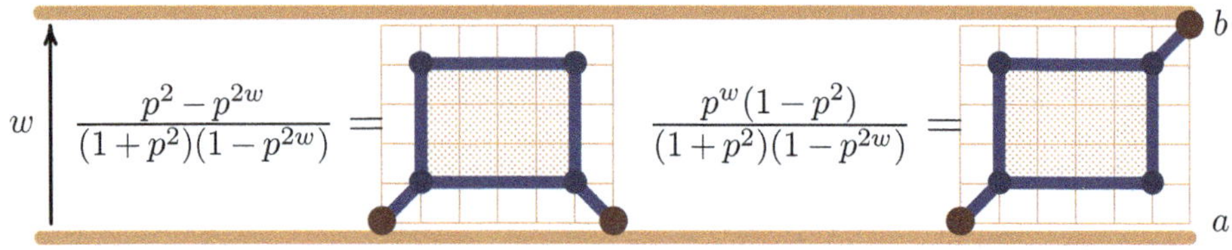

FIG. 5.31. The term on the left generates paths from the origin and terminating at their first return to the bottom wall without intersecting the top wall. Similarly, the term on the right generates paths from the origin and terminating on their first visit to the top wall without intersecting the bottom wall after they depart from the origin.

On the other hand, $c_0 = c_0^{(1)} + c_0^{(2)}$, where

$$c_0^{(1)} = \frac{p^w(1-p^2)}{(1+p^2)(1-p^{2w})}, \text{ and } c_0^{(2)} = \frac{q^u(1-q^2)}{(1+q^2)(1-q^{2u})}. \tag{5.204}$$

The generating function $c_0^{(1)}$ generates directed paths from $\vec{0}$ into the α layer and which terminate in the first visit to the defect line $\mathfrak{B}$ (without intersecting $\mathfrak{A}$ again). Similarly, $c_0^{(2)}$ generates directed paths from $\vec{0}$ into the β layer and which terminate in first visit to the defect line $\mathfrak{B}$ (without intersecting $\mathfrak{A}$ again); see figure 5.31.

The generating functions c_s and c_0 are functions of (w, u) and it is not hard to see that

$$c_s(w,u) - c_0(w,u) = c_s(\tfrac{1}{2}w, \tfrac{1}{2}u) \tag{5.205}$$

see, for example, reference [6] for a proof-by-picture.

The periodic layered environment in figure 5.30 is reduced to a slit $\mathbb{S}_w$ of width w (figure 5.29) by putting $\beta = 0$ and putting $\alpha = 1$ in the above.

Since $q = 0$ if $\beta = 0$, the generating function g_0 in theorem 5.1 reduces to the generating function of directed paths in $\mathbb{S}_w$. Comparing equation (5.202) to equation (5.3) shows that $p = t\,D(t)$. Thus, define $p_+ = t\,D^+(t)$, and $p_- = t\,D^-(t)$ (see equation (5.4)). Then $p_+ p_- = 1$, and $t(1+p^2) = p$ for both $p = p_+$ and $p = p_-$.

The generating function g_0 in $\mathbb{S}_w$ may be computed from theorem 5.1 by putting $q = B = 0$, and $p = p_+$. This gives

$$g_0 = \frac{a(1-bp_+t)p_-^w - a(1-bp_-t)p_+^w + ab(p_-^2 - p_+^2)t^2}{(1-ap_+t)(1-bp_+t)p_-^w - (1-ap_-t)(1-bp_-t)p_+^w}. \tag{5.206}$$

This generating function includes paths which start and end in the defect line $\mathfrak{A}$ (these are 'loops'), as well as paths starting in the defect line $\mathfrak{B}$ and crossing over to end in $\mathfrak{A}$ (these are 'bridges'). The generating function of loops is given by

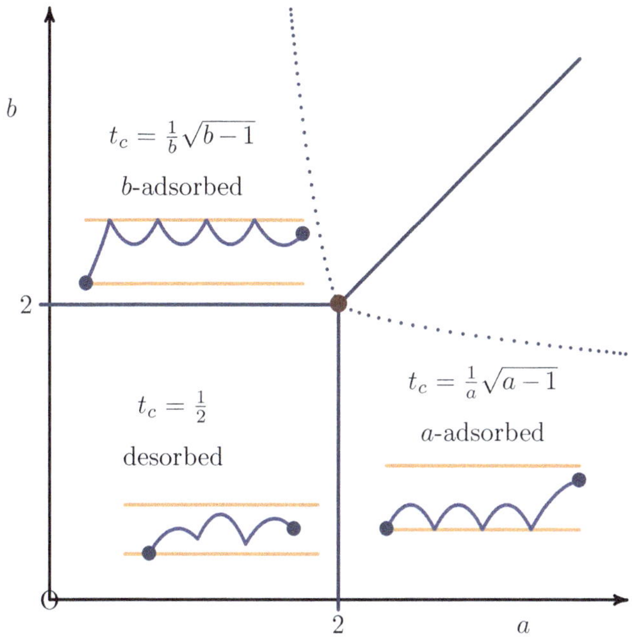

FIG. 5.32. The phase diagram of paths in $\mathbb{S}_w$ in the limit $w \to \infty$. The zero force curve is denoted by the dotted curve in the adsorbed phases. The force is short-ranged in the adsorbed phases (either attractive for large a and b, or repulsive for small a and b), and long-ranged repulsive in the desorbed phase.

$$L_0 = \frac{a(1-bp_+t)p_-^w - a(1-bp_-t)p_+^w}{(1-ap_+t)(1-bp_+t)p_-^w - (1-ap_-t)(1-bp_-t)p_+^w}, \tag{5.207}$$

and of bridges is given by

$$B_0 = \frac{ab(p_-^2 - p_+^2)t^2}{(1-ap_+t)(1-bp_+t)p_-^w - (1-ap_-t)(1-bp_-t)p_+^w}. \tag{5.208}$$

Thus, $g_0 = L_0 + B_0$.

The free energy $\mathcal{D}_w$ can be calculated from singularities in $p_\pm$ as well as from the zeros of the denominator of g_0 given by solutions of

$$(1-ap_+t)(1-bp_+t)p_-^w = (1-ap_-t)(1-bp_-t)p_+^w \tag{5.209}$$

in the t-plane. An asymptotic approximation for $\mathcal{D}_w$ was determined in reference [74]. It is given by

$$\mathcal{D}_w = \begin{cases} \log\left(\frac{a}{\sqrt{a-1}}\right) + \frac{(a-2)^2(ab-a-b)}{2(a-b)(a-1)^{w+1}} + O(w(a-1)^{-2w}), & a > 2, \text{ and } a > b; \\ \log\left(\frac{a}{\sqrt{a-1}}\right) + \frac{(a-2)^2}{2(a-1)^{w/2+1}} + O(w(a-1)^{-w}), & a > 2, \text{ and } a = b; \\ \log 2 - \frac{\pi^2}{8w^2} + \frac{\pi^2 a}{4(2-a)w^3} + O(w^{-4}), & a < 2, \text{ and } b = 2; \\ \log 2 - \frac{\pi^2}{2w^2} - \frac{2\pi^2(ab-a-b)}{(2-a)(2-b)w^3} + O(w^{-4}), & a < 2, \text{ and } b < 2. \end{cases}$$

Taking $w \to \infty$ gives the limiting free energy of the model (in the limit that the lengths of the paths first goes to infinity, followed by $w \to \infty$). Three phases are apparent: a desorbed phase with $\mathcal{D}_\infty = \log 2$, a phase where $\mathcal{D}_\infty$ is a function of a only (the path is adsorbed onto $\mathfrak{A}$ if $a > 2$, and $a > b$), and a phase where $\mathcal{D}_\infty$ is a function of b only (the path is adsorbed onto $\mathfrak{B}$ if $b > 2$, and $b > a$); see figure 5.32.

Paths confined to $\mathbb{S}_w$ lose entropy. This induces a repulsive entropic force on the walls of $\mathbb{S}_w$. If both a and b are large, then the path will adsorb on the walls of $\mathbb{S}_w$ and thus should instead result in a net attractive force between the walls (as a result of the path bridging between the walls of $\mathbb{S}_w$).

The force is given by a discrete derivative of $\mathcal{D}_w$ to w; this may approximated by treating w as a continuous variable and then taking the derivative of $\mathcal{D}_w$. This gives the force f_w asymptotically:

$$f_w = \begin{cases} -\frac{(a-2)^2(ab-a-b)\log(a-1)}{2(a-b)(a-1)^{w+1}} + O((a-1)^{-2w}), & a > 2, \text{ and } a > b; \\ -\frac{(a-2)^2\log(a-1)}{4(a-1)^{w/2+1}} + O((a-1)^{-w}), & a \geq 2, \text{ and } a = b; \\ \frac{\pi^2}{4w^3} - \frac{3\pi^2 a}{4(2-a)w^4} + O(w^{-5}), & a < 2, \text{ and } b = 2; \\ \frac{\pi^2}{w^3} + \frac{6\pi^2(ab-a-b)}{(2-a)(2-b)w^4} + O(w^{-5}), & a < 2, \text{ and } b < 2. \end{cases}$$

In the adsorbed phases the force decays exponentially with w; this is a *short-ranged* force. In this regime the force may be both positive or negative (depending on the sign of the factor $(ab-a-b)$ in the asymptotic regime). This gives a *zero force curve* which is asymptotic to

$$ab = a + b \tag{5.210}$$

in the phase diagram (this is indicated by the dotted curve in figure 5.32).

In the desorbed phase (when $a \leq 2$, and $b \leq 2$, but excluding the point $a = b = 2$), the force is repulsive and proportional to w^{-3}. That is, it decays with w as an inverse power. This is a *long-ranged* repulsive force (see references [6, 74] for additional results).

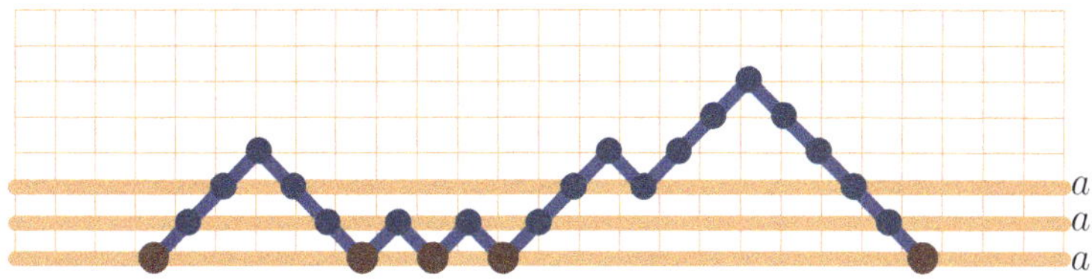

FIG. 5.33. A Dyck path adsorbing into a thick boundary layer.

5.7.2 *Adsorption of Dyck paths into a thick boundary*

The choice of $a_i = a$ for $0 \leq i \leq w$ and of $a_i = 1$ for $i > w$ in figure 5.28 gives a model of Dyck paths adsorbing into a layer of thickness w (see reference [6] and figure 5.33).

By equations (5.198) and (5.199), the generating function of this model is (simplify the notation by putting $g = D^+$):

$$L_w(a,t) = \cfrac{1}{1 - \cfrac{a^2t^2}{1 - \cfrac{a^2t^2}{1 - \cfrac{\cdots}{1 - \cfrac{a^2t^2}{1 - at^2 g}}}}}, \tag{5.211}$$

where there are w divisions in the stacked fraction. This is a rational function in the variables (a, t, g). Compare this to equation (5.201) to see that it may be recovered from equation (5.207) by first putting $a = 1$ and then taking $t \to at$, and $b \to \frac{1}{a} g$. Then $p_+ = tD^+(t) \to (at)D^+(at)$, with $D^+(t)$ given in equation (5.4). Simplify the notation again by putting $r_- = D^+(at)$, and $r_+ = D^-(at)$. This gives

$$L_w = \frac{(1 - at^2 r_- g) r_+^{N+1} - (1 - at^2 r_+ g) r_-^{N+1}}{(1 - a^2t^2 r_-)(1 - at^2 r_- g) r_+^{N+1} - (1 - a^2t^2 r_+)(1 - at^2 r_+ g) r_-^{N+1}}. \tag{5.212}$$

Singularities in L_w include the branch points in g (at $t = \pm\frac{1}{2}$) and in $r_\pm$ (at $t = \pm\frac{1}{2a}$), as well as a pole when the denominator vanishes. It may be verified that $r_\pm = 2$ when $t = \pm\frac{1}{2a}$. Note that L_w is not singular when $t = \frac{1}{2a}$, and $a > 1$. In fact, putting $g^* = g(\frac{1}{2a})$ gives

$$L_w = \frac{2[wg^* - 2a(w+1)]}{(w+1)g^* - 2a(w+2)} \tag{5.213}$$

when $t = \frac{1}{2a}$, and $a > 1$. To see that L_w is not singular, consider the continued fraction in equation (5.211). Suppose that $t > \frac{1}{2a}$, and $t < \frac{1}{2}$; say that $at = \frac{1}{2}(1+\epsilon)$ for some small positive ϵ. Since $t < \frac{1}{2}$, it follows that $gt < 1$. If $w = 1$, then $\frac{1}{1-at^2 g} < \frac{2}{1-\epsilon}$. That is, $D_1 < \infty$.

In the case $w = 2$,

$$\frac{1}{1-\frac{a^2t^2}{1-at^2g}} < \frac{1}{1-\frac{2(1+\epsilon)^2}{4(1-\epsilon)}} < \frac{2}{1-4\epsilon} \tag{5.214}$$

for small enough values of $\epsilon > 0$. This shows that $L_2 < \infty$ for some $t > \frac{1}{2a}$ as claimed if $a > 1$. Proceed recursively in this way by showing that there is a finite integer m such that $L_w < \frac{2}{1-m\epsilon}$.

Consequently, the radius of convergence of L_w is determined by the branch point at $t = \frac{1}{2}$ and by the zero of the denominator of the generating function. These singularities meet at the critical adsorption point in the model, the location of which can be determined by putting $t = \frac{1}{2}$ in the denominator of the generating function and then solving for the critical value of a. For $w = 1$, the critical point is located at $a_c = \sqrt{5}-1$, and more complicated expressions are obtained for larger values of w.

The critical point a_c can be determined asymptotically as a function of w by expanding the terms in the denominator of the generating function asymptotically in w and then solving for a_c. The result is

$$a_c = 1 + \frac{\pi^2}{8w^2} - \frac{3\pi^2}{8w^3} + \frac{(324+5\pi^2)\pi^2}{384w^4} + O(w^{-5}). \tag{5.215}$$

5.7.3 *Dyck paths with short-ranged interactions*

Put $a_0 = a$ and put $a_i = q^i$ for $i \geq 1$ in equation (5.198). Then $a_0 a_1 = aq$, and $a_i a_{i+1} = q^{2i+1}$ for $i \geq 1$. This is a model of a Dyck path in $\mathbb{J}^2_+$ attracted to $\partial\mathbb{J}^2_+$ by a force-inducing field decaying exponentially with distance. There is also a binding energy term, represented by a, for visits to the adsorbing line.

The generating function of this model is given by equation (5.198), namely,

$$D_q(a,q,t) = \cfrac{1}{1-\cfrac{aqt^2}{1-\cfrac{q^3t^2}{1-\cfrac{q^5t^2}{1-\cdots}}}}. \tag{5.216}$$

This is related to the Rogers-Ramanujan continued fraction [463] and to the *fountain-of-coins* problem [442]. The function D_q also can be expressed as the generating function of 132-pattern-avoiding permutations. That is, if $a = 1$, then

$$D_q(1,q,t) = \sum_{\pi \in S(132)} q^{|\pi|+2e_1(\pi)} t^{2|\pi|}, \tag{5.217}$$

where $|\pi|$ is the length of a permutation π in the symmetric group $S(132)$ avoiding the pattern 132, and $e_1(\pi)$ is the number of non-inversions in π; see for example reference [75]. Define

$$E_q(q,t) = \cfrac{1}{1 - \cfrac{q^3 t^2}{1 - \cfrac{q^5 t^2}{1 - \cfrac{q^7 t^2}{1 - \cdots}}}}. \tag{5.218}$$

Then D_q is given by

$$D_q(a,q,t) = \frac{1}{1 - aqt^2 E_q}. \tag{5.219}$$

Observe that $E_q(q,t) = D_q(1,q,tq)$. This shows that, if $a = 1$, then

$$t^2 q\, D_q(1,q,tq)\, D_q(1,q,t) - D_q(1,q,t) + 1 = 0. \tag{5.220}$$

Recurrences of this type can be solved by assuming that $D_q(1,q,t) = \frac{H(tq)}{H(t)}$, where $H(t)$ is to be determined. Substitution and simplification shows that

$$qt^2\, H(tq^2) - H(tq) + H(t) = 0. \tag{5.221}$$

Suppose that

$$H(t) = \sum_{n=0}^{\infty} \alpha_n\, t^n. \tag{5.222}$$

Since $D_q(1,q,t)$ is a function of t^2, and $D_q(1,q,t) = 1 + O(t^2)$, it is the case that $\alpha_0 = 1$, and $\alpha_n = 0$, if n is odd.

Substituting equation (5.222) into equation (5.221) gives a recurrence for α_n with $\alpha_0 = 1$, and $\alpha_1 = 0$. Iterating the recurrence gives an expression for α_n. This shows that

$$H(t) = \sum_{n=0}^{\infty} \frac{(-1)^n\, q^{n(2n-1)}\, t^{2n}}{(q^2;q^2)_n}, \tag{5.223}$$

where $(t;q)_n$ is the q-Pochhammer function; see equation (D.41) in appendix D. If $a = 1$, then

$$D_q(1,q,t) = \frac{H(tq)}{H(t)} = \frac{\sum_{n=0}^{\infty} \frac{(-1)^n\, q^{n(2n+1)}\, t^{2n}}{(q^2;q^2)_n}}{\sum_{n=0}^{\infty} \frac{(-1)^n\, q^{n(2n-1)}\, t^{2n}}{(q^2;q^2)_n}}, \tag{5.224}$$

and comparison to equation (5.218) gives an identity closely related to the Rogers-Ramanujan identities; see for example references [36, 378].

Note that $E_q(q,t) = D_q(1,q,tq)$ and that $D_q(a,q,t)$ may be obtained from equation (5.219):

$$D_q(a,q,t) = \frac{H(tq)}{H(tq) - aqt^2\, H(tq^2)}. \tag{5.225}$$

The generating variable q is conjugate to the area enclosed under the path (in units of $\frac{1}{2}$ unit squares).

The function $H(tq)$ is absolutely convergent if $|q| < 1$, and it is an analytic function inside the unit disk in the q-plane. This shows that $D_q(a,q,t)$ is a meromorphic function inside the unit disk $|q| < 1$, since it is the ratio of holomorphic functions. Its singularities on $|q| < 1$ are given by the roots of the denominator.

By Worpitzky's theorem (see theorem D.20 in appendix D [572]), $D_q(a,q,t)$ is convergent if $|aqt^2| \leq \frac{1}{4}$, and $\sup_n |q^{2n+1}t^2| \leq \frac{1}{4}$, for $n \geq 1$. The theorem shows that $D_q(a,q,t)$ is defined in a disk centred at $\frac{4}{3}$ and of radius $\frac{2}{3}$ in the complex t-plane and, if $D_q(a,q,t)$ is real, then $1 \leq D_q(a,q,t) \leq 2$. Moreover, if $a = q = 1$, and $t = \frac{1}{2}$, then $D_q(1,1,t) = 2$; thus, Worpitzky's theorem shows that the critical curve at $a = q = 1$ is $t_c = \frac{1}{2}$.

Applying Worpitzky's theorem to equation (5.218) shows that if $q = 1$, then $E_q(1,t) = 2$, and the radius of convergence of $E_q(1,t)$ is $t_e = \frac{1}{2}$.

Equation (5.224) shows that $D_q(a,q,t)$ is singular for $|q^{2n}| = 1$ for any values of t and a, and $n \in \mathbb{N}$. The accumulation of singularities on the boundary of the unit disk in the q-plane shows that $|q| = 1$ is an essential singularity in $D_q(a,q,t)$. Thus, if $|q| > 1$, then the radius of convergence in the t-plane is $t_c = 0$.

This leaves the situation when $0 \leq q < 1$. By Worpitzky's theorem, $E_q(q,t)$ is convergent if $|q^3t^2| \leq \frac{1}{4}$. This shows that the radius of convergence of $E_q(q,t)$ is $t_e \geq \frac{1}{2}q^{3/2}$.

Note that $E_q(q,t) \geq \frac{1}{1-q^3t^2}$, and thus $t_e \leq q^{-3/2}$. This shows that

$$\frac{1}{2q^{3/2}} \leq t_e \leq \frac{1}{q^{3/2}} \tag{5.226}$$

when $0 \leq q < 1$.

The function $D_q(a,q,t)$ has a simple pole when $1 - aqt^2\, E_q(q,t) = 0$. But $D_q \geq \frac{1}{1-aqt^2}$ and thus $t_c \leq \frac{1}{\sqrt{aq}}$, unless $\sqrt{aq} \geq 2|q|^{3/2}$ (that is, for $q \in (0,1)$ and for large values of q, $t_c < t_e$).

Putting $q = 1$ shows that $E_q(1,t) = \frac{1}{2t^2}\left(1 - \sqrt{1-4t^2}\right) \leq 2$. This proves that $1 \leq E_q(q,t) \leq 2$ if $t \leq \frac{1}{2}$, and $q \in (0,1)$. Hence, $D_q(a,q,t) \leq \frac{1}{1-2aqt^2}$ and so $t_c \geq \frac{1}{\sqrt{2aq}}$ for large values of a.

These arguments give the following bounds on t_c when a is large:

$$\frac{1}{\sqrt{2\,aq}} \leq t_c \leq \frac{1}{\sqrt{aq}}. \tag{5.227}$$

This shows that t_c increases without bound as $a \to 0$; however, this increase would violate the bound $t_c \leq t_e$ for small a. Since t_e is finite, there exists an a_c such that $t_c = t_e$ for $a < a_c$. Comparing the lower bound of the above with the upper bound in equation (5.226) shows that $a_c \geq \frac{1}{2}q^2$. By comparing the upper bound in the above with the lower bound in equation (5.226), it follows that $a_c \leq 4q^2$.

Putting $q = 1$ gives $a_c = 2$, and hence $a_c \leq 2$ for $q \in (0,1)$, since a_c is an increasing function of q.

Taken together, the following bounds have been determined on the critical adsorption point a_c:

$$a_c \begin{cases} = \infty, & \text{if } q > 1; \\ = 2, & \text{if } q = 1; \\ \leq \min\{2, 4q^2\}, & \text{if } q \in (0,1); \\ \geq \frac{1}{4}q^2, & \text{if } q \in (0,1). \end{cases} \tag{5.228}$$

5.7.4 *Dyck paths in a long-ranged field*

A Dyck path in a long-ranged force-inducing field is obtained by putting $a_i = i^{-m}$ in equation (5.198) for $i \geq 1$. This gives the generating function

$$D_p(a, m, t) = \cfrac{1}{1 - \cfrac{at^2}{1 - \cfrac{2^{-m}t^2}{1 - \cfrac{2^{-m}3^{-m}t^2}{1 - \cfrac{3^{-m}4^{-m}t^2}{1 - \cdots}}}}}. \tag{5.229}$$

By Worpitzky's theorem (theorem D.20 in appendix D) D_p is convergent if $\sup_{j\geq 0}\{|j^{-m}(j+1)^{-m}t^2|\} \leq \frac{1}{4}$, and $at^2 \leq \frac{1}{4}$ [572].

Define the continued fraction

$$E_p(m, t) = \cfrac{1}{1 - \cfrac{2^{-m}t^2}{1 - \cfrac{2^{-m}3^{-m}t^2}{1 - \cfrac{3^{-m}4^{-m}t^2}{1 - \cdots}}}}. \tag{5.230}$$

Then $D_p(a,m,t) = \frac{1}{1 - at\, E_p(m,t)}$. Let t_e be the radius of convergence of E_p.

Let e_j be the j-finite fraction approximation to E_p. That is, $e_1 = \frac{1}{1-2^{-m}t^2}$, and so on. Then e_j converges to E_p as $j \to \infty$, and $t < t_e$. Each e_j is a rational function in t, so suppose that

$$e_j = \frac{p_j}{q_j}. \tag{5.231}$$

It may be shown that p_j and q_j satisfy the recurrences

$$p_j = p_{j-1} - \zeta_j\, p_{j-2}, \quad \text{and } q_j = q_{j-1} - \zeta_j\, q_{j-2}, \tag{5.232}$$

where $\zeta_j = t^2 j^{-m}(j+1)^{-m}$; $p_0 = 1$, $p_{-1} = 0$; $q_0 = 1$, and $q_{-1} = 1$.

The generating function of paths avoiding the adsorbing line and with vertices at height j weighted by j^{-m} is E_p. Hence, E_p is a formal power series in t with non-negative coefficients for every finite value of $m > 0$ and so its radius of convergence t_e is determined by a singularity on the positive real axis. Upper bounds on t_e can be determined by examining the finite fraction approximations e_j. Since $E_p \geq e_1$, it follows that $E_p \geq \frac{1}{1-2^{-m}t^2}$ and thus $t_e \leq \sqrt{2^m}$.

Since $E_p > e_2 > e_1$, it similarly follows that a bound on t_e can be obtained by locating the zeros of the denominator of e_2. That is, solve for $q_2 = 0$ to see that $t_e \leq \sqrt{2^m/(1+3^{-m})}$. This bound may be improved by considering e_3 and so on.

A lower bound on t_e can be determined for $m > 0$. It follows from Worpitzky's theorem that $t_e \geq \sqrt{2^{m-2}}$ if $m \geq 0$. This can be improved by using Pringsheim's theorem [572]. By an equivalence transformation of E_p,

$$E_p = \cfrac{1}{1 - \cfrac{1}{2^m/t^2 - \cfrac{1}{3^m - \cfrac{1}{4^m/t^2 - \cfrac{1}{5^m - \cdots}}}}}. \tag{5.233}$$

By Pringsheim's theorem (theorem D.19 in appendix D), this is convergent if both $(2j)^m t^{-2} \geq 2$ and $(2j+1)^m \geq 2$ for $j = 1, 2, \ldots$; that is, if both $m \geq \frac{\log 2}{\log 3}$ and $t^2 \leq 2^{m-1}$ [474]. This gives the improved lower bound on t_e for larger values of m: $t_e \geq \sqrt{2^{m-1}}$ if $m \geq \frac{\log 2}{\log 3}$.

In the event that $m < 0$, Van Vleck's theorem can be used (see theorem D.21 in appendix D [554]). Choose $\theta_n = \pi$ for all n in theorem D.21. Put $r_2 = t^2 2^{-m}$. Then

$$r_j(1 - r_{j-1}) = t^2 (j-1)^{-m} j^{-m}. \tag{5.234}$$

It follows that $E_p = K$ in theorem D.21.

Obviously, $r_2 > 0$, and $r_j(1 - r_{j-1}) \geq 0$. It follows inductively that $r_j \in [0, 1]$ for all j if $r_2 \in (0, 1)$. Assume that t is small enough that $r_2 \in (0, 1)$, so that $r_j \in (0, 1)$. If the root test is applied to the series in theorem D.21 in appendix D, then it follows that the radius of convergence t_e in the t-plane is given by that real value of t such that:

$$\limsup_{j \to \infty} \left(\prod_{\ell=2}^{j} \frac{r_\ell}{1 - r_\ell} \right)^{1/j} = 1. \tag{5.235}$$

The critical value t_e can be bounded by noting from equation (5.234) that

$$\frac{r_j}{1 - r_j} = \frac{r_{j-1} r_j (j-1)^m j^m}{t^2}. \tag{5.236}$$

The product evaluates to

$$\prod_{\ell=2}^{j} \frac{r_\ell}{1-r_\ell} = \left(\tfrac{1}{2} j!(j+1)!\right)^m t^{-2(j-1)} \prod_{\ell=2}^{j} r_\ell r_{\ell+1}. \tag{5.237}$$

By taking the power $\frac{1}{j}$ and then the limit superior as $j \to \infty$, using the Stirling approximation for the factorials, the critical point t_e is given by

$$t_e^2 = \limsup_{j\to\infty} \left(j^m (j+1)^m e^{-2m}\right) \left(\prod_{\ell=2}^{j} r_\ell r_{\ell+1}\right)^{1/j}, \tag{5.238}$$

where r_ℓ is also a function of t. Since $r_\ell \in [0,1]$, this shows that $t_e = 0$ if $m < 0$.

5.8 Paths in wedges and the kernel method

Paths in wedges in $\mathbb{J}^2$ or $\mathbb{L}^2$ are not translationally invariant and so do not as a general rule renew themselves. A consequence is that the generating functions cannot be determined using a factorisation method. Instead, recurrences are obtained using a Temperley method, and they are generally difficult to solve. The *kernel method* (see for example references [19, 51, 56, 186, 423, 424]), first encountered in section 5.1.4, is often useful in these models.

The kernel method is generally implemented as follows. Suppose that $g(a,b,t)$ is a generating function satisfying a recurrence of the generic form

$$g(a,b,t) = f(a,b) + F(a,b,t)\, g(a,b,t) + B(a,b,t). \tag{5.239}$$

The coefficient $F(a,b,t)$ is the *bulk coefficient*, and it describes the next step of the path in the absence of boundary conditions. The term $B(a,b,t)$ is a *boundary term*, which introduces corrections in $g(a,b,t)$ due to boundary conditions. The terms in the boundary term are not explicitly known in general and must be solved for at the same time that $g(a,b,t)$ is determined. The term $f(a,b)$ is a *source term*, which seeds the recurrence.

Write the recurrence in equation (5.239) in *kernel form*:

$$K(a,b,t)\, g(a,b,t) = f(a,b) + B(a,b,t). \tag{5.240}$$

The coefficient $K(a,b,t) = 1 - F(a,b,t)$ is the *kernel*. Notice that if $B(a,b,t)$ is known, then $g(a,b,t)$ is solved for. If $B(a,b,t)$ is not known, then choose a catalytic variable (say b) and suppose $K(b,t) \equiv K(a,b,t)$ (so that a is just a parameter).

Solve $K(b,t) = 0$ for b to find the roots of the kernel $\{b_j^*\}$ (such that $K(b_j^*,t) = 0$). One root is a 'physical root' (it expands as a series with non-negative coefficients in t); let this be b_0^*.

Substituting $b = b_0^*$ in equation (5.240) gives $f(a,b_0^*) + B(a,b_0^*,t) = 0$. In some cases this is enough to solve for $g(a,b,t)$.

More generally, the above is not enough. In that case observe that all the roots $\{b_j^*\}$ give $f(a,b_j^*) + B(a,b_j^*,t) = 0$, namely, a set of equations in the boundary

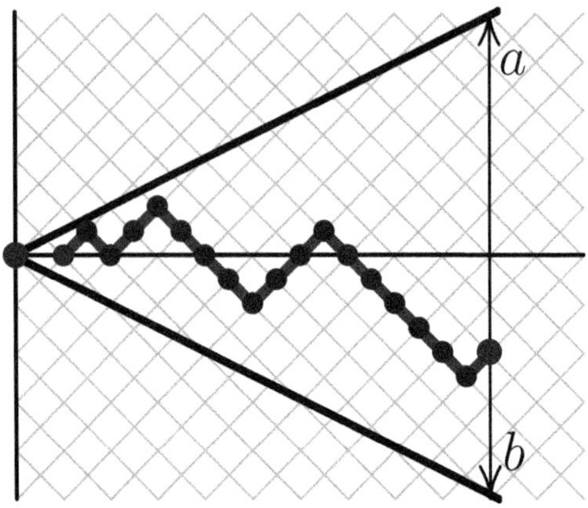

FIG. 5.34. A directed path giving north-east and south-east steps in the wedge V_2. The starting vertex is at $(2,0)$. The generating variables (a,b) are conjugate to the vertical distance between the endpoint of the path and the walls of V_2.

---

terms which must be solved. Solving these equations may pose special challenges; in that case the composition of roots of $K(b,t)$ should be examined. Notice that if $K(b_j^*,t)=0$, then $K(b_j^*\circ b_l^*, b_l^*)=0$ (obtained by $t\to b_l^*$). By systematically examining the properties of the composition of roots and the transformation of the kernel under these compositions, enough information is obtained in order to solve for $B(a,b,t)$ and thus for $g(a,b,t)$; this is the *obstinate kernel method* (see reference [53]).

5.8.1 *Directed paths in a wedges*

Define a wedge-shaped region in $\mathbb{J}^2$ by

$$V_p = \{(x,y)\in\mathbb{Z}^2 \mid \text{where } -\tfrac{x}{p}\le y\le \tfrac{x}{p}, \text{ and } x>0\}, \tag{5.241}$$

and let $\partial V_p = \{(x,y)\in V_p \mid y=\pm\lceil\frac{1}{p}x\rceil\}$ be the boundary of V_p.

A directed path is confined to V_p if all its vertices are points in V_p. In figure 5.34 an example of a directed path confined to the wedge V_2 is illustrated. The path gives north-east and south-east steps onto vertices of V_2. In this particular example the starting vertex for the path has coordinates $(2,0)$.

Let $c_n^{(p)}$ denote the number of paths of length n in V_p (with starting point $(j,0)$ and where j is taken as the least integer such that $c_n^{(p)}$ is non-negative for an infinite set of n). Let $c_n^{(p)}(u,v)$ be the number of paths of length n in V_p, with endpoints a vertical distance u from the top wall of ∂V_p (given by vertices (x,y) with $y=\lceil\frac{x}{p}\rceil$), and a vertical distance v from the bottom wall of ∂V_p (vertices (x,y) with $y=-\lceil\frac{x}{p}\rceil$).

The generating function of the path is

$$g_p(a,b) = \sum_{n,u,v\geq 0} c_n^{(p)}(u,v)\, a^u b^v t^n, \tag{5.242}$$

where t is the generating variable conjugate to the length of the path. Consider the case that $p = 2$ (generally it is possible to show, using the methods of reference [261], that the coefficients $c_n^{(2)}$ of t^n in $g_p(1,1)$ grow at the rate $C_0\, 2^{n+o(n)}$, for some constant C_0 which depends on p).

If $p = 2$, then u and v can only be 0 if $n = 2+4m$ for some $m \in \mathbb{N}$. This gives rise to parity effects in $g_2(a,b)$; therefore, define h_0 to be the generating function of paths of even length and let h_1 be the generating function of paths of odd length. Then

$$g_2(a,b) = h_0(a,b) + h_1(a,b). \tag{5.243}$$

To determine g_2, a set of recurrences will be determined for h_0 and h_1 by using the Temperley method.

Each path generated by h_0 is either the single vertex at $(2,0)$ (generated by ab) or is obtained by appending a north-east step onto a path of odd length (that is, $t\frac{b}{a}h_1$) or is obtained by appending a south-east step onto a path of odd length (that is, $t\frac{a}{b}h_1$).

It remains to subtract paths stepping over ∂V_2: paths of odd length with a final vertex in the top boundary of V_2 have generating function $h_1(0,b)$. Hence, subtract these paths if they give a north-east step outside V_2 (subtract $t\frac{b}{a}h_1(0,b)$ for paths stepping over the upper boundary of V_2).

Similarly, subtract $t\frac{a}{b}h_1(a,0)$ for paths stepping over the lower boundary of V_2.

This gives the recurrence

$$h_0(a,b) = ab + t\left(\tfrac{b}{a} + \tfrac{a}{b}\right) h_1(a,b) - t\tfrac{b}{a}h_1(0,b) - t\tfrac{a}{b}h_1(a,0). \tag{5.244}$$

A similar argument gives a recurrence for h_1 in terms of h_0:

$$h_1(a,b) = t\left(a^2+b^2\right) h_0(a,b) - tb^2 h_0(0,b) - ta^2 h_0(a,0). \tag{5.245}$$

In other words, a set of coupled recurrences in h_0 and h_1 is obtained.

Define $P = P(a,b) = t(\frac{b}{a}+\frac{a}{b})$, and $Q = Q(a,b) = t(a^2+b^2)$. Put $K(a,b) = 1-PQ = 1-t^2(\frac{b}{a}+\frac{a}{b})(a^2+b^2)$ and substitute the equations in one another. Simplification gives

$$\begin{aligned} K(a,b)\,h_0(a,b) &= ab - ta^2P\,h_0(a,0) - tb^2P\,h_0(b,0) \\ &\qquad - t\tfrac{a}{b}\,h_1(a,0) - t\tfrac{b}{a}\,h_1(b,0), \text{ and} \\ K(a,b)\,h_1(a,b) &= ab\,Q - ta^2\,h_0(a,0) - tb^2\,h_0(b,0) \\ &\qquad - t\tfrac{a}{b}Q\,h_1(a,0) - t\tfrac{b}{a}Q\,h_1(b,0), \end{aligned} \tag{5.246}$$

where it should be noted that $h_0(a,0) = h_0(0,a)$ and $h_1(a,0) = h_1(0,a)$ (and $h_1(a,0) = h_1(0,b) = 0$, since paths of odd length cannot end in ∂V_2).

The above is a set of coupled functional recurrences for h_0 and h_1 in kernel form, and the kernel is $K(a,b)$. Simplifying $K(a,b)$ gives

$$K(a,b) = \tfrac{1}{ab}\left(ab - t^2(a^2+b^2)^2\right). \tag{5.247}$$

The kernel is a quartic in b. Pairs $(a,\beta(a))$ which kill the kernel ($K(a,\beta(a))=0$) can be used in equation (5.246) to determine the generating function. The kernel $K(a,b)$ has four roots which may be useful. However, they will on their own not be enough, and an infinite collection of pairs $(a,\beta(a))$ will be generated in order to solve for h_0 and h_1. This is the *iterated kernel method*, and it uses iterated composition of the roots of the kernel; see for example references [51, 52, 59].

5.8.1.1 *Determination of $g_2(a,b)$ by the iterated kernel method:* The four roots of the kernel have the following properties. The first is a real 'physical root' denoted by $\beta_0(a)$ and it is a power series in t:

$$\beta_0(a) = a^3t^2 + 2\,a^7t^6 + 9\,a^{11}t^{10} + 52\,a^{15}t^{14} + \cdots. \tag{5.248}$$

The remaining roots are all singular at $t=0$. There is a real root amongst them denoted $\beta_1(a)$ with Laurent expansion

$$\beta_1(a) = a^{1/3}t^{-2/3} - \tfrac{2}{3}\,a^{5/3}t^{2/3} - \tfrac{28}{81}\,a^{13/3}t^{10/3} + \cdots. \tag{5.249}$$

The remaining two roots are a complex conjugate pair denoted by

$$\beta_\pm(a) = -\tfrac{1}{2}(1\mp i\sqrt{3})a^{1/3}t^{-2/3} + \tfrac{1}{3}(1\pm i\sqrt{3})a^{5/3}t^{2/3} - \tfrac{1}{3}a^3t^2 + \cdots. \tag{5.250}$$

Implement the kernel method by substituting $b=\beta_0(a)\equiv\beta_0$ in equations (5.246). This gives a system of two linear equations in $h_0(a,0)$, $h_0(b,0)$, $h_1(a,0)$ and $h_1(b,0)$. These equations may be simplified to

$$ta^2\,h_0(a,0) + t\,\beta_0^2\,h_0(\beta_0,0) = Qa\,\beta_0, \quad \text{and} \tag{5.251}$$

$$ta^2\,h_1(a,0) + t\,\beta_0^2\,h_1(\beta_0,0) = 0. \tag{5.252}$$

The second equation shows that $\beta_0^2\,h_1(\beta_0(a),0) = -a^2\,h_1(a,0)$. This shows that $h_1(\beta_0(a),0)$ and $h_1(a,0)$ have opposite signs if they are not 0. The conclusion is that $h_1(a,0)=0$ (since $h_1(a,b)$ is a series in t with non-negative coefficients which do not all vanish).

For h_0, the above gives the result

$$h_0(a,0) = \tfrac{1}{ta}Q(a,\beta_0)\,\beta_0 - \tfrac{1}{a^2}\beta_0^2\,h_0(\beta_0,0). \tag{5.253}$$

Observe that this is a recurrence for $h_0(a,0)$. Define $\beta^{(n)} = (\beta_0\circ\beta_0\circ\cdots\circ\beta_0)$ to be the composition of β_0 n times with itself. Put $\beta^{(0)}(a)=a$. Then the recurrence for $h_0(a,0)$ becomes

$$h_0(\beta^{(n-1)},0) = \frac{Q(\beta_0^{(n-1)},\beta_0^{(n)})\,\beta_0^{(n)}}{t\,\beta_0^{(n-1)}} - \left(\frac{\beta_0^{(n)}}{\beta^{(n-1)}}\right)^2 h_0(\beta^{(n)},0). \tag{5.254}$$

Iterating this gives a formal solution for $h_0(a,0)$. That is,

$$h_0(a,0) = \tfrac{1}{ta^2}\sum_{n=0}^{\infty}(-1)^n\, Q(\beta_0^{(n)},\beta_0^{(n+1)})\,\beta_0^{(n)}\,\beta_0^{(n+1)}, \tag{5.255}$$

where $Q(a,b) = t(a^2+b^2)$. Since $h_1(a,0) = 0$, this result also gives $h_0(a,b)$ by equation (5.246). If $h_1(a,b)$ is computed from equation (5.246), then the full generating function $g_2(a,b)$ for this model is computed from equation (5.243). So far, and in particular,

$$h_0(a,b) = \frac{ab - tP(a,b)\left(a^2h_0(a,0)+b^2h_0(0,b)\right)}{K(a,b)}. \tag{5.256}$$

The radius of convergence is given by the dominant root of the kernel in equation (5.247) (if $h_0(a,0)$ does not have a singularity which cancels this root).

5.8.1.2 *The roots of the kernel:* The physical root has the series expansion

$$\beta_0(a) = 2\,t^2a^3\sum_{n=0}^{\infty}\frac{1}{3n+2}\binom{4n+1}{n}(at)^{4n}. \tag{5.257}$$

The radius of convergence of this series is the solution of $|at|^4 = \frac{3^3}{4^4}$. The physical root counts directed paths with first step in the south-east direction, above the line $Y = -\frac{1}{2}X$ and with last vertex in the line $Y = -\frac{1}{2}X$.

The real root $\beta_1(a)$ (see equation (5.249)) is the inverse function of $\beta_0(a)$; direct calculation shows that $\beta_0\circ\beta_1(a) = \beta_1\circ\beta_0(a) = a$. The roots $\beta_+(a)$ and $\beta_-(a)$ are not independent, it may be shown that $\beta_+\left(\frac{i}{8}x^3\right) = -\beta_-\left(\frac{1}{8\,i}x^3\right)$, and

$$\beta_+\left(\tfrac{i}{8}x^3\right) = -\tfrac{i}{2}xt^{-2/3} + \tfrac{i}{48}x^5t^{2/3} + \tfrac{i}{1536}x^9t^2 + \tfrac{7i}{165888}x^{13}t^{10/3} + \cdots.$$

The roots of the kernel can be obtained by solving

$$r^2 = (a^2+b^2), \;\; \text{and } t^2r^4 = ab \tag{5.258}$$

There are four solutions. If

$$a_1(r) = \frac{\sqrt{2}\,r^3t^2}{\sqrt{1+\sqrt{1-4\,t^4r^4}}} = \frac{r\sqrt{1-\sqrt{1-4\,t^4r^4}}}{\sqrt{2}}; \;\; \text{and} \tag{5.259}$$

$$b_1(r) = \frac{\sqrt{2}\,r^3t^2}{\sqrt{1-\sqrt{1-4\,t^4r^4}}} = \frac{r\sqrt{1+\sqrt{1-4\,t^4r^4}}}{\sqrt{2}}, \tag{5.260}$$

then the solutions are $(a_1(r),b_1(r))$, $(-a_1(r),-b_1(r))$, $(b_1(r),a_1(r))$ and $(-b_1(r),-a_1(r))$.

Compute $r^2 = a^2 + b^2$; then it follows that

$$a_1(r) = t^2 r^3 + t^6 r^7 \sum_{n=0}^{\infty} \binom{4n+3}{2n} \frac{(tr)^{4n}}{(2n+1)4^{n+1/2}}, \quad \text{and} \tag{5.261}$$

$$b_1(r) = r - t^4 r^5 \sum_{n=0}^{\infty} \binom{4n+1}{2n} \frac{(tr)^{4n}}{(2n+2)4^n}. \tag{5.262}$$

The roots of the quartic are found by inverting a_1 to obtain a function $r_a(a)$. This gives the identity map $r_a \circ a_1 = a_1 \circ r_a = 1$. It follows that $\beta_0(a) = b_1 \circ r_a$. Inverting b_1 to obtain $r_b(b)$ similarly gives a second root by the composition $\beta_1 = a_1 \circ r_b$.

This gives for example $\beta_0 \circ \beta_1 = b_1 \circ (r_a \circ a_1) \circ r_b = b_1 \circ r_b = 1$ since $a_1^{-1} = r_a$ and $b_1^{-1} = r_b$. Hence, $\beta_0 \circ \beta_1 = \beta_1 \circ \beta_0 = 1$ as claimed above.

The roots $\beta_\pm$ are given by the compositions $b_1 \circ r_b$, and $a_1 \circ r_a$.

Explicit expressions for r_a and r_b can be obtained, but they are large. Since both r_a^2 and r_b^2 are roots of the quartic $t^4 x^4 - c^2 x + c^4$ (where $c = a$ for r_a, and $c = b$ for r_b), the first terms in the series expansions of r_a^2 and r_b^2 can be obtained. Comparing the resulting coefficients to sequences of integers shows that

$$r_a^2(a) = a^2 \sum_{n=0}^{\infty} \frac{1}{3n+1} \binom{4n}{n} (at)^{4n}. \tag{5.263}$$

Other roots of the quartic are unphysical. This series expansion for $r_a^2(a)$ can be substituted in $b_1(r)$ to obtain an expression for the physical root $\beta_0(a)$ of the kernel:

$$\beta_0(a) = \tfrac{1}{\sqrt{2}} r_a(a) \sqrt{1 + \sqrt{1 - 4\,t^4 r_a^4(a)}}. \tag{5.264}$$

Compositions of this root with itself gives the series for $h_0(a,0)$ in equation (5.255).

The composition of $a_1(r)$ and $r_a(a)$ must yield the identity, which is given by

$$\tfrac{1}{\sqrt{2}} r_a(a) \sqrt{1 - \sqrt{1 - 4t^4 r_a^4(a)}} = a. \tag{5.265}$$

Putting $b = \beta_0(a)$ in equation (5.258) gives

$$a\,\beta_0(a) = t^2 r_a(a)^4. \tag{5.266}$$

By equation (5.257), this gives the combinatorial identity

$$\left(\sum_{n=0}^{\infty} \frac{1}{3n+1} \binom{4n}{n} (at)^{4n} \right)^2 = 2 \sum_{n=0}^{\infty} \frac{1}{3n+2} \binom{4n+1}{n} (at)^{4n}. \tag{5.267}$$

5.8.1.3 *The growth of $c_n^{(2)}$:* It was argued above that $c_n^{(2)} = C_0\, 2^{n+o(n)}$. It is possible to compute C_0 accurately by examining the residues of $\frac{g_2(1,1)}{t^{n+1}}$ at $t = \pm\frac{1}{2}$.

Notice that $\beta_0(a) = a^3t^2 + O\left(a^7t^6\right)$, so compositions $\beta_0^{(n)}(a)$ quickly approache 0 with increasing n at $|t| = \frac{1}{2}$ and at $a = 1$. That is, the recurrence $x_{n+1} = \beta_0(x_n)\big|_{t=\frac{1}{2}}$ is a fixed point recurrence of order 3 (with a fixed point at 0). This fast convergence at $a = 1$ allows the very accurate numerical estimation of $\beta_0^{(n)}(1)$ at $t = \frac{1}{2}$.

Using a symbolic computations programme (Maple) gives

$$\begin{aligned}
\beta_0^{(0)}(1) &= 1, \\
\beta_0^{(1)}(1) &= 2.9559774252208477098\ldots \times 10^{-1}, \\
\beta_0^{(2)}(1) &= 6.4633625443847777820\ldots \times 10^{-3}, \\
\beta_0^{(3)}(1) &= 6.7501832073150278963\ldots \times 10^{-8}, \\
\beta_0^{(4)}(1) &= 7.6892979457392165146\ldots \times 10^{-23}, \\
\beta_0^{(5)}(1) &= 1.1365801752937162161\ldots \times 10^{-67}, \quad \text{and} \\
\beta_0^{(6)}(1) &= 3.6706268625677440729\ldots \times 10^{-202}.
\end{aligned}$$

The expression of $g_2(1,1)$ in terms of $h_0(1,0)$ and $h_0(0,1)$ is given by

$$g_2(1,1) = \frac{1 - 2\,t^2(h_0(1,0) + h_0(0,1))}{1 - 4t^2}, \qquad (5.268)$$

where, by equation (5.255),

$$h_0(1,0) = h_0(0,1) = \sum_{n=0}^{\infty} (-1)^n\, Q(\beta_0^{(n)}, \beta_0^{(n+1)})\; \beta_0^{(n)}\; \beta_0^{(n+1)},$$

and where $Q(a, \beta_0) = a^2 + \beta_0^2$.

Numerical evaluation of the residue at $t = \frac{1}{2}$ using the calculated values of $\beta^{(n)}(1)$ above, gives the leading order behaviour of the number of paths of length n:

$$c_n^{(2)} = 2^{n-1} \times 0.67874053079810945741723 27\ldots + \text{parity term} + \cdots. \qquad (5.269)$$

The seven values of $\beta_0^{(n)}(1)$ listed above give C_0 accurately at least to $O(10^{-401})$.

Parity effects can be estimated as well by determining the residue of $\frac{g_2(1,1)}{t^{n+1}}$ at $t = -\frac{1}{2}$. This gives the same results as above, only with a factor of $(-1)^n$. That is,

$$c_n^{(2)} = 2^{n-1}(1 + (-1)^n) \times (0.678740530798109457417232 7\ldots) + \cdots. \qquad (5.270)$$

Notice that $c_{2n+1}^{(2)} = 0$ in this expression, which cannot be true. This result follows because only even values of n were used in the above. Thus, $(-1)^n = 1$, so

$$c_n^{(2)} = 2^n \times (0.67874053079810945741723 27\ldots) + \cdots . \tag{5.271}$$

In the case that n is odd notice that $h_1(a,0) = h_1(0,b) = 0$. Thus, by equation (5.244), $h_0(a,b) = ab + t(\frac{b}{a} + \frac{a}{b})\, h_1(a,b)$. This implies that $c_{2n}^{(2)} = 2\, c_{2n-1}^{(2)}$. That is, an asymptotic expression for c_n for odd values of n is obtained by inserting a factor of one-half in the expression for c_n in equation (5.271).

Corrections to the asymptotic expression for $c_n^{(2)}$ arise from singularities in equation (5.256). There are, for example, branch points and possibly other singularities in $h_0(1,0)$ and $h_0(0,1)$, and these are due to branch points in β_0.

Put $a = 1$ in equations (5.263) and (5.266). By equation (5.263), $r_1(1)$ is convergent on the disk $|t| \leq \frac{1}{4}3^{3/4}$ in the t-plane. Since $\frac{1}{4}3^{3/4} > \frac{1}{2}$, this proves that the simple poles at $t = \pm 1/2$ are within the radius of convergence of $\beta_0(1)$.

The function $\beta_0(1)$ is a complicated expression of nested radicals which explicitly contain radicals of the form $\sqrt{81 - 768t^4}$. This shows that there are branch points at $t = \frac{1}{4}3^{3/4}\omega$, where ω is a fourth root of unity. There may be more branch points on the circle $|t| = \frac{1}{4}3^{3/4}$.

Since $r_1(t)$ is a positive term power series, application of the triangle inequality in equation (5.263) for (complex) t such that $|t| \leq \frac{1}{4}3^{3/4}$ gives

$$|r_1^2(1)| \leq \sum_{n=0}^{\infty} \frac{1}{3n+1} \binom{4n}{n} \left(\frac{3^{3/4}}{4}\right)^{4n} = \frac{4}{3}. \tag{5.272}$$

Thus, by equation (5.266),

$$|\beta_0(1)| \leq \frac{16}{9} \left(\frac{3^{3/4}}{4}\right)^2 = \frac{1}{\sqrt{3}}, \quad \text{if } 4|t| \leq 3^{3/4}. \tag{5.273}$$

Since $\beta_0(a)$ is a power series with positive coefficients in both a and t, $|\beta_0(a)| \leq |\beta_0(1)|$ for any $|t| \leq \frac{1}{4}3^{3/4}$, and $|a| \leq 1$, and, by the triangle inequality, $\beta_0(a)$ is a maximum when $t = \frac{1}{4}3^{3/4}$ for a fixed value of a. Thus, for fixed values of $|a| \leq 1$, $\beta_0(a)$ is a maximum on the closed disk $|t| \leq \frac{1}{4}3^{3/4}$ when $t = \frac{1}{4}3^{3/4}$.

Thus, it follows that, for $t = \frac{1}{4}3^{3/4}$, $|\beta_0(a)| \leq \frac{1}{\sqrt{3}}$ for $|a| \leq 1$. By equation (5.257), $|\beta_0(a)| \leq \frac{|a|}{\sqrt{3}}$ for $|a| \leq 1$ when $t = \frac{1}{4}3^{3/4}$. In other words, taking compositions of β_0 at $a = 1$,

$$|\beta_0(\beta_0(1))| \leq \left| \frac{4\,(3^{3/4})}{4\sqrt{3}\,(\sqrt{3}\,3^{3/4})} \right| = \frac{1}{\sqrt{3^2}}. \tag{5.274}$$

It follows inductively that

$$\left|\beta_0^{(n)}(1)\right| \leq \sqrt{3^{-n}}. \tag{5.275}$$

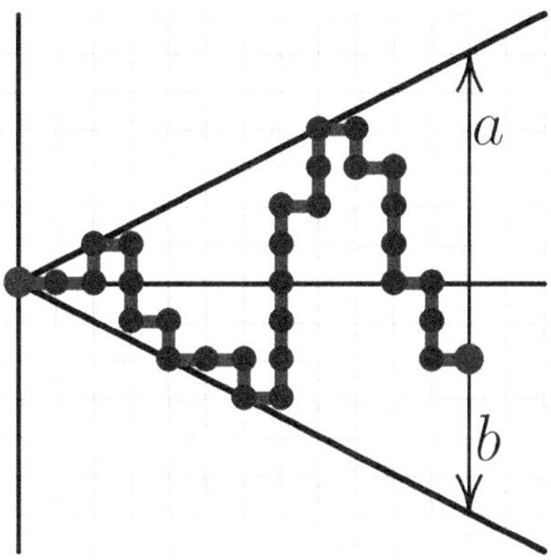

FIG. 5.35. A partially directed path from the origin in the wedge U_p. The generating variables (a, b) are conjugate to the vertical distance of the endpoints of the wedge.

Branch points in $\beta_0^{(n)}(1)$ will occur at values of t where $|\beta_0^{(n-1)}(1)| = \frac{1}{4}3^{3/4}$. However, $|\beta_0^{(n-1)}(1)| \leq \sqrt{3^{1-n}} < \frac{1}{4}3^{3/4}$ for $a = 1$ and for $n > 2$. This shows that the branch points in $\beta^{(n)}(1)$ for $n > 1$ lie outside the circle with radius $|t| = \frac{1}{4}3^{3/4}$, and contributions of these branch points to c_n are dominated by the contributions to c_n, which is a result of the branch points in $\beta_0(1)$ itself.

The radius of convergence of $\beta_0(a)$ is $|at| = \frac{1}{4}3^{3/4}$, while $|\beta_0^{(n-1)}(1)| \leq \sqrt{3^{1-n}}$. The radius of convergence of $\beta_0^{(n)}(1)$ in the t-plane is on or outside the circle with radius $|t| = \frac{1}{4}\,3^{3/4}\sqrt{3^{n-1}}$, for $n > 2$. This shows that the generating function $h_0(a, 0)$ has infinitely many singularities in the t-plane for $a = 1$. Hence, $h_0(1, 0)$ (and $g_2(1, 1)$) cannot be holonomic [57].

Observe that the bound in equation (5.275) proves that $h_0(1, 0)$ is an absolutely convergent series on the open disk with radius $|t| = 3\sqrt{3}$ and which includes the simple poles of $h_0(1, 1)$ at $t = \pm\frac{1}{2}$ and the branch points on the circle $|t| = \frac{1}{4}3^{3/4}$ in its interior. The asymptotic behaviour of c_n is given by equation (5.271), with corrections due to the branch points. These corrections grow at an exponential rate $\left(4 \cdot 3^{-3/4}\right)^{n+o(n)}$. Since this rate of growth is less than 2, the effects of the correction terms will disappear quickly with increasing n.

5.8.2 *Partially directed paths in wedges*

A partially directed path from the origin in a wedge U_p, defined by

$$U_p = \left\{ (x, y) \in \mathbb{Z}^2 \;\middle|\; \text{where } -px \leq y \leq px, \text{ and } x \geq 0 \right\}, \tag{5.276}$$

is illustrated in figure 5.35. The boundary of U_p is $\partial U_p = \{(x, y) \in U_p \mid y = \pm px\}$.

Introduce generating variables x for horizontal steps, and y for vertical steps. Let the generation function of these paths be $\overline{g}_p(a,b)$, where a and b are generating variables of the vertical distances between the endpoint of the path and ∂U_p (see figure 5.35).

Consider paths in V_p which are either the empty path or which end in a horizontal step. Denote the generating function of these paths by $f_p(a,b)$. Then

$$f_p(a,b) = 1 + x\,(ab)^p\,\overline{g}_p(a,b). \tag{5.277}$$

Notice that $f_p(by,b)$ is the generating function of paths ending in a vertex (x,y) in the top wall of U_p ($y = \lfloor px \rfloor$), while $f_p(a,ay)$ is the generating function of paths ending in a vertex (x,y) in the bottom wall of U_p ($y = -\lfloor px \rfloor$).

A recurrence for $f_p(a,b)$ is obtained by using the Temperley method. This is implemented by adding edges to paths ending in a horizontal step. That is, either add vertical steps up followed by a horizontal step (a $\ulcorner$-conformation), or add just one horizontal step or add vertical steps down followed by a horizontal step (a $\llcorner$-conformation). These cases are given on the left in figure 5.36.

Adding edges in a $\ulcorner$-conformation may take the path outside U_p, and such paths must be subtracted. Since $f_p(by,b)$ is the generating function of paths ending in a vertex in the top wall of U_p, adding one more vertical step to such a path will take it outside U_p. Paths stepping over the top wall of ∂U_p and ending with a horizontal step are generated by $\frac{x(ab)^p(yb/a)}{1-yb/a}\,f_p(by,b)$ and must be subtracted out.

Similarly, paths stepping over the bottom wall of ∂U_p and ending with a horizontal step are generated by $\frac{x(ab)^p(ya/b)}{1-ya/b}\,f_p(a,ay)$, and these paths must be subtracted out.

Putting the above together gives the recurrence

$$\begin{aligned} f_p(a,b) = 1 + x(ab)^p\,f_p(a,b) &+ \frac{x\,(ab)^p\,(yb/a)}{1-yb/a}\,f_p(a,b) + \frac{x\,(ab)^p\,(ya/b)}{1-ya/b}\,f_p(a,b) \\ &- \frac{x(ab)^p(yb/a)}{1-yb/a}\,f_p(by,b) - \frac{x(ab)^p(ya/b)}{1-ya/b}\,f_p(a,ay). \end{aligned}$$

This may be put into kernel form:

$$K(a,b)\,f_p(a,b) = X(a,b) + Y(a,b)\,f_p(a,ya) + Z(a,b)\,f_p(yb,b) \tag{5.278}$$

with

$$\begin{aligned} K(a,b) &= (a-yb)(b-ya)(1-x(ab)^p) - xy(ab)^p(a^2+b^2-2yab); \\ X(a,b) &= (a-yb)(b-ya); \\ Y(a,b) &= -xya^{p+1}b^p(a-yb); \\ Z(a,b) &= -xya^pb^{p+1}(b-ya). \end{aligned}$$

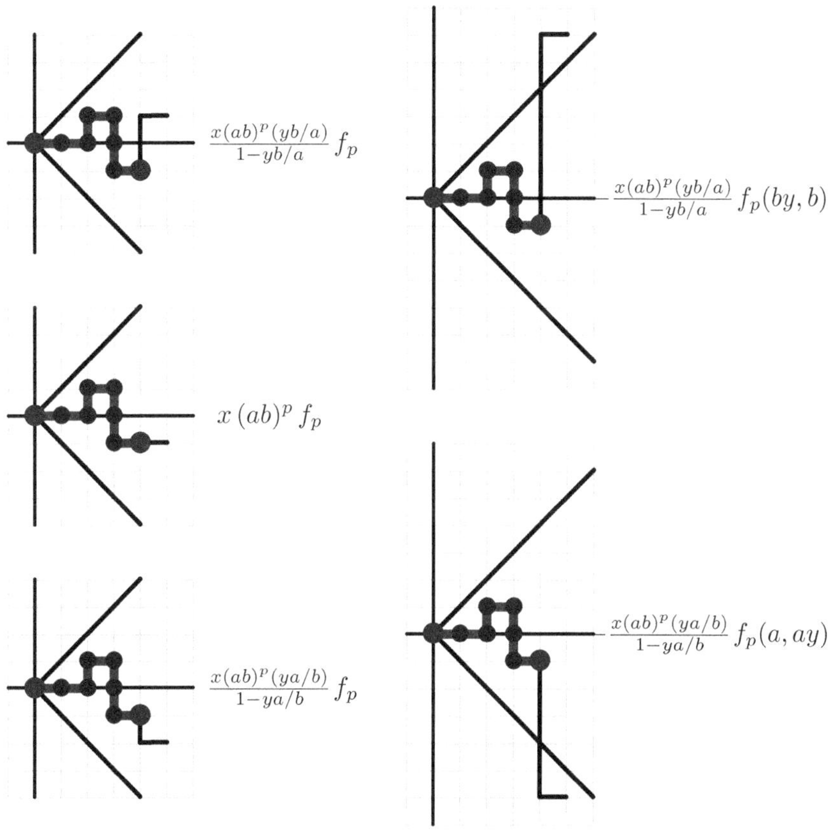

FIG. 5.36. The Temperley method can be used to find a recurrence for $f_p(a,b)$.

5.8.2.1 *The iterated kernel method for the case $p = 1$:* For $p = 1$, the kernel is a quadratic in b, and it has two roots $b = \beta_\pm(a)$. One of these roots is the 'physical root', and it will be denoted β_1. The other root is denoted by β_{-1}.

Then $K(a,\beta_1) = K(a,\beta_{-1}) = 0$. The roots of the kernel can be determined explicitly:

$$\beta_\pm(a) = \frac{a}{2}\left(\frac{1+y^2 \pm \sqrt{(1-y^2)(1-4xya^2-y^2)}}{y+xa^2-xy^2a^2}\right). \tag{5.279}$$

Examination of the two roots show that

$$\beta_1(a) \equiv \beta_-(a) = ya + O(xy^2a^3), \quad \text{and} \tag{5.280}$$

$$\beta_{-1}(a) \equiv \beta_+(a) = \tfrac{1}{y}a + O(xy^{-2}a). \tag{5.281}$$

Repeated mapping of $a \to \beta_1(a)$ generates a sequence of pairs $(\beta_n(a), \beta_{n+1}(a))$, where $\beta_n = \beta_1 \circ \beta_1 \circ \cdots \circ \beta_1 = \beta_1^{(n)}(a)$ is the composition of β_1 n times with itself (where $\beta_0(a) = a$). For each such pair, $K(\beta_n(a), \beta_{n+1}(a)) = 0$. Obviously, β_0 is the identity, so $K(\beta_0(a), \beta_1(a)) = K(a, \beta_1(a)) = 0$.

A similar approach for β_{-1} generates a sequence of pairs $(\beta_{-n}(a), \beta_{-n-1}(a))$, where $\beta_{-n} = \beta_{-1} \circ \beta_{-1} \circ \cdots \circ \beta_{-1} = \beta_{-1}^{(n)}(a)$ is the composition of β_{-1} n times with itself. For each such pair $K(\beta_{-n}(a), \beta_{-n-1}(a)) = 0$.

It can be shown that $\beta_1 \circ \beta_{-1} = \beta_{-1} \circ \beta_1 = \beta_0$. That is, the set of functions $\{\beta_n \mid n \in \mathbb{Z}\}$ is a presentation of the infinite group $(\mathbb{Z}, +)$ with identity β_0 and inverses $\beta_n \circ \beta_{-n} = \beta_0$. Set $b = \beta_1(a)$ in equation (5.278) and set $a = \beta_n(a)$ for any finite $n \geq 0$. Since $K(\beta_n(a), \beta_{n+1}(a)) = 0$,

$$\begin{aligned} X(\beta_n(a), \beta_{n+1}(a)) + Y(\beta_n(a), \beta_{n+1}(a))\, f_1(\beta_n(a), y\beta_n(a)) \\ + Z(\beta_n(a), \beta_{n+1}(a))\, f_1(y\beta_{n+1}(a), \beta_{n+1}(a)) = 0. \end{aligned} \tag{5.282}$$

Solve this equation for $f_1(\beta_n(a), y\beta_n(a))$:

$$\begin{aligned} f_1(\beta_n(a), y\beta_n(a)) = &-\left(\frac{X(\beta_n(a), \beta_{n+1}(a))}{Y(\beta_n(a), \beta_{n+1}(a))}\right) \\ &-\left(\frac{Z(\beta_n(a), \beta_{n+1}(a))}{Y(\beta_n(a), \beta_{n+1}(a))}\right) f_1(y\beta_{n+1}(a), \beta_{n+1}(a)) \end{aligned} \tag{5.283}$$

Simplify the above by making use of $f_1(a, b) = f_1(b, a)$ and defining the following:

$$\begin{aligned} &\mathfrak{F}_n(a) = f_1(\beta_n(a), y\beta_n(a)) = f_1(y\beta_n(a), \beta_n(a)); \\ &\mathfrak{X}_n(a) = -\left(\frac{X(\beta_n(a), \beta_{n+1}(a))}{Y(\beta_n(a), \beta_{n+1}(a))}\right), \quad \text{and } \mathfrak{Z}_n(a) = -\left(\frac{Z(\beta_n(a), \beta_{n+1}(a))}{Y(\beta_n(a), \beta_{n+1}(a))}\right). \end{aligned}$$

Equation (5.283) may be written as

$$\mathfrak{F}_n(a) = \mathfrak{X}_n(a) + \mathfrak{Z}_n(a)\, \mathfrak{F}_{n+1}(a). \tag{5.284}$$

Starting at $n = 0$, iterate this to get a formal series solution for $\mathfrak{F}_0(a)$:

$$f_1(a, ya) = \mathfrak{F}_0(a) = \sum_{n=0}^{\infty} \mathfrak{X}_n(a) \prod_{k=0}^{n-1} \mathfrak{Z}_k(a). \tag{5.285}$$

This also gives $F(yb, b)$:

$$f_1(yb, b) = f_1(b, yb) = \mathfrak{F}_0(b) = \sum_{n=0}^{\infty} \mathfrak{X}_n(b) \prod_{k=0}^{n-1} \mathfrak{Z}_k(b) \tag{5.286}$$

by symmetry. Thus, the formal solution for the generating function is

$$f_1(a, b) = \frac{X}{K} + \frac{Y}{K} \sum_{n=0}^{\infty} \mathfrak{X}_n(a) \prod_{k=0}^{n-1} \mathfrak{Z}_k(a) + \frac{Z}{K} \sum_{n=0}^{\infty} \mathfrak{X}_n(b) \prod_{k=0}^{n-1} \mathfrak{Z}_k(b), \tag{5.287}$$

where $X \equiv X(a, b)$; $Y \equiv Y(a, b)$; $Z \equiv Z(a, b)$; and $K \equiv K(a, b)$.

5.8.2.2 *Determining $\beta_n(a)$:* It follows directly from equation (5.279) that

$$\frac{1}{\beta_1} + \frac{1}{\beta_{-1}} = \frac{1+y^2}{ya}. \tag{5.288}$$

Substituting $a = \beta_{n-1}$ gives the second order recurrence

$$\frac{1}{\beta_n} = \frac{1+y^2}{y}\frac{1}{\beta_{n-1}} - \frac{1}{\beta_{n-2}}, \tag{5.289}$$

which may be iterated to obtain

$$\frac{1}{\beta_n} = \frac{y(1-y^{2n})}{y^n(1-y^2)}\frac{1}{\beta_1} - \frac{y^2(1-y^{2n-2})}{y^n(1-y^2)}\frac{1}{a}. \tag{5.290}$$

By simplifying the expressions for $\mathfrak{X}(a,b)$ and $\mathfrak{Z}(a,b)$, equation (5.285) becomes

$$f_1(a,ya) = \sum_{n=0}^{\infty}(-1)^n\left(\frac{\beta_{n+1}-y\beta_n}{xya\beta_n\beta_{n+1}}\right)\prod_{k=0}^{n-1}\left(\frac{\beta_{k+1}-y\beta_k}{\beta_k-y\beta_{k+1}}\right). \tag{5.291}$$

Substituting the expression for β_n and simplifying gives:

$$\frac{\beta_{n+1}-y\beta_n}{xya\beta_n\beta_{n+1}} = \frac{y^n}{a} - \frac{y}{\beta_1}, \tag{5.292}$$

$$\frac{\beta_{k+1}-y\,\beta_k}{\beta_k-y\beta_{k+1}} = \frac{y^{2k+1}}{xya^2} - \frac{1}{xa\beta_1} - 1. \tag{5.293}$$

These results give the following expression for $f_1(a,ya)$:

$$f_1(a,ya) = \left(\frac{1}{xya^2} - \frac{1}{xa\beta_1}\right)\sum_{n=0}^{\infty}(-1)^n y^{n(n+1)}\left(\frac{1}{xya^2} - \frac{1}{xa\beta_1} - 1\right)^n. \tag{5.294}$$

Define the function $Q(a)$ by

$$Q(a) \equiv Q(a;x,y) = \frac{1}{xa^2} - \frac{y}{xa\beta_1} - y. \tag{5.295}$$

This simplifies the expression for $f_1(a,ya)$:

$$f_1(a,ya) = \left(1+\tfrac{1}{y}Q(a)\right)\sum_{n=0}^{\infty}(-1)^n y^{n^2}Q(a)^n. \tag{5.296}$$

The $a \leftrightarrow b$ symmetry of $f_1(a,b)$ gives a similar expression for $f_1(yb,b)$. By equation (5.278), the full solution for $f_1(a,b)$ is

$$f_1(a,b) = \frac{X}{K} + \frac{Y}{K}\left(1+\tfrac{1}{y}Q(a)\right)\sum_{n\geq 0}(-1)^n Q(a)^n y^{n^2} \\ + \frac{Z}{K}\left(1+\tfrac{1}{y}Q(b)\right)\sum_{n\geq 0}(-1)^n Q(b)^n y^{n^2}. \quad (5.297)$$

Putting $a = b = 1$ and $x = y = t$ (so that t is conjugate to the number of edges) gives the generating function of partially directed walks ending in a horizontal step in V_1:

$$f_1(1,1) = \frac{1-t}{1-2t-t^2} - \frac{1-t^2-\sqrt{(1-t^2)(1-5t^2)}}{1-2t-t^2}\sum_{n=0}^{\infty}(-1)^n t^{n^2} Q(1;t,t)^n,$$

where

$$Q(1;t,t) = \tfrac{1}{2t}\left(1-3t^2-\sqrt{(1-t^2)(1-5t^2)}\right). \quad (5.298)$$

The full generating function of all paths in V_1 follows from equation (5.277):

$$\overline{g}_1(1,1) = \frac{1+t}{1-2t-t^2} \quad (5.299) \\ - \frac{1-t^2-\sqrt{(1-t^2)(1-5t^2)}}{t(1-2t-t^2)}\sum_{n=0}^{\infty}(-1)^n t^{n^2} Q(1;t,t)^n.$$

Both $f_1(1,1)$ and $g_1(1,1)$ are θ-functions with natural boundaries, so they are not holonomic [57].

5.8.2.3 *The asymptotics of the number of paths in V_1:* Denote the number of paths in V_1 of length n by v_n. An asymptotic approximation to v_n can be determined from the singularities of $\overline{g}_1(1,1)$; these singularities (see equation (5.299)) are either zeros of the factor $1-2t-t^2$, branch points in $\sqrt{(1-t^2)(1-5t^2)}$ or singularities in the function defined by the series $\sum_{n=0}^{\infty}(-1)^n t^{n^2} Q(1;t,t)^n$.

The generating function $\overline{g}_1(1,1)$ has simple poles at the roots of $1-2t-t^2$ ($t = -1\pm\sqrt{2}$). In addition, there are branch points at $t = \pm 1$ and again at $t = \pm\frac{1}{\sqrt{5}}$ in $\sqrt{(1-t^2)(1-5t^2)}$.

The series $\sum_{n=0}^{\infty}(-1)^n t^{n^2} Q(1;t,t)^n$ is a Jacobi θ-function, and it is convergent inside the unit circle except at singularities of $Q(1;t,t)$ (when $t = \pm\frac{1}{\sqrt{5}}$). This shows that the dominant singularity is the simple pole at $\sqrt{2}-1$ in $\overline{g}_1(1,1)$. There are subdominant contributions to the asymptotics at the singularities $t = \pm\frac{1}{\sqrt{5}}$ (which also give rise to parity effects). Extracting the residues and combining the results give

$$v_n = A_0\left(1+\sqrt{2}\right)^n + \frac{5^{n/2}}{(n+1)^{3/2}}\left(A_1 + (-1)^n A_2 + O(\tfrac{1}{n})\right), \quad (5.300)$$

where the constants are

$$A_0 = 0.27730985348603118827\ldots,$$
$$A_1 = 3.71410486533662324953\ldots, \quad \text{and} \tag{5.301}$$
$$A_2 = 0.20697997020804157910\ldots.$$

These constants were obtained by expanding the expression for $g_1(1,1)$ about $t = \sqrt{2}-1$, and about $t = \pm\frac{1}{\sqrt{5}}$ (using the first forty terms).

5.9 Spiral walks

A self-avoiding walk from the origin in $\mathbb{L}^2$ is a *spiral walk* [409, 475] if every step it gives is either in the same direction as the last step or takes a 90° right turn. An example of a spiral walk is given in figure 5.37. This is an *out-spiral* since it spirals outwards from the origin.

An out-spiral starts at the origin, gives a *train* of j_1 north steps, followed by a train of i_1 east steps. This is followed by trains of j_2 south steps and i_2 west steps. The walk continues in this way, alternating trains of $\{j_k\}$ north steps or south steps with trains of $\{i_k\}$ east steps or west steps. The $\{j_k, i_k\}$ are the lengths of *trains* of the out-spiral. By convention, j_1 is the number of north steps steps from the origin, followed by a train of i_1 east steps.

Assume that all out-spirals start with a north step ($j_1 > 0$). Let the trains of the out-spiral have lengths $\{j_1, i_1, j_2, i_2, \ldots, j_m, i_m\}$. If $i_m > 0$, then the number of trains is $2m$, and the number of right turns is $2m-1$. If $i_m > 0$ is the last train, then $j_{m-1} < j_m$, and, if j_m is the last train, then $i_{m-2} < i_{m-1}$.

In an out-spiral, if $i_m = i_{m-1} > 0$ or if $j_m = j_{m-1} > 0$, and $i_m = 0$, then the spiral is a *stopped spiral*. The out-spiral in figure 5.37 is a stopped spiral. The number of trains in a stopped spiral is at least 3.

The generating function of a stopped spiral can be computed by summing over the lengths of its trains. Introduce generating variable t conjugate to length, and z conjugate to the number of trains. Then the generating function of stopped spirals with the first step a north step and with m sides is

$$G_\perp(m;z,t) = z^m \sum_{j_1=1}^{\infty} \sum_{i_1=1}^{\infty} \sum_{j_2=j_1+1}^{\infty} \cdots \sum_{j_{\lceil m/2\rceil}=j_{\lceil m/2\rceil-1}+1}^{\infty} \sum_{i_{\lfloor m/2\rfloor}=i_{\lfloor m/2\rfloor-1}+1}^{\infty} t^K,$$

where $K = i_{\lfloor m/2\rfloor} + \sum_k (i_k + j_k)$ if m is even, and $K = j_{\lceil m/2\rceil} + \sum_k (i_k + j_k)$ if m is odd. For $m \geq 3$, this can be evaluated to give

$$G_\perp(m;z,t) = \frac{z^m(1-t)}{zt} \prod_{j=1}^{\lceil m/2\rceil} \left(\frac{t^j}{1-t^j}\right) \prod_{i=1}^{\lfloor m/2\rfloor} \left(\frac{t^i}{1-t^i}\right). \tag{5.302}$$

Summing from $m=3$ gives the generating function of stopped spirals with first steps in the north direction. If $(t;q)_n$ is the q-Pochhammer function, then

$$G_\perp(z,t) = \frac{(1-t)}{zt} \sum_{m=3}^{\infty} \frac{z^m t^{\lceil m/2\rceil\,(\lfloor m/2\rfloor+1)}}{(t;t)_{\lceil m/2\rceil}\,(t;t)_{\lfloor m/2\rfloor}}. \tag{5.303}$$

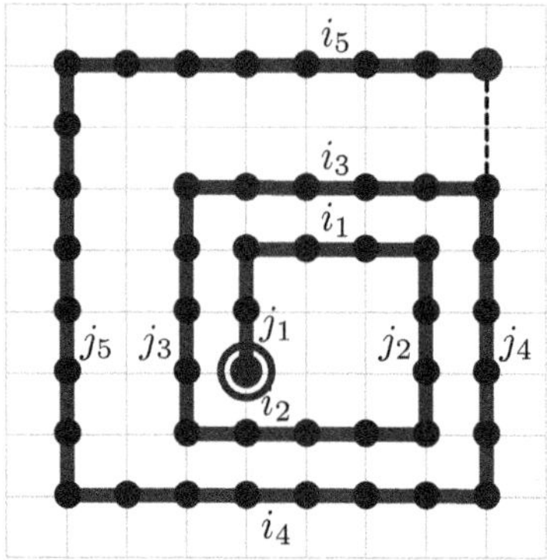

FIG. 5.37. An *out-spiral* is a self-avoiding walk from the origin in the square lattice. It is convention that the spiral rotates clockwise from the origin; that is, north steps are followed by east steps, which in turn are followed by south steps and then by west steps. This out-spiral steps north for j_1 steps, then east for i_1 steps, then south for j_2 steps, then west for i_2 steps, then north for j_3 steps, and so on. It is convention that $j_1 > 0$ so that the walk starts with a north step. In this out-spiral $i_5 = i_4$; this is an example of a *stopped spiral.*

The full generating function of out-spirals with first steps in the north direction is obtained by adding in the generating functions of out-spirals of 0, 1 and 2 sides and by adding edges to the endpoints of stopped spirals.

The infinite series defining $G_\perp(z,t)$ is absolutely convergent in the unit disk with a natural boundary $|t| = 1$. It is holonomic inside the unit disk in the t-plane.

An asymptotic approximation can be determined for $G_\perp(z,t)$ by using a saddle point approximation. Use equation (D.41) from appendix D to note that $(t;t)_n\,(t^{n+1};t)_\infty = (t;t)_\infty$. In the above this gives

$$G_\perp(z,t) = \frac{(1-t)}{zt\,(t;t)_\infty^2} \sum_{m=0}^{\infty} z^m\, t^{\lceil m/2\rceil}\, t^{\lceil m/2\rceil \lfloor m/2\rfloor}\, \frac{(t^{\lceil m/2\rceil};t)_\infty (t^{\lfloor m/2\rfloor};t)_\infty}{(1-t^{\lceil m/2\rceil})(1-t^{\lfloor m/2\rfloor})}.$$

Approximate the summation by an integral and use the approximation in equation (D.76) in appendix D for the q-Pochhammer functions. The summation is approximated by the integral

$$\int_0^\infty \frac{z^m\, t^{\lceil m/2\rceil + \lceil m/2\rceil \lfloor m/2\rfloor}}{\sqrt{(1-t^{\lfloor m/2\rfloor})(1-t^{\lceil m/2\rceil})}}\, e^{-\frac{1}{|\log t|}\left(\mathcal{L}i_2(t^{\lfloor m/2\rfloor}) + \mathcal{L}i_2(t^{\lceil m/2\rceil}) - \frac{\pi^2}{3}\right)}\, dm. \tag{5.304}$$

This resolves differently for even and odd values of m because of the ceiling and floor functions. Approximate the above by only integrating through even m.

That is, approximate $\lceil \frac{m}{2} \rceil \approx \lfloor \frac{m}{2} \rfloor \approx \frac{m}{2}$, substitute $m = \frac{2\log s}{\log t}$ and simplify. The integral becomes

$$\frac{2}{|\log t|} \int_0^1 \frac{s^{\frac{2\log z}{\log t}}}{1-s} \, e^{-\frac{1}{|\log t|}\left(2\mathcal{L}i_2(s) + \log^2 s - \frac{\pi^2}{3}\right)} \, ds. \tag{5.305}$$

Use equation (D.39) from appendix D to approximate this integral. Put $N = \frac{1}{|\log t|}$, put

$$N f(s) = \frac{2\log z \cdot \log s}{\log t} - \frac{1}{|\log t|}\left(2\mathcal{L}i_2(s) + \log^2 s - \frac{\pi^2}{3}\right)$$

and let $g(s) = \frac{2}{|\log t|\,(1-s)}$.

The saddle point is located at $s_0 = \frac{1}{1+z} \in (0,1)$. Substituting s_0 and using equation (D.39) from appendix D to estimate the integral. This gives

$$\frac{2\sqrt{\pi}}{\sqrt{z(1+z)\,|\log t|}} \, e^{\frac{2\mathcal{L}i_2(\frac{z}{1+z}) + \log^2(1+z)}{|\log t|}}. \tag{5.306}$$

Lastly, approximate $(t;t)_\infty$ by equation (D.74) in appendix D. Taken together, this gives the approximation

$$G_\perp(z,t) \approx \frac{(1-t)\sqrt{|\log t|}}{zt\sqrt{\pi\, z(1+z)}} \, e^{\frac{2\mathcal{L}i_2(z/(1+z)) + \log^2(1+z) + \frac{\pi^2}{3}}{|\log t|}}. \tag{5.307}$$

Putting $z = 1$ gives the simpler formula for t approaching 1:

$$G_\perp(1,t) \approx \frac{(1-t)\sqrt{|\log t|}}{\sqrt{2\pi}\; t} \, e^{\frac{\pi^2}{2|\log t|}} \approx \frac{|\log^{3/2} t|}{\sqrt{2\pi}\; t} \, e^{\frac{\pi^2}{2|\log t|}}. \tag{5.308}$$

The generating functions of spiral walks were studied in references [42, 250, 253]. An asymptotic formula for the number of spiral walks of length n is due to Joyce [342]; the number of spiral walks of length n grows asymptotically as $\frac{1}{4} \frac{1}{3^{5/4}} \frac{\pi}{n^{7/4}} \, e^{2\pi\sqrt{n}/\sqrt{3}}$. Related results on these models can be found in references [381, 383, 386, 392, 580].

6

CONVEX LATTICE VESICLES AND DIRECTED ANIMALS

Convex lattice vesicles in $\mathbb{L}^2$ are directed lattice polygon models of two-dimensional, fluid-filled, vesicles which undergo pressure-induced changes in shape. These models are related to models of directed animals and directed percolation. There are many different models of convex polygons, including partitions [8, 272, 276, 471] (see figure 3.9).

Pressure-induced shape changes in vesicles are (usually) first order phase transitions between deflated and inflated phases, and the transition is like a condensation transition of droplets. If the condensation is near a boundary, then it is a two-dimensional model of a wetting transition along a surface. This wetting transition also can be modelled by two-dimensional directed percolation along a boundary.

Models of convex lattice polygons in $\mathbb{L}^2$ are classical models in enumerative combinatorics [31, 353–356, 486]. Examples of convex polygons include partitions, stacks and staircase polygons, or other lattice polygons with the property that the intersection of the polygon with a horizontal or vertical line has at most two connected components. For recent work, see references [48–51, 55, 60, 66] as well as references [54, 292, 482, 493].

6.1 Partitions

Partitions can be represented by bottom-adjusted Ferrer's diagrams in $\mathbb{L}^2$ (see figure 6.1). The perimeter of the Ferrer's diagram is a *partition polygon* enclosing a subset of $\mathbb{R}^2$ which is its *area*. The area is tiled with unit lattice squares called *plaquettes*.

The area-perimeter generating function of partition polygons can be obtained by cutting it into strips as shown in figure 6.1(a). Introduce the generating variables q, conjugate to plaquettes, and t, conjugate to perimeter edges. If there are n such strips and the j-th strip has length i_j (see figure 6.1(a)), then the area-perimeter generating function is

$$V_p(t,q) = \sum_{n=1}^{\infty} t^{2n} \sum_{i_n=1}^{\infty} \cdots \sum_{i_2=i_3}^{\infty} \sum_{i_1=i_2}^{\infty} t^{2i_1} q^{i_1+i_2+\cdots+i_n} = \sum_{n=1}^{\infty} \frac{t^{2n+2} q^n}{(t^2 q; q)_n}, \tag{6.1}$$

where t is the perimeter generating variable (conjugate to perimeter edges), and q is the area generating variable (conjugate to area plaquettes). Alternatively, a Ferrer's diagram can be cut into $\mathbb{L}$-shaped strips. Putting $L = 2(i_1 + j_1)$ and putting $\alpha = \sum i_k + \sum j_\ell$ give the generating function

The Statistical Mechanics of Interacting Walks, Polygons, Animals and Vesicles, 2nd edition, E.J. Janse van Rensburg. © E.J. Janse van Rensburg. Published in 2015 by Oxford University Press.

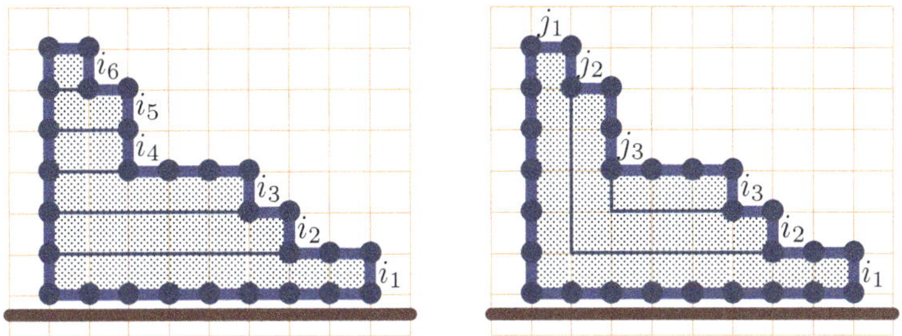

(a) (b)

FIG. 6.1. The area-perimeter generating function of partition polygons can be computed by (a) cutting the partition into n strips of lengths $\langle i_1, i_2, \ldots, i_n \rangle$ or (b) cutting it into m L-shaped strips with legs of lengths (i_k, j_k), where $i_k \geq i_{k+1} + 1$, and $j_k \geq j_{k+1} + 1$.

$$V_p(t,q) = \sum_{m=1}^{\infty} t^4 q^m \sum_{\substack{i_m = 0 \\ j_m = 0}}^{\infty} \cdots \sum_{\substack{i_2 = i_3 + 1 \\ j_2 = j_3 + 1}}^{\infty} \sum_{\substack{i_1 = i_2 + 1 \\ j_1 = j_2 + 1}}^{\infty} t^L q^\alpha = \sum_{m=1}^{\infty} \frac{t^{4m} q^{m^2}}{(t^2 q; q)_m^2}. \tag{6.2}$$

Equations (6.1) and (6.2) give a partition identity [531].

If x generates horizontal perimeter edges, and y generates vertical perimeter edges, then the generating function of partitions of height h is

$$V_h(x,y,q) = \frac{x^2 y^{2h} q^h}{(x^2 q; q)_h} \tag{6.3}$$

by equation (6.1) and figure 6.1(a). Putting $x = y = t$ and summing over n gives equation (6.1).

If $N_{(w+h)w\alpha}$ is the number of partitions of area α, width w and height h, then the generating function of partition vesicles of width w and height h is

$$V_{wh}(x,y,q) = x^{2w} y^{2h} \sum_{\alpha=0}^{wh} N_{(w+h)w\alpha} q^\alpha = x^{2w} y^{2h} \begin{bmatrix} w+h \\ w \end{bmatrix} \tag{6.4}$$

by lemma D.9 in appendix D.

Singularities in $V_p(t,q)$ are seen in equations (6.1) and (6.2). Poles located at $(1 - t^2 q^n) = 0$ for $n \in \mathbb{N}$ accumulate on $q = 1$, which is a line of essential singularities in the q-plane. The radius of convergence $q_c(t)$ of $V_p(t,q)$ is given by a singularity on the real axis. For $t \geq 1$, $q_c(t) = \frac{1}{t^2}$, and, for $t < 1$, $q_c(t) = 1$.

The microcanonical density function of the perimeter is $\mathcal{V}(\epsilon) = 1$ for $\epsilon \in [0, 2]$. This shows that the partition undergoes a first order transition, from a phase of deflated ⊾-shaped partitions to a phase of inflated □-shaped partitions.

The radius of convergence is plotted in figure 6.2 (notice that $t_c = 1$). The critical line $q_c(t)$ separates a phase of finite partitions from a phase of infinite

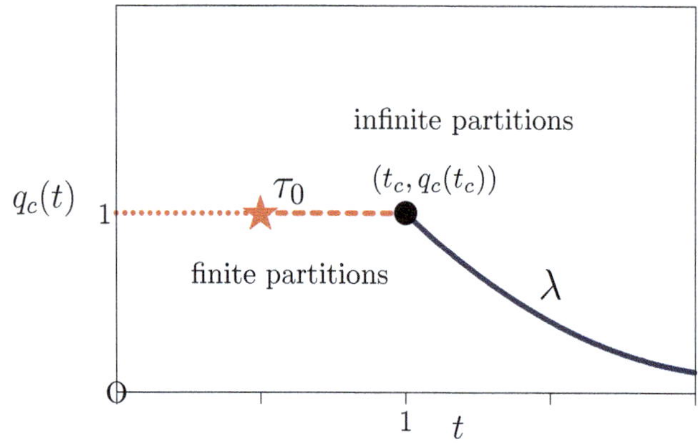

FIG. 6.2. The singularity or phase diagram of inflating partitions. A τ_0-line meets a λ-isotherm at a tricritical point at $(t_c, q_c) = (1, 1)$. The point on the τ_0-line marked by $\star$ is a special point where the generating function diverges; to the left of this point, $V_p(t, 1)$ is finite (and dominated by partitions of finite size), and, to the right of this point, $V_p(t, 1)$ is divergent. This special point is located at $(t, q) = (1/\sqrt{2}, 1)$.

---

partitions. The tricritical point is located at $(t_c, q_c(t_c)) = (1, 1)$, separating the critical curve into a τ_0-line and a λ-isotherm.

D'Alembert's ratio test shows that $V_p(t, 1) < \infty$ if $t^2 < \frac{1}{2}$, and $V_p(t, 1) = \infty$ if $t^2 \geq \frac{1}{2}$. This shows that, on the τ_0-line, there is a special point where finite partitions give way to infinite partitions as $t^2 \to \frac{1}{2}^-$ with $q = 1$.

The limiting free energy of the model is given by $\mathfrak{V}(t) = -\log q_c(t)$. That is, $\mathfrak{V}(t) = 0$ if $t \leq 1$, and $\mathfrak{V}(t) = 2 \log t$ if $t > 1$. The first derivative of $\mathfrak{V}(t)$ is the energy of the model, and this is given by $\mathfrak{E}(t) = 0$ if $t < 1$, and $\mathfrak{E}(t) = 2$ if $t > 1$. The jump discontinuity in $\mathfrak{E}(t)$ is a latent heat in the model and is consistent with a first order transition between the deflated and inflated phases.

6.1.1 *Asymptotics of the partition generating function*

There are three scaling regimes at $t < 1$, $t = 1$, and $t > 1$, respectively [306].

If $t > 1$, then the radius of convergence of $V_q(t, q)$ is $q_c(t) < 1$. Noting that $(t^2q; q)_n \geq (t^2q; q)_\infty$ if $q < q_c(t) < 1$, it follows from equation (6.2) that

$$\frac{t^4 q}{(1-t^2q)^2} \leq V_p \leq \frac{t^4 q}{(1-t^2q)^2} \left(1 + \frac{1}{(t^2q^2; q)^2_\infty} \sum_{n=1}^{\infty} t^{4n} q^{n^2+2n} \right). \tag{6.5}$$

The terms in square brackets on the right-hand side are finite along the λ-isotherm. Thus, on approach to the λ-isotherm, $V_p(t, q) \sim \frac{t^4 q}{(1-t^2q)^2}$ (see figure 6.2). Comparison to equation (4.19) shows that $\gamma_+ = 2$ in this model.

When $t = 1$, the tricritical point is approached as $q \to 1^-$. By equation (6.1),

$$V_p(1,q) = \sum_{n=1}^{\infty} \frac{q^n}{(q;q)_n} = \sum_{n=1}^{\infty} \frac{q^n}{1-q^n} \frac{(q^n;q)_\infty}{(q;q)_\infty}. \tag{6.6}$$

Approximate $(q;q)_\infty$ and $(q^n;q)_\infty$ by equations (D.74) and (D.77) from appendix D. This gives

$$V_p(1,q) \approx \sqrt{\frac{|\log q|}{2\pi}}\, e^{\frac{\pi^2}{3|\log q|}} \sum_{n=1}^{\infty} \frac{q^n}{\sqrt{1-q^n}} e^{-\frac{\mathcal{L}i_2(q^n)}{|\log q|}}. \tag{6.7}$$

Approximate the series in equation (6.7) by an integral. Change variables by putting $n = \frac{\log s}{\log q}$. The range of integration is $s \in (0,q)$, so

$$V_p(1,q) \approx \frac{1}{\sqrt{2\pi|\log q|}}\, e^{\frac{\pi^2}{3|\log q|}} \int_0^q \frac{s}{\sqrt{1-s}} e^{-\frac{\mathcal{L}i_2(s)}{|\log q|}}\, ds. \tag{6.8}$$

Notice that

$$\int_0^q \frac{1}{\sqrt{1-s}} e^{-\frac{\mathcal{L}i_2(s)}{|\log q|}}\, ds = \int_0^q \frac{1}{s\sqrt{1-s}} e^{-\frac{(\mathcal{L}i_2(s)+\log s\cdot\log q)}{|\log q|}}\, ds. \tag{6.9}$$

Use a saddle point approximation to approximate the integral. Put $N = -\frac{1}{|\log q|}$, put $f(s) = \mathcal{L}i_2(s) + \log s \cdot \log q$ and put $g(s) = \frac{1}{s\sqrt{1-s}}$ in equation (D.39) in appendix D. The saddle point is at $s_0 = 1-q$. This is in the region of integration if $1-q < q$ or if $q > \frac{1}{2}$. Substituting the results into equation (D.39) and simplifying gives

$$\int_0^q \frac{1}{\sqrt{1-s}} e^{-\frac{\mathcal{L}i_2(s)}{|\log q|}}\, ds \approx \sqrt{2\pi}\,|\log q|\, e^{-\frac{\mathcal{L}i_2(1-q)}{|\log q|}}, \tag{6.10}$$

for q close to 1.

Approximate $|\log q| \approx 1-q$ as $q \to 1^-$. Use the identity $\mathcal{L}i_2(q) + \mathcal{L}i_2(1-q) = \frac{\pi^2}{6} - \log q \cdot \log(1-q)$. Then the approximation in equation (6.8) becomes

$$V_p(1,q) \approx \frac{1}{\sqrt{|\log q|}}\, e^{\frac{\mathcal{L}i_2(q)-\frac{\pi^2}{6}}{|\log q|}} \quad \text{as } q \to 1^-. \tag{6.11}$$

It remains to consider the case that $t < 1$. By equation (6.1),

$$V_p(t;q) = \sum_{n=1}^{\infty} \frac{1}{1-t^2q^n} \frac{(t^2q^n;q)_\infty}{(t^2q;q)_\infty}\, t^{2n+2}q^n. \tag{6.12}$$

Approximate $(t^2q^n;q)_\infty$ and $(t^2q;q)_\infty$ by equation (D.76). Approximate the summation by an integral and change variables $n = \frac{\log s}{\log q}$. This gives the approximation

$$V_p(t,q) \approx \frac{t^2 e^{\frac{\mathcal{L}i_2(t^2 q)}{|\log q|}}}{|\log q|} \int_0^1 \frac{e^{-\frac{1}{|\log q|}(\mathcal{L}i_2(t^2 s)+2\log s\cdot \log t)}}{\sqrt{1-t^2 s}}\, ds. \tag{6.13}$$

Approximate the integral by using the saddle point approximation in equation (D.39): put $N = -\frac{1}{|\log q|}$, put $f(s) = \mathcal{L}i_2(t^2 s) + 2\log s \cdot \log t$ and put $g(s) = \frac{1}{\sqrt{1-t^2 s}}$. The saddle point is at $s_0 = \frac{1-t^2}{t^2} \in (0,1)$ if $\frac{1}{2} < t^2 < 1$. That is, the approximation is only valid between the special point and the tricritical point along the τ_0-line in figure 6.2. The approximation simplifies to

$$V_p(t,q) \approx \sqrt{\frac{2\pi(1-t^2)}{|\log q|(1-t^2 q)}}\, e^{\frac{\mathcal{L}i_2(t^2 q)+\mathcal{L}i_2(t^2)-\frac{\pi^2}{6}+\log^2 t^2}{|\log q|}} \tag{6.14}$$

for $q \to 1^-$, and $\frac{1}{2} < t^2 < 1$.

6.1.2 *Scaling of partition vesicles*

Tricritical coordinates (τ, g) are introduced in the model by placing the τ-axis along the line $q = 1$ (this is the τ_0-line). The λ-isotherm is given by $qt^2 = 1$, and the g-axis is placed normal to the τ_0-line through the tricritical point so that the origin in the $g\tau$-plane is at the tricritical point. In these coordinates the λ-isotherm has the formula

$$g = \frac{\tau(2+\tau)}{(1+\tau)^2}. \tag{6.15}$$

Taking $\tau \to 0^+$, this shows by equation (4.8) that the shift-exponent is $\psi_\lambda = 1$. By equation (4.17), the crossover exponent is $\phi = 1$, and, by equation (4.46), the specific heat exponent is $\alpha = 1$. These values are consistent with a first order inflation or condensation transition in the model.

The scaling of the generating function is uncovered by using scaling fields $\sigma = |\log q| \approx 1-q$ (for $q<1$, and q close to 1), and $\rho = 1 - t^2 q$. The field σ is a measure of the distance from the τ_0-line, and ρ is a measure of the distance from the λ-isotherm. Since $\mathcal{L}i_2(x)$ is singular at $x=1$, define $L(1-x) = \mathcal{L}i_2(x)$. The scaling of the generating function of partition polygons is

$$V_p(t,q) \approx \begin{cases} \frac{t^4 q}{\rho^2}, & \text{if } \tau > 0; \\ \frac{1}{\sqrt{\sigma}}\, e^{\frac{1}{\sigma}\left(L(\sigma)-\frac{\pi^2}{6}\right)}, & \text{if } \tau = 0; \\ \sqrt{\frac{2\pi(1-t^2)}{\sigma\rho}}\, e^{\frac{1}{\sigma}\left(L(\rho)-\frac{\pi^2}{6}+\mathcal{L}i_2(t^2)+\log^2 t^2\right)}, & \text{if } \tau < 0 \end{cases} \tag{6.16}$$

by equations (6.5), (6.11) and (6.14), and where $L(x) = \mathcal{L}i_2(1-x)$ is singular when $x = 0$.

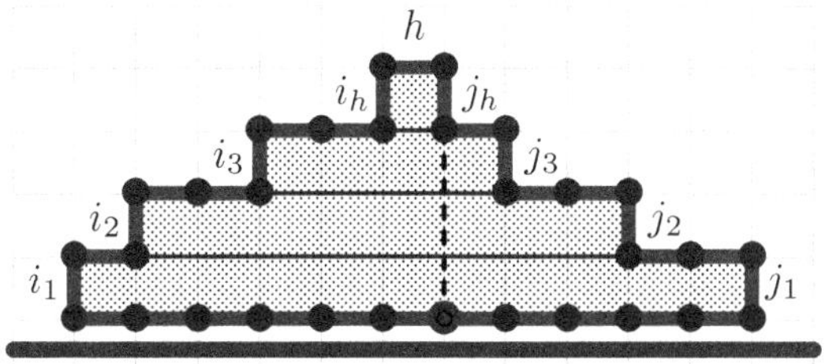

FIG. 6.3. A stack of height h, area $\sum_k i_k + \sum_l j_l$ and perimeter $2h + 2i_1 + 2j_1$. The stack can be cut into two partitions along the dashed line.

6.2 Stacks

Stacks (figure 6.3) are a generalisation of partitions and were first counted in references [596, 597] (see reference [471] as well).

The area-perimeter generating function of stacks may be determined by summing over $\{i_k, j_l\}$ and h in figure 6.3. Let q be conjugate to area and let t be conjugate to perimeter. Let x be conjugate to horizontal perimeter edges, and y to vertical perimeter edges. Then the generating function of stacks is

$$S(x,y,q) = \sum_{h=1}^{\infty} \frac{y^{2h} x^2 q^h}{(x^2q;q)_h (x^2q;q)_{h-1}}. \tag{6.17}$$

Each term in $S(x,y,q)$ is the generating function of stacks of height h.

Expanding the q-factorials in the denominator using equation (D.55) and then selecting the coefficient of x^w gives the generating function of stacks of base width w and height h:

$$S_{wh}(x,y,q) = x^{2w} y^{2h} q^{w+h-1} \sum_{i=0}^{w-1} \begin{bmatrix} h+i-1 \\ i \end{bmatrix} \begin{bmatrix} w+h-3-i \\ w-1-i \end{bmatrix}. \tag{6.18}$$

The generating function of stacks of base width w is

$$S_w(x,y,q) = (x^2q)^w \sum_{h=1}^{\infty} y^{2h} q^{h-1} \sum_{i=0}^{w-1} \begin{bmatrix} h+i-1 \\ i \end{bmatrix} \begin{bmatrix} w+h-3-i \\ w-1-i \end{bmatrix}. \tag{6.19}$$

See reference [382] for related results.

The area-perimeter generating functions are obtained by putting $x = y = t$ in the above. For example, the full generating function is

$$S(t,q) = \sum_{h=1}^{\infty} \frac{t^{2h+2} q^h}{(t^2q;q)_h (t^2q;q)_{h-1}}. \tag{6.20}$$

The radius of convergence $q_c(t)$ of $S(t,q)$ can be determined by D'Alembert's ratio test. For $t > 1$, there is a curve of double poles along $t^2q = 1$. For $t < 1$, there a line of essential singularities along $q = 1$. In addition, $S(t,1) < \infty$ for $t < \frac{1}{2}(\sqrt{5}-1)$ but $S(t,1) = \infty$ if $t > \frac{1}{2}(\sqrt{5}-1)$. That is, the phase diagram is similar to the phase diagram of partitions, with a special point on the τ_0-line (see figure 6.2). There is a tricritical point at $(t,q) = (1,1)$, and the λ-isotherm is given by $t^2 q = 1$, which is the same shape as for partition polygons. This shows that the shift-exponent is $\psi_\lambda = 1$, and the crossover exponent is $\phi = 1$.

6.2.1 *Scaling of the stack polygon generating function*

Put $x = y = t$ in the above. If $t > 1$, then $(t^2q;q)_1 \geq (t^2q;q)_n \geq (t^2q;q)_\infty$ for $t^2 q \leq 1$, so

$$\frac{1}{(t^2q;q)_1^2}\sum_{h=1}^{\infty} t^{2h+2}q^h(1-t^2q^h) \leq S(t,q) \leq \frac{1}{(t^2q;q)_\infty^2}\sum_{h=1}^{\infty} t^{2h+2}q^h. \tag{6.21}$$

This shows that

$$\frac{t^2}{(1-t^2x)^3}\left(1-\frac{t^2(1-t^2q)}{(1-t^2q^2)}\right) \leq S(t,q) \leq \frac{t^2}{(1-t^2q)^3\,(t^2q^2;q)_\infty^2}. \tag{6.22}$$

Along the λ-isotherm

$$S(t,q) \simeq \frac{t^2}{(1-t^2q)^3} \quad \text{as } q \nearrow t^{-2}, \text{ and } t > 1. \tag{6.23}$$

By equation (4.19), $\gamma_+ = 3$. The divergence along the λ-isotherm is consistent with deflated stacks which are 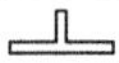-shaped in the limit; each 'leg' of the deflated stack contributes a factor $\frac{1}{1-t^2q}$.

Putting $t = 1$ in the stack generating function gives

$$S(1,q) = \sum_{h=1}^{\infty} \frac{q^h}{(q;q)_h(q;q)_{h-1}} = \sum_{h=1}^{\infty} \frac{q^h}{1-q^h}\,\frac{(q^h;q)_\infty^2}{(q;q)_\infty^2}. \tag{6.24}$$

Approximate $(q;q)_\infty$ and $(q^h;q)_\infty$ by equations (D.74) and (D.77) from appendix D. This gives

$$S(1,q) \approx \frac{|\log q|}{2\pi}\sum_{h=1}^{\infty} q^h e^{-\frac{2\mathcal{L}i_2(q^h)}{|\log q|}}. \tag{6.25}$$

Approximate the summation by an integral. Change variables by putting $h = \frac{\log s}{\log q}$ and simplify. Then the range of integration is $s \in (0,q)$, and

$$S(1,q) \approx \frac{1}{2\pi}\int_0^q e^{-\frac{2\mathcal{L}i_2(s)}{|\log q|}}\, ds. \tag{6.26}$$

This integral can be approximated using the saddle point approximation in equation (D.39). Put $N = -\frac{1}{|\log q|}$, put $f(s) = 2\mathcal{L}i_2(s) + 2\log s \cdot \log q$ and put

$g(s) = \frac{1}{2\pi s^2}$. The saddle point $s_0 = 1-q$, and $s_0 \in (0,q)$ if $q \in (\frac{1}{2},1)$. Substituting the results into equation (D.39) and simplifying gives

$$S(1,q) \approx \frac{\sqrt{q}\,|\log q|}{\sqrt{4\pi}}\, e^{-\frac{2\,\mathcal{L}i_2(1-q)}{|\log q|}} \approx \frac{\sqrt{q}}{\sqrt{4\pi}\,|\log q|}\, e^{\frac{2}{|\log q|}\left(\mathcal{L}i_2(q)-\frac{\pi^2}{6}\right)} \tag{6.27}$$

for q close to 1.

The case $t < 1$ remains. Replace $(t^2q;q)_{h-1}(t^2q^h;q)_\infty = (t^2q;q)_\infty$ in equation (6.20). Approximate $(t^2q^h;q)_\infty$ and $(t^2q;q)_\infty$ by equation (D.76) and simplify:

$$S(t,q) \approx \sum_{h=1}^{\infty} \frac{t^{2h+2}q^h\, e^{-\frac{2\,\mathcal{L}i_2(t^2q^h)}{|\log q|}}\, e^{\frac{2\,\mathcal{L}i_2(t^2q)}{|\log q|}}}{1-t^2q}. \tag{6.28}$$

Approximate the sum by an integral and change variables $h = \frac{\log s}{\log q}$. This gives

$$S(t,q) \approx \frac{t^2 e^{\frac{2\,\mathcal{L}i_2(t^2q)}{|\log q|}}}{(1-t^2q)|\log q|} \int_0^1 s^{-\frac{2\log t}{|\log q|}}\, e^{-\frac{2\,\mathcal{L}i_2(t^2s)}{|\log q|}}\, ds. \tag{6.29}$$

Use a saddle point method (equation (D.39) in appendix D) to approximate the integral. Put $f(s) = 2\mathcal{L}i_2(t^2s) + 2\log s \cdot \log t$, put $g(s) = 1$ and put $N = \frac{-1}{|\log q|}$. The saddle point is $s_0 = \frac{1-t}{t^2}$. Note that $s_0 \in (0,1)$ if $\frac{1}{2}(\sqrt{5}-1) < t < 1$. Simplify the result to find that

$$S(t,q) \approx \frac{\sqrt{\pi t(1-t)}}{(1-t^2q)\sqrt{|\log q|}}\, e^{\frac{2\,\mathcal{L}i_2(t^2q)+2\,\mathcal{L}i_2(t^2)-\frac{\pi^2}{3}+\log^2 t^2}{|\log q|}} \tag{6.30}$$

as $q \to 1^-$, and $\frac{1}{2}(\sqrt{5}-1) < t < 1$.

6.2.2 *Scaling of stack vesicles*

The critical curve of stack vesicles is the same as for partition vesicles (see figure 6.2). The shift-exponent is $\psi_\lambda = 1$, and the crossover exponent is $\phi = 1$ by equation (4.17) so, by equation (4.46), the specific heat exponent is $\alpha = 1$.

Introduce scaling fields $\sigma = |\log q| \approx 1-q$ (for $q < 1$, and q close to 1), and $\rho = 1 - t^2q$. Since $\mathcal{L}i_2(x)$ is singular at $x=1$, define $L(1-x) = \mathcal{L}i_2(x)$. Then in terms of these scaling fields and the function L, the scaling of the stack vesicle generating function becomes

$$S(t,q) \approx \begin{cases} \frac{t^2}{\rho^3}, & \text{if } \tau > 0; \\ \frac{\sqrt{q}}{\sqrt{4\pi\sigma}}\, e^{\frac{2}{\sigma}\left(L(\sigma)-\frac{\pi^2}{6}\right)}, & \text{if } \tau = 0; \\ \frac{\sqrt{\pi t(1-t)}}{\rho\sqrt{\sigma}}\, e^{\frac{1}{\sigma}\left(2L(\rho)-\frac{\pi^2}{3}+2\,\mathcal{L}i_2(t^2)+\log^2 t^2\right)}, & \text{if } \tau < 0 \end{cases} \tag{6.31}$$

by equations (6.23), (6.27) and (6.30).

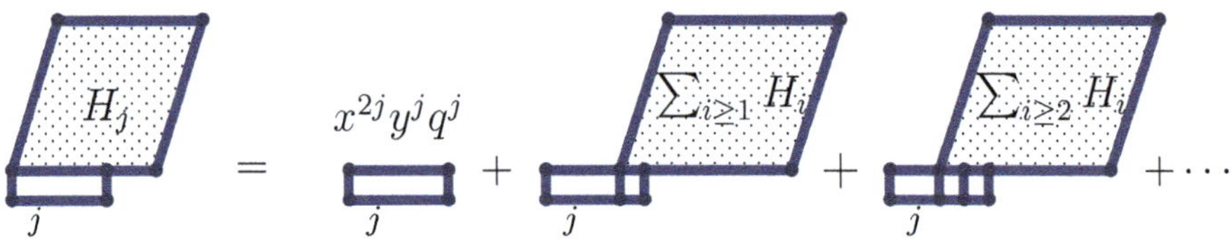

FIG. 6.4. A staircase polygon with a base of width j can be decomposed in the classes illustrated here. This gives a recursion relation for H_j.

6.3 Staircase vesicles

Staircase polygons (see section 5.6) counted by area and perimeter are *staircase vesicles*. The generating function of this model has been determined in several classical ways, which are reviewed below.

Denote the area-perimeter generating function by $H(t,q) \equiv H$ (q is the area generating variable, and t is the perimeter generating variable). Introduce generating variables x, conjugate to horizontal perimeter edges, and y, conjugate to vertical perimeter edges.

6.3.1 *Staircase vesicles by a functional recurrence*

The Temperley method [544] can be used to find a functional recurrence for $H(t,q)$. The implementation is illustrated in figure 6.4 and proceeds by constructing $H(t,q)$ row by row (or layer by layer) [66].

Let $H_j \equiv H_j(t,q)$ be the generating function of staircase polygons with a *bottom row* (or *base*) of width j. Then each vesicle is either a row of j plaquettes or a row of j plaquettes forming the base of a staircase vesicle. The base can overlap the second row by up to $j-1$ edges (see figure 6.4).

The expansion in figure 6.4 gives the recurrence

$$H_j = y^2q^j\left(x^{2j} + x^{2j-2}\sum_{i\geq 1} H_i + x^{2j-4}\sum_{i\geq 2} H_i + \cdots + x^2 \sum_{i\geq j-1} H_i + \sum_{i\geq j} H_i\right).$$

Inspecting the above shows that

$$H_{j+2} - q(1+x^2-y^2q^{j+1})H_{j+1} + x^2q^2H_j = 0. \tag{6.32}$$

This is a homogeneous linear recurrence with non-constant coefficients. Recurrences of this particular type can be solved (see reference [478]) by assuming that

$$H_j = A\lambda^j \sum_{m=0}^{\infty} r_m(q)q^{mj}. \tag{6.33}$$

The normalising coefficient A is a function of q and cancels on substitution in equation (6.32). Simplification gives

$$\sum_{m=1}^{\infty} \left(\lambda y^2 q^{m+1} r_{m-1}(q) + (\lambda^2 q^{2m} - \lambda(1+x^2)q^{m+1} + x^2q^2) r_m(q)\right) q^{mj} \\ + \left(\lambda^2 - (1+x^2)q\lambda + x^2q^2\right) = 0. \quad (6.34)$$

Consequently, λ and $r_m(q)$ are solutions of

$$\lambda^2 - (1+x^2)q\lambda + x^2q^2 = 0, \text{ and} \quad (6.35)$$

$$\lambda y^2 q^{m+1} r_{m-1}(q) + \left(\lambda^2 q^{2m} - \lambda(1+x^2)q^{m+1} + x^2q^2\right) r_m(q) = 0. \quad (6.36)$$

Equation (6.35) has two solutions: $\lambda = q$, or the physical solution $\lambda = x^2 q$, while $r_m(q)$ can be found from the first order recursion in equation (6.36). That is,

$$r_m(q) = \left(\frac{-\lambda y^2 q^{m+1}}{(1-\lambda q^{m-1})(x^2q^2 - \lambda q^{m+1})}\right) r_{m-1}(q). \quad (6.37)$$

Iterating this and putting $\lambda = x^2 q$ gives

$$r_m(q) = \frac{(-y^2)^m q^{\binom{m+1}{2}}}{(q;q)_m (x^2q;q)_m}. \quad (6.38)$$

This gives the generating function of staircase polygons of base width j:

$$H_j(x,y,q) = A(x^2q)^j \sum_{m=0}^{\infty} \frac{(-y^2)^m q^{\binom{m+1}{2}+mj}}{(q;q)_m (x^2q;q)_m}. \quad (6.39)$$

To determine A, put $j = 0$ in the above; then $H_0(x,y,q) = y^2$. This shows that

$$A = \frac{y^2}{\sum_{m=0}^{\infty} \frac{(-y^2)^m q^{m(m+1)/2}}{(q;q)_m (x^2q;q)_m}}. \quad (6.40)$$

This gives the full generating function for staircase polygons of base width j (see references [66, 384]).

Putting $x = y = t$ and comparing the result to equation (D.48) gives the area-perimeter generating function of staircase vesicles in terms of q-Bessel functions:

$$H(t;q) = \frac{\sum_{j=1}^{\infty} t^2 (t^2q)^j J(t^2q, q, t^2q^{j+1})}{J(t^2q, q, t^2q)}. \quad (6.41)$$

There is an essential singularity in $H(t,q)$ at $q = 1$. It will be shown that $t^2q = 1$ is a curve of singularities in A and in $\sum_{j \geq 1} H_j(t,q)$. A numerical study of $H(t,q)$ was done in reference [66].

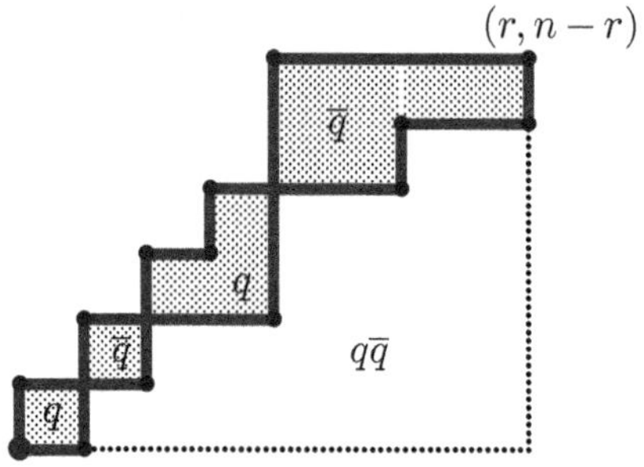

FIG. 6.5. A chain of staircase polygons with area fugacity alternating between q and $\overline{q}$.

The generating function $H_{j,k}$ of staircase polygons of base width j and of top row width k was also determined in reference [382]. It satisfies the recurrence

$$H_{j,k}(x,y,q) = y^2(x^2q)^n\,\delta_{m,n} + \sum_{k=1}^{n} x^{2(n-k)} \sum_{\ell=0}^{\infty} H_{m,k+\ell}. \tag{6.42}$$

Obviously, $H_j = \sum_{k=1}^{\infty} H_{j,k}$.

$H_{j,k} \equiv H_{j,k}(x,y,q)$ can be shown to satisfy the recurrence

$$\begin{aligned} H_{j,k+2} &- q(1+x^2-y^2q^{k+1})H_{j,k+1} + x^2q^2H_{j,k} \\ &= y^2q(x^2q)^{n+1}\left(x^2\delta_{j,k+2} - (1-x^2)\delta_{j,k+1} + \delta_{j,k}\right). \end{aligned} \tag{6.43}$$

This is an inhomogeneous linear recurrence which can be solved by putting

$$K_j(x,y,q) = \frac{y^2q^n\,H(y^2q^j)}{H(y^2)}, \text{ where } H(t) = \sum_{m=0}^{\infty} \frac{(-\overline{x}^2)^m\,q^{\binom{m+1}{2}}\,t^m}{(q;q)_m\,(\overline{x}^2q;q)_m} \tag{6.44}$$

with $\overline{x} = \frac{1}{x}$. Then

$$H_{j,k}(x,y,q) = \frac{y^2q\,H_j\,(K_k - H_k)}{K_1 - H_1} + y^2(x^2q)^n\,\delta_{j,k} \text{ for } j \geq k, \tag{6.45}$$

where $K_k \equiv K_k(x,y,t)$, and $H_k \equiv H_k(x,y,t)$ (see reference [66]).

6.3.2 *Pólya's method for staircase vesicles*

Pólya's method [467] for staircase vesicles is inspired by lemma D.9 in appendix D. Let $N_{nk\alpha}$ be the number of directed paths giving east and north steps from $\vec{0}$ to the vertex $(k, n-k)$ in $\mathbb{L}^2$, enclosing an area α underneath the path in the first quadrant (see lemma D.10). The path is said to *overstep* area α.

By lemma D.9, the area generating function of such paths from $\vec{0}$ to $(k, n-k)$ is $\left[\begin{smallmatrix} n \\ k \end{smallmatrix}\right]_q$, with q the area generating variable. Let $\overline{q} = \frac{1}{q}$. Then the product of q-factorials $\left[\begin{smallmatrix} n \\ k \end{smallmatrix}\right]_q \left[\begin{smallmatrix} n \\ k \end{smallmatrix}\right]_{\overline{q}}$ (see equation (D.49) and figure 6.5) generates two paths from

$\vec{0}$ to $(k, n-k)$. The first path oversteps the area generated by q, and the second path oversteps the area generated by $\overline{q}$. Summing over k gives all such pairs of paths of length n. This is illustrated in figure 6.5, and a chain of staircase polygons is obtained. Some of these could be the empty polygon. The generating function of this is

$$P_n(q) = \sum_{k=0}^{n} \begin{bmatrix} n \\ k \end{bmatrix}_q \begin{bmatrix} n \\ k \end{bmatrix}_{\overline{q}} = \sum_{k=0}^{n} \begin{bmatrix} n \\ k \end{bmatrix}_q^2 q^{-k(n-k)}. \tag{6.46}$$

Each staircase polygon in the chain has area once generated by q or once generated by $\overline{q}$. Thus, $P_n(q) = P_n(\overline{q})$. If the perimeter is generated by t, then

$$W(t,q) = \tfrac{1}{2} \sum_{n=1}^{\infty} \left(P_n(q) + P_n(\overline{q})\right) t^{2n} = \sum_{n=1}^{\infty} P_n(q) t^{2n} \tag{6.47}$$

generates a string of staircase polygons, some of which are degenerate (of area 0 and length 2). This gives

$$\begin{aligned} W(t,q) &= \left(2t^2 + H(t,q) - 1\right) + \left(2t^2 + H(t,q) - 1\right)^2 + \cdots \\ &= \frac{1 - 2t^2 - H(t,q)}{2t^2 + H(t,q)}. \end{aligned} \tag{6.48}$$

Solving for $H(t,q)$ gives

$$H(t,q) = \frac{1}{1 + W(t,q)} - 2t^2. \tag{6.49}$$

6.3.3 *Staircase polygons and the wasp-waist method*

This solution is due to Prellberg and Brak [471]. Introduce generating variables (x, y), with x conjugate to the width of the polygon, and y conjugate to its height. As before, let q be the area generating variable. Denote the generating function by $g(x,y,q)$. Then $H(t,q) = 1 + g(t^2, t^2, q)$ is the area-perimeter generating function.

A functional recurrence for the generation function $g(x,y,q)$ is obtained by the wasp-waist factorisation given in figure 6.6. The catalytic variable is x, so denote $G(x) = g(x,y,q)$. Figure 6.6 gives the following non-linear functional recurrence for $G(x)$:

$$G(x) = xyq + xqG(x) + yG(xq) + G(x)G(xq). \tag{6.50}$$

This may be linearised by putting

$$G(x) = \alpha \frac{F(xq)}{F(x)} - y. \tag{6.51}$$

Simplifying gives the homogeneous linear recurrence

$$A^2 F(xq^2) + A(xq - y - 1) F(xq) + yF(x) = 0. \tag{6.52}$$

where A is a normalising function.

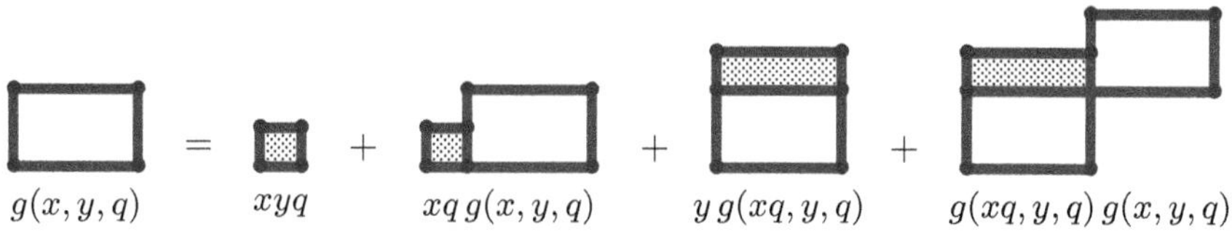

FIG. 6.6. The factorisation of the staircase polygon generating function. A staircase polygon either is a single square, or has a single square as its first column, or it has no wasp-waist (a column of height 1), or it can be factored at its first wasp-waist.

The recurrence in equation (6.52) is of the general form

$$\sum_{k=0}^{N} \alpha_k H(xq^k) + x H(xq) = 0 \quad \text{with} \sum_{k=0}^{N} \alpha_k = 0 \tag{6.53}$$

where the α_k are constants independent of x. Functional recurrence of this type has a general solution, which is regular at $x = 0$, given by

$$H(x) = \sum_{n=0}^{\infty} \frac{q^{\binom{n}{2}} (-x)^n}{\prod_{m=1}^{n} \Lambda(q^m)} \quad \text{where } \Lambda(s) = \sum_{k=0}^{N} \alpha_k s^k. \tag{6.54}$$

This can be verified by substituting equation (6.54) into equation (6.53); see for example reference [469].

To use the above to find $F(x)$ in equation (6.52), it is necessary to fix A. Assuming that $F(x) = 1 + \cdots$ so that $F(x)$ is regular at the origin shows that A is a solution of

$$A^2 + (xq - y - 1) A + y = 0. \tag{6.55}$$

Terms in y are cancelled if $A = y$ (but notice that higher order terms will cancel the terms in x). Thus, choosing $A = y$ gives

$$\sum_{k=0}^{2} \alpha_k F(xq^k) + x F(xq) = 0. \tag{6.56}$$

It may be verified that $\alpha_2 = \frac{y}{q}$; that $\alpha_1 = -\frac{1+y}{q}$ and that $\alpha_0 = \frac{1}{q}$. Note that $\alpha_0 + \alpha_1 + \alpha_2 = 0$. It follows that $\Lambda(s) = \frac{1}{q}(1-s)(1-ys)$.

Substituting these into equation (6.54) gives the following result fot $F(x)$:

$$F(x) = \sum_{n=0}^{\infty} \frac{q^{\binom{n}{2}} (-xq)^n}{(q;q)_n (yq;q)_n} = J(yq, q, xq), \tag{6.57}$$

where the q-deformed Bessel function is defined in equation (D.48) in appendix D. Thus, the full solution for $G(x) = g(x, y, q)$ is given by

TABLE 6.1. Staircase polygon tricritical exponents

ϕ	α	$2-\alpha_t$	$2-\alpha_u$	y_t	ν_t
$\frac{2}{3}$	$\frac{1}{2}$	$\frac{1}{3}$	$\frac{1}{2}$	4	$\frac{1}{6}$

$$G(x) = \frac{y\, J(yq, q, xq^2)}{J(yq, q, xq)} - y. \tag{6.58}$$

Let t be the perimeter generating variable and put $y = x = t^2$. Then the generating function is

$$H(t, q) = \frac{t^2\, J(t^2 q, q, t^2 q^2)}{J(t^2 q, q, t^2 q)} - t^2. \tag{6.59}$$

The function $J(t^2 q, q, t^2 q^2)$ is absolutely convergent and analytic inside the unit disk $|q| < 1$ (and so $H(t, q)$ is a meromorphic function in $|q| < 1$ since it is the ratio of two holomorphic functions). This implies that singularities in $H(t, q)$ on $|q| < 1$ are given by roots of $J(t^2 q, q, t^2 q)$.

6.3.4 *Tricritical scaling of staircase vesicles*

The perimeter generating function of staircase polygons is recovered when $q = 1$ in equation (5.160). That is,

$$H(t, 1) = \frac{2t^2}{1 - 2t^2 + \sqrt{1 - 4t^2}}. \tag{6.60}$$

Comparison to equation (4.11) shows that $2 - \alpha_u = \frac{1}{2}$.

Assume that $G(x)$ has asymptotic expansion in $|\log q|$ for $q \lessapprox 1$:

$$G(x) = \sum_{n=0}^{\infty} S_n(x, y)\, |\log q|^n. \tag{6.61}$$

Insert this into equation (6.50) and determine

$$S_0(x, y) = \tfrac{1}{2}\left(1 - x - y - \sqrt{1 - 2x - 2y + y^2 - 2xy + x^2}\right), \tag{6.62}$$

and

$$\begin{aligned} S_n(x, y) = {} & \frac{(y + S_0(x, y)) \sum_{m=1}^{n} S_{n-m}^{(m)}(x, y)\, \frac{1}{m!} q^m}{1 - x - y - 2 S_0(x, y)} \\ & + \frac{\sum_{k=1}^{n-1} S_k(x, y) \sum_{m=0}^{n-k} S_{n-m-k}^{(m)}(x, y)\, \frac{1}{m!} q^m}{1 - x - y - 2 S_0(x, y)}. \end{aligned} \tag{6.63}$$

Examining $S_1(x, y)$ shows that $S_1(t^2, t^2) \simeq \frac{1}{1-4t^2}$. Since $S_0(t^2, t^2) \simeq \sqrt{1 - 4t^2}$, the gap exponent $\Delta = \frac{3}{2}$ (see equations (4.16) and (4.17)). By equation (4.17), the crossover exponent is $\phi = \frac{2}{3}$ and by equation (4.14) it follows that $2 - \alpha_t = \frac{1}{3}$.

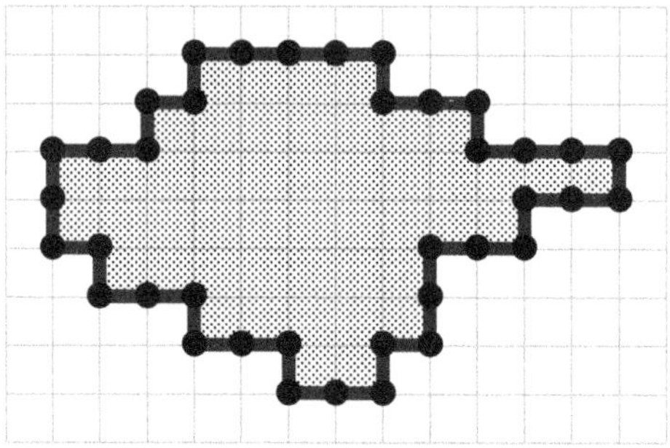

FIG. 6.7. A convex lattice polygon. The intersection of the interior of this polygon with a horizontal or vertical line is always an interval.

The remaining exponents can be found by using equations (4.29), (4.46), (4.53) and (4.54). This gives $\nu_t = \frac{1}{6}$, and $y_g = 6$, while $y_t = 4$, and $\alpha = \frac{1}{2}$ (see table 6.1).

The τ_0-line is a line of essential singularities in the generating function. Along the τ_0-line, $H(t, 1)$ is finite (and the essential singularities are condensation transitions where 'inflated' droplets condense as $q \nearrow 1^-$ [7, 289]).

The λ-isotherm is a curve of simple poles in the phase diagram. This follows from equation (6.59): the smallest solution of $J(t^2q, q, t^2q) = 0$ for $q > 1$ is the dominant singularity in $H(t, q)$ [468].

6.4 Convex polygons

A lattice vesicle in $\mathbb{L}^2$ is (fully) convex if its intersection with any line parallel to the lattice axes is an interval. The perimeter of a fully convex lattice vesicle is a *convex polygon*. A convex polygon is illustrated in figure 6.7.

Let c_n^c be the number of all convex polygons in $\mathbb{L}^2$ (equivalent under translations) of perimeter length n. It is known that [127, 352]

$$c_{2n+8}^c = (2n+1)4^n - 2(2n+1)\binom{2n}{n}. \tag{6.64}$$

The perimeter generating function is

$$C^c(t) = \frac{t^4(1-6t^2+11t^4-4t^6)}{(1-4t^2)^2} - \frac{4t^8}{\sqrt{1-4t^2}^3}. \tag{6.65}$$

The radius of convergence of $C^c(t)$ is $t_c = \frac{1}{2}$. Both terms are singular at this point, but the first term dominates. The second term is a *confluent correction*.

The (known) area-perimeter generating function of convex square lattice polygons is a complicated expression which was determined in reference [382]. This may be given as follows. Let q be the area generating variable, let x generate horizontal perimeter edges and let y generate vertical perimeter edges. Let S be

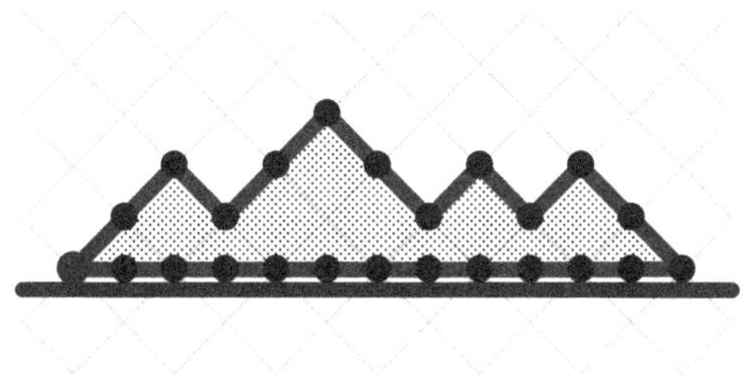

FIG. 6.8. A Dyck path vesicle has as its perimeter the horizontal line and the path itself. Its interior is the region underneath the path in the half-space above the horizontal axis.

---

the stack generating function in equation (6.17) and let S_w be the generating function of stacks of base width w in equation (6.19). Let H_w be the generating function of staircase polygons of base width w in equation (6.39) and let K_w be the related generating function in equation (6.44). Define the functions

$$U_w = \sum_{v=1}^{w-2} (w-v-1) S_v \overline{x}^{2v}, \quad \text{and } V_w = \sum_{v=1}^{w-1} \overline{x}^{2v} \sum_{p=0}^{\infty} H_{v+p}, \tag{6.66}$$

where $\overline{x} = \frac{1}{x}$. Put $A = \frac{y^2 q}{K_1 - H_1}$. The area-perimeter generating function for convex vesicles is given by [382]:

$$\begin{aligned} C^c(x,y,q) &= S + \sum_{w=3}^{\infty} S_w U_w + 2y^2 \sum_{w=2}^{\infty} S_w \sum_{v=1}^{w-1} \overline{x}^{2v} \sum_{u=v}^{\infty} (x^2 q)^u U_{u+1} \\ &\quad + 2 \left[\sum_{w=2}^{\infty} S_w V_w \right] \left[1 + A \sum_{w=2}^{\infty} U_{w+1} (K_w - H_w) \right] \\ &\quad + 2A \sum_{w=2}^{\infty} S_w \sum_{v=1}^{w-1} \overline{x}^{2v} \sum_{u=v+1}^{\infty} U_{u+1} \sum_{s=0}^{u-v-1} (H_v K_{u+s} - K_v H_{u+s}). \end{aligned} \tag{6.67}$$

It is not known if a simpler expression exists.

6.5 Dyck path vesicles

A Dyck path from $\vec{0}$ in $\mathbb{J}_+^2$ (see equation (5.2)) and with north-east and south-east steps is shown in figure 6.8. The path is completed into a vesicle by joining its endpoints in $\partial\mathbb{J}_+^d$ with a straight line. The total area enclosed underneath the path and above $\partial\mathbb{J}_+^d$ is the *area of the path.* If every second vertex of the path is a *visit*, then the area of the path is a minimum of $\frac{1}{4}n$ (unit squares). The maximum area of the path is $\frac{1}{8}n^2$. Define the area function of a Dyck path ω by $\text{Area}(\omega) = [\text{Area of } \omega] - \frac{n}{4}$. The path illustrated in figure 6.8 has Area $= 6$.

The generating function $D(t,q)$ (with q the area generating variable, and t the perimeter generating variable) of Dyck path vesicles can be determined from the factorisation illustrated in figure 5.2:

$$D(t,q) = 1 + t^2 D(t\sqrt{q}, q) D(t,q). \tag{6.68}$$

Iterating gives an infinite fraction representation of the generating function:

$$D(t,q) = \frac{1}{1 - t^2 D(t\sqrt{q}, q)} = \cfrac{1}{1 - \cfrac{t^2}{1 - \cfrac{qt^2}{1 - \cfrac{q^2t^2}{1 - \cdots}}}}. \tag{6.69}$$

For larger values of t, the generating function $D(t\sqrt{q}, q)$ will converge for small q, and there is a simple pole in the generating function when $1 - t^2 D(t\sqrt{q}, q) = 0$. That is,

$$D(t,q) > 1 + t^2 D(t\sqrt{q}, q), \tag{6.70}$$

so, if $1 - t^2 D(t\sqrt{q}, q) \to 0$ for large enough t, then $D(t,q) \to \infty$.

Equation (6.68) may be solved by assuming that for a function A,

$$D(t,q) = A\left(\frac{H(t^2q)}{H(t^2)}\right). \tag{6.71}$$

Substitution into equation (6.68) and simplification give the homogeneous linear second order functional recurrence

$$A^2t^2 H(t^2q^2) - AH(t^2q) + H(t^2) = 0. \tag{6.72}$$

Assume that $H(t^2) = 1 + O(t^2)$ is regular as $t \to 0$. Then $A^2t^2 - A + 1 + O(t^2) = 0$. The only choice for A which cancels the constant terms above is $A = 1$; this reduces equation (6.72) to the following functional recurrence for $H(t^2)$:

$$t^2 H(t^2q^2) - H(t^2q) + H(t^2) = 0. \tag{6.73}$$

Equation (6.73) can be solved by assuming an infinite series solution. Substitute $\sigma = t^2$ in the above to find $\sigma H(\sigma q^2) - H(\sigma q) + H(\sigma) = 0$. Assume that $H(\sigma) = \sum_{n \geq 0} \alpha_n \sigma^n$. Substitution into the recurrence shows that

$$\alpha_n = \frac{q^{2(n-1)}}{q^n - 1} \alpha_{n-1}. \tag{6.74}$$

Putting $\alpha_0 = 1$ and iterating this recurrence gives an expression for α_n. Simplifying gives an expression for $D(t,q)$ as a ratio of two infinite series:

$$D(t,q) = \frac{H(t\sqrt{q})}{H(t)} \quad \text{where } H(t) = \sum_{n=0}^{\infty} \frac{(-1)^n q^{n(n-1)} t^{2n}}{(q;q)_n}. \tag{6.75}$$

Since $D(t,q)$ was also represented as a continued fraction (equation (6.69)), this gives a non-trivial identity involving a ratio of infinite series on one side, and an infinite fraction on the other side. Define the function

$$E(t^2,q) = \sum_{n=0}^{\infty} \frac{(-1)^n q^{n^2} t^{2n}}{(q;q)_n} \tag{6.76}$$

so that, if $\overline{q} = \frac{1}{q}$, then $D(t,q)$ is given by

$$D(t,q) = \frac{E(t^2,q)}{E(t^2\overline{q},q)}. \tag{6.77}$$

See reference [197] and note that this is related to Rogers-Ramanujan identities and the fountain-of-coins problem [442].

By Worpitzky's theorem [572], the continued fraction in equation (6.69) converges absolutely if $|q^n t^2| \leq \frac{1}{4}$ for all n (see theorem D.20 in appendix D). Since $D(t,q)$ has a natural boundary at $|q| = 1$, this shows that $q_c(t) = 1$ if $t^2 \leq \frac{1}{4}$.

Setting $q = 1$ in equation (6.75) and solving explicitly gives $D^+(t)$ in equation (5.4). This shows that there is a critical point at $(t,q) = (\frac{1}{2}, 1)$.

Putting $t = \frac{1}{2}$ in the infinite fraction representation of $D(t,q)$ gives

$$C\left(\tfrac{1}{4},q\right) = \cfrac{1}{1 - \cfrac{1/4}{1 - \cfrac{q/4}{1 - \cfrac{q^2/4}{1 - \cdots}}}}, \tag{6.78}$$

which has an essential singularity at $q = 1$ (since $(q;q)_n = 0$ for all n if $q = 1$).

$E(t^2,q)$ in equation (6.76) is absolutely convergent if $|q| < 1$. Thus, it is analytic inside the unit disk $|q| < 1$. It follows that $D(t,q)$ is a meromorphic function in the unit disk $|q| < 1$ since it is the ratio of two holomorphic functions. The singularities in $D(t,q)$ on $|q| < 1$ are given by roots of $E(t^2q^{-1},q)$.

If $t^2 > \frac{1}{4}$ then, by equation (6.68), $D(t,q)$ is singular along curves $t^2 D(t\sqrt{q},q) = 1$. This shows that $q_c(t) < 1$ if $t > \frac{1}{4}$ and the singular points are roots of $E(t^2q^{-1},q)$. This gives a curve of simple poles on the real axis in the q-plane. Thus, $q_c(t) = 1$ if $t \leq \frac{1}{2}$, and $q_c(t) < 1$ if $t > \frac{1}{2}$.

6.6 Bargraph and column convex vesicles

A bargraph polygon is given by the area underneath a partially directed path in $\mathbb{L}^2_+$ (see section 5.5). The path starts in $\vec{0}$ and ends in a vertex in $\partial\mathbb{L}^2_+$, but otherwise avoids the boundary of $\mathbb{L}^2_+$. By connecting the endpoints of the

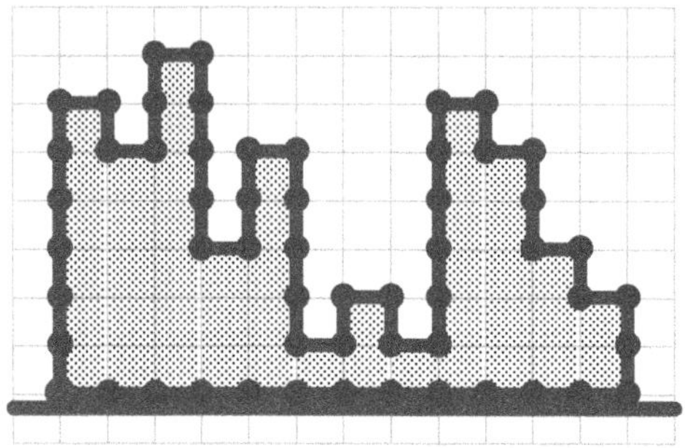

FIG. 6.9. A bargraph vesicle is the area underneath a partially directed path from $\vec{0}$ in $\mathbb{L}^2_+$ with its endpoint in $\partial\mathbb{L}^2_+$. The perimeter of a bargraph vesicle is a bargraph polygon which is partially convex because the intersection of the bargraph vesicle and a vertical line is an interval.

partially directed path, a bargraph polygon or vesicle is obtained. Notice that these vesicles are only partially convex; see figure 6.9 for an example.

Introduce generating variables t conjugate to the perimeter, x conjugate to the width, y conjugate to one-half the number of vertical edges, and q conjugate to the area. Then the generating function of bargraph vesicles is denoted by $B(x, y, q)$; therefore, $B(t^2, t^2, q)$ is the area-perimeter generating function. The catalytic variable will be x (put $B(x) \equiv B(x, y, q)$).

The factorisation of bargraphs is given in figure 6.10 (see reference [469]). This gives the non-linear inhomogeneous recurrence

$$B(x) = xq(1 + B(xq))(y + B(x)) + yB(xq). \tag{6.79}$$

A lattice vesicle in $\mathbb{L}^2$ with the property that its intersection with any vertical line is an interval (possibly empty) is a *column convex vesicle.* The perimeter of a column convex vesicle is a column convex polygon. If ω is a column convex polygon with bottom vertex $\vec{b}$ and top vertex $\vec{t}$, then these vertices partition ω into two mutually avoiding partially directed paths from $\vec{b}$ to $\vec{t}$.

A subclass of column convex polygons is the class of *directed column convex polygons.* These are formed by a partially directed path from $\vec{b}$ to $\vec{t}$ *above* a fully directed path (giving north and east steps) from $\vec{b}$ to $\vec{t}$ (see figure 6.11).

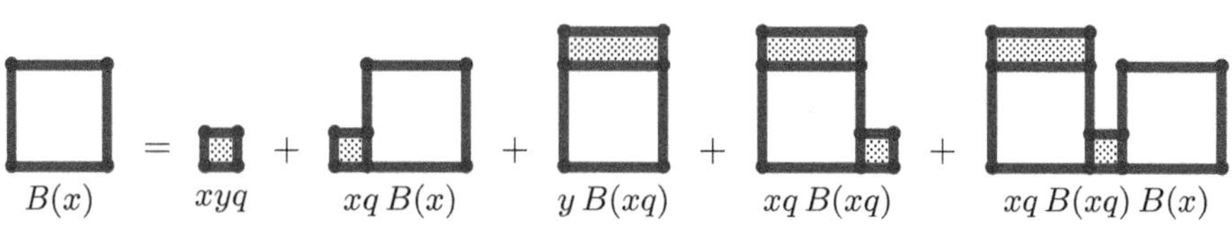

FIG. 6.10. The factorisation of bargraph polygons.

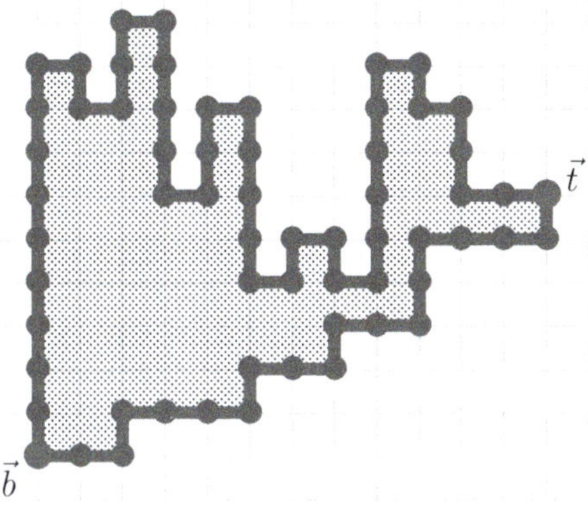

FIG. 6.11. A directed column convex polygon with bottom vertex $\vec{b}$ and top vertex $\vec{t}$. The polygon is composed of a partially directed path from $\vec{b}$ to $\vec{t}$ above a fully directed path from $\vec{b}$ to $\vec{t}$. This is a subclass of column convex polygons (which are formed by two partially directed paths from $\vec{b}$ to $\vec{t}$).

Let the generating function of directed column convex polygons be $F(x, y, q, \mu)$ with $\{x, y, q\}$ defined as above, and with μ conjugate to the height of the last column. There are two catalytic variables x and μ in this model. Define $F(x, y, q, \mu) \equiv F(x, \mu)$. The factorisation in figure 6.12 (see reference [469]) gives the recurrence

$$\begin{aligned} F(x,\mu) &= xyq\mu + xqF(x,\mu) + y\mu F(xq,\mu) + xyq\mu F_\mu(xq,1) \\ &\quad + xqF_\mu(xq,1)F(x,\mu) + F(xq,1)F(x,\mu), \end{aligned} \tag{6.80}$$

where $F_\mu(x, 1) = \frac{\partial F}{d\mu} \mid_{\mu=1}$. This is a non-linear and inhomogeneous functional-differential equation. Denote $F(x) \equiv F(x, 1) \equiv F(x, y, q, 1)$ and let $F_\mu(x) \equiv F_\mu(x, 1) \equiv \frac{\partial}{\partial \mu} F(x, y, q, \mu) \mid_{\mu=1}$.

Putting $\mu = 1$ at first, or first taking the partial derivative to μ and then putting $\mu = 1$ in the recurrence, gives the two equations

$$F(x) = qx(1+F_\mu(xq))(y+F(x)) + F(xq)(y+F(x)), \quad \text{and} \tag{6.81}$$

$$F_\mu(x) = qx(1+F_\mu(xq))(y+F_\mu(x)) + F(xq)(y+F_\mu(x)) + yF_\mu(xq). \tag{6.82}$$

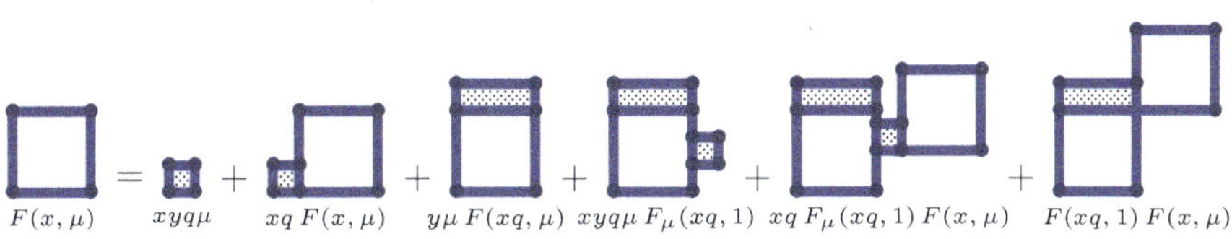

FIG. 6.12. The factorisation of directed column convex polygons.

Let $x \to xq$ to obtain two more equations, and eliminate $F_\mu(x)$, $F_\mu(xq)$ and $F_\mu(xq^2)$ between the four equations. This gives the functional recurrence for directed column convex polygons:

$$\begin{aligned} F(xq^2)F(xq)F(x) + yF(xq^2)F(xq) + yF(xq^2)F(x) \\ - (1+q)F(xq)F(x) + y^2F(xq^2) - y(1+q)F(xq) \\ + q(1 - xq(1-y))F(x) - xyq^2(1-y) = 0. \end{aligned} \tag{6.83}$$

Solving equations (6.79) and (6.83) follows the same general method seen in section 5.7 (and also for staircase polygons earlier in this section; see equations (6.53) and (6.54)). Notice that the substitution

$$B(x) = y\left(\frac{H(xq)}{H(x)} - 1\right) \tag{6.84}$$

turns equation (6.79) into

$$yH(xq^2) + (xq - 1 - y)H(xq) + H(x) = 0 \tag{6.85}$$

giving $\Lambda(s) = \frac{1}{q} - \frac{(1+y)s}{q} + \frac{ys^2}{q}$; thus, $\Lambda(1) = 0$. Therefore, the solution is given in terms of q-Bessel functions (see equation (D.48) in appendix D) by

$$B(x) = y\left(\frac{H(xq)}{H(x)} - 1\right), \quad \text{where } H(x) = \sum_{n=0}^{\infty} \frac{q^{\binom{n}{2}}(-xq)^n}{(q;q)_n(yq;q)_n} = J(yq, q, xq).$$

Similarly, substituting $F(x) = y\left(\frac{T(xq)}{T(x)} - 1\right)$ into equation (6.83) gives the the third order functional recurrence

$$y^2T(xq^3) - y(y+q+1)T(xq^2) + (x(y-1)q^2 + y(q+1) + q)T(xq) - qT(x) = 0.$$

Then $\Lambda(s) = \frac{-q}{(y-1)q^2} + \frac{(q+y(1+q))s}{(y-1)q^2} - \frac{y(1+y+q)s^2}{(y-1)q^2} + \frac{y^2s^3}{(y-1)q^2}$; therefore, $\Lambda(1) = 0$. This gives the generating function for directed column convex polygons

$$F(x) = y\left(\frac{T(xq)}{T(x)} - 1\right), \quad \text{where } T(x) = \sum_{n=0}^{\infty} \frac{q^{\binom{n}{2}}(-x(1-y)q)^n}{(q;q)_n(y;q)_n(yq;q)_n}. \tag{6.86}$$

See reference [469] for additional results. Generating functions for column convex polygons was studied in references [71, 72, 151, 391] as well. In reference [66] an explicit formula for the area-perimeter generating function of column convex polygons is given in terms of q-Bessel functions.

6.7 Heaps of dimers, and directed animals

Directed animals on $\mathbb{L}^2$ were enumerated using a variety of methods [139, 140, 232, 254]. These methods can be difficult to implement, but simpler proofs can be obtained using a heap structure. The simplest proofs rely on the monoid structure of heaps and pyramids [38]. This general approach can be used to determine the generation functions for even more general models of multidirectional lattice animals [564].

Let $\mathbb{Z}_e^2 = \{\vec{v} \in \mathbb{Z}^2 \mid \vec{v}(1) + \vec{v}(2) \text{ is even}\}$ be the set of vertices with even parity in $\mathbb{Z}^2$, where $\vec{v}(j)$ is the j-th Cartesian component of $\vec{v}$. Define the directed square lattice

$$\mathbb{K}^2 = \{\langle \vec{u} \rightsquigarrow \vec{v} \rangle \mid \vec{u}, \vec{v} \in \mathbb{Z}_e^2, \|\vec{u} - \vec{v}\|_2 = \sqrt{2}, \vec{v}(2) > \vec{u}(2) \geq 0\} \tag{6.87}$$

Then $\mathbb{K}^2$ is a directed half-lattice with vertices of even parity and edges directed in the north-east or north-west directions. The boundary of $\mathbb{K}^2$ is denoted by $\partial\mathbb{K}^2$ and is the set of all vertices $\vec{v} \in \mathbb{K}^2$ with $\vec{v}(2) = 0$.

An animal α in $\mathbb{K}^2$ is a *directed animal* if every vertex $\vec{v} \in \alpha$ is the endpoint of a directed path $\vec{0} \rightsquigarrow \vec{v}$ in α. The *source vertex* of the directed animal is said to be at $\vec{0}$, and it is grown recursively on even vertices by appending directed edges in the north-east or north-west directions. Directed animals may also be grown from other source vertices.

Let a_v be the number of directed animals of v vertices or *sites*. Then $a_1 = 1$; $a_2 = 2$; $a_3 = 5$; and so on. The generating function is

$$\mathfrak{a}(t) = \sum_{v=1}^{\infty} a_v t^v, \tag{6.88}$$

and it can be determined using *heaps of dimers* [562].

Heaps of dimers in $\mathbb{K}^2$ are formed when horizontally unit length oriented dimers (with midpoints having integer Cartesian coordinates) are dropped vertically down columns onto $\partial\mathbb{K}^2$. Dimers are long enough in the horizontal direction so that those in adjacent columns overlap, but those in next-nearest-neighbouring columns do not.

The dimers are released at a great height and drop until they hit $\partial\mathbb{K}^2$ or until they land on top of dimers already in the heap. Dimers falling into $\partial\mathbb{K}^2$ are *minimal*. If a heap has one minimal dimer, then it is a *pyramid*. A pyramid (of dimers) is illustrated in figure 6.13.

A heap of dimers is *strict* if no one dimer lies exactly on another (but is offset one lattice step to the left or right side). A heap is connected if its projection on $\partial\mathbb{K}^2$ is connected.

There is a bijection between directed animals and strict pyramids. This is also seen in figure 6.13. Orient the animal as on the left and then replace each vertex with a dimer. This gives a unique strict pyramid, since the animal has one source vertex at $\vec{0}$. Conversely, every strict pyramid corresponds to a unique directed animal.

Define the directed half-lattice

$$\mathbb{K}_+^2 = \{\langle \vec{u} \rightsquigarrow \vec{v} \rangle \in \mathbb{K}^2 \mid \vec{v}(1) \leq 0,\ \vec{u}(1) \leq 0\}. \tag{6.89}$$

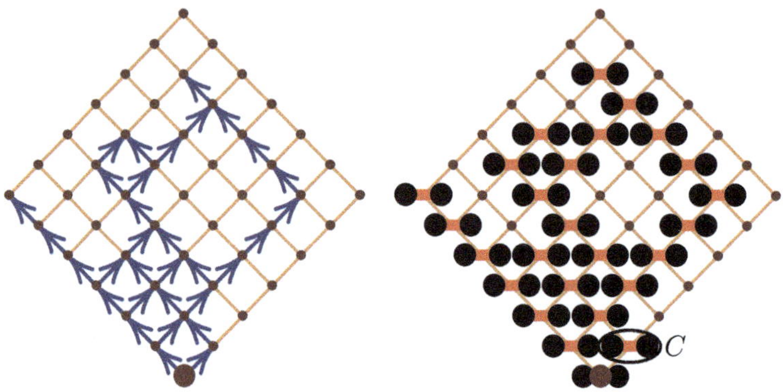

FIG. 6.13. The directed animal on the left was grown from the single source vertex at the bottom. It may be represented as a heap of horizontal dimers falling vertically down into an (inverted) heap of dimers which form a strict pyramid supported at its nadir (which is at the origin).

Then $\vec{0} \in \mathbb{K}^2$, and directed animals with source $\vec{0}$ can be grown in $\mathbb{K}^2_+$. One such directed animal is illustrated in figure 6.14.

Denote the number of directed animals in $\mathbb{K}^2_+$ with source at $\vec{0}$ and with v vertices by a_v^+. Then $a_1^+ = 1$; $a_2^+ = 1$; and $a_3^+ = 2$. The generating function is

$$\mathfrak{a}^+(t) = \sum_{v=1}^{\infty} a_v^+ t^n. \qquad (6.90)$$

A pyramid is a *strict half-pyramid* if it is a strict pyramid with source $\vec{0}$ in $\mathbb{K}^2_+$. There is a bijection from directed animals in $\mathbb{K}^2_+$ to strict half-pyramids (see figure 6.14).

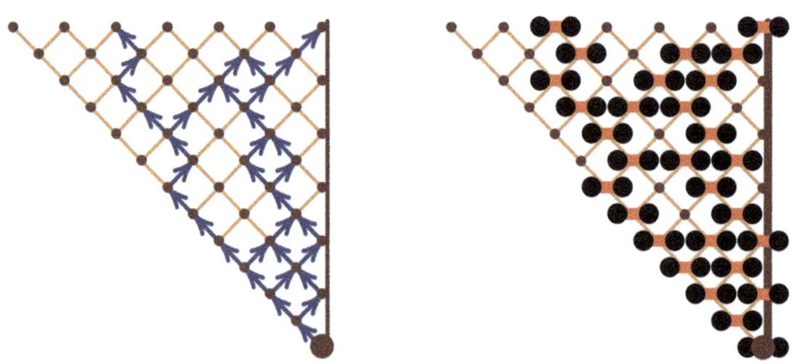

FIG. 6.14. A directed animal in a half-space on the left may be represented by a strict half-pyramid on the right. Vertices where the animal returns to the boundary are visits. Visits correspond to dimers in the right-most column.

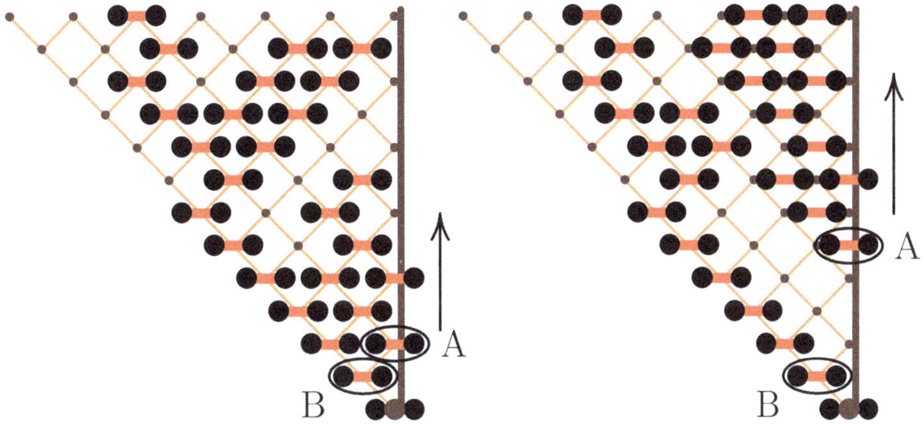

FIG. 6.15. The factorisation of a strict half-pyramid. Lifting the dimer A takes with it a strict half-pyramid of dimers with A as their source dimer. Lifting B next will take the remaining dimers, also in a strict half-pyramid, except for the original source dimer at $\vec{0}$. Dimers in the right-most column are weighted by a in a model of adsorbing directed animals.

Strict half-pyramids can be decomposed as shown in figure 6.15. Mark the lowest dimer in the same column and above the dimer at $\vec{0}$ by A, and let B be the dimer next to and offset to the north-west of $\vec{0}$.

Lifting the dimer A vertically captures dimers above A and lifts them out of the half-pyramid as shown. This necessarily forms a half-pyramid with A as its source dimer. Dimers remaining behind form a half-pyramid with B as source dimer resting on the original source dimer at $\vec{0}$. This is seen by lifting the dimer B vertically and noticing that it lifts all remaining dimers (except the source dimer at $\vec{0}$).

The decomposition of strict half-pyramids in figure 6.15 gives a factorisation for strict half-pyramids in figure 6.16. Every strict half-pyramid is either a single dimer, has no dimer at position A (but one at B) or has dimers at both A and B. This shows that the generating functions of directed animals in a half-space has recurrence

$$\mathfrak{a}^+(t) = t + t\mathfrak{a}^+(t) + t\left(\mathfrak{a}^+(t)\right)^2. \tag{6.91}$$

Finding the physical root gives the generating function

$$\mathfrak{a}^+(t) = \frac{2t}{1 - t + \sqrt{(1+t)(1-3t)}}. \tag{6.92}$$

The radius of convergence is $t_c = \frac{1}{3}$, so $a_v^+ = 3^{v+o(v)}$.

Strict pyramids can be factored into strict half-pyramids. Consider the pyramid in figure 6.13. Either this is a half-pyramid or, if it is not, then it has a

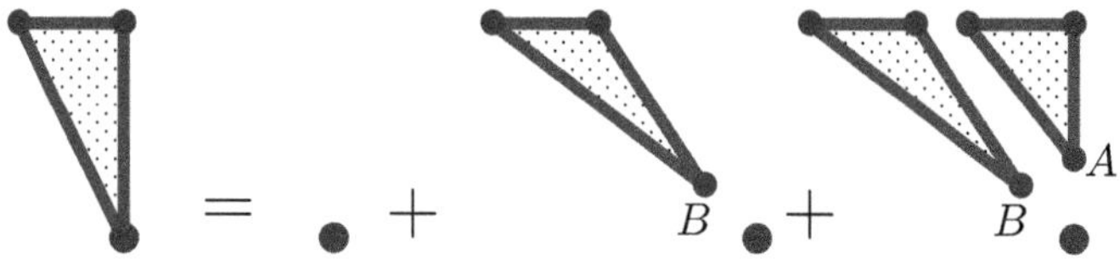

FIG. 6.16. The factorisation of strict half-pyramids.

lowest dimer in the first column to the east (right) of the column containing the source dimer at $\vec{0}$. In the pyramid in figure 6.13 this lowest dimer in the first column to the right is marked by C. Lifting the dimer at C vertically pushes a strict pyramid with C as its minimal dimer from the original pyramid. A strict half-pyramid is left behind.

Thus, every strict pyramid is either a strict half-pyramid or, if not, then it can be decomposed into a strict half-pyramid and a strict pyramid as shown schematically in figure 6.17. In terms of generating functions,

$$\mathfrak{a}(t) = \mathfrak{a}^+(t) + \mathfrak{a}^+(t)\,\mathfrak{a}(t). \tag{6.93}$$

Solving for $\mathfrak{a}(t)$ gives

$$\mathfrak{a}(t) = \frac{2t}{1 - 3t + \sqrt{(1+t)(1-3t)}}. \tag{6.94}$$

This shows that $a_v = 3^{v+o(v)}$. More careful asymptotic analysis shows that $a_v = \frac{1}{\sqrt{3\pi v}} 3^v \left(1 - \frac{1}{16v} + O(\frac{1}{v^2})\right)$ (see, for example, [197]).

The boundary of $\mathbb{K}^2_+$ is $\partial\mathbb{K}^2_+ = \{\vec{v} \in \mathbb{K}^2 \mid \vec{v}(1) = 0\}$ (these are the vertices in the right most column of $\mathbb{K}^2_+$). A *visit* in a directed animal α in $\mathbb{K}^2_+$ is a vertex $\vec{v} \in \alpha$ such that $\vec{v} \in \partial\mathbb{K}^2_+$. Although $\vec{0} \in \partial\mathbb{K}^2_+$, the convention is that $\vec{0}$ is not a visit. In the corresponding half-pyramid the visits correspond to dimers located in the right-most column (see figure 6.14).

Let a be the generating variable of visits in directed animals in $\mathbb{K}^2_+$. Denote the generating function of adsorbing directed animals by $\mathfrak{a}^+(a,t)$. Then the decomposition in figure 6.15 gives

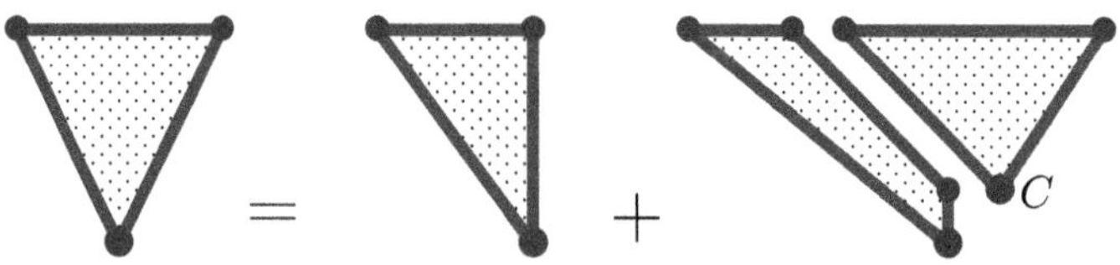

FIG. 6.17. The factorisation of a strict pyramid into a strict half-pyramid and a strict pyramid.

TABLE 6.2. Directed animal tricritical exponents

ϕ	α	γ_t	γ_u	y_t	ν_t	γ_+	γ_-
$\frac{1}{2}$	0	$\frac{1}{2}$	1	2	$\frac{1}{4}$	1	$-\frac{1}{2}$

$$\mathfrak{a}^+(a,t) = at + at\,\mathfrak{a}^+(1,t) + at\,\mathfrak{a}^+(1,t)\,\mathfrak{a}^+(a,t). \tag{6.95}$$

This gives equation (6.91) if $a = 1$. Solving for $\mathfrak{a}^+(a,t)$ gives

$$\begin{aligned}\mathfrak{a}^+(a,t) &= \frac{at\,(1+\mathfrak{a}^+(1,t))}{1-at\,\mathfrak{a}^+(1,t)} \\ &= \frac{2at(1+t)}{(1+t)(1-at)+(1+at)\sqrt{(1+t)(1-3t)}}.\end{aligned} \tag{6.96}$$

The critical curve can be determined from $\mathfrak{a}^+(a,t)$:

$$t_c(a) = \begin{cases} \frac{1}{3}, & \text{if } a \leq 3; \\ \frac{1}{2a}\left(\sqrt{4a-3}-1\right), & \text{if } a > 3. \end{cases} \tag{6.97}$$

The free energy is $\mathcal{A}(a) = -\log t_c(a)$, and the critical adsorption point is $a_c = 3$. The generating function has the asymptotic behaviour $\mathfrak{a}^+(3,t) \sim (1-3t)^{-1/2}$, so the tricritical exponent is $\gamma_t = \frac{1}{2}$ by equation (4.23). Since $\frac{1}{3} - t_c(a) \sim (a-3)^2$ as $a \to 3^+$ the crossover exponent $\phi = \frac{1}{2}$ by equations (4.5) and (4.17). The remaining tricritical exponents can be computed from equation (4.25); thus, $\phi\gamma_u = \gamma_t = \frac{1}{2}$, and $\gamma_u = 1$. The exponents along the λ-isotherm and the τ_0-line can also be determined: $\gamma_+ = 1$, and $\gamma_- = -\frac{1}{2}$, by equations (4.19) and (4.21).

6.8 Directed percolation

Directed percolation clusters are directed animals grown from source vertices along open edges and unblocked vertices. Let the set of vertices with even parity in $\mathbb{Z}^2$ be $\mathbb{Z}^2_e = \{\vec{v} \in \mathbb{Z}^2 \mid \vec{v}(1) + \vec{v}(2) \text{ is even}\}$, where $\vec{v}(j)$ is the j-th Cartesian component of $\vec{v}$. Define the directed square lattice

$$\mathbb{D}^2 = \{\langle \vec{u} \rightsquigarrow \vec{v} \rangle \mid \vec{u}, \vec{v} \in \mathbb{Z}^2_e,\ \|\vec{u}-\vec{v}\|_2 = \sqrt{2},\ \vec{v}(1) > \vec{u}(1) \geq 0\}. \tag{6.98}$$

Directed clusters can be grown from source vertices along open directed edges (directed in the north-east and south-east directions) in $\mathbb{D}^2$, as illustrated in figure 6.18.

Define a directed percolation process on sites in $\mathbb{D}^2$ as follows [141]. Let $\vec{0}$ be *unblocked* and let every other vertex in $\mathbb{D}^2$ be unblocked with probability p_s (and *blocked* with probability $q_s = 1 - p_s$). A (directed) edge in $\mathbb{D}^2$ (in the south-east or north-east direction) is *open* with probability p_b and *closed* with probability $q_b = 1 - p_b$. This gives a model with two parameters: p_s and p_b.

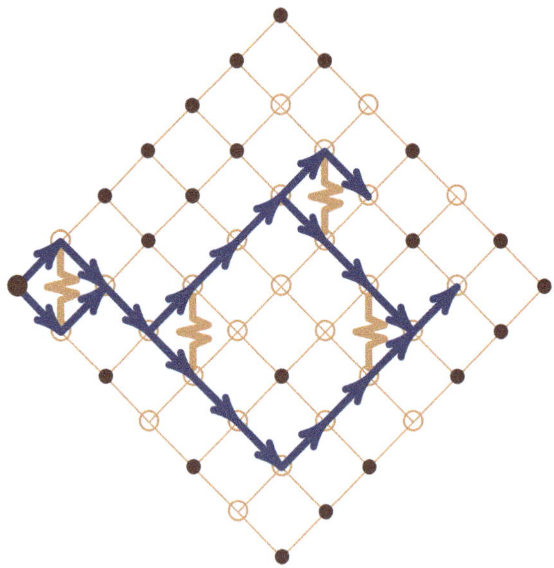

FIG. 6.18. A directed animal of size twenty edges and with four contacts oriented in the east direction is grown along unblocked sites and open directed edges from the source vertex at $\vec{0}$ on the left. Blocked vertices are indicated by •'s, while unblocked vertices are represented by ∘'s. Contacts are indicated by squiggly lines connecting adjacent occupied vertices.

A site $\vec{w}$ is a *successor* of the site $\vec{v}$ if there is a directed edge $\langle \vec{v} \rightsquigarrow \vec{w} \rangle$ *from* $\vec{v}$ *to* $\vec{w}$ in $\mathbb{D}^2$. The site $\vec{v}$ is a *predecessor* of $\vec{w}$. An *open path* $\vec{v}_0 \rightsquigarrow \vec{v}_n$ in directed percolation is a sequence of sites $\langle \vec{v}_0, \vec{v}_1, \ldots, \vec{v}_n \rangle$ such that, for each j, the directed edge $\langle \vec{v}_{j-1} \rightsquigarrow \vec{v}_j \rangle$ is open, and $\vec{v}_j$ is an unblocked site. The open path is *occupied* if first $\vec{v}_0$ is occupied, and then recursively all directed edges $\langle \vec{v}_{j-1} \rightsquigarrow \vec{v}_j \rangle$ and sites $\vec{v}_j$ are occupied.

Directed percolation clusters rooted at $\vec{0}$ are grown recursively by occupying first $\vec{0}$ (which is unblocked) and then occupying recursively only those unblocked sites which are successors of occupied sites along open directed edges. If this recursive process terminates with probability 1, then the process is subcritical. If it does not terminate with positive probability, then it is supercritical.

A typical directed percolation cluster is shown in figure 6.18. The cluster with source $\vec{0}$ is a directed animal oriented in the east direction. *Perimeter sites* of the animal are sites which have predecessors in the animal, but they are not in the animal. Each directed percolation cluster (or directed animal) has a weight, which will be determined below.

6.8.1 *The collapse transition in directed percolation clusters*

Let α be a directed animal grown from $\vec{0}$. The number of vertices in α is denoted v, and the number of edges is n. A *contact* between vertices in the animal is a pair of vertices $\vec{v}, \vec{u} \in \alpha$ such that $\|\vec{v} - \vec{u}\|_2 = \sqrt{2}$, and $\vec{v}(1) = \vec{u}(1)$. Denote the number of contacts in α by k. The animal in figure 6.18 has four contacts.

The animal α is a simple, planar, directed and connected graph. Its cyclomatic index c is given by Euler's formula $c = n - v + 1$, where c is the number of independent cycles. This shows that the model may be examined in either the $\{v, n, k\}$- or the $\{v, c, k\}$-ensemble.

Denote the number of directed animals of v vertices, n edges and k contacts by $a_v^+(n, k)$. The partition function of directed site animals in the edge-contact ensemble is

$$A_v(b, b_k) = \sum_{k \geq 0; n \geq 0} a_v^+(n, k) b^n b_k^k, \tag{6.99}$$

where b_k is conjugate to the number of contacts, and b is conjugate to the number of independent cycles. This is a model of *collapsing* or self-interacting directed site animals which undergo a *collapse transition* at a critical value of the activities b or b_k. The limiting free energy of the model is defined by

$$\mathcal{A}(b, b_k) = \lim_{v \to \infty} \tfrac{1}{v} \log A_v(b, b_k), \tag{6.100}$$

and the existence of this limit (the thermodynamic limit) is shown by concatenating animals as in figure 2.3 and following the arguments leading to theorem 3.2.

Let v_1 be the total number of occupied sites with one open edge from an occupied predecessor, let v_2 be the total number of occupied sites with two open edges from occupied predecessors, let v_3 be the number of perimeter sites adjacent to the cluster with one occupied predecessor and let v_4 be the number of perimeter sites with two occupied predecessors. It follows (by counting edges and vertices in the animal and in its perimeter) that

$$\begin{aligned} v_1 &= 2v - n - 2, & v_2 &= n - v + 1, \\ v_3 &= n - 2k + 2, & v_4 &= v + k - n - 1. \end{aligned} \tag{6.101}$$

The probability that $\vec{0}$ is the root of an animal α of v vertices is given by

$$\text{P}_r(\alpha) = p_s^{v-1} p_b^{v_1} (1 - q_b^2)^{v_2} (1 - p_b p_s)^{v_3} (q_s + p_s q_b^2)^{v_4}. \tag{6.102}$$

This may be seen by noting that $v - 1$ sites are unblocked and v_1 of these sites have one open edge from a predecessor with probability p_b, while v_2 have two open edges from two predecessors with probability $1 - q_b^2$. In addition, a site is in the perimeter with one predecessor with probability $(1 - p_b p_s)$, and in the perimeter with two predecessors with probability $q_s + p_s q_s^2$.

Using the relations in equations (6.101),

$$P_r(\alpha) = \mathfrak{X} \left(\frac{p_s p_b^2 (q_s + p_s q_b^2)}{1 - q_b^2} \right)^v \left(\frac{(1-q_b^2)(1-p_b p_s)}{p_b(q_s + p_s q_b^2)} \right)^n \left(\frac{(q_s + p_s q_b^2)}{(1-p_b p_s)^2} \right)^k,$$

where $\mathfrak{X} = \frac{(1-q_b^2)(1-p_b p_s)}{p_s p_b (q_s + p_s q_b^2)}$. Summing this over $\{n, k\}$ for animals with v vertices gives the probability that the directed animal has v vertices

$$P_r(|\alpha|{=}v) = \sum_{|\alpha|=v} P_r(\alpha) = \mathfrak{X} \left(\frac{p_s p_b^2 (q_s + p_s q_b^2)}{1 - q_b^2} \right)^v A_v(b, b_k), \tag{6.103}$$

where

$$b = \frac{(1-q_b^2)(1-p_b p_s)}{p_b(q_s + p_s q_b^2)}, \quad \text{and } b_k = \frac{q_s + p_s q_b^2}{(1-p_b p_s)^2}. \tag{6.104}$$

Explicit solutions for p_b and p_s in these equations in terms of b and b_k can be found. Taking the logarithm of the above, dividing by v and letting $v \to \infty$ gives

$$\mathcal{A}(b, b_k) = -\log\left(\frac{p_s p_b^2 (q_s + p_s q_b^2)}{1 - q_b^2} \right) + \lim_{v\to\infty} \tfrac{1}{v} \log P_r(|\alpha|{=}v). \tag{6.105}$$

If $p_b > 0$ is small enough, then $P_r(|\alpha|{=}v) \leq e^{-\gamma v}$ for some fixed positive number γ (see theorem E.7 in appendix E, for a proof of this fact in the case of ordinary bond percolation).

On the other hand, if $p_b < 1$ is close enough to 1, and p_s is large enough, then percolation occurs in the directed model. Then $\limsup_{v\to\infty} \frac{1}{v} \log P_r(|\alpha|{=}v) = 0$; see appendix E for the similar result in ordinary bond percolation (and, in particular, theorem E.13). In particular, define

$$\zeta(p_s, p_b) = -\lim_{v\to\infty} \tfrac{1}{v} \log P_r(|\alpha|{=}v). \tag{6.106}$$

This limit exists and $\zeta(p_s, p_b) = 0$ in the supercritical phase while $\zeta(p_s, p_b) > 0$ in the subcritical phase.

Theorem 6.1 *If both $p_s < 1$ and $p_b < 1$ are close to 1, then $\zeta(p_s, p_b) = 0$ in the supercritical phase. Otherwise, $\zeta(p_s, p_b) > 0$, and the system is in the subcritical phase.* □

That is, defining $\mathfrak{Y} = \left(\frac{p_s p_b^2 (q_s + p_s q_b^2)}{1 - q_b^2} \right)$ in equation (6.105), there exists a $\gamma > 0$ such that

$$\mathcal{A}(b, b_k) \begin{cases} = -\log \mathfrak{Y}, & \text{if } p_b \text{ or } p_s \text{ are large enough;} \\ \leq -\log \mathfrak{Y} - \gamma, & \text{if } p_b \text{ and } p_s \text{ are small enough.} \end{cases} \tag{6.107}$$

This shows that there is a non-analyticity in $\mathcal{A}(b, b_k)$ corresponding to a phase transition in the model. At this transition the model undergoes a transition from

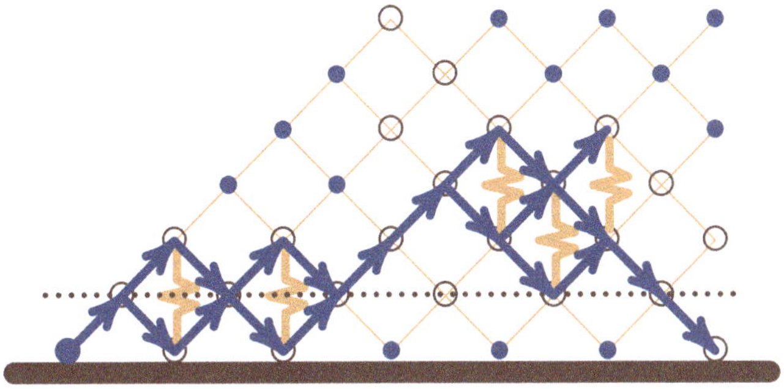

FIG. 6.19. A directed percolation cluster in $\mathbb{D}^2_+$. Blocked vertices are indicated by $\bullet$, while unblocked vertices are represented by $\circ$. Directed edges that return to the boundary are visits. The dotted line is parallel to the adsorbing line and is the super-diagonal. Occupied vertices in the super-diagonal are s-visits. The cluster is a directed animal of size nineteen edges, with three visits, five s-visits and five contacts.

---

an extended phase poor in contacts (and in edges or cycles) to a contact- and edge-rich *collapsed* phase.

Directed percolation was reviewed in reference [172]. Critical exponents for directed percolation clusters have been estimated numerically: the entropic exponent has estimated value $\gamma = 2.2783(7)$. Divergence of the correlation length normal to the preferred direction to leading order is proportional to $(p_c - p)^{-\nu_\perp}$, and along the preferred direction to leading order is proportional to $(p_c - p)^{-\nu_\parallel}$. Numerical estimates of these metric exponents are $\nu_\perp = 1.0969(3)$, and $\nu_\parallel = 1.7339(3)$ [182, 183].

The collapse transition in directed site-animals was examined in references [140, 141]. Directed site-animals are also equivalent to Baxter's directed hard-square lattice-gas model with anisotropic next-nearest-neighbour interactions [27, 139]. A three-dimensional model of directed site-animals was solved in reference [140]. See references [232, 561] for additional results on directed animals.

Compact directed percolation or *compact directed animals* are models equivalent to staircase vesicles (see section 6.3); in these models all lattice sites are unblocked.

6.8.2 *Collapse in an adsorbing directed animal*

Let $\mathbb{D}^2$ be given by equation (6.98) and define the half-directed lattice

$$\mathbb{D}^2_+ = \{\langle \vec{u} \rightsquigarrow \vec{v} \rangle \in \mathbb{D}^2 \mid \vec{v}(2) \geq 0,\ \vec{u}(2) \geq 0\}. \tag{6.108}$$

The boundary of $\mathbb{D}^2_+$ is $\partial\mathbb{D}^2_+ = \{\vec{v} \in \mathbb{Z}^2 \mid \vec{v}(2) = 0\}$ (see figure 6.19).

A directed percolation cluster grown from $\vec{0}$ in $\mathbb{D}^2_+$ is illustrated in figure 6.19. This percolation cluster is a directed animal in $\mathbb{D}^2_+$ but making *visits* to $\partial\mathbb{D}^2_+$ (this is also called the *adsorbing line*).

The *super-diagonal* in $\mathbb{D}^2_+$ is composed of the vertices $S\mathbb{D}^2_+ = \{\vec{v} \in \mathbb{Z}^2 \mid \vec{v}(2) = 1\}$ (see figure 6.19), and it is parallel to $\partial\mathbb{D}^2_+$ (see figure 6.19). A vertex of a directed animal in the super-diagonal is an *s-visit*.

As above, edges in $\mathbb{D}^2_+$ are directed in the north-east and south-east directions. Edges are *open* with probability p_b and *closed* with probability $q_b = 1 - p_b$. The origin $\vec{0}$ is *unblocked*, and sites above and disjoint with $\partial\mathbb{D}^2_+$ are unblocked with probability p_s. If a site is not unblocked, then it is *blocked* (with probability $q_s = 1 - p_s$). Sites in the adsorbing line are unblocked with probability p_w and blocked with probability $q_w = 1 - p_w$.

A directed percolation cluster α is grown from $\vec{0}$ in $\mathbb{D}^2_+$ as before: a site $\vec{w}$ is recursively included in the cluster if $\vec{w}$ is unblocked and it has a predecessor site $\vec{v} \in \alpha$ with $\langle \vec{v} \rightsquigarrow \vec{w} \rangle$ open. This defines a directed percolation process in $\mathbb{D}^2_+$, and α is a directed site animal rooted at $\vec{0}$ in this half-lattice. A site in α which is not $\vec{0}$ or a *visit* to $\partial\mathbb{D}^2_+$ has one occupied predecessor with probability p_b and two occupied predecessors with probability $1 - q_b^2$.

A perimeter site in $\partial\mathbb{D}^2_+$ has probability $(1 - p_b p_w)$ (it will be in α if it is both unblocked and the edge from its predecessor is open). Perimeter sites not in $\partial\mathbb{D}^2_+$ have one occupied predecessor with probability $(1 - p_s p_b)$ or have two occupied predecessors in α with probability $(q_s + p_s q_b^2)$.

Suppose that α has v sites, n edges, k contacts, w visits to the adsorbing line and w_s *s*-visits to the super-diagonal. Let the number of independent cycles in α be denoted by c.

Furthermore, let v_1 be the number of vertices in α with one predecessor (excluding visits to $\partial\mathbb{D}^2_+$), and let v_2 be the number of vertices of α with two predecessors.

Continue by denoting the number of perimeter sites in $\partial\mathbb{D}^2_+$ by s_d, the number of perimeter sites with one predecessor in α by s_1, and the number of perimeter sites with two predecessors in α by s_2. It follows that $s_d + s_1 + 2s_2 = q$, if q is the number of *perimeter edges* (the number of edges with one endpoint in α, since each perimeter site in $\partial\mathbb{D}^2_+$ is incident with exactly one perimeter edge).

Each site $\vec{u}_1$ (with the exception of $\vec{0}$) in α has two directed lattice edges of the form $\langle \vec{u}_1 \rightsquigarrow \vec{u}_2 \rangle$ pointing from it. These edges are either edges in the animal, perimeter edges, or edges from visits pointing to a site *below* $\partial\mathbb{D}^2_+$. This shows that

$$2v - 1 = n + w + q, \tag{6.109}$$

in particular since $\vec{0}$ is not a visit (and so the one edge from it below $\partial\mathbb{D}^2_+$ should be accounted for).

The following relations are also valid (these include Euler's relation $n = v + c - 1$ for planar graphs):

$$v_1 = v - w - c - 1, \quad v_2 = c, \quad n = v + c - 1;$$
$$s_1 = v - 2k + c - w_s, \quad s_2 = k - c, \text{ and } s_d = w_s - w.$$

These can be seen as follows. It is the case that $v_2 = c$ since each site with two predecessors closes an independent cycle in the animal. By counting sites in the animal, $c = 1 + v_1 + v_2 + w$. This shows that $v_1 = v - w - c - 1$.

Every nearest-neighbour contact in the animal corresponds to two occupied sites which are predecessors to a common site in the lattice. If this site is occupied, then it closes a cycle. If it is not occupied, then it is a perimeter site which has two predecessors in α. This shows that $k = c + s_2$.

The vertex at $\vec{0}$ has no predecessors; thus, $w_s = s_d + w$. Finally, counting perimeter edges gives $q = s_1 + s_d + 2s_2$. Since $n + w + q = 2v - 1$, it follows that $2v - 1 = n + w + s_1 + s_d + 2s_2$. Using Euler's relation gives $s_1 = v - w - c - s_d - 2s_2$. Substitution of s_d and s_2 gives the required expression for s_1.

The probability of the cluster α is

$$\mathrm{P}_r^+(\alpha) = p_s^{v-1-w} p_w^w p_b^{v_1} (1 - q_b^2)^{v_2} (1 - p_s p_b)^{s_1} (q_s + p_s q_b^2)^{s_2} (1 - p_b p_w)^{s_d}, \quad (6.110)$$

where P_r^+ is a probability measure on the space of directed half-space clusters rooted at $\vec{0}$. Substituting and simplifying,

$$\mathrm{P}_r^+(\alpha) = \frac{(p_b p_s (1 - p_b p_s))^v}{p_b p_s} \left(\frac{(1 - q_b^2)(1 - p_b p_s)}{p_b (q_s + p_s q_b^2)} \right)^c \times$$
$$\left(\frac{q_s + p_s q_b^2}{(1 - p_b p_s)^2} \right)^k \left(\frac{p_w}{p_b p_s (1 - p_b p_w)} \right)^w \left(\frac{1 - p_b p_w}{1 - p_b p_s} \right)^{w_s}. \quad (6.111)$$

Define b and b_k as in equation (6.104), and a and a_s by

$$a = \frac{p_w}{p_s p_b (1 - p_b p_w)}, \quad \text{and } a_s = \frac{1 - p_b p_w}{1 - p_b p_s}. \quad (6.112)$$

Summing equation (6.111) over all α with v vertices gives the probability that the cluster at the origin has v vertices. Put $\mathfrak{Z}(v) = \frac{[p_b p_s (1 - p_b p_s)]^v}{p_b p_s}$ to see that

$$\mathrm{P}_r^+(|\alpha| {=} v) = \sum_{|\alpha| = v} \mathrm{P}_r^+(\alpha) = \mathfrak{Z}(v) \sum_{c,k,w,w_s} a_v(c,k,w,w_s) a^w a_s^{w_s} b^k b_c^c. \quad (6.113)$$

The existence of the limit

$$\lim_{v \to \infty} \tfrac{1}{v} \log \mathrm{P}_r^+(|\alpha| {=} v) = -\zeta^+(p_s, p_b, p_w) \leq 0 \quad (6.114)$$

can be shown using the methods of section 11.3; thus, the limiting free energy

$$\mathcal{A}(a, a_s, b, b_c) = \lim_{v \to \infty} \tfrac{1}{v} \log \sum_{c,k,w,w_s} a_v(c,k,w,w_s) a^w a_s^{w_s} b^k b_c^c$$
$$= -\log (p_b p_s (1 - p_b p_s)) - \zeta^+(p_s, p_b, p_w) \quad (6.115)$$

exists. This is the free energy of a model of collapsing and adsorbing directed animals; the collapse transition is driven by the activities b_c and b conjugate to the number of cycles and to the number of contacts, respectively. The adsorption transition is driven by activities a and a_s conjugate to the number of visits to the adsorbing plane and to the super-diagonal.

It remains to show that $\mathcal{A}(a, a_s, b, b_c)$ is non-analytic. This can be done by examining the function $\zeta^+(p_s, p_b, p_w)$.

6.8.3 *The function* $\zeta^+(p_s, p_b, p_w)$

Set up a directed percolation model in $\mathbb{D}^2_+$ with directed edges as above (in north-east and south-east directions) and with densities (p_s, p_b, p_w) (see figure 6.19). Grow directed percolation clusters from the source at $\vec{0}$ along open directed edges (open with probability p_b) and unblocked sites (bulk sites are unblocked with probability p_s, and sites in $\partial\mathbb{D}^2_+$ are unblocked with probability p_w).

Let $\vec{1} = (1,1)$ and let $\theta(p_s, p_b, p_w; \vec{v})$ be the probability that the vertex $\vec{v}$ is in an infinite cluster. By the fundamental theorem of percolation, $\theta(p_s, p_b, p_w; \vec{0}) = 0$ in the *subcritical* phase, and $\theta(p_s, p_b, p_w; \vec{0}) > 0$ in the *supercritical* phase. (Notice that the critical values of p_s and p_b are not the trivial values 0 or 1 and that $\theta(p_s, p_b, p_w; \vec{0})$ is monotone non-decreasing in p_s, p_b, or p_w.)

Lemma 6.2 *The probability* $\theta(p_s, p_b, p_s; \vec{0}) = \theta(p_s, p_b, 0; \vec{1})$.

Proof If $p_w = 0$, then all vertices in $\partial\mathbb{D}^2_+$ are blocked and the model reduces to directed percolation in the directed half-lattice $\mathbb{D}^2_+ + \vec{1}$ with site and bond densities p_s and p_b. Choose the origin at $\vec{1}$ to complete the proof. □

Lemma 6.3 *If* $0 < p_w \leq p_s$, *then* $\theta(p_s, p_b, p_w; \vec{0}) = 0 \Leftrightarrow \theta(p_s, p_b, p_s; \vec{0}) = 0$.

Proof Notice that any (directed) path from $\vec{0}$ to ∞ must pass through $\vec{1}$. Thus, since the directed edge $\langle \vec{0} \rightsquigarrow \vec{1} \rangle$ is open with probability p_b,

$$\begin{aligned}
\theta(p_s, p_b, p_s; \vec{0}) &\geq \theta(p_s, p_b, p_w; \vec{0}), \quad \text{since } p_s \geq p_w; \\
&= p_w p_b \theta(p_s, p_b, p_w; \vec{1}), \quad \text{since } \vec{0} \text{ is unblocked with probability } p_w; \\
&\geq p_w p_b \theta(p_s, p_b, 0; \vec{1}), \quad \text{since } p_w > 0; \\
&= p_w p_b \theta(p_s, p_b, p_s; \vec{0}), \quad \text{by lemma 6.2.}
\end{aligned} \tag{6.116}$$

This completes the proof. □

To see that $\zeta^+(p_s, p_b, p_w) = 0$ in the supercritical phase, the general method for ordinary percolation (see section E.4.1) can be used, but some technical difficulties due to directedness must be overcome.

Theorem 6.4 *If* $\theta(p_s, p_b, p_w; \vec{0}) > 0$, *then* $\zeta^+(p_s, p_b, p_w) = 0$.

Proof Put $\vec{p} \equiv \langle p_s, p_b, p_w \rangle$ and put $\theta(\vec{p}; \vec{0}) \equiv \theta(p_s, p_b, p_w; \vec{0})$. Let $m \in 2\mathbb{N}$ be even and let T_m be the triangle in $\mathbb{D}^2_+$ with corners $(0,0)$, $(m,0)$ and (m,m). The number of vertices in T_m is $(\frac{m}{2}+1)^2$. The boundary of T_m is denoted by $\partial T_m = \{(2i,0), (m,2i), (2i,2i) | i = 0,1,\ldots,\frac{m}{2}\} \cup \{(2i+1,1), (m-1,2i+1) | i = 0,1,\ldots,\frac{m}{2}-1\}$.

Let r_m be the number of unblocked sites in T_m and which can be the source of an infinite cluster. If $\vec{v} \in T_m$, then $\vec{v}$ is a site counted by r_m with probability $\theta(\vec{p};\vec{v}) \geq \theta(\vec{p};\vec{0})$ (since every directed path from $\vec{v}$ to ∞ avoids $\vec{0}$ but not vice versa).

An upper bound on the expectation of r_m will be found, and this will give a lower bound on the probability that r_m exceeds $\frac{1}{2}\theta(\vec{p};\vec{0})\times(\frac{m}{2}+1)^2$.

The expected value of r_m is bounded above

$$\begin{aligned}
\mathbb{E} r_m &= \sum_{n=0}^{(\frac{m}{2}+1)^2} n\,\mathrm{P}_r(r_m{=}n) \\
&\leq (\tfrac{m}{2}+1)^2\,\mathrm{P}_r\Big(r_m{\geq}\tfrac{1}{2}\theta(\vec{p};\vec{0})\times(\tfrac{m}{2}+1)^2\Big) \\
&\quad + \tfrac{1}{2}\theta(\vec{p};\vec{0})\times(\tfrac{m}{2}+1)^2\,\mathrm{P}_r\Big(r_m{<}\tfrac{1}{2}\theta(\vec{p};\vec{0})\times(\tfrac{m}{2}+1)^2\Big) \qquad (6.117) \\
&\leq (\tfrac{m}{2}+1)^2\,\mathrm{P}_r\Big(r_m{\geq}\tfrac{1}{2}\theta(\vec{p};\vec{0})(\tfrac{m}{2}+1)^2\Big) + \tfrac{1}{2}\theta(\vec{p};\vec{0})\times(\tfrac{m}{2}+1)^2.
\end{aligned}$$

Denote the number of vertices of T_m which can be reached from ∂T_m by an open directed path by u_m. Denote the number of vertices in T_m which is the source of an open path to ∂T_m by v_m. Suppose that $\vec{v}$ is counted by v_m, and that V_p is the open directed path from ∂T_m to $\vec{v}$.

Reflect T_m through its symmetric anti-diagonal (which passes through the sites $(\frac{m}{2},\frac{m}{2})$ and $(m,0)$), and let T'_m be the image of T_m, let $\vec{v}'$ be the image of $\vec{v}$ and let V'_p be the image of V_p.

Under this reflection, all open edges are reflected to open edges, closed edges to closed edges, unblocked vertices to unblocked vertices, and blocked vertices to blocked vertices. If the direction of edges in T'_m is reversed, then V'_p is a directed path of open edges from $\vec{v}'$ to ∂T_m. Thus, $\vec{v}'$ contributes to u_m in T'_m.

Similarly, it may be shown that vertices counted by u_m in T'_m are reflected into vertices counted by v_m in T_m. Averaging over all realisations of open and closed edges, and unblocked and blocked vertices in T_m, this shows that the most likely values of u_m and v_m are the same; let this number be M.

Observe that $v_m \geq r_m$, since each vertex counted by r_m is the source of an open directed path of edges to ∂T_m. Thus,

$$\mathrm{P}_r\Big(v_m{\geq}\tfrac{1}{2}\theta(\vec{p};\vec{0})\times(\tfrac{m}{2}+1)^2\Big) \geq \mathrm{P}_r\Big(r_m{\geq}\tfrac{1}{2}\theta(\vec{p};\vec{0})\times(\tfrac{m}{2}+1)^2\Big) \geq \tfrac{1}{2}\theta(\vec{p};\vec{0}).$$

Thus, $M = \Theta(m^2)$, provided that $\theta(p;\vec{0}) > 0$.

The event K_m that all vertices in ∂T_m are unblocked, that all directed edges between vertices of ∂T_m are open, and all directed edges in the perimeter of T_m are closed has probability $p_w^{m/2+1} p_s^{5m/2-4} p_b^{3m-3}(1-p_b)^{m+1}$

Thus, the probability that both K_m occurs and that $(\frac{m}{2}+1)^2 \geq v_m \geq \frac{1}{2}\theta(\vec{p};\vec{0})\times(\frac{m}{2}+1)^2$ is at least

$$\mathrm{P}_r\Big(r_m{\geq}\tfrac{1}{2}\theta(\vec{p};\vec{0})\times\tfrac{m(m+1)}{2}\Big)\mathrm{P}_r(K_m) \geq \tfrac{1}{2}\theta(\vec{p};\vec{0})\times p_w^{m/2+1} p_s^{5m/2-4} p_b^{3m-3}(1-p_b)^{m+1}.$$

The most likely value of v_m is M, and so it follows that $M \geq r_m$ as well. The number M is also the most likely value of u_m (this is the number of vertices in T_m which can be reached by an open path from ∂T_m and thus from $\vec{0}$).

Let $\mathrm{P}_r(|C|{=}M)$ be the probability that the cluster C at $\vec{0}$ has size M vertices. Since v_m takes on at most $(\frac{m}{2}+1)^2$ values, of which the most likely value equals M, it follows that $(\frac{m}{2}+1)^2 \mathrm{P}_r(|C|{=}M) \geq \mathrm{P}_r(v_m{\geq}\frac{1}{2}\theta(\vec{p};\vec{0}){\times}(\frac{m}{2}+1)^2)$. Thus

$$\begin{aligned}\mathrm{P}_r(|C|{=}M) &\geq \tfrac{4}{(m+2)^2}\,\mathrm{P}_r\Big((\tfrac{m}{2}+1)^2{\geq}r_m{\geq}\tfrac{1}{2}\theta(\vec{p};\vec{0}){\times}(\tfrac{m}{2}+1)^2\Big)\,\mathrm{P}_r(K_m)\\ &\geq \tfrac{4}{(m+2)^2}\,\theta(\vec{p};\vec{0})\,p_w^{m/2+1}p_s^{5m/2-4}p_b^{3m-3}(1-p_b)^{m+1}.\end{aligned}$$

Since $M=\Theta(m^2)$ if $\theta(\vec{p};\vec{0})>0$, it follows that $\lim_{M\to\infty}\mathrm{P}_r(|C|{=}M)^{1/M}=1$ in the supercritical phase. □

Corollary 6.5 *If* $\theta(p_s,p_b,p_s;\vec{0})>0$ *then* $\zeta^+(p_s,p_b,p_w)=0$ *for all* $p_w\in(0,1]$.

Proof If $\theta(p_s,p_b,p_s;\vec{0})>0$, then $\theta(p_s,p_b,p_w;\vec{0})>0$, for all $p_w\in(0,1]$. This follows from lemma 6.3 if $p_w\leq p_s$ and since $\theta(p_s,p_b,p_w;\vec{0})$ is non-decreasing with p_w if $p_w\geq p_s$. The corollary now follows by theorem 6.4 . □

To prove that $\zeta^+(p_s,p_b,p_w)>0$ if $\theta(p_s,p_b,p_w;\vec{0})=0$ requires careful arguments, and only an outline is given here.

Consider a directed percolation in $\mathbb{D}^2$ (see equation 6.98). Define vertices with coordinates $\{(2m,0)\mid m\in\mathbb{N}\}$ as a *defect line* in the model. Let the densities of open directed edges be p_b, of unblocked vertices in the defect line be p_w, and of the remaining vertices be p_s.

Grow directed clusters from a vertex $\vec{v}$ recursively as before. Let $\mathrm{P}_r^{\,s}(|C|{=}v)$ be the probability that $\vec{0}$ is in a cluster C of size v vertices. The probability that $\vec{0}$ is in an infinite cluster is $\theta^s(p_s,p_b,p_w;\vec{0})$. The function $\zeta^s(p_s,p_b,p_w)$ is defined as before (see equation (6.106)). By theorem 6.1, it follows that in the subcritical phase

$$\zeta^s(p_s,p_b,p_w)\geq\zeta^s(p_s,p_b,p_s)>0,\ \text{ for all } p_w\leq p_s, \tag{6.118}$$

since directed percolation is an increasing process with p_w.

Partition the process above into two half-lattices by cutting it parallel to and just above $\partial\mathbb{D}^2_+$. This gives a directed half-lattice $\mathbb{L}^2_+$ on and above the defect line with densities (p_s,p_b,p_w), and a directed half-lattice $\mathbb{L}^2_-$ on and below the defect line with densities (p_s,p_b,p_w).

A probability measure $\mathrm{P}_r^{\,+}$ can be defined on clusters in $\mathbb{L}^2_+$ and $\mathbb{L}^2_-$ (see for example equation (6.110)). The probability that $\vec{0}$ is the root of a cluster C of v sites in $\mathbb{L}^2_-$ is $\mathrm{P}_r^{\,+}(p_s,p_b,p_w;|C|{=}v)$ (by reflection symmetry). Similarly, the probability that $\vec{0}$ is the root of a cluster C of v sites in $\mathbb{L}^2_+$ is $\mathrm{P}_r^{\,+}(p_s,p_b,p_w;|C|{=}v)$.

Joining the clusters at $\vec{0}$ shows that

$$\mathrm{P}_r^{\,s}(|C|{=}2v)\geq p_b\big(\mathrm{P}_r^{\,+}(p_s,p_b,p_w;|C|{=}v)\big)^2. \tag{6.119}$$

Taking logarithms, dividing by $2v$ and letting $v \to \infty$ show that $\zeta^+(p_s, p_b, p_w) > 0$ in the subcritical phase (see equation (6.106) and theorem 6.1 for similar results, and theorem E.13 in appendix E for ordinary percolation).

By lemma 6.3, $\theta(p_s, p_b, p_w; \vec{0}) = 0$ if and only if $\theta(p_s, p_b, p_s; \vec{0}) = 0$. This shows that, in the subcritical phase defined by $\theta(p_s, p_b, p_w; \vec{0}) = 0$, it is the case that $\zeta^+(p_s, p_b, p_w) > 0$. This gives the following theorem.

Theorem 6.6 *The probability* $\theta(p_s, p_b, p_w; \vec{0}) = 0 \;\Rightarrow\; \zeta^+(p_s, p_b, p_w) > 0$*; moreover, if* $0 < p_w \leq p_s$*, then* $\zeta^+(p_s, p_b, p_s) > 0 \;\Leftrightarrow\; \zeta^+(p_s, p_b, p_w) > 0$.

Proof That $\theta(p_s, p_b, p_w; \vec{0}) = 0 \;\Rightarrow\; \zeta^+(p_s, p_b, p_w) > 0$ follows from the arguments above.

To prove the rest of the theorem, consider the following:

$$\begin{aligned} \zeta^+(p_s, p_b, p_w) > 0 &\Rightarrow \theta(p_s, p_b, p_s; \vec{0}) = 0, \quad \text{by corollary 6.5;} \\ &\Rightarrow \zeta^+(p_s, p_b, p_s) > 0, \quad \text{by the above.} \end{aligned}$$

On the other hand,

$$\begin{aligned} \zeta^+(p_s, p_b, p_w) = 0 &\Rightarrow \theta(p_s, p_b, p_w; \vec{0}) > 0, \quad \text{by the above;} \\ &\Rightarrow \theta(p_s, p_b, p_s; \vec{0}) > 0, \quad \text{by theorem 6.3, since } 0 < p_w \leq p_s; \\ &\Rightarrow \zeta^+(p_s, p_b, p_s) = 0, \quad \text{by theorem 6.4.} \end{aligned}$$

This completes the proof. □

Notice that, if there exists a directed path $\vec{0} \rightsquigarrow \vec{v}$, then $\theta(p_s, p_b, p_w; \vec{0}) > 0$ if and only if $\theta(p_s, p_b, p_w; \vec{v}) > 0$ in the above. That is, $\theta(p_s, p_b, p_w; \vec{0})$ may be replaced by $\theta(p_s, p_b, p_w; \vec{v})$ in the above.

7

SELF-AVOIDING WALKS AND POLYGONS

Concatenation of self-avoiding walks and polygons in section 1.1 produced a subadditive inequality for c_n in equation (1.2) from which the existence of a connective constant $\kappa_d = \log \mu_d$ for the self-avoiding walk in the lattice $\mathbb{L}^d$ is proven (see equation (1.1)).

In this chapter the growth constant is examined for other classes of walks, including half-space and positive walks (see equation (2.20)) in the half-lattice

$$\mathbb{L}_+^d = \{\langle \vec{u} \sim \vec{v} \rangle \in \mathbb{L}^d \mid \vec{u}(d) \geq 0 \text{ and } \vec{v}(d) \geq 0\},$$

where $\vec{u}(d)$ is the d-th Cartesian coordinate of the vector $\vec{u}$ and where the boundary of $\mathbb{L}_+^d$ is

$$\partial \mathbb{L}_+^d = \{\langle \vec{u} \sim \vec{v} \rangle \in \mathbb{L}^d \mid \vec{u}(d) = 0 \text{ and } \vec{v}(d) = 0\}.$$

These results lead to a proof that $\mu_p = \mu_d$ (see theorem 1.4). A second consequence is a pattern theorem for self-avoiding walks.

7.1 Walks, bridges, polygons and pattern theorems

Additional results for c_n and related functions require additional constructive approaches, including the notions of *unfolding*, and the Hammersley-Welsh bridge decomposition of half-space walks [263]. These constructions relate walks to bridges and ultimately to polygons and also give rise to pattern theorems and limit ratio theorems for c_n and related functions (see equations (1.3) and (1.8) as well as references [348, 349, 394, 399]).

An unrelated and different proof of a pattern theorem for the self-avoiding walk by Kesten [348] is reproduced in appendix C.

7.1.1 *Half-space walks and bridges*

An n-step self-avoiding walk $\omega = \langle \omega_0, \omega_1, \ldots, \omega_n \rangle$ from the origin $\vec{0} \in \mathbb{L}^d$ has vertices ω_j for $j = 0, 1, \ldots, n$. Let $\omega_j(k)$ be the k-th Cartesian coordinate of the vertex ω_j in the self-avoiding walk ω.

A self-avoiding walk ω is a *positive walk* if $\omega_0(d) \leq \omega_j(d)$ for all $0 \leq j \leq n$. If, instead, $\omega_0(1) \leq \omega_j(1)$ for all $0 \leq j \leq n$, then it is a positive walk oriented in the $\vec{e}_1$ direction (where $\langle \vec{e}_1, \vec{e}_2, \cdots, \vec{e}_d \rangle$ is the canonical orthonormal basis of $\mathbb{R}^d$).

A self-avoiding walk ω is a *half-space walk* if $\omega_0(1) < \omega_j(1)$ for $1 \leq j \leq n$. The number of half-space walks from $\vec{0}$ of length n is denoted h_n, and the number of positive walks of length n is denoted by c_n^+. By removing the first edge of a half-space walk and by rotating it appropriately, it follows that $c_n^+ = h_{n+1}$.

The Statistical Mechanics of Interacting Walks, Polygons, Animals and Vesicles, 2nd edition, E.J. Janse van Rensburg. © E.J. Janse van Rensburg. Published in 2015 by Oxford University Press.

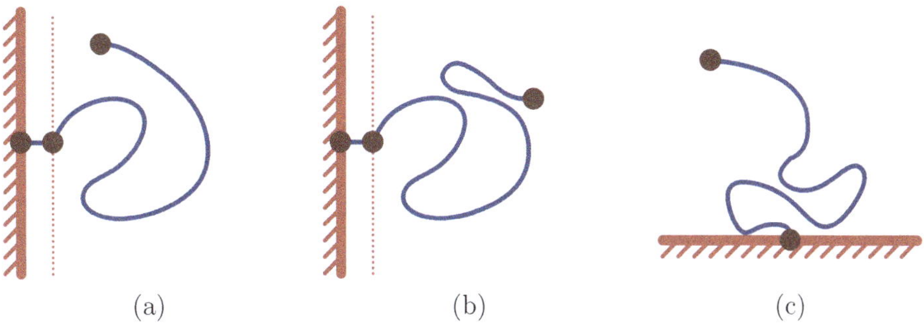

FIG. 7.1. Schematic illustrations of (a) a half-space walk, (b) a bridge and (c) a positive walk. Notice that the first step of the walks in (a) and (b) is in the positive $\vec{e}_1$ direction from the origin. Half-space walks, bridges and positive walks oriented in other directions are obtained by appropriately rotating these walks.

Notice that $h_n \le h_{n+1}$ (this is seen by prepending an edge $\langle(\omega_0 - \vec{e}_1) \sim \omega_0\rangle$ on the first vertex ω_0 of each half-space walk). The walk of length 0 is by convention a half-space walk. This shows that $h_0 = 1$, and $h_1 = 1$.

A half-space walk is a *bridge* if $\omega_0(1) < \omega_i(1) \le \omega_n(1)$ for $1 \le i \le n$. The number of bridges of length n from the origin is b_n. It is the case that $b_n \le b_{n+1}$ (this follows by appending a horizontal edge $\langle \omega_n \sim (\omega_n + \vec{e}_1)\rangle$ on the last vertex ω_n of the each bridge). By convention, the walk of length 0 (the vertex at the origin) is a bridge. That is, $b_0 = 1$, and $b_1 = 1$. A bridge is schematically illustrated in figure 7.1(b).

A *doubly unfolded* walk is a bridge ω such that both $\omega_0(1) < \omega_i(1) \le \omega_n(1)$ and $\omega_0(2) \le \omega_i(2) \le \omega_n(2)$ for $1 \le i \le n$. A doubly unfolded walk is illustrated in figure 7.2(c).

The two bridges $\phi = \langle \phi_0, \phi_1, \ldots, \phi_n\rangle$ and $\psi = \langle \psi_0, \psi_1, \ldots, \psi_m\rangle$ of lengths n and m, respectively, are concatenated by placing ψ such that $\phi_n = \psi_0$. Since there are b_n choices for ϕ, and b_m choices for ψ, this shows that

$$b_n b_m \le b_{n+m}. \tag{7.1}$$

Since $b_n \le c_n$, it follows by theorem A.1 (in appendix A), and theorem 1.1 that

$$\lim_{n\to\infty} \tfrac{1}{n} \log b_n = \sup_{n>0} \tfrac{1}{n} \log b_n = \log \mu_b, \tag{7.2}$$

where $\mu_b \le \mu_d$ is the growth constant of bridges in d dimensions.

A self-avoiding walk can be cut into two half-space walks. To see this, let $\omega = \langle \omega_0, \omega_1, \omega_2, \ldots, \omega_n\rangle$ be a self-avoiding walk with $\omega_0 = \vec{0}$. Denote the *subwalk* from ω_i to ω_j by $\omega[i,j] = \langle \omega_i, \omega_{j+1}, \ldots, \omega_j\rangle$.

The *bottom vertex* $\vec{b}$ of ω is its lexicographic bottom vertex with j-th Cartesian coordinate $\vec{b}(j)$ (the lexicographic top vertex of ω is its *top vertex* $\vec{t}$). Let

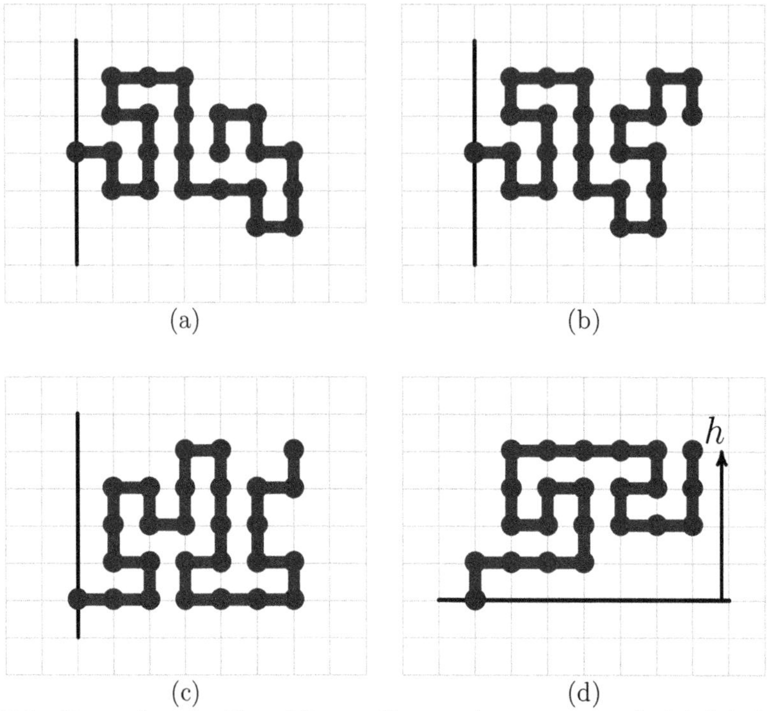

FIG. 7.2. Examples of self-avoiding walks $\omega = \langle \omega_0, \omega_1, \ldots, \omega_n \rangle$. (a) A half-space walk oriented in the $\vec{e}_1$ direction (such that $\omega_0(1) < \omega_i(1)$ for $1 \leq i \leq n$). (b) If $\omega_0(1) < \omega_i(1) \leq \omega_n(1)$, then ω is a bridge. (c) If ω is a bridge, and $\omega_0(2) \leq \omega_i(2) \leq \omega_n(2)$, then ω is a doubly unfolded walk. (d) A doubly unfolded walk (or a bridge or a half-space walk) may be oriented in the $\vec{e}_2$ direction (or in other directions). In this example the doubly unfolded walk is a bridge in the $\vec{e}_2$ direction and it has an endpoint at height h.

$Q_\omega = \{\vec{v} \in \mathbb{L}^d \mid \vec{v}(1) = \vec{b}(1)\}$ be the $(d-1)$-dimensional lattice $\mathbb{L}_b^{d-1}$ with origin at $\vec{b}$ and orthogonal to $\vec{e}_1$. Note that Q_ω is the *bottom plane* of ω, and ω has a last vertex $\vec{b}_1 \in Q_\omega$. Let $\vec{b}_1 = \omega_k$.

Then $\omega[k, n]$ is the subwalk from $\vec{b}_1$ to ω_n. It is necessarily a half-space walk, since its first step $\omega[k, k+1]$ is in the $\vec{e}_1$ direction and it never returns to Q_ω. The number of such half-space walks is at most h_{n-k}. Denote this half-space walk by ψ_2.

The subwalk of ω from $\vec{0}$ to $\vec{b}_1$ is $\omega[0, k]$. It is a positive walk oriented in the $\vec{e}_1$ direction from its bottom vertex $\vec{b}_1$. By *prepending* the edge $\langle (\vec{b}_1 - \vec{e}_1) \sim \vec{b}_1 \rangle$ and fixing the origin at $\vec{b}_1 - \vec{e}_1$, this becomes a half-space walk of length $k+1$. Denote this half-space walk by ψ_1. Notice that ω can be uniquely recovered from ψ_1 and ψ_2.

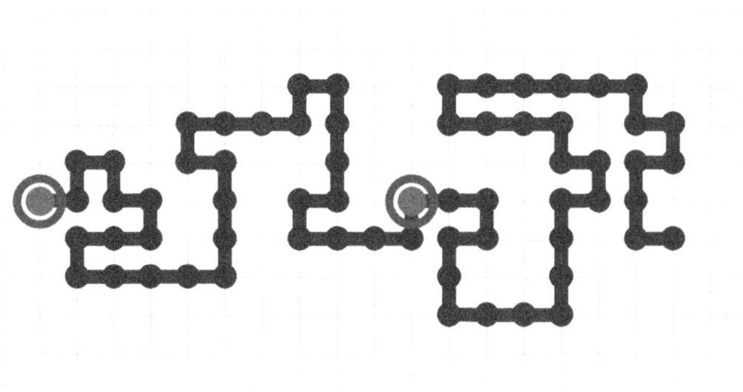

FIG. 7.3. Concatenating two bridges by placing the first vertex of the second on the last vertex of the first.

If ω had length n, and $\omega[0,k]$ and $\omega[k,n]$ have lengths $(k, n-k)$, then (ψ_1, ψ_2) have lengths $(k+1, n-k)$. There are c_n choices for ω, and at most $\sum_{k=0}^{n} h_{k+1} h_{n-k}$ choices for (ψ_1, ψ_2). This shows that

$$b_n \leq h_n \leq c_n^+ \leq c_n \leq \sum_{k=0}^{n} h_{k+1} h_{n-k}, \tag{7.3}$$

since every bridge is also a half-space walk.

7.1.2 *The Hammersley-Welsh construction*

Half-space walks can be decomposed into sets of bridges using the Hammersley-Welsh construction [263].

Let $\omega = \langle \omega_0, \omega_1, \ldots, \omega_n \rangle$ be a half-space walk of length n, with lexicographic bottom vertex $\vec{b}$ and with lexicographic top vertex $\vec{t}$.

Let $P_\omega = \{\vec{v} \in \mathbb{L}^d \mid \vec{v}(1) = \vec{t}(1)\}$ be the $(d-1)$ dimensional sub-lattice $\mathbb{L}_t^{d-1}$ containing $\vec{t}$ and orthogonal to $\vec{e}_1$; P_ω is the *top plane* of ω. The *bottom plane* is defined by $Q_\omega = \{\vec{v} \in \mathbb{L}^d \mid \vec{v}(1) = \vec{b}(1)\}$.

The *span* of a collection of vertices $\{\vec{v}_i\}$ in the $\vec{e}_1$ direction is defined by

$$S_1(\{\vec{v}_i\}) = \max_{0 \leq i \leq n} \vec{v}_i(1) - \min_{0 \leq i \leq n} \vec{v}_i(1). \tag{7.4}$$

This defines the span $S_1(\omega)$ of a walk ω in the $\vec{e}_1$ direction (see equation 1.19).

Denote the number of self-avoiding walks of length n from $\vec{0}$ and with span ℓ in the $\vec{e}_1$ direction by $c_{n,\ell}$. Similarly, define $h_{n,\ell}$ (the number of half-space walks of length n from $\vec{0}$ with span ℓ in the $\vec{e}_1$ direction) and $b_{n,\ell}$ (the number of bridges of length n with span ℓ in the $\vec{e}_1$ direction).

The Hammersley-Welsh construction [263] (or a *bridge decomposition*) cuts bridges recursively from a half-space walk. Eventually the entire half-space walk is decomposed as a set of bridges of (strictly) decreasing spans. The bridges can

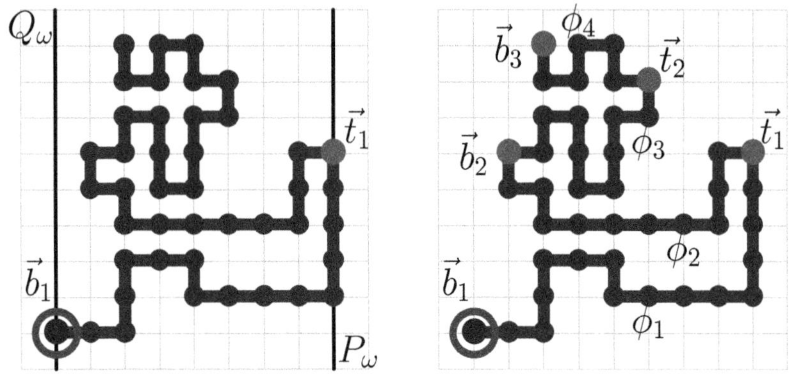

FIG. 7.4. The bridge decomposition of a half-space walk ω. The walk ω has bottom vertex $\vec{b_1} = \vec{0}$ in its bottom plane Q_ω. The top vertex of ω is located in its top plane P_ω which is orthogonal to $\vec{e}_1$. There exists a last vertex $\vec{t_1} = \omega_k$ of ω in P_ω. Cutting ω in $\vec{t_1}$ gives a bridge $\phi_1 = \omega[0,k]$ and a remaining subwalk $\omega[k,n]$. The subwalk $\omega[k,n]$ becomes a half-space walk $\omega^\dagger[k,n]$ with bottom vertex ω_k if it is reflected in P_ω, and the decomposition can be applied to it recursively, cutting a bridge ${\phi_2}^\dagger$ from it. This gives a sequence of bridges $\langle \phi_1, {\phi_2}^\dagger, \ldots, \phi_{2i-1}, \phi_{2i}^\dagger, \ldots \rangle$, such that $\omega = \phi_1\phi_2\phi_3\cdots$ where $\phi_{2i}^\dagger$ is the reflection of ϕ_{2i} in P_ω. This is a bridge decomposition of the half-space walk.

---

be concatenated into an *unfolded walk* (the half-space walk is said to have been *unfolded*).

Let $\omega = \langle \omega_0, \omega_1, \ldots, \omega_n \rangle$ be a half-space walk of length n from $\vec{0}$ with span $\ell = S_1(\omega)$. The bottom vertex of ω is $\vec{b} = \vec{b_1} = \omega_0 = \vec{0}$, and the top vertex is $\vec{t}$. The walk is confined to a *slab* W with boundary $\partial W = Q_\omega \cup P_\omega$ (see figure 7.4).

There is a last vertex $\vec{t_1} = \omega_k$ of ω in P_ω. The subwalk $\omega[0,k]$ is a bridge from $\vec{0}$ to $\vec{t_1}$ of span ℓ in the $\vec{e}_1$ direction. The number of conformations of $\omega[0,k]$ is less than or equal to $b_{k,\ell}$, the number of bridges of length k. Put $\phi_1 = \omega[0,k]$ and let its span be $S_1(\phi_1)$.

Delete the subwalk $\omega[0,k]$ to obtain a walk ψ_1 which has exactly one vertex (namely $\vec{t_1}$) in P_ω. Place the origin at $\vec{t_1}$ and reflect ψ through P_ω. This gives a half-space walk ψ_1^*.

A bridge ${\phi_i}^* = \omega[0,k']$ may be cut recursively from ψ_i^* for $i = 1, 2, \ldots$ by repeating the construction above. This gives a finite sequence of remaining walks $\langle \psi_1, \psi_2, \ldots, \psi_N \rangle$, which must be finite.

Eventually ω is partitioned $\omega = \phi_1\phi_2\phi_3\cdots\phi_N$, with each ϕ_{2i+1} and ϕ_{2i}^* a bridge. By construction, the spans of the bridges are strictly decreasing: $S_1(\phi_{i-1}) > S_1(\phi_i)$ for $i = 2, 3, \ldots, N$. If the length of ϕ_i is n_i, then $\sum_{i=1}^N n_i = n$, and $\sum_{i=1}^N S_1(\phi_i) \leq n$.

The decomposition gives a mapping from half-space walks to sets of bridges $\langle \phi_1, \phi_2^*, \ldots, \phi_N \rangle$ with strictly decreasing spans $S_1(\phi_1) > S_1(\phi_2) > \cdots > S_1(\phi_N) > 0$ such that $\sum_{i=1}^{N} S_1(\phi_i) = s \leq n$.

For each $s \leq n$, the number of choices for $\{S_1(\phi_1), S_1(\phi_2), \ldots, S_1(\phi_N)\}$ is bounded from above by $p(s)$, the number of partitions of s into distinct parts.

The bridges $\langle \phi_1, \phi_2^*, \ldots, \phi_N \rangle$ can be concatenated into a single bridge ϕ of span $S_1(\phi) = s \leq n$ (see figure 7.3). The total number of such bridges is $b_{n,s}$, namely, the number of bridges of length n and span s. This shows that

$$h_n \leq \sum_{s=1}^{N} p(s) b_{n,s}. \tag{7.5}$$

This gives the following lemma.

Lemma 7.1 *If $\gamma_0 > \pi\sqrt{\frac{2}{3}}$, then there exist $\epsilon > 0$, an $N_0(\epsilon) \in \mathbb{N}$, and a constant K such that*

$$h_n \leq K e^{(\gamma_0 - \epsilon)\sqrt{n}}\, b_n, \text{ for all } n \geq N_0(\epsilon).$$

This shows that $\lim_{n\to\infty} \frac{1}{n} \log h_n = \mu_b$.

Proof Since $p(s) \leq p(s+1)$, it follows from equation (7.5) that $h_n \leq p(n) b_n$. It is known that $\log p(n) = \pi\sqrt{\frac{2n}{3}}(1+o(1))$ [275]. For any $\gamma_0 > \pi\sqrt{\frac{2}{3}}$, there exist an $\epsilon > 0$, a $K > 0$, and an $N_0(\epsilon)$ so that for all $n \geq N_0(\epsilon)$, $\log p(n) \leq \log K + (\gamma_0 - \epsilon)\sqrt{n}$. This shows that $h_n \leq K e^{(\gamma_0-\epsilon)\sqrt{n}} b_n$ for all $n \geq N_0(\epsilon)$.

If $\gamma_0 > \pi\sqrt{\frac{2}{3}}$, then, by equation (7.2), there exist $\epsilon > 0$ and an $N_0(\epsilon)$ such that

$$b_n \leq h_n \leq K e^{(\gamma_0-\epsilon)\sqrt{n}}\, b_n. \tag{7.6}$$

This shows $\lim_{n\to\infty} \frac{1}{n} \log h_n = \mu_b$. □

Notice that, by increasing γ_0, it may be assumed that $K = 1$ in lemma 7.1. That is, there exist a $\gamma_1 > 0$, and an $N_1 \in \mathbb{N}$ such that $b_n \leq h_n \leq e^{\gamma_1\sqrt{n}} b_n$ for all $n \geq N_1$.

Lemma 7.1 has the following implication for c_n.

Theorem 7.2 *If $\gamma_0 > \frac{2\pi}{\sqrt{3}}$, then there exist $\epsilon > 0$ and an N_ϵ such that $c_n \leq e^{(\gamma_0-\epsilon)\sqrt{n}} \mu_d^n$ for all $n \geq N_\epsilon$ whenever $n \geq N_\epsilon$.*

Proof For given and fixed γ_0, and $\epsilon > 0$ (such that there exists a fixed N_ϵ in lemma 7.1), the sum in equation (7.3) is partitioned into three parts: (1) $m \in \{0, 1, \ldots, N_\epsilon\}$, (2) $m \in \{N_\epsilon + 1, N_\epsilon + 2, \ldots, n - N_\epsilon - 1\}$ and (3) $m \in \{n - N_\epsilon, n - N_\epsilon + 1, \ldots, n\}$. Since $h_m \leq h_{m+1}$, this gives

$$c_n \leq (N_\epsilon + 1) h_{N_\epsilon+1} h_n + \sum_{m=N_\epsilon+1}^{n-N_\epsilon-1} h_{m+1} h_{n-m} + (N_\epsilon + 1) h_n h_{N_\epsilon}.$$

For values of $n \geq N_\epsilon$ use equation (7.6) on h_n in the first and third terms on the right-hand side and then on h_{m+1} and h_{n-m} in the summation in the middle, because $m + 1 \geq N_\epsilon$, and $n - m \geq N_\epsilon$. This shows that, for all $n \geq N_\epsilon$,

$$c_n \le 2K(N_\epsilon+1)h_{N_\epsilon+1}e^{(\gamma_0-\epsilon)\sqrt{n}}b_n+K^2\sum_{m=N_\epsilon+1}^{n-N_\epsilon-1}e^{(\gamma_0-\epsilon)(\sqrt{m+1}+\sqrt{n-m})}b_{m+1}b_{n-m}.$$

Since $\sqrt{m+1}+\sqrt{n-m} \le \sqrt{2(n+1)}$, and $b_{m+1}b_{n-m} \le b_{n+1}$, the result is

$$c_n \le 2K(N_\epsilon+1)h_{N_\epsilon+1}e^{(\gamma_0-\epsilon)\sqrt{n}}b_n + K^2(n-2N_\epsilon-2)e^{\sqrt{2}(\gamma_0-\epsilon)\sqrt{n+1}}b_{n+1}.$$

The last term grows faster than the first, and, for large n, it is strictly bigger. Thus, K may be chosen so that

$$b_n \le c_n \le K^2 n e^{\sqrt{2}(\gamma_0-\epsilon)\sqrt{n+1}}b_{n+1}.$$

Now replace $\sqrt{2}\gamma_0 \to \gamma_0$, and $\sqrt{2}\epsilon \to \epsilon$. Then the above is true for $n \ge N_\epsilon$. Since $b_n \le \mu_b^n$ by equation (7.2), the result is that

$$b_n \le c_n \le K^2 n e^{(\gamma-\epsilon)\sqrt{n+1}}\mu_b^{n+1}.$$

Taking the power $\frac{1}{n}$ and letting $n \to \infty$ shows that $\mu_b = \mu_d$, by theorem 1.1.

By slightly increasing γ_0 and N_ϵ, replace the above by $c_n \le e^{(\gamma_0-\epsilon)\sqrt{n}}\mu_d^n$. This completes the proof. □

The total number of bridges in the Hammersley-Welsh construction is N (see equation (7.5)). Put $s_i = S_1(\phi_i)$ in the bridge decomposition; then the order of N may be estimated as follows. The number N is a maximum if $s_{i+1} = s_i - 1$ for each i and for each $s_N = 1$ (that is, if the bridges have minimal spans). This shows that $\min\left\{\sum_{i=1}^N s_i\right\} = \frac{1}{2}N(N+1) \le n$, and, hence, $N \le \frac{3\sqrt{n}}{2}$.

By equations (7.1), (7.3) and (7.5),

$$b_n \le h_n \le c_n^+ \le c_n \le b_{n+1}\sum_{k=0}^{n} p(k)p(n-k). \tag{7.7}$$

Thus, a corollary of lemma 7.1, theorem 7.2 and equation (7.3) is the following:

Corollary 7.3 $\lim_{n\to\infty} b_n^{1/n} = \lim_{n\to\infty} h_n^{1/n} = \lim_{n\to\infty} (c_n^+)^{1/n} = \lim_{n\to\infty} c_n^{1/n} = \mu_d.$ □

The Hammersley-Welsh bound was improved to $c_n \le e^{\gamma n^{2/(d+2)}\log n}\mu_d^n$ for every $n \ge 2$ by Kesten; see reference [349].

7.1.3 *Doubly unfolded walks and tails*

A self-avoiding walk $\omega = \langle \omega_0, \omega_1, \ldots, \omega_n \rangle$ is an *unfolded walk* if $\omega_0(1) \le \omega_i(1) \le \omega_n(1)$ for $i = 0, 1, \ldots, n$. The number of unfolded walks of length n from $\vec{0}$ is denoted by $c_n^\dagger$. By prepending an edge in the $\vec{e}_1$ direction on ω_0, an unfolded walk becomes a unique bridge. Similarly, removing the first edge in a bridge gives a unique unfolded walk. Thus, $c_n^\dagger = b_{n+1}$.

A bridge $v = \langle v_0, v_1, \ldots, v_n \rangle$ is a *doubly unfolded walk* if $v_0(2) \leq v_i(2) \leq v_n(2)$, where $v(2)$ is the second Cartesian coordinate of the vertex v. This doubly unfolded walk is a bridge in the $\vec{e}_1$ direction and unfolded in the $\vec{e}_2$ direction. A rotation or a reflection of a doubly unfolded walk gives doubly unfolded walks oriented in other coordinate directions. Doubly unfolded walks are illustrated in figures 7.2(c) and (d) and schematically illustrated in figure 7.5(a). The number of doubly unfolded walks of length n from $\vec{0}$ is denoted by $c_n^{\ddagger}$. Clearly, $c_n^{\ddagger} \leq b_n$.

The walk $\tau = \langle \tau_0, \tau_1, \tau_2, \ldots, \tau_n \rangle$ is a *tail* if $\tau_0(1) \leq \tau_i(1) \leq \tau_n(1)$ for $i = 1, 2, \ldots, n$ and if $\tau_0(d) < \tau_i(d)$ for $i = 1, 2, \ldots, n$. That is, τ is unfolded in the $\vec{e}_1$ direction and is a half-space walk in the $\vec{e}_d$ direction.

The number of tails of length n from $\vec{0}$ is c_n^t. Every doubly unfolded walk which is a bridge in the $\vec{e}_d$ direction and unfolded in the $\vec{e}_1$ direction is also a tail. Thus, $c_n^{\ddagger} \leq c_n^t$.

The next result has a proof similar to the proofs of lemma 7.1 and theorem 7.2.

Lemma 7.4 *If $\gamma_0 > \frac{2\pi}{\sqrt{3}}$, then there exist $\epsilon > 0$, and an N_ϵ, such that $b_n \leq e^{(\gamma_0 - \epsilon)\sqrt{n}} c_n^{\ddagger}$, for all $n \geq N_\epsilon$.*

Proof Consider a bridge $v = \langle v_0, v_1, \ldots, v_n \rangle$. There is a first vertex v_i such that $v_i(2) \leq v_j(2)$ for $j = 0, 1, \ldots, n$. Define $Q = \{\vec{v} \in \mathbb{L}^d \mid \vec{v}(2) = v_i(2)\}$

Cut v into two subwalks (ϕ_1, ϕ_2) at v_i. Then ϕ_1 is a half-space walk oriented in the $\vec{e}_2$ direction and, by adding the edge $\langle (v_i - \vec{e}_2) \sim v_i \rangle$ to ϕ_2, it becomes a half-space walk.

Apply the Hammersley-Welsh construction to each of ϕ_1 and ϕ_2 to obtain bridges ψ_1 and ψ_2 oriented in the $\vec{e}_2$ direction.

Reflect ψ_1 through Q to obtain ψ_1^*. Form the walk $v' = \psi_1^* \psi_2$. Then $v' = \langle v_0', v_1', \ldots, v_n' \rangle$, and $v_0'(2) \leq v_j'(2) \leq v_n'(2)$. Thus, v' is unfolded in the $\vec{e}_2$ direction. The Hammersley-Welsh construction did not change any first coordinate. Thus, v' is a bridge in the $\vec{e}_1$ direction. This shows that v' is doubly unfolded.

It only remains to account for the number of bridges v which may be unfolded to the same doubly unfolded walk. This is bounded in a way similar to the method in the proof of theorem 7.2. □

Every doubly unfolded walk is a half-space walk, and every half-space walk is a positive walk in the half-lattice $\mathbb{L}_+^d$ (see equation (2.20)). Thus, $c_n^{\ddagger} \leq h_n \leq c_n^+$. Since $c_n^{\ddagger} \leq b_n$, and $c_n^{\ddagger} \leq c_n^t$, all these functions grow at the same exponential rate (see corollary 7.3).

Corollary 7.5 *If ψ_n is any of $\{c_n^{\ddagger}, c_n^{\dagger}, c_n^t, b_n, h_n, c_n^+, c_n\}$, then $\lim_{n \to \infty} \frac{1}{n} \log \psi_n = \log \mu_d$.* □

7.1.4 *Loops, hoops and the most popular class argument*

A walk $\omega = \langle \omega_0, \omega_1, \ldots, \omega_n \rangle$ in $\mathbb{L}^d$ is a *hoop* if $\omega_0(d) = \omega_n(d) \leq \omega_i(d)$ for $i = 1, 2, \ldots, n-1$. That is, a hoop is a positive walk with its endpoints in the

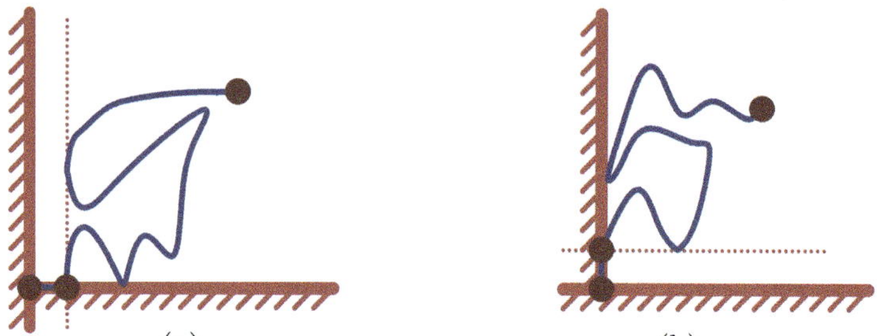

(a) (b)

FIG. 7.5. (a) A schematic diagram of a doubly unfolded walk. The walk is a bridge in the horizontal direction and is unfolded in the vertical direction. (b) A schematic representation of a tail. A tail is unfolded in the horizontal direction and is a half-space walk in the vertical direction.

---

sub-lattice $\mathbb{L}^{d-1}$ orthogonal to $\vec{e}_d$. The number of hoops of length n is denoted by ℓ_n^o.

A hoop $\omega = \langle \omega_0, \omega_1, \ldots, \omega_n \rangle$ in $\mathbb{L}^d$ is a *loop* if both $\omega_0(d) = \omega_n(d) < \omega_i(d)$ and $\omega_0(1) \leq \omega_i(1) \leq \omega_n(1)$ for $i = 1, 2, \ldots, n-1$. The number of loops of length n is denoted by ℓ_n. Since each loop is also a hoop, it follows that $\ell_n \leq \ell_n^o \leq c_n$.

The endpoints of a hoop or a loop are in the *base plane*; this is normally the $(d-1)$-dimensional sub-lattice $\mathbb{B}_d \equiv \mathbb{L}^{d-1}$ containing $\vec{0}$ and orthogonal to $\vec{e}_d$.

A loop is illustrated in figure 7.7(a). Observe that $\ell_n \leq \ell_{n+1}$, since the last edge of a loop is right-most and may be replaced by two edges in a $\neg$-conformation.

Loops can be constructed by putting together tails in the $\vec{e}_1$ and $-\vec{e}_1$ directions, as shown in figure 7.7(a) (delete the edge marked by X to obtain a tail and a reflected tail). Two tails with endpoints at the same height can be put together in this way to find a loop.

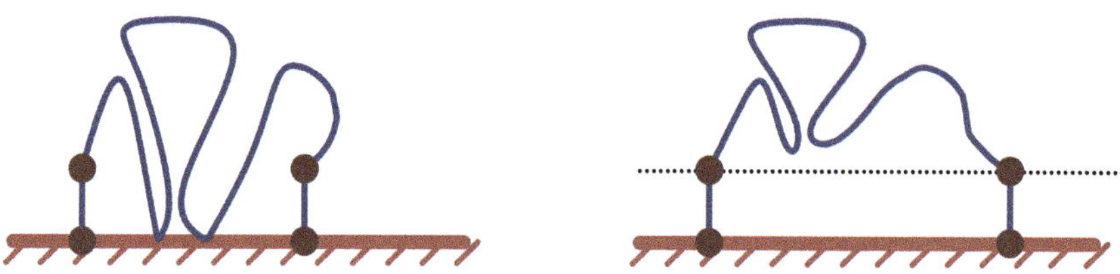

(a) (b)

FIG. 7.6. Schematic representations of (a) a hoop and (b) a loop. A hoop is a positive walk from $\vec{0}$ with its endpoints in the sub-lattice $\mathbb{L}^{d-1}$ of $\mathbb{L}^d$. A loop is a hoop which is unfolded in the $\vec{e}_1$ direction and with all vertices (except endpoints) disjoint with the sub-lattice $\mathbb{L}^{d-1}$.

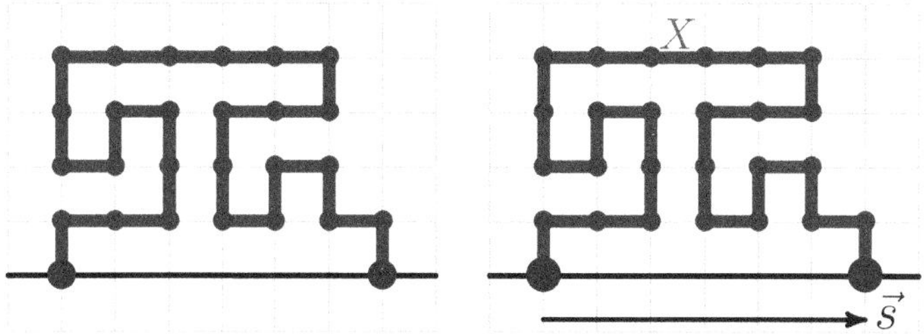

(a) (b)

FIG. 7.7. A loop is illustrated in (a). This loop may be cut into two tails by deleting the edge marked by X in (b). The vector between the endpoints of the loop is indicated by $\vec{s}$.

Denote the number of tails of length n and with an endpoint at height h (above the base plane $\mathbb{B}_d$ orthogonal to $\vec{e}_d$) by $c_n^t(h)$. Then figure 7.7(b) gives

$$c_{2n+2} \geq \ell_{2n+2} \geq \ell_{2n+1} \geq \sum_{n=0}^{n} c_n^t(h) c_n^t(h). \tag{7.8}$$

Since $c_n^t = \sum_{h=0}^n c_n^t(h)$, there is a value h^* (a function of n) which maximises $c_n^t(h)$. That is, $c_n^t(h^*) \leq c_n^t \leq (n+1)c_n^t(h^*)$. This value of h^* defines a *most popular class* of tails of length n.

By corollary 7.5, $\lim_{n\to\infty} \frac{1}{n} \log c_n^t(h^*) = \lim_{n\to\infty} \frac{1}{n} \log c_n^t = \log \mu_d$.

By replacing the $c_n^t(h)$ on the right-hand side of equation (7.8) with the most popular classes, it follows that

$$c_{2n+2} \geq \ell_{2n+2} \geq \ell_{2n+1} \geq \left(c_n^t(h^*)\right)^2. \tag{7.9}$$

By taking logarithms, dividing by n and taking $n \to \infty$, it follows (in particular because $\ell_n \leq \ell_n^o \leq c_n$) that the growth constant of loops and hoops is equal to μ_d.

Theorem 7.6 *The limit* $\lim\limits_{n\to\infty} \frac{1}{n} \log \ell_n = \lim\limits_{n\to\infty} \frac{1}{n} \log \ell_n^o = \log \mu_d$. □

The *end-to-end displacement vector* $\vec{s}$ of a loop is the vector pointing from its lexicographic least endpoint to its lexicographic most endpoint; see figure 7.7(b).

Denote the number of loops of length n and denote the end-to-end displacement vector $\vec{s}$ by $\ell_n(\vec{s})$. Then there is (at least) one most popular end-to-end displacement vector $\vec{s}^*$ (which is a function of n), so that

$$\ell_n(\vec{s}^*) \leq \ell_n = \sum_{\vec{s}} \ell_n(\vec{s}) \leq (n+1)^d \ell_n(\vec{s}^*), \tag{7.10}$$

since each component of $\vec{s}^*$ may have up to $(n+1)$ values. This most popular class argument shows that $\lim_{n\to\infty} \frac{1}{n} \log \ell_n(\vec{s}^*) = \log \mu_d$ by theorem 7.6.

Two loops with the same length and most popular span can be put together by reflecting one through its base plane and then translating it until the endpoints of the two are identified. That is, $\frown + \smile \to \circ$. The result is an unrooted lattice polygon so $(\ell_n(s^*))^2 \leq p_{2n} \leq c_{2n-1}$. Take logarithms, divide by $2n$ and let $n \to \infty$. This gives the following theorem.

Theorem 7.7 *The limit* $\lim_{n\to\infty} \frac{1}{2n} \log p_{2n} = \lim_{n\to\infty} \frac{1}{n} \log c_n = \log \mu_d$. □

This proves the claim that $\mu_p = \mu_d$ (see theorem 1.4). By considering equation (1.11), a corollary of the theorem above is the following.

Corollary 7.8 *For any vertex* $\vec{v} \in \mathbb{Z}^d$,

$$\lim_{n\to\infty} \frac{1}{2n} \log p_{2n} = \lim_{n\to\infty} \frac{1}{n} \log c_n(\vec{v}) = \log \mu_d,$$

where the limit $\lim_{n\to\infty} \frac{1}{n} \log c_n(\vec{v})$ *is taken through all* $n \in \mathbb{N}$ *with parity equal to that of* $\|\vec{v}\|_\infty$. □

7.1.5 *Walk, bridge and polygon generating functions*

The self-avoiding walk generating function $C(t)$ (equations (1.56) and (1.14)) has the lower bound

$$C(t) = \sum_{n=0}^{\infty} c_n t^n \geq \frac{1}{1-\mu_d t} \tag{7.11}$$

by theorem 1.1. This shows that $C(\mu_d^{-1}) = \infty$ (and its radius of convergence is $t_c = \mu_d^{-1}$ by theorem 1.1).

The half-space walk, bridge, loop and polygon generating functions are

$$H(t) = \sum_{n=0}^{\infty} h_n t^n; \; B(t) = \sum_{n=0}^{\infty} b_n t^n; \; L(t) = \sum_{n=0}^{\infty} \ell_n t^n; \text{ and } P(t) = \sum_{n=0}^{\infty} p_{2n} t^{2n},$$

respectively. These are formal power series with integer coefficients with dominant singularity on the real axis in the complex t-plane. By theorems 1.1 and 1.4, corollary 7.3 and theorem 7.6, the radii of convergence of these series are at $t_c = \mu_d^{-1}$.

The two-point correlation function $C(\vec{v};t)$ (see equation (1.37)) is the generating function of $c_n(\vec{v})$. Multiplying equation (1.10) with t^n and summing over n,

$$P(t) \leq 1 + (d-1)t^{2-\|\vec{v}\|_1} C(\vec{v};t), \text{ and } C(\vec{v};t) \leq C(t). \tag{7.12}$$

This shows that the radius of convergence of $C(\vec{v};t)$ is also at $t_c = \mu_d^{-1}$.

Since $\ell_{n-1} \leq b_n \leq h_n \leq c_n$, and $p_{2n} \leq c_{2n-1}$, it follows that $tL(t) \leq B(t) \leq H(t) \leq C(t)$, and $P(t) \leq tC(t)$, for $t \in [0, t_c]$.

Theorem 7.9 *The generating functions $C(t)$, $H(t)$ and $B(t)$ are related by*

$$H(t) \leq e^{B(t)-1}, \quad \text{and} \quad tC(t) \leq (H(t))^2.$$

Since $C(t)$ is divergent at $t_c = \mu_d^{-1}$, it follows that $H(t)$ and $B(t)$ are divergent at $t = t_c$. That is, $\lim_{t\to t_c^-} C(t) = \lim_{t\to t_c^-} H(t) = \lim_{t\to t_c^-} B(t) = \infty$.

The bridge generating function has the lower bound

$$B(t) \geq 1 + \tfrac{1}{2}\log\left(\frac{t}{1-\mu_d t}\right).$$

Moreover, if $B_s(t)$ is the generating function of bridges of span s, then

$$H(t) \leq \prod_{s=1}^{\infty} (1 + B_s(t)).$$

Hence, $\prod_{s=1}^{\infty} (1 + B_s(t_c)) = \infty$.

Proof Denote the number of bridges of length n with span s by $b_{n,s}$. The Hammersley-Welsh decomposition of half-space walks into sets of bridges provides the basic construction.

Each half-space walk ω of length n may be decomposed into a finite sequence of bridges $\langle \phi_1, \phi_2, \ldots, \phi_N \rangle$ with spans s_i in the $\vec{e}_1$ direction and lengths n_i, where $s_{i+1} > s_i$ and $\sum n_i = n$. This set of bridges can be uniquely reassembled to recover ω. There are h_n choices for ω, and at most b_{n_i,s_i} choices for each bridge ϕ_i of span s_i. Thus,

$$h_n \leq \sum_{\{n_i\},\{s_i\},N} \prod_{i=1}^{N} b_{n_i,s_i}$$

where the summation is over all $\{n_i\}$, $\{s_i\}$ and N, subject to the constraints on n_i and s_i; that is, $\sum_i n_i = n$, and $s_i \leq n_i$.

Multiplying by t^n and summing over n gives

$$H(t) \leq \prod_{s=1}^{\infty} \left(1 + \sum_{m=1}^{\infty} b_{m,s} t^m\right) \leq e^{\sum_{s=1}^{\infty}\left(\sum_{m=1}^{\infty} b_{m,s} t^m\right)} = e^{B(t)-1}. \tag{7.13}$$

Hence, if $H(t)$ is divergent, then $B(t)$ is divergent.

On the other hand, by equation (7.3),

$$tC(t) \leq \sum_{n=0}^{\infty} \left(\sum_{m=0}^{\infty} h_{m+1} h_{n-m}\right) t^{n+1} = (H(t))^2. \tag{7.14}$$

However, $C(t)$ is divergent at t_c. Thus, $H(t)$ is divergent if $t = t_c$. The lower bound on $B(t)$ is obtained from equation (7.11) and equations (7.13) and (7.14).

The generating function of bridges of span s is $B_s(t) = \sum_{m=1}^{\infty} b_{m,s} t^m$. By equation (7.13), it follows that $H(t) \leq \prod_{s=1}^{\infty} (1 + B_s(t))$.

This completes the proof. □

7.1.6 *A pattern theorem for walks and polygons*

A *pattern* P in a self-avoiding walk $\omega = \langle \omega_0, \omega_2, \ldots, \omega_n \rangle$ is a subwalk. A pattern $P = \langle p_0, p_1, \ldots, p_N \rangle$ *occurs* in ω at a vertex ω_j if there is a vector $\vec{v} \in \mathbb{Z}^d$ such that $\omega_{j+\ell} = p_\ell + \vec{v}$ for $\ell = 0, 1, 2, \ldots, N$. If there are k distinct vectors $\vec{v}_m$ such that P occurs, then P occurs k times in ω.

The pattern P is a *proper pattern* if, for every $k \in \mathbb{N}$, there is a walk on which P occurs at least k times. If P can occur three times in ω, then by theorem C.2 P is a proper pattern [261]. In addition, if P is a proper pattern, then by theorem C.3 there exist a hypercube Q and a self-avoiding walk ω such that P occurs at least once in ω, $\omega \subseteq Q$, and the endpoints of ω are opposite corners of Q. If P is a bridge, then P is a proper pattern. By theorem C.3, if P is a proper pattern, then there is a bridge on which P occurs at least once. A proper pattern is also called a *Kesten pattern.*

A bridge is *decomposable* if it can be cut in a vertex into two (non-trivial) bridges. If a bridge is not decomposable, then it is a *prime bridge.* The number of prime bridges of length n is denoted q_n. The generating function of prime bridges is $Q(t) = \sum_{n=0}^{\infty} q_n t^n$. By convention, $q_0 = 0$.

If ϕ is a bridge of length at least 1, then either it is a prime bridge or there is a first vertex $\vec{v}$ where ϕ can be cut into a prime bridge π and a bridge ψ. That is, $\phi = \pi\psi$ (where π is a prime bridge, and ψ is possibly the trivial bridge of length 0). This renewal of ϕ gives the recurrence

$$b_n = \sum_{m=1}^{n} q_m b_{n-m}, \quad \text{for } n \geq 1. \tag{7.15}$$

Multiply this by t^n and sum over n. Since $b_0 = 1$, and $q_0 = 0$, the result is

$$B(t) - 1 = Q(t) B(t), \quad \text{which gives } B(t) = \frac{1}{1 - Q(t)}. \tag{7.16}$$

A corollary of theorem 7.9 is that $\lim_{t \to t_c^-} Q(t) = 1$. Since $Q(t)$ is continuous within its radius of convergence, and $B(t) \geq 1$ while $Q(t) \leq 1$ for $t \in [0, t_c]$, it follows that $Q(t_c) = 1$.

Corollary 7.10 *The limits* $\lim_{t \to t_c^-} B(t) = \infty$, *and* $\lim_{t \to t_c^-} Q(t) = Q(t_c) = 1$.

Proof By the Perron-Frobenius theorem, $B(t)$ and $Q(t)$ have dominant singularities on the positive real axis in the t-plane. The function $B(t)$ is a continuous function of $t \in (0, t_c)$, and $B(t) < \infty$ if and only if $Q(t) < 1$ for $t \in \mathbb{R}^+$. By theorem 7.9, $B(t) \to \infty$ as $t \to t_c^-$. Thus, by equation (7.16), it follows that $\lim_{t \to t_c^-} Q(t) = 1$ (and that $B(t) < \infty$ if $Q(t) < 1$). □

Let P be a proper pattern, let $b_n[\leq k, P]$ be the number of bridges of length n, let $q_n[\leq k, P]$ be the number of prime bridges such that P occurs no more than k times and let $c_n[\leq k, P]$ be the number of self-avoiding walks on which P occurs no more than k times.

Let $c_n[k, P]$ be the number of self-avoiding walks of length n from $\vec{0}$ on which P occurs exactly at k different vertices and let $c_n[\geq k, P]$ be the number of self-avoiding walks from the origin of length n on which P occurs at least at k different vertices.

Denote by $C(t; [\leq k, P])$, $B(t; [\leq k, P])$ and $Q(t; [\leq k, P])$ the generating functions $c_n[\leq k, P]$, $b_n[\leq k, P]$ and $q_n[\leq k, P]$, respectively. The following lemma is obvious since, for any proper pattern P, there is a bridge containing P at least once. Theorem 7.12 is a corollary of lemma 7.11.

Lemma 7.11 *If P is a proper pattern, then for any $k \in \mathbb{N}$, and $t \leq t_c$,*

$$B(t; [\leq k, P]) < B(t), \quad \textit{and}\, Q(t; [\leq k, P]) < Q(t).$$

In particular, $Q(t_c; [\leq k, P]) < 1$, and $B(t_c; [\leq k, P]) < \infty$. □

Theorem 7.12 *If P is a proper pattern, then there exist an $\epsilon > 0$ and an $N_\epsilon \in \mathbb{N}$ such that*

$$c_n[0, P] < \mu_d^n (1 - \epsilon)^n$$

for all $n \geq N_\epsilon$.

Proof It may be assumed, without loss of generality, that the pattern P is a prime bridge (see theorem C.3 in appendix C). Define the generating functions of walks, bridges and prime bridges which do not contain P by $C(t; [0, P])$, $B(t; [0, P])$, and $Q(t; [0, P])$, respectively (see section (7.1.5)).

Since P is a prime bridge, an occurrence of P cannot be created by concatenating bridges in the arguments leading to equations (7.15) and (7.16). This shows that

$$B(t; [0, P]) = \frac{1}{1 - Q(t; [0, P])}.$$

It follows mutatus mutandis from theorem 7.9 that there is a critical point $t = t_P$ such that

$$\lim_{t \to t_P^-} C(t; [0, P]) = \lim_{t \to t_P^-} B(t; [0, P]) = \infty,$$

where t_P is the radius of convergence of $C(t; [0, P])$ and $B(t; [0, P])$. Thus, it is the case that $\lim_{t \to t_P^-} Q(t; [0, P]) = Q(t_P; [0, P]) = 1$.

Since $Q(t) > Q(t; [0, P])$ for all $0 \leq t \leq t_c$, it follows that $Q(t_c; [0, P]) < 1$. This shows that $t_P > t_c = \frac{1}{\mu_d}$, since $Q(t)$ and $Q(t; [0, P])$ are continuous functions within their radii of convergence. Since $C(t; [0, P])$ is the generating function of $c_n[0, P]$, it follows that $\limsup_{n \to \infty} (c_n[0, P])^{1/n} = \frac{1}{t_P} < \mu_d$. Thus, there is an $\epsilon > 0$ such that $\limsup_{n \to \infty} (c_n[0, P])^{1/n} < \mu_d(1 - \epsilon)$. This shows that there exists an $N_\epsilon \in \mathbb{N}$ such that, for all $n \geq N_\epsilon$, $c_n[0, P] < (\mu_d(1 - \epsilon))^n$. This completes the proof. □

Similar to the above, define the functions $p_n[\leq k, P]$, $\ell_n[\leq k, P]$, $\ell_n^o[\leq k, P]$, $b_n[\leq k, P]$, $h_n[\leq k, P]$, $c_n^t[\leq k, P]$, $c_n^+[\leq k, P]$, $c_n^\dagger[\leq k, P]$ or $c_n^\ddagger[\leq k, P]$ (where p_n, ℓ_n,

ℓ_n^o, b_n, h_n, c_n^t, c_n^+, $c_n^\dagger$ and $c_n^\ddagger$ are as defined before). Define $c_n[\leq k, P](\vec{v})$ to be the number of self-avoiding walks from $\vec{0}$ to $\vec{v}$ on which the pattern P occurs no more than k times. Then

$$\begin{aligned} b_n[\leq \epsilon n, P] \leq h_n[\leq \epsilon n, P] \leq c_n^+[\leq \epsilon n, P] &\leq c_n[\leq \epsilon n, P], \\ \ell_n[\leq \epsilon n, P] \leq \ell_n^o[\leq \epsilon n, P] &\leq c_n[\leq \epsilon n, P], \\ c_n[\leq \epsilon n, P](\vec{v}) &\leq c_n[\leq \epsilon n, P], \\ c_n^\ddagger[\leq \epsilon n, P] \leq c_n^t[\leq \epsilon n, P] &\leq c_n[\leq \epsilon n, P]. \end{aligned} \tag{7.17}$$

The pattern theorem follows from these inequalities and the last theorem.

Theorem 7.13. (The pattern theorem for walks) *If P is a proper pattern, then there exists an $\epsilon_0 > 0$ such that, for all non-negative $\epsilon < \epsilon_0$,*

$$\limsup_{n\to\infty} \left(c_n[\leq \epsilon n, P]\right)^{1/n} < \mu_d.$$

Similarly, for polygons, there exists an $\epsilon_1 > 0$ such that, for all non-negative $\epsilon < \epsilon_1$,

$$\limsup_{n\to\infty} \left(p_n[\leq \epsilon n, P]\right)^{1/n} < \mu_d.$$

It follows from equation (7.17) that, for all $\epsilon < \epsilon_0$, $\limsup_{n\to\infty} \left(\psi_n[\leq \epsilon n, P]\right)^{1/n} < \mu_d$, where ψ_n is any one of ℓ_n, ℓ_n^o, b_n, h_n, c_n^t, c_n^+, $c_n^\dagger$, $c_n^\ddagger$ or $c_n(\vec{v})$.

Proof The proofs for walks and polygons are similar.

The proofs for $\psi_n \in \{\ell_n, \ell_n^o, b_n, h_n, c_n^\dagger, c_n^\ddagger, c_n^t, c_n(\vec{v})\}$ follow because these quantities are less than or equal to c_n.

By theorems 1.1 and 7.12, there exist an $\epsilon > 0$ and an $N_\epsilon \in \mathbb{N}$ such that, for all $n \geq N_\epsilon$,

$$c_n[0, P] < (\mu_d(1-\epsilon))^n, \quad \text{and } c_n < (\mu_d(1+\epsilon))^n.$$

Let ω be a walk of length n on which P occurs no more than $\lfloor \epsilon n \rfloor$ times. By starting at the origin, colour subwalks of length m in ω by $1, 2, 3, \ldots$ where $m \geq |P|$ and let $M = \lfloor \frac{n}{m} \rfloor$ be the number of colours used.

The pattern P will occur on at most $\lfloor \epsilon n \rfloor$ monochromatic subwalks.

Let $r = n - mM$ and suppose that $\lfloor \epsilon n \rfloor \leq \frac{M}{2}$. Then

$$\begin{aligned} c_n[\leq \epsilon n, P] &\leq \sum_{j=0}^{\lfloor \epsilon n \rfloor} \binom{M}{j} (c_m)^j (c_m[0, P])^{M-j} c_r \\ &\leq (\lfloor \epsilon n \rfloor + 1) \binom{M}{\lfloor \epsilon n \rfloor} \mu_d^{Mm} c_r (1+\epsilon)^{m\lfloor \epsilon n \rfloor} (1-\epsilon)^{mM - m\lfloor \epsilon n \rfloor}. \end{aligned}$$

Take the power $\frac{1}{n}$ and let $n \to \infty$. This shows that

$$\limsup_{n\to\infty} \left(c_n[\leq \epsilon n, P]\right)^{1/n} \leq \left(\frac{\mu_d(1-\epsilon)}{(\epsilon m)^\epsilon (1-\epsilon m)^{1/m - \epsilon}} \right) \frac{(1+\epsilon)^{\epsilon m}}{(1-\epsilon)^{\epsilon m}} < \mu_d$$

for $\epsilon > 0$ small enough. □

If P is a prime bridge pattern, then P cannot be created by concatenating walks or polygons. In addition, P cannot be created in the Hammersley-Welsh construction, since cutting half-space walks into bridges cannot create prime bridge patterns. Consequently, the entire set of arguments leading to theorem 7.7 can be used on walks and polygons with at most $\lfloor \epsilon n \rfloor$ occurrences of P. For example, by concatenating polygons (see figure 1.5), it follows that

$$p_n[\leq \epsilon n, P]\, p_m[\leq \epsilon m, P] \leq (d-1)\, p_{n+m}[\leq \epsilon(n+m), P].$$

Therefore, the limit

$$\mu_\epsilon = \lim_{n\to\infty} \left(p_n[\leq \epsilon n, P]\right)^{1/n} \tag{7.18}$$

exists. Similar arguments can be made for loops and bridges. For walks and half-space walks, the methods of theorem 7.7 may be used.

Concatenating polygons as in illustrated in figure 1.5 shows that if P is a proper pattern, then

$$p_{n_1}[m_1, P]\, p_{n_2}[m_2, P] \leq (d-1) \sum_{j=0}^{2} p_{n_1+n_2}[m_1+m_2+j, P]. \tag{7.19}$$

This follows because at most two new copies of the proper pattern P may be created when the polygons are concatenated. Put $\widetilde{p}_n[m, P] = \frac{1}{d-1} p_n[m, P]$ and compare this to equation (3.70). This shows that $\widetilde{p}_n[m, P]$ satisfies the supermultiplicative relation of section (3.4). By corollary 3.14, there are the following integrated density functions:

$$\mathcal{P}_P(\leq \epsilon) = \lim_{n\to\infty} \left(p_n[\leq \epsilon n, P]\right)^{1/n}, \quad \text{and } \mathcal{P}_P(\geq \epsilon) = \lim_{n\to\infty} \left(p_n[\geq \epsilon n, P]\right)^{1/n}. \tag{7.20}$$

By theorem 3.17, there is a sequence $\langle \delta_n \rangle$ such that $\delta_n = \lfloor \epsilon n \rfloor + o(n)$, and

$$\log \mathcal{P}_P(\epsilon) = \min\{\log \mathcal{P}_P(\leq \epsilon), \log \mathcal{P}_P(\geq \epsilon)\} = \lim_{n\to\infty} \tfrac{1}{n} \log p_n[\delta_n, P]. \tag{7.21}$$

A similar approach may be followed for bridges.

The following corollary is true not only for c_n but also for polygons, bridges, loops and unfolded walks, amongst the models considered above. Let ϵ_0 be as it is defined in corollary 7.13.

Corollary 7.14 *Let P be a proper pattern and suppose that $0 < \epsilon < \epsilon_0$. Then there exist a $k > 0$ (possibly dependent on P and ϵ) and an $N_\epsilon \in \mathbb{N}$ such that*

$$c_n[\leq \epsilon n, P] < e^{-kn} c_n$$

for all $n \geq N_\epsilon$. This is similarly the case if c_n is replaced by p_n, ℓ_n, ℓ_n^o, b_n, h_n, c_n^t, c_n^+, $c_n^\dagger$, $c_n^\ddagger$ or $c_n(\vec{v})$.

Proof By theorem 7.13,

$$\limsup_{n\to\infty} \tfrac{1}{n} \log\Big(\tfrac{1}{c_n} \left(c_n[\leq \epsilon n, P]\right)\Big) = \log \tfrac{\mu_\epsilon}{\mu} < 0.$$

Put $k = \log \frac{\mu_d}{\mu_\epsilon} > 0$. By the definition of the limit, there is an $N_\epsilon \in \mathbb{N}$ such that

$$\tfrac{1}{n} \log\Big(\tfrac{1}{c_n} \left(c_n[\leq \epsilon n, P]\right)\Big) < -k.$$

This completes the proof. □

7.1.7 *Limit ratio theorems*

Limit ratio theorems for walks and polygons (see equations (1.3) and (1.8)) are consequences of the pattern theorem. Define the *cube* Q with bottom corner $\vec{a} = (a_1, a_2, \ldots, a_d)$ and with side-length $b \in \mathbb{N}$ by

$$Q = \{\vec{x} \in \mathbb{Z}^d \mid \text{for } a_i \leq x_i \leq a_i + b, \text{ and for all } i = 1, 2, \ldots, d\}. \tag{7.22}$$

If $P = \langle p_0, p_1, \ldots, p_N\rangle$ is proper pattern, then by theorem C.3 in appendix C there exist a cube Q and a self-avoiding walk $\phi = \langle \phi_0, \phi_1, \ldots, \phi_n\rangle$ such that $\phi \subseteq Q$, the endpoints of ϕ are opposite corners of Q, and P occurs at least once in ϕ.

Let $\omega = \langle \omega_0, \omega_1, \ldots, \omega_n\rangle$ be a self-avoiding walk in $\mathbb{L}^d$. The *event* (P, Q) occurs at ω_j in ω if (1) there exist a vector $\vec{v} \in \mathbb{Z}^d$ and a j such that $p_k + \vec{v} = \omega_{j+k}$ for $k = 0, 1, \ldots, N$, and (2) $\omega \cap Q = P + \vec{v}$.

In other words, the self-avoiding walk ω passes through Q exactly once, tracing out the pattern P, and then avoids Q otherwise. The pair (P, Q) is a *proper pattern-cube pair.*

The proof of the following lemma is mutatis mutandis to the proofs of theorems 7.12 and 7.13.

Lemma 7.15 *If (P, Q) is a proper pattern-cube pair, then*

- *there exist an $\epsilon > 0$ and an $N_\epsilon \in \mathbb{N}$ such that*

$$c_n[0, (P,Q)] < (\mu_d(1-\epsilon))^n \text{ for all } n \geq N_\epsilon;$$

- *there exists an $\epsilon_0 > 0$ such that*

$$\limsup_{n\to\infty} \left(c_n[\leq \epsilon n, (P,Q)]\right)^{1/n} < \mu_d \text{ for all non-negative } \epsilon < \epsilon_0;$$

- *there exist an $\epsilon > 0$ and an $N_\epsilon \in \mathbb{N}$ such that*

$$p_n[0, (P,Q)] < (\mu_d(1-\epsilon))^n \text{ for all } n \geq N_\epsilon;$$

- *there exists an $\epsilon_0 > 0$ such that*

$$\limsup_{n\to\infty} \left(p_n[\leq \epsilon n, (P,Q)]\right)^{1/n} < \mu_d \text{ for all non-negative } \epsilon < \epsilon_0.$$

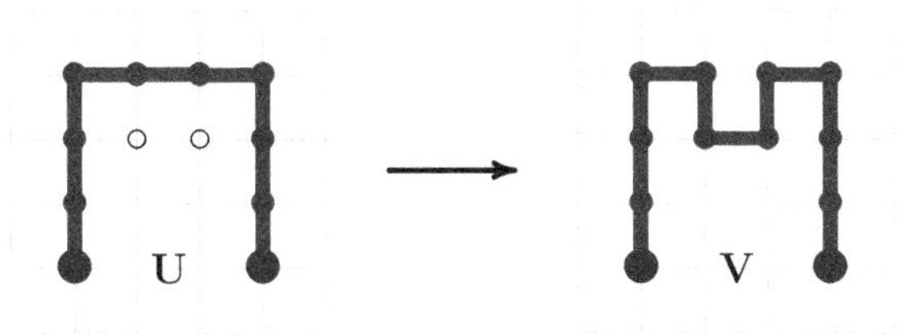

FIG. 7.8. The pattern U on the left can be changed to V to increase the length of a walk by 2.

By replacing P by (P,Q) in the inequalities in equation (7.17), it follows that

$$\limsup_{n\to\infty} \left(\psi_n[\leq \epsilon n,(P,Q)]\right)^{1/n} < \mu_d,$$

where $\psi_n \in \{p_n, \ell_n, \ell_n^o, b_n, h_n, c_n^t, c_n^+, c_n^\dagger, c_n^\ddagger, c_n(\vec{v})\}$. □

The pattern U in figure 7.8 is a proper pattern which can be changed into the pattern V as illustrated. By making such changes and by carefully accounting the outcome, lemma 7.16 follows. The limit ratio theorem is a corollary of lemma 7.16.

Lemma 7.16 *Define ϕ_n to be any of $\frac{c_{n+2}}{c_n}$, $\frac{p_{n+2}}{p_n}$ (for even n), $\frac{c_{n+2}(\vec{v})}{c_n(\vec{v})}$ (where n has the same parity as $\|\vec{v}\|_1$), $\frac{\ell_{n+2}}{\ell_n}$, $\frac{\ell^o_{n+2}}{\ell^o_n}$, $\frac{b_{n+2}}{b_n}$, $\frac{h_{n+2}}{h_n}$, $\frac{c^t_{n+2}}{c^t_n}$, $\frac{c^+_{n+2}}{c^+_n}$, $\frac{c^\dagger_{n+2}}{c^\dagger_n}$ or $\frac{c^\ddagger_{n+2}}{c^\ddagger_n}$. Then there exists a constant $D > 0$ such that*

$$\phi_n \phi_{n+2} - \phi_n^2 \geq -\tfrac{1}{n} D$$

for sufficiently large n.

Proof Let Q be a cube of dimension 3^d and consider the two patterns U and V in figure 7.8. Both U and V occur in Q with their endpoints on two corners of Q. Denote the proper pattern-cube pairs $U \equiv (U,Q)$ and $V \equiv (V,Q)$ such that only the vertices of U and V inside Q are occupied, and none other.

The two vertices marked by $\circ$ in U are not occupied by the rest of the walk. An occurrence of U can be replaced freely by V in a given walk. Similarly, V may be replaced freely by U.

First consider the proof for walks (that is, for c_n).

For $n > 0$, $i > 0$, and $j > 0$, let $\mathcal{C}_n\{i,j\}$ be the set of walks from $\vec{0}$, of length n and on which (U,Q) occurs at exactly i vertices, and (V,Q) occurs at exactly j vertices. Notice that these occurrences are necessarily disjoint.

Denote the cardinality of $\mathcal{C}_n\{i,j\}$ by $c_n\{i,j\}$. Similarly, define $c_n\{\geq i, \geq j\}$ to be the number of walks of length n from $\vec{0}$ on which (U,Q) and (V,Q) occur at least i and j times, respectively.

CLAIM: $ic_n\{i,j\} = (j+1)c_{n+2}\{i-1,j+1\}$.
PROOF OF CLAIM: Consider the pairs of walks (ω, ω') such that $\omega \in \mathcal{C}_n\{i,j\}$ and ω' can be obtained from ω by changing an occurrence of (U,Q) to (V,Q). In particular, $\omega' \in \mathcal{C}_{n+2}\{i-1,j+1\}$.

There are i places in ω where an ω' can be created. Thus, there are $ic_n\{i,j\}$ pairs (ω, ω').

On the other hand, there are $j+1$ places in ω' where an ω can be recovered. Thus, there are $(j+1)c_{n+2}\{i-1,j+1\}$ pairs (ω,ω'). This proves the claim $\triangle$

A consequence of the claim above is that

$$c_{n+2}\{\geq 0, \geq 1\} = \sum_{\substack{i\geq 1\\ j\geq 0}} c_{n+2}\{i-1,j+1\} = \sum_{\substack{i\geq 1\\ j\geq 0}} \frac{i}{j+1} c_n\{i,j\}.$$

Square the left-hand side and use the Schwarz inequality to see that

$$(c_{n+2}\{\geq 0, \geq 1\})^2 \leq \sum_{i\geq 1}\sum_{j\geq 0}\left(\frac{i}{j+1}\right)^2 c_n\{i,j\} \sum_{k\geq 1}\sum_{\ell\geq 0} c_n\{k,\ell\}. \tag{7.23}$$

In addition,

$$\begin{aligned} c_{n+4}\{\geq 0, \geq 2\} &= \sum_{i\geq 2}\sum_{j\geq 0} c_{n+4}\{i-2,j+2\} \\ &= \sum_{i\geq 2}\sum_{j\geq 0} \frac{i(i-1)}{(j+1)(j+2)} c_n\{i,j\}. \end{aligned} \tag{7.24}$$

Define

$$Z_n = \tfrac{1}{c_n} c_{n+4}\{\geq 0, \geq 2\} - \tfrac{1}{c_n^2} c_{n+2}^2\{\geq 0, \geq 1\}.$$

Let $\phi_n = \frac{c_{n+2}}{c_n}$ and put $\mathcal{E}_n = \phi_n\phi_{n+2} - \phi_n^2 - Z_n$.

CLAIM: The function $\mathcal{E}_n \to 0$ exponentially fast with increasing n.
PROOF OF CLAIM: Use the definition of ϕ_n and the triangle inequality:

$$|\mathcal{E}_n| \leq \left|\tfrac{1}{c_n}\left(c_{n+4} - c_{n+4}\{\geq 0, \geq 2\}\right)\right| + \left|\tfrac{1}{c_n^2}\left(c_{n+2}^2 - c_{n+2}^2\{\geq 0, \geq 1\}\right)^2\right|.$$

This shows that $|\mathcal{E}_n| \leq \frac{1}{c_n} c_{n+2}\{1,(V,Q)\} + \frac{1}{c_n^2} c_{n+2}c_{n+2}\{0,(V,Q)\}$. By lemma 7.15, both terms on the right-hand side decay to 0 exponentially fast with n. $\triangle$

It remains to find a lower bound of the form $-\frac{C}{n}$ on Z_n, for some constant C and n large enough. By equations (7.23) and (7.24),

$$\begin{aligned} Z_n c_n &\geq \sum_{i\geq 0}\sum_{j\geq 0} \frac{i(i-1)}{(j+1)(j+2)} c_n\{i,j\} - \sum_{i\geq 0}\sum_{j\geq 0} \frac{i^2}{(j+1)^2} c_n\{i,j\} \\ &= \sum_{i\geq 0}\sum_{j\geq 0} \frac{(-i^2 - ij - i)}{(j+1)^2(j+2)} c_n\{i,j\}. \end{aligned}$$

However, $c_n\{i,j\} = 0$ if $i > n$ or $j > n$ – so use the bound $(-i^2 - ij - i) \geq -3n^2$. Split the summation over j into $0 \leq j < \epsilon n$, and $\epsilon n \leq j \leq n$, for some small $\epsilon > 0$. This shows that

$$Z_n \geq \tfrac{1}{c_n}\left(-3n^2(c_n - c_n\{\geq 0, \geq \epsilon n\}\right) + \tfrac{1}{(\epsilon n)^3 c_n}\left(-3n^2 c_n\{\geq 0, \geq \epsilon n\}\right).$$

The first term decays to 0 exponentially fast by lemma 7.15, and the second term is asymptotic to $-3/(\epsilon^3 n)$. This completes the proof. □

If Kesten's limit ratio theorem in appendix A (see theorem A.5) is used, then a corollary of lemma 7.16 is the following ratio limit theorem (see reference [399] for more details).

Theorem 7.17 (Limit ratio theorems) *If $\vec{v} \neq \vec{0}$, then the following limits exist:*

(a) $\lim_{n\to\infty} \frac{c_{n+2}}{c_n} = \mu_d^2$;
(b) $\lim_{n\to\infty} \frac{p_{n+2}}{p_n} = \mu_d^2$, *taken through even n;*
(c) $\lim_{n\to\infty} \frac{c_{n+2}(\vec{v})}{c_n(\vec{v})} = \mu_d^2$, *taken through n with the same parity as $\|\vec{v}\|_1$;*
(d) $\lim_{n\to\infty} \frac{b_{n+1}}{b_n} = \mu_d$; *and*
(e) $\lim_{n\to\infty} \frac{\psi_{n+2}}{\psi_n} = \mu_d^2$,

for any $\psi_n \in \{h_n, \ell_n, \ell_n^o, n_n^t, c_n^+, c_n^\dagger, c_n^\ddagger\}$.

Proof Observe that $c_{n+2} \geq c_n$, and similarly for all the other cases (use lemma 1.7 for $c_n(\vec{v})$). Hence, all these results, except for (d), follow from theorem A.5 and lemma 7.16.

It remains to prove (d). Since $b_{n+2} \geq b_n$,

$$\lim_{n\to\infty} \frac{b_{n+2}}{b_n} = \mu_d^2. \tag{7.25}$$

Proceed as in Madras and Slade [399] by defining $L_j = \liminf_{n\to\infty} \frac{b_{n-j}}{b_n}$. It remains to show that $L_1 = \frac{1}{\mu_d}$ and that this limit inferior is a limit.

By equation (7.25), $L_{j+2} = \mu_d^{-2} L_j$ for every $j \geq 1$. Thus, $L_j = \mu_d^{-j}$ for even values of j, and $L_j = \mu_d^{1-j} L_1$ for odd values of j. By equation (7.15),

$$\frac{b_{n-j}}{b_n} = \sum_{m=1}^{n-j} \frac{q_m b_{n-j-m}}{b_n}.$$

Fatou's lemma gives

$$L_j \geq \sum_{m=1}^{\infty} q_m L_{j+m} \tag{7.26}$$

Define the odd and even sums: $S_o = \sum_{\substack{m \geq 1 \\ m \text{ odd}}} q_m \mu_d^{-m}$, and $S_e = \sum_{\substack{m \geq 1 \\ m \text{ even}}} q_m \mu_d^{-m}$. By corollary 7.10, $S_o + S_e = 1$. By equation (7.26) with $j = 0$,

$$1 \geq L_1 \mu_d S_o + S_e.$$

This shows that $L_1 \leq \mu_d^{-1}$. Using equation (7.26) again with $j = 1$,

$$L_1 \geq \mu_d^{-1} S_o + L_1 S_e.$$

This shows that $L_1 \geq \mu_d^{-1}$. Thus, $L_1 = \mu_d^{-1}$, and $\limsup_{n\to\infty} \frac{b_{n+1}}{b_n} = \mu_d$. However, from equation (7.25),

$$\liminf_{n\to\infty} \frac{b_{n+1}}{b_n} = \liminf_{n\to\infty} \frac{b_{n+1}}{b_{n-1}} \frac{b_{n-1}}{b_n} = \mu_d^2 L_1 = \mu_d,$$

as required. □

7.2 Patterns in interacting models of walks and polygons

The Hammersley-Welsh decomposition of a half-space walk ω into bridges was used in section 7.1.6 to prove a pattern theorem for walks and polygons. In an interacting model this decomposition may change the energy of the walk, giving rise to complications in the proof of a pattern theorem. Thus, the discussion is limited to models with an energy which is additive over concatenation and decomposition: if two walks ω_1 and ω_2 of energies m_1 and m_2, respectively, are concatenated to obtain a walk $\omega = \omega_1\omega_2$, then the energy of ω is $m_1 + m_2$. Similarly, if a walk ω of energy m is decomposed into two subwalks ω_1 and ω_2 of energies m_1 and m_2, respectively, by cutting it in a vertex, then $m = m_1 + m_2$.

Consider an interacting model where $c_n(m)$ is the number of self-avoiding walks in $\mathbb{L}^d$ from $\vec{0}$ of length n and with energy m, and $c_n^h(m)$ is the number of half-space walks similarly defined. Define the partition functions

$$Z_n(z) = \sum_{m\geq 0} c_n(m)\, z^m, \quad \text{and } Z_n^h(z) = \sum_{m\geq 0} c_n^h(m)\, z^m \tag{7.27}$$

of walks and half-space walks similarly. The partition function of bridges of length n is defined in the same way and denoted by $Z_n^b(z)$.

Suppose that the half-space walk ω with energy m is decomposed into a set of bridges $(\phi_1, \phi_2, \ldots, \phi_N)$ such that bridge ϕ_j has (1) energy $\vec{e}_2$, (2) span in the $\vec{e}_1$ direction $S_1(\phi_j)$, and (3) length n_j. Assume that $\sum_j \vec{e}_2 = m$ (the energies of the bridges add up to the energy of ω).

Let $h_n(m)$ be the number of half-space walks of energy m and length n from the origin. Denote the number of bridges ϕ of length n, span $s = S_1(\phi)$ and energy m by $b_{n,s}(m)$ and let $Z_{n,s}^b(z)$ be the partition function of this class of bridges.

There are at most $b_{n_j,s_j}(\vec{e}_2)$ choices for ϕ_j in the bridge decomposition of ω if ϕ_j has length n_j, energy $\vec{e}_2$ and span $s_j = S_1(\phi_j)$. Similar to equation (7.5), this bridge decomposition of half-space walks gives

$$h_n(m) \leq \sum \prod_{j=1}^{N} b_{n_j,s_j}(\vec{e}_2), \tag{7.28}$$

where the summation is over all $\{n_j\}$, $\{s_j\}$, $\{\vec{e}_2\}$ and N, constrained such that energies add up to m, lengths to n, spans to $s \leq n$, and $N \leq n$. Multiplying this by z^m and summing over m gives

$$Z_n^h(z) \leq \sum \prod_{j=1}^{N} Z_{n_j,s_j}^b(z), \tag{7.29}$$

where the remaining summation is over all $\{n_j\}$, $\{s_j\}$, and N and where $Z_{n,s}^b(z)$ is the partition function of bridges of length n and span s in the $\vec{e}_1$ direction.

7.2.1 *Generating functions of interacting walks*

The assumption that the energy of a model of walks is additive under concatenation and decomposition implies that the results of sections 7.1.1, 7.1.2, 7.1.3 and 7.1.4 generalise to interacting models (but with $\{p_n, \ell_n, b_n, h_n, c_n(\vec{v}), c_n^+, c_n^\ddagger, c_n\}$ replaced by the corresponding partition functions). A consequence of corollaries 7.3 and 7.5 and of theorems 7.6 and 7.8 is the following theorem.

Theorem 7.18 *Let $p_n(m)$, $\ell_n(m)$, $b_n(m)$, $h_n(m)$, $c_n(m;\vec{v})$, $c_n^+(m)$, $c_n^\ddagger(m)$ and $c_n(m)$ be the number of polygons, loops, bridges, half-space walks, walks from $\vec{0}$ to $\vec{v}$, positive walks, doubly unfolded walks or self-avoiding walks, respectively, of energy m and length n.*

Let $Z_n^a(z)$ be the partition function of any of these models, and be defined by

$$Z_n^a(z) = \sum_{m\geq 0} a_n(m)\, z^n,$$

where a is any one of $\{p, \ell, b, h, c(\vec{v}), c^+, c^\ddagger, c\}$. Then the limiting free energy is given by

$$\mathcal{F}(z) = \lim_{n\to\infty} \tfrac{1}{n} \log Z_n^a(z)$$

and is independent of the choice of a. Existence of the limit is guaranteed by the supermultiplicativity of the bridge partition function:

$$Z_n^b(z)\, Z_m^b(z) \leq Z_{n+m}^b(z)$$

(which follows from the concatenation of the bridges, as in figure 7.3 and from equation (7.2)). □

The generating functions of these partition functions are defined in the usual way. For example, the bridge generating function is

$$B_z(t) = \sum_{n=0}^{\infty} Z_n^b(z) t^n. \tag{7.30}$$

Similarly, the half-space walk, polygon, loop, and self-avoiding walk generating functions can be defined.

By the construction in figure 1.3, a self-avoiding walk can be decomposed into two subwalks by cutting it in a chosen vertex. If a walk of length n_1+n_2 and energy m is cut in its n_1-th vertex into subwalks of energy m_1 and energy $m-m_1$, then there are at most $c_{n_1}(m_1)$ choices for the first subwalk, and at most $c_{n_2}(m-m_1)$ choices for the second subwalk. This shows that $c_{n_1+n_2}(m) \leq \sum_{m_1=0}^{m} c_{n_1}(m_1) c_{n_2}(m-m_1)$. Multiplying by z^m and summing over m gives in terms of partition functions of self-avoiding walks that

$$Z_{n_1}^c(z) Z_{n_2}^c(z) \geq Z_{n_1+n_2}(z). \tag{7.31}$$

In other words, $\log Z_n^c(z)$ is a subadditive function, and, by theorem A.1 in appendix A, this also proves the existence of a limiting free energy (see theorem 7.18). However, there is an additional consequence of equation (7.31) for $Z_n^c(z)$, namely, that

$$\mathcal{F}(z) = \lim_{n\to\infty} \tfrac{1}{n} \log Z_n^c(z) = \inf_{n>0} \tfrac{1}{n} \log Z_n^c(z). \tag{7.32}$$

This shows that

$$Z_n^c(z) \geq e^{n\mathcal{F}(z)} = \mu_z^n, \tag{7.33}$$

where $\mu_z = e^{\mathcal{F}(z)}$. That is, the self-avoiding walk generating function has the lower bound

$$C_z(t) \geq \frac{1}{1-t\mu_z}. \tag{7.34}$$

This gives the following theorem.

Theorem 7.19 *The radius of convergence of $C_z(t)$ is $t_c(z) = \mu_z^{-1}$. Moreover, $\lim_{t\nearrow\mu_z^{-1}} C_z(t) = \infty$.* □

The proof of a pattern theorem follows the basic outline for the non-interacting case. The bridge generating function is shown to be divergent, and it is related to the prime bridge generating function.

7.2.2 *The bridge generating function*

The number of bridges of length n and energy m is denoted $b_n(m)$ and has the partition function and generating function

$$Z_n^b(z) = \sum_{m\geq 0} b_n(m) z^m, \quad \text{and } B_z(t) = \sum_{n\geq 0} Z_n^b(z) t^n, \quad \text{respectively.} \tag{7.35}$$

Equation (7.29) gives. after multiplication by t^n and summing over n,

$$H_z(t) \leq \prod_{s=1}^{\infty} \left(1 + \sum_{m=1}^{\infty} Z_{m,s}(z) t^m \right) \leq e^{\sum_{s=1}^{\infty} \sum_{m=1}^{\infty} Z_{m,s}^b(z) t^m} = e^{B_z(t)-1}, \quad (7.36)$$

where $H_z(t)$ is the half-space walk generating function. If $H_z(t)$ is divergent as $t \nearrow \mu_z^{-1}$, then $B_z(t)$ is also divergent at μ_z^{-1}.

Theorem 7.20 *The generating functions $C_z(t)$, $H_z(t)$ and $B_z(t)$ have the critical point and radius of convergence $t_c = \mu_z^{-1}$ (where $\mu_z = e^{\mathcal{F}(z)}$). Moreover,*

$$\lim_{t \to t_c^-} C_z(t) = \lim_{t \to t_c^-} H_z(t) = \lim_{t \to t_c^-} B_z(t) = \infty.$$

The bridge generating function has the lower bound

$$B_z(t) \geq 1 + \tfrac{1}{2} \log \left(\frac{t}{1 - \mu_z t} \right).$$

Proof By equation (7.36), it is enough to show that $H_z(t)$ is divergent.

By equation (7.3), a self-avoiding walk can be decomposed into two half-space walks. Accounting for the energy in this decomposition gives

$$c_n(m) \leq \sum_{n_1=0}^{n} \sum_{m_1=0}^{m} h_{n_1+1}(m_1) h_{n-n_1}(m - m_1).$$

Multiply by z^m and sum over m to obtain

$$Z_n^c(z) \leq \sum_{n_1=0}^{n} Z_{n_1+1}^h(z) Z_{n-n_1}^h(z).$$

Multiplying by t^{n+1} and summing over n give

$$t C_z(t) \leq (H_z(t))^2.$$

By theorem 7.19, $C_z(t)$ approaches infinity as $t \nearrow \mu_z^{-1}$. This shows that $H_z(t)$ and (by virtue of equation (7.36)) $B_z(t)$ diverge at their dominant singularities on the real axis. The lower bound on $B_z(t)$ is obtained in the same way as the bound on $B(t)$ was obtained in theorem 7.9. □

7.2.3 *A pattern theorem for interacting models*

The results in theorem 7.20 make it possible to prove a pattern theorem for interacting models in a similar way as for walks in section 7.1.6. The proof is almost unchanged, except that the corresponding partition functions replace the functions $\{b_n, q_n, p_n, c_n\}$.

Patterns, proper patterns, prime patterns, prime bridges and prime bridge patterns are defined as in section 7.1.6. Let $q_n(m)$ be the number of prime bridges of length n and with energy m. The prime bridge partition function is defined by

$$Z_n^q(z) = \sum_{m\geq 0} q_n(m)\, z^m \tag{7.37}$$

and its generating function is $Q_z(t)$.

If ϕ is a bridge, then it is either a prime bridge or it may be decomposed into prime bridges (see equation (7.15)). If ϕ has energy m, then this shows that

$$b_n(m) = \sum_{n_1=1}^{n} \sum_{m_1=0}^{m} q_{n_1}(m_1)\, b_{n-n_1}(m-m_1). \tag{7.38}$$

Multiply by z^m and sum over m. This gives a renewal equation for the partition functions:

$$Z_n^b(z) = \sum_{n_1=1}^{n} Z_{n_1}^q(z)\, Z_{n-n_1}^b(z), \tag{7.39}$$

where $Z_0^q(z) = 0$, and $Z_0^b(z) = 1$. Multiplying by t^n and summing over n, this becomes

$$B_z(t) - 1 = Q_z(t)\, B_z(t), \quad \text{with solution } B_z(t) = \frac{1}{1-Q_z(t)}. \tag{7.40}$$

Let $t_c = \mu_z^{-1}$. By theorem 7.20, this gives the following theorem (the proof is similar to that of corollary 7.10).

Theorem 7.21 *The left-limits*

$$\lim_{t\nearrow \mu_z^{-1}} B_z(t) = \infty, \quad \textit{and} \ \lim_{t\nearrow t_c} Q_z(t) = Q_z(t_c) = 1. \qquad \square$$

Denote by $B_z(t; [\leq k, P])$ and $Q_z(t; [\leq k, P])$ the generating functions for bridges and prime bridges, respectively, on which the proper pattern P occurs at most k times. Theorem 7.21 gives the following lemma.

Lemma 7.22 *If P is a proper pattern, then, for any $k \in \mathbb{N}$, and $t \leq \mu_z^{-1}$,*

$$B_z(t; [\leq k, P]) < B_z(t), \quad \textit{and}\ Q_z(t; [\leq k, P]) < Q_z(t).$$

In particular, $Q_z(t_c; [\leq k, P]) < 1$, and $B_z(t_c; [\leq k, P]) < \infty$. $\square$

Let $Z_n^c(z; [k, P])$ be the partition function of walks which contain the proper pattern P exactly k times. Then lemma 7.22 may be used to prove theorem 7.23. The proof is similar to that of theorem 7.12.

Theorem 7.23 *If P is a proper pattern, then there exist an $\epsilon > 0$ and an $N_\epsilon \in \mathbb{N}$ such that*

$$Z_n^c(z; [0, P]) < (\mu_z(1-\epsilon))^n \quad \textit{for all } n \geq N_\epsilon. \qquad \square$$

It remains to prove the equivalent of theorem 7.13 for the interacting models in this section. As in theorem 7.18, let $Z_n^a(z)$ be the partition function of an interacting model, where a is any one of $\{\ell, \ell^o, b, h, c^t, c^+, c^\dagger, c^\ddagger, c(\vec{v}), p, c\}$. In addition, let $Z_n^a(z; [\leq k, P])$ be the partition function of a model which contains the pattern P at most k times. Then the pattern theorem is proven similarly to the proof of theorem 7.13.

Theorem 7.24 *Suppose that $a \in \{\ell, \ell^o, b, h, c^t, c^+, c^\dagger, c^\ddagger, c(\vec{v}), p, c\}$ and let P be a proper pattern. Then there exists an $\epsilon_0 > 0$ (dependent on a) such that, for all non-negative $\epsilon < \epsilon_0$,*

$$\limsup_{n\to\infty} (Z_n^a(z; [\leq \epsilon n, P]))^{1/n} < \mu_z = e^{\mathcal{F}(z)}. \qquad \square$$

Corollary 7.25 *Suppose that $a \in \{\ell, \ell^o, b, h, c^t, c^+, c^\dagger, c^\ddagger, c(\vec{v}), p, c\}$, let P be a proper pattern and let $0 < \epsilon < \epsilon_0$ be as defined in theorem 7.24. Then there exist a $k_z > 0$ and an $N_\epsilon \in \mathbb{N}$ such that*

$$Z_n^a(z; [\leq \epsilon n, P]) < e^{-k_z n} Z_n^a(z) \quad \textit{for all } n \geq N_\epsilon. \qquad \square$$

7.3 Patterns, curvature and knotting in stiff lattice polygons

A lattice polygon with a bending energy is a model of a semi flexible ring polymer [203, 204]. The bending energy is proportional to the polygon's total curvature, given by $s\frac{\pi}{2}$, where s is the total number right angles between adjacent edges along the polygon.

Define $p_n(s)$ to be the number of lattice polygons in $\mathbb{L}^3$ and of length n with s right angles between adjacent edges. Concatenating two lattice polygons in $\mathbb{L}^3$ (see figure 1.5 and equation 1.7) and accounting for changes in the number of right angles,

$$p_{n_1}(s_1) p_{n_2}(s_2) \leq 2 \sum_{j=-4}^{4} p_{n_1+n_2}(s_1 + s_2 + j). \tag{7.41}$$

Thus, $p_n(s)$ satisfies equation (3.70). It may also be shown that $\epsilon_m = 1$ in equation (3.71). By corollary 3.14, the integrated density function exists, and a density function for right angles may be defined on $(0, 1)$ by

$$\log \mathcal{P}_s(\epsilon) = \lim_{n\to\infty} \tfrac{1}{n} \log p_n(\lfloor \epsilon n \rfloor + \sigma_n) = \min\{\log \mathcal{P}_s(\leq\epsilon), \log \mathcal{P}_s(\geq\epsilon)\}, \tag{7.42}$$

where $\sigma_n = o(n)$ is a sequence of integers. By theorem 3.7, there exists a limiting free energy $\mathcal{S}_s(z)$ which is convex in $\log z$ for $z \in (0, \infty)$.

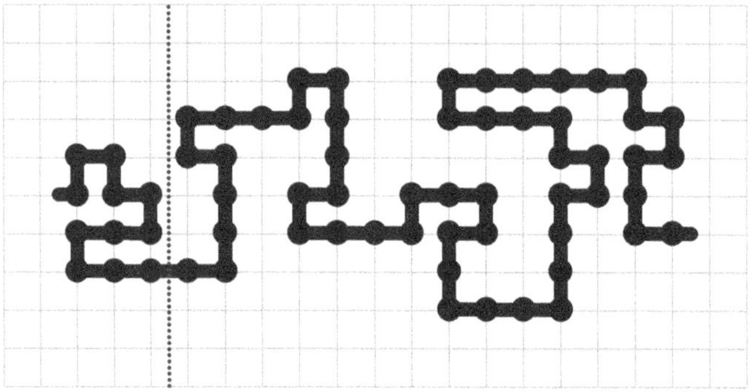

FIG. 7.9. A bridge in the hypercubic lattice which starts in the midpoint of an edge, and ends in the midpoint of an edge. The first and last (half)-steps of the bridge is in the $\vec{e}_1$-direction. This bridge can be decomposed by cutting a prime bridge from it where the vertical dotted line cuts an edge in its midpoint.

Note that $\mathcal{P}_s(0) = 1$, and, since $p_n(s) \leq \binom{n-1}{s}(2d)^s$, it follows that $\mathcal{P}_s(\epsilon) \leq \frac{(2d)^\epsilon}{\epsilon^\epsilon(1-\epsilon)^{1-\epsilon}}$. Taking $\epsilon \to 0^+$ establishes the right-continuity of $\mathcal{P}_s(\epsilon)$ at $\epsilon = 0$. The bounds $2^{(n-2)/4} \leq p_n(n) \leq p_n$ give $2^{1/4} \leq \mathcal{P}_s(1) \leq \mu_d$.

Theorem 7.26 *The function $\mathcal{P}(\epsilon)$ is continuous in $[0,1)$, and the right-derivative $\frac{d^+}{d\epsilon}\mathcal{P}(\epsilon) \mid_{\epsilon=0} = \infty$.*

Proof Continuity in $(0,1)$ follows from theorem 3.17. Right-continuity at $\epsilon = 0^+$ was shown above.

The number of polygons of length n with s right angles is bounded from below by the number of strict partition polygons with exactly s right angles (see section 6.1). Partition polygons of height l and width $\frac{n}{2} - l$ have $2 + 2l$ right angles, and the number of such polygons is $\binom{n/2-l-2}{l-1}$. Let $l = \lfloor \delta n \rfloor$ for some small $\delta > 0$. Then

$$\mathcal{P}(2\delta) \geq \lim_{n\to\infty} \binom{\frac{1}{2}n - \lfloor \delta n \rfloor - 2}{\lfloor \delta n \rfloor - 1}^{1/n} = \frac{(\frac{1}{2} - \delta)^{\frac{1}{2}-\delta}}{\delta^\delta(\frac{1}{2} - 2\delta)^{\frac{1}{2}-2\delta}}.$$

The limit of the ratio on the right is 1 as $\delta \to 0^+$, and its right-derivative is infinite as $\delta \to 0^+$. Since $\mathcal{P}(\epsilon)$ has a right-derivative everywhere in $[0,1)$, its right-derivative at $\epsilon = 0$ is infinite. □

7.3.1 *A pattern theorem for stiff polygons*

A pattern theorem for stiff polygons and walks is proven using generating functions and the bridge decomposition in the Hammersley-Welsh construction (see section 7.1.2). Some minor modification in the definition of bridges is needed in this model, and this is illustrated in figure 7.9.

Any bridge starts in an edge in the $\vec{e}_1$ direction and ends in a vertex with a maximal coordinate in the $\vec{e}_1$ direction. Delete half of the first edge and append this onto the last vertex. The resulting walk is a modified bridge which starts in the midpoint of an edge and ends in a midpoint of an edge. The total energy of the modified bridge is the total number of right angles.

A bridge is decomposable if it can be cut in the midpoint of an edge into two bridges, as illustrated in figure 7.9. Otherwise, it is a prime bridge.

Denote the number of bridges of length n and with m right angles by $b_n(m)$, and the number of prime bridges of length n and with m right angles by $q_n(m)$. The generating function of bridges is $B_z(t)$, and that of prime bridges of length n is $Q_z(t)$.

The prime decomposition of bridges proceeds as in section 7.1.2 but with prime bridges cut at midpoints of edges. The number of right angles is additive over this decomposition, so equation (7.40) becomes

$$B_z(t) = \frac{1}{1 - Q_z(t)}. \tag{7.43}$$

The bridge decomposition of half-space walks and of walks into half-space walks (see figure 7.4 and equation 7.28) proceeds as before but one right angle is lost when a walk is cut in its bottom vertex into half-space walks. By removing and adding back half-edges, the number of right angles in the resulting walks and bridges can be recovered. This gives the following relations for the half-space and walk generating functions, respectively:

$$tC_z(t) \leq z\,(H_z(t))^2, \quad \text{and } H_z(t) \leq e^{B_z(t)-1}. \tag{7.44}$$

The existence of a thermodynamic limit (for stiff self-avoiding walks) follows from the construction in figure 1.3. This gives

$$\sum_{j=-1}^{0} \sum_{m_1=0}^{n} c_{n_1}(m_1 + j)\, c_{n_2}(m - m_1) \geq c_{n_1+n_2}(m). \tag{7.45}$$

Multiply this by z^m and sum over m to see that

$$(1+z)\, Z^c_{n_1}(z)\, Z^c_{n_2}(z) \geq Z^c_{n_1+n_2}(z), \tag{7.46}$$

where $Z^c_n(z)$ is the partition function of stiff walks. Thus, $\log((1+z)\,Z^c_n(z))$ is subadditive in n, and, by theorem (A.1) in appendix A, the limit

$$\mathcal{F}_c(z) = \lim_{n\to\infty} \tfrac{1}{n} \log Z^c_n(z) = \inf_{n\geq 1} \tfrac{1}{n} \log((1+z) Z^c_n(z)) \tag{7.47}$$

exists. Moreover, $(1+z)\,Z^c_n(z) \geq e^{n\mathcal{F}_c(z)} = \mu_z^n$. Multiplying by t^n and summing over t gives $(1+z)C_z(t) \geq \frac{1}{1-t\mu_z}$, and thus $\lim_{t \nearrow \mu_z^{-1}} C_z(t) = \infty$ (see theorem 7.19). This completes all the ingredients in the proof of the pattern theorem in theorem 7.24 and corollary 7.25.

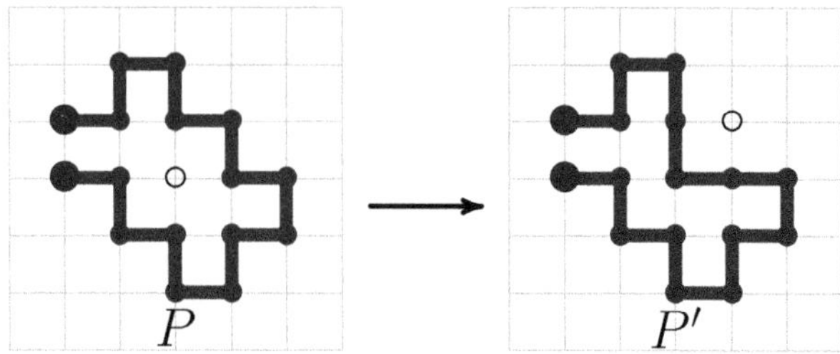

(a) (b)

FIG. 7.10. (a) A proper pattern P which contains fourteen right angles between its endpoints can be changed to the proper pattern P' in (b), which contains twelve right angles as shown.

Let $Z_n^a(z)$ be the partition function of stiff half-space walks when $a = h$, of stiff bridges if $a = b$, of polygons if $a = p$ (and of stiff walks if $a = c$). Then the pattern theorem for these models is the following.

Theorem 7.27 *Let $a \in \{c, h, b, p\}$ and let P be a proper pattern. Then there exists a small $\epsilon_0 > 0$ such that, for all $\epsilon \in [0, \epsilon_0)$, there exist a k_z and an $N_\epsilon \in \mathbb{N}$ such that*

$$Z_n^a(z; [\leq \epsilon n, P]) \leq e^{-k_z n} Z_n^a(z) \quad \textit{for all } n \geq N_\epsilon. \qquad \square$$

Theorem 7.27 is particularly true for polygons and bridges (and in fact also for loops, hoops, positive walks, unfolded walks and doubly unfolded walks) because the unfolding of bridges into doubly unfolded walks (see section 7.1.3) does not change the number of right angles (and concatenating these into loops and polygons may create or remove at most a constant number of right angles). The generating functions of doubly unfolded walks, loops and polygons have the same dominant singularity as $Z_n^c(z)$, and so the result in equation (7.27) applies to polygons in particular (that is, when the superscript a is p).

7.3.2 *The density function of stiff polygons*

The existence of the density function $\mathcal{P}(\epsilon)$ for stiff polygons was shown in theorem 7.26, and the derivative of $\mathcal{P}(\epsilon)$ at $\epsilon \to 0^+$ was shown to be infinite. In this section it is shown that the left-derivative $\frac{d^-}{d\epsilon}\mathcal{P}(\epsilon) \mid_{\epsilon=1} = -\infty$.

Let P be the proper pattern in figure 7.10(a). By equation (7.27), there exist a k_z and an $\epsilon > 0$ (both dependent on P) for every finite $z > 0$ such that $Z_n^p(z; [\leq \lfloor \epsilon n \rfloor, P]) \leq e^{-k_z n} Z_n^p(z)$. Without loss of generality, assume $\epsilon < \frac{1}{14}$. Then

$$\left(1 - e^{-k_z n}\right) Z_n^p(z) \leq Z_n^p(z; [\geq \lfloor \epsilon n \rfloor, P]) \leq Z_n^p(z). \qquad (7.48)$$

This shows that $\lim_{n\to\infty} \frac{1}{n} \log Z_n^p(z; [\geq \lfloor \epsilon n \rfloor, P]) = \mathcal{F}_c(z)$.

For polygons with at least $\lfloor \epsilon n \rfloor$ occurrences of P,

$$Z_n^p(z;[\geq \lfloor \epsilon n \rfloor, P]) = \sum_{m \geq 14\lfloor \epsilon n \rfloor}^{n} p_n(m;[\geq \lfloor \epsilon n \rfloor, P])\, z^m, \tag{7.49}$$

since these polygons have at least $14\lfloor \epsilon n \rfloor$ right angles. There is a most popular value of m, say m_z, which maximises the summation on the right-hand side. Obviously, $m_z \geq \lfloor \alpha n \rfloor$ where $\alpha = 14\epsilon < 1$. By theorems 3.16 and 3.17, there exists a $\kappa_z \geq \alpha$ such that $m_z = \lfloor \kappa_z n \rfloor + o(n)$. Notice that $\kappa_z \to 1$ as $z \to \infty$, since $p_n(m;[\geq \lfloor \epsilon n \rfloor, P])$ grows at most exponentially with n. Thus, it follows that

$$\begin{aligned}\left(1 - e^{-k_z n}\right) Z_n^p(z) &\leq Z_n^p(z;[\geq \lfloor \epsilon n \rfloor, P])\, z^{-m_z} \\ &\leq (n+1) p_n(m_z;[\geq \lfloor \epsilon n \rfloor, P]).\end{aligned} \tag{7.50}$$

Let $0 < \delta < \epsilon$. Since the pattern P occurs at least $\lfloor \epsilon n \rfloor$ times in each polygon, select $\lfloor \delta n \rfloor$ occurrences of P and perform the construction in figure 7.10(b) on each to obtain P'. This reduces the number of right angles by $2\lfloor \delta n \rfloor$ in each polygon, while leaving all other edges unchanged. This shows that

$$\binom{\lfloor \epsilon n \rfloor}{\lfloor \delta n \rfloor} p_n(m_z;[\geq \lfloor \epsilon n \rfloor, P]) \leq p_n(m_z - 2\lfloor \delta n \rfloor;[\geq \lfloor \epsilon n \rfloor - \lfloor \delta n \rfloor, P]), \tag{7.51}$$

while $p_n(m_z - 2\lfloor \delta n \rfloor;[\geq \lfloor \epsilon n \rfloor - \lfloor \delta n \rfloor, P]) \leq p_n(\geq (m_z - 2\lfloor \delta n \rfloor))$. Since $m_z = \lfloor \kappa_z n \rfloor + o(n)$, it follows from theorem 3.13 that the integrated density function $\mathcal{P}(\geq \epsilon)$ exists and that

$$\log \mathcal{P}(\geq(\kappa_z - 2\delta)) = \lim_{n\to\infty} \tfrac{1}{n} \log p_n(\geq(m_z - 2\lfloor \delta n \rfloor)). \tag{7.52}$$

Multiply both sides of equation (7.51) by z^{-m_z} and use the lower bound in equation (7.50). Take the power $\frac{1}{n}$ and take $n \to \infty$. This gives

$$\frac{\epsilon^\epsilon}{\delta^\delta (\epsilon - \delta)^{\epsilon - \delta}}\, e^{\mathcal{F}_c(z) - \kappa_z \log z} \leq \mathcal{P}(\geq(\kappa_z - 2\delta)). \tag{7.53}$$

This is true for every $z > 0$. Since $\log \mathcal{P}(\kappa_z) \leq \mathcal{F}(z) - \kappa_z \log z$, the above may be cast in the form

$$\frac{\epsilon^\epsilon}{\delta^\delta (\epsilon - \delta)^{\epsilon - \delta}} \mathcal{P}(\kappa_z) \leq \mathcal{P}(\geq(\kappa_z - 2\delta)) = \mathcal{P}(\kappa_z - 2\delta) \tag{7.54}$$

for values of z such that κ_z is close to 1 and for $\delta > 0$ small enough. Take $z \to \infty$ so that $\kappa_z \to 1^-$. By left-continuity of $\mathcal{P}(z)$, this gives

$$\left(\frac{\epsilon^\epsilon}{\delta^\delta (\epsilon - \delta)^{\epsilon - \delta}} - 1\right) \mathcal{P}(1^-) \leq \mathcal{P}(1 - 2\delta) - \mathcal{P}(1^-). \tag{7.55}$$

Divide by 2δ and take $\delta \to 0^+$. The left-hand side is divergent and so the left-derivative of $\mathcal{P}(\epsilon)$ at $\epsilon = 1$ is infinite. This enhances theorem 7.26 to the following:

Theorem 7.28 *The function $\mathcal{P}(\epsilon)$ is continuous in $[0,1)$ and has an infinite right-derivative at $\epsilon = 0$, and an infinite left-derivative at $\epsilon = 1$.* □

FIG. 7.11. An example of a proper pattern Q in $\mathbb{L}^3$. The union of the unit dual 3-cells centred at the vertices is a topological 3-ball C. The pair (C, Q) is a knotted ball-pair.

7.3.3 *Knotted polygons*

Let Q be the proper pattern in figure 7.11. The union of the unit dual 3-cells centred at the interior vertices of Q is a (topological) ball C, and the pair (C, Q) is a *knotted ball-pair* in 3-space. The pair (C, Q) is said to occur in a polygon ω if $\omega \cap \{\vec{v} + (C, Q)\} = \vec{v} + Q$, where $\vec{v} + (C, Q)$ is the translation of (C, Q) by $\vec{v}$, and $\vec{v} + Q$ is the translation of Q by $\vec{v}$. The pattern Q is a *knotted arc* since the occurrence of (C, Q) necessarily implies that ω will have a non-trivial knot type.

The number of occurrences of (C, Q) in a polygon ω is given by the number of distinct vectors $\vec{v}$ for which $\omega \cap \{\vec{v} + (C, Q)\} = \vec{v} + Q$.

Let $P_n(z; k)$ be the partition function of polygons of length n, with an activity z conjugate to curvature and which contains the knotted ball-pair (C, Q) exactly k times. By the arguments preceding and following equation (7.27), there exist a small $\epsilon > 0$, an N_ϵ and a $k_z > 0$ such that

$$P_n(z; [\geq \lfloor \epsilon n \rfloor, (C, Q)]) \geq (1 - e^{-k_z n}) P_n(z) \tag{7.56}$$

for all $n \geq N_\epsilon$. This gives the following theorem.

Theorem 7.29 *There exists an $\epsilon_c(z) > 0$ such that, for all $0 \leq \epsilon < \epsilon_c(z)$, and $z > 0$, $\lim_{n\to\infty} [P_n(z; [\lfloor \epsilon n \rfloor, (C, Q)])/P_n(z)] = 0$. Moreover, the rate of approach of the ratio to zero is exponential.*

Proof (C, Q) is a proper pattern; the result follows from equation (7.27). □

Corollary 7.30 *The ratio of the partition function $P_n^{\emptyset}(z)$ of stiff unknotted polygons, and of stiff polygons $P_n(z)$, approaches zero exponentially fast with increasing length. That is, $\lim_{n\to\infty} [P_n^{\emptyset}(z)/P_n(z)] = 0$ for all $z \in (0, \infty)$. Moreover, the rate of approach of the ratio to 0 is exponential.* □

The corollary follows by the pattern theorem, since an unknotted polygon cannot contain the ball-pair (C, Q). Thus, there exist an $N_\epsilon \in \mathbb{N}$ and a $k_z > 0$ such that, for all $n \geq N_\epsilon$, $P_n^{\emptyset}(z) \leq P_n(z; [0, (C, Q)]) \leq e^{-k_z n} P_n(z)$.

A corollary of the above is a proof of the Frisch-Wasserman-Delbruck conjecture [126, 212] for cubic lattice polygons: lattice polygons are knotted with a probability which approaches 1 as their lengths increase. This conjecture was proven for uniformly weighted cubic lattice polygons in references [466, 539]. In the case of stiff polygons a similar result may be stated [450] as follows.

Theorem 7.31 *The limiting free energy $\mathcal{F}_s(z)$ of polygons in $\mathbb{L}^3$ (with z conjugate to the curvature of the polygons) is completely determined by knotted polygons. That is, for every $z > 0$, there is an $\epsilon_c > 0$ such that*

$$\mathcal{F}_s(z) = \lim_{n\to\infty} \tfrac{1}{n} \log P_n(z) = \lim_{n\to\infty} \tfrac{1}{n} \log P_n(z; [\geq \lfloor \epsilon_c n \rfloor, (C,Q)]),$$

where (C,Q) is the knotted ball-pair with Q (illustrated in figure 7.11). □

The knot complexity (measured by the number of prime knot factors) of stiff polygons becomes arbitrarily large for finite values of z. For a similar result involving adsorbing lattice polygons in $\mathbb{Z}^3$, see section 9.2.1 and reference [557].

By concatenating two polygons (see figure 1.5),

$$\sum_{m_1=0}^{n} p_{n_1}(m-m_1,[k_1,R])p_{n_2}(m_1,[k_2,R]) \leq 2\sum_{j=-4}^{4} p_{n_1+n_2}(m+j,[k_1+k_2,R]),$$

where $R = (C,Q)$. Multiply by z^m and sum over m:

$$P_{n_1}(z,[k_1,(C,Q)])P_{n_2}(z,[k_2,(C,Q)]) \leq 10 P_{n_1+n_2}(z,[k_1+k_2,(C,Q)]) \quad (7.57)$$

for the partition functions of stiff polygons with k occurrences of the pattern (C,Q). This may be compared to equation (3.46) to see that $\frac{1}{10}P_n(z,[k,(C,Q)])$ satisfies the supermultiplicative inequality in equation (3.25) with the result that, by theorem 3.9, the limit

$$\log \mathcal{P}_z(\epsilon) = \lim_{n\to\infty} \tfrac{1}{n} \log P_n(z,[\lfloor \epsilon n \rfloor,(C,Q)]) \quad (7.58)$$

exists. The function $\mathcal{P}_z(\epsilon)$ is the microcanonical density function of the knotted ball-pair (C,Q) parameterised by z. By theorem 7.29, there exists an $\epsilon_c(z) > 0$ such that, for all $\epsilon < \epsilon_c(z)$, $\log \mathcal{P}_z(\epsilon) < \sup_\epsilon \log \mathcal{P}_z(\epsilon) = \mathcal{F}(z)$.

Put $\mathcal{F}_\epsilon(z) = \log \mathcal{P}_z(\epsilon)$; this is the limiting free energy of stiff polygons which contain (C,Q) at a density ϵ. The joint density function of right angles and (C,Q) is defined by

$$\log \mathcal{P}(\delta,\epsilon) = \inf_{z>0} \left(\log \mathcal{P}_z(\epsilon) - z \log \delta\right), \quad (7.59)$$

as may be seen from equation (3.48).

A second corollary of theorem 7.29 is the following:

Corollary 7.32 *Let $\mu_\emptyset$ be the growth constant of unknotted lattice polygons. Then $\mu_\emptyset < \mu_3$.*

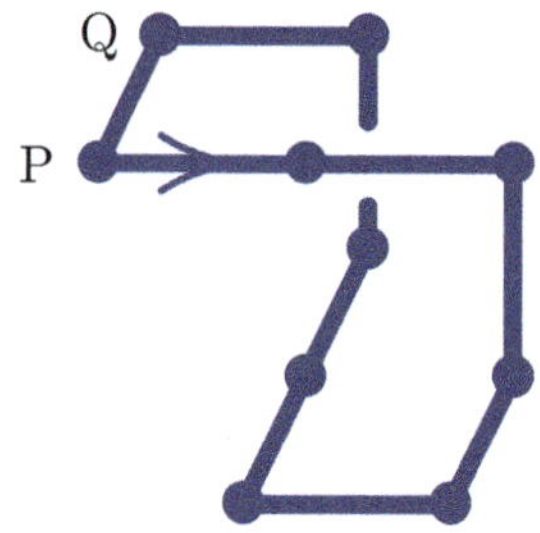

FIG. 7.12. $B = \{\vec{e}_1, \vec{e}_1, -\vec{e}_3, -\vec{e}_2, -\vec{e}_1, \vec{e}_2, \vec{e}_2, \vec{e}_3, -\vec{e}_1, -\vec{e}_2\}$ (starting at P) is a curlicue polygon of total writhe $W(B) = +\frac{1}{2}$.

Proof The pattern (C, Q) illustrated in figure 7.11 cannot occur in polygons counted by $p_n(\emptyset)$. This shows that there exist small $\alpha > 0$ and an $N \in \mathbb{N}_\epsilon$ such that $p_n(\emptyset) < e^{-\alpha n} p_n$ for all $n \geq N_\epsilon$ (this follows, for example, by putting $z = 1$ in theorem 7.29).

By theorem 1.8, $p_n(\emptyset) = \mu_\emptyset^{n+o(n)}$, and, by corollary 7.8, $p_n = \mu_d^{n+o(n)}$. Hence, by the above, it follows that $\mu_\emptyset < \mu_3$. □

More generally, theorem 7.29 implies that $\frac{p_n(\emptyset)}{p_n} = e^{-\alpha_0 n + o(n)}$ for some $\alpha_0 > 0$. If $p_n(\emptyset) = \mu_\emptyset^{n+o(n)}$, and $p_n = \mu_3^{n+o(n)}$, then $\alpha_0 = \log \frac{\mu_3}{\mu_\emptyset}$. A numerical estimate for α_0 is given in equation (1.81).

There are numerous studies of knotted walks and polygons; see, for example, references [17, 143, 146, 328, 360, 408, 420, 421, 448, 449, 452, 456].

7.4 Writhe in stiff polygons

Singular points in a regular projection of a directed knotted curve into a geometric plane are oriented double points called *crossings* (see section 1.7). Crossings are either left-handed (or negative) or right-handed (or positive); this is illustrated in figure 1.9.

The *writhing number* [213] of a regular projection of a simple closed curve C is the sum of the signed crossings in the projection. The *writhe* of C, denoted $\text{wr}(C)$, is the average of the writhing numbers over all directions $\widehat{\xi}$ (with the uniform measure) normal to planes into which C is projected. This defines the writhe of a lattice polygon in $\mathbb{L}^3$ as well.

Let B be the *curlicue polygon* illustrated in figure 7.12. Denote the writhe of B by $W(B)$. Then $W(B)$ can be computed by using the Lacher-Sumners theorem [366], which states that the writhe of a lattice polygon in the cubic lattice is the average of the four *linking numbers* of the polygon with a push-off (of itself) into four non-antipodal octants. This may be applied to B.

Lemma 7.33. ([327]) *The writhe of B is $W(B) = +\frac{1}{2}$.*

Proof By the Lacher-Sumners theorem [366], only the linking numbers of B with its four push-offs into non-antipodal octants must be computed. Construct the push-offs by adding the following four vectors to the coordinates of the vertices of B: $\vec{v}_1 = \frac{1}{2}(\vec{e}_1 + \vec{e}_2 + \vec{e}_3)$; $\vec{v}_2 = \frac{1}{2}(-\vec{e}_1 + \vec{e}_2 + \vec{e}_3)$; $\vec{v}_3 = \frac{1}{2}(-\vec{e}_1 - \vec{e}_2 + \vec{e}_3)$; and $\vec{v}_4 = \frac{1}{2}(\vec{e}_1 - \vec{e}_2 + \vec{e}_3)$. Let B_i be the curve defined by $B + \vec{v}_i$. Then the linking numbers can be computed explicitly: $L(B, B_1) = L(B, B_3) = L(B, B_4) = 1$, and $L(B, B_2) = -1$. Thus, $W(B) = +\frac{1}{2}$.

For additional results, see references [310, 311, 455]. □

Let A be a polygon which intersects the curlicue polygon B in a self-avoiding walk B', which starts at P and ends at Q in figure 7.12. A projection of B' and one of its push-offs, B'_1, is illustrated in figure 7.13. The pattern B' is the *curlicue pattern.*

The polygon A traverses all the edges in B, except for the single edge joining P and Q. The walk B' is a proper or Kesten pattern in A, and it can be truncated from A by deleting it between the vertices P and Q and then reconnecting A by adding the single edge from P to Q. Suppose this truncation leaves the polygon A'.

Lemma 7.34 $W(A) = W(A') + W(B)$.

Proof Consider A and its push-off A_1. Assume that A contains the subwalk $\{\vec{e}_1, \vec{e}_1, \vec{e}_1, -\vec{e}_3, -\vec{e}_2, -\vec{e}_1, \vec{e}_2, \vec{e}_2, \vec{e}_3, -\vec{e}_1, -\vec{e}_1\}$, which intersects B in B'.

In figure 7.12 a projection of this subwalk in a lattice plane is plotted, with a projection of the push-off A_1. If B' and B'_1 are truncated along the dotted lines in figure 7.13, then the polygons A' and A'_1 are obtained.

Consider the triplet (U, B', B'_1) of the ball C (the union of 3-cells dual to B'), and the two arcs B' and B'_1.

By a small deformation of B' inside U, the $(+)$-crossing in the dotted circle in figure 7.13 can be changed into a $(-)$-crossing. The resulting pair of curves is isotopic (by an isotopy in U) to the pair (A', A'_1). (This is most easily seen by using Reidemeister move I (see figure 1.10) inside U to convert A to A' and A_1 to A'_1 [82].)

The change in the sign of a single crossing between A and A_1 shows that $L(A, A_1) = L(A', A'_1) + 1$. However, $L(B, B_1) = 1$. Therefore,

$$L(A, A_1) = L(A', A'_1) + L(B, B_1).$$

Similar calculations show that

$$\begin{aligned} L(A, A_2) &= L(A', A'_2) + L(B, B_2); \\ L(A, A_3) &= L(A', A'_3) + L(B, B_3); \\ L(A, A_4) &= L(A', A'_4) + L(B, B_4). \end{aligned}$$

Use the Lacher-Sumners theorem; the average of these four equations gives the writhe of the polygons. □

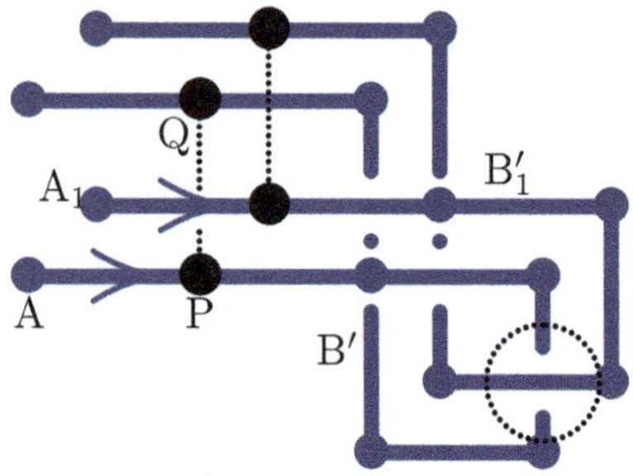

FIG. 7.13. The projections of a subwalk and its push-off in a coordinate plane in $\mathbb{L}^3$. The subwalks B' and B'_1 can be truncated by adding edges along the dotted lines and then removing the resulting polygons. Alternatively, the same can be achieved by reversing the positive crossing in the dotted circle and then using Reidemeister moves in the knot diagram.

The additivity of writhe over the curlicue pattern in figure 7.12 may be used to prove that the expected value of the writhe should increase almost surely as $O(\sqrt{n})$ with the length of the polygon.

Let $p_n(s)$ be the number of polygons with s right angles and let $Z_n^p(z)$ be the partition function of this model. This partition function defines a probability distribution $\Pi_n(z,s)$ on the set of polygons of length n:

$$\Pi_n(z,s) = \frac{p_n(s)\, z^s}{Z_n^p(z)}. \tag{7.60}$$

Compute the expected value of the absolute writhe with respect to $\Pi_n(z,s)$. For small values of z, polygons with low curvature have larger measure and contribute more to the absolute writhe, and, for large values of z, polygons with larger curvature will contribute more.

Theorem 7.35 *Let A be a polygon of length n sampled from the distribution $\Pi_n(z,s)$ for any $z>0$. Then, for every function $f(n)=o(\sqrt{n})$, the probability that the absolute value of the writhe of A is less than $f(n)$ approaches 0 as $n\to\infty$.*

Proof Let $z>0$ and denote the ball-pair of B' and its dual 3-cells in figure 7.13 by $P=(U,B')$. Reflect P in the xy-plane to find its mirror image $P^*=(U,B^{*\prime})$. Notice that P and P^* are Kesten patterns.

Let $Z_n^p(z;\leq l)$ be the partition function of polygons which contains P or P^* at most l times. By corollary 7.25, if $\epsilon>0$ is small enough, there exist a $k(z)>0$ and an N_0 such that

$$Z_n^p(z;\leq\lfloor \epsilon n\rfloor) < e^{-k(z)n}\, Z_n^p(z)$$

for all $n\geq N_0$.

In other words, the probability that A contains the patterns P or P^* at least $\lfloor \epsilon n \rfloor$ times exceeds $(1 - e^{-k(z)n})$ for $n \geq N_0$.

The distribution of the patterns P and P^* along A is binomial. Consider $\lfloor \epsilon n \rfloor$ occurrences of P or P^* along A. The probability that B occurs exactly k times amongst these $\lfloor \epsilon n \rfloor$ occurrences is less than $\lfloor \epsilon n \rfloor^{-1/2}$ (this follows from Stirling's formula).

By lemma 7.34, the writhe of A is the sum over two terms; the first term is obtained by truncating the $\lfloor \epsilon n \rfloor$ occurrences of P or P^* from A, while the second term is obtained from the $\lfloor \epsilon n \rfloor$ copies of B or B^*.

Suppose that the absolute writhe of A is less than $f(n)$; then the contribution from the copies of B or B^* to the writhe of A is at most one of $\lceil 2f(n)+1 \rceil$ different values (whatever the writhe of the truncated polygon, if the writhe due to B or B^* is added to it, then a result with absolute value less than $f(n)$ must be obtained).

In other words,

$$\mathrm{P}_r(|W(A)|<f(n)) \leq Q_n(z) \frac{\lceil 2f(n)+1 \rceil}{\sqrt{\lfloor \epsilon n \rfloor}} + (1 - Q_n(z))R_n,$$

where R_n is the contribution from polygons which contain fewer than $\lfloor \epsilon n \rfloor$ occurrences of B or B^*, and where $Q_n(z)$ is the probability that a polygon of length n will contain at least $\lfloor \epsilon n \rfloor$ occurrences of B or B^*.

However, the probability that a polygon has at least $\lfloor \epsilon n \rfloor$ occurrences of B or B^* is at least $(1 - e^{-k(z)n})$, so $Q_n(z) \geq (1 - e^{-k(z)n})$. Thus, $(1 - Q_n(z)) \leq e^{-k(z)n}$, and this gives

$$\mathrm{P}_r(|W(A)|<f(n)) \leq Q_n(z) \frac{\lceil 2f(n)+1 \rceil}{\sqrt{\lfloor \epsilon n \rfloor}} + e^{-k(z)n} R_n.$$

Since $Q_n(z) \to 1$ as $n \to \infty$, it follows that $\mathrm{P}_r(|W(A)|<f(n)) \to 0$ as $n \to \infty$ if $f(n) = o(\sqrt{n})$. This completes the proof. □

The above is a generalisation of the results in reference [311]; see references [351, 455] and [558] as well.

The square root bound in theorem 7.35 was tested numerically in reference [311]. If it is assumed that the absolute writhe of a polygon satisfies a power law relation with n, say $\langle |W| \rangle \simeq n^{\omega}$, then the numerical simulations give the estimate $\omega = 0.522 \pm 0.004$.

The geometry and topology of polymers are further discussed in references [350, 558, 583] and reference [456].

7.5 Torsion in polygons

The entanglement complexity of walks and polygons may be measured by *torsion*, which captures information about planarity [311, 548, 565]. (Writhe, in contrast, may be interpreted as a global measure of supercoiling.) A polygon with high

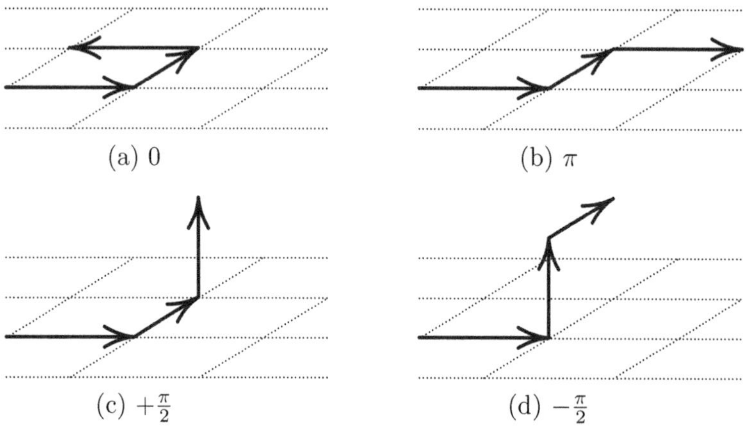

FIG. 7.14. Dihedral angles in self-avoiding walks or in lattice polygons are composed of three consecutive runs or trains of edges. The size of the dihedral angle is independent of the length of the runs but dependent on their changes in direction. In (a) and (b), two planar dihedral angles of size 0 and π, respectively, are illustrated. These angles do not contribute to the torsion. In (c), a positive dihedral angle is illustrated (the sign is determined by a right-hand rule). It contributes $+1$ to the torsion. The negative dihedral angle in (d) contributes -1 to the torsion.

---

plectonemic or solenoidal supercoiling will have high torsion. In this section the discussion is limited to polygons in $\mathbb{L}^3$ (but the results may be generalised to $\mathbb{L}^d$ with $d \geq 3$).

A polygon is a piecewise linear embedding of the circle and consists of *runs* or *trains* of collinear edges forming line segments. A sequence of three successive runs which are mutually orthogonal is a *dihedral angle.*

Consider three successive runs $\{\ell_1, \ell_2, \ell_3\}$ so that $\ell_1\ell_2\ell_3$ is a subwalk consisting of three line segments in the polygon. The sign of the dihedral angle between ℓ_1 and ℓ_3 is obtained by a right-hand rule, as illustrated in figure 7.14. A dihedral angle of size $+\frac{\pi}{2}$ is a *positive dihedral angle*, and a dihedral angle of size $-\frac{\pi}{2}$ is a *negative dihedral angle.* In the cubic lattice the dihedral angle may have size 0 or π (in which case the three runs are coplanar), or it may have sizes $\pm\frac{\pi}{2}$ (in which case the three runs are not planar).

Dihedral angles in a polygon ω can be labelled by τ_i for $i = 1, 2, \ldots, m$, and their signs can be denoted by $\sigma(\tau_i)$, where $\sigma(\tau) = 0$ if the dihedral angle τ consists of coplanar line segments, and $\sigma(\tau) = \pm 1$ if the dihedral angle τ has size $\pm\frac{\pi}{2}$.

The *torsion t* and *excess torsion t_e* of a polygon ω is defined by

$$t(\omega) = \sum_{i=1}^{N} \sigma(\tau_i), \quad \text{and } t_e(\omega) = \left| \sum_{i=0}^{N} \sigma(\tau_i) \right|. \tag{7.61}$$

7.5.1 *The free energy of polygons with torsion*

Let $p_n(t)$ be the number of polygons with torsion t, and $\overline{p}_n(t)$ be the number of polygons with excess torsion t. Then $\overline{p}_n(0) = p_n(0)$, and $\overline{p}_n(t) = 2p_n(t)$ if $t > 0$ (and less than or equal to n). Notice that there are no polygons with negative excess torsion. The partition functions of these models are given by

$$Z_n^t(z) = \sum_{t=-n}^{n} p_n(t)\, z^t, \quad \text{and } \overline{Z^t}_n(z) = \sum_{t=0}^{n} \overline{p}_n(t)\, z^t. \tag{7.62}$$

Since $p_n(t) = p_n(-t)$, it follows that $Z_n^t(z) = Z_n^t\left(\frac{1}{z}\right)$.

The existence of a thermodynamic limit in these models is shown by concatenating two polygons (see figure 1.5). By equation (1.7),

$$\sum_{s=0}^{t} p_n(t-s)\, p_m(s) \leq 2 \sum_{i=-6}^{6} p_{n+m}(t+i), \tag{7.63}$$

where it is assumed that $n \geq m$ without loss of generality (notice that the removal of any edge in a polygon can reduce the number of dihedral angles by as many as three, and that the addition of new edges to concatenate the polygons can create as many as six new dihedral angles). Multiplying this by z^t and summing over t gives

$$Z_n^t(z)\, Z_m^t(z) \leq 2\psi\, Z_{n+m}^t(z), \tag{7.64}$$

where $\psi = \sum_{i=-6}^{6} z^i$. Thus, $-\log\left(Z_n^t(z)/(2\psi)\right)$ satisfies a subadditive inequality of the type in equation (A.1) in appendix A. This gives the following theorem.

Theorem 7.36 *The limiting free energy of polygons with torsion is given by*

$$\mathcal{F}^t(z) = \lim_{n\to\infty} \tfrac{1}{n} \log Z_n^t(z) = \sup_{n>0} \tfrac{1}{n} \log\Big(\tfrac{1}{2\psi} Z_n^t(z) \Big),$$

where $\psi = \sum_{i=-6}^{6} z^i$. *The microcanonical density of the torsion is*

$$\mathcal{P}^t(\epsilon) = \lim_{n\to\infty} \left[p_n(\lfloor \epsilon n \rfloor + \sigma_n) \right]^{1/n} = \min\left\{ \mathcal{P}^t(\leq\epsilon), \mathcal{P}^t(\geq\epsilon) \right\},$$

where $\sigma_n = o(n)$.

Proof It only remains to prove existence of $\mathcal{P}^t(\epsilon)$. But this follows from theorem 3.6, while theorem 3.13, corollary 3.14 and theorem 3.16 give the result that $\mathcal{P}^t(\epsilon) = \min\{\mathcal{P}^t(\leq\epsilon), \mathcal{P}^t(\geq\epsilon)\}$. It may be shown that this model is regular (see remark 3.3.1.3); therefore, existence of the density function also follows from this fact and the supermultiplicative inequality in equation (7.63). □

More work is needed to show that there exist a limiting free energy and a density function in the model of polygons with excess torsion. The argument has two steps [548]. First, suppose that $z \geq 1$. Then

$$\sum_{t=-n}^{n} p_n(t)\, z^t \leq \sum_{t=-n}^{n} p_n(t)\, z^{|t|} \leq 2 \sum_{t=0}^{n} p_n(t)\, z^t \leq 2 \sum_{t=-n}^{n} p_n(t)\, z^t. \qquad (7.65)$$

Take logarithms, divide by n and let $n \to \infty$. This gives existence of a limiting free energy for $z \geq 1$:

$$\overline{\mathcal{F}^t}(z) = \lim_{n\to\infty} \tfrac{1}{n} \log \overline{p}_n(z) = \mathcal{F}^t(z) \text{ if } z \geq 1. \qquad (7.66)$$

Second, if $z < 1$, then equation (7.63) is used to see that

$$\sum_{t=-n}^{n} \sum_{s=-m}^{m} p_n(t-s) p_m(s)\, z^{|t-s|+|s|} \leq 2 \sum_{i=-6}^{6} \sum_{t=-(n+m)}^{n+m} p_{n+m}(t+i)\, z^{|t|}, \qquad (7.67)$$

where the triangle inequality $|t| \leq |t-s| + |s|$ was used. Since $||t+i| - |i|| \leq |t|$, it follows from equation (7.67) that

$$\overline{Z^t}_n(z) \overline{Z^t}_m(z) \leq 2\psi \overline{Z^t}_{n+m}(z), \qquad (7.68)$$

where $\psi = \sum_{i=-6}^{6} z^i$. Thus, the limiting free energy exists by theorem A.1, and the density function exists by theorem 3.6. This gives the following theorem.

Theorem 7.37 *The limiting free energy of polygons with excess torsion is given by*

$$\overline{\mathcal{F}^t}(z) = \lim_{n\to\infty} \tfrac{1}{n} \log \overline{Z^t}_n(z) = \sup_{n>0} \tfrac{1}{n} \log\Big(\tfrac{1}{2\psi} \overline{Z^t}_n(z)\Big).$$

The density function of excess torsion is

$$\overline{\mathcal{P}^t}(\epsilon) = \lim_{n\to\infty} \left[\overline{p}_n(\lfloor \epsilon n \rfloor + \tau_n)\right]^{1/n},$$

where $\tau_n = o(n)$. □

Observe that $\mathcal{F}^t(1) = \overline{\mathcal{F}^t}(1) = \log \mu_3$ and that $\mathcal{F}^t(z) = \overline{\mathcal{F}^t}(z) \geq \log \mu_3$.

Consider the polygon of length 24 and with torsion 14 in figure 7.15. By increasing the length in an obvious way, a class of polygons with length n and torsion $n-10$ is found. Thus, $Z_n^t(z) \geq z^{n-10} p_n(n-10)$. By taking logarithms, letting $n \to \infty$ and noting that $Z_n^t(z)$ is non-increasing with $z \geq 1$, it follows that

$$\overline{\mathcal{F}^t}(z) = \mathcal{F}^t(z) = \mathcal{F}^t(\tfrac{1}{z}) \geq \max\{\log z, \log \mu_3\} \text{ if } z \geq 1. \qquad (7.69)$$

If $z \leq 1$, then $\overline{\mathcal{F}^t}(z) = \log \mu_3$. This is proven in the next theorem.

Theorem 7.38 *The free energy* $\overline{\mathcal{F}^t}(z) = \log \mu_3$, *for all* $z \leq 1$.

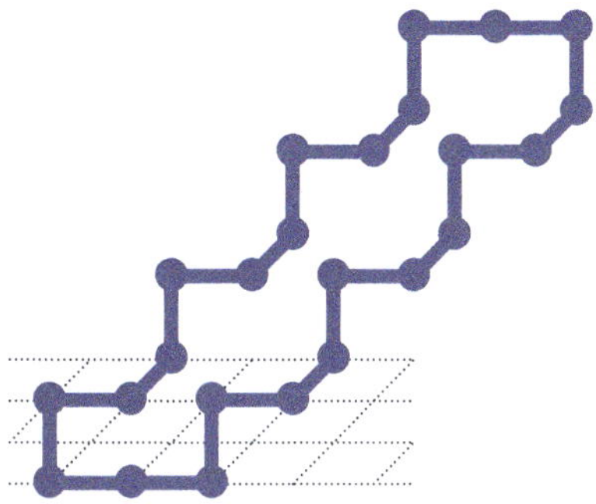

FIG. 7.15. A polygon of torsion 14 and length 24. By lengthening this polygon in the obvious way, a family of polygons of torsion $n-10$ and length n is found.

Proof Note that $\overline{p}_n(0) = p_n(0)$ and that $\overline{p}_n(1) = p_n$. Moreover, $\overline{p}_n(0) \leq \overline{p}_n(1)$.

Choose $m = n$, choose $s = -k$ and choose $t = 2k$ on the left-hand side of equation (7.63). This gives

$$\sum_{k=-n}^{n} p_n(k) p_n(-k) \leq 2 \sum_{l=-6}^{6} p_{2n}(l).$$

Let $\pm k^*$ be the most popular torsion in polygons of length n. That is, $p_n(-k^*) = p_n(k^*) \geq p_n(k)$ for all $|k| \leq n$. Then $p_n(k^*) \geq \frac{1}{2n+1} p_n$. Since $p_n(k^*) = p_n(-k^*)$, this gives

$$\left(\tfrac{1}{2n+1}\right)^2 p_n^2 \leq p_n(k^*) p_n(-k^*) \leq \sum_{k=-n}^{n} p_n(k) p_n(-k) \leq 2 \sum_{l=-6}^{6} p_{2n}(l).$$

Let l^* be the most popular value of l in the summation on the right-hand side. Then $-6 \leq l^* \leq 6$, and $p_{2n}(l^*) \geq p_{2n}(l)$ for $-6 \leq l \leq 6$. This shows that $\left(\frac{1}{2n+1}\right)^2 p_n^2 \leq 26 p_{2n}(l^*)$. Thus, if $z \leq 1$, then

$$p_{2n} \geq \overline{p}_{2n}(z) \geq p_{2n}(l^*) z^{|l^*|} \geq \tfrac{1}{26} \left(\tfrac{1}{2n+1}\right)^2 p_n^2 z^{|l^*|}.$$

Take logarithms, divide by $2n$ and let $n \to \infty$ to see that $\lim_{n\to\infty} \frac{1}{n} \log \overline{p}_n(z) = \log \mu_3$ for all $z \leq 1$. $\square$

By equation (7.69) and theorem 7.38, there is a non-analyticity in $\overline{\mathcal{F}^t}(z)$ at some critical value of z. This critical value must be at the point $z_t = \inf\{z \mid \mathcal{F}^t(z) > \log \mu_3\} \geq 1$.

On the other hand, let $z_d = \inf\{z \mid \overline{\mathcal{F}^t}(z) = \mathcal{F}^t(z)\}$. Clearly, $z_d \leq z_t$, and $z_d \leq 1$. If $z_d < 1$, then $\mathcal{F}^t(z)$ is non-analytic at z_d and at $\frac{1}{z_d}$.

FIG. 7.16. A Kesten pattern with one negative and two positive dihedral angles.

A corollary of the next theorem (theorem 7.39) is that $\mathcal{F}^t(z)$ and $\overline{\mathcal{F}^t}(z)$ are both strictly increasing for $z > 1$. Since $\overline{\mathcal{F}^t}(z) = \log \mu_d$ for $z < 1$, this proves that there is a non-analyticity at $z_t = 0$, that $\mathcal{F}^t(z) > \overline{\mathcal{F}^t}(z)$ if $z < 1$ and that $z_d = 0$.

7.5.2 *The expected torsion of polygons*

The expected torsion per edge is defined by

$$\langle t \rangle_n = \tfrac{z}{n} \tfrac{d}{dz} \log Z_n^t(z). \tag{7.70}$$

Since $\langle \frac{1}{n} \log Z_n^t(z) \rangle$ is a sequence of convex functions pointwise convergent to $\mathcal{F}^t(z)$, which is itself convex, it follows from theorem B.9 in appendix B that the derivative of $\mathcal{F}^t(z)$ exists almost everywhere and

$$\tau_z = \lim_{n \to \infty} \langle t \rangle_n = \lim_{n \to \infty} \tfrac{z}{n} \tfrac{d}{dz} \log Z_n^t(z) = z \tfrac{d}{dz} \mathcal{F}^t(z) \text{ for almost all } z. \tag{7.71}$$

The expected excess torsion $\overline{\tau}_z$ may similarly be defined and $\tau_z = \overline{\tau}_z$ for almost all $z > 1$. Moreover, by theorem 7.38, $\overline{\tau}_z = 0$ for all $z < 1$.

Theorem 7.39 *For almost all $z > 1$, both $\tau_z > 0$ and $\overline{\tau}_z > 0$.*

Proof The proof relies on the tossing of a biased coin.

The expected value of the torsion at finite values of n is

$$\langle t \rangle_n = \frac{\sum_{t=-n}^{n} t\, p_n(t)\, z^t}{n \sum_{t=-n}^{n} p_n(t)\, z^t}. \tag{7.72}$$

Since $\mathcal{F}^t(z) = \overline{\mathcal{F}^t}(z)$ for all $z \geq 1$, only the expected torsion τ_z for $z > 1$ need be considered. Since $\mathcal{F}^t(z)$ is convex, it is also differentiable almost everywhere. Thus, τ_z exists almost everywhere.

It will be shown for almost all $z > 1$ that there exists an $\epsilon > 0$ such that

$$\tau_z \geq \left(\frac{z - \overline{z}}{z + \overline{z}} \right) \epsilon,$$

where $\overline{z} = \frac{1}{z}$.

Let $\vec{e}_1$, $\vec{e}_2$ and $\vec{e}_3$ be the canonical orthonormal basis vectors of $\mathbb{Z}^3$. Let P be the walk illustrated in figure 7.16. The union of dual 3-cubes centred on the vertices of this walk is a topological 3-ball C, and P contains one negative dihedral angle and two positive dihedral angles.

A reflection of P (to obtain the walk P^*) through its centre of mass leaves its endpoints and C unchanged. The walk P^* contains two negative dihedral angles and one positive dihedral angle.

The pattern P is also a Kesten pattern (see section 7.1.6) and, since the endpoints of P and P^* are the same points in the boundary of C, either P or P^* can be inserted into C in any given polygon containing (C, P) or (C, P^*).

Let α_n be a class of polygons with at least $\lfloor \epsilon n \rfloor$ occurrences of (C, P) or (C, P^*) and with $\lfloor \epsilon n \rfloor$ of the (C, P) or (C, P^*) fixed. By corollary 8.28, the number of such classes of polygons is at least $(1 - e^{-kn}) 2^{-\lfloor \epsilon n \rfloor} p_n$, where $k > 0$ is a small number, and p_n is the number of polygons of length n. The factor $2^{-\lfloor \epsilon n \rfloor}$ appears because each class α_n represents $2^{\lfloor \epsilon n \rfloor}$ polygons. (This is a consequence of the binomial choice of either P or P^* in each C.)

Suppose that the contribution to the torsion of a polygon in the class α_n from the fixed part outside the union of the C's is t, and let there be, in the C's, t_+ occurrences of the pattern P, and t_- occurrences of the pattern P^* (thus, $t_+ + t_- = \lfloor \epsilon n \rfloor$). In the partition function, the weight of each P is z, and of P^* is $z^{-1} = \overline{z}$. Since P and P^* are independent and binomially distributed, the normalised contribution of the class α_n to the partition function is

$$z^t \sum_{t_+=0}^{\lfloor \epsilon n \rfloor} \binom{\lfloor \epsilon n \rfloor}{t_+} \left(\frac{z}{z + \overline{z}} \right)^{t_+} \left(\frac{\overline{z}}{z + \overline{z}} \right)^{\lfloor \epsilon n \rfloor - t_+} = z^t.$$

Similarly, the normalised contribution to the numerator in equation (7.72) is

$$z^t \sum_{t_+=0}^{\lfloor \epsilon n \rfloor} (t_+ - t_- + t) \binom{\lfloor \epsilon n \rfloor}{t_+} \left(\frac{z}{z + \overline{z}} \right)^{t_+} \left(\frac{\overline{z}}{z + \overline{z}} \right)^{\lfloor \epsilon n \rfloor - t_+} = z^t \left(t + \tfrac{2z}{z+\overline{z}} - 1 \right) \lfloor \epsilon n \rfloor .$$

The contribution from the class α_n^* (the mirror image of α_n) is similarly obtained by replacing t by $-t$: this is $z^{-t}(-t + \frac{2z}{z+\overline{z}} - 1) \lfloor \epsilon n \rfloor$.

Without loss of generality, assume that $t \geq 0$. In this case the combined contribution to the numerator in equation (7.72) and its mirror image is

$$t(z^t - \overline{z}^t) + \left(\frac{2z}{z + \overline{z}} - 1 \right) \lfloor \epsilon n \rfloor (z^t + \overline{z}^t).$$

For $z > 1$, the first term is always non-negative and so a lower bound on the numerator in equation (7.72) can be obtained by ignoring it. This gives a lower bound for each of the classes α_n.

Let Ψ_n be the set of polygons with fewer than $\lfloor \epsilon n \rfloor$ occurrences of P or P^* in C and fixed outside the union of the C's. Let the number of such polygons be q_n. Then q_n still grows exponentially since it is supermultiplicative under concatenation, as shown in figure 1.5. Corollary 8.28 implies that there exists an

$\epsilon_0 > 0$ such that, for every positive $\epsilon < \epsilon_0$, q_n/p_n is exponentially small. That is, q_n is bounded from above and below by $e^{-k_1 n} p_n \geq q_n \geq e^{-k_2 n} p_n$ if n is large, where $0 < k_1 < k_2$, and for every positive $\epsilon < \epsilon_0$.

Assume that $\epsilon < \epsilon_0$ is fixed. Let R be the minimum contribution to the numerator in equation (7.72) by a polygon of length n, and S be the maximum contribution to the partition function by a polygon of length n. Using these bounds and the contributions from the classes α_n, the expected torsion for n large enough is bounded from below as follows:

$$\langle t \rangle_n \geq \frac{(1 - e^{-kn}) 2^{-\lfloor \epsilon n \rfloor} \left(\frac{2z}{z + \overline{z}} - 1 \right) \left(\frac{\lfloor \epsilon n \rfloor}{n} \right) (z^t + \overline{z}^t) p_n + e^{-k_2 n} p_n R}{2^{-\lfloor \epsilon n \rfloor} (z^t + \overline{z}^t) p_n + e^{-k_1 n} p_n S}, \tag{7.73}$$

since each class α_n contains $2^{\lfloor \epsilon n \rfloor}$ polygons. Now take $\epsilon < k_1$; since decreasing ϵ also decreases q_n, this is always possible.

Let $n \to \infty$ to obtain

$$\lim_{n \to \infty} \langle t \rangle_n \geq \left(\frac{2z}{z + \overline{z}} - 1 \right) \epsilon = \left(\frac{z - \overline{z}}{z + \overline{z}} \right) \epsilon. \tag{7.74}$$

This completes the proof. □

By theorem 7.38, $\overline{\mathcal{F}^t}(z) = \log \mu_3$ if $z \leq 1$. This implies that the expected excess torsion per edge is 0 if $z < 1$. The left-derivative of $\overline{\mathcal{F}^t}(z)$ is also zero at $z = 1$, and, by theorem 7.72, the right-derivative at z is either 0 or positive.

If the right-derivative is positive, then there is a jump discontinuity at $z = 1$ in the first derivative of $\overline{\mathcal{F}^t}(z)$, which indicates a *first order* transition in this model. In addition, since $\mathcal{F}^t(z) = \overline{\mathcal{F}^t}(z)$ if $z \geq 1$, and $\mathcal{F}^t(z) = \mathcal{F}^t(\frac{1}{z})$, this would indicate a first order transition in the expected torsion as well.

If the right-derivative at $z = 1$ is 0, on the other hand, then there is a continuous transition in the excess torsion model.

For $z < 1$, the excess torsion model should have critical exponents given by dilute polymer exponents (see table 1.4). If $z > 1$, then there is a transition to a phase of polygons with excess torsion. The conformations of a polygon in this phase should be dominated by branched structures called *cruciforms* (made up of supercoiled segments of the polygon [112]). The branched nature of these conformations suggest that branched polymer exponents may be seen in this phase (see table 2.5).

8

SELF-AVOIDING WALKS IN SLABS AND WEDGES

Walks and polygons confined to slabs or wedges in $\mathbb{L}^d$ are models of polymers in confined spaces [148] (for example, polymers confined in capillaries or in the interstitial spaces between particles in a colloid). A confined polymer loses conformational entropy and this induces a repulsive entropic force on the walls of the confining space. This repulsive force underlies the phenomenon of steric stabilisation of colloidal particles [430]. A polymer may also adsorb on particles in a colloid and even bridge gaps between neighbouring particles. This may cause attractive forces between particles. This may explain the phenomenon of sensitised flocculation of the colloid [113].

Lattice walks and polygons in slabs and wedges in $\mathbb{L}^d$ pose unique questions which require different approaches from those in section 7.1 (see, for example, references [529, 588, 589]). In this chapter, the existence of growth constants and free energies, their properties and a pattern theorem for walks in slabs and in wedges will be examined.

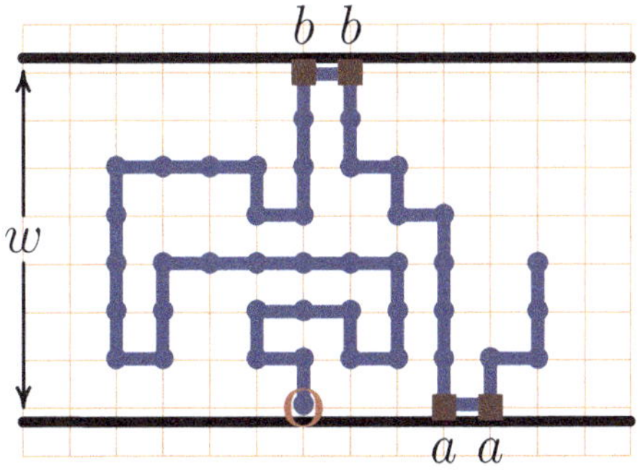

FIG. 8.1. A self-avoiding walk in a slab $\mathbb{S}_w$. The walk starts from the origin in the bottom wall and makes two top visits to $T(\mathbb{S}_w)$ and two bottom visits to $B(\mathbb{S}_w)$. By convention, the vertex at the origin is not a bottom visit.

The slab $\mathbb{S}_w$ in $\mathbb{L}^d$ is defined by

$$\mathbb{S}_w = \{\langle \vec{v} \sim \vec{u} \rangle \in \mathbb{L}^d \mid 0 \leq \vec{v}(d), \vec{u}(d) \leq w\}, \tag{8.1}$$

where $\vec{v}(d)$ is the d-th Cartesian component of the vector $\vec{v}$, and $\mathbb{S}_w$ is a slab of *width* or *height* w.

The *bottom wall* $B(\mathbb{S}_w)$ and *top wall* $T(\mathbb{S}w)$ of a slab $\mathbb{S}_w$ are given by

$$B(\mathbb{S}_w) = \{\vec{v} \in \mathbb{L}^d \mid \vec{v}(d) = 0\}, \quad \text{and } T(\mathbb{S}_w) = \{\vec{v} \in \mathbb{L}^d \mid \vec{v}(d) = w\}, \tag{8.2}$$

The Statistical Mechanics of Interacting Walks, Polygons, Animals and Vesicles, 2nd edition, E.J. Janse van Rensburg. © E.J. Janse van Rensburg. Published in 2015 by Oxford University Press.

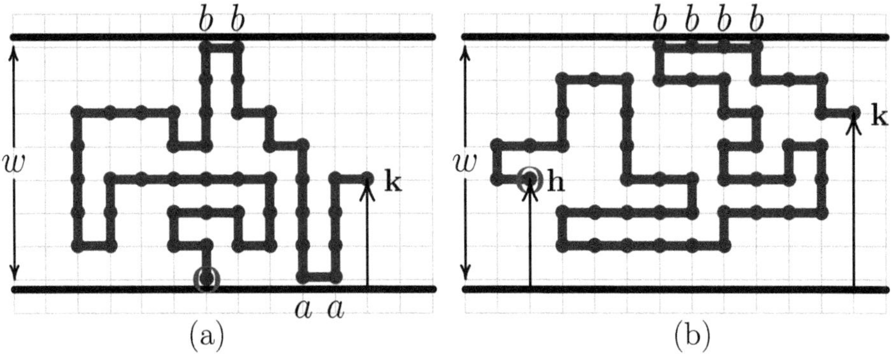

FIG. 8.2. (a) A self-avoiding walk from the origin in $\mathbb{S}_w$ and ending in a vertex at height k. Bottom visits in the walk are weighted by a, and top visits are weighted by b. (b) A self-avoiding walk in $\mathbb{S}_w$ from a vertex at height h and terminating in a vertex at height k.

---

respectively. The *boundary* of $\mathbb{S}_w$ is defined by $\partial\mathbb{S}_w = B(\mathbb{S}_w) \cup T(\mathbb{S}_w)$.

It is a convention that the origin $\vec{0} \in B(\mathbb{S}_w)$. In figure 8.1(a) a self-avoiding walk from the $\vec{0}$ in $\mathbb{S}_w$ is illustrated. The walk is oriented from $\vec{0}$ and is *confined* to $\mathbb{S}_w$, making *visits* to $\partial\mathbb{S}_w$. Visits of the walk to $B(\mathbb{S}_w)$ are *bottom visits*, and visits to $T(\mathbb{S}_w)$ are *top visits*. By convention, the vertex at $\vec{0} \in B(\mathbb{S}_w)$ is not a bottom visit.

8.1 Self-avoiding walks in slabs

Denote the number of self-avoiding walks from $\vec{0}$ in $\mathbb{S}_w$ of length n by $c_n(w)$. Let $c_n(v_1, v_2; w)$ be the number of walks from $\vec{0}$ in $\mathbb{S}_2$ making v_1 bottom visits and v_2 top visits. Introduce activities a, conjugate to bottom visits, and b, conjugate to top visits, and define the partition function by

$$S_n(a, b; w) = \sum_{v_1, v_2 \geq 0} c_n(v_1, v_2; w)\, a^{v_1} b^{v_2}. \tag{8.3}$$

In figure 8.2 the *heights* of the endpoints of walks in $\mathbb{S}_w$ are indicated. In figure 8.2(a), a walk from $\vec{0}$ in the bottom wall ends in a vertex that is height k above the bottom wall; the height of this endpoint is k. Denote the number of walks from $\vec{0}$ in $\mathbb{S}_w$ ending in a vertex at height k and with v_1 bottom visits and v_2 top visits as $c_n([0k], v_1, v_2; w)$. The partition function of this model is defined similarly to equation (8.3) by introducing visit activities a and b and is denoted by $S_n([0k], a, b; w)$.

Denote the number of walks in $\mathbb{S}_w$ which start in a vertex $\vec{v}$ of height h (above $B(\mathbb{S}_w)$) and end in a vertex $\vec{w}$ of height k as $c_n([hk], v_1, v_2; w)$. These walks have v_1 bottom visits and v_2 top visits and are distinct up to equivalence under translation parallel to $B(\mathbb{S}_w)$. The partition function of this model is $S_n([hk], a, b; w)$, defined similarly to equation (8.3).

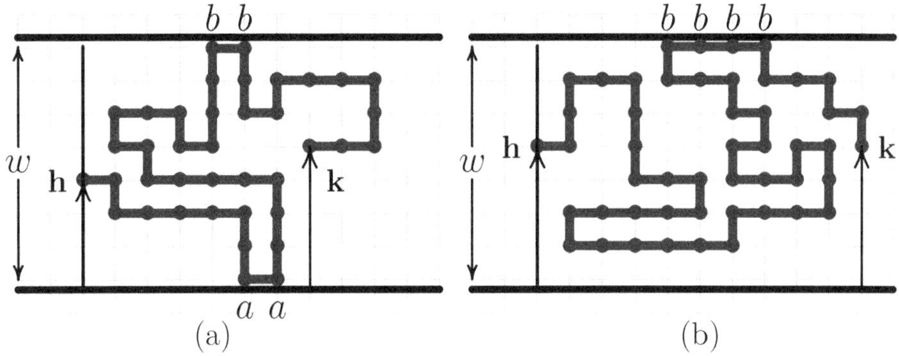

FIG. 8.3. (a) A w-half-space walk in $\mathbb{S}_w$. The endpoints of the walk are at heights h and k. (b) A w-bridge in $\mathbb{S}_w$. Notice that the w-half-space walk and the w-bridge are oriented in the $\vec{e}_1$ direction, parallel to the $B(\mathbb{S}_w)$ and $T(\mathbb{S}_w)$.

Summing over the heights of the endpoints give the partition functions

$$S_n([h*], a, b; w) = \sum_{k=0}^{w} S_n([hk], a, b; w), \quad \text{and}$$
$$S_n([*k], a, b; w) = \sum_{h=0}^{w} S_n([hk], a, b; w).$$

The partition function of walks with endpoints at any height is $S_n([**], a, b; w) = \sum_{k=0}^{w} S_n([*k], a, b; w)$. Notice that equation (8.3) is recovered by $S_n(a, b; w) \equiv S_n([0*], a, b; w)$.

8.1.1 *Half-space walks, and bridges in a slab*

Let $\omega = \langle \omega_0, \omega_1, \ldots, \omega_n \rangle$ be a walk in $\mathbb{S}_w$. Denote the k-th Cartesian coordinate of ω_j by $\omega_j(k)$. The heights of the endpoints of ω are given by $\omega_0(d)$ and $\omega_n(d)$. The walk ω is a *w-half-space walk* if $\omega_0(1) < \omega_i(1)$ for $1 \leq i \leq n$. If $\omega_0(1) < \omega_i(1) \leq \omega_n(1)$ for $1 \leq i \leq n$, then ω is a *w-bridge*. A w-half-space walk and a w-bridge are illustrated in figure 8.3.

The number of w-half-space walks of length n from $\vec{0}$ in $\mathbb{S}_w$, with v_1 bottom visits and v_2 top visits, is denoted by $h_n(v_1, v_2; w)$. The number of w-half-space walks of length n from a vertex at height h to a vertex at height k (in $\mathbb{S}_w$), is denoted by $h_n([hk], v_1, v_2; w)$. Summing over the heights of the endpoints gives $h_n([h*], v_1, v_2; w) = \sum_{k=0}^{w} h_n([hk], v_1, v_2; w)$, and $h_n([*k], v_1, v_2; w) = \sum_{h=0}^{w} h_n([hk], v_1, v_2; w)$, and, if summed over both endpoints, $h_n([**], v_1, v_2; w) = \sum_{k=0}^{w} h_n([*k], v_1, v_2; w)$. Notice that $h_n(v_1, v_2; w) = h_n([0*], v_1, v_2; w)$.

By prepending $w+1$ edges in a $\ulcorner$-conformation on a w-half-space walk, the height of the initial vertex can be changed. This shows that, for $h, k \notin \{0, w\}$,

$$h_n([kk], v_1, v_2; w) \leq h_{n+w+1}([hk], v_1, v_2; w). \tag{8.4}$$

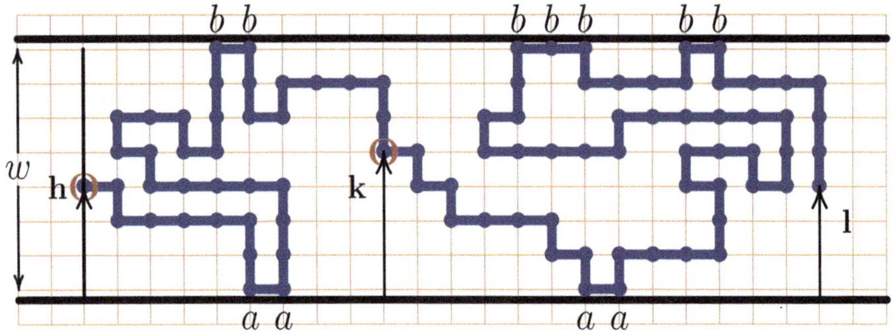

FIG. 8.4. Concatenating two w-bridges. Notice that the numbers of bottom visits and top visits are additive under the construction, since the starting vertex of the w-bridge on the right is by convention never a visit, even when $k = 0$, or $k = w$.

If either h or k is equal to 0 or w, then the number of visits may change.

The number of w-bridges in $\mathbb{S}_w$ is defined similarly to the above. There are $b_n(v_1, v_2; w)$ w-bridges of length n from $\vec{0}$ with v_1 bottom visits and v_2 top visits. Similarly, if the endpoints are at height h or k, then define $b_n([hk], v_1, v_2; w)$, $b_n([*k], v_1, v_2; w)$, $b_n([h*], v_1, v_2; w)$ and $b_n([**], v_1, v_2; w)$. By reflecting w-bridges in the $\vec{e}_1$ direction and then placing the horizontal step on the first vertex,

$$b_n([hk], v_1, v_2; w) = b_n([kh], v_1, v_2; w) \text{ if } 0 < h, \text{ and } k < w. \tag{8.5}$$

Two w-bridges of which the first ends in a vertex at height k and the second starts in a vertex at height k can be concatenated (see figure 8.4). This shows that

$$\sum_{k=0}^{w} \sum_{v_1, u_1 \geq 0} b_n([hk], v - v_1, u - u_1; w) b_m([kl], v_1, u_1; w) \leq b_{n+m}([hl], v, u; w).$$

Let $S_n^b([hk], a, b; w) = \sum_{v,u} b_n([hk], a, b; w) a^v b^u$ be the partition function of w-bridges. Multiply the inequality above by $a^v b^u$ on both sides and sum over all values of v and u. This shows that

$$\sum_{k=0}^{w} S_n^b([hk], a, b; w) S_m^b([kl], a, b; w) \leq S_{n+m}^b([hl], a, b; w). \tag{8.6}$$

Define the matrix $\mathbf{S}_n^b$ with the hk-th element given by $[\mathbf{S}_n^b]_{hk} = S_n^b([hk], a, b; w)$. Then the last inequality may be cast in matrix form:

$$\mathbf{S}_n^b \mathbf{S}_m^b \leq \mathbf{S}_{n+m}^b, \tag{8.7}$$

where for matrices $\mathbf{A} = [a_{ij}]$ and $\mathbf{B} = [b_{ij}]$, $\mathbf{A} \leq \mathbf{B}$ if $a_{ij} \leq b_{ij}$ for all i and j.

First, put $h = k = l$ in equation (8.6) to obtain

$$S_n^b([hh], a, b; w)\, S_m^b([hh], a, b; w) \leq S_{n+m}^b([hh], a, b; w). \tag{8.8}$$

The consequence is the following theorem (which is a corollary of theorem A.1 in appendix A).

Theorem 8.1 *There exists a free energy $\mathcal{S}_w(a,b)$ for adsorbing bridges in $\mathbb{S}_w$:*

$$\mathcal{S}_w(a,b) = \sup_{n>0} \tfrac{1}{n} \log S_n^b([hh], a, b; w) = \lim_{n\to\infty} \tfrac{1}{n} \log S_n^b([hh], a, b; w);$$

$\mathcal{S}_w(a,b)$ is convex in both $\log a$ and $\log b$. □

The convexity of $\mathcal{S}(a,b)$ follows from equation (3.10).

A w-bridge decomposition of w-half-space walks in $\mathbb{S}_w$ proceeds similarly as in figure 7.4 and as discussed in section 7.1.2. The heights of vertices above $B(\mathbb{S}_w)$ are preserved in this decomposition. The result is a lemma similar to lemma 7.1.

Lemma 8.2 *If $\gamma_0 > \pi\sqrt{\frac{2}{3}}$, then there exist $\epsilon > 0$, an $N_\epsilon \in \mathbb{N}$ and a constant K such that*

$$h_n([hk], v_1, v_2; w) \leq K\, e^{(\gamma_0-\epsilon)\sqrt{n}}\, b_n([hk], v_1, v_2; w) \ \textit{ for all } n \geq N_\epsilon. \qquad \square$$

Denote the partition functions of w-half-space walks by $S_n^h([hk], a, b; w)$. By multiplying the inequality in lemma (8.2) by $a^{v_1} b^{v_2}$, summing over v_1 and v_2 and then also summing over k and h, the following lemma is obtained.

Lemma 8.3 *If $\gamma_0 > \pi\sqrt{\frac{2}{3}}$, then there exist $\epsilon > 0$, an N_ϵ and a constant K such that*

$$S_n^h([A], a, b; w) \leq K\, e^{(\gamma_0-\epsilon)\sqrt{n}}\, S_n^b([A], a, b; w) \quad \textit{for all } n \geq N_\epsilon$$

and where $A \in \{hk, h*, *k, **\}$. Putting $A = 0*$ shows that

$$S_n^h(a, b; w) \leq K\, e^{(\gamma_0-\epsilon)\sqrt{n}}\, S_n^b(a, b; w), \ \textit{ for all } n \geq N_\epsilon. \qquad \square$$

The existence of a limiting free energy for w-half-space walks is a corollary of lemma 8.3.

Corollary 8.4 *The free energy $\mathcal{S}_w(a,b) = \lim_{n\to\infty} \frac{1}{n} \log S_n^h([hh], a, b; w)$.*

Proof Observe that

$$S_n^b([hk], a, b; w) \leq S_n^h([hk], a, b; w), \tag{8.9}$$

since every w-bridge is also a w-half-space walk. By lemma 8.3, if $\gamma_0 > \pi\sqrt{\frac{2}{3}}$, there exist $\epsilon > 0$, an N_ϵ and a constant K such that

$$S_n^h([hk], a, b; w) \leq K\, e^{(\gamma_0-\epsilon)\sqrt{n}}\, S_n^b([hk], a, b; w) \ \text{ for all } n \geq N_\epsilon. \tag{8.10}$$

Putting $h = k$ and taking logarithms on both sides of inequalities (8.9) and (8.10), then dividing by n and taking the limit inferior on the right-hand side of

(8.9) and the limit superior on the left-hand side of (8.10) prove existence of the limit

$$\mathcal{S}_w(a,b) = \lim_{n\to\infty} \tfrac{1}{n} \log S_n^h([hh],a,b;w) = \lim_{n\to\infty} \tfrac{1}{n} \log S_n^b([hh],a,b;w)$$

by corollary 8.5. This completes the proof. □

There is a *most popular* value $h^* \in \{0,1,\ldots,w\}$ (which is a function of $\{a,b,w,n\}$ such that

$$S_n^b([hh],a,b;w) \le S_n^b([h^*h^*],a,b;w), \quad \text{for all } h \in \{0,1,\ldots,w\}. \tag{8.11}$$

By prepending and appending edges (see equation (8.4)), the heights of the first and final vertices of a w-bridge can be changed. By accounting for the (possible) creation of new visits, this gives

$$S_n^b([h^*h^*],a,b;w) \le z(a,b)\, S_{n+2(w+1)}^b([hh],a,b;w) \ \text{ for } 0 \le h \le w, \tag{8.12}$$

where $z(a,b) = \max\{1, \frac{1}{a}, \frac{1}{b}, \frac{1}{b^2}, \frac{1}{ab}, \frac{1}{ab^2}\}$.

Comparison of equations (8.11) and (8.12) with theorem 8.1 gives the following corollary.

Corollary 8.5 *Suppose that h^* is as defined in equation (8.11). Then*

$$\mathcal{S}_w(a,b) = \lim_{n\to\infty} \tfrac{1}{n} \log S_n^b([h^*h^*],a,b;w) = \lim_{n\to\infty} \tfrac{1}{n} \log S_n^b([hh],a,b;w).$$

Similarly, by lemma 8.2,

$$\mathcal{S}_w(a,b) = \lim_{n\to\infty} \tfrac{1}{n} \log S_n^h([h^*h^*],a,b;w) = \lim_{n\to\infty} \tfrac{1}{n} \log S_n^h([hh],a,b;w).$$

Proof It only remains to prove the case for w-half-space walks. By lemma 8.3, if $\gamma_0 > \pi\sqrt{\frac{2}{3}}$, then there exist $\epsilon > 0$, an $N_\epsilon \in \mathbb{N}$ and a constant K such that

$$S_n^h([hh],a,b;w) \le K e^{(\gamma_0-\epsilon)\sqrt{n}} S_n^b([hh],a,b;w) \le S_n^b([h^*h^*],a,b;w) \ \text{ for } n \ge N_\epsilon.$$

By equation (8.12),

$$S_{n-2(w+1)}^b([h^*h^*],a,b;w) \le z(a,b)\, S_n^h([hh],a,b;w),$$

since $S_n^b([hh],a,b;w) \le S_n^h([hh],a,b;w)$. Take logarithms and divide the last two inequalities by n. Take $n \to \infty$ to complete the proof. □

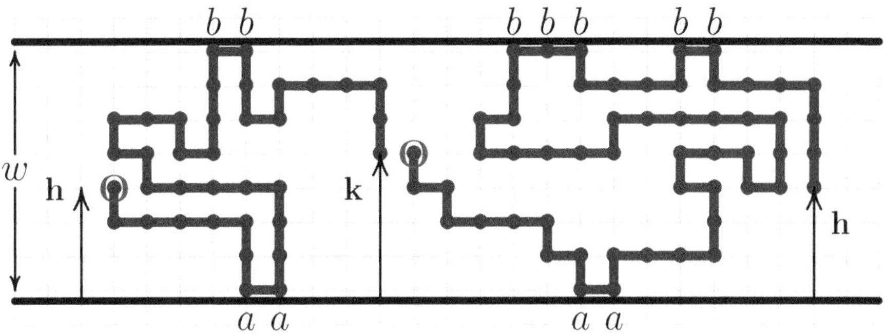

FIG. 8.5. By removing the first horizontal step of two w-bridges, reflecting the second in the $\vec{e}_1$ direction and then placing them as shown, they can be put together into a new w-bridge by placing an edge between them and by prepending a horizontal step on the left-most endpoint.

8.1.2 *The w-bridge free energy*

The most popular values of the heights $[hk]$ in the w-bridge partition function $S_n^b([hk], a, b; w)$ are $[h^*k^*]$. Consider two w-bridges of length $n+1$ each and with heights $[hk]$. Delete the first horizontal steps in each, reflect the second in the $\vec{e}_1$ direction, and then place them as illustrated in figure 8.5. Rejoin them by putting a horizontal edge between them and prepend a horizontal step on the left-most endpoint. Putting $[hk] = [h^*k^*]$ gives

$$\left(S_{n+1}^b([h^*k^*], a, b; w)\right)^2 \leq S_{2n+2}^b([h^*h^*], a, b; w) \leq S_{2n+2}^b([h^*k^*], a, b; w).$$

Take logarithms, divide by n and let $n \to \infty$. By theorem 8.1 and corollary 8.5, the following lemma is proven.

Lemma 8.6 *Let $[h^*k^*]$ be the most popular heights of the endpoints of a w-bridge. Then $\mathcal{S}_w(a,b) = \lim_{n\to\infty} \frac{1}{n} \log S_n^b([h^*k^*], a, b; w)$.* □

The partition function of w-bridges from $\vec{0} \in B(\mathbb{S}_w)$ is explicitly defined by

$$S_n^b(a, b; w) = \sum_k \sum_{v_1, v_2 \geq 0} b_n([0k], v_1, v_2; w)\, a^{v_1} b^{v_2}. \tag{8.13}$$

There is a most popular value of k, and the height of the starting point may also be relaxed from the origin to its most popular value. This shows that

$$S_n^b([00], a, b; w) \leq S_n^b(a, b; w) \leq (w+1)\, S_n^b([h^*k^*], a, b; w). \tag{8.14}$$

Take logarithms of the above and divide it by n. Take $n \to \infty$ and compare this to corollary 8.5 and lemma 8.6. This gives the following theorem.

Theorem 8.7 *The free energy* $\mathcal{S}_w(a,b) = \lim_{n\to\infty} \frac{1}{n} \log S_n^b(a,b;w)$. *Moreover,* $\mathcal{S}_w(a,b)$ *is finite for finite values of* a *and* b, *and convex in both* $\log a$ *and* $\log b$. *It is also given by corollary 8.4.* □

If $\gamma_0 > \pi\sqrt{\frac{2}{3}}$, then, by equation (8.4) and lemma 8.3, there exist $\epsilon > 0$, an N_ϵ and a constant K such that, for any $h,k \in \{0,1,\ldots,w\}$ and for all $n \geq N_\epsilon$,

$$\begin{aligned} S_{n-2w}^h([kk],a,b;w) &\leq S_{n-w}^h([hk],a,b;w) \leq S_{n-w}^h([h*],a,b;w) \\ &\leq S_{n-w}^h([**],a,b;w) \leq S_n^h(a,b;w) \leq K e^{(\gamma_0-\epsilon)\sqrt{n}} S_n^b(a,b;w). \end{aligned} \tag{8.15}$$

This is particularly so because the height of the first vertex of a half-space walk can be changed by prepending w edges. Notice that $S_n^h(a,b;w) \leq K e^{(\gamma_0-\epsilon)\sqrt{n}} S_n^b(a,b;w)$ by unfolding half-space walks into bridges (similar to equation (8.15)). A similar set of inequalities is obtained by replacing $S_{n-w}^h([h*],a,b;w)$ by $S_{n-w}^h([*k],a,b;w)$ in the above.

Similarly, note that, if $\gamma_0 > \pi\sqrt{\frac{2}{3}}$, then there exist $\epsilon > 0$, an N_ϵ and a constant K such that, for all $n > N_\epsilon$, $S_{n-w}^h([kk],a,b;w) \leq K e^{(\gamma_0-\epsilon)\sqrt{n}} S_n^b([hk],a,b;w)$. Thus, $S_n^b([hk],a,b;w) \leq S_n^b([h*],a,b;w) \leq S_n^b([**],a,b;w)$. Taking this together with equation (8.15) and noting corollary 8.4 give the following theorem.

Theorem 8.8 *Let* $S_n^*(a,b;w)$ *be any of the following:*

- $S_n^h([hh],a,b;w)$ *or* $S_n^h([hk],a,b;w)$;
- $S_n^h([h*],a,b;w)$, $S_n^h([*k],a,b;w)$ *or* $S_n^h([**],a,b;w)$;
- $S_n^h(a,b;w)$;
- $S_n^b([hh],a,b;w)$ *or* $S_n^b([hk],a,b;w)$;
- $S_n^b([h*],a,b;w)$, $S_n^b([*k],a,b;w)$ *or* $S_n^b([**],a,b;w)$;
- $S_n^b(a,b;w)$

for fixed $h,k \in \{0,1,\ldots,w\}$. *Then* $\lim_{n\to\infty} \frac{1}{n} \log S_n^*(a,b;w) = \mathcal{S}_w(a,b)$. □

By exploiting the matrix notation introduced in equation (8.7) and by defining the matrix $[\mathcal{S}_w]$ with its hk-th elements $[\mathcal{S}_w]_{hk} = \mathcal{S}_w(a,b)$, theorem 8.8 gives

$$\lim_{n\to\infty} \tfrac{1}{n} \log \mathbf{S}_n^b(a,b) = [\mathcal{S}_w]. \tag{8.16}$$

On the left-hand side, $\log \mathbf{S}_n^b(a,b)$ is the matrix $\mathbf{S}_n^b(a,b)$ with each element replaced by its logarithm.

8.1.3 *The free energy of walks in* $\mathbb{S}_w$

The existence of a limiting free energy for w-bridges and w-half-space walks in theorem 8.8 can be used to show that a w-walk from $\vec{0}$ in $\mathbb{S}_w$ has a limiting free energy, which is $\mathcal{S}_w(a,b)$.

Each half-space walk from the origin is also a self-avoiding walk from the origin. This shows that

$$S_n^h([0*], a, b; w) \leq S_n(a, b; w) \tag{8.17}$$

(see equation (8.3)). By theorem 8.8,

$$\mathcal{S}_w(a, b) \leq \liminf_{n\to\infty} \tfrac{1}{n} \log S_n(a, b; w). \tag{8.18}$$

Arguments similar to the arguments leading to equation (7.3) and theorem 7.2 will be made.

Theorem 8.9 *For all* $a, b \in (0, \infty)$*, and* $w \in \mathbb{N}$*,* $\lim_{n\to\infty} \frac{1}{n} \log S_n(a, b; w) = \mathcal{S}_w(a, b)$*.*

Proof Let ω be a walk from $\vec{0} \in \mathbb{S}_w$ of length n, with v_1 visits in $B(\mathbb{S}_w)$, and v_2 visits in $T(\mathbb{S}_w)$. Define the *top vertex* $\vec{t}$ of ω as its lexicographic top vertex, located in a $(d-1)$-dimensional hyperplane $\mathbb{B}_1$ normal to the $\vec{e}_1$ direction.

The walk ω may enter and exit $\mathbb{B}_1$ several times, but it does have a *last* vertex $\vec{t}_1$ in $\mathbb{B}_1$ at height h. Cutting ω in $\vec{t}_1$ gives two subwalks (one may be empty), namely, ω_1, from $\vec{0}$ to $\vec{t}_1$, and ω_2, from $\vec{t}_1$ (at height h in $\mathbb{B}_1$) to the end of ω.

Reflecting ω_2 through $\mathbb{B}_1$ gives a half-space walk ψ_2 from $\vec{t}_1$ at height h (since the first edge in ω_2 is normal to $\mathbb{B}_1$). Adding one edge in the $\vec{e}_1$ direction to ω_1 and then reflecting it in the $\vec{e}_1$ direction gives a second half-space walk ψ_1 from a vertex at height h.

Notice that, if ψ_1 has length $m+1$, then ψ_2 has length $n-m$.

It remains to account for the number of visits in walks ψ_1 and ψ_2. Reflecting ω_1 to ψ_1 removes the $\vec{0}$ as a root, and it becomes a visit of ψ_1 in $B(\mathbb{S}_w)$. Adding one extra edge in the $\vec{e}_1$ direction at height h to the end of ω_1 may create at most one more visit in either $B(\mathbb{S}_w)$ (if $h = 0$) or $T(\mathbb{S}_w)$ (if $h = w$).

Assume that ψ_2 has length $n-m$, $v_1 - u_1$ bottom visits and $v_2 - u_2$ top visits. Then the number of choices for ψ_2 is at most $h_{n-m}([h*], v_1 - u_1, v_2 - u_2; w)$.

For each choice of ψ_2, the walk ψ_1 has length $m+1$, and either (1) $u_1 + 1$ visits to $B(\mathbb{S}_w)$, and u_2 visits to $T(\mathbb{S}_w)$; (2) $u_1 + 2$ visits to $B(\mathbb{S}_w)$, and u_2 visits to $T(\mathbb{S}_w)$; or (3) $u_1 + 1$ visits to $B(\mathbb{S}_w)$, and $u_2 + 1$ visits to $T(\mathbb{S}_w)$. This shows that the number of choices for ψ_1 is at most the sum of $h_{m+1}([h*], u_1 + 1, u_2; w)$, $h_{m+1}([h*], u_1 + 2, u_2; w)$ and $h_{m+1}([h*], u_1 + 1, u_2 + 1; w)$.

In terms of partition functions, this gives the following upper bound

$$\begin{aligned} S_n(a, b; w) \leq{}& \left(\tfrac{1}{a} + \tfrac{1}{a^2} + \tfrac{1}{ab}\right) \\ &\times \sum_{m=0}^{n} \sum_{h=0}^{w} S_{n-m}^h([h*], a, b; w)\, S_{m+1}^h([h*], a, b; w). \end{aligned} \tag{8.19}$$

By theorem (8.8), $S_n^h([h*], a, b; w) = e^{n\mathcal{S}_w(a,b)+o(n)}$. This shows that the upper bound cannot grow at an exponential rate faster than this. Take logarithms, divide by n and let $n \to \infty$ to obtain $\limsup_{n\to\infty} \frac{1}{n} \log S_n(a, b; w) \leq \mathcal{S}_w(a, b; w)$. Comparison with equation (8.18) completes the proof. □

Consider walks counted by the partition function $S_n([**],a,b;w)$ and with first and last vertices at any height. By unfolding these walks towards the left and prepending a horizontal edge, it follows that, for $\gamma_0 > \pi\sqrt{\frac{2}{3}}$, there exist a small $\epsilon > 0$ and an N_ϵ such that, for all $n \geq N_\epsilon$, $e^{-(\gamma_0-\epsilon)\sqrt{n}} S_{n-1}([**],a,b;w) \leq (1+\frac{1}{a}+\frac{1}{b})\, S_n^h([**],a,b;w)$. In addition, $S_{n-w}^h([**],a,b;w) \leq (1+a)S_n^h([0,*],a,b;w) \leq (1+a)S_n(a,b;w)$, while $S_n^h([0,*],a,b;w) \leq S_n([**],a,b;w)$. This, together with theorem 8.9, gives the following corollary.

Corollary 8.10 *The limit* $\lim_{n\to\infty} \frac{1}{n} \log S_n([**],a,b;w) = \mathcal{S}_w(a,b)$. □

If $a=b=1$, then the *growth constant* μ_w for walks in $\mathbb{S}_w$ is defined by

$$\log \mu_w = \mathcal{S}_w(1,1) = \lim_{n\to\infty} \tfrac{1}{n} \log c_n(w), \tag{8.20}$$

where $c_n(w) = S_n([0*],1,1;w)$ is the number of self-avoiding walks of length n from $\vec{0}$ in $\mathbb{S}_w$. Notice that $\mathcal{S}_w(0,1) = \mathcal{S}_w(1,0) = \log \mu_{w-1}$, and $\mathcal{S}_w(0,0) = \log \mu_{w-2}$. By inclusion, $\mu_w \leq \mu_{w+1} \leq \mu_d$, and $\mu_0 = \mu_{d-1}$. Since $\mathbb{S}_w \subseteq \mathbb{S}_{w+1}$, $\mu_w \leq \mu_{w+1} \leq \mu_d$, where μ_d is the growth constant of self-avoiding walks.

8.2 Generating functions of walks in slabs

Define the generating function $G_w(a,b,t) = \sum_{n=0}^{\infty} S_n([0*],a,b;w)t^n$ of walks from $\vec{0}$ in $\mathbb{S}_w$ and define

$$G_w([**],a,b,t) = \sum_{n=0}^{\infty} S_n([**],a,b;w)t^n. \tag{8.21}$$

The radii of convergence of $G_w(a,b,t)$ and $G_w([**],a,b,t)$ are given by a singularity at t_c on the positive real axis in the t-plane. By corollary 8.10, $\log t_c = -\mathcal{S}_w(a,b)$.

Suppose $w>0$ and let ω be a walk in $\mathbb{S}_w$ of length $n+m$ (and endpoints at any height) and with v_1 visits in $B(\mathbb{S}_w)$, and v_2 visits in $T(\mathbb{S}_w)$. Let ω_n be the n-th vertex in ω, let $\omega[0,n] = \langle \omega_0,\omega_1,\ldots,\omega_n\rangle$ be the subwalk of length n, and let $\omega[n,n+m] = \langle \omega_n,\omega_{n+1},\ldots,\omega_{n+m}\rangle$ be the subwalk of length m. Suppose that ω_n has height h. The number of conformations of $\omega[0,n]$, with u_1 visits to $B(\mathbb{S}_w)$ and with u_2 visits to $T(\mathbb{S}_w)$ is at most $c_n([*h],u_1,u_2;w)$. Similarly, the number of conformations of $\omega[n,n+m]$ with v_1-u_1 visits to $B(\mathbb{S}_w)$ and with v_2-u_2 visits to $T(\mathbb{S}_w)$ is at most $c_m([h*],v_1-u_1,v_2-u_2;w)$, since $\omega[0,n]\cap\omega[n,n+m] = \omega_n$.

Summing over $\{u_1,u_2,h\}$ gives the following:

$$\begin{aligned} c_{n+m}([**], v_1, v_2, w) \leq & \sum_{u_1,u_2} c_n([*0], u_1+1, u_2; w)\, c_m([0*], v_1-u_1, v_2-u_2; w) \\ & + \sum_{h=1}^{w-1} \sum_{u_1,u_2} c_n([*h], u_1, u_2; w)\, c_m([h*], v_1-u_1, v_2-u_2; w) \\ & + \sum_{u_1,u_2} c_n([*w], u_1, u_2+1; w)\, c_m([w*], u_1, u_2; w). \quad (8.22) \end{aligned}$$

Notice that, in the case that $h = 0$, or $h = w$, an extra visit is needed since one is lost when the two subwalks are put together. Multiply the above by $a^{v_1} b^{v_2}$ and sum over v_1 and v_2. This gives in terms of partition functions

$$\begin{aligned} S_{n+m}([**], a, b; w) \leq & \tfrac{1}{a} S_n([*0], a, b; w)\, S_m([0*], a, b; w) \\ & + \sum_{h=1}^{w-1} S_n([*h], a, b; w)\, S_m([h*], a, b; w) \\ & + \tfrac{1}{b} S_n([*w], a, b; w)\, S_m([w*], a, b; w) \\ \leq & \left(1 + \tfrac{1}{a} + \tfrac{1}{b}\right) \sum_{h=0}^{w} S_n([*h], a, b; w)\, S_m([h*], a, b; w) \\ \leq & \left(1 + \tfrac{1}{a} + \tfrac{1}{b}\right) S_n([**], a, b; w)\, S_m([**], a, b; w). \quad (8.23) \end{aligned}$$

If $w = 0$, then a similar construction shows that

$$S_{n+m}([**], a, b; w) \leq \tfrac{1}{ab} S_n([**], a, b; w)\, S_m([**], a, b; w). \quad (8.24)$$

The last two inequalities can be combined in

$$S_{n+m}([**], a, b; w) \leq \left(1 + \tfrac{1}{a} + \tfrac{1}{b} + \tfrac{1}{ab}\right) S_n([**], a, b; w)\, S_m([**], a, b; w). \quad (8.25)$$

Thus, $(1 + \frac{1}{a} + \frac{1}{b} + \frac{1}{ab}) S_n([**], a, b; w)$ satisfies a submultiplicative inequality of the sort in theorem A.1 and corollary 8.10. Thus,

$$\log S_n([**], a, b; w) \geq n \mathcal{S}_w(a, b) - \log(1 + \tfrac{1}{a} + \tfrac{1}{b} + \tfrac{1}{ab}). \quad (8.26)$$

This gives the following theorem.

Theorem 8.11 *Suppose that $a, b > 0$. Then*

$$(1 + \tfrac{1}{a} + \tfrac{1}{b} + \tfrac{1}{ab}) G_w([**], a, b, t) \geq \frac{1}{1 - t e^{\mathcal{S}_w(a,b)}}.$$

Moreover, $G_w([**], a, b, t_c) = \infty$ if $\log t = -\mathcal{S}_w(a, b)$ (and so $t = t_c$).

Proof Exponentiate both sides of equation (8.26). Multiply both sides of the resulting equation by t^n and sum over n to obtain the claimed lower bound. Since $\log t_c = -\mathcal{S}_w(a, b)$, this lower bound shows that $G_w([**], a, b, t_c) = \infty$.

This completes the proof. □

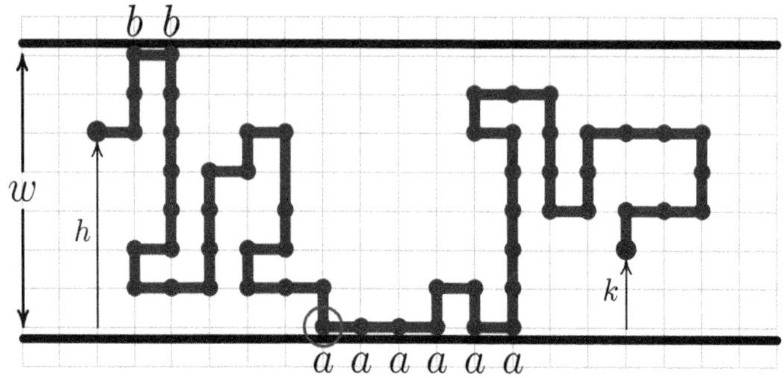

FIG. 8.6. A walk in $\mathbb{S}_w$ with root in its lexicographic first vertex in the bottom hard wall $B(\mathbb{S}_w)$. The number of such walks of length n, with endpoints at heights $[hk]$, and with v_1 visits in $B(\mathbb{S}_w)$ and v_2 visits in the top hard wall $T(\mathbb{S}_w)$ is $c_n^\ddagger([hk], v_1, v_2; w)$.

The divergence of the generating function $G_w([**], a, b, t)$ at its critical point implies that the generating function of bridges confined to $\mathbb{S}_w$ is also divergent at its critical point. To see this requires proof.

Define $c_n^\ddagger([hk]; v_1, v_2; w)$ to be the number of walks with (1) at least one vertex in the bottom wall $B(\mathbb{S}_w)$ and (2) rooted at their lexicographic bottom *visit*, and (3) of length n and with endpoints at heights h and k respectively (see figure 8.6). Define the partition function

$$V_n^\ddagger([hk], a, b; w) = \sum_{v_1, v_2 \geq 0} c_n^\ddagger([hk]; v_1, v_2; w) a^{v_1} b^{v_2}. \tag{8.27}$$

The generating function of this class of walks is

$$G_w^\ddagger([hk], a, b, t) = \sum_{n=0}^{\infty} V_n^\ddagger([hk], a, b; w) t^n. \tag{8.28}$$

Put $V_n^\ddagger([**], a, b; w) = \sum_{h,k} V_n^\ddagger([hk], a, b; w)$ and define the generating function $G_w^\ddagger([**], a, b, t)$ if the endpoints are not fixed at heights h and k. The following lemma shows that $G_w^\ddagger([**], a, b, t)$ is divergent at its critical point t_c.

Lemma 8.12 *The radius of convergence of $G_w^\ddagger([**], a, b, t)$ is $t_c = e^{-\mathcal{S}_w(a,b)}$, and $G_w^\ddagger([**], a, b, t) \to \infty$ as $t \nearrow e^{-\mathcal{S}_w(a,b)}$ for all $a, b > 0$.*

Proof Consider first the case that $d = 2$, and $w = 0$.

Then $c_n^\ddagger([**], n, n+1; w) = 1$, and $G_w^\ddagger([**], a, b, t) = \frac{b}{1-tab}$. This shows that $G_w^\ddagger([**], a, b, t)$ is divergent as $t \nearrow (ab)^{-1}$.

Next, consider the case that $d > 2$, and $w = 0$. Then $c_n^\ddagger([**], n, n+1; 0)$ is the number of walks of length n in $\mathbb{L}^{d-1}$. Each vertex (apart from the first vertex at $\vec{0}$) is weighted by ab. Hence $c_n^\ddagger([**], n, n+1; 0) = c_n^{(d-1)} a^n b^{n+1}$. By theorem 1.1 and

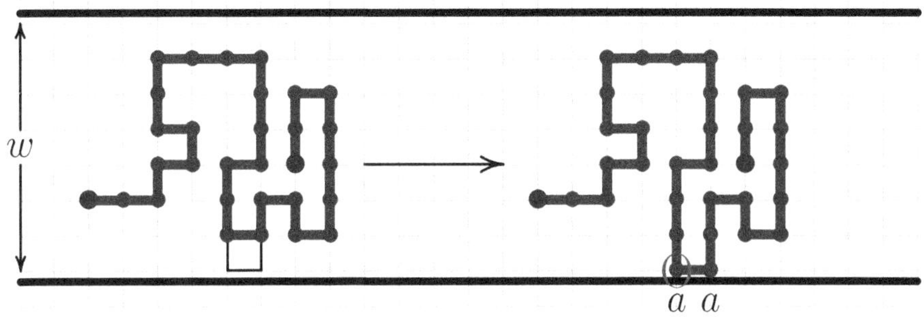

FIG. 8.7. A walk in $\mathbb{S}_w$ which is not rooted in $B(\mathbb{S}_w)$ may be rooted by finding the lexicographic bottom edge in the set of edges with midpoints of minimum height. By adding $2w$ edges towards $B(\mathbb{S}_w)$, two bottom visits are created in $B(\mathbb{S}_w)$, and one of these visits (the bottom visit) can be chosen as the root of the walk.

---

equation (1.56), it follows that $G_w^\ddagger([**],a,b,t) = \frac{1}{a}C(tab)$, where $C(t)$ is the self-avoiding walk generating function. It is divergent if t approaches $t_c = \frac{1}{ab}\mu_{d-1}^{-1}$. This shows that $\mathcal{S}_0(a,b) = \log(ab) + \log\mu_{d-1}$.

This completes the proof for $w = 0$.

Let $d \geq 2$ and let $w > 0$. Notice that $V_n^\ddagger([hk],a,b;w) \leq S_n([hk],a,b;w)$ since every walk counted by $V_n^\ddagger([hk],a,b;w)$ is also counted by $S_n([hk],a,b;w)$ with the same weight. This shows that

$$G_w^\ddagger([hk],a,b,t) \leq G_w([**],a,b,t).$$

There are two cases to consider.

Case 1: Walks counted by $c_n([hk];v_1,v_2;w)$ which have visits to $B(\mathbb{S}_w)$. These walks may be rooted in their lexicographic bottom visit.

This shows that $c_n([hk];v_1,v_2;w) \leq c_n^\ddagger([hk];v_1,v_2;w)$ if $v_2 > 0$.

Case 2: Walks counted by $c_n([hk];v_1,v_2;w)$ which have no visits to $B(\mathbb{S}_w)$ (so that $v_1 = 0$). Continue as illustrated in figure 8.7. Suppose that ω is a walk with $v_1 = 0$ (0 visits to $B(\mathbb{S}_w)$). Let E be the set of edges of ω with midpoints of minimal height $h > 0$. Choose the bottom edge of E, say $\langle \vec{s}\sim\vec{t}\rangle$. Then $\langle \vec{s}\sim\vec{t}\rangle$ is perpendicular to $\vec{e}_d$ and parallel to $\vec{e}_j$ for a value of $j \in \{1,2,\ldots,d-1\}$. Suppose that $j = 1$ (the other cases are dealt with similarly). Then the sites $\{\vec{s}-k\vec{e}_2, \vec{t}-k\vec{e}_2\}$ and $\{\vec{s}-k\vec{e}_2-\ell\vec{e}_d, \vec{t}-k\vec{e}_2-\ell\vec{e}_d\}$ are not occupied by the walk, for $k,\ell \in \mathbb{N}$.

Create a root in $B(\mathbb{S}_w)$ as follows. If $w = 1$, then replace $\langle \vec{s}\sim\vec{t}\rangle$ by the sequence of steps $\langle \vec{s}\sim(\vec{s}-\vec{e}_d)\sim(\vec{t}-\vec{e}_d)\sim\vec{t}\rangle$. This increases the length by $2w$ and creates two visits in $B(\mathbb{S}_w)$.

This process generalises to $w > 1$ as follows: suppose the height of $\langle \vec{s}\sim\vec{t}\rangle$ is $h > 0$ (and $h \leq w$). Starting at $\vec{s}$, drop a sequence of $h-1$ steps in the $-\vec{e}_d$

direction to the site $\vec{s} - (h-1)\vec{e}_d$ at height $h = 1$ above $B(\mathbb{S}_w)$. Continue from this site by appending $w - h$ edges in the $-\vec{e}_2$ direction, and then in the $-\vec{e}_d$ direction into a site $\vec{b} \in B(\mathbb{S}_w)$. Repeat this, but starting at $\vec{t}$ to a site $\vec{c} \in B(\mathbb{S}_w)$ adjacent to $\vec{b}$. This completes the construction, since one of $\vec{b}$ and $\vec{c}$ is the bottom visit and can be chosen as the new root. This increases the length by $2w$ and creates two visits in $B(\mathbb{S}_w)$ as well.

This construction shows that $c_n([hk]; 0, v_2; w) \leq c^\ddagger_{n+2w}([hk]; 2, v_2; w)$.

Summing the inequalities for cases 1 and 2, multiplying with $a^{v_1} b^{v_2}$ and then summing over $\{v_1, v_2\}$ gives

$$\sum_{v_1,v_2} c_n([hk], v_1, v_2; w) a^{v_1} b^{v_2} \leq \sum_{v_1,v_2} c^\ddagger_n([hk], v_1, v_2; w) a^{v_1} b^{v_2} + \max\left\{1, \tfrac{w}{a^2}\right\} \sum_{v_1,v_2} c^\ddagger_{n+2w}([hk], v_1, v_2; w) a^{v_1} b^{v_2}.$$

Multiplying this with t^n and then summing over n give

$$G^\ddagger_w([hk], a, b, t) \leq G_w([hk], a, b, t) \leq \left(1 + \max\left\{1, \tfrac{w}{a^2}\right\} t^{-2w}\right) G^\ddagger_w([hk], a, b, t). \tag{8.29}$$

Summing over h and k gives

$$G^\ddagger_w([**], a, b, t) \leq G_w([**], a, b, t) \leq \left(1 + \max\left\{1, \tfrac{w}{a^2}\right\} t^{-2w}\right) G^\ddagger_w([**], a, b, t). \tag{8.30}$$

By theorem 8.11, $G_w([**], a, b, t_c) = \infty$ for $t_c = e^{-\mathcal{S}_w(a,b)}$. This completes the proof for the case that $w > 0$. □

The number of walks of length n, from $\vec{0}$ in $\mathbb{S}_w$ to a vertex at height k, with v_1 visits in $B(\mathbb{S}_w)$ and with v_2 visits in $T(\mathbb{S}_w)$ is denoted by $c_n([0k], v_1, v_2; w)$. Summing over k gives $c_n([0*], v_1, v_2; w)$, and the generating function of these walks is $G(a, b, t)$ (see equation (8.21)).

A consequence of lemma 8.12 is that $G_w([0*], a, b, t) \equiv G_w(a, b, t)$ is divergent at $t = t_c = e^{-\mathcal{S}_w(a,b)}$ (and has radius of convergence equal to t_c). This is proven in theorem 8.13.

Theorem 8.13 *The radius of convergence of $G_w(a, b, t)$ in the t-plane is $t_c = e^{-\mathcal{S}_w(a,b)}$. Moreover, $G_w(a, b, t_c) = \infty$.*

Proof Each walk in $\mathbb{S}_w$ counted by $c^\ddagger_n([hk], v_1, v_2; w)$ (see figure 8.6) is rooted at its lexicographic bottom visit in $B(\mathbb{S}_w)$. By cutting these walks in their roots into two subwalks, a pair of walks in $\mathbb{S}$ from the origin, one of length ℓ, and the other of length $n - \ell$, is obtained. This shows that

$$c^\ddagger_n([**], v_1, v_2; w) \leq \sum_{\ell=0}^{n} \sum_{u_1,u_2} c_\ell([0*], v_1 - u_1, v_2 - u_2; w) c_{n-\ell}([0*], u_1, u_2; w).$$

Multiply by $a^{v_1} b^{v_2} t^n$ and sum over $\{v_1, v_2, n\}$. This gives in terms of generating functions

$$G_w^{\ddagger}([**], a, b, t) \leq (G_w([0*], a, b, t))^2 = (G_w(a, b, t))^2 .$$

Since $G_w([0*], a, b, t) \leq G_w([**], a, b, t)$, it follows from theorem 8.11 and lemma 8.12 that $t_c = e^{-\mathcal{S}_w(a,b)}$ and $G_w(a, b, t_c) = \infty$. □

Denote the generating function of half-space walks in $\mathbb{S}_w$ by

$$H_w([hk], a, b, t) = \sum_{n=0}^{\infty} S_n^h([hk], a, b; w) t^n. \tag{8.31}$$

Similarly, define the bridge generating function in $\mathbb{S}_w$ by

$$B_w([hk], a, b, t) = \sum_{n=0}^{\infty} S_n^b([hk], a, b; w) t^n. \tag{8.32}$$

A bridge decomposition of half-space walks in $\mathbb{S}_w$ may be done similarly to equations (7.28) and (7.29). This shows that

$$S_n^h([**], a, b; w) \leq \sum_{\{n_i, s_i\}} \left(\prod_{i \geq 1} S_{n_i, s_i}^b([**], a, b; w) \right), \tag{8.33}$$

where $S_{n_i,s_i}^b([**], a, b; w)$ is the partition function of bridges in $\mathbb{S}_w$ of length n_i with span in the $\vec{e}_1$ direction equal to s_i (where the summation is over $\{n_i, s_i\}$ is such that $\sum n_i = n$, and $s_{i+1} > s_i$). Multiplying by t^n and summing over n for each realisation of $\{s_i\}$ shows that

$$\begin{aligned} H_w([**], a, b, t) &\leq \prod_{s=1}^{\infty} \left(1 + \sum_{m=1}^{\infty} S_{m,s}^b([**], a, b; w) t^m \right) \\ &\leq e^{\sum_{s\geq 1} \sum_{m=1}^{\infty} S_{m,s}^b([**],a,b;w)\, t^m} = e^{B_w([**],a,b,t)-1}. \end{aligned} \tag{8.34}$$

This shows that $B(w([**], a, b, t) = \infty$ if and only if $H_w([**], a, b, t) = \infty$.

A consequence of lemma 8.13 is that the half-space walk and bridge generating functions of walks in $\mathbb{S}_w$ are also divergent at the critical point $t_c = e^{-\mathcal{S}_w(a,b)}$. The proof of this is similar to the proofs in sections 7.1.4 and 7.2.

Theorem 8.14 *The radius of convergence of $H_w([**], a, b, t)$ and $B_w([**], a, b, t)$ in the t-plane is $t_c = e^{-\mathcal{S}_w(a,b)}$.*

Moreover, $H_w([**], a, b, t_c) = B_w([**], a, b, t_c) = \infty$.

Proof The generating function of half-space walks in $\mathbb{S}_w$ may be related to the generating function $G_w([0*], a, b, t)$ as follows: multiply equation (8.19) by t^n and sum it over n. This gives

$$G_w([0*],a,b,t) \leq \left(\tfrac{1}{a}+\tfrac{1}{a^2}+\tfrac{1}{ab}\right)\sum_{h=0}^{w} H_w([h*],a,b,t) \times \left(H_w([h*],a,b,t) - S_0^h([h*],a,b;w)\right)$$
$$\leq \left(\tfrac{1}{a}+\tfrac{1}{a^2}+\tfrac{1}{ab}\right)\sum_{h=0}^{w} \left(H_w([h*],a,b,t)\right)^2.$$

Thus, $\sum_{h=0}^{w}\left(H_w([h*],a,b,t)\right)^2 = \infty$ if and only if $G_w([0*],a,b,t) = \infty$. By lemma 8.13, $G_w([0*],a,b,t)$ is finite if $t < t_c$, and infinite if $t = t_c$.

Observe that $H_w([h*],a,b,t) \leq H_w([**],a,b,t)$.

The height of the first vertex in a half-space walk can be changed by prepending $w+1$ edges in a $\ulcorner$-pattern on each such walk. This shows that $S_n^h([**],a,b;w) \leq \max\left\{1,\frac{1}{a},\frac{1}{b},\frac{1}{ab}\right\} S_{n+w+1}^h([h*],a,b;w)$, where $\max\left\{1,\frac{1}{a},\frac{1}{b},\frac{1}{ab}\right\}$ accounts for the cases when new visits in $\partial\mathbb{S}_w$ are created due to the prepending set of edges.

Thus, $H_w([**],a,b,t) = \infty$ if and only if $G_w([0*],a,b,t) = \infty$.

By equation (8.34), $B_w([**],a,b,t) = \infty$ if and only if $H_w([**],a,b,t) = \infty$.

The proof is completed by noting $G_w([0*],a,b,t) = \infty$ if and only if $t = t_c$, where $t_c = e^{-\mathcal{S}_w(a,b)}$, by theorem 8.13. □

The heights of bridge endpoints may be changed by appending and prepending sequences of $w+1$ edges. This shows that $B_w([**],a,b,t) = \infty$ if and only if $B_w([hk],a,b,t) = \infty$. A corollary of theorem 8.14 is

Corollary 8.15 *The radius of convergence of $H_w([hk],a,b,t)$ and $B_w([hk],a,b,t)$ in the t-plane is $t_c = e^{-\mathcal{S}_w(a,b)}$. Moreover, $H_w([hk],a,b,t_c) = B_w([hk],a,b,t_c) = \infty$. In addition, the radius of convergence of $G_w([hk],a,b,t)$ is $t_c = e^{-\mathcal{S}_w(a,b)}$, and $G_w([hk],a,b,t_c) = \infty$.* □

8.3 A pattern theorem for walks in $\mathbb{S}_w$

A pattern theorem for self-avoiding walks in slabs will be proven by following methods similar to the methods for walks in slits and prisms pioneered in reference [522] (by using the Perron-Frobenius theorem).

A *proper pattern* P in a walk in $\mathbb{S}_w$ is defined in the same way as before (see section 7.1.6) [316]. Theorems C.2 and C.3 (in appendix C) for walks in $\mathbb{S}_w$ can also be proven without much difficulty. That is, if P is a proper pattern in $\mathbb{S}_w$, then there is a w-bridge ϕ in which P occurs at least once (bridges and prime bridges are defined as in section 7.1.6). In what follows, use the convention that the first vertex of a bridge or a prime bridge is not weighted by a or b if it is a visit to $\partial\mathbb{S}_w$.

The decomposition of bridges into prime bridges proceeds in the same way as in section 7.1.6 but the confining space $\mathbb{S}_w$ causes complications. For example, two prime bridges can only be concatenated if the height of the endpoint of the first is the same as the height of the first vertex of the second (in $\mathbb{S}_w$).

Define the matrix $\mathbf{B}$ so its hk-th element is the generating function of bridges from a vertex at height h to a vertex at height k:

$$[\mathbf{B}]_{hk} = [B_w([hk], a, b, t)] \,. \tag{8.35}$$

The product $\mathbf{BB}$ has elements which are the concatenation of bridges with end-points of appropriate heights. That is, the hk-th element of the product is the sum of the products of generating functions given by

$$[\mathbf{BB}]_{hk} = \sum_{\ell=0}^{w} B_w([h\ell], a, b, t) B_w([\ell k], a, b, t). \tag{8.36}$$

Since $B_w([hk], a, b, t) > 0$ for all $a, b > 0$, $\mathbf{B}$ is an irreducible matrix.

Next, define a matrix $\mathbf{Q}$ with its hk-th element given by the prime bridge generation function

$$[\mathbf{Q}]_{hk} = [Q_w([hk], a, b, t)] \,. \tag{8.37}$$

Here, $Q_w([hk], a, b, t)$ is the generating function of prime bridges in $\mathbb{S}_w$ and with the first vertex at height h and the last vertex at height k. Since $Q_w([hk], a, b, t) > 0$ for all $a, b > 0$, $\mathbf{Q}$ is an irreducible matrix.

The prime bridge decomposition of bridges in $\mathbb{S}_w$ is given by

$$\mathbf{B} - \mathbf{I} = \mathbf{QB}, \;\; \text{so that } \mathbf{B} = (\mathbf{I} - \mathbf{Q})^{-1}, \tag{8.38}$$

where $\mathbf{I}$ is the identity matrix.

Notice the following about $\mathbf{Q}$: for each $t \in (0, t_c)$, $\mathbf{Q}$ is a positive matrix with finite elements. If the spectral radius of $\mathbf{Q}$ is strictly less than 1, then the spectral radius of $\mathbf{B}$ is positive (on the real axis) and is associated with an eigenvector with all positive elements (by the Perron-Frobenius theorem).

By corollary 8.15, $\mathbf{B}$ is a finite positive matrix for all $0 < t < t_c$ but, at $t = t_c$, some (or all) elements of $\mathbf{B}$ are infinite. This is only possible if $\mathbf{I} - \mathbf{Q}$ is a singular matrix at $t = t_c$. That is, when $\det(\mathbf{I} - \mathbf{Q}) = 0$, and the spectral radius of $\mathbf{Q}$ is $\rho(\mathbf{Q}) = 1$ by the Perron-Frobenius theorem (this is when $t = t_c$ where $\rho(\mathbf{Q})$ is a simple root of the characteristic polynomial of $\mathbf{Q}$). Moreover, if $t < t_c$, then $\rho(\mathbf{Q}) < 1$ (since $\mathbf{Q}$ is a positive $(w+1) \times (w+1)$-matrix with characteristic polynomial of degree $w+1$ and since all the elements of $\mathbf{B}$ are finite if $\rho(\mathbf{Q}) < 1$).

Let P be a prime bridge pattern, let $\mathbf{B}[\widetilde{P}]$ be the matrix of generating functions of bridges not containing P and let $\mathbf{Q}[\widetilde{P}]$ be the matrix of generating functions of prime bridges not containing P. The prime decomposition of bridges not containing P gives, similar to the above,

$$\mathbf{B}[\widetilde{P}] - \mathbf{I} = \mathbf{Q}[\widetilde{P}]\mathbf{B}[\widetilde{P}], \;\; \text{so that } \mathbf{B}[\widetilde{P}] = (\mathbf{I} - \mathbf{Q}[\widetilde{P}])^{-1}. \tag{8.39}$$

Similar to $\mathbf{B}$ and $\mathbf{Q}$, the matrices $\mathbf{B}[\widetilde{P}]$ and $\mathbf{Q}[\widetilde{Q}]$ are positive and irreducible for $t > 0$, and for t less than the spectral radius of $\mathbf{B}[\widetilde{P}]$.

Arguing as above, there is a critical value t_P such that $\mathbf{B}[\widetilde{P}]$ is not singular for all $t \in (0, t_P)$ (where $\rho(\mathbf{Q}[\widetilde{P}]) < 1$). By the Perron-Frobenius theorem, $\rho(\mathbf{Q}[\widetilde{P}])$ is a simple root of the characteristic polynomial of $\mathbf{Q}[\widetilde{P}]$. As before, $\mathbf{B}[\widetilde{P}]$ is singular when $\rho(\mathbf{Q}[\widetilde{P}]) = 1$ (and when $t = t_P$).

Notice that all the elements of $\mathbf{B}$ and $\mathbf{B}[\widetilde{P}]$ are continuous functions of t within their radii of convergence, and the determinants $\det(\mathbf{I} - \mathbf{Q})$ and $\det(\mathbf{I} - \mathbf{Q}[\widetilde{P}])$ are polynomials in the Q_w of degree $w+1$.

Since each element of $\mathbf{Q}[\widetilde{P}]$ is strictly less than the corresponding element of $\mathbf{Q}$, and since $\mathbf{Q}$ and $\mathbf{Q}[\widetilde{P}]$ are irreducible matrices, $\rho(\mathbf{Q}[\widetilde{P}]) < \rho(\mathbf{Q})$ (for all $t \in (0, t_c)$); this is a corollary of Wielandt's theorem (see for example theorem 6.4 and its corollary in reference [501]). Thus, when $t = t_c$, then $\rho(\mathbf{Q}[\widetilde{P}]) < \rho(\mathbf{Q}) = 1$, so $\rho(\mathbf{Q}[\widetilde{P}]) < 1$, and $\mathbf{B}[\widetilde{P}]$ is not singular. Instead, $\mathbf{B}[\widetilde{P}]$ is singular when $\rho(\mathbf{Q}[\widetilde{P}]) = 1$ at $t = t_P$, which implies that $t_P > t_c$.

By theorem 8.8 and corollary 8.15, this shows that

$$\limsup_{n\to\infty} \tfrac{1}{n} \log S_n^b([hk], a, b; w; [0, P]) < \lim_{n\to\infty} \tfrac{1}{n} \log S_n^b([hk], a, b; w) = \mathcal{S}_w(a, b),$$

since $-\log t_P < -\log t_c = \mathcal{S}_w(a, b)$. This gives the following theorem, which is analogous to theorem 7.12.

Theorem 8.16 *If P is a prime w-pattern, then there exist an $\epsilon > 0$ and an $N_\epsilon \in \mathbb{N}$ such that, for all $n \geq N_\epsilon$, $S_n^b([hk], a, b; w; [0, P]) < (e^{\mathcal{S}_w(a,b)}(1 - \epsilon))^n$.* □

This theorem is the basis of a pattern theorem for w-walks.

Denote the number of w-walks of length n from $\vec{0}$, with endpoint in $B(\mathbb{S}_w)$, with v_1 visits to $B(\mathbb{S}_w)$, with v_2 visits to $T(\mathbb{S}_w)$ and which contains the prime pattern P at most k times by $c_n([00], v_1, v_2; w; [\leq k, P])$. Let the partition function be $S_n([00], a, v_2; w; [\leq k, P])$. Since the height of the endpoints of unfolded walks can be changed by adding sequences of edges to endpoints and since the unfolding of a walk changes the partition function by a multiplicative factor which is subexponential and does not change the number of visits to $B(\mathbb{S}_w)$ or $T(\mathbb{S}_w)$, it follows that the methods leading to theorem 7.13 and corollary 7.14 apply to the case here with little change. The result is the following pattern theorem for walks in $\mathbb{S}_w$.

Theorem 8.17 *If P is a proper w-pattern, then there exists an $\epsilon_0 > 0$ such that, for all $\epsilon < \epsilon_0$,*

$$\limsup_{n\to\infty} \tfrac{1}{n} \log S_n([**], a, b; w; [\leq \epsilon n, P]) < \mathcal{S}_w(a, b).$$

Thus, for walks from $\vec{0} \in B(\mathbb{S}_w)$,

$$\limsup_{n\to\infty} \tfrac{1}{n} \log S_n([0*], a, b; w; [\leq \epsilon n, P]) < \mathcal{S}_w(a, b).$$

The same theorem applies if S_n is replaced by S_n^h or S_n^b and if $[**]$ is replaced by $[h*]$, $[*k]$ or $[hk]$.

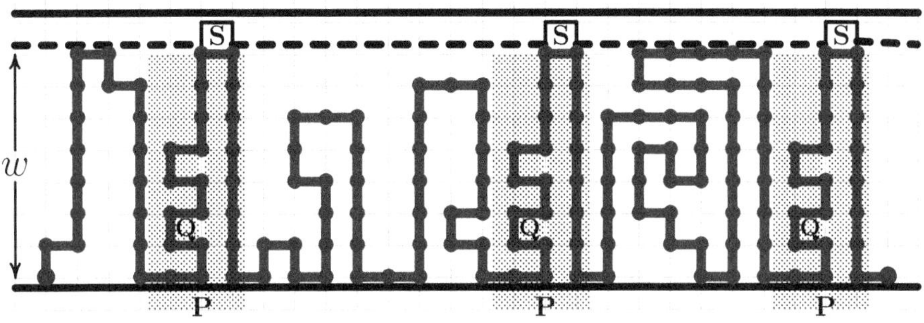

FIG. 8.8. A walk in $\mathbb{S}_w$ can be placed in $\mathbb{S}_{w+1}$ by adding steps in the locations marked by S. If this is done in prime patterns P by removing edges from the squares labelled by Q, then the walks do not change length.

Proof Since every proper pattern is a subpattern of a prime pattern, it is sufficient to assume that P is a prime pattern.

The case for $S_n^b([hk], a, b; w; [\leq \epsilon n, P])$ is a corollary of theorem 8.16. By summing over the heights of endpoints, the theorem is proven for bridges. It only remains to verify that S_n^b can be replaced by S_n^h or S_n. But these are consequences of theorems 8.13 and 8.14 and corollary 8.15. □

8.4 Growth constants and free energies of walks in slabs

8.4.1 *The growth constant of walks in slabs*

Consider the case that $a = b = 1$ so that $\mathcal{S}_w(1,1) = \log \mu_w$, where μ_w is the growth constant of walks in $\mathbb{S}_w$. By theorem 8.8, equation (8.20) and corollary 8.15, it is sufficient to consider w-bridges with both endpoints in $B(\mathbb{S}_w)$ in order to examine μ_w.

Suppose that P is the prime w-pattern illustrated in figure 8.8. Let ϕ be a w-bridge contributing to the count $b_n([00]; v_1, v_2, [> \lfloor \epsilon n \rfloor, P])$ so that P occurs at least $\lfloor \epsilon n \rfloor + 1$ times, for a small $\epsilon > 0$.

Choose any m of the P in ω. Delete a $\sqsubset$-conformation of three edges (marked by Q in figure 8.8) and place a $\sqcap$-conformation of three edges in P at the position marked by S in figure 8.8. This gives a w-bridge in $\mathbb{S}_{w+1}$.

There are $\sum_{v_1,v_2} b_n([00]; v_1, v_2, w; [> \lfloor \epsilon n \rfloor, P])$ choices for ω and, for each such bridge, at least $\binom{\lfloor \epsilon n \rfloor}{m}$ ways to choose m places where the construction can be performed. Hence,

$$\binom{\lfloor \epsilon n \rfloor}{m} \sum_{v_1,v_2} b_n([00]; v_1, v_2, w; [> \lfloor \epsilon n \rfloor, P]) \leq \sum_{v_1,v_2} b_n([00]; v_1, 2m, w). \qquad (8.40)$$

By theorem 8.17, $\epsilon > 0$ can be chosen small so that

$$\lim_{n\to\infty} \tfrac{1}{n} \log \sum_{v_1,v_2} b_n([00]; v_1, v_2, w; [>\lfloor \epsilon n \rfloor, P]) = \mathcal{S}_w(1,1) = \log \mu_w. \qquad (8.41)$$

Let $\delta < \epsilon$ and put $m = \lfloor \delta n \rfloor$ in equation (8.40). Take the power $\frac{1}{n}$ on both sides and let $n \to \infty$. Then

$$\left(\frac{\epsilon^\epsilon}{\delta^\delta (\epsilon - \delta)^{\epsilon-\delta}} \right) \mu_w \leq \mu_{w+1}. \qquad (8.42)$$

In particular, if $\delta = \frac{1}{2}\epsilon$, then $\mathcal{S}_w(1,1) + \epsilon \log 2 \leq \mathcal{S}_{w+1}(1,1)$. Thus, $\mathcal{S}_w(1,1) < \mathcal{S}_{w+1}(1,1)$ and so $\mu_w < \mu_{w+1}$. This gives the following lemma.

Lemma 8.18 *The growth constant $\mu_w < \mu_{w+1}$ and $\mu_w < \mu_d$.* □

Since $\mu_w < \mu_d$, and $\langle \mu_w \rangle$ is a monotonic sequence, $\lim_{w\to\infty} \mu_w \leq \mu_d$ exists.

Theorem 8.19 *The limit $\lim_{w\to\infty} \mu_w = \mu_d$.*

Proof Let $\partial \mathbb{L}^d_+$ be the boundary of the half-lattice $\mathbb{L}^d_+$ (so that $\partial \mathbb{L}^d_+$ is orthogonal to $\vec{e}_d$). Then $\partial \mathbb{L}^d_+$ is the base plane of loops in $\mathbb{L}^d_+$ (see equation (2.20) and section 7.1.4). Let ℓ_n be the number of loops of length n from $\vec{0}$ in $\mathbb{L}^d_+$. By theorem 7.6, $\lim_{n\to\infty} \frac{1}{n} \log \ell_n = \log \mu_d$.

By doubly unfolding loops in the $\vec{e}_1$ direction (see section 7.1.3) and by prepending a horizontal edge, bridges which are also loops in $\mathbb{L}^d_+$ are obtained. Let the number of such bridges of length n from $\vec{0}$ be $\ell^\ddagger_n$. Then $\lim_{n\to\infty} \frac{1}{n} \log \ell^\ddagger_n = \log \mu_d$, by the constructions in section 7.1.3.

Two bridges which are also loops can be concatenated in a way similar to the concatenation in figure 8.4. This shows that $\ell^\ddagger_n \ell^\ddagger_m \leq \ell^\ddagger_{n+m}$; therefore, $\ell^\ddagger_n$ is a supermultiplicative function. Let $\epsilon > 0$ then there exists an $N > 0$ such that $\frac{1}{n} \log \ell^\ddagger_n > \log \mu_d - \epsilon$ for all $n \geq N$. For this value of N, all loops counted by $\ell^\ddagger_N$ fit into $\mathbb{S}_w$, with $w \geq \lceil \frac{N}{2} \rceil = w_0(N)$.

Concatenate p bridges which are also loops. This shows that

$$\left(\ell^\ddagger_N \right)^p \leq \sum_{v_1,v_2} c_{pN}([00], v_1, v_2; w) \leq c_{pN}(w)$$

for all $w \geq w_0(N)$, where $c_n(w)$ is the number of self-avoiding walks of length n from $\vec{0}$ in $\mathbb{S}_w$ (see equation (8.20)). Take logarithms of the above, divide by pN and let $p \to \infty$. This shows that $\log \mu_w \geq \frac{1}{N} \log \ell^\ddagger_N \geq \log \mu_d - \epsilon$. Thus, for any $w \geq w_0(N)$, $\log \mu_w \geq \log \mu_d - \epsilon$. □

8.4.2 *The free energy $\mathcal{S}_w(a,1)$*

The free energy $\mathcal{S}_w(a,1)$ is the free energy of walks that are in $\mathbb{S}_w$ and which adsorb onto $B(\mathbb{S}_w)$. By theorem 8.1, $\mathcal{S}_w(a,1)$ is convex in $\log a$, and $\mathcal{S}_w(a,1) = \mathcal{S}_w(1,a)$.

The construction in figure 8.8 changes the number of visits in $T(\mathbb{S}_w)$ but not in $B(\mathbb{S}_w)$. Multiply equation (8.40) by a^{v_1} and sum over v_1 and v_2. This gives in terms of partition functions

$$\binom{\lfloor \epsilon n \rfloor}{m} S_n^b([00]; a, 1, w; [>\lfloor \epsilon n \rfloor, P]) \leq n\, S_n^b([00]; a, 1, w). \tag{8.43}$$

Fix $\epsilon > 0$ small so that $\limsup_{n\to\infty} \frac{1}{n} \log S_n([00], a, 1, w; [>\lfloor \epsilon n \rfloor, P]) = \mathcal{S}_w(a, 1)$ by theorem 8.17. Let $\delta \in (0, \epsilon)$ and put $m = \lfloor \delta n \rfloor$. Take the power $\frac{1}{n}$ and let $n \to \infty$. This gives the following theorem.

Theorem 8.20 *For any $a > 0$, $\mathcal{S}_w(a, 1)$ is a strictly increasing function of w. There exists small $\epsilon > \delta > 0$ such that*

$$\left(\frac{\epsilon^\epsilon}{\delta^\delta (\epsilon - \delta)^{\epsilon - \delta}} \right) e^{\mathcal{S}_w(a,1)} \leq e^{\mathcal{S}_{w+1}(a,1)}.$$

If $\delta = \frac{\epsilon}{2}$, then $\mathcal{S}_w(a, 1) + \epsilon \log 2 \leq \mathcal{S}_{w+1}(a, 1)$. That is, $\mathcal{S}_w(a, 1) < \mathcal{S}_{w+1}(a, 1)$. Similarly, $\mathcal{S}_w(1, b) < \mathcal{S}_{w+1}(1, b)$. □

Since $S_n(a, b; w) \leq a^n b^n c_n(w)$ if $a, b \geq 1$ (see equation (8.3)), it is the case that $\mathcal{S}_w(a, 1) \leq \log a + \log \mu_w$. Similarly, if $a < 1$, then $S_n(a, 1; w) \leq c_n(w)$. Thus, the limit $\lim_{w\to\infty} \mathcal{S}_w(a, 1)$ exists (as does the limit $\lim_{w\to\infty} \mathcal{S}(1, b)$).

8.5 Polygons and walks in wedges

Let $\vec{v} \in \mathbb{L}^d$ and let $\vec{v}(j)$ be the j-th Cartesian coordinate of $\vec{v}$. Define the vector function $\vec{f} = \langle f_2, f_3, \ldots, f_d \rangle$ so that $\vec{f}$ is a sequence of non-negative real-valued functions f_j on $\mathbb{R}$. The f-*wedge* is defined by

$$W_f = \left\{ \langle \vec{v} \sim \vec{u} \rangle \in \mathbb{L}^d \;\middle|\; 0 \leq \vec{v}(i) \leq f_i(\vec{v}(i)), 0 \leq \vec{u}(i) \leq f_i(\vec{u}(i)), 2 \leq i \leq d \right\}. \tag{8.44}$$

Assume that $f_2(v) \geq 1$ for $v \in \mathbb{N}_0$, so the lattice sites $\vec{v}_0 = (m, 0, \ldots, 0) \in W_f$ and $\vec{v}_1 = (m, 1, 0, \ldots, 0) \in W_f$ for all $m \in \mathbb{N}_0$.

A self-avoiding walk or a polygon is in W_f if every vertex in the walk or polygon is contained in W_f.

Let $c_n(f)$ be the number of self-avoiding walks of length n from $\vec{0}$ in W_f and let $p_n(f)$ be the number of polygons of length n which pass through $\vec{0}$ in W_f. Then $c_n(f) > 0$ for $n \geq 0$, and $p_n(f) > 0$ for even $n \geq 4$.

Since the removal of one or two edges incident with $\vec{0}$ in a polygon in W_f gives a walk from $\vec{0}$,

$$\liminf_{n\to\infty} \tfrac{1}{n} \log p_n(f) \leq \liminf_{n\to\infty} \tfrac{1}{n} \log c_n(f) \leq \log \mu_d, \tag{8.45}$$

where $\liminf_{n\to\infty} \frac{1}{n} \log p_n(f)$ is taken through even values of $n \in \mathbb{N}$.

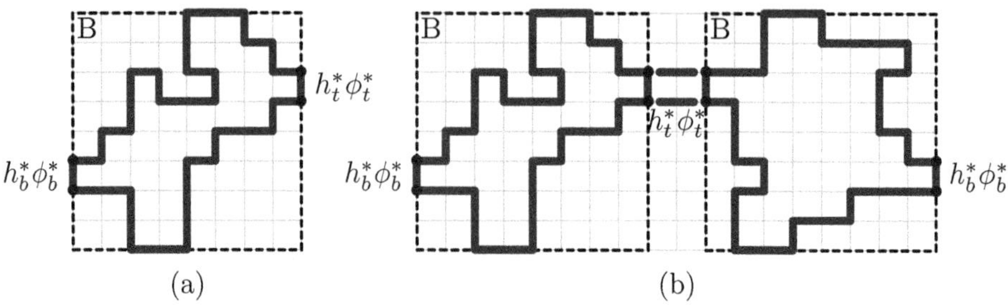

FIG. 8.9. Polygons in boxes B. (a) For a given n, there are most popular sizes of the boxes, and most popular locations and orientations of the bottom and top edges of the polygons. This gives the most popular class of polygons. These polygons can be concatenated as shown in (b).

8.5.1 *Polygons in wedges*

Let $\mathbb{L}_+^d$ be positive half-lattice with boundary $\partial\mathbb{L}_+^d$ (see equation (2.20)). Then $W_f \subseteq \mathbb{L}_+^d$. Let ω be a positive lattice polygon rooted at $\vec{0}$ in $\mathbb{L}_+^d$. Order the edges of ω lexicographically by the coordinates of their midpoints The bottom edge $\vec{b}(\omega)$ of ω is the lexicographic bottom edge in the ordering. Notice that $\vec{b}(\omega) \perp \vec{e}_1$ so that it is parallel to a unit vector $\phi_b \in \{\vec{e}_2, \vec{e}_3, \ldots, \vec{e}_d\}$.

There is a smallest d-dimensional rectangular box $B \subseteq \mathbb{L}_+^d$ containing ω. Denote the boundary of B by ∂B. The midpoint of $\vec{b}(\omega)$ is a point h_b in ∂B. Denote the position and orientation of $\vec{b}(\omega)$ by $h_b\phi_b$.

Similarly, define the top edge $\vec{t}(\omega)$ of ω by a lexicographic ordering of edges of ω by their midpoints. Denote the orientation and position of $\vec{t}(\omega)$ by $h_t\phi_t$.

Let $p_n[h_b\phi_b, h_t\phi_t]$ be the number of polygons in a box B, with their bottom and top edges of orientation and position in ∂B given by $[h_b\phi_b, h_t\phi_t]$.

Notice that there are at most n^d choices for the dimensions of B, that h_b and h_t have at most n^{d-1} values each and that ϕ_b and ϕ_t have at most $d-1$ possible orientations each. There is a most popular box size B, denoted B^*, for which there are most popular values of $[h_b\phi_b, h_t\phi_t]$ denoted by $[h_b^*\phi_b^*, h_t^*\phi_t^*]$ (see figure 8.9). Since there are at most n^d choices for B and at most $(d-1)^2 n^{2(d-1)}$ for $[h_b\phi_b, h_t\phi_t]$,

$$p_n[h_b^*\phi_b^*, h_t^*\phi_t^*] \le p_n \le (d-1)^2 n^{3d-2} p_n[h_b^*\phi_b^*, h_t^*\phi_t^*]. \qquad (8.46)$$

Notice that, by reflection symmetry, $p_n[h_b^*\phi_b^*, h_t^*\phi_t^*] = p_n[h_t^*\phi_t^*, h_b^*\phi_b^*]$. This shows that

$$\lim_{n\to\infty} \tfrac{1}{n} \log p_n[h_b^*\phi_b^*, h_t^*\phi_t^*] = \lim_{n\to\infty} \tfrac{1}{n} \log p_n[h_t^*\phi_t^*, h_b^*\phi_b^*] = \log \mu_d. \qquad (8.47)$$

These most popular classes of polygons, with the construction in section 3.2.2 and figure 3.4, give the following theorem.

Theorem 8.21 *Let W_f be an f-wedge and suppose that, for at least one value of i, $f_i(v) \geq 1$ for $v \in \mathbb{N}_0$. Then $\liminf_{n\to\infty} \frac{1}{n} \log p_n(f) \geq \log \mu_d$ if and only if $\lim_{x\to\infty} f_i(x) = \infty$ for all $2 \leq i \leq d$.*

Proof Define

$$a_n = \min \Big\{ m \;\Big|\; f_i(k) \geq n+1, \text{ for all } k \geq m \text{ and } m,k \in \mathbb{N}; \text{ for all } i = 2,\ldots,d \Big\}.$$

Choose polygons of length n in a box B^* of most popular size ℓ_*. Let B_1^* and B_2^* be two such boxes and place them in $\mathbb{L}^d$ to rest on $\partial\mathbb{L}^d$ such that $B_2^* = B_1^* + (\ell_* + 1)\vec{e}_1$.

Choose a polygon ω_1 of length n and of most popular class $[h_b^*\phi_b^*, h_t^*\phi_t^*]$ in B_1^*, and a polygon ω_2 of length n and of most popular class $[h_t^*\phi_t^*, h_b^*\phi_b^*]$ in B_2^*. Suppose that ω_1 has top edge $\vec{t}(\omega_1)$ and that ω_2 has bottom edge $\vec{b}(\omega_2)$. By construction, $\vec{b}(\omega_2) = \vec{t}(\omega_1) + \vec{e}_1$. This placement of B_1^* and B_2^* and choice of polygons are such that ω_1 and ω_2 can be concatenated, using the construction in figure 8.9. This gives a polygon of length $2n$ (see figure 8.9).

Notice that B_1^* and B_2^* may be placed in this way inside the wedge W_f if B_1^* is placed such that its corners have $\vec{e}_1$-coordinates bigger than or equal to a_n.

Generalise this construction as follows.

Concatenate N polygons in boxes $\langle B_1^*, B_2^*, \ldots, B_N^* \rangle$ (of most popular sizes) by placing the boxes as above in sequence. In each box B_{2j-1}^* choose a polygon of length n of most popular class $[h_b^*\phi_b^*, h_t^*\phi_t^*]$, and, in each box B_{2j}^*, a polygon of length n and most popular class $[h_t^*\phi_t^*, h_b^*\phi_b^*]$. This gives a polygon of length nN in the wedge W_f. This polygon may be rooted in $\vec{0}$ by running two walks of $2a_n + K$ edges, one from $\vec{0}$, and the other from $\vec{1} = (0,1,0,\ldots,0)$, to the box B_1^* and then to the endpoints of the bottom edge of the polygon in B_1^*. Translate B_1^* in the $\vec{e}_1$ direction and increase K, if necessary, to accommodate this (see figure 3.4 and section 3.2.2).

The resulting polygon has length $2a_n + K + nN$ in W_f. For each polygon in a box B_i^*, there are $p_n[h_b^*\phi_b^*, h_t^*\phi_t^*]$ choices. This shows that

$$\left(p_n[h_b^*\phi_b^*, h_t^*\phi_t^*]\right)^N \leq p_{2a_n+K+nN}(f).$$

Take logarithms and divide by nN to obtain

$$\tfrac{1}{n} \log p_n[h_b^*\phi_b^*, h_t^*\phi_t^*] \leq \tfrac{1}{nN} \log p_{2a_n+K+nN}(f).$$

Define the sequence $\langle m_i \rangle$ by

$$\liminf_{n\to\infty} \tfrac{1}{n} \log p_n(f) = \lim_{i\to\infty} \tfrac{1}{m_i} \log p_{m_i}(f).$$

For fixed n and for each m_i, there is a largest N_i such that $2a_n + K + nN_i \leq m_i \leq 2a_n + K + n(N_i+1)$. Since $p_n(f) \leq p_{n+2}(f)$,

$$\tfrac{1}{n} \log p_n[h_b^*\phi_b^*, h_t^*\phi_t^*] \leq \tfrac{1}{nN_i} \log p_{2a_n+K+nN_i}(f) \leq \tfrac{1}{nN_i} \log p_{m_i}(f).$$

Take $i \to \infty$ on the right-hand side. Then $N_i \to \infty$, and $m_i \to \infty$. Since $\frac{m_i}{nN_i} \to 1$, this gives

$$\tfrac{1}{n} \log p_n[h_b^*\phi_b^*, h_t^*\phi_t^*] \leq \liminf_{m\to\infty} \tfrac{1}{m} \log p_m(f).$$

Finally, take $n \to \infty$ on the left-hand side. It follows from equation (8.47) that $\log \mu_d \leq \liminf_{m\to\infty} \frac{1}{m} \log p_m(f)$. □

A corollary of theorem 8.21 is that the growth constant of walks confined to an f-wedge is μ_d if and only if $\lim_{x\to\infty} f_i(x) = \infty$ for all $2 \leq i \leq d$.

Theorem 8.22 *Let W_f be an f-wedge. If $c_n(f)$ is the number of walks from the origin in W_f of length n, then $\lim_{n\to\infty} \frac{1}{n} \log c_n(f) = \log \mu_d$ if and only if $\lim_{x\to\infty} f_i(x) = \infty$ for all $2 \leq i \leq d$.*

Proof Since $\lim_{x\to\infty} f_i(x) = \infty$ for all $2 \leq i \leq d$, there is an x_0 such that $f_2(x) \geq 1$ for all $x \geq x_0$. Let $p_n(f, x_0)$ be the number of polygons of length n in W_f and rooted at $(x_0, 0, 0, \ldots, 0)$.

By theorem 8.21, $\lim_{n\to\infty} \frac{1}{n} \log p_n(f, x_0) = \log \mu_d$.

Let ω be a polygon of length n, contributing to $p_n(f, x_0)$. Delete one edge incident with x_0 in ω and run a sequence of x_0 edges along the $\vec{e}_1$ direction from $\vec{0}$ to the vertex $(x_0, 0, 0, \ldots, 0)$. The result is a walk from the origin of length $x_0 + n - 1$ in W_f. This shows that $p_n(f, x_0) \leq c_{x_0+n-1}(f) \leq c_{x_0+n-1}$. Similarly, by deleting two edges in ω, $p_n(f, x_0) \leq 2dc_{x_0+n-2}(f) \leq 2dc_{x_0+n-2}$. Take logarithms, divide by n and let $n \to \infty$ (through even numbers). This gives $\lim_{n\to\infty} \frac{1}{n} \log c_n(f) = \log \mu_d$. This completes the proof. □

8.5.2 *Polygons and walks in general wedges*

Let $\vec{v}$ be a vertex in $\mathbb{L}^d$ and denote the j-th Cartesian component of $\vec{v}$ by $\vec{v}(j)$. Let both $f = \langle f_2, f_3, \ldots, f_d\rangle$ and $g = \langle g_2, g_3, \ldots, g_d\rangle$ be sequences of $(d-1)$ non-negative real-valued functions on $\mathbb{Z}$. The *fg-wedge* is defined by

$$\begin{aligned} W_{fg} = \{\langle \vec{v} \sim \vec{u}\rangle \in \mathbb{L}^d \mid\ & g_i(\vec{v}(i)) \leq \vec{v}(i) \leq f_i(\vec{v}(i)), \text{ and} \\ & g_i(\vec{u}(i)) \leq \vec{u}(i) \leq f_i(\vec{u}(i)), \text{ for all } 2 \leq i \leq d\}. \end{aligned} \tag{8.48}$$

Assume that, for at least one value of i, $f_i(x) \geq 1 + g_i(x)$ for all $x \in \mathbb{N}_0$. Then a polygon of even length $n \geq 4$ can be placed in W_{fg} such that it is rooted at $\vec{0}$.

Define $p_n(fg)$ to be the number of polygons in W_{fg}, rooted at $\vec{0}$. Similarly, let $c_n(fg)$ be the number of walks from $\vec{0}$ in W_{fg}.

Similar to the discussion in section 8.5.1, let ω_1 and ω_2 be polygons of length n in the most popular classes given by $[h_b^*\phi_b^*, h_t^*\phi_t^*]$ and $[h_t^*\phi_t^*, h_b^*\phi_b^*]$, respectively (and in boxes B^* of most popular size). Concatenate ω_1 and ω_2 (see figure 8.9 and use reflection symmetry) to find that

$$(p_n[h_b^*\phi_b^*, h_t^*\phi_t^*])^2 = p_n[h_b^*\phi_b^*, h_t^*\phi_t^*]\, p_n[h_t^*\phi_t^*, h_b^*\phi_b^*] \leq p_{2n}[h_b^*\phi_b^*, h_b^*\phi_b^*]. \tag{8.49}$$

By equation (8.47) (and since $p_n[h_b^*\phi_b^*, h_b^*\phi_b^*] \leq p_{n+2}[h_b^*\phi_b^*, h_b^*\phi_b^*]$), the following lemma is established.

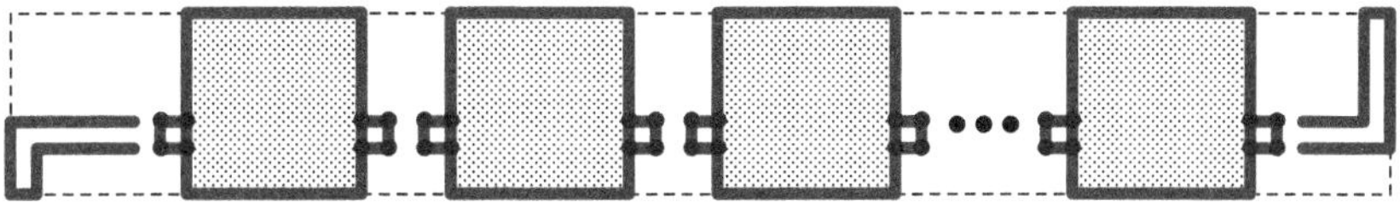

FIG. 8.10. Concatenating most popular polygons in a row.

Lemma 8.23 *The limit* $\lim_{n\to\infty} \frac{1}{n} \log p_n[h_b^*\phi_b^*, h_b^*\phi_b^*] = \log \mu_d$. □

Polygons in the class $[h_b^*\phi_b^*, h_b^*\phi_b^*]$ can be concatenated in a string as shown in figure 8.10. More generally, in d dimensions, let ω be a polygon of length n, and let B be the smallest rectangular box containing ω. The boundary ∂B of B has $2d$ corners, and these may or may not be vertices in ω.

Let the lexicographic bottom corner of ∂B be denoted $\vec{v}_b$, and the lexicographic top corner of ∂B be $\vec{v}_t$. If $\vec{v}_b \in \omega$, then the bottom edge of ω is incident with $\vec{v}_b$. Similarly, if $\vec{v}_t \in \omega$, then the top edge of ω is incident with $\vec{v}_t$. Define the edge $\vec{b}_2 = \langle \vec{v}_b \sim (\vec{v}_b + \vec{e}_2)\rangle$. Then $\vec{b}_2$ is incident with $\vec{v}_b$ and $\vec{b}_2 \perp \vec{e}_1$.

Similarly, define the edge $\vec{t}_2 = \langle (\vec{v}_t - \vec{e}_2) \sim \vec{v}_t\rangle$. Then $\vec{t}_2$ is incident with $\vec{v}_t$, and with $\vec{t}_2 \perp \vec{e}_1$. Notice that $\vec{b}_2 \| \vec{t}_2$.

Let ω be a polygon in B such that $\vec{b}_2 \in \omega$, and $\vec{t}_2 \in \omega$. Then the bottom edge of ω is $\vec{b}_2$, and the top edge of ω is $\vec{t}_2$. Let $p_n[\vec{b}_2, \vec{t}_2]$ be the number of polygons of length n which contain the edges $\vec{b}_2$ and $\vec{t}_2$ in the smallest rectangular box B containing ω. These polygons are in the class $[\vec{b}_2, \vec{t}_2]$.

Some polygons in the class $[\vec{b}_2\vec{t}_2]$ can be created by using the construction illustrated in figure 8.10. This is done as follows. A polygon v contributing to $p_n[h_b^*\phi_b^*, h_b^*\phi_b^*]$ fits into the most popular box B_0 for polygons of this size. Let the side-lengths of B_0 be $\langle h_1, h_2, \ldots, h_d\rangle$.

Concatenate a string of N polygons contributing to $p_n[h_b^*\phi_b^*, h_b^*\phi_b^*]$ and then append $2dn$ edges as illustrated in figure 8.10 so that the polygon has length $nN + 2dn$ and is in the class $[\vec{b}_2, \vec{t}_2]$ in a box B. Since each of the N polygons fits into the most popular rectangular box B_0, the rectangular box B has side-lengths $\langle n + Nh_1, h_2, h_3, \ldots, h_d\rangle$.

This shows that

$$(p_n[h_b^*\phi_b^*, h_b^*\phi_b^*])^N \le p_{nN+2dn}[B, \vec{b}_2, \vec{t}_2], \tag{8.50}$$

where $p_{nN+2dn}[B, \vec{b}_2, \vec{t}_2]$ is the number of polygons of length $nN + 2dn$ in B in the class $[\vec{b}_2, \vec{t}_2]$.

There are most popular values for polygons in rectangular boxes. In particular, for each n, there is a most popular box B^* such that $p_n[B, \vec{b}_2, \vec{t}_2] \le p_n[B^*, \vec{b}_2, \vec{t}_2]$. Define $p_n[b_2^*, t_2^*] \equiv p_n[B^*, \vec{b}_2, \vec{t}_2]$. Then, by equation (8.50),

$$(p_n[h_b^*\phi_b^*, h_b^*\phi_b^*])^N \le p_{nN+2dn}[b_2^*, t_2^*]. \tag{8.51}$$

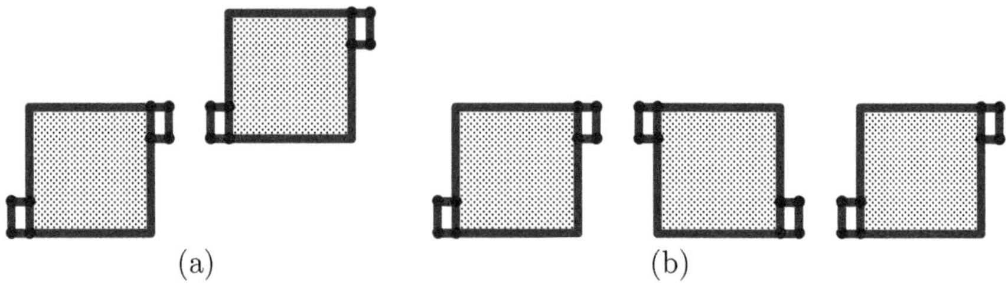

FIG. 8.11. Polygons in the class counted by $p_n[b_2^*, t_2^*]$ can be concatenated into strings as shown in (a), or as shown in (b).

Take logarithms, divide by nN and let $N \to \infty$. This gives

$$\tfrac{1}{n} \log p_n[h_b^* \phi_b^*, h_b^* \phi_b^*] \leq \limsup_{n\to\infty} \tfrac{1}{n} \log p_n[b_2^*, t_2^*]. \tag{8.52}$$

Concatenating two polygons contributing to $p_n[b_2^*, t_2^*]$ and $p_m[b_2^*, t_2^*]$ as shown in figure 8.11(a) shows that $p_n[b_2^*, t_2^*] p_m[b_2^*, t_2^*] \leq p_{n+m}[b_2^*, t_2^*]$, since the concatenated polygon also passes through the edges $\vec{b}_2$ and $\vec{t}_2$ in the smallest box containing the pair. By lemma A.1 in appendix A this shows that the limit superior above is a limit; therefore,

$$\tfrac{1}{n} \log p_n[h_b^* \phi_b^*, h_b^* \phi_b^*] \leq \lim_{n\to\infty} \tfrac{1}{n} \log p_n[b_2^*, t_2^*] \leq \log \mu_d. \tag{8.53}$$

Take $n \to \infty$ on the left-hand side and use lemma 8.23 to finish the proof of the following lemma.

Lemma 8.24 *The limit* $\lim_{n\to\infty} \frac{1}{n} \log p_n[b_2^*, t_2^*] = \log \mu_d$. □

Polygons in the class $[b_2^*, t_2^*]$ are useful in the study of polygons confined to W_{fg}-wedges. Notice that such polygons can be concatenated as shown in figure 8.11, and, if this is recursively done, then sequences of concatenated polygons which are horizontal or follow the profile of a curve in $\mathbb{L}^d$ may be obtained. In fact, since the concatenation can be done in the vertical direction instead (by reflecting polygons through the main diagonal), it is possible to concatenate polygons such that they stay close to (within a distance n, say) from the trajectory of any non-decreasing function h in the first quadrant. In particular, it is possible to concatenate polygons between two functions f and g, as illustrated in figure 8.13.

Notice that, if f_i and g_i are strictly increasing functions of $\mathbb{R}$, then their inverse functions f_i^{-1} and g_i^{-1} are defined.

This gives the following theorem.

Theorem 8.25 *Let W_{fg} be an fg-wedge. Suppose that f_i and g_i are strictly increasing functions on $\mathbb{R}$ such that $f_i \geq g_i \geq 0$. Moreover, suppose that $f_i(x) \geq 1 + g_i(x)$ for all $x \in \mathbb{N}_0$, for at least one value of $i \in \{2, 3, \ldots, d\}$.*

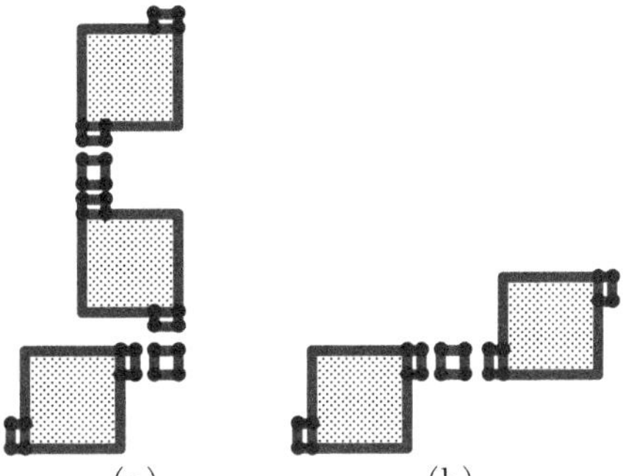

FIG. 8.12. Polygons in the class $[b_2^*, t_2^*]$ can be concatenated in either (a) a vertical direction or (b) a horizontal direction.

If $p_n(fg)$ is the number of polygons of length n in W_{fg} and rooted at $\vec{0}$, then $\lim_{n\to\infty} \frac{1}{n} \log p_n(fg) = \log \mu_d$ if and only if $\lim_{x\to\infty} (f_i(x) - g_i(x)) = \infty$, and $\lim_{x\to\infty} \left(g_i^{-1}(x) - f_i^{-1}(x)\right) = \infty$, for all $2 \leq i \leq d$.

Proof If either $\lim_{x\to\infty} (f_i(x) - g_i(x)) < \infty$, or $\lim_{x\to\infty} \left(g_i^{-1}(x) - f_i^{-1}(x)\right) < \infty$, then $\lim_{n\to\infty} \frac{1}{n} \log p_n(fg) < \log \mu_d$ by the pattern theorem for polygons (see corollary 7.14), since large patterns cannot be placed inside W_{fg}.

Hence, choose a (large) $n \in \mathbb{N}$ and choose x and y such that $f_i(x) - g_i(x) > 8n$, and $g_i^{-1}(y) - f_i^{-1}(y) > 8n$, for all $i \in \{2, 3, \ldots, d\}$. Then, for all $\overline{x} > x$, it is the case that $f_i(\overline{x}) - g_i(\overline{x}) > 8n$ and, for all $\overline{y} > y$, it is the case that $g_i^{-1}(\overline{y}) - f_i^{-1}(\overline{y}) > 8n$. Thus, for $\overline{x} > x$, a hypercube of side-length n can be placed in W_{fg} without intersecting f or g.

Let ω_j be polygons of length n in the class $[b_2^*, t_2^*]$ and contained in the most popular rectangular box B. Then the side-lengths of B are less than n.

Suppose that ω_1 was placed in W_{fg} such that the bottom corner of B has coordinates $(x_1, y_2, y_3, \ldots, y_d)$ with $x_1 \geq x$, and $y_i \geq y$. Suppose the top corner of B has coordinates $(\overline{x}_1, \overline{y}_2, \overline{y}_3, \ldots, \overline{y}_d)$.

Either a polygon ω_2 in the class $[b_2^*, t_2^*]$ can be concatenated on ω_2 as illustrated in figure 8.11, or this cannot be done. If this cannot be done, then the vertex $(\overline{x}_1 + n + 1, \overline{y}_2 - 1, \overline{y}_3, \ldots, \overline{y}_d)$ is not in W_{fg} (in fact, there is a last vertex $(\overline{x}_1 + j, \overline{y}_2 - 1, \overline{y}_3, \ldots, \overline{y}_d)$ for $1 \leq j \leq n$ which is in W_{fg}). This vertex is a corner of a hypercube C of side-lengths $4n$ which is in W_{fg} and has bottom vertex $(\overline{x}_1 + j - 4n, \overline{y}_2 - 1, \overline{y}_3, \ldots, \overline{y}_d)$.

Thus, the top edge of ω_1 is inside C. Since the side-length of C is $4n$, this means that the hypercube D of side-length $2n$ and with a bottom vertex with the coordinates $(\overline{x}_1 - n + 1, \overline{y}_2 - 1, \ldots, \overline{y}_d)$ is in C and so inside W_{fg}. By placing a suitably reflected polygon ω_2 of length n in the class $[b_2^*, t_2^*]$ as shown in figure 8.12, it can be concatenated onto the top edge of ω_1.

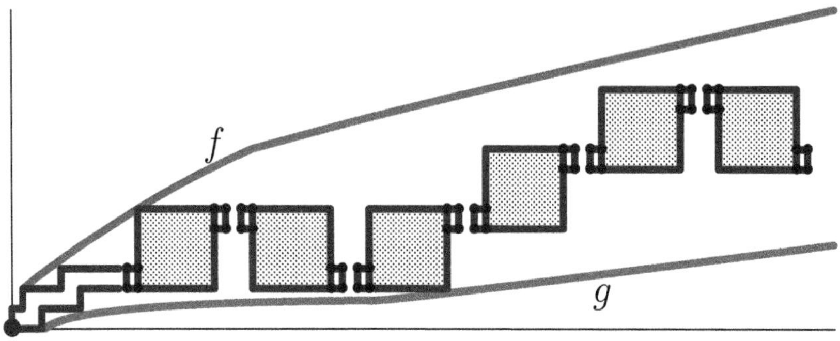

FIG. 8.13. A string of most popular polygons in an fg-wedge.

Since C has size $4n$, a third polygon ω_3, suitably reflected, can be concatenated onto ω_2 as illustrated in figure 8.12 so the top edge of ω_3 is again incident on the top vertex of the entire polygon.

This process may be repeated recursively, each time concatenating another polygon until a string of polygons similar to that illustrated in figure 8.13 is obtained.

The first polygon ω_1 can be joined to $\vec{0}$ by $2m$ edges in W_{fg}, where $m = x_1 + y_2 + y_3 + \cdots + y_d$. If N polygons are concatenated, then this shows that, for each even $r \geq 0$,

$$(p_n[b_2^*, t_2^*])^N \leq p_{nN+2m+r}(fg),$$

since each polygon can be chosen in $p_n[b_2^*, t_2^*]$ ways. Choose an even $r_N \geq 0$ for each N such that

$$\liminf_{n\to\infty} \tfrac{1}{n} \log p_n(fg) = \liminf_{N\to\infty} \tfrac{1}{nN} \log p_{nN+2m+r_N}(fg).$$

This shows that

$$\tfrac{1}{n} \log p_n[b_2^*, t_2^*] \leq \liminf_{N\to\infty} \tfrac{1}{nN} \log p_{nN+2m+r_N}(fg) = \liminf_{n\to\infty} \tfrac{1}{n} \log p_n(fg) \leq \log \mu_d.$$

Finally, take $n \to \infty$ on the left-hand side to establish existence of the limit by lemma 8.24. □

A corollary of theorem 8.25 is a similar theorem for walks from the origin in W_{fg}. Since it is only necessary to join a walk with a single path to the origin, the conditions on W_{fg} can be slightly relaxed.

Theorem 8.26 *Let W_{fg} be an fg-wedge. Suppose that f_i and g_i are strictly increasing functions on $\mathbb{R}$ such that $f_i \geq g_i \geq 0$ for $i \in \{2, 3, \ldots, d\}$. Moreover, suppose that, for each vertex $\vec{v} \in W_{fg}$, there is a walk from the origin to $\vec{v}$.*

If $c_n(fg)$ is the number of walks of length n from $\vec{0}$ in W_{fg}, then the limit $\lim_{n\to\infty} \frac{1}{n} \log c_n(fg) = \log \mu_d$ exists if and only if $\lim_{x\to\infty}(f_i(x) - g_i(x)) = \infty$, and $\lim_{x\to\infty}(g_i^{-1}(x) - f_i^{-1}(x)) = \infty$, for all $2 \le i \le d$.

Proof That $\lim_{x\to\infty}(f_i(x) - g_i(x)) = \infty$, and $\lim_{x\to\infty}(g_i^{-1}(x) - f_i^{-1}(x)) = \infty$, for all $2 \le i \le d$, are necessary conditions follows again from the pattern theorem for walks (see corollary 7.14).

Conversely, concatenate N polygons of length n in the class $[b_2^*, t_2^*]$ as in the proof of theorem 8.25. Join the bottom vertex of the resulting polygon with a walk of length $m = x_1 + y_2 + y_3 + \cdots + y_d$ to $\vec{0}$ (and delete the bottom edge of the polygon). This gives a walk of length $nN + m - 1$ and shows that $(p_n[b_2^*, t_2^*])^N \le c_{nN+m+r}(fg)$ for even $r \ge 0$. By deleting two edges instead of just the bottom edge, $(p_n[b_2^*, t_2^*])^N \le 2d c_{nN+m+r-1}(fg)$ for even $r \ge 0$. This shows that the last two inequalities apply for both odd and even $nN + m + r$. Taking logarithms, dividing by nN and letting $N \to \infty$ while choosing r such that the limit inferior is realised (similar to the proof of theorem 8.25), it follows that

$$\tfrac{1}{n} \log p_n[b_2^*, t_2^*] \le \liminf_{n\to\infty} \tfrac{1}{n} \log c_n(fg).$$

Take $n \to \infty$ on the left-hand side to complete the proof. □

Corollary 7.5, theorems 7.6 and 7.7 and corollary 7.8 generalise to models of walks in wedges; therefore, a corollary of theorems 8.25 and 8.26 is the following.

Corollary 8.27 *Let W_{fg} be an fg-wedge. Suppose that f_i and g_i are strictly increasing functions on $\mathbb{R}$ such that $f_i \ge g_i \ge 0$ for $i \in \{2, 3, \ldots, d\}$. Moreover, suppose that, for each $\vec{v} \in W_{fg}$, there is a walk from $\vec{0}$ to $\vec{v}$.*

If $c_n(fg)$ is the number of walks of length n from $\vec{0}$ in W_{fg}, then the limit $\lim_{n\to\infty} \frac{1}{n} \log c_n(fg) = \log \mu_d$ exists if and only if $\lim_{x\to\infty}(f_i(x) - g_i(x)) = \infty$, and $\lim_{x\to\infty}(g_i^{-1}(x) - f_i^{-1}(x)) = \infty$, for all $2 \le i \le d$.

This is also true if c_n is replaced by any of b_n, h_n, c_n^t, c_n^+, $c_n^\dagger$, $c_n^\ddagger$ or $c_n(\vec{v})$. Moreover, if $f_i(x) \ge 1 + g_i(x)$ for all $x \in \mathbb{N}_0$, for at least one value of $i \in \{2, 3, \ldots, d\}$, then c_n can also be replaced by p_n. □

A second corollary is a pattern theorem for walks in W_{fg}. This follows directly from the last corollary and from corollary 7.14. Let $c_n[\le\epsilon n, P, fg]$ be the number of self-avoiding walks of length n in W_{fg} from $\vec{0}$ and containing the proper pattern P at most ϵn times. Then corollary 7.14 becomes the following in W_{fg}.

Corollary 8.28 *Let P be a proper pattern and suppose that $0 < \epsilon < \epsilon_0$ (where ϵ_0 is defined in corollary 7.13). Then there exist a $k > 0$ (possibly dependent on P and ϵ) and an $N_\epsilon \in \mathbb{N}$ such that*

$$c_n[\le\epsilon n, P, fg] < e^{-kn} c_n(fg)$$

for all $n \ge N_\epsilon$. This is similarly the case if c_n is replaced by b_n, h_n, c_n^t, c_n^+, $c_n^\dagger$, $c_n^\ddagger$ or $c_n(\vec{v})$. Moreover, if $f_i(x) \ge 1 + g_i(x)$ for all $x \in \mathbb{N}_0$, for at least one value of $i \in \{2, 3, \ldots, d\}$, then c_n can also be replaced by p_n. □

9

INTERACTION MODELS OF SELF-AVOIDING WALKS

Interacting models of self-avoiding walks and polygons have partition functions similar to equation (3.1). Existence of a limiting free energy is generally shown by concatenating clusters (walks or polygons) to find supermultiplicative inequalities for the partition function. In this chapter, the limiting free energies and phase diagrams of adsorbing, collapsing and pulled self-avoiding walks are examined.

9.1 Adsorbing self-avoiding walks and polygons

A polymer grafted to a surface may adsorb if its monomers are strongly attracted to molecules in the surface. The adsorption is a conformational rearrangement of the polymer which is a phase transition [39, 117].

An adsorbing polymer (see figure 9.1) can be modelled by a *positive walk* from the origin in the half-lattice $\mathbb{L}^d_+$ (see equation (2.20)). The *adsorbing wall* is the boundary $\partial\mathbb{L}^d_+$ of $\mathbb{L}^d_+$ (see section 7.1.1 and figure 2.8). The model is also illustrated in figure 3.1(a).

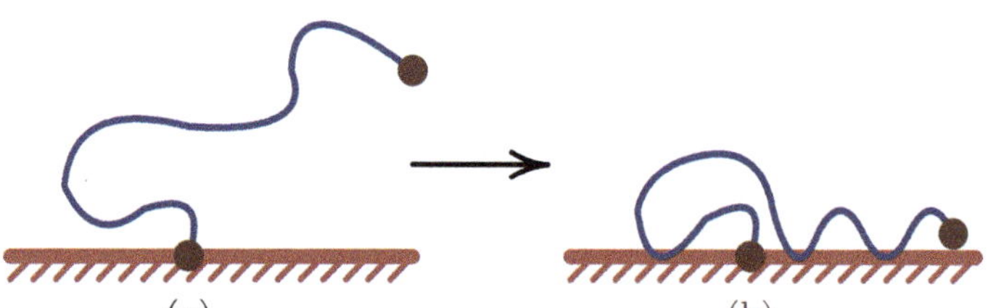

FIG. 9.1. The adsorption transition in a self-avoiding walk. (a) A positive walk grafted to an adsorbing wall. If the walk does not interact with the wall, then it explores conformations in the bulk solvent. This is the desorbed or coiled phase. (b) If the walk is strongly attracted to the wall, then it adsorbs onto the wall and explores conformations which are close to the wall. This is the adsorbed phase.

Let $\omega = \langle \omega_0, \omega_1, \ldots, \omega_n \rangle$ be a positive walk with vertices ω_j (so $\omega_0 = \vec{0}$). Denote the i-th Cartesian coordinate of the vertex ω_j by $\omega_j(i)$. If $\omega_j(d) = 0$, then $\omega_j \in \partial\mathbb{L}^d_+$ and therefore is called a *visit* in the adsorbing wall. By convention, ω_0 is not a visit, even though it is a vertex in the adsorbing wall.

The Statistical Mechanics of Interacting Walks, Polygons, Animals and Vesicles, 2nd edition, E.J. Janse van Rensburg. © E.J. Janse van Rensburg. Published in 2015 by Oxford University Press.

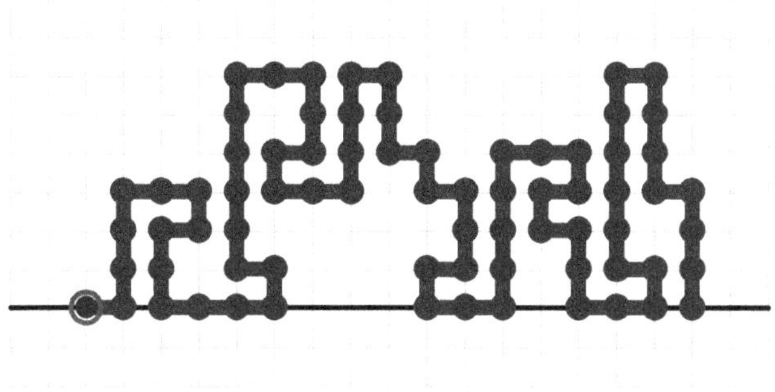

FIG. 9.2. A positive bridge in $\mathbb{L}^2_+$ with four excursions, five incursions (including the final vertex) and twelve visits to the adsorbing boundary $\partial\mathbb{L}^2_+$.

Let $c_n^+(v)$ be the number of positive walks of length n with v visits. Introduce the activity a conjugate to the number of visits. By equation (3.1), the partition function of adsorbing walks is given by

$$A_n^+(a) = \sum_{v \geq 0} c_n^+(v)\, a^v. \tag{9.1}$$

To show that a limiting free energy exists in this model requires the introduction of various subclasses of adsorbing walks, namely, loops, hoops, positive bridges and unfolded positive walks.

9.1.1 *The free energy of adsorbing hoops*

Recall that *hoops* and *loops* were introduced in section 7.1.4 (see figure 7.6). A positive walk $\omega = \langle \omega_0, \omega_1, \ldots, \omega_n \rangle$ is a *hoop* if $0 = \omega_0(d) = \omega_n(d) \leq \omega_j(d)$ for $i = 1, 2, \ldots, n-1$ and is a *loop* if both $0 = \omega_0(d) = \omega_n(d) < \omega_j(d)$ and $0 = \omega_0(1) \leq \omega_i(1) \leq \omega_n(1)$. The number of hoops with v visits and length n is denoted by $\ell_n^o(v)$ so that $\ell_n^o(v) \leq c_n^+(v)$ for each n and v. Denote the partition function of adsorbing hoops by $L_n^o(a)$.

The number of *loops* of length n is denoted by ℓ_n (see figure 7.6). The only vertices of a loop in $\partial\mathbb{L}^d_+$ are its end-vertices. Notice that the end-vertices of loops are also left-most and right-most in the $\vec{e}_1$ direction (see, for example, figures 2.8(b), 3.1(a) and 7.7(a)).

A hoop $\omega = (\omega_0, \omega_1, \ldots, \omega_n)$ is said to be $\vec{e}_1$-*unfolded* (or just *unfolded*) if $\omega_0(1) < \omega_i(1)$ for all $i = 1, 2, \ldots, n$. It is a *positive bridge* if $\omega_0(1) < \omega_i(1) \leq \omega_n(1)$ for all $i = 1, 2, \ldots, n$ (see figure 9.2; positive bridges are a subclass of hoops).

Denote the number of $\vec{e}_1$-unfolded hoops of length n from $\vec{0}$ by $\ell_n^\dagger(v)$, and the number of positive bridges of length n from $\vec{0}$ by $\ell_n^b(v)$. Observe that the first vertex of a positive bridge is at $\vec{0}$, and the last vertex is necessarily a visit, because it is also a hoop. Let the partition functions of $\vec{e}_1$-unfolded hoops and of positive bridges be $L_n^\dagger(a)$ and $L_n^b(a)$, respectively.

An $\vec{e}_1$-unfolded hoop may be unfolded into a positive bridge by using the Hammersley-Welsh construction of section 7.1.2. The unfolding does not remove or create new visits to $\partial\mathbb{L}^d_+$. Similar to lemma 7.1, there exist constants $\gamma_1 > 0$ and $N_1 \in \mathbb{N}$ such that, for all $n \geq N_1$,

$$\ell^b_n(v) \leq \ell^\dagger_n(v) \leq e^{\gamma_1\sqrt{n}}\,\ell^b_n(v). \tag{9.2}$$

This shows that $L^b_n(a) \leq L^\dagger_n(a) \leq e^{\gamma_1\sqrt{n}} L^b_n(a)$.

Two positive bridges can be concatenated by placing the first vertex of the second on the last vertex of the first. This shows that

$$\sum_{w=0}^{v} \ell^b_n(w)\,\ell^b_m(v-w) \leq \ell^b_{n+m}(v). \tag{9.3}$$

Multiply this by a^v and sum over v. This shows that

$$L^b_n(a)\,L^b_m(a) \leq L^b_{n+m}(a). \tag{9.4}$$

By lemma A.1 in appendix A and equation (9.2), the result is existence of the limit

$$\mathcal{A}^+(a) = \lim_{n\to\infty} \tfrac{1}{n} \log L^b_n(a) = \lim_{n\to\infty} \tfrac{1}{n} \log L^\dagger_n(a). \tag{9.5}$$

A hoop can be $\vec{e}_1$-unfolded using the Hammersley-Welsh unfolding in section 7.1.2. This shows that there exist constants γ_2 and N_2 such that, for all $n \geq N_2$,

$$\ell^\dagger_n(v) \leq \ell^o_n(v) \leq e^{\gamma_2\sqrt{n}} \left(\ell^\dagger_{n+1}(v+1) + \ell^\dagger_n(v)\right)$$

since no visits are created in the unfolding (but one horizontal edge may have to be added to achieve the strict inequality $\omega_0(1) < \omega_1(1)$; this gives $\ell^\dagger_{n+1}(v+1)$ on the right-hand side since this increases the number of visits by 1). Comparison with equation (9.5) proves the existence of a limiting free energy for adsorbing hoops:

$$\mathcal{A}^+(a) = \lim_{n\to\infty} \tfrac{1}{n} \log L^o_n(a). \tag{9.6}$$

Similar to the above, a hoop may be unfolded and then lifted one step in the $\vec{e}_d$ direction and reconnected to $\partial\mathbb{L}^d_+$ by inserting two new edges to find that there exist constants γ_3, N_3 and K_3 such that, for all $n \geq N_3$,

$$\ell^o_n(v) \leq e^{\gamma_3\sqrt{n}}\,\ell_{n+2} \leq e^{\gamma_3\sqrt{n}}\,\ell^o_n. \tag{9.7}$$

The last inequality follows because $\ell_{n+2} \leq \ell^o_n$ (a hoop is obtained if the first and last edges of a loop are deleted).

Suppose that $a \in (0,1]$; then it follows from the above that

$$a\ell_n \le a\ell_n^o(1) \le L_n^o(a) = \sum_{v=0}^{n} \ell_n^o(v)\, a^v \le e^{\gamma_3\sqrt{n}}\ell_{n+2}\sum_{v=0}^{n} a^v.$$

By taking logarithms, dividing by n and letting $n \to \infty$, it follows that (see section 7.1.4 and theorem 7.6)

$$\mathcal{A}^+(a) = \log \mu_d \ \text{ if } a \in (0,1]. \tag{9.8}$$

Notice that $\frac{1}{a}L_n^o(a) \ge \ell_n$, so $\lim_{a\to 0^+}\mathcal{A}^+(a) \ge \log\mu_d$. Thus, define $\mathcal{A}^+(0) = \log\mu_d$ so that $\mathcal{A}^+(a) = \log\mu_d$ for all $a \in [0,1]$.

9.1.2 *The free energy of adsorbing walks*

Let $\omega = \langle \omega_0, \omega_1, \ldots, \omega_n\rangle$ be a positive walk. Unfolding ω with the Hammersley-Welsh construction (see section 7.1.2) does not change the number of visits, and the result is a *doubly unfolded positive walk* with $\omega_0(1) \le \omega_i(1) \le \omega_n(1)$ for all $0 \le i \le n$ (see section 7.1.3).

Denote the number of doubly unfolded positive walks of length n and with v visits in $\partial\mathbb{L}_+^d$ by $c_n^{+\ddagger}(v)$. Let the partition function of doubly unfolded positive walks be

$$A_n^{+\ddagger}(a) = \sum_{v\ge 0} c_n^{+\ddagger}(v)\, a^v. \tag{9.9}$$

By theorem 7.2, there exist constants γ_4 and N_4 such that, for all $n \ge N_4$, $c_n^{+\ddagger}(v) \le c_n^+(v) \le e^{\gamma_4\sqrt{n}} c_n^{+\ddagger}(v)$. In terms of partition functions,

$$A_n^{+\ddagger}(a) \le A_n^+(a) \le e^{\gamma_4\sqrt{n}}\, A_n^{+\ddagger}(a) \ \text{ for all } n \ge N_4. \tag{9.10}$$

This gives the following theorem.

Theorem 9.1 *The limiting free energy of adsorbing walks is given by*

$$\mathcal{A}^+(a) = \lim_{n\to\infty} \tfrac{1}{n}\log A_n^+(a) = \lim_{n\to\infty}\tfrac{1}{n}\log L_n(a) = \lim_{n\to\infty}\tfrac{1}{n}\log L_n^b(a).$$

Moreover, $\mathcal{A}^+(a) = \log\mu_d$ *for all* $0 \le a \le 1$ *and it is a convex function of* $\log a$ *on its domain.*

Proof Let $c_n^{+\ddagger}(v,h)$ be the number of doubly unfolded positive walks of length n with v visits and with endpoints at height h above $\partial\mathbb{L}_+^d$. By equation (9.9),

$$A_n^{+\ddagger}(a) = \sum_{h\ge 0}\sum_{v\ge 0} c_n^{+\ddagger}(v,h)\, a^v.$$

For given (a,n), there are most popular values of h in the summation on the right-hand side; denote this by h^*. This shows that

$$\sum_{v\ge 0} c_n^{+\ddagger}(v,h^*)\, a^v \le A_n^{+\ddagger}(a) \le (n+1)\sum_{v\ge 0} c_n^{+\ddagger}(v,h^*)\, a^v, \tag{9.11}$$

since h^* has at most $n+1$ different possible values.

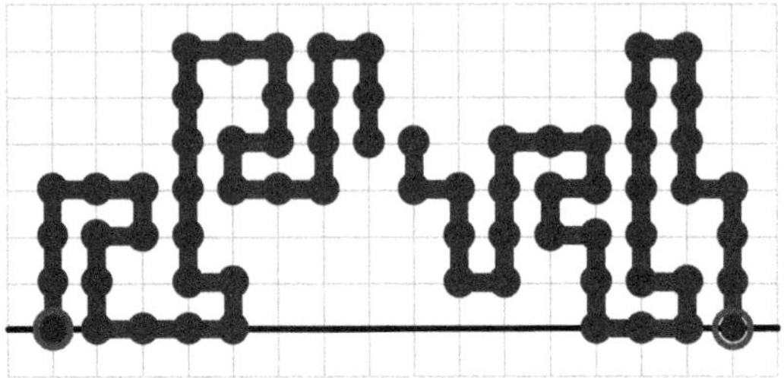

FIG. 9.3. On the left is a doubly unfolded positive walk, and on the right is a reflected (in the $\vec{e}_1$ direction) doubly unfolded positive walk. Both walks have endpoints at the same height. They can be concatenated into a hoop by inserting a single edge between their endpoints. By adding another edge onto the root on the left in the obvious way, a positive bridge is obtained.

Let ω_1 and ω_2 be two doubly unfolded positive walks of length n and with their endpoints at height h^*. Reflect ω_2 in the $\vec{e}_1$ direction and concatenate the image with ω_1 by placing them as illustrated in figure 9.3. Joining them into a single walk by adding a single edge between their endpoints at height h^* gives a hoop of length $2n+1$ and with end-vertices left-most and right-most in the $\vec{e}_1$ direction. This hoop is turned into a positive bridge by adding a step in the $\vec{e}_1$ direction on its first vertex and then placing this new endpoint at $\vec{0}$. This creates one new visit, so

$$\sum_{w=0}^{v} c_n^{+\ddagger}(w,h^*)c_n^{+\ddagger}(v-w,h^*) \leq \ell_{2n+2}^b(v+1) \leq c_{2n+2}^{+\ddagger}(v+1).$$

The last inequality follows because every positive bridge is a doubly unfolded positive walk. Multiply this inequality by a^v and sum over v. If $a > 0$, then this gives

$$\Big(\sum_{v\geq 0} c_n^{+\ddagger}(v,h^*)\,a^v\Big)^2 \leq \tfrac{1}{a}L_{2n+2}^b(a) \leq \tfrac{1}{a}\sum_{v\geq 0} c_{2n+2}^{+\ddagger}(v,h^*)\,a^v \leq \tfrac{1}{a}A_{2n+2}^{+\ddagger}(a).$$

Take logarithms, divide by $2n$ and let $n\to\infty$. By equations (9.5) and (9.11), this shows that

$$\mathcal{A}^+(a) = \lim_{n\to\infty}\tfrac{1}{n}\log\sum_{v\geq 0} c_n^{+\ddagger}(v,h^*)\,a^v = \lim_{n\to\infty}\tfrac{1}{n}\log A_n^{+\ddagger}(a).$$

Comparison of this result to equations (9.1), (9.5), (9.6), (9.8) and (9.10) completes the proof for the case $a > 0$.

If $a = 0$, then $A_n^+(0) = c_{n-1}^+$, since the first step in the walk must be away from the adsorbing hyperplane, and the rest of the walk is in the half-lattice $\mathbb{L}_+^d + \vec{e}_d$. By corollary 7.3, $\mathcal{A}^+(0) = \log\mu_d$.

In addition, $\mathcal{A}^+(a)$ is a convex function of $\log a$ by equation (3.10) and theorem 3.3. □

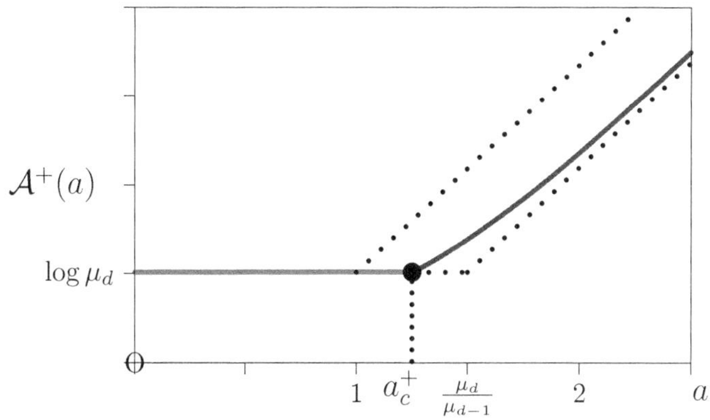

FIG. 9.4. The free energy $\mathcal{A}^+(a)$ of adsorbing positive walks is equal to $\log \mu_d$ for $a < a_c^+$ and is strictly increasing for $a > a_c^+$. This shows that the density of visits for $a > a_c^+$ is strictly positive and is an adsorbed phase in the model. Bounds on the critical curve are given by dotted lines. By theorem 9.2, $\log \mu_{d-1} + \log a \leq \mathcal{A}^+(a) \leq \log \mu_d + \log a$, and $\mathcal{A}^+(a) \geq \log \mu_d$. By corollary 9.3, $1 \leq a_c^+ < \frac{\mu_d}{\mu_{d-1}}$.

Theorem 9.2 *For all $a \leq 1$, $\mathcal{A}^+(a) = \log \mu_d$. If $a > 1$, then $\mathcal{A}^+(a) \geq \log \mu_d$ and*

$$\log \mu_{d-1} + \log a \leq \mathcal{A}^+(a) \leq \log \mu_d + \log a,$$

where $\mu_1 = 1$.

Proof By theorem 9.1, it only remains to show the bounds for $a \geq 1$. Since $v \leq n$ in equation (9.1), choose $v = n$ to see that $c_n^+(n)a^n \leq A_n^+(a) \leq c_n^+ a^n$ for $a \geq 1$. The number of positive walks from $\vec{0}$ confined to $\partial\mathbb{L}_+^d$ is $c_n^+(n)$ and this is also the number of self-avoiding walks from $\vec{0}$ in $\mathbb{L}^{d-1}$: $c_n^+(n) = c_n^{(d-1)}$. Take logarithms, divide by n and let $n \to \infty$ to find the claimed bounds. □

It will be seen that $\mathcal{A}^+(a) > \log \mu_{d-1} + \log a$ for $a \geq 0$ (see theorem 9.7).

Define $a_c^+ = \sup\{a \mid \mathcal{A}^+(a) = \log \mu_d\}$. Then $\mathcal{A}^+(a) > \log \mu_d$ if $a > a_c^+$, and $\mathcal{A}^+(a) = \log \mu_d$ if $a \leq a_c^+$.

Corollary 9.3 *There is a non-analyticity in the free energy of adsorbing positive walks $\mathcal{A}^+(a)$ in $\mathbb{L}_+^d$ at a critical value $a_c^+ \in [1, \frac{\mu_d}{\mu_{d-1}})$ (where $\mu_1 = 1$).*

Proof It only remains to prove that $a_c^+ < \frac{\mu_d}{\mu_{d-1}}$.

Denote the number of walks from $\vec{0}$ in $\mathbb{L}^{d-1}$ by $c_n^{(d-1)}$. Let ω be a walk of length n from $\vec{0}$ in the $(d-1)$-dimensional lattice $\partial\mathbb{L}_+^d \simeq \mathbb{L}^{d-1}$. Then ω is a positive walk in $\mathbb{L}_+^d$ with n visits.

Label the edges along ω from $\vec{0}$ by $1, 2, \ldots, n$ and let $\alpha > 0$ be small. Proceed by replacing $\lfloor \alpha n \rfloor$ edges with even labels with three edges in a $\sqcap$-conformation

TABLE 9.1. Surface exponents for adsorbing walks

	$d = 2$	$d = 3$	Mean Field
a_c^+	1.77564 [30]	1.334(27) [321]	−−
ϕ	$\frac{1}{2}$ [25, 34, 83]	0.5005 [278, 321]	$\frac{1}{2}$ [83]
γ_1	$\frac{61}{64}$ [95]	0.697(2) [278]	$\frac{1}{2}$
γ_{11}	$-\frac{3}{16}$ [168]	−0.383(5) [278]	$-\frac{1}{2}$
γ_s	$\frac{93}{64}$ [83]	1.304(16) [416]	−−
γ_{ss}	0.860(26) [412]	0.806(15) [416]	−−

into $\mathbb{L}_+^d$. This gives a walk in $\mathbb{L}_+^d$ of length $n + 2\lfloor \alpha n \rfloor$. Multiplying by a^n shows that

$$A_{n+2\lfloor\alpha n\rfloor}^+(a) \geq \binom{\lfloor \frac{n}{2} \rfloor}{\lfloor \alpha n \rfloor} c_n^{(d-1)} \, a^n. \tag{9.12}$$

Put $a = \frac{\mu_d}{\mu_{d-1}}$, take logarithms and divide by n. Taking $n \to \infty$ shows that

$$\mathcal{A}^+(\tfrac{\mu_d}{\mu_{d-1}}) \geq -\log\Big(\sqrt{2}\alpha^\alpha \left(\tfrac{1}{2} - \alpha\right)^{1/2-\alpha}\Big) - 2\alpha \mathcal{A}^+(\tfrac{\mu_d}{\mu_{d-1}}) + \log \mu_d.$$

The right-hand is strictly larger than $\log \mu_d$ for small $\alpha > 0$. By theorem 9.1, $\mathcal{A}^+(a)$ is a convex function of $\log a$. It is therefore continuous on $(0, \infty)$ and this implies that there exists an $\epsilon > 0$ such that $\mathcal{A}^+(\frac{\mu_d}{\mu_{d-1}} - \epsilon) > \log \mu_d$. Hence, it follows that $a_c^+ < \frac{\mu_d}{\mu_{d-1}}$. □

9.1.3 *The adsorption transition of self-avoiding walks*

The free energy $\mathcal{A}^+(a)$ of adsorbing walks is a non-decreasing convex function of $\log a$ (see figure 9.4). The order parameter of the model is the expected density of visits, given by

$$\langle v \rangle = \frac{d}{d \log a} \mathcal{A}^+(a). \tag{9.13}$$

The density of visits $\langle v \rangle$ characterises two phases, namely, $a < a_c^+$ (where $\langle v \rangle = 0$ (the desorbed phase)), and $a > a_c^+$ (where $\langle v \rangle > 0$ (the adsorbed phase)). Numerical estimates of the location of the critical point gives $a_c^+ = 1.759(18)$ in $\mathbb{L}^2$ [321]. Exact enumeration studies gave the estimate $a_c^+ = 1.77564$ if $d = 2$ [30]. In $\mathbb{L}^3$ the critical point is located at $a_c^+ = 1.338(5)$ [416], and $a_c^+ = 1.334(27)$ if $d = 3$ [321].

In the desorbed phase, the walk has critical exponents of linear polymers in a good solvent. The partition function scales as

$$A_n^+(a) \simeq B_- n^{\gamma_1 - 1} \mu_d^n \text{ if } a < a_c^+ \tag{9.14}$$

(see, for example, equation (2.35)). The entropic exponent γ_1 is that of positive walks (attached to a hard wall; see section 2.2.2). The partition function of

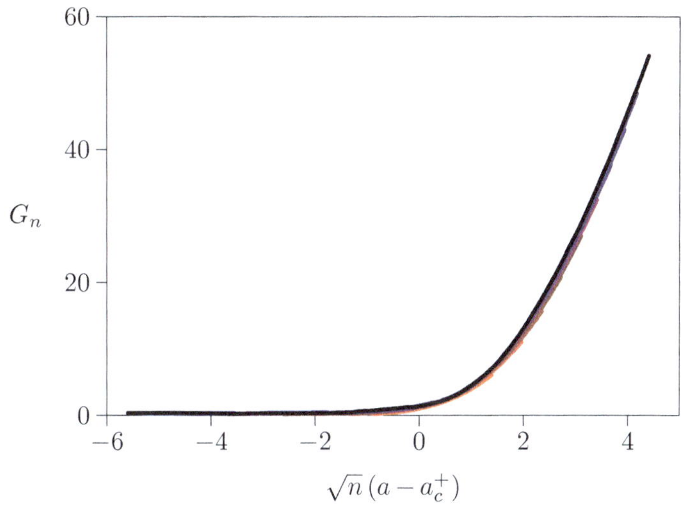

FIG. 9.5. The scaling function f in equation (9.16) for adsorbing walks in $\mathbb{L}^3_+$ can be uncovered by defining the finite size approximation $F_n^+ = \frac{1}{n}\log A_n^+(a)$ of the free energy and then plotting $G_n = n(F_n^+(a) - \log\mu)$ against $n^\phi(a - a_c^+)$ (where $\phi = \frac{1}{2}$). The values of n ranged from 10 to 100 in steps of 10 [321].

---

adsorbing hoops (or adsorbing bridges) will have a similar scaling form but with γ_1 replaced by γ_{11} (see equation (2.39)).

In the adsorbed phase the exponents will be those of a self-avoiding walk in one dimension less, with the scaling form

$$A_n^+(a) \simeq B_+ n^{\gamma^{(d-1)}-1} e^{n\mathcal{A}^+(z)}, \text{ if } a > a_c^+, \tag{9.15}$$

where $\gamma^{(d-1)}$ is the self-avoiding walk entropic exponent in one dimension lower.

At the critical point $a = a_c^+$, scaling similar to equation (9.14) is expected but with γ_1 replaced by a *surface exponent* γ_s. In the case of adsorbing hoops at the critical point, the exponent γ_{ss} should be seen instead. The exponents γ_s and γ_{ss} are *special point surface critical exponents*, or entropic exponents at the *special point*.

Conformal invariance calculations give $\gamma_s = \frac{93}{64}$ in $d = 2$ [83]. Monte Carlo simulations at the critical point in $\mathbb{L}^2$ give the estimates $\gamma_s = 1.478(20)$ and $\gamma_{ss} = 0.860(26)$ [412] in two dimensions. Simulations in $\mathbb{L}^3$ give the estimates $\gamma_s = 1.304(16)$ and $\gamma_{ss} = 0.806(15)$ [387, 416] in three dimensions.

The adsorption crossover exponent ϕ (as $a \searrow a_c^+$) is thought to have exact value $\phi = \frac{1}{2}$ in two dimensions [25, 34, 83]. Numerical estimates in two dimensions give $\phi = 0.562(20)$ [412], and $\phi = 0.501(14)$ [321]. In three dimensions, estimates of the crossover exponent include $\phi = 0.530(7)$ [387, 416], $\phi = 0.483(3)$ [358], and $\phi = 0.5005(36)$ [321]. These results suggest that $\phi = \frac{1}{2}$ in three dimensions

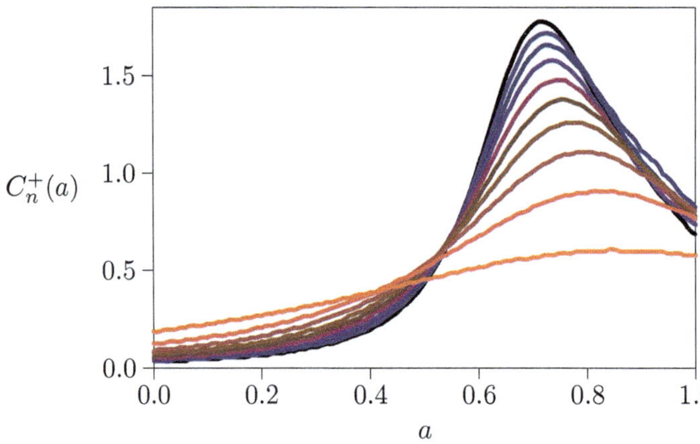

FIG. 9.6. The finite size specific heat of adsorbing walks. The values of n range from 10 to 100 in steps of 10. [321]

---

as well and it is feasible that the crossover exponent for polymer adsorption is equal to $\frac{1}{2}$ in all dimensions.

At the critical point the adsorbing walk is deformed and the ratio of the normal and parallel components of the radius of gyration approaches a limiting value of $\frac{R_\perp^2}{R_\parallel^2} = 0.320(3)$ [358].

Finite size approximations to $\mathcal{A}^+(a)$ have a scaling form given by equation (4.27). The bulk entropy of the walk in the desorbed phase is $\log \mu_d$. Thus,

$$\mathcal{A}^+(a) \approx \log \mu_d + \tfrac{1}{n} f(n^\phi(a - a_c^+)) \tag{9.16}$$

for some scaling function f. This may be tested by plotting $n\,(\mathcal{A}^+(a) - \log \mu_d)$ against $n^\phi(a - a_c^+)$ (with $\phi = \frac{1}{2}$). This is done in figure 9.5 for values of n between 10 and 100 in $\mathbb{L}_+^3$, where $F_n^+(a) = \frac{1}{n} \log A_n^+$. The data collapse to a single curve, revealing f (see reference [321]).

The finite size specific heat $C_n^+(a)$ is the second derivative of the finite size free energy. For adsorbing walks this is plotted in figure 9.6. Since $\phi = \frac{1}{2}$, it follows from equation (4.29) that $\alpha = 0$. By equation (4.3), this shows that the peak in $C_n^+(a)$ should approach a constant or grow sublinear in n. Normalising $C_n^+(a)$ to have peak height equal to 1, and plotting it against $n^\phi(a - a_c^+)$ collapses the specific heat curves to reveal the underlying scaling function in figure 9.7 [321].

9.1.4 *The density of visits in adsorbing walks*

In equation (9.3) put $w = v_1$ and put $v - w = v_2$:

$$\ell_{n_1}^b(v_1)\,\ell_{n_2}^b(v_2) \le \ell_{n_1+n_2}^b(v_1 + v_2). \tag{9.17}$$

Since $\ell_n^b = \mu_d^{n+o(n)}$, and $\ell_n^b(v) > 0$ for $0 < v \le n$, it follows from theorem 3.9 that the limit

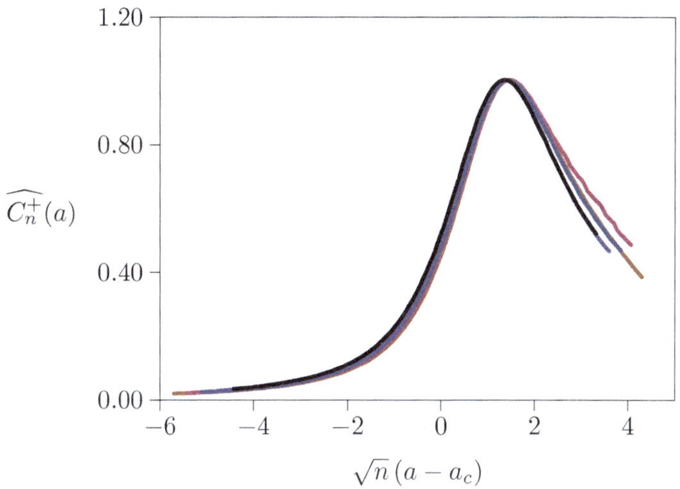

FIG. 9.7. Rescaling of the data in figure 9.6. The curves are normalised to have height 1 and are plotted against $n^\phi(a-a_c^+)$ (where $\phi=\frac{1}{2}$). These data are for values of n between 50 and 100 [321].

$$\mathcal{P}^+(\epsilon) = \lim_{n\to\infty} \left(\ell_n^b(\lfloor \epsilon n \rfloor)\right)^{1/n} \tag{9.18}$$

exists and is concave for $\epsilon \in [0,1]$. In addition, $\ell_n^b(\lfloor \epsilon n\rfloor + \eta_n) \leq [\mathcal{P}^+(\epsilon)]^n$ (where $\eta_n \in \{0,1\}$).

By the proof of theorem 9.1, $\left(c_n^{+\ddagger}(v)\right)^2 \leq \ell_{2n+2}^b(v) \leq c_{2n+2}^{+\ddagger}(v)$, and, by equation (9.10), $c_n^{+\ddagger}(v) \leq c_n^+(v) \leq K_0 e^{\gamma_0\sqrt{n}} c_n^{+\ddagger}(v)$. This shows that

$$\mathcal{P}^+(\epsilon) = \lim_{n\to\infty} \left(c_n^+(\lfloor \epsilon n \rfloor)\right)^{1/n} \tag{9.19}$$

exists, and is the density of visits in a positive walk. Notice that $\mathcal{P}^+(0) = \mu_d$, that $\mathcal{P}^+(\epsilon) \geq \mu_{d-1}$ and that $\mathcal{P}^+(1) = \mu_{d-1}$. The following properties of $\mathcal{P}^+(\epsilon)$ were shown in reference [557].

Lemma 9.4 *For the density function $\mathcal{P}^+(\epsilon)$, $\mathcal{P}^+(0) = \mu_d$; $\mathcal{P}^+(\epsilon) > \mu_{d-1}$ if $\epsilon < 1$; $\mathcal{P}^+(1) = \mu_{d-1}$; and $D^-\mathcal{P}^+(1) = \lim_{\epsilon\to 1^-} \frac{d}{d\epsilon}\mathcal{P}^+(\epsilon) = -\infty$.*

Proof A positive walk $\omega = \langle \omega_0, \omega_1, \ldots, \omega_n\rangle$ of length n and with n visits is entirely in the adsorbing wall $\partial\mathbb{L}_+^d \simeq \mathbb{L}^{d-1}$.

The edges of ω are $\langle \omega_j \sim \omega_{j+1}\rangle$ for $j = 0,1,\ldots,n-1$. Choose k of these n edges. Some of these will be isolated but, more generally, there may be runs of chosen edges of length ℓ forming subwalks or *trains* along the walk. Let $T = \langle\langle \vec{v}_0 \sim \vec{v}_1\rangle, \langle \vec{v}_1 \sim \vec{v}_2\rangle, \ldots, \langle \vec{v}_{m-1} \sim \vec{v}_m\rangle\rangle$ be a train of m chosen edges. Add m edges $\{\langle(\vec{v}_0 + j\vec{e}_d) \sim (\vec{v}_0 + (j+1)\vec{e}_d)\rangle \mid j = 0,1,\ldots,m-1\}$ in the $\vec{e}_d$ direction on $\vec{v}_0$. For $j = 1,2,\ldots,m$, add another m edges $\langle(\vec{v}_j + (m-j+1)\vec{e}_d) \sim (\vec{v}_j + (m-j)\vec{e}_d)\rangle$ to each of the (translated) vertices $\vec{v}_j$ for $j = 1, 2, \ldots, m$. This reconnects the walk, while $\langle \vec{v}_0 \sim \vec{v}_1 \rangle$ has been lifted to a height of m above $\partial \mathbb{L}^d_+$, and $\langle \vec{v}_1 \sim \vec{v}_2 \rangle$ to a height $m-1$, and so on.

Select k edges in ω, with some possibly isolated, and some forming trains of edges. Perform the above construction on each train and isolated edge. There are $c_n^+(n)$ choices for ω, and $\binom{n}{k}$ choices for the edges. Thus,

$$c_{n+2k}^+(n) \geq \binom{n}{k} c_n^+(n).$$

Put $k = \lfloor \alpha N \rfloor$, and $n + 2k = N$. Take the power $\frac{1}{N}$ of the above and then $N \to \infty$. By corollary D.2 in appendix D, theorem 3.9 and the comments above, this gives

$$\mathcal{P}^+(1-2\alpha) \geq \left(\frac{1-2\alpha}{\alpha}\right)^\alpha \left(\frac{1-2\alpha}{1-3\alpha}\right)^{1-3\alpha} \left(\mathcal{P}^+(1)\right)^{1-2\alpha}.$$

Take logarithms, rearrange and divide by 2α to obtain

$$\tfrac{1}{2\alpha}\big(\log \mathcal{P}^+(1) - \mathcal{P}^+(1-2\alpha)\big) \leq \tfrac{1}{2}\log\!\left(\tfrac{\alpha}{1-2\alpha}\right) + \tfrac{1-3\alpha}{2\alpha}\log\!\left(\tfrac{1-3\alpha}{1-2\alpha}\right) + \log \mathcal{P}^+(1).$$

Take the limit $\alpha \to 0^+$. Since $\lim_{\alpha\to 0^+} \frac{1-3\alpha}{2\alpha} \log(\frac{1-3\alpha}{1-2\alpha}) = -\frac{1}{2}$, it follows that

$$\lim_{\alpha\to 0^+} \tfrac{1}{2\alpha}\big(\log \mathcal{P}^+(1) - \mathcal{P}^+(1-2\alpha)\big) \leq -\infty.$$

This also shows that $\mathcal{P}^+(1-2\alpha) > \mathcal{P}^+(1)$ for small $\alpha > 0$.

This completes the proof. □

Theorem 9.5 *There exists an* $\epsilon^* < 1$ *such that*

$$\mathcal{A}^+(a) = \sup_{0\leq\epsilon\leq 1} \{\log \mathcal{P}^+(\epsilon) + \epsilon \log a\} = \log \mathcal{P}^+(\epsilon^*) + \epsilon^* \log a.$$

Moreover, $\epsilon^* = 0$ *if* $a < a_c^+$, *and* $\epsilon^* > 0$ *if* $a > a_c^+$. *In addition* $\epsilon^* = \frac{d^-}{d\log a}\mathcal{A}^+(a)$ *for almost all* $a \in [0, \infty)$.

Proof By lemma 9.4 and theorem B.11 in appendix B, $\mathcal{A}^+(a)$ is the Legendre transform of $\mathcal{P}^+(\epsilon)$. For $a < a_c^+$, $\mathcal{A}^+(a) = \log \mu_d$ by theorem 9.2, so $\mathcal{A}^+(a) = \log \mathcal{P}^+(0)$, and $\epsilon^* = 0$.

It remains to prove that $\epsilon^* < 1$ and that $\epsilon^* > 0$ for $a > a_c^+$. If $a > a_c^+$, then $\mathcal{A}^+(a) > \log \mu_d$; that is, $\epsilon^* > 0$.

To see that $\epsilon^* < 1$, consider $\log \mathcal{P}^+(\epsilon) + \epsilon \log a$. This a bounded function for $\epsilon \in [0,1]$ (for any finite $a \in [1,\infty)$) and so its supremum is achieved at some $\epsilon^* \in [0,1]$. If $\epsilon^* = 1$, then $\log \mathcal{P}^+(1) + \log a > \log \mathcal{P}^+(\epsilon) + \epsilon \log a$ for all $\epsilon \in [0,1)$. Rearrange this to obtain $\frac{\log \mathcal{P}^+(1) - \log \mathcal{P}^+(\epsilon)}{1-\epsilon} > -\log a$. Take $\epsilon \to 1^-$ to see that $D^-\mathcal{P}^+(1) \geq -\log a > -\infty$. This is a contradiction with lemma 9.4.

Finally, by using the arguments in section 3.3.1.4; $\epsilon^* = \frac{d^-}{d\log a}\mathcal{A}^+(a)$ almost everywhere. □

The function ϵ^* is the density of visits and, for almost every $a > 0$, it is equal to the left-derivative $D^-\mathcal{A}^+(a) = \frac{d^-}{d\log a}\mathcal{A}^+(a)$ (note that $D^-\mathcal{A}^+(a)$ is a lower semicontinuous function of a).

Since $\mathcal{P}^+(\epsilon) \geq \mu_{d-1}$, it follows from the above that $\mathcal{A}^+(a) > \log\mu_{d-1} + \epsilon\log a$ for each $\epsilon \in [0,1)$, so $\mathcal{A}^+(a) \geq \log\mu_{d-1} + \log a$. This recovers the lower bound in theorem 9.2. This theorem also gives the following bounds for $\mathcal{P}^+(\epsilon)$:

$$\log\mathcal{P}^+(\epsilon) = \inf_{a\in[0,\infty)}\{\mathcal{A}^+(a) - \epsilon\log a\}\begin{cases} = \log\mu_d, & \text{if } \epsilon = 0; \\ < \log\mu_d, & \text{if } \epsilon \in (0,1]. \end{cases} \tag{9.20}$$

By equation (3.62), $D^+\log\mathcal{P}^+(0) = -\log a_c^+ \leq 0$. This lower bound on a_c^+ is strict, as will be shown in the next section. Taken together, the above gives the following theorem.

Theorem 9.6 *The limit $\mathcal{P}^+(\epsilon) = \lim_{n\to\infty}(c_n^+(\lfloor\epsilon n\rfloor))^{1/n}$ is a concave function on $\epsilon \in [0,1]$. In addition, $\log\mathcal{P}^+(0) = \log\mu_d$; $\log\mathcal{P}^+(\epsilon) < \log\mu_d$ if $\epsilon \in (0,1]$; and $\mathcal{P}^+(\epsilon)$ is right-continuous at $\epsilon = 0$ so that $\lim_{\epsilon\to 0^+}\mathcal{P}^+(\epsilon) = \log\mu_d$.*

Proof It only remains to show that $\mathcal{P}^+(\epsilon)$ is right-continuous at $\epsilon = 0$. Positive walks in $\mathbb{L}_+^d$ from $\vec{0}$ with v visits include a class of walks with the first v steps in $\partial\mathbb{L}_+^d$ and with the remaining $n-v$ steps disjoint with $\partial\mathbb{L}_+^d$. By deleting the $(v+1)$-th edge in each such walk, it follows that $c_n^+(\lfloor\epsilon n\rfloor) \geq c_{\lfloor\epsilon n\rfloor}^+(\lfloor\epsilon n\rfloor)c_{n-\lfloor\epsilon n\rfloor}^+$. Take the $\frac{1}{n}$-th power and let $n \to \infty$. By noting that $c_v^+(v)$ is the number of walks of length v in $\mathbb{L}^{d-1}$, it follows that $\mathcal{P}^+(\epsilon) \geq \mu_{d-1}^{\epsilon}\mu_d^{1-\epsilon}$. Take $\epsilon \to 0^+$ to see that $\mathcal{P}^+(0^+) \geq \mu_d$.

However, by equation (9.19), $\mathcal{P}^+(0^+) \leq \mu_d$. This completes the proof. □

Consider positive walks, as above, of length n, with the first $\lfloor\epsilon^* n\rfloor$ steps in $\partial\mathbb{L}_+^d$ and with the remaining $n - \lfloor\epsilon^* n\rfloor$ steps disjoint with $\partial\mathbb{L}_+^d$. This shows that

$$A_n^+(a) \geq c_{\lfloor\epsilon^* n\rfloor}^+(\lfloor\epsilon^* n\rfloor)c_{n-\lfloor\epsilon^* n\rfloor}^+ a^{\lfloor\epsilon^* n\rfloor}. \tag{9.21}$$

Note that $c_v^+(v)$ is the number of walks of length v in $\mathbb{L}^{d-1}$. Take logarithms, divide by n, and take $n \to \infty$. This gives

$$\mathcal{A}^+(a) \geq \epsilon^*\log\mu_{d-1} + (1-\epsilon^*)\log\mu_d + \epsilon^*\log a. \tag{9.22}$$

Use theorem 9.5, replace ϵ^* by $\frac{d^-}{d\log a}\mathcal{A}^+(a)$, and rearrange to obtain the differential inequality

$$\tfrac{d^-}{d\log a}\mathcal{A}^+(a) \leq \frac{\mathcal{A}^+(a) - \log\mu_d}{\log\mu_{d-1} - \log\mu_d + \log a} \quad \text{for almost all } a > 0. \tag{9.23}$$

Since $\mathcal{A}^+(a)$ is convex and finite, it is measurable. Choose $a > z > 1$, with z large enough that $\mathcal{A}^+(z) > \log\mu_d$. Separate variables and integrate:

$$\int_z^a \frac{d\,\mathcal{A}^+(a)}{\mathcal{A}^+(a) - \log \mu_d} \leq \int_z^a \frac{d \log a}{\log \mu_{d-1} - \log \mu_d + \log a}. \tag{9.24}$$

This gives, after simplification,

$$\frac{\mathcal{A}^+(a) - \log \mu_d}{\log a - \log(\mu_d/\mu_{d-1})} \leq \frac{\mathcal{A}^+(z) - \log \mu_d}{\log z - \log(\mu_d/\mu_{d-1})} \quad \text{for } a > z > \tfrac{\mu_d}{\mu_{d-1}}. \tag{9.25}$$

This result gives the following theorem.

Theorem 9.7 *For every $\delta > 0$, there exists an a_δ, such that*

$$\log \mu_{d-1} + \log a < \mathcal{A}^+(a) \leq \log \mu_{d-1} + (1+\delta) \log a$$

for all $a > a_\delta$.

Proof To see that $\log \mu_{d-1} + \log a < \mathcal{A}^+(a)$, consider the inequality in equation (9.12). Take logarithms and divide by n. Take $n \to \infty$ to see that

$$(1+2\alpha)\mathcal{A}^+(a) \geq -\log\Big(\sqrt{2}\alpha^\alpha (\tfrac{1}{2} - \alpha)^{1/2-\alpha}\Big) + \log \mu_{d-1} + \log a$$

for $\alpha \in [0, \frac{1}{2})$. Rearrange this to

$$\mathcal{A}^+(a) \geq -\Big(\log\Big(\sqrt{2}\alpha^\alpha (\tfrac{1}{2} - \alpha)^{1/2-\alpha}\Big) + 2\alpha\,\mathcal{A}^+(a)\Big) + \log \mu_{d-1} + \log a.$$

Since $\mathcal{A}^+(a)$ is finite for finite values of a, there is a small value of $\alpha \in (0, \frac{1}{2})$ for any $a > 0$ such that the factor in square brackets is strictly negative. This shows that $\log \mu_{d-1} + \log a < \mathcal{A}^+(a)$.

It remains to determine the upper bound.

By theorem 9.2, $\mathcal{A}^+(z) \leq \log \mu_d + \log z$. Use this bound on the right-hand side of equation (9.25):

$$\mathcal{A}^+(a) - \log \mu_d \leq \Big(\log a - \log(\tfrac{\mu_d}{\mu_{d-1}})\Big)\left(\frac{\log z}{\log z - \log(\mu_d/\mu_{d-1})}\right). \tag{9.26}$$

Let $\delta > 0$ and choose z so large that

$$1 \leq \frac{\log z}{\log z - \log(\mu_d/\mu_{d-1})} \leq 1 + \delta.$$

For any $a > z$, equation (9.26) becomes

$$\log \mu_{d-1} + \log a \leq \mathcal{A}^+(a) \leq \log \mu_d + (1+\delta)\Big(\log a - \log(\tfrac{\mu_d}{\mu_{d-1}})\Big).$$

Expand and simplify the right-hand side to obtain

$$\log \mu_{d-1} + \log a \leq \mathcal{A}^+(a) \leq \log \mu_{d-1} + (1+\delta) \log a + \delta \log(\tfrac{\mu_{d-1}}{\mu_d}).$$

Since $\frac{\mu_d}{\mu_{d-1}} > 1$, the last term may be taken away. □

9.1.5 *Location of the critical point* a_c^+

A positive walk from $\vec{0}$ in $\mathbb{L}_+^d$ consists of subwalks or trains which are in $\partial\mathbb{L}_+^d$ (*incursions*), or *loops* with all vertices, except for the two endpoints, disjoint with $\partial\mathbb{L}_+^d$ (*excursions*), and then terminates in a *tail* with only one endpoint in $\partial\mathbb{L}_+^d$.

Denote the number of positive walks in $\mathbb{L}_+^d$ of length n and with w excursions and v visits as $c_n^+(w,v)$. The partition function is

$$A_n^+(w;a) = \sum_{v=0}^{n} c_n^+(w,v)\, a^n. \tag{9.27}$$

Let $\ell_n^b(w,v)$ be the number of positive bridges of length n from $\vec{0}$ in $\mathbb{L}_+^d$ and with w distinct excursions and v visits. An adsorbing bridge with four excursions, five incursions (including the final vertex) and twelve visits is illustrated in figure 9.2.

Two positive bridges can be concatenated by placing the first vertex of the second on the last vertex of the first. If the first has length n, and the second has length m, then

$$\sum_{v_1=0}^{v} \ell_n^b(w_1, v-v_1)\,\ell_m^b(w_2,v_1) \leq \ell_{n+m}^b(w_1+w_2,v). \tag{9.28}$$

Multiple by a^v and sum over v. Denote the partition function by $A_n^b(w;a)$. Then

$$A_n^b(w_1;a)\,A_m^b(w_2;a) \leq A_{n+m}^b(w_1+w_2;a) \ \text{ for fixed } z \in (0,\infty). \tag{9.29}$$

Comparison of this to equation (3.25), lemma 3.8 and theorem 3.9 establishes the existence of the limit

$$\mathcal{P}_w^+(\delta;a) = \lim_{n\to\infty} \left(Z_n^b(\lfloor \delta n \rfloor;a)\right)^{1/n}, \ \text{for fixed } a \in (0,\infty). \tag{9.30}$$

Notice that $\delta \in [0,\frac{1}{4}]$ (since each excursion has length at least three edges followed by a step in $\partial\mathbb{L}_+^d$).

By theorem 3.9, $\mathcal{P}_w^+(\delta;a)$ is log-concave in δ. By following the arguments of section 9.1.2 mutatis mutandis, that is, arguing through equations (9.9) and (9.10) via the proof of theorem 9.1, the following theorem is obtained.

Theorem 9.8 *The density of excursions in a model of adsorbing positive walks with activity* $a \in (0,\infty)$ *is given by*

$$\mathcal{P}_w^+(\delta;a) = \lim_{n\to\infty} \left(A_n^b(\lfloor \delta n \rfloor;a)\right)^{1/n} = \lim_{n\to\infty} \left(A_n^+(\lfloor \delta n \rfloor;a)\right)^{1/n}.$$

Moreover, $\mathcal{P}_w^+(\delta;a)$ *is log-concave for* $\delta \in [0,\frac{1}{4}]$ *and is a convex function of* $\log a$ *for fixed values of* $\delta \in [0,\frac{1}{4}]$.

Finally, $\mathcal{P}_w^+(0;a) = a\,\mu_{d-1}$, *and* $\mathcal{P}_w^+(0^+;a) \geq \max\{\mu_d, a\,\mu_{d-1}\}$. *This shows that* $\mathcal{P}_w^+(\epsilon;a)$ *is not right-continuous at* $\epsilon = 0$ *for* $a < 1$.

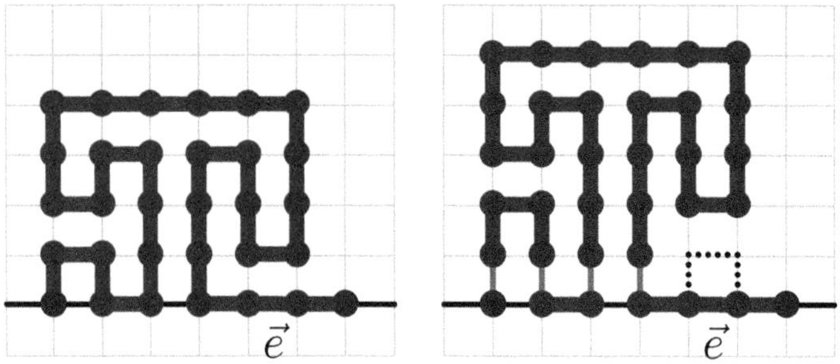

FIG. 9.8. By translating the excursions in the positive walk on the left one step away from the adsorbing boundary, a new excursion can be created on the edge-visit marked by $\vec{e}$ as shown on the right.

Proof Notice that $c_n^+(0,n) = c_n^{(d-1)}$. This shows that $\mathcal{P}_w^+(0;a) = a\mu_{d-1}$. On the other hand, $A_n^+(1;a) \geq a\ell_{n-2}$, where ℓ_n is the number of loops of length n. Thus, $\mathcal{P}_w^+(0^+;a) \geq \max\{\mu_d, a\mu_{d-1}\}$. □

The density of visits and excursions is

$$\log \mathcal{P}_w^+(\delta,\epsilon) = \inf_{0<a<\infty}\{\mathcal{P}_w^+(\delta;a) - \epsilon\log a\} = \lim_{n\to\infty}\left(c_n^+(\lfloor\delta n\rfloor,\epsilon_n)\right)^{1/n} \tag{9.31}$$

by theorem 3.6, where $\langle\epsilon_n\rangle$ is a sequence of integers such that $\frac{1}{n}\epsilon_n \to \epsilon$ as $n\to\infty$ (see reference [262]).

If $\mathcal{P}^+(\epsilon)$ is the density of adsorbing walks (see theorem 9.14), then $\mathcal{P}^+(\epsilon) = \sup_{0<\delta<1/4}\mathcal{P}_w^+(\delta,\epsilon)$. To see this, suppose that this is not the case but that there exists $\kappa>0$ such that $\mathcal{P}^+(\epsilon) - \mathcal{P}_w^+(\delta,\epsilon) \geq \kappa > 0$, for all $\delta\in(0,\frac{1}{4})$. By the definition of $\mathcal{P}_w^+(\delta,\epsilon)$, this shows that there exists an $N_0\in\mathbb{N}$ such that

$$c_n^+(\lfloor\delta n\rfloor,\epsilon_n) \leq \left(\mathcal{P}^+(\epsilon) - \tfrac{\kappa}{2}\right)^n \quad \text{for all } n\geq N_0 \text{ and } \delta\in(0,\tfrac{1}{4}). \tag{9.32}$$

In other words, for all $k\in\mathbb{N}$, it is the case that $c_n^+(k,\epsilon_n) \leq \left(\mathcal{P}^+(\epsilon)-\frac{\kappa}{2}\right)^n$. Sum over k to obtain

$$c_n^+(\epsilon_n) \leq \tfrac{n}{4}\left(\mathcal{P}^+(\epsilon)-\tfrac{\kappa}{2}\right)^n. \tag{9.33}$$

Take the power $\frac{1}{n}$ and $n\to\infty$ on the left-hand side, since $\frac{1}{n}\epsilon_n\to\epsilon$; this shows that $\mathcal{P}^+(\epsilon) < \mathcal{P}^+(\epsilon) - \frac{1}{2}\kappa$, which is a contradiction.

The function $\mathcal{P}_w^+(\delta,\epsilon)$ is defined for all $\epsilon\in(0,1)$ and for $0\leq\delta<\min\{\frac{1}{4},\frac{1}{2}\epsilon\}$. Denote the domain of $\mathcal{P}_w^+(\delta,\epsilon)$ in the $\delta\epsilon$-plane by Δ; then $\mathcal{P}_w^+(\delta,\epsilon)$ satisfies the following inequality.

Lemma 9.9 *Let $\mathcal{P}^+(\delta,\epsilon)>0$ for all $(\delta,\epsilon)\in\Delta$, where Δ is the domain of $\mathcal{P}^+(\delta,\epsilon)$. Then, for every $(\delta,\epsilon)\in\Delta$, there exists a $\delta_*>0$ such that*

$$(1+\mu_d^{-2})^{(\epsilon-3\gamma)/2}\mu_d^{-2\gamma}\mathcal{P}_w^+(\gamma,\epsilon) \leq \mathcal{P}_w^+\left(\tfrac{\gamma+\delta_*}{1+2\gamma+2\delta_*},\tfrac{\epsilon}{1+2\gamma+2\delta_*}\right).$$

A suitable choice for δ^* is $\delta_* = \frac{\epsilon-3\gamma}{2(1+\mu_d^2)}$.

Proof An *edge-visit* $\vec{e}$ in a positive walk ω in $\mathbb{L}^d$ is an edge in $\partial\mathbb{L}^d$ (see figure 9.8, where an edge-visit is marked by $\vec{e}$). If an edge-visit is disjoint with excursions in the walk, then a new excursion may be created by replacing the edge-visit by three edges in a $\sqcap$-conformation. This is only possible if there is space available adjacent to execute this construction.

Suppose that ω is a positive walk with w excursions and v visits. Translate all excursions in ω by $\vec{e}_d$ and reconnect ω by adding two edges (for each translated excursion), as illustrated in figure 9.8. This gives a new walk of length $n+2w$ with at least $v-w$ edge-visits.

It is possible to create a new excursion on an edge-visit as shown in figure 9.8 (if the edge-visit is the middle edge of a train of three edge-visits). Subtracting the edge-visits which are incident with excursions leaves at least $v-3w$ edge-visits, of which at least half can be used to create new excursions. Hence, the maximum number of new excursions which can be created is at least $\lfloor \frac{1}{2}(v-3w) \rfloor$.

If h locations for new excursions are chosen, then

$$\binom{\lfloor \frac{1}{2}(v-3w)\rfloor}{h} c_n^+(w,v) \le c_{n+2w+2h}^+(w+h,v).$$

Choose $v = \lfloor \epsilon n \rfloor$, and $w = \lfloor \gamma n \rfloor$. Let $h = \lfloor \delta n \rfloor$ and substitute these into the above. Take the $\frac{1}{n}$-th power and then let $n \to \infty$. This gives

$$\begin{aligned}\left(\frac{(\frac{1}{2}(\epsilon-3\gamma))^{(\epsilon-3\gamma)/2}}{\delta^\delta(\frac{1}{2}(\epsilon-3\gamma)-\delta)^{(\epsilon-3\gamma)/2-\delta}}\right)\mathcal{P}_w^+(\gamma,\epsilon) &\le \left(\mathcal{P}_w^+\left(\tfrac{\gamma+\delta}{1+2\gamma+2\delta}, \tfrac{\epsilon}{1+2\gamma+2\delta}\right)\right)^{1+2\gamma+2\delta} \\ &\le \mu_d^{2(\gamma+\delta)}\,\mathcal{P}_w^+\left(\tfrac{\gamma+\delta}{1+2\gamma+2\delta}, \tfrac{\epsilon}{1+2\gamma+2\delta}\right),\end{aligned}$$

since $\mathcal{P}_w^+(\gamma,\epsilon) \le \mathcal{P}^+(\epsilon) \le \mu_d$. Rearrange to obtain

$$\left(\frac{(\frac{1}{2}(\epsilon-3\gamma))^{(\epsilon-3\gamma)/2}\mu_d^{-2(\gamma+\delta)}}{\delta^\delta(\frac{1}{2}(\epsilon-3\gamma)-\delta)^{(\epsilon-3\gamma)/2-\delta}}\right)\mathcal{P}_w^+(\gamma,\epsilon) \le \mathcal{P}_w^+\left(\tfrac{\gamma+\delta}{1+2\gamma+2\delta}, \tfrac{\epsilon}{1+2\gamma+2\delta}\right).$$

The factor on the left in brackets is a maximum if $\delta = \delta^* = \frac{\epsilon-3\gamma}{2(1+\mu_d^2)}$. This completes the proof. □

A corollary of lemma 9.9 is that $a_c^+ > 1$.

Theorem 9.10 *The critical value* $\log a_c^+ \ge \frac{3}{2}\log(1+\mu_d^{-2}) > 0$.

Proof Take the limit $\gamma \to 0^+$ in lemma 9.9. Since $\mathcal{P}_w^+(\gamma,\epsilon)$ is log-concave, this shows that

$$(1+\mu_d^{-2})^{\epsilon/2}\mathcal{P}_w^+(0^+,\epsilon) \le \mathcal{P}_w^+\left(\tfrac{\delta^*}{1+2\delta^*}, \tfrac{\epsilon}{1+2\delta^*}\right), \tag{9.34}$$

where $\delta_* = \frac{\epsilon}{2(1+\mu_d^2)}$.

The free energy of adsorbing walks is the Legendre transform of $\log \mathcal{P}^+(\epsilon)$. If $a > a_c^+$, then there is a positive density of visits $\epsilon^* > 0$, and, by theorem 9.6,

$$\mathcal{A}^+(a) = \log \mathcal{P}^+(\epsilon^*) + \epsilon^* \log a \;\; \text{almost surely.}$$

Suppose that the free energy is obtained for a zero density of excursions for some values of a. In this case

$$\mathcal{A}^+(a) = \log \mathcal{P}_w^+(0^+, \epsilon^*) + \epsilon^* \log a \;\; \text{almost surely}$$

for some $\epsilon^* \geq 0$ where $\mathcal{P}_w^+(0^+, \epsilon^*) = \lim_{\gamma \to 0^+} \mathcal{P}_w^+(\gamma, \epsilon^*)$. If $\epsilon^* > 0$, then, by equation (9.34), this shows that

$$\log \mathcal{P}_w^+(0^+, \epsilon^*) + \epsilon^* \log a$$
$$\leq \log \mathcal{P}_w^+\left(\tfrac{\delta^*}{1+2\delta^*}, \tfrac{\epsilon^*}{1+2\delta^*}\right) + \tfrac{\epsilon^*}{1+2\delta^*} \log a + \log\left(\frac{a^{2\delta^*/(1+2\delta^*)}}{\sqrt{1+\mu_d^{-2}}}\right)^{\epsilon^*}.$$

That is, if $a > 1$ is chosen such that the factor in square brackets is less than 1, then this is gives a contradiction, *unless* $\epsilon^* = 0$ (and hence $\delta^* \to 0$). For these small values of a, the free energy is given by $\mathcal{A}^+(a) = \log \mathcal{P}_w^+(0,0) = \log \mu_d$, since $\epsilon^* = 0$, and the model is in the desorbed phase. This provides a lower bound on the critical value of a when the right-most factor in brackets is equal to 1. Simplification gives

$$\log a_c^+ \geq \inf_{0 \leq \delta < 1/4} \left\{ \tfrac{1+2\delta}{2\delta} \left(\log \sqrt{1+\mu_d^{-2}} \right) \right\} = \tfrac{3}{2} \log(1+\mu_d^{-2}).$$

This completes the proof. □

If $\epsilon > 0$ in lemma 9.9, then $\delta^* > 0$ so the adsorbed phase is characterised by a dense phase of excursions from the adsorbing line.

Let ω be a positive self-avoiding walk in $\mathbb{L}_+^d$ and let V be the set of visits by ω to $\partial \mathbb{L}_+^d$. If $V \neq \emptyset$, then it has a lexicographic top visit $\vec{t} \in V$; this is the *top visit* of ω. It is possible to create new visits in ω near its top visit. This is shown in the next lemma.

Lemma 9.11 *Let ω be a positive walk in $\mathbb{L}_+^d$. By adding at most four edges to ω, it is possible to create at least one and as many as three visits in $\mathbb{L}_+^d$. The resulting walk has at least one edge incident with its top visit and which is normal to both $\vec{e}_1$ and $\vec{e}_d$.*

If there is an edge of ω in $\partial \mathbb{L}_+^d$ incident with the top visit $\vec{t}$ and normal to $\vec{e}_1$ and $\vec{e}_d$, then the number of visits can be increased without increasing the number of excursions.

Proof Without loss of generality, assume that $\vec{0}$ is also a visit. Then the set of visits V of ω is not empty and there is a top visit $\vec{t}$. The proof is a case analysis.

Notice that the vertices $\vec{t}+\vec{e}_j \not\in \omega$ for $j = 1, \ldots, d-1$. It is possible that $\vec{t}+\vec{e}_d \in \omega$.

Case 1: If $\vec{t}$ is an endpoint of ω, then new visits can be created by adding two edges $\langle \vec{t} \sim (\vec{t}+\vec{e}_1)\rangle$, and $\langle (\vec{t}+\vec{e}_1) \sim (\vec{t}+\vec{e}_1+\vec{e}_2)\rangle$, starting from $\vec{t}$. This increases the length of the path by 2 and creates two new visits. The new top visit is $\vec{t}+\vec{e}_1+\vec{e}_2$, and the edge incident on it is perpendicular to both $\vec{e}_1$ and $\vec{e}_d$. This does not increase the number of excursions in the path.

Next, assume that $\vec{t}$ is not an endpoint of ω, in which case there are two edges incident with $\vec{t}$. Let the vertex preceding $\vec{t}$ in ω be $\vec{t}_a$ and let the vertex following it be $\vec{t}_b$.

Case 2: If the edge $\langle \vec{t}_a \sim \vec{t}\rangle$ preceding $\vec{t}$ is perpendicular to $\vec{e}_1$ and to $\vec{e}_d$, then the vertices $\vec{t}_a+\vec{e}_1$ and $\vec{t}+\vec{e}_1$ are not occupied. Thus, $\langle \vec{t}_a \sim \vec{t}\rangle$ may be replaced by the three edges $\langle \vec{t}_a \sim (\vec{t}_a+\vec{e}_1)\rangle$, $\langle (\vec{t}_a+\vec{e}_1) \sim (\vec{t}+\vec{e}_1)\rangle$, and $\langle (\vec{t}+\vec{e}_1) \sim \vec{t}\rangle$ in a $\sqcup$-conformation. The new top visit is $\vec{t}+\vec{e}_1$, and the edge $\langle (\vec{t}_a+\vec{e}_1) \sim (\vec{t}+\vec{e}_1)\rangle$ is perpendicular to both $\vec{e}_1$ and $\vec{e}_d$.

Similarly, if $\langle \vec{t} \sim \vec{t}_b\rangle$ is perpendicular to $\vec{e}_1$ and $\vec{e}_d$, then pairs of visits may be created by adding three edges in a $\sqcup$-conformation in the $\vec{e}_1$ direction. This construction does not increase the number of excursions in the path.

Case 3: This leaves the case that the edges $\langle \vec{t}_a \sim \vec{t}\rangle$ and $\langle \vec{t} \sim \vec{t}_b\rangle$ are both not perpendicular to $\vec{e}_1$ and $\vec{e}_d$. Without loss of generality, assume that $\langle \vec{t}_a \sim \vec{t}\rangle$ is parallel to $\vec{e}_1$ and that $\langle \vec{t} \sim \vec{t}_b\rangle$ is parallel to $\vec{e}_d$. Then the vertex $\vec{t}_b$ is not a visit but $\vec{t}_a \in V$ is. If the vertex $\vec{t}+\vec{e}_1+\vec{e}_d$ is not occupied, then new visits may be created by replacing $\langle \vec{t} \sim \vec{t}_b\rangle$ by the three edges $\langle \vec{t} \sim (\vec{t}+\vec{e}_2)\rangle$, $\langle (\vec{t}+\vec{e}_2) \sim (\vec{t}+\vec{e}_1+\vec{e}_2)\rangle$ and $\langle (\vec{t}+\vec{e}_1+\vec{e}_2) \sim (\vec{t}+\vec{e}_1+\vec{e}_d)\rangle$, where $\vec{t}+\vec{e}_1+\vec{e}_d = \vec{t}_b$. This increases the length by four and creates three new visits. The new top visit is $\vec{t}+\vec{e}_1+\vec{e}_2$ and the edge $\langle (\vec{t}+\vec{e}_1) \sim (\vec{t}+\vec{e}_1+\vec{e}_2)\rangle$ is perpendicular to both $\vec{e}_1$ and $\vec{e}_d$. This construction does not increase the number of excursions in the path.

Case 4: The last case is when the two edges $\langle \vec{t}_a \sim \vec{t}\rangle$ and $\langle \vec{t} \sim \vec{t}_b\rangle$ are both not perpendicular to $\vec{e}_1$ and $\vec{e}_d$, and the vertex $\vec{t}+\vec{e}_1+\vec{e}_d = \vec{t}_b+\vec{e}_1$ is occupied. If $\langle \vec{t}_b \sim (\vec{t}_b+\vec{e}_1)\rangle$ is an edge, then new visits may be created by replacing the two edges $\langle \vec{t} \sim \vec{t}_b\rangle$ and $\langle \vec{t}_b \sim (\vec{t}_b+\vec{e}_1)\rangle$ by the four edges $\langle \vec{t} \sim (\vec{t}+\vec{e}_2)\rangle$, $\langle (\vec{t}+\vec{e}_2) \sim (\vec{t}+\vec{e}_1+\vec{e}_2)\rangle$, $\langle (\vec{t}+\vec{e}_1+\vec{e}_2) \sim (\vec{t}+\vec{e}_1)\rangle$ and $\langle (\vec{t}+\vec{e}_1) \sim (\vec{t}_b+\vec{e}_1)\rangle$. This increases the length by two edges and the number of visits by 3. The new top visit is $\vec{t}+\vec{e}_1+\vec{e}_2$, and the edge $\langle (\vec{t}+\vec{e}_1) \sim (t+\vec{e}_1+\vec{e}_2)\rangle$ is normal to both $\vec{e}_1$ and $\vec{e}_d$.

Otherwise, if $\langle \vec{t}_b \sim (\vec{t}_b+\vec{e}_1)\rangle$ is not in the walk, then there is an edge incident with $\vec{t}_b+\vec{e}_1$ which is perpendicular to $\vec{e}_1$ and $\vec{e}_d$. Let this edge be $\langle (\vec{t}_b+\vec{e}_1) \sim \vec{c}\rangle$. The vertex $\vec{c}-\vec{e}_d$ is in $\mathbb{L}^d_+$ and is not a visit because it is lexicogaphically larger than $\vec{t}$. This is similarly the case for the vertex $\vec{t}_b+\vec{e}_1-\vec{e}_d$. Replace $\langle (\vec{t}_b+\vec{e}_1) \sim \vec{c}\rangle$ by the three edges $\langle (\vec{t}_b+\vec{e}_1) \sim (\vec{t}_b+\vec{e}_1-\vec{e}_d)\rangle$, $\langle (\vec{t}_b+\vec{e}_1-\vec{e}_d) \sim (\vec{c}-\vec{e}_d)\rangle$ and $\langle (\vec{c}-\vec{e}_d) \sim \vec{c}\rangle$ to create two new vertices and to increase the length of the walk by 2. The new top visit is one of $\vec{t}_b+\vec{e}_1-\vec{e}_d$ and $\vec{c}-\vec{e}_d$ and the edge $\langle (\vec{t}_b+\vec{e}_1-\vec{e}_d) \sim (\vec{c}-\vec{e}_d)\rangle$ is normal to the $\vec{e}_1$ direction.

Observe that this is the only case in which the number of excursions in ω is increased by 1.

Finally, observe that, if there is an edge in $\partial\mathbb{L}^d_+$ incident with $\vec{t}$ and perpendicular to both $\vec{e}_1$ and $\vec{e}_d$, then it is in one of cases 1 or 2. In this case the number of visits can be increased in pairs, each time increasing the length by 2 without increasing the number of excursions. □

Once the construction in lemma 9.11 has been applied once to a positive walk, then any subsequent application of the construction only uses cases 1 and 2 so that pairs of visits are created each time by only increasing the length by 2.

Let ω be a positive walk of length n with $\lfloor \gamma n \rfloor$ excursions and $\lfloor \epsilon n \rfloor$ visits. By repeated application of the construction in lemma 9.11, at least $2\lfloor \frac{1}{2}\alpha n \rfloor$ and at most $2\lfloor \frac{1}{2}\alpha n \rfloor + 3$ new visits are created, while the length of the walk increases by at least $2\lfloor \frac{1}{2}\alpha n \rfloor$ and at most $2\lfloor \frac{1}{2}\alpha n \rfloor + 4$. The number of excursions increases by at most 1. This shows that

$$c_n^+(\lfloor \gamma n \rfloor, \lfloor \epsilon n \rfloor) \leq \sum_{i=0}^{4}\sum_{j=0}^{3}\sum_{k=0}^{1} c^+_{n+2\lfloor \alpha n/2 \rfloor + i}(\lfloor \gamma n \rfloor + k, \lfloor \epsilon n \rfloor + 2\lfloor \tfrac{1}{2}\alpha n \rfloor + j). \quad (9.35)$$

Take the power $\frac{1}{n}$ of the above and let $n \to \infty$. This gives

$$\mathcal{P}^+_w(\gamma, \epsilon) \leq \left(\mathcal{P}^+_w\left(\tfrac{\gamma}{1+\alpha}, \tfrac{\epsilon+\alpha}{1+\alpha}\right)\right)^{1+\alpha}. \quad (9.36)$$

Take $\gamma \to 0^+$ in lemma 9.9 and use the bound in equation (9.36) to obtain

$$\begin{aligned}(1+\mu_d^{-2})^{\epsilon/2}\mathcal{P}^+_w(0^+, \epsilon) &\leq \mathcal{P}^+_w\left(\tfrac{\delta^*}{1+2\delta^*}, \tfrac{\epsilon}{1+2\delta^*}\right) \\ &\leq \left(\mathcal{P}^+_w\left(\tfrac{\delta^*}{(1+2\delta^*)(1+\alpha)}, \tfrac{\epsilon}{(1+2\delta^*)(1+\alpha)} + \tfrac{\alpha}{1+\alpha}\right)\right)^{1+\alpha}. \quad (9.37)\end{aligned}$$

Choose

$$\alpha = \frac{2\delta^*\epsilon}{(1+2\delta^*)(1-\epsilon)} \quad (9.38)$$

in the above and note that $\delta^* = \frac{\epsilon}{2(1+\mu_d^2)}$. Simplification gives

$$(1+\mu_d^{-2})^{\epsilon/2}\mathcal{P}^+_w(0^+, \epsilon) \leq \mu_d^\alpha \mathcal{P}^+_w\left(\tfrac{\delta^*}{(1+2\delta^*)(1+\alpha)}, \epsilon\right). \quad (9.39)$$

The choice of α in equation (9.38) gives $\alpha \leq \frac{\epsilon^2}{1-\epsilon}$ so that $\mu_d^\alpha < \mu_d^\epsilon \leq (1+\mu_d^{-2})^{\epsilon/2}$ if $\frac{\epsilon}{1-\epsilon} < 1$; that is, if $\epsilon \in (0, \frac{1}{2})$. Choose $\epsilon \in (0, \frac{1}{2})$ to see that

$$(1+\mu_d^{-2})^{\epsilon/2}\mathcal{P}^+_w(0^+, \epsilon) \leq \mu_d^{\epsilon^2/(1-\epsilon)} \mathcal{P}^+_w\left(\tfrac{\delta^*}{(1+2\delta^*)(1+\alpha)}, \epsilon\right). \quad (9.40)$$

Finally, observe that $(1+\mu_d^{-2}) > \mu_d^{2\epsilon/(1-\epsilon)}$ for $\epsilon > 0$ and small enough. This proves that there exists small $\epsilon > 0$ such that $\mathcal{P}^+_w(0^+, \epsilon) < \mathcal{P}^+_w\left(\frac{\delta^*}{(1+2\delta^*)(1+\alpha)}, \epsilon\right)$, where $\frac{\delta^*}{(1+2\delta^*)(1+\alpha)} > 0$. This gives the following theorem, which shows that, if $\epsilon > 0$, then there is an density of excursions in the walk.

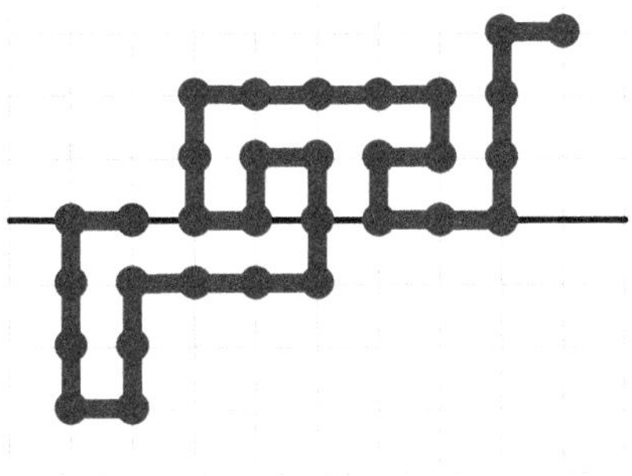

FIG. 9.9. A self-avoiding walk adsorbing in a defect place in $\mathbb{L}^d$.

Theorem 9.12 *If $\epsilon > 0$ is small enough, then there exists a $\delta > 0$ such that $\mathcal{P}^+(0^+, \epsilon) < \mathcal{P}(\delta, \epsilon)$ and where $\delta \leq \epsilon$.* □

9.1.6 *Walks adsorbing at a defect plane*

A walk in $\mathbb{L}^d$ adsorbing at a *defect plane* $\mathbb{L}^{d-1} = \partial\mathbb{L}^d_+$ is shown in figure 9.9. The walk starts in $\vec{0}$ and is weighted by the number of *visits* it makes to $\mathbb{L}^{d-1}$ (with the convention that the vertex at $\vec{0}$ is not a visit). Let $c_n^>(v)$ be the number of walks from $\vec{0}$ of length n and with v visits in the defect plane. The partition function of this model is $A_n^>(a) = \sum_{v\geq 0} c_n^>(v)\, a^v$. Since $c_n^+(v) \leq c_n^>(v) \leq c_n$, it follows that $A_n^+(a) \leq A_n^>(a) \leq (n+1)c_n$ for $a \in [0,1]$, since there are at most n visits. This shows that $\lim_{n\to\infty} \frac{1}{n} \log A_n^>(a) = \log \mu_d$ if $a \leq 1$ (see theorem 9.2).

Theorem 9.13 *The limiting free energy of walks adsorbing at a defect plane in $\mathbb{L}^d$ is given by $\mathcal{A}^>(a) = \lim\limits_{n\to\infty} \frac{1}{n} \log A_n^>(a)$. The function $\mathcal{A}^>(a)$ is also a convex function of $\log a$. For all $a \leq 1$, $\mathcal{A}^>(a) = \log \mu_d$, and, if $a > 1$, then $\mathcal{A}^>(a) \geq \log \mu_d$, and*

$$\log \mu_{d-1} + \log a < \mathcal{A}^>(a) \leq \log \mu_d + \log a.$$

$\mathcal{A}^>(a)$ is a non-analytic function at $a = a_c^>$, where $a_c^> \in [1, a_c^+]$.

Proof If $a > 1$, then the existence of $\mathcal{A}^>(a)$ and the upper bound $\mathcal{A}^>(a) \leq \log \mu_d + \log a$ for $a \geq 1$ are proven using arguments similar to those leading to theorems 9.1 and 9.2. The limiting free energy $\mathcal{A}^<(a)$ is not analytic at $a_c^> = \sup\{a \mid \mathcal{A}^>(a) = \log \mu_d\}$; this is the adsorption critical point. Since $A_n^>(a) \geq A_n^+(a) > \log \mu_{d-1} + \log a$ for all $a \geq 0$, the lower bound follows (see corollary 9.7). This gives $a_c^> \leq a_c^+$. □

The generating function of this model has a singularity structure similar to figure 3.3. The order parameter is the expected density of visits, given by

$$\langle v \rangle = \tfrac{d}{d\log a} \mathcal{A}^>(a). \tag{9.41}$$

This gives two phases: a desorbed phase, with $\langle v \rangle = 0$ when $a < a_c^>$, and an adsorbed phase, with $\langle v \rangle > 0$ when $a > a_c^>$. It is believed (but remains unproven) that $a_c^> = 1$. In section 9.1.7 it will be shown that $1 \le a_c^> < a_c^+$.

The partition function $A_n^>(z)$ has scaling similar to equation (9.14); namely,

$$A_n^>(a) \simeq B_- n^{\gamma-1} \mu_d^n \ \text{ if } a \le a_c^>, \tag{9.42}$$

where the entropic exponent γ is that of self-avoiding walks (see equation (1.12)). In the adsorbed phase the walk stays close to the defect plane, and the entropic exponent γ takes its value in $(d-1)$ dimensions.

$$A_n^>(a) \simeq B_+ n^{\gamma^{(d-1)}-1} e^{n\mathcal{A}^>(a)} \ \text{ if } a > a_c^>. \tag{9.43}$$

The crossover exponent of the transition is expected to be $\phi = \frac{1}{2}$.

9.1.7 *Proving that* $a_c^> < a_c^+$

By corollary 9.3 and theorem 9.10, $a_c^+ \in [\frac{3}{2} \log(1+\mu_d^{-2}), \mu_d/\mu_{d-1}]$. It will be shown that $1 \le a_c^> < a_c^+$ by comparing positive walks in $\mathbb{L}_+^d$ with walks adsorbing in the defect plane $\mathbb{L}^{d-1} = \partial\mathbb{L}_+^d$.

Define the density of function of visits $\mathcal{P}^>(\epsilon)$ for walks adsorbing in a defect plane using arguments similar to those leading to equation (9.19).

Theorem 9.14 *The limit* $\mathcal{P}^>(\epsilon) = \lim_{n\to\infty} \left(c_n^>(\lfloor \epsilon n \rfloor)\right)^{1/n}$ *exists and is concave on* $[0,1]$. *Moreover,* $\log \mathcal{P}^>(0) = \log \mu_d$, *and* $\log \mathcal{P}^>(\epsilon) < \log \mu_d$ *if* $\epsilon \in (0,1]$. *In addition,* $\lim_{\epsilon\to 0^+} \mathcal{P}^>(\epsilon) = \log \mu_d$ *(*$\mathcal{P}^>(\epsilon)$ *is right-continuous at* $\epsilon = 0$*).* □

By equation (3.62), $\log a_c^+ = -D^+ \log \mathcal{P}^+(0)$, and $\log a_c^> = -D^+ \log \mathcal{P}^>(0)$. Since $\mathcal{P}^+(\epsilon)$ and $\mathcal{P}^>(\epsilon)$ are concave functions, this shows that

$$\log \mathcal{P}^+(\epsilon) \le \log \mu_d - \epsilon \log a_c^+, \ \text{ and } \log \mathcal{P}^>(\epsilon) \le \log \mu_d - \epsilon \log a_c^>. \tag{9.44}$$

Since $\mathcal{P}(0) = \mathcal{P}^+(0) = \mu_d$, this gives the following result.

Lemma 9.15 *The difference* $\log a_c^+ - \log a_c^> = \frac{1}{\mu_d} \lim_{\epsilon\to 0^+} \frac{1}{\epsilon}\left(\mathcal{P}^>(\epsilon) - \mathcal{P}^+(\epsilon)\right)$.

Proof By the above, $\log a_c^+ - \log a_c^> = D^+ \log \mathcal{P}^>(0) - D^+ \log \mathcal{P}^+(0)$. Use the definition of the right-derivative to complete the proof. □

Since the defect plane $\mathbb{L}^{d-1}$ is a $d-1$ dimensional sub-lattice of $\mathbb{L}^d$, each visit of a walk to $\partial\mathbb{L}_+^d$ is incident with an *edge-visit* (an edge of the walk in $\mathbb{L}^{d-1}$). If there are v visits, then there are at least $\lfloor \frac{1}{2} v \rfloor$ edge-visits. A positive walk can be used to create walks adsorbing in a defect plane by replacing edge-visits by ⊔-conformations in the negative half-lattice $\mathbb{L}_-^d$. This is illustrated in figure 9.10.

Theorem 9.16 $\mathcal{P}^+$ *and* $\mathcal{P}^>$ *are related by* $(1+\mu_d^{-2})^{\epsilon/4} \mathcal{P}^+(\epsilon) \le \mathcal{P}^>\left(\frac{2(1+\mu_d^2)\epsilon}{2(1+\mu_d^2)+\epsilon}\right)$.

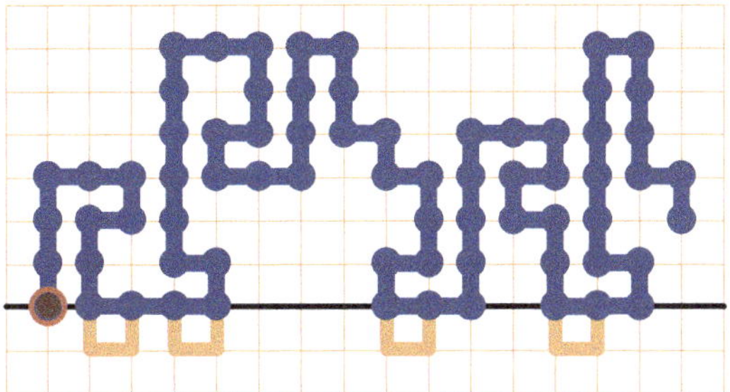

FIG. 9.10. A positive walks with edge-visits. This walk can be turned into a walk adsorbing in the defect plane $\partial\mathbb{L}^d_+$ by replacing edge-visits by ⊔-conformations as shown.

Proof Let ω be a positive walk in $\mathbb{L}^d_+$ of length n and with $\lfloor \epsilon n \rfloor$ visits in $\partial\mathbb{L}^d_+$. Then ω has at least $\lfloor \frac{1}{2}\epsilon n \rfloor$ edge-visits in $\partial\mathbb{L}^d_+$, not all of which are necessarily disjoint (see figure 9.10).

Since each edge-visit is incident with at most two other edge-visits, there is a subset E of at least $\lfloor \frac{1}{2} \lfloor \frac{1}{2}\epsilon n \rfloor \rfloor$ edge-visits which are disjoint. Choose m of these disjoint edge-visits in E and replace them with three edges in a ⊔-conformation in the negative half-lattice $\partial\mathbb{L}^d_-$ (see figure 9.10).

This creates a walk at the defect plane $\mathbb{L}^{d-1}$ of length $n+2m$. By accounting for the different ways in which the above construction can be completed,

$$\binom{\lfloor \lfloor \epsilon n/2 \rfloor /2 \rfloor}{m} c_n^+(\lfloor \epsilon n \rfloor) \le c_{n+m}^>(\lfloor \epsilon n \rfloor).$$

Let $m = \lfloor \delta n \rfloor$, where $\delta < \frac{1}{4}\epsilon$. Take the power $\frac{1}{n}$ of the above and let $n \to \infty$. The result is

$$\left(\frac{(\frac{1}{4}\epsilon)^{\epsilon/4}}{\delta^\delta (\frac{1}{4}\epsilon - \delta)^{\epsilon/4-\delta}} \right) \mathcal{P}^+(\epsilon) \le \left(\mathcal{P}^> \left(\tfrac{\epsilon}{1+2\delta} \right) \right)^{1+2\delta}.$$

But $\mathcal{P}^>(\epsilon) \le \mathcal{P}^>(0) = \mu_d$ (see theorem 9.14); therefore,

$$\left(\frac{(\frac{1}{4}\epsilon)^{\epsilon/4} \mu_d^{-2\delta}}{\delta^\delta (\frac{1}{4}\epsilon - \delta)^{\epsilon/4-\delta}} \right) \mathcal{P}^+(\epsilon) \le \mathcal{P}^> \left(\tfrac{\epsilon}{1+2\delta} \right).$$

The factor in square brackets on the left-hand side is a maximum if $\delta = \frac{\epsilon}{4(1+\mu_d^2)}$ (see for example equation (D.15) in appendix D). Simplify to complete the proof. □

A corollary of this theorem is

Corollary 9.17 *The difference* $\log a_c^+ - \log a_c^> \ge \frac{1}{4} \log(1 + \mu_d^{-2}) > 0.$

Proof By theorem 9.14, $\mathcal{P}^>(\epsilon)$ is right-continuous at $\epsilon = 0$ and, since it is concave, it is right-differentiable at $\epsilon = 0$. That is, $D^+\mathcal{P}^>(0) = \frac{d\mathcal{P}^>}{d\epsilon}\mid_{\epsilon=0^+}$ exists. By equation (3.62), $D^+\mathcal{P}^>(0) \leq 0$, since $a_c^> \geq 1$ (by theorem 9.13).

Eliminate $\mathcal{P}^+(\epsilon)$ between lemma 9.15 and theorem 9.16. This gives

$$\log a_c^+ - \log a_c^> \geq \lim_{\epsilon\to 0^+} \tfrac{1}{\epsilon}\left(1 - \tfrac{(1+\mu_d^{-2})^{-\epsilon/4}}{\mathcal{P}^>(\epsilon)}\mathcal{P}^>\left(\tfrac{2(1+\mu_d^2)\epsilon}{2(1+\mu_d^2)+\epsilon}\right)\right). \tag{9.45}$$

Assume that $\epsilon > 0$ is small. There are two cases to consider.

Case 1: Suppose that $D^+\mathcal{P}^>(0) = 0$. Then there exists an $\eta > 0$ such that

$$\mathcal{P}^>(0) - \epsilon\eta \leq \mathcal{P}^>(\epsilon) \leq \mathcal{P}^>(0) + \epsilon\eta.$$

Since $\mathcal{P}^>(\epsilon)$ is right-continuous at $\epsilon = 0$, η may be taken arbitrary small if $\epsilon \to 0^+$ in the above. Using this on the terms in the right-hand side of equation (9.45) shows that $(\mathcal{P}^>(\epsilon))^{-1} \leq (\mathcal{P}^>(0) - \epsilon\eta)^{-1} \leq \mu_d^{-1}(1 + O(\eta\epsilon))$, and

$$\mathcal{P}^>\left(\tfrac{2(1+\mu_d^2)\epsilon}{2(1+\mu_d^2)+\epsilon}\right) \leq \mathcal{P}^>(0) + \eta\left(\tfrac{2(1+\mu_d^2)\epsilon}{2(1+\mu_d^2)+\epsilon}\right) = \mu_d + \eta\left(\tfrac{2(1+\mu_d^2)\epsilon}{2(1+\mu_d^2)+\epsilon}\right).$$

Substitute these upper bounds on the right-hand side of equation (9.45) and simplify:

$$\log a_c^+ - \log a_c^> \geq \lim_{\epsilon\to 0^+} \tfrac{1}{\epsilon}\left(1 - \left(1 + \tfrac{\eta}{\mu_d}\left(\tfrac{2(1+\mu_d^2)\epsilon}{2(1+\mu_d^2)+\epsilon}\right) + O(\eta^2\epsilon^2)\right)(1+\mu_d^{-2})^{-\epsilon/4}\right).$$

Expand $(1+\mu_d^{-2})^{-\epsilon/4} = 1 - \frac{\epsilon}{4}\log(1+\mu_d^{-2}) + O(\epsilon^2)$. Substitute, simplify and take $\epsilon \to 0^+$. The result is that

$$\log a_c^+ - \log a_c^> \geq \lim_{\epsilon\to 0^+} \left(\tfrac{1}{4}\log(1+\mu_d^{-2}) - \tfrac{\eta}{\mu_d}\left(1 + \tfrac{2(1+\mu_d^2)}{2(1+\mu_d^2)+\epsilon}\right)\right).$$

Evaluating the limit gives

$$\log a_c^+ - \log a_c^> \geq \tfrac{1}{4}\log(1+\mu_d^{-2}) - \tfrac{2}{\mu_d}\eta.$$

Take $\eta \to 0^+$ to obtain the claimed lower bound.

Case 2: It only remains to consider the case that $D^+\mathcal{P}^>(0) < 0$. Take $\epsilon > 0$ very small and note that there exists an $\eta > 0$ such that

$$\mathcal{P}^>(0) + \epsilon(1-\eta)D^+\mathcal{P}^>(0) \geq \mathcal{P}^>(\epsilon) \geq \mathcal{P}^>(0) + \epsilon(1+\eta)D^+\mathcal{P}^>(0).$$

Observe as well that if $\epsilon \to 0^+$, then η may be taken arbitrary small. Use these bounds on $\mathcal{P}^>(\epsilon)$ on the factors on the right-hand side of equation (9.45):

$$\begin{aligned}\frac{\mathcal{P}^>\left(\frac{2(1+\mu_d^2)\epsilon}{2(1+\mu_d^2)+\epsilon}\right)}{\mathcal{P}^>(\epsilon)} &\leq \frac{\mathcal{P}^>(0) + \left(\frac{2(1+\mu_d^2)(1-\eta)\epsilon}{2(1+\mu_d^2)+\epsilon}\right)D^+\mathcal{P}^>(0)}{\mathcal{P}^>(0) + \epsilon(1+\eta)D^+\mathcal{P}^>(0)} \\ &\leq \left(1 + \left(\tfrac{2(1+\mu_d^2)(1-\eta)\epsilon}{2(1+\mu_d^2)+\epsilon}\right)\left(\tfrac{D^+\mathcal{P}^>(0)}{\mathcal{P}^>(0)}\right)\right)\left(1 - \epsilon(1+\eta)\left(\tfrac{D^+\mathcal{P}^>(0)}{\mathcal{P}^>(0)}\right) + \eta_0\epsilon^2\right),\end{aligned}$$

provided that $\epsilon < -\mathcal{P}^>(0)/(2(1+\eta)D^+\mathcal{P}^>(0))$. The factor η_0 is given by $\eta_0 = (1+2(1+\eta)^2)(D^+\mathcal{P}^>(0)/\mathcal{P}^>(0))^2$. Further simplification gives

$$\frac{1}{\mathcal{P}^>(\epsilon)}\mathcal{P}^>\left(\frac{2(1+\mu_d^2)\epsilon}{2(1+\mu_d^2)+\epsilon}\right) \leq 1 - 2\eta\epsilon\left(\frac{D^+\mathcal{P}^>(0)}{\mathcal{P}^>(0)}\right) + \eta_1\epsilon^2,$$

where $\eta_1 > 0$ is finite. Substitution of this into equation (9.45) and again expanding $(1+\mu_d^{-2})^{-\epsilon/4} = 1 - \frac{\epsilon}{4}\log(1+\mu_d^{-2}) + O(\epsilon^2)$ gives

$$\log a_c^+ - \log a_c^>$$
$$\geq \lim_{\epsilon\to 0^+} \tfrac{1}{\epsilon}\Big(1 - \Big(1 - 2\epsilon\eta\Big(\tfrac{D^+\mathcal{P}^>(0)}{\mathcal{P}^>(0)}\Big) + \eta_1\epsilon^2\Big)\big(1 - \tfrac{\epsilon}{4}\log(1+\mu_d^{-2}) + O(\epsilon^2)\big)\Big)$$

after simplification. Expanding the right-hand side and taking $\epsilon \to 0^+$ gives

$$\log a_c^+ - \log a_c^> \geq \tfrac{1}{4}\log(1+\mu_d^{-2}) + 2\eta\frac{D^+\mathcal{P}^>(0)}{\mathcal{P}^>(0)}.$$

Taking $\eta \to 0^+$ completes the proof for this case as well. □

Since $a_c \geq 1$, a corollary of corollary 9.17 is that

$$a_c^+ \geq \left(1+\mu_d^{-2}\right)^{1/4}. \tag{9.46}$$

In $\mathbb{L}^2$ this shows that $a_c^+ \geq 1.034\ldots$, and in $\mathbb{L}^3$ $a_c^+ \geq 1.011\ldots$; these lower bounds are far smaller than the numerical estimates in table 9.1.

9.1.8 *Adsorbing walks in a slab*

In section 8.1 a model of walks in a slab $\mathbb{S}_w$ was examined. Denote the *bottom wall* of $\mathbb{S}_w$ by $B(\mathbb{S}_w)$, and the *top wall* of $\mathbb{S}_w$ by $T(\mathbb{S}_w)$. Visits of the walk to $B(\mathbb{S}_w)$ are weighted by a, and to $T(\mathbb{S}_w)$ are weighted by b. The partition function of the model is given by $S_n(a,b;w)$ in equation (8.3).

Existence of a limiting free energy is shown in theorem 8.9. Since all walks in $\mathbb{S}_w$ also contributes to the partition function of walks in $\mathbb{L}^d_+$, it follows that $\mathcal{S}_w(a,1) \leq \mathcal{A}^+(a)$ for (see theorem 9.1). A corollary of this and of lemma 8.20 is that $\lim_{w\to\infty}\mathcal{S}_w(a,1) \leq \mathcal{A}^+(a)$ exists.

Arguments similar to those explained in the proof of theorem 8.19 show that $\lim_{w\to\infty}\mathcal{S}_w(a,1) = \mathcal{A}^+(a)$. The arguments are almost identical here but with $\ell_n^\dagger$ (the number of loops) replaced by the partition function of adsorbing loops $L_n^\dagger(a)$ (see equation (9.5)).

By theorem 8.17, there exist small $\epsilon > 0$, and $N_\epsilon \in \mathbb{N}$, such that, for all $n \geq N_\epsilon$, $\limsup_{n\to\infty}\frac{1}{n}\log S_n([00],a,1,w;[>\lfloor\epsilon n\rfloor,P]) \geq \mathcal{S}_w(a,1) - \epsilon$. These results give the following theorem.

Theorem 9.18 *Let $a > 0$ and let $0 \leq b \leq 1$; then $\lim_{w\to\infty}\mathcal{S}_w(a,b) = \mathcal{A}^+(a)$. Since $\mathcal{S}_w(a,b) = \mathcal{S}_w(b,a)$, the theorem is also true if a and b are interchanged in the above.*

Proof The theorem is true if $b = 1$, by the arguments above. For $0 \leq b \leq 1$, $\mathcal{S}_{w-1}(a,1) = \mathcal{S}_w(a,0) \leq \mathcal{S}_w(a,b) \leq \mathcal{S}_w(a,1)$. Take $w \to \infty$ to complete the proof. □

Note that $\mathcal{A}^+(a) = \log \mu_d$ if $a < a_c^+$ (see corollary 9.3). A corollary of theorem 9.18 is the following.

Corollary 9.19 *If $a \leq 1$, and $b \leq a_c^+$ then $\lim_{w\to\infty} \mathcal{S}_w(a,b) = \log \mu_d$. Similarly, if $b \leq 1$, and $a \leq a_c^+$, then $\mathcal{S}_w(a,b) = \log \mu_d$.* □

Since $\mathcal{S}_w(a,b)$ is convex in both $\log a$ and $\log b$, it follows that

$$\mathcal{S}_w(\sqrt{a_1 a_2}, \sqrt{b_1 b_2}) \leq \tfrac{1}{2}(\mathcal{S}_w(a_1,b_1) + \mathcal{S}_w(a_2,b_2)) \,. \tag{9.47}$$

Put $b_1 = a_2 = a$ and put $a_1 = b_2 = a$ in the above to see that

$$\mathcal{S}_w(\sqrt{a}, \sqrt{a}) \leq \tfrac{1}{2}(\mathcal{S}_w(a,1) + \mathcal{S}_w(1,a)) \leq \tfrac{1}{2}\big(\mathcal{A}^+(a) + \mathcal{A}^+(a)\big) = \mathcal{A}^+(a). \tag{9.48}$$

Since $\mathcal{A}^+(a) = \log \mu_d$ if $a \leq a_c^+$ and $\liminf_{w\to\infty} \mathcal{S}_w(\sqrt{a}, \sqrt{a}) \geq \log \mu_d$, the result is the existence of the limit

$$\lim_{w\to\infty} \mathcal{S}_w(\sqrt{a}, \sqrt{a}) = \log \mu_d \text{ for all } a \leq a_c^+. \tag{9.49}$$

Continue by choosing $a_1 = b_1 = \sqrt{a}$ and choosing $a_2 = a$ with $b_2 = 1$ in equation (9.47). This gives

$$\mathcal{S}_w(a^{3/4}, a^{1/2}) \leq \tfrac{1}{2}\Big(\mathcal{S}_w(a^{1/4}, a^{1/2}) + \mathcal{S}_w(a,1)\Big). \tag{9.50}$$

The result, using the above, and the same arguments as in the above, is that

$$\lim_{w\to\infty} \mathcal{S}_w(a^{3/4}, a^{1/4}) = \log \mu_d \text{ for all } a \leq a_c^+. \tag{9.51}$$

By repeated application of this, it may be proven that $\lim_{w\to\infty} \mathcal{S}_w(a,b) = \log \mu_d$ on a dense subset of points on the curve $ab = c$ for $1 < c < a_c^+$. Since $\lim_{w\to\infty} \mathcal{S}_w(a,b)$ is the limit of a sequence of convex functions, it is itself convex and thus continuous. This shows that $\lim_{w\to\infty} \mathcal{S}_w(a,b) = \log \mu_d$ for all points (a,b), with $ab < a_c^+$ and with both $a < a_c^+$, and $b < a_c^+$.

On the other hand, if $b \geq 1$, then $\mathcal{S}_w(a,b) \geq \mathcal{S}_w(a,1)$, $\lim_{w\to\infty} \mathcal{S}_w(a,1) = \mathcal{A}^+(a)$, and $\mathcal{A}^+(a) > \log \mu_d$, for all $a > a_c^+$. This shows that $\lim_{w\to\infty} \mathcal{S}_w(a,b) > \log \mu_d$ if either $a > a_c^+$ or if $b > a_c^+$. Taken together, the following theorem is obtained.

Theorem 9.20 *If both $a < a_c^+$ and $b < a_c^+$ so that $ab < a_c^+$, then $\lim_{w\to\infty} \mathcal{S}_w(a,b) = \log \mu_d$. On the other hand, if either $a > a_c^+$ or $b > a_c^+$, then $\lim_{w\to\infty} \mathcal{S}_w(a,b) > \log \mu_d$. Finally, if both $a > a_c^+$ and $b > a_c^+$, then $\lim_{w\to\infty} \mathcal{S}_w(a,b) \geq \max\{\mathcal{A}^+(a), \mathcal{A}^+(b)\}$.*

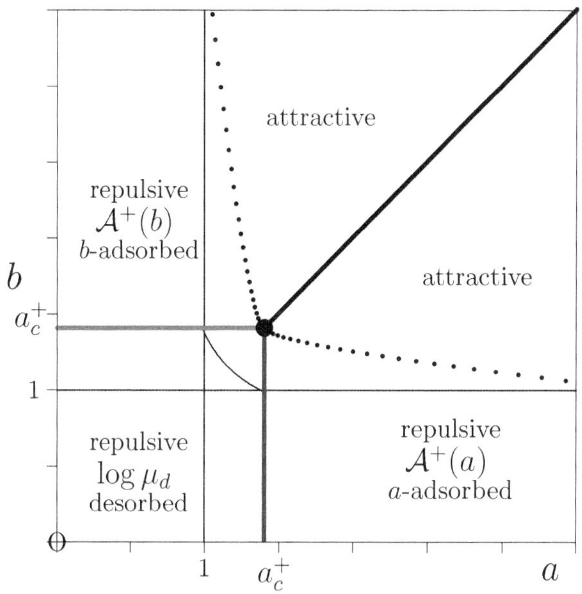

FIG. 9.11. The phase diagram of walks adsorbing in $\mathbb{S}_w$ in the $w \to \infty$ limit.

Proof It only remains to prove that $\lim_{w\to\infty} \mathcal{S}_w(a,b) \geq \max\{\mathcal{A}^+(a), \mathcal{A}^+(b)\}$. This follows because $\mathcal{S}_w(a,b) \geq \mathcal{S}_w(a,1)$ if $b \geq 1$. Then use theorem 9.18. □

The phase diagram of walks adsorbing in the walls of $\mathbb{S}_w$ (in the limit $w \to \infty$) is illustrated in figure 9.11. When both $a < a_c^+$ and $b < a_c^+$ so that $ab < a_c^+$, $\lim_{w\to\infty} \mathcal{S}_w(a,b) = \log \mu_d$. If $b < 1$, then $\lim_{w\to\infty} \mathcal{S}_w(a,b) = \mathcal{A}^+(a)$. There is an adsorption transition of the walk onto the bottom hard wall $B(\mathbb{S}_w)$ at $a = a_c^+$, in the limit $w \to \infty$.

Similarly, there is an adsorption transition of the walk onto $T(\mathbb{S}_w)$ if $a < 1$ at a critical point $b = a_c^+$, in the limit $w \to \infty$.

For both $a > 1$ and $b > 1$, the situation is more complex. There are points (a,b) in the phase diagram with $ab < a_c^+$, where $\lim_{w\to\infty} \mathcal{S}_w(a,b) = \log \mu_d$, and the model is in a desorbed or free phase in this regime. Increasing either a or b causes the model to pass through an adsorption transition onto $B(\mathbb{S}_w)$ or $T(\mathbb{S}_w)$ in the limit $w \to \infty$.

It is conjectured in reference [316] that $\lim_{w\to\infty} \mathcal{S}_w(a,b) = \mathcal{A}^+(a)$ for all $a > b > a_c^+$ (and $\lim_{w\to\infty} \mathcal{S}_w(a,b) = \mathcal{A}^+(b)$ for all $b > a > a_c^+$). This conjecture implies that there is a line of first order transitions in the model along the line $a = b$ for $a > a_c^+$ where the path switches its adsorbed state from $B(\mathbb{S}_w)$ to $T(\mathbb{S}_w)$ or vice versa. It is a conjecture that the three critical curves meet in the point $(a,b) = (a_c^+, a_c^+)$ [316].

The three critical curves divide the phase diagram as shown in figure 9.11 into a-adsorbed, b-adsorbed and desorbed phases.

A walk in $\mathbb{S}_w$ exerts net forces on $B(\mathbb{S}_w)$ and $T(\mathbb{S}_w)$. These forces may be examined by using free energy differences. The force is defined by the discrete derivative

$$F_w(a,b) = \mathcal{S}_w(a,b) - \mathcal{S}_{w-1}(a,b). \tag{9.52}$$

For all values of $b \leq 1$, the force is strictly positive for finite w. This follows because

$$F_w(a,b) = \mathcal{S}_w(a,b) - \mathcal{S}_{w-1}(a,b) > \mathcal{S}_{w-1}(a,1) - \mathcal{S}_{w-1}(a,1) = 0 \tag{9.53}$$

by theorem 9.18 and lemma 8.20. There is a net repulsive force between $B(\mathbb{S}_w)$ and $T(\mathbb{S}_w)$ in this regime. In the desorbed regime (for both $a < a_c^+$ and $b < a_c^+$) the force is expected to be short ranged (decay exponentially fast with w).

On the other hand, in the adsorbed regimes the force may be long ranged [316].

If both $a > 1$ and $b > 1$, then the walk is attracted to both hard walls. It is expected that if a and b are large enough, then the walk will bridge the gap between the walls, resulting in a net attractive force between them. This is conjectured and not known. It is also conjectured that the attractive forces in these adsorbed phases will be long ranged (that is, decay slower than exponential with w) [316].

Numerical results in reference [309] and the results obtained from a directed path version of this model strongly support these expectations [74].

The regime of attractive forces is separated from the regime of repulsive forces by a curve of zero force. The presumed limit of this curve in the $w \to \infty$ limit is indicated by a dotted line in figure 9.11. It is conjectured that the limiting zero-force curve passes through the point $(a,b) = (a_c^+, a_c^+)$ and that it is asymptotic to both $a = 1$ and $b = 1$ in the $w \to \infty$ limit.

9.2 Adsorbing polygons

Let $\mathbb{L}_+^d$ be the positive half-lattice with boundary $\partial\mathbb{L}_+^d$ (see equation (2.20)). Lattice polygons in $\mathbb{L}_+^d$ which intersect $\partial\mathbb{L}_+^d$ are *positive polygons*. The two positive polygons ω_1 and ω_2 are equivalent if $\omega_2 = \omega_1 + \vec{v}$ in $\mathbb{L}_+^d$, where $\vec{v} \in \mathbb{Z}^d$ is a vector orthogonal to $\vec{e}_d$.

Denote the number of *positive polygons* of length n by p_n^+. Denote the number of positive polygons of length n and with v visits in $\partial\mathbb{L}_+^d$ as $p_n^+(v)$. It is convenient to assume that $p_n^+(0) = 0$. Since $\mathbb{L}^d$ is bipartite, $p_n^+ = 0$ for all odd values of n. Thus, assume that n is even and that limits are taken through even numbers.

A lattice polygon in $\mathbb{L}^d$ is *attached* to $\partial\mathbb{L}_+^d$ if it contains a vertex $\vec{v} \in \partial\mathbb{L}_+^d$ (that is, ω intersects $\partial\mathbb{L}_+^d$). Similar to positive polygons, two attached polygons are equivalent if the second is a translation of the first parallel to $\partial\mathbb{L}_+^d$. Let $p_n^>(v)$ be the number of attached polygons in $\mathbb{L}^d$ of length n and with v visits in $\partial\mathbb{L}_+^d$. Put $p_n^>(0) = 0$.

Positive polygons with exactly two visits to $\partial\mathbb{L}_+^d$ are attached to $\partial\mathbb{L}_+^d$ by a $\sqcup$-conformation of edges. By deleting these edges and replacing them with a single edge, $p_n^+(2) \geq p_{n-2}$, where p_n is the number of lattice polygons of length n. In addition, $\sum_{v\geq 0} p_n^+(v) = p_n^+ \leq p_n$, and $\sum_{v\geq 0} p_n^>(v) = p_n^> \leq n p_n$.

Let a be the generating variable of visits in positive and attached polygons. Let $P_n^+(a)$ be the partition function of *adsorbing positive polygons*, and $P_n^>(a)$ be the partition function of *adsorbing attached polygons*.

By removing one edge incident with $\partial\mathbb{L}_+^d$ from attached and positive polygons, it follows that

$$P_n^+(a) \leq P_n^>(a) \leq A_{n-1}^>(a) \;\text{ for } a \in (0,\infty), \tag{9.54}$$

where $A_n^>(z)$ is the partition function of walks adsorbing at a defect plane (see section 9.1.6).

Existence of a limiting free energy for attached and positive walks is shown using most popular class arguments (see the proofs of theorem 9.2 and corollary 9.3) [262].

Theorem 9.21 *The limiting free energies of attached and positive polygons are*

$$\mathcal{A}_p^+(a) = \lim_{n\to\infty} \tfrac{1}{n} \log P_n^+(a), \quad \textit{and}\; \mathcal{A}_p^>(a) = \lim_{n\to\infty} \tfrac{1}{n} \log P_n^>(a),$$

respectively, for $a \in (0,\infty)$. *In addition,* $\mathcal{A}_p^+(a)$ *and* $\mathcal{A}_p^>(a)$ *are convex functions of* $\log a$.

Proof Let ω be a polygon in $\mathbb{L}^d$. Find the *top edge* $\vec{e}_t$ and the *bottom edge* $\vec{e}_b$ of ω by a lexicographic ordering of the edges via the coordinates of their midpoints. Then $\vec{e}_t \perp \vec{e}_1$, and $\vec{e}_b \perp \vec{e}_1$, while $\vec{e}_t$ is parallel to $\widehat{\psi} \in \{\vec{e}_2,\ldots,\vec{e}_d\}$, and $\vec{e}_b$ is parallel to $\widehat{\phi} \in \{\vec{e}_2,\ldots,\vec{e}_d\}$.

Let h_t be the d-th coordinate of $\vec{e}_t$, and h_b be the d-th coordinate of $\vec{e}_b$. Then the *height* and orientation of $\vec{e}_t$ are denoted by $[h_t\widehat{\psi}]$, and thos of $\vec{e}_b$ are denoted by $[h_b\widehat{\phi}]$. Let $p_n^+(v;[h_b\widehat{\phi}],[h_t\widehat{\psi}])$ be the number of positive polygons with a bottom edge $\vec{e}_b$ and a top edge $\vec{e}_t$ and of height and orientation $[h_b\widehat{\phi}]$ and $[h_t\widehat{\psi}]$, respectively. Define the partition function

$$P_n^+(a;[h_b\widehat{\phi}],[h_t\widehat{\psi}]) = \sum_{v\geq 0} p_n^+(v;[h_b\widehat{\phi}],[h_t\widehat{\psi}])\, a^v.$$

Summing over $\{[h_b\widehat{\phi}],[h_t\widehat{\psi}]\}$ recovers $P_n^+(a)$ in equation (9.54).

There are most popular values of $\{[h_b\widehat{\phi}],[h_t\widehat{\psi}]\}$ in $P_n^+(a;[h_b\widehat{\phi}],[h_t\widehat{\psi}])$, denoted by $\{[h_b^*\widehat{\phi}^*],[h_t^*\widehat{\psi}^*]\}$; this defines the most popular class of polygons. Thus,

$$P_n^+(a;[h_b^*\widehat{\phi}^*],[h_t^*\widehat{\psi}^*]) \leq P_n^+(a) \leq n^2(d-1)^2\, P_n^+(a;[h_b^*\widehat{\phi}^*],[h_t^*\widehat{\psi}^*]) \tag{9.55}$$

since h_b and h_t have at most n possible values each, while $\widehat{\phi}^*$ and $\widehat{\psi}^*$ have at most $d-1$ possible orientations (note that these are functions of n and a).

Positive polygons in the class with bottom and top edges of height and orientation $\{[h\widehat{\phi}],[h\widehat{\phi}]\}$ can be concatenated because their top and bottom edges have the same height and are parallel. If a polygon of length n and with $v-u$ visits is concatenated with a polygon of length m and with u visits (see figure 1.5), then

$$\sum_{u=0}^{v} p_n^+(v-u;[h\widehat{\phi}],[h\widehat{\phi}])\,p_m^+(u;[h\widehat{\phi}],[h\widehat{\phi}]) \leq p_{n+m}^+(v;[h\widehat{\phi}],[h\widehat{\phi}]). \tag{9.56}$$

Multiply by a^v and sum over v. This gives

$$P_n^+(a;[h\widehat{\phi}],[h\widehat{\phi}])\,P_m^+(a;[h\widehat{\phi}],[h\widehat{\phi}]) \leq P_{n+m}^+(a;[h\widehat{\phi}],[h\widehat{\phi}]). \tag{9.57}$$

Consider next the most popular values of $[h\widehat{\phi}]$ in the above for positive polygons of length n; denote these by $[h^*\widehat{\phi}^*]_n$ (generally a function of n). Put $N = nk+r$ for some fixed value of n and use the inequality in equation (9.57) repeatedly to see that

$$\left(P_n^+(a;[h^*\widehat{\phi}^*]_n,[h^*\widehat{\phi}^*]_n)\right)^k P_r^+(a;[h^*\widehat{\phi}^*]_n,[h^*\widehat{\phi}^*]_n) \leq P_N^+(a;[h^*\widehat{\phi}^*]_n,[h^*\widehat{\phi}^*]_n).$$

The right-hand side may be bounded from above by replacing $[h^*\widehat{\phi}^*]_n$ by $[h^*\widehat{\phi}^*]_N$. Increase r (and decrease k) if necessary so that the left-hand side is not 0. This gives

$$\left(P_n^+(a;[h^*\widehat{\phi}^*]_n,[h^*\widehat{\phi}^*]_n)\right)^k P_r^+(a;[h^*\widehat{\phi}^*]_n,[h^*\widehat{\phi}^*]_n) \leq P_N^+(a;[h^*\widehat{\phi}^*]_N,[h^*\widehat{\phi}^*]_N).$$

Take logarithms, divide by N and take the limit inferior of the right-hand side with n fixed. Then $k\to\infty$ and the result is that

$$\tfrac{1}{n}\log P_n^+(z;[h^*\widehat{\phi}^*]_n,[h^*\widehat{\phi}^*]_n) \leq \liminf_{N\to\infty} \tfrac{1}{N}\log P_N^+(z;[h^*\widehat{\phi}^*]_N,[h^*\widehat{\phi}^*]_N).$$

Take the limit superior of the left-hand side to show that the limit

$$\mathcal{A}_p^+(a) = \lim_{n\to\infty} \tfrac{1}{n}\log P_n^+(a;[h^*\widehat{\phi}^*]_n,[h^*\widehat{\phi}^*]_n) \tag{9.58}$$

exists. Continue by suppressing the explicit dependence on n of $[h^*\widehat{\phi}^*]_n$ so that $[h^*\widehat{\phi}^*]_n \equiv [h^*\widehat{\phi}^*]$.

By reflection symmetry, $P_n^+(a;[h_b^*\widehat{\phi}^*],[h_t^*\widehat{\psi}^*]) = P_n^+(a;[h_t^*\widehat{\psi}^*],[h_b^*\widehat{\phi}^*])$. By concatenating two polygons as in figure 1.5,

$$\sum_{u=0}^{v} p_n^+(v-u;[h_b^*\widehat{\phi}^*],[h_t^*\widehat{\psi}^*])\,p_n^+(u;[h_b^*\widehat{\phi}^*],[h_t^*\widehat{\psi}^*]) \leq p_{2n}^+(v;[h_b^*\widehat{\phi}^*],[h_b^*\widehat{\psi}^*]).$$

Multiply this by a^v and sum over v. This gives

$$\left(P_n^+(a;[h_b^*\widehat{\phi}^*],[h_t^*\widehat{\psi}^*])\right)^2 \leq P_{2n}^+(a;[h_b^*\widehat{\phi}^*],[h_b^*\widehat{\phi}^*]) \leq P_{2n}^+(a;[h^*\widehat{\phi}^*],[h^*\widehat{\phi}^*]).$$

Since $P_n^+(a;[h^*\widehat{\phi}^*],[h^*\widehat{\phi}^*]) \leq P_n^+(a;[h_b^*\widehat{\phi}^*],[h_t^*\widehat{\psi}^*])$, it follows by comparison to equation (9.58) that the limit

$$\mathcal{A}_p^+(a) = \lim_{n\to\infty} \tfrac{1}{n} \log P_n^+(a;[h_b^*\widehat{\phi}^*],[h_t^*\widehat{\psi}^*])$$

exists as claimed. It is a convex function of $\log a$ by theorem (3.3).

Comparison to equation (9.55) shows that this is equal to $\lim_{n\to\infty} \frac{1}{n} \log P_n^+(a)$.

Similar arguments show that

$$\mathcal{A}_p^>(a) = \lim_{n\to\infty} \tfrac{1}{n} \log P_n^>(a;[h_b^*\widehat{\phi}^*],[h_t^*\widehat{\psi}^*])$$

exists and that this is equal to $\lim_{n\to\infty} \frac{1}{n} \log P_n^>(a)$. □

Theorem 9.22 *For all values of $a \leq 1$, $\mathcal{A}_p^+(a) = \mathcal{A}_p^>(a) = \log \mu_d$. If $a > 1$, then*

$$\begin{aligned} \log\mu_{d-1} + \log a \leq \mathcal{A}_p^+(a) \leq \mathcal{A}_p^>(A) \leq \log \mu_d + \log a, &\qquad \textit{if } d > 2; \\ \tfrac{1}{2}\log a \leq \mathcal{A}_p^+(a) \leq \mathcal{A}_p^>(a) \leq \log\mu_2 + \tfrac{1}{2}\log a, &\qquad \textit{if } d = 2. \end{aligned}$$

Proof Let ω be a positive polygon of length n and with v visits. Then $\omega + \vec{e}_d$ is a polygon with zero visits. This polygon can be reattached to $\partial\mathbb{L}_+^d$ by replacing an edge with its midpoint at height 1 by three edges in a ⊔-conformation. This shows that $p_n^+(v) \leq p_{n+2}^+(2)$ or that $p_n^+ \leq n p_{n+2}^+(2)$. Consequently, if $a < 1$, then $\frac{1}{n-2} p_{n-2}\, a^2 \leq p_n^+(2)\, a^2 \leq P_n^+(a) \leq p_{n+2} \sum_{v\geq 0} a^v$. Take logarithms, divide by n and let $n \to \infty$ to see that $\mathcal{A}_p^+(a) = \log \mu_d$.

Since $P_n^+(a) \leq P_n^>(a) \leq n p_n$ if $a \leq 1$, it follows that $\mathcal{A}_p^>(a) = \log\mu_d$ if $a \leq 1$.

Next, if $a > 1$, and $d > 2$, then $p_n^+(n)\, a^n \leq P_n^+(a) \leq P_n^>(a) \leq n p_n\, a^n$. Since $p_n^+(n)$ is the number of polygons of length n in $\partial\mathbb{L}_+^d \equiv \mathbb{L}^{d-1}$, the bounds follow by taking logarithms, dividing by n and taking $n \to \infty$.

If $d = 2$, then the maximum number of visits of a positive polygon to $\partial\mathbb{L}_+^d$ is $\frac{n}{2}$, and $p_n^+(\frac{n}{2}) = 1$. Thus, $a^{n/2} \leq P_n^+(a) \leq P_n^>(a) \leq n\, a^{n/2} p_n$. Take logarithms, divide by n and take $n \to \infty$ to complete the proof. □

Comparison of theorem 9.22 with theorem 9.2 shows that the limiting free energy of adsorbing positive walks is not equal to the limiting free energy of adsorbing positive polygons in $\mathbb{L}_+^2$: $\mathcal{A}_p^+(a) < \mathcal{A}^+(a)$ for sufficiently large values of a. If $d \geq 3$, then $\mathcal{A}^+(a) = \mathcal{A}_p^+(a)$. The proof of this uses the methods of section 7.1.4.

Theorem 9.23 *The limiting free energy of adsorbing polygons $\mathcal{A}_p^+(a)$ is non-analytic at a critical point $a = a_p^+$. If $d = 2$, then $a_p^+ \in [1, \mu_d^2]$, and, if $d \geq 3$, then $a_p^+ \in [1, \frac{\mu_d}{\mu_{d-1}}]$.*

If $d = 2$ then $\mathcal{A}_p^+(a) < \mathcal{A}^+(a)$ for large values of a. Thus, $a_c^+ \leq a_p^+$ in $\mathbb{L}_+^2$.

If $d \geq 3$, then $\mathcal{A}_p^+(a) = \mathcal{A}^+(a)$; thus, adsorbing polygons and walks have the same adsorption critical point. That is, $a_p^+ = a_c^+$. □

A similar theorem is obtained for attached polygons adsorbing in a defect plane.

Theorem 9.24 *The limiting free energy of adsorbing polygons $\mathcal{A}_p^>(a)$ is non-analytic at a critical point $a = a_p^>$. If $d = 2$, then $a_p^> \in [1, \mu_d^2]$, and, if $d \geq 3$, then $a_p^> \in [1, \frac{\mu_d}{\mu_{d-1}}]$.*

If $d = 2$, then $\mathcal{A}_p^>(a) < \mathcal{A}^>(a)$ for large values of a. Thus, $a_c^> \leq a_p^>$ in $\mathbb{L}^2$.

If $d \geq 3$, then $\mathcal{A}_p^>(a) = \mathcal{A}^>(a)$, so adsorbing polygons and walks have the same adsorption critical point; that is, $a_p^> = a_c^>$.

Moreover, since $\mathcal{A}^>(a) \geq \mathcal{A}^+(a)$, it follows that $a_p^> \leq a_p^+$. □

Similar to the case of walks adsorbing at a defect plane, it is conjectured that $a_p^> = 1$ (and that $a_p^> < a_p^+$).

9.2.1 *Knotting in adsorbing polygons*

Knotting in adsorbing polygons in $\mathbb{L}_+^3$ was examined in reference [557]. Let $p_n^+(v)$ be the number of positive polygons in $\mathbb{L}_+^3$ of length n and with v visits (and with the convention that $p_n^+(0) = 0$).

The number of positive polygons with v visits, length n and knot type K is denoted by $p_n^+(K, v)$. For example, the number of unknotted positive polygons of length n with v visits is $p_n^+(\emptyset, v)$.

Let $P_n^+(a)$ be the partition function of adsorbing positive polygons, and let $P_n^\emptyset(a)$ be the partition function of unknotted positive polygons. By theorem 9.21, $\mathcal{A}_p^+(a)$ is the limiting free energy of adsorbing positive polygons.

If two unknotted polygons are concatenated, then the resulting polygon is also an unknot. Thus, by considering unknotted polygons in the proof of theorem 9.21, the existence of the limiting free energy of adsorbing unknotted positive polygons, $\mathcal{A}_p^\emptyset(a)$, follows. This gives the following theorem (the convexity follows from the general arguments in section 3.2).

Theorem 9.25 *There exists a limiting free energy in models of adsorbing positive polygons and of adsorbing unknotted positive polygons defined by*

$$\mathcal{A}_p^+(a) = \lim_{n\to\infty} \tfrac{1}{n} \log P_n^+(a), \quad \textit{and}\ \mathcal{A}_p^\emptyset(a) = \lim_{n\to\infty} \tfrac{1}{n} \log P_n^\emptyset(a)$$

for $a \in (0, \infty)$. Moreover, $\mathcal{A}_p^\emptyset(a) \leq \mathcal{A}_p^+(a)$ for all $a \in (0, \infty)$, and both $\mathcal{A}_p^+(a)$ and $\mathcal{A}_p^\emptyset(a)$ are convex functions of $\log a$. □

The conclusions of theorem 9.22 are similarly true for the models here. This gives the following theorem (see theorem 9.2).

Theorem 9.26 *The limiting free energy $\mathcal{A}_p^\emptyset(a) \leq \mathcal{A}_p^+(a)$ for all $a \in (0, \infty)$. For all values of $a \leq 1$, $\mathcal{A}_p^\emptyset(a) = \log \mu_\emptyset$, and $\mathcal{A}_p^+(a) = \log \mu_3$, where μ_3 is the growth constant of polygons in $\mathbb{L}^3$, and $\mu_\emptyset$ is the growth constant of unknotted lattice polygons in $\mathbb{L}^3$. Thus, $\mathcal{A}_p^\emptyset(a) < \mathcal{A}_p^+(a)$ for $a \leq 1$.*

If $a > 1$, then $\log \mu_2 + \log a \leq \mathcal{A}_p^\emptyset(a) \leq \mathcal{A}_p^+(a) \leq \log \mu_3 + \log a$. □

By theorem 9.26, there exist critical values of a for the adsorption of polygons and knotted polygons.

Denote the critical point for adsorbing polygons by a_p^+ (see theorem 9.23) and for adsorbing unknotted polygons by $a_p^\emptyset$ (these are defined by $\mathcal{A}_p^+(a) = \log \mu_3$ for all $a < a_p^+$, and $\mathcal{A}_p^+(a) > \log \mu_3$ for $a > a_p^+$, and similarly for $a_p^\emptyset$).

By theorem 9.23, $a_p^+ = a_c^+$ and it is conjectured that $a_p^+ < a_p^\emptyset$ [557]. By theorem 9.23, it follows that $a_p^+ \in [1, \frac{\mu_3}{\mu_2}]$, and $a_p^\emptyset \in [1, \frac{\mu_\emptyset}{\mu_2}]$. Use equation (3.19) to define

$$\log \mathcal{P}_p^+(\epsilon) = \inf_{a>0} \mathcal{A}_p^+(a) - \epsilon \log a, \quad \text{and } \log \mathcal{P}_p^\emptyset(\epsilon) = \inf_{a>0} \mathcal{A}_p^\emptyset(a) - \epsilon \log a.$$

In addition, $\mathcal{P}_p^+(\epsilon)$ and $\mathcal{P}_p^\emptyset(\epsilon)$ have domain $\epsilon \in [0, 1)$.

By theorem 9.22, $\mathcal{P}_p^+(0) = \lim_{\epsilon \to 0^+} \mathcal{P}_p^+(\epsilon) = \mu_3$, and a similar argument shows that $\mathcal{P}_p^\emptyset(0) = \lim_{\epsilon \to 0^+} \mathcal{P}_p^\emptyset(\epsilon) = \mu_\emptyset$ (where $\mu_\emptyset$ is defined in theorem 1.8). By corollary 7.32, $\mathcal{P}_p^\emptyset(0) < \mathcal{P}_p^+(0)$.

By the bounds in theorem 9.26, $\lim_{\epsilon \to 1^-} \mathcal{P}_p^+(\epsilon) \geq \mu_2$, and similarly for $\mathcal{P}_p^\emptyset(\epsilon)$. Hence, define $\mathcal{P}_p^+(1) = \lim_{\epsilon \to 1^-} \mathcal{P}_p^+(\epsilon)$, and $\mathcal{P}_p^\emptyset(1) = \lim_{\epsilon \to 1^-} \mathcal{P}_p^\emptyset(\epsilon)$, so these are defined on $[0, 1]$.

The following lemma follows mutatis mutandis from the proof of lemma 9.4.

Lemma 9.27 *The limit* $\lim_{\epsilon \to 1^-} \frac{d}{d\epsilon} \mathcal{P}_p^+(\epsilon) = \lim_{\epsilon \to 1^-} \frac{d}{d\epsilon} \mathcal{P}_p^\emptyset(\epsilon) = -\infty$. □

A consequence of this lemma is the following theorem.

Theorem 9.28 *There exists an* $\epsilon^* < 1$ *such that*

$$\mathcal{A}_p^+(a) = \sup_{0 \leq \epsilon \leq 1} \left\{ \log \mathcal{P}_p^+(\epsilon) + \epsilon \log a \right\} = \log \mathcal{P}_p^+(\epsilon^*) + \epsilon^* \log a.$$

Moreover, if $a > a_p^+$, *then* $\epsilon^* > 0$.

There exists an $\epsilon^\dagger < 1$ *such that*

$$\mathcal{A}_p^\emptyset(a) = \sup_{0 \leq \epsilon \leq 1} \left\{ \log \mathcal{P}_p^\emptyset(\epsilon) + \epsilon \log a \right\} = \log \mathcal{P}_p^\emptyset(\epsilon^\dagger) + \epsilon^\dagger \log a.$$

Moreover, if $a > a_p^\emptyset$, *then* $\epsilon^\dagger > 0$. □

The proof of theorem 9.28 is similar to the proof of theorem 9.5.

The free energies and density functions of adsorbing polygons and adsorbing unknotted polygons are related in the following way.

Theorem 9.29 *The free energy* $\mathcal{A}_p^+(a) > \mathcal{A}_p^\emptyset(a)$ *for all* $a \in (0, \infty)$ *if and only if* $\mathcal{P}_p^+(\epsilon) > \mathcal{P}_p^\emptyset(\epsilon)$ *for all* $\epsilon \in (0, 1)$.

Proof Define the function $f(a) = \mathcal{A}_p^+(a) - \epsilon \log a$ for a given $\epsilon \in (0, 1)$. By theorem 9.26, $f'(a) < 0$ for small $a < a_p^+$, and $f'(a) > 0$ for $a \gg 0$. Since $f(a)$

is convex in $\log a$, there exists a finite value of $a > 0$, say $a = a^*$, such that $f(a^*) = \inf_{a\in(0,\infty)} f(a)$ (by the intermediate value theorem).

By the definition of $\mathcal{P}_p^+(\epsilon)$, this shows that

$$\mathcal{P}_p^+(\epsilon) = \mathcal{A}_p^+(a^*) - \epsilon \log a^*.$$

It may similarly be shown that there exists a finite $a^\dagger > 0$ such that, for every $\epsilon \in (0,1)$,

$$\mathcal{P}_p^\emptyset(\epsilon) = \mathcal{A}_p^\emptyset(a^\dagger) - \epsilon \log a^\dagger.$$

If $\mathcal{A}_p^+(z) > \mathcal{A}_p^\emptyset(a)$ for all $a \in (0,\infty)$, then

$$\mathcal{P}_p^+(\epsilon) = \mathcal{A}_p^+(a^*) - \epsilon \log a^* > \mathcal{A}_p^\emptyset(a^*) - \epsilon \log a^* \geq \mathcal{P}_p^\emptyset(\epsilon),$$

since $\mathcal{A}_p^\emptyset(a^*) - \epsilon \log a^*$ is an upper bound on $\mathcal{P}_p^\emptyset(\epsilon)$.

Conversely, if $\mathcal{P}_p^+(\epsilon) > \mathcal{P}_p^\emptyset(\epsilon)$ for all $\epsilon \in (0,1)$, then, by theorem 9.28,

$$\mathcal{A}_p^+(a) \geq \mathcal{P}_p^+(\epsilon^\dagger) + \epsilon^\dagger \log a > \mathcal{P}_p^\emptyset(\epsilon^\dagger) + \epsilon^\dagger \log a = \mathcal{A}_p^\emptyset(a)$$

for some $\epsilon^\dagger \in (0,1)$ if $a \in (0,1)$ and since $\mathcal{P}_p^+(\epsilon^\dagger) + \epsilon^\dagger \log a$ is a lower bound on $\mathcal{A}_p^+(a)$. This completes the proof. □

Thus, to show that $\mathcal{A}_p^+(a) > \mathcal{A}_p^\emptyset(a)$ for all $a \in (0,\infty)$ it is only necessary to prove that $\mathcal{P}_p^+(\epsilon) > \mathcal{P}_p^\emptyset(\epsilon)$ for all $\epsilon \in (0,1)$.

Theorem 9.30 *The density function $\mathcal{P}_p^+(\epsilon) > \mathcal{P}_p^\emptyset(\epsilon)$ for all $\epsilon \in (0,1)$. Hence, $\mathcal{A}_p^+(a) > \mathcal{A}_p^\emptyset(a)$ for all $a \in (0,\infty)$.*

Proof The proof consists of calculating bounds separating $\mathcal{P}_p^+(\epsilon)$ and $\mathcal{P}_p^\emptyset(\epsilon)$.

CLAIM: The density function $\mathcal{P}_p^\emptyset(\epsilon) \leq \epsilon \log \mu_2 + (1-\epsilon)\log \mu_T$, where $\mu_T < \mu_3$.
PROOF OF CLAIM: Each unknotted positive polygon of length n consists of subwalks in $\partial\mathbb{L}_+^3$ (these are called *trains*), separated by subwalks with only their endpoints in $\partial\mathbb{L}_+^3$ (these are *loops*; see figure 7.6).

Each train has a length and an end-to-end displacement vector. Let $T_n(\vec{v})$ be the number of trains from $\vec{0}$ of length n and with end-to-end vector $\vec{v}$ and let $T_n = \sum_{\vec{v}} T_n(\vec{v})$. Since a train is a self-avoiding walk in $\mathbb{L}^2$, it follows that $T_n = \mu_2^{n+o(n)}$.

Loops in unknotted positive polygons cannot contain the knotted ball-pair (C,P) shown in figure 7.11, since these polygons are unknotted. Let $\bar{\ell}_n(\vec{v})$ be the number of loops not containing (C,P) of length n and with end-to-end displacement vector $\vec{v}$, and put $\bar{\ell}_n = \sum_{\vec{v}} \bar{\ell}_n(\vec{v})$.

It follows that $\bar{\ell}_n \leq c_n^+[0,(C,P)] \leq c_n[0,(C,P)]$, since each such loop is a self-avoiding walk in $\mathbb{L}^3$ not containing (C,P). By theorem 7.13, there are a $\mu_T < \mu_3$ and an $N \in \mathbb{N}$ such that $\bar{\ell}_n \leq c_n[0,(C,P)] < \mu_T^n$ for all $n \geq N$.

An adsorbing polygon can be decomposed into a sequence of t alternating trains and loops of $2t$ lengths $\langle n_j, m_j\rangle$ and $2t$ displacement vectors $\langle \vec{v}_j, \vec{w}_j\rangle$. Observe that $\sum_j (\vec{v}_j + \vec{w}_j) = \vec{0}$ and that $\sum_j (n_j + m_j) = n$. This shows that

$$p_n(\emptyset, \lfloor \epsilon n \rfloor) \leq \sum_{t=1}^{n/2} \sum_{\langle \vec{v}_j \rangle}{}' \sum_{\langle \vec{w}_j \rangle}{}' \sum_{\langle n_i \rangle} \sum_{\langle m_j \rangle} \prod_{k=1}^{t} T_{n_k}(\vec{v}_k) \prod_{\ell=1}^{t} \overline{\ell}_{m_\ell}(\vec{w}_\ell),$$

where the primed summations are done only if the intersections between trains and loops are empty. The number of non-zero terms on the right-hand side must be of order $e^{o(n)}$, since $\sum_k (n_k + m_k) = n$ and the right-hand side grows more slowly than μ_3^n by corollary 7.32. In addition, there are most popular values for t, $\vec{v}_j$ and $\vec{w}_j$, so this may be bounded by

$$p_n^{\emptyset}(\lfloor \epsilon n \rfloor) \leq e^{o(n)} \mu_2^{\lfloor \epsilon n \rfloor + o(n)} \mu_T^{n - \lfloor \epsilon n \rfloor + o(n)}$$

since $\sum n_k = \lfloor \epsilon n \rfloor$; $T_n = \mu_2^{n+o(n)}$; $\sum m_\ell = n - \lfloor \epsilon n \rfloor$; and $\overline{\ell}_n \leq c_n[0, (C,P)] = \mu_T^{n+o(n)}$. By taking the power $\frac{1}{n}$ and letting $n \to \infty$, the result is that

$$\log \mathcal{P}_p^{\emptyset}(\epsilon) \leq \epsilon \log \mu_2 + (1-\epsilon) \log \mu_T.$$

This completes the proof of the claim. $\triangle$

Claim: The density function $\mathcal{P}_p^{+}(\epsilon) \geq \epsilon \log \mu_2 + (1-\epsilon)\log \mu_3$.
Proof of claim: Let ω_1 be a positive polygon of length $\lfloor \epsilon n \rfloor$ placed in $\partial \mathbb{L}_+^3$. Then ω_1 has $\lfloor \epsilon n \rfloor$ visits. Let $\vec{e}_t$ be the top edge of ω_1. Since ω_1 is a polygon in $\mathbb{L}^2$, it is necessarily the case that $\vec{e}_t \| \vec{e}_2$.

Suppose ω_2 is a polygon of length $n + 2 - \lfloor \epsilon n \rfloor$ and with exactly two visits in $\partial \mathbb{L}_+^3$. Then ω_2 has exactly one edge in $\partial \mathbb{L}_+^3$ and which is also the middle edge in a $\sqcup$-conformation. Let this edge be $\vec{e}_b$.

The polygons ω_1 and ω_2 may be concatenated into a single polygon of length n by translating ω_2 such that $\vec{e}_t = \vec{e}_b$.

There are $p_{\lfloor \epsilon n \rfloor}^{(2)}$ choices for ω_1 where $p_n^{(2)}$ is the number of polygons of length n in $\mathbb{L}^2$ (counted up to equivalence under translation in $\mathbb{L}^2$).

The number of choices for ω_2 is slightly more complicated. Since the bottom edge is parallel to $\vec{e}_2$ and since the $\sqcup$-conformation is arbitrarily placed otherwise, there are at least $\frac{1}{2n} p_{n-\lfloor \epsilon n \rfloor}^{(3)}$ choices for ω_2 if the $\sqcup$-conformation is removed.

Since the result of the concatenation is an adsorbing polygon of length n and with $\lfloor \epsilon n \rfloor$ visits, the above shows that

$$p_{\lfloor \epsilon n \rfloor}^{(2)} p_{n-\lfloor \epsilon n \rfloor}^{(3)} \leq 2n p_n^{+}(\lfloor \epsilon n \rfloor).$$

Take logarithms, divide by n and let $n \to \infty$ to complete the proof. $\triangle$

Compare the two claims and note that $\mu_T < \mu_3$. This shows that

$$\mathcal{P}_p^{\emptyset}(\epsilon) \leq \epsilon \log \mu_2 + (1-\epsilon)\log \mu_T < \epsilon \log \mu_2 + (1-\epsilon) \log \mu_3 \leq \mathcal{P}_p^{+}(\epsilon)$$

for every $\epsilon \in (0,1)$. This completes the proof. $\square$

By theorem 9.30, it follows that

$$\alpha(a) = \mathcal{A}_p^+(a) - \mathcal{A}_p^\emptyset(a) > 0, \text{ for all } a \in (0,\infty). \tag{9.59}$$

Since the partition functions in these models are related to the free energies by $P_n^+(a) = e^{n\mathcal{A}_p^+(a)+o(n)}$ and by $P_n^\emptyset(a) = e^{n\mathcal{A}_p^\emptyset(a)+o(n)}$, the corollary of the above is stated as follows.

Corollary 9.31 *Let $a \in (0,\infty)$. Then the ratio of partition functions of unknotted adsorbing polygons and adsorbing polygons approaches 0 exponentially fast with increasing length:*

$$\lim_{n\to\infty} \frac{P_n^\emptyset(a)}{P_n^+(a)} = \lim_{n\to\infty} e^{-\alpha(a)n+o(n)} = 0. \qquad \square$$

This generalises corollary 7.30 to adsorbing polygons.

9.3 Copolymer adsorption

Let $\omega = \langle \omega_0, \omega_1, \ldots, \omega_n \rangle$ be a positive walk from $\vec{0}$ in in the half-lattice $\mathbb{L}_+^d$ (see equation (2.20)) with boundary $\partial\mathbb{L}_+^d$. Colour the vertices along ω by a sequence of colours $\chi = \langle \chi(1), \chi(2), \ldots \rangle$ by assigning colour $\chi(i)$ to vertex ω_i. This is an unambiguous and fixed assignment of colours for walks, bridges, half-space walks and unfolded walks. The colouring is called a *quench* since it is fixed along the walk.

A *random colouring* of a walk is obtained as follows. Let $\chi(i)$ be a random variable in the probability space Υ for each i and define the random sequence $\chi = \langle \chi(1), \chi(2), \ldots \rangle$ in $\Omega = \Upsilon \times \Upsilon \times \cdots$. A random colouring $\chi \in \Omega$ of a walk ω from $\vec{0}$ in $\mathbb{L}_+^d$ is an assignment of the colours from χ to the vertices of ω starting in $\vec{0}$. If $\Upsilon = \{0,1\}$ is endowed with a distribution from which colours are selected, then χ is a random binary sequence. If the colours are selected identically and independently, then Ω is endowed with a probability measure (so that expectation values can be computed over Ω). Let $c_n^+(i_1, i_2, \ldots, i_v)$ be the number of positive walks from $\vec{0}$ of length n and with v visits to $\partial\mathbb{L}_+^d$ (these are the i_1-th, i_2-th, … , i_v-th vertices of ω). Define the partition function

$$A_n^+(a_0, a_1 \mid \chi) = \sum_{v=0}^{n} \sum_{\{i_k\}} c_n^+(i_1, i_2, \ldots, i_v)\, {a_0}^{v-\sum_{i=1}^{v}\chi_{i_k}}\, {a_1}^{\sum_{i=1}^{v}\chi_{i_k}}, \tag{9.60}$$

where visits of colour 0 are weighted by a_1, and visits of colour 1 are weighted by a_1. Since χ is random and fixed, this is a *quenched partition function* of a walk with (fixed and random) colouring χ.

Taking the average of the partition function $A_n^+(a_0, a_1 \mid \chi)$ over Ω defines the *annealed ensemble* in the model, with the partition function $\langle A_n^+(a_0, a_1 \mid \chi)\rangle_\chi$. In systems such as polyelectrolytes, changes in randomness are quick, and the annealed ensemble is used [46]. The *annealed free energy* per monomer is given by

$F_n^a(a_0, a_1) = \frac{1}{n} \log \langle A_n^+(a_0, a_1 \mid \chi) \rangle_\chi$. The annealed model is useful for modelling homopolymer adsorption onto a random heterogeneous surface if the surface is mobile and changes character on a time scale that is shorter than the time scale of conformational changes in the polymer [76, 426, 427, 538].

Quenched models with simple periodic colourings (for example, the alternating colouring $\chi = \langle 0, 1, 0, 1, \ldots, 0, 1, \ldots \rangle$) may be analysed in some cases using the methods for adsorbing positive walks, although the statements and proofs of theorems change in minor ways. The situation is more difficult for longer period or aperiodic sequences.

The *quenched averaged free energy* per monomer is given by the average of the quenched free energy, namely $F_n^c(a_0, a_1) = \frac{1}{n} \langle \log A_n^+(a_0, a_1 \mid \chi) \rangle_\chi$. The annealed average free energy may be a good approximation of the quenched average free energy for weakly interacting models, but this is not generally the case. Better approximations of the quenched average free energy can be determined by using the *Morita approximation* [428]; see, for example, references [363, 409] and, in particular, reference [528] for a comprehensive review of this method for models of adsorbing quenched polymers.

9.3.1 *Annealed copolymer adsorption*

Let $c_n^+(v)$ be the number of positive walks from $\vec{0}$ of length n with v visits. Choose $\Upsilon = \{0, 1\}$ and $\chi \in \Omega$ with $\chi(i)$ selected independently and binomially so that $\chi(i) = 0$ with probability p, and $\chi(i) = 1$ with probability $q = 1 - p$, for each $i = 0, 1, \ldots$. The average of the quenched partition function over Ω simplifies to

$$A_n^a(a_0, a_1) = \sum_{v=0}^{n} c_n^+(v) \sum_{w=0}^{v} \binom{v}{w} (a_0 p)^{v-w} (a_1 q)^w = \sum_{v=0}^{n} c_n^+(v) \, (a_0 p + a_1 q)^v.$$

By taking logarithms, dividing by n and letting $n \to \infty$, it follows from theorem 9.1 that

$$\mathcal{A}_a^+(a_0, a_1) = \lim_{n \to \infty} \tfrac{1}{n} \log A_n^a(a_0, a_1) = \mathcal{A}^+(a_0 p + a_1 q). \tag{9.61}$$

The annealed copolymer free energy $\mathcal{A}^a(a_0, a_1)$ inherits all the properties of $\mathcal{A}^+(a)$. For example, by corollary 9.3 and equation (9.46), in the phase diagram there is a critical curve $a_0 p + a_1 q = a_c^+ > 1$ that corresponds to an adsorption transition in the model.

Theorem 9.32 *The annealed free energy of a model of adsorbing walks is given by*

$$\mathcal{A}_a^+(a_0, a_1) = \lim_{n \to \infty} \tfrac{1}{n} \log A_n^a(a_0, a_1),$$

where a_0 is conjugate to visits of colour 0, and a_1 is conjugate to visits of colour 1.

There is an adsorption critical point in the model along the critical curve $a_0 p + a_1 q = a_c^+ > 1$ (where a_c^+ is the adsorption critical point of adsorbing walks;

see corollary 9.3 and theorem 9.10). If $a_0p + a_1q \le a_c^+$*, then* $\mathcal{A}_a^+(a_0, a_1) = \log \mu_d$*, and, if* $a_0p + a_1q > a_c^+$*, then*

$$\log \mu_{d-1} + (a_0p + a_1q) \le \mathcal{A}_a^+(a_0, a_1) \le \log \mu_{d-1} + (a_0p + a_1q)\,. \qquad \square$$

Put $a_1 = 1$, and put $a_0 = a$. Then the critical point in the annealed model of copolymer adsorption is given by

$$a_c^a = 1 + \tfrac{1}{p}(a_c^+ - 1). \tag{9.62}$$

Note that $a_c^a > a_c^+ > 1$ if $p < 1$.

9.3.2 *Averaged quenched copolymer adsorption*

Choose $\Upsilon = \{0, 1\}$ and let Ω and χ be as above. Suppose that $a_0 = a$ and that $a_1 = 1$ and let $A_n^+(a \mid \chi) \equiv A_n^+(a, 1 \mid \chi)$ be as defined in equation (9.60). For various realisations of χ, this is a model of quenched copolymer adsorption [453], which includes alternating or block copolymers [242, 427, 582], and models of random quenches [41].

The average quenched model is obtained by taking the average of the quenched free energy over all quenches in the probability space Ω. Existence of a thermodynamic limit in this model can be shown using arguments similar to those leading to theorem 9.1 and starting with positive bridges, as in section 9.1.1.

Let $\ell_n^b(i_1, i_2, \ldots, i_v)$ be the number of positive bridges in $\mathbb{L}_+^d$, with v visits to $\partial\mathbb{L}_+^d$ and with its i_j-th vertices (for $j = 1, \ldots, v$) as visits in $\partial\mathbb{L}_+^d$. The first and last vertices of positive bridges are necessarily in the adsorbing plane (by definition). Two positive bridges can be concatenated by inserting an edge in the $\vec{e}_1$ direction between the last vertex of the first, and the first vertex of the second (see figure 9.2). This shows that

$$\ell_n^b(i_1, i_2, \ldots, i_v)\,\ell_m^b(j_1, j_2, \ldots, j_w) \le \ell_{n+m+1}^b(i_1, i_2, \ldots, i_v, i_{v+1}, \ldots, i_{v+w}), \tag{9.63}$$

where $i_{v+k} = n + 1 + j_k$ for $k = 1, 2, \ldots, w$.

Choose the two random and independent colourings χ_1 and χ_2 for the positive bridges and multiply the above by $a^{(v+w) - \sum_{k=0}^{v} \chi_{i_k} - \sum_{k=0}^{w} \chi_{j_k}}$. Define χ_{12} to be the sequence composed of the first $n+1$ colours of χ_1 concatenated with the first $m+1$ colours of χ_2. Sum the resulting inequality over all i_k and j_k and then over all v and w. Denote the partition function of positive bridges by L_n^b. This shows that

$$L_n^b(a \mid \chi_1)\,L_m^b(a \mid \chi_2) \le L_{n+m+1}^b(a \mid \chi_{12}). \tag{9.64}$$

Take logarithms of the above and then the average over χ_1 and χ_2 to see that

$$\left\langle \log L_n^b(a \mid \chi)\right\rangle_\chi + \left\langle \log L_m^b(a \mid \chi)\right\rangle_\chi \le \left\langle \log L_{n+m+1}^b(a \mid \chi)\right\rangle_\chi, \tag{9.65}$$

where $\langle \cdot \rangle_\chi$ is the uniform average with respect to all $\chi \in \Omega$.

Theorem 9.33 *Suppose that $\chi \in \Omega$. The averaged quenched free energy of a model of adsorbing bridges is given by*

$$\mathcal{A}^+_{qu}(a) = \sup_{n\geq 1} \tfrac{1}{n} \left\langle \log L^b_{n-1}(a \mid \chi)\right\rangle_\chi = \lim_{n\to\infty} \tfrac{1}{n} \left\langle \log L^b_n(a \mid \chi)\right\rangle_\chi .$$

Moreover, $\mathcal{A}^+_{qu}(a) = \log \mu_d$ if $a \leq 1$, and $\log \mu_{d-1} + \frac{1}{2}\log a \leq \mathcal{A}^+_{qu}(a) \leq \mathcal{A}^+_a(\frac{1}{2}, 1, a)$, where $\mathcal{A}^+_a(p, a_0, a_1)$ is the annealed free energy of a model of adsorbing copolymers defined in equation (9.61).

Proof Existence of the limit follows from the superadditive relation (9.65) and theorem A.1 in appendix A.

Let $\vec{1} = (1, 0, 0, \ldots, 0)$. Every positive bridge from $\vec{0}$ contains the edge $\langle \vec{0} \sim \vec{1}\rangle$, so $\vec{1}$ is necessarily a visit to $\partial\mathbb{L}^d_+$.

If $0 < a \leq 1$, then a lower bound on $\left\langle \log L^b_n(a \mid \chi)\right\rangle_\chi$ is obtained by counting positive bridges from $\vec{0}$ with only two visits (at $\vec{1}$ and at its endpoint).

Let ω be an positive bridge coloured by χ with only two visits (these must be at $\vec{1}$ and at its endpoint). Then ω is composed of the edge $\langle \vec{0} \sim \vec{1}\rangle$ followed by an unfolded hoop from $\vec{1}$ to its endpoint in $\partial\mathbb{L}^d_+$. The number of such unfolded hoops is $\ell^\dagger_{n-1}$ (see equation (9.2)), and the two visits have weights a or 1, depending on the colouring χ.

If $a \leq 1$, the average over χ gives a lower bound $\log(\ell^\dagger_{n-2}\, a^2) \leq \left\langle \log L^b_n(a \mid \chi)\right\rangle_\chi$. Divide by n and take $n \to \infty$ to see that $\log \mu_d \leq \mathcal{A}^+_{qu}(a)$. An upper bound for $a \leq 1$ is obtained by noting that $\left\langle \log L^b_n(a \mid \chi)\right\rangle_\chi \leq \left\langle \log L^b_n(1 \mid \chi)\right\rangle_\chi = \log \ell^b_n$. This shows that $\mathcal{A}^+_{qu}(a) \leq \log \mu_d$, by equation (9.5) and theorem 9.1. Thus $\mathcal{A}^+_{qu}(a) = \log \mu_d$ if $0 \leq a \leq 1$.

If $a > 1$, then a lower bound is obtained by considering positive bridges with all vertices in $\partial\mathbb{L}^d_+ \simeq \mathbb{L}^{d-1}$. The number of such bridges is at least $b^{(d-1)}_n$, that is, the number of bridges from the origin in $\mathbb{L}^{d-1}$. This shows that $L^b_n(a \mid \chi) \geq a^{v(\chi)} b^{(d-1)}_n(\chi)$, where $v(\chi)$ is the number of times the colour 0 appears in χ, and $b^{(d-1)}_n(\chi)$ is the number of bridges coloured by χ. Hence,

$$\left\langle \log L^b_n(a \mid \chi)\right\rangle_\chi \geq \langle v(\chi)\rangle_\chi \log a + \langle \log b^{(d-1)}_n(\chi)\rangle_\chi .$$

Since χ is selected uniformly in Ω, $\langle v(\chi)\rangle_\chi = \frac{1}{2}n$, and $\langle \log b^{(d-1)}_n(\chi)\rangle_\chi = \log b^{(d-1)}_n$, since $b^{(d-1)}_n(\chi) = b^{(d-1)}_n$ for each realisation of χ. Divide by n and let $n \to \infty$ to see that $\mathcal{A}^+_{qu}(a) \geq \log \mu_{d-1} + \frac{1}{2}\log a$.

An upper bound for $a > 1$ is obtained by using Jensen's inequality: observe that $\left\langle \log L^b_n(a \mid \chi)\right\rangle_\chi \leq \log \left\langle L^b_n(a \mid \chi)\right\rangle_\chi$. The average $\left\langle L^b_n(a \mid \chi)\right\rangle_\chi$ is the annealed free energy with $a_0 = a$, and $a_1 = 1$, as in section 9.3.1. This is so because positive bridges have the same free energy as adsorbing walks. Using this when dividing by n and letting $n \to \infty$ gives the claimed upper bound. This completes the proof. □

This model undergoes an adsorption transition.

Theorem 9.34 *There exists a critical point* $a_c^q \geq 1$ *such that* $\mathcal{A}_{qu}^+(a) = \log \mu_d$ *for all* $a < a_c^q$, *and* $\mathcal{A}_{qu}^+(a) > \log \mu_d$ *if* $a > a_c^q$. □

It remains to prove that $\mathcal{A}_{qu}^+(a)$ is the limiting free energy of the averaged quenched model. This is done by using most popular class arguments with positive walks, unfolded walks, loops and positive bridges in the same way as was done in section 9.1.1 (or for polygons in section 9.2). Thus, a corollary of those sections and theorem 9.33, is the existence of a limiting free energy.

Theorem 9.35 *Suppose that* $\chi \in \Omega$. *Then the averaged quenched free energy of a model of adsorbing positive walks is given by*

$$\mathcal{A}_{qu}^+(a) = \lim_{n\to\infty} \tfrac{1}{n} \left\langle \log A_n^+(a \mid \chi) \right\rangle_\chi .$$

Moreover, $\mathcal{A}_{qu}^+(a) = \log \mu_d$ *if* $a \leq 1$, *and* $\log \mu_{d-1} + \frac{1}{2} \log a \leq \mathcal{A}_{qu}^+(a) \leq \mathcal{A}_a^+(\frac{1}{2}, 1, a)$ *(where* $\mathcal{A}_a^+(p, a_0, a_1)$ *is the annealed free energy of a model of adsorbing copolymers defined in equation (9.61)).*

The averaged quenched free energy of a model of adsorbing polygons is similarly

$$\mathcal{A}_{p,qu}^+(a) = \lim_{n\to\infty} \tfrac{1}{n} \left\langle \log P_n^+(a \mid \chi) \right\rangle_\chi .$$

Observe that $\mathcal{A}_{p,qu}^+(a) = \mathcal{A}_{qu}^+(a)$ *if* $d \geq 3$ *but that* $\mathcal{A}_{p,qu}^+(a) \neq \mathcal{A}_{qu}^+(a)$ *if* $d = 2$ *in the adsorbed phase.* □

The location of the averaged quenched critical point a_c^q may be related to the critical point a_c^a of the corresponding annealed model (see section 9.3.1). By the arithmetic-geometric mean inequality, $\log \langle A_n^+(a \mid \chi)\rangle_\chi \geq \langle \log A_n^+(a \mid \chi)\rangle$. Dividing by n and taking $n \to \infty$ gives $\mathcal{A}_a^+(a, 1) \geq \mathcal{A}_{qu}^+(a)$. By equation (9.62),

$$a_c^q \geq a_c^a > a_c^+ > 1. \tag{9.66}$$

9.3.3 *Self-averaging in quenched copolymer adsorption*

Let Ω and χ be as above and let $A_n^+(a \mid \chi)$ be as defined in equation (9.60). Let the averaged quenched free energy be given by $\mathcal{A}_{qu}^+(a)$ as in theorem 9.35. Self-averaging in this model is shown in the next theorem (see references [291, 453, 454, 603]).

Theorem 9.36 *Let* χ *be an infinite random sequence in* Ω. *Then*

$$\lim_{n\to\infty} \tfrac{1}{n} \log A_n^+(a \mid \chi) = \mathcal{A}_{qu}^+(a) \quad \textit{almost surely.}$$

Proof Let $n = Nm + r$ for some fixed value of m and where $0 \leq r \leq m-1$. Divide an n-edge positive walk into subwalks of length m vertices by deleting N edges and let the remainder of the walk be of length r.

Let the first m colours in χ be $\chi^{(1)}$, the second m be $\chi^{(2)}$ and so on, until the remainder has colouring $\chi^{(N+1)}$. Then χ has prefix $\chi^{(1)}\chi^{(2)}\ldots\chi^{(N+1)}$.

Each subwalk of length $m-1$ has at most c_{m-1} conformations. If the subwalk has at least one visit to $\partial\mathbb{L}_+^d$, then its partition function is denoted by $Y_{m-1}^+(a \mid \chi)$. Obviously, if walks are coloured by ψ, then

$$Y_{m-1}(a \mid \psi) \leq n c_{m-1} + \sum_{k=0}^{m-1} A_{m-1-k}^+(a \mid \psi^{(1)}) A_k^+(a \mid \psi^{(2)}), \tag{9.67}$$

since each walk contributing to $Y_{m-1}(a \mid \psi)$ is either disjoint with $\partial\mathbb{L}_+^d$ or has one visit in $\partial\mathbb{L}_+^d$ (in which case the walk is cut into two positive walks in the visit). In this case $\psi^{(1)}$ denotes the first $m-k$ colours of ψ, and $\psi^{(2)}$ denotes the last $k+1$ colours of ψ.

In equation (9.67) choose $\psi = \chi^{(i)}$. Take logarithms, divide by m, sum over $i = 1, 2, \ldots, N$ and divide by N. By the strong law of large numbers,

$$\limsup_{N\to\infty} \tfrac{1}{N} \sum_{i=1}^{N} \tfrac{1}{m} \log Y_{m-1}(a|\chi^{(i)}) = \left\langle \tfrac{1}{m} \log Y_{m-1}(a|\chi) \right\rangle_\chi \leq \mathcal{A}_{qu}^+(a) \tag{9.68}$$

almost surely.

The decomposition of positive walks of colour χ as above into subwalks of length $m-1$ gives

$$A_n^+(a \mid \chi) \leq (2d)^N \left(\prod_{i=1}^{N} Y_{m-1}(a \mid \chi^{(i)}) \right) Y_{r-1}(a \mid \chi^{(N+1)}).$$

Take logarithms, divide by n and take the lim sup of the left-hand side as $n \to \infty$. If m is kept fixed, then $N \to \infty$ while $r \in \{0, 1, \ldots, m-1\}$. By equation (9.68), the result is that

$$\limsup_{n\to\infty} \tfrac{1}{n} \log A_n^+(a \mid \chi) \leq \limsup_{N\to\infty} \tfrac{1}{N} \sum_{i=1}^{N} \tfrac{1}{m} \log Y_{m-1}(a \mid \chi^{(i)}) \leq \mathcal{A}_{qu}^+(a).$$

Next, a lower bound is constructed by using positive bridges with the partition function $L_n^b(a \mid \chi)$. Let $n = Nm - r$ and concatenate positive bridges on $\partial\mathbb{L}_+^d$ in the $\vec{e}_1$ direction by placing the left-most endpoint of the second bridge one step above the right-most endpoint of the first bridge and then by inserting an edge between these vertices. This gives

$$A_n^+(a \mid \chi) \geq \left(\prod_{i=1}^{N} L_{m-1}^b(a \mid \chi^{(i)}) \right) L_r^b(a \mid \chi^{(N+1)}).$$

Take logarithms and the limit inferior of the left-hand side as $n \to \infty$ with m fixed. Argue as above on the right-hand side using the strong law of large numbers. That is, for almost every colouring χ,

$$\liminf_{n\to\infty} \tfrac{1}{n} \log A_n^+(z \mid \chi) \geq \liminf_{N\to\infty} \tfrac{1}{N} \sum_{i=1}^{N} \tfrac{1}{m} \log L_{m-1}^b(a \mid \chi^{(i)})$$
$$= \left\langle \tfrac{1}{m} \log L_{m-1}^b(a \mid \chi) \right\rangle_\chi .$$

The right-hand side converges to $\mathcal{A}_{qu}^+(z)$ as $m \to \infty$ by theorem 9.33. This completes the proof. □

Existence of the limit $\lim_{n\to\infty} \frac{1}{n} \log A_n^+(a \mid \chi)$ (almost surely) is also a consequence of the local superadditive ergodic theorem [4].

9.4 Collapsing self-avoiding walks

A collapsing self-avoiding walk (see figure 3.1(d)) is a model of the *collapse transition* of a linear polymer in a poor solvent. The collapse transition occurs at the θ-point, which is a tricritical point separating a phase of the linear polymer in a good solvent (this is the *coiled phase*) from a collapsed or *globule phase* when the polymer is in a poor solvent.

Let $\omega = \langle \omega_0, \omega_1, \ldots, \omega_n \rangle$ be a self-avoiding walk from $\vec{0}$ in $\mathbb{L}^d$. A *contact* in ω is an edge $\langle \omega_j \sim \omega_k \rangle$ in $\mathbb{L}^d$ such that $\|\omega_j - \omega_k\|_2 = 1$, and $|j-k| > 1$. That is, the vertices ω_j and ω_k are adjacent in $\mathbb{L}^d$ but not in ω.

Denote by $c_n(k)$ the number of walks from $\vec{0}$ of length n and with k contacts. The partition function of this model is

$$C_n(x) = \sum_{k\geq 0} c_n(k)\, x^k, \tag{9.69}$$

where x is conjugate to the number of contacts. The partition function of collapsing bridges is

$$C_n^b(x) = \sum_{k\geq 0} b_n(k)\, x^k, \tag{9.70}$$

where $b_n(k)$ is the number of bridges of length n and with k contacts. Two bridges may be concatenated similarly to the case illustrated in figure 7.3. If an extra edge in the $\vec{e}_1$ direction is inserted between the two bridges, then no new contacts are created in this construction, and the result is that

$$\sum_{\ell=0}^{k} b_n(k-\ell)\, b_m(\ell) \leq b_{n+m+1}(k). \tag{9.71}$$

Multiply by x^k and sum over k to obtain

$$C_n^b(x) C_m^b(x) \leq C_{n+m+1}^b(x). \tag{9.72}$$

Since $C_n^b(x) \leq (2d-1)^n$ if $x \leq 1$, and $C_n^b(x) \leq (2d-1)^n x^n$ if $x > 1$, theorem A.1 in appendix A gives existence of a limiting free energy for collapsing bridges:

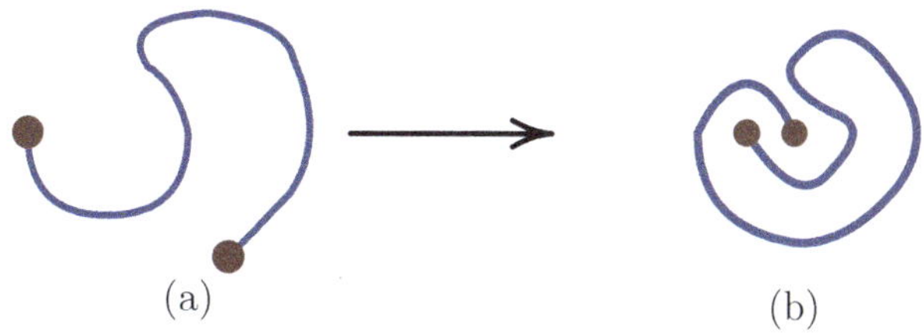

FIG. 9.12. A schematic diagram of a collapsing walk. (a) A walk in the coiled phase. It collapses at the θ-point to a compact walk in the globule phase (b).

$$\mathcal{C}_k(x) = \sup_{n>0} \tfrac{1}{n} \log C_n^b(x) = \lim_{n\to\infty} \tfrac{1}{n} \log C_n^b(x). \tag{9.73}$$

The partition function $C_n^h(x)$ for half-space walks (see section 7.1.1) is similarly defined. Cutting a self-avoiding walk in its bottom vertex (see equation (7.3)) to create two half-space walks destroys contacts, so that

$$C_n^b(x) \le C_n(x) \le \sum_{m=0}^{n} C_{m+1}^h(x) C_{n-m}^h(x) \;\text{ if } 0 \le x \le 1. \tag{9.74}$$

Similarly, half-space walks may be partitioned into bridges (see figure 7.4 and lemma 7.1). This also destroys nearest-neighbour contacts, so that if $\gamma_0 > \pi\sqrt{\frac{2}{3}}$, then there exist an $\epsilon > 0$ and an N_0 such that

$$C_n^h(x) \le K\, e^{(\gamma_0-\epsilon)\sqrt{n}} \sum_{j=0}^{k} C_{n+j}^b(x) \;\text{ if } x \le 1 \text{ and where } k = O(\sqrt{n}), \tag{9.75}$$

for all $n \ge N_0$ and where equation (9.72) was used.

By comparing equations (9.73) with (9.74) and (9.75), the following theorem is obtained.

Theorem 9.37 *If $0 \le x \le 1$, then there exists a limiting free energy in a model of collapsing self-avoiding walks defined by*

$$\mathcal{C}_k(x) = \lim_{n\to\infty} \tfrac{1}{n} \log C_n^b(x) = \lim_{n\to\infty} \tfrac{1}{n} \log C_n(x). \qquad \square$$

For values $x > 1$, the existence of the free energy is not known. Define

$$\mathcal{C}_k(x) = \limsup_{n\to\infty} \tfrac{1}{n} \log C_n(x) \;\text{ if } x > 1 \tag{9.76}$$

as a working definition and then proceed with analysing the model.

The free energy of *collapsing polygons* is shown to exist as follows. Let $p_n(k)$ be the number of polygons of length n with k contacts. Then the partition function is $C^p(x) = \sum_k p_n(k)\, x^k$. Concatenating polygons as shown in figure 1.5 gives

$$\sum_{\ell=0}^{k} p_n(k-\ell)\,p_m(\ell) \leq (d-1)\,p_{n+m}(k+2). \tag{9.77}$$

Multiply by x^k and sum over k to obtain

$$C_n^p(x)\,C_m^p(x) \leq x^{-2} C_{n+m}^p(x). \tag{9.78}$$

Since $C_n^p(x) \leq x C_{n-1}(x)$, theorem A.1 from appendix A gives the following result.

Theorem 9.38 *For all $x \geq 0$, there exists a limiting free energy in a model of collapsing polygons defined by*

$$\mathcal{C}_k^p(x) = \lim_{n\to\infty} \tfrac{1}{n} \log C_n^p(x).$$

Moreover, $\mathcal{C}_k^p(x) = \mathcal{C}_k(x)$ for all $0 \leq x \leq 1$, and $\mathcal{C}_k^p(x) \leq \mathcal{C}_k(x)$ for all $x > 1$, where $\mathcal{C}_k(x)$ is the limiting free energy of collapsing walks.

Proof Existence of $\mathcal{C}_k^p(x)$ for $x > 0$ follows from equation (9.78). *Define* $\mathcal{C}_k^p(0) = \mathcal{C}_k(0)$. Then it only remains to prove that $\mathcal{C}_k^p(x) = \mathcal{C}_k(x)$ for all $0 < x \leq 1$.

Delete the bottom edge of each collapsing polygon of length n to see that $C_n(x)^p \leq \max\{C_{n-1}(x), xC_{n-1}(x)\}$. Thus, $\mathcal{C}_k^p(x) \leq \mathcal{C}_k(x)$ for $0 \leq x \leq 1$. It only remains to show that $\mathcal{C}_k^p(x) \geq \mathcal{C}_k(x)$ for $0 < x \leq 1$. Only an outline of the proof is given here.

The double unfolding of bridges (see section 7.1.3) reduces the number of nearest-neighbour contacts. Since $0 < x \leq 1$, this shows that $C_n^b(x) \leq e^{O(\sqrt{n})} C_n^\ddagger(x)$, where $C_n^\ddagger(x)$ is the partition function of doubly unfolded collapsing walks (see lemma 7.4).

It only remains to use most popular class arguments (section 7.1.4) to put doubly unfolded walks together into loops and polygons (without creating new nearest-neighbour contacts). This is not difficult to do. □

9.4.1 *The collapse transition in self-avoiding walks*

The collapse of self-avoiding walks is thought to be a phase transition from an expanded phase (the *coiled* phase) to a *collapsed* walk or a *globule* in the large x regime (see figure 9.12). The collapsed phase is characterised by a non-negative density of edges in boxes centred about the origin, and the metric exponent in this phase is $\nu = \frac{1}{d}$ in d dimensions.

In the coiled phase the usual self-avoiding walk exponents are found. The critical point for collapse is the *θ-point* at a critical activity x_c. The θ-point has tricritical scaling [116] (see reference [189] as well).

The location of the θ-point has been estimated in numerous simulations. In $\mathbb{L}^2$ the estimate $x_c = 1.4986(12)$ was made in reference [88]. Earlier estimates are $x_c = 1.5037(23)$ [434], and $x_c = 1.941(47)$ [509].

TABLE 9.2. Tricritical θ-exponents for collapsing walks

	$d = 2$	$d = 3$	Mean Field
x_c	1.4986(12) [88]	1.318(29) [547]	--
ϕ_θ	$\frac{3}{7}$ [164]	$\frac{1}{2}$	$\frac{1}{2}$
ν_θ	$\frac{4}{7}$ [436, 438]	$\frac{1}{2}$	$\frac{1}{2}$
γ_θ	$\frac{8}{7}$	$\frac{3}{2}$	$\frac{3}{2}$
γ_u	$\frac{8}{3}$	3	3
α	$-\frac{1}{3}$	0	0

There are several estimates for the location of the θ-point in $\mathbb{L}^3$. In reference [546] it was found that $x_c = 1.320(20)$ for collapsing walks, and $x_c = 1.321(34)$ for collapsing polygons. These results compare well with $x_c = 1.318(29)$ [547] and with the earlier result $x_c = 1.319(43)$ in reference [413].

The tricritical crossover exponent of the collapse transition of linear polymers in two dimensions has the exact value $\phi_\theta = \frac{3}{7}$ [164, 165, 167, 170]. The metric exponent at the θ-point has the exact value $\nu_\theta = \frac{4}{7}$ [436, 438]. The exact tricritical exponents can be computed from these results. The tricritical metric exponent ν_t in equation (4.48) is the same as $\nu_\theta = \frac{4}{7}$, so $\nu_t = \frac{4}{7}$. By equation (4.50), this gives $y_g = \frac{7}{4}$ and, by equation (4.57), it follows that $y_t = y_g \phi_\theta = \frac{3}{4}$. This implies in turn that $\gamma_t = \frac{8}{7}$, and $\gamma_u = \frac{8}{3}$. The exponent γ_t is usually denoted by γ_θ.

By equation (4.46), the specific heat exponent associated with the collapse transition is $\alpha = -\frac{1}{3}$ in two dimensions. Numerical tests of scaling of two dimensional collapsing walks were carried out in reference [434], where a value consistent with $\gamma_- \approx 1$ (see equation (4.21)) was obtained along the τ_0-curve. These results were also checked by ϵ-expansions in $3-\epsilon$ dimensions [159, 162, 536].

The amplitude ratios of metric quantities in two dimensions are related as in equation (1.25). Series enumeration suggests that this relationship is also valid at the θ-point [458].

The upper critical dimension of the collapse transition is $d = 3$ [116, 162, 163, 536], so mean field exponents are found for $d \geq 3$ (with logarithmic corrections to scaling laws if $d = 3$). The mean field values are given in table 9.2. These are determined by noting that mean field θ-polymers have random walk exponents, with the metric exponent $\nu_\theta = \frac{1}{2}$, which describes the metric scaling of the tricritical clusters at the θ-point. The mean field crossover exponent is $\phi_\theta = \frac{1}{2}$, and the exponent ν_t in equation (4.48) is the same as ν_θ $(=\frac{1}{2})$. By equation (4.50), $y_g = 2$. These values show by equation (4.57) that $y_\theta \equiv y_t = \phi_\theta y_g = 1$, with the result that $\gamma_u = \frac{3}{y_\theta} = 3$, and $\gamma_\theta = 3\nu_\theta = \frac{3}{2}$.

The specific heat of collapsing polygons in three dimensions is plotted in figure 9.13 for lengths $n = 200$, 400 and 800 against $\beta = \log x$. The height in the

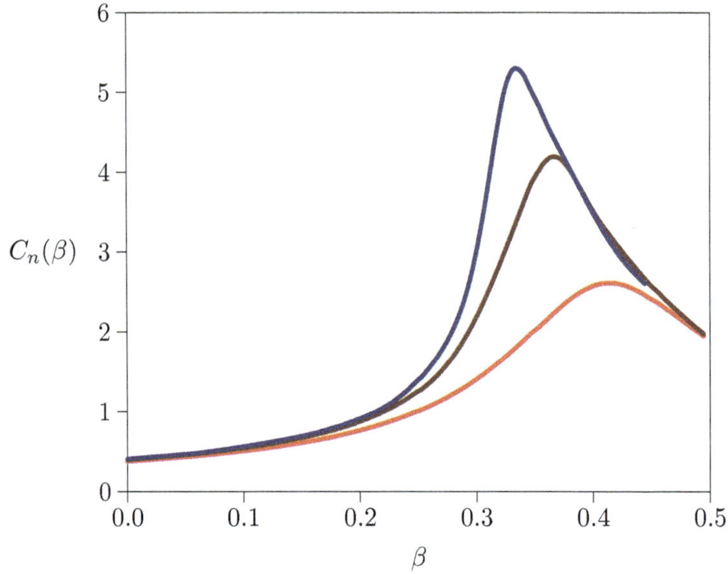

FIG. 9.13. The specific heat for collapsing polygons of lengths 200, 400 and 800 in $d = 3$ dimensions against $\beta = \log x$ [546].

---

peak grows proportional to $n^{\alpha\phi}$ (where α is the specific heat exponent, and ϕ the crossover exponent; see equation (4.31)). Since $\alpha = 2 - \frac{1}{\phi_\theta} = 0$ is the mean field value of α, this shows that the height in the specific heat peak will increase slower than any power law in three dimensions, or will approach a constant. In figure 9.14, the data in figure 9.13 is rescaled to have height 1 and a crossover exponent equal to $\frac{1}{2}$. This rescaling uncovers the scaling function of the specific heat.

A renormalisation group approach to θ-polymers in three dimensions was developed in references [103, 104, 153, 154], while an approach using the Edwards model [149, 173] was developed in reference [162]. Computer simulations of collapsing walks give results consistent with these results [238, 413, 509, 546, 547].

The physical nature of the collapse transition was also discussed by Douglas and Freed [154], while experimental work can be found in references [540, 541].

The collapse of stiff walks (with a curvature activity) was studied in references [24, 155]; in these studies there is the suggestion of a curvature inducing freezing transition (which is first order) separating two different collapsed phases. The issue of first order collapse transitions in models of homopolymers was also discussed in reference [483]. More numerical work in $\mathbb{L}^2$ was done in references [100, 237, 404, 413, 414], and in $\mathbb{L}^3$ in references [238, 415]. Exact enumeration studies were done in references [32, 33, 290, 458, 477]. Partition function zeros were studied in reference [480].

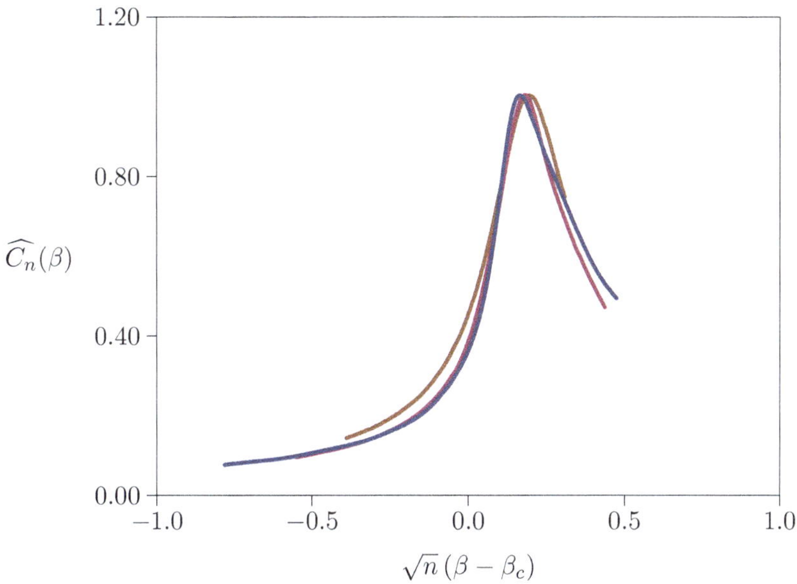

FIG. 9.14. The rescaled specific heat for collapsing polygons of lengths 200, 400 and 800 in $d = 3$ dimensions against $\beta = \log x$. The data are the same as in figure 9.13 but rescaled so that the peak height is equal to 1, the crossover exponent is equal to $\frac{1}{2}$, and the critical point at $\beta_c \approx \log(1.318)$.

9.5 Collapsing and adsorbing polygons

Let $p_n^+(k,v)$ be the number of positive polygons in the half-lattice $\mathbb{L}_+^d$ (see equation (2.20)), attached to the adsorbing plane $\partial\mathbb{L}_+^d$ (that is, each polygon has at least one vertex in $\partial\mathbb{L}_+^d$), with v *visits* to $\partial\mathbb{L}_+^d$ and with k nearest-neighbour *contacts* between vertices in the polygon.

The partition function of this model is

$$P_n^+(x,a) = \sum_{v\geq 0}\sum_{k\geq 0} p_n^+(k,v)\, x^k a^v, \tag{9.79}$$

where x is conjugate to the number of contacts, and a is conjugate to the number of visits.

The partition function of attached polygons at a defect plane (see section (7.1.4)) may be found as follows. Let $p_n^>(k,v)$ be the number of lattice polygons attached to the adsorbing plane $\partial\mathbb{L}_+^d$ (that is, each polygon has at least one vertex in $\partial\mathbb{L}_+^d$), with v *visits* to $\partial\mathbb{L}_+^d$ and with k nearest-neighbour *contacts* between vertices in the polygon.

The partition function of this model is

$$P_n^>(x,a) = \sum_{v\geq 0}\sum_{k\geq 0} p_n^>(k,v)\, x^k a^v. \tag{9.80}$$

The self-avoiding walk version of this model was studied in references [567–569]. In this section the results are reviewed in $\mathbb{L}^d$ with $d \geq 3$ dimensions. The results in $\mathbb{L}^2$ are different, but not in important ways.

The existence of limiting free energies is shown using the same arguments presented in section 9.2 in the proof of theorem 9.21. Those methods give the following theorem.

Theorem 9.39 *The limiting free energy of adsorbing and collapsing polygons is given by*

$$\mathcal{C}_p^+(x,a) = \lim_{n\to\infty} \tfrac{1}{n} \log P_n^+(x,a), \quad \textit{and}\, \mathcal{C}_p^>(x,a) = \lim_{n\to\infty} \tfrac{1}{n} \log P_n^>(x,a),$$

for all $x, a \in (0,\infty)$. □

Bounds on the limiting free energies are obtained using arguments similar to those in the proof of theorem 9.22.

Theorem 9.40 *Let* $\mathcal{C}_k^p(x) = \mathcal{C}_p^>(x,1)$ *be the free energy of a model of collapsing polygons with contact activity* x. *Then*

$$\mathcal{C}_p^+(x,a) = \mathcal{C}_p^>(x,a) = \mathcal{C}_k^p(x) \textit{ for all } a \leq 1.$$

If $a > 1$, *and* $\mathcal{C}_k^{(d-1)}(x)$ *is the limiting free energy of collapsing polygons in* $d-1$ *dimensions, then*

$$\mathcal{C}_k^{(d-1)}(x) + \log a \leq \mathcal{C}_p^+(x,a) \leq \mathcal{C}_p^>(x,a) \leq \mathcal{C}_k^p(x) + \log a.$$

Proof Let $a < 1$. Note that $p_n^+(k,v) \leq p_n^>(k,v)$. Translate attached polygons $p_n^>(k,v)$ in the $\vec{e}_d$ direction until they are disjoint with $\partial\mathbb{L}_+^d$. This shows that $p_n^>(k,v) \leq n p_n(k)$ (where $p_n(k)$ is the number of polygons with k nearest-neighbour contacts). Thus, $P_n^+(x,a) \leq P_n^>(x,a) \leq \frac{n}{1-a} P_n(x)$.

Let ω be a polygon of length n and with k contacts. Find the bottom vertex $\vec{b}$ of ω by a lexicographic ordering of vertices in the order $\langle \vec{e}_d, \vec{e}_1, \vec{e}_2, \ldots, e_{d-1}\rangle$ (that is, $\vec{b}$ is the bottom-most and left-most vertex in ω). Place ω in $\mathbb{L}_+^d$ such that $\vec{b} = \vec{e}_d$. Let $\langle \vec{b} \sim \vec{c}\rangle$ be the edge with lexicographic least midpoint incident with $\vec{b}$. Replace this edge with three edges in the ⊔-conformation $\vec{b} \sim (\vec{b} - \vec{e}_d) \sim (\vec{c} - \vec{e}_d) \sim \vec{c}$. Then ω is a positive polygon with two visits and length $n+2$ (and with $k+1$ contacts). Since there are $p_n(k)$ choices for ω, this shows that $p_n(k) \leq p_{n+2}^+(k+1,2)$. That is, in terms of partition functions, $P_n(x) \leq \sum_k p_{n+2}^+(k+1,2)x^k \leq x^{-1}a^{-2}\sum_k p_{n+2}^+(k,v)\, x^k a^v = x^{-1}a^{-2} P_{n+2}^+(x,a)$.

By comparing the inequalities of the last two paragraphs,

$$xa^2 P_{n-2}(x) \leq P_n^+(x,a) \leq P_n^>(x,a) \leq \tfrac{n}{1-a} P_n(x). \tag{9.81}$$

By taking logarithms, dividing by n and letting $n \to \infty$, the results claimed for $a < 1$ are obtained.

If $a = 1$, then a similar argument shows that $p_n(k) = p_n^+(k) \leq p_n^>(k) \leq n p_n(k)$. This shows that $\mathcal{C}_p^+(x,1) = \mathcal{C}_p^>(x,1) = \mathcal{C}_k^p(x)$.

If $a > 1$, then

$$\sum_{k \geq 0} p_n^+(k,n)\, x^k a^n \leq p_n^+(x,a) \leq p_n^>(x,a) \leq a^n \sum_{k \geq 0} p_n^+(k)\, x^k.$$

Since $p_n^+(k,n)$ is also the number of polygons with k contacts in $\mathbb{L}^d$ dimensions, the claimed inequalities are found after taking logarithms, dividing by n and letting $n \to \infty$. □

An immediate consequence of theorem 9.40 is that both $\mathcal{C}_p^+(x,a)$ and $\mathcal{C}_p(x,a)$ are non-analytic functions of a, for any finite value of x. The non-analyticities are located at

$$\begin{aligned} a_c^>(x) &= \sup\{a \mid \mathcal{C}_p^>(x,a) = \mathcal{C}_k^p(x)\} \text{ and} \\ a_c^+(x) &= \sup\{a \mid \mathcal{C}_p^+(x,a) = \mathcal{C}_k^p(x)\}. \end{aligned} \tag{9.82}$$

Corollary 9.41 *The free energies $\mathcal{C}_P^+(x,a)$ and $\mathcal{C}_p^>(x,a)$ are non-analytic functions of a for each $0 \leq x < \infty$ at $a_c^+(x)$ and $a_c^>(x)$, respectively.*

Moreover,

$$\begin{aligned} &1 \leq a_c^>(x) \leq a_c^+(x) \leq \tfrac{\mu_d}{\sqrt{\mu_{d-1}}}, && \textit{if } x < 1; \\ &1 \leq a_c^>(x) \leq a_c^+(x) \leq x^{d-1} \tfrac{\mu_d}{\mu_{d-1}}, && \textit{if } x \geq 1, \end{aligned}$$

where μ_d is the growth constant of walks in d dimensions.

Proof Observe that $P_n^>(x,a) \geq P_n^+(x,a)$ for all (x,a). By equation (9.82), this shows that $a_c^>(x) \leq a_c^+(x)$. By theorem 9.40, $1 \leq a_n^>$.

It remains to prove the upper bounds.

If $0 \leq x < 1$, then $p_n \geq p_n(x) \geq p_n(0)$. By subdividing every edge in polygons of length n, polygons are obtained with zero contacts, so that $p_{2n}(0) \geq p_n$. Hence, $\log \mu_d \geq \mathcal{C}_k^p(x) \geq \log \sqrt{\mu_d}$. These bounds are enough to find the claimed inequalities for $0 \leq x < 1$: for $a \leq a_c^+(x)$, $\mathcal{C}_p^+(x,a) = \mathcal{C}_k^p(x) \geq \mathcal{C}_k^{(d-1)}(x) + \log a$. The upper bound on $\log a$ is found when this is an equality. Thus, $\log a \leq \mathcal{C}_k^p(x) - \mathcal{C}_k^{(d-1)}(x) \leq \log \mu_d - \log \sqrt{\mu_{d-1}}$.

Next, consider $x \geq 1$. By theorem 9.40, the model is in the adsorbed phase if $\mathcal{C}_p^+(x,a) > \mathcal{C}_k^p(x)$. The upper bound on $a_c^+(x)$ is found when the lower bound in theorem 9.40 is an equality, that is, when $\log a_c^+(x) \leq \mathcal{C}_k^p(x) - \mathcal{C}_k^{(d-1)}(x)$.

In d dimensions, $p_n(x) = \sum_{k \geq 0} p_n(k)\, x^k \leq x^{(d-1)n} p_n$, since the maximum number of contacts in a polygon in d dimensions is at most $(d-1)n$ and since $x \geq 1$. Thus, $\mathcal{C}_k^{(d-1)}(x) \leq \log(x^{d-1}\mu_d)$.

On the other hand, $p_n(x) \geq p_n$ if $x \geq 1$; therefore, $\mathcal{C}_k^{(d-1)} \geq \log \mu_{d-1}$. Substitution of these bounds gives $\log a_c^+(x) \leq \log(x^{d-1}\mu_d) - \log \mu_{d-1}$.

This completes the proof. □

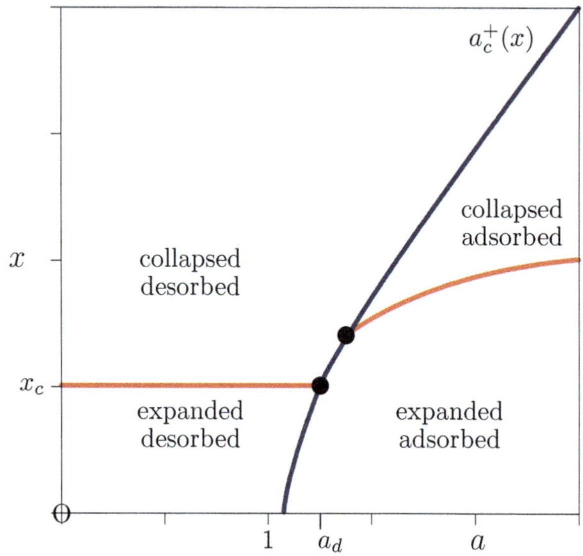

FIG. 9.15. The phase diagram of adsorbing and collapsing polygons may have four phases, as shown. Critical curves meet in at most two multicritical points (which may coincide).

By corollary 9.41, a desorbed phase is found for small values of a, and an adsorbed phase is found if $a > a_c^+(x)$.

It is not known whether there is an expanded phase at small enough values of x or a collapsed phase at large values of x, although there is a tremendous amount of numerical evidence which suggests this [569]. Such a collapse transition will go through a tricritical θ-point at a critical point $x = x_c(z)$ and, moreover, may depend on whether the transition occurs in an adsorbed or desorbed phase.

By making assumptions, it may be shown that $x_c(a)$ is independent of a for all $a < a_c^+(x_c)$ (where $x_c = x_c(1)$ is the critical activity for the collapse transition in polygons).

Theorem 9.42 . *Suppose that $\mathcal{C}_p^+(x,1) = \mathcal{C}_k^p(x)$ is non-analytic at $x = x_c$; then $\mathcal{C}_p^+(x,a)$ and $\mathcal{C}_p^>(x,a)$ are non-analytic functions of x at $x = x_c$ for all $a \leq 1$.*

Moreover, if $x_c^+(a)$ is a continuous function of x at $x = x_c$, then $\mathcal{C}_p^+(x,a)$ is a non-analytic function of x at $x = x_c$ for all $a < a_c^+(x_c)$.

The same statement is true for $\mathcal{C}_p^>(x,a)$ but for $a < a_c^>(x_c)$ instead.

Proof Since $\mathcal{C}_p^+(x,a) = \mathcal{C}_p^>(x,a) = \mathcal{C}_p(x,1) = \mathcal{C}_k^p(x)$ if $a \leq 1$, it follows that $x_c(a) = x_c$ if $a \leq 1$. Let $\epsilon > 0$ and find a_d by

$$a_d < \inf \left\{ a_c^+(x) \;\middle|\; x \in [x_c - \epsilon, x_c + \epsilon] \right\}.$$

The free energy $\mathcal{C}_p^+(x,a)$ is an analytic function of a for all such $a < a_d$, and $x \in [x_c - \epsilon, x_c + \epsilon]$. Hence, $\mathcal{C}_p^+(x,a) = \mathcal{C}_k^p(x)$ for all $x \in [x_c - \epsilon, x_c + \epsilon]$, and

TABLE 9.3. θ-Exponents in two dimensions

Exponent	Exact [558]	Numerical [209]
$\gamma_{1,\theta}$	$\frac{4}{7}$	0.57(2)
$\gamma_{11,\theta}$	$-\frac{4}{7}$	−0.55(5)
$\gamma_{s,\theta}$	$\frac{8}{7}$	1.14(5)
$\gamma_{ss,\theta}$	−−	0.51(8)
$\phi_{s,\theta}$	$\frac{8}{21}$	0.40(5)

$a \leq a_d$; if this were not the case, then there would be a phase boundary at a critical value of a for some $x_d \in [x_c - \epsilon, x_c + \epsilon]$, which is a contradiction.

However, $\mathcal{C}_k^p(x)$ was assumed to be non-analytic at $x = x_c$; thus, $\mathcal{C}_p^+(x, a)$ is non-analytic at x_c for each $a < a_d$.

Notice that, if ϵ is small enough, then a_d can be taken arbitrarily close to $a_c^+(x_c)$. Hence, $\mathcal{C}_p^+(x, a)$ is non-analytic at x_c for each $a < a_c^+(x_c)$.

If $\mathcal{C}_p^>(x, a)$ is considered instead, then the same result may be shown using similar arguments. □

The phase diagram of adsorbing and collapsing walks or polygons is illustrated in figure 9.15, based on numerical simulations of the model in references [568, 569]. Notice that the adsorption critical curve $a_c^+(x)$ is drawn with $a_c^+(x) \geq 1$; it can be shown that $a_c^+(x) > 1$ for all $x \geq 0$.

Similar results are obtained for collapsing walks or polygons adsorbing in a defect plane. It is conjectured that the adsorption critical point in this model is $a_c^>(x) = 1$ for all $x > 0$, based on the numerical data [568].

Numerical simulations suggest that there is no distinct collapsed and adsorbed phase in two dimensions.

In higher dimensions the phase diagram is illustrated with two triple points (figure 9.15) but these may coincide in a single multicritical point where four phases meet.

The collapse transition at $x = x_c$ between the desorbed expanded and desorbed collapsed phase should be the usual θ-transition, with the self-avoiding walk θ-exponents given in table 9.2.

In addition, θ-conditions affect the surface exponents of self-avoiding walks. For example, there are new distinct θ-values for the exponents γ_1 and γ_{11}, denoted by $\gamma_{1,\theta}$ and $\gamma_{11,\theta}$, in table 9.1. These exponents have the exact values $\gamma_{1,\theta} = \frac{4}{7}$, and $\gamma_{11,\theta} = -\frac{4}{7}$, in two dimensions [558].

A distinct set of θ-surface-exponents is obtained at the θ-point. For example, the surface exponent $\gamma_{s,\theta}$ for a positive walk adsorbing at the θ-point has the exact value $\frac{8}{7}$, and the crossover exponent associated with the adsorption transition at the θ-point $\phi_{s,\theta}$, is $\frac{8}{21}$ [558]. These exponents are listed in table 9.3.

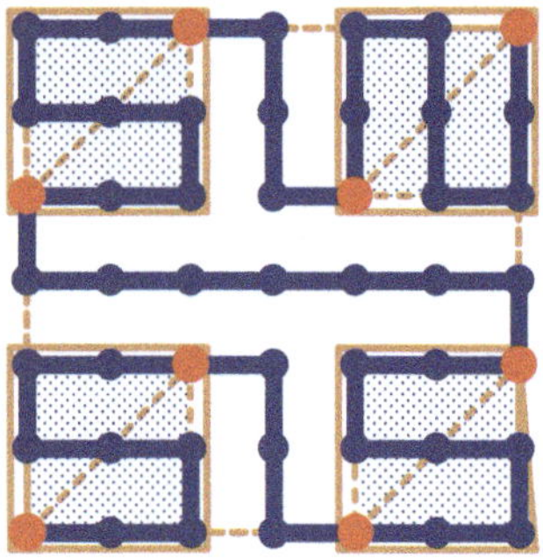

FIG. 9.16. There are at least 2^4 walks which cross L_6^2; each of the walks in the smaller shaded squares can be chosen in two ways (by reflection through the diagonal). Thus, $\varsigma_6^c(48) \geq 2^4$.

9.6 Walks crossing a square as a model of the θ-transition

The θ-transition may be modelled as a dilute-dense transition from the coiled phase (dilute) to the globule phase (dense). A suitable model for this is a self-avoiding walk crossing a square [584]. Several conjectures about this model were proven in reference [395].

9.6.1 *Walks crossing a square*

Let $L_n^d = \{\langle \vec{v} \sim \vec{w} \rangle \in \mathbb{L}^d \bigm| -\lceil \frac{1}{2}n \rceil \leq \vec{v}(i), \text{and } \vec{w}(i) \leq \lceil \frac{1}{2}n \rceil \text{ for } i = 1, 2, \ldots, d\}$ be a d-dimensional hypercube centred at $\vec{0}$ in $\mathbb{L}^d$ and with side-length n (and where $\vec{v}(i)$ is the i-th Cartesian coordinate of the vector $\vec{v}$). The boundary of L_n^d is defined by $\partial L_n^d = \{\langle \vec{v} \sim \vec{w} \rangle \in \mathbb{L}^d \bigm| \text{where } \|\vec{v}\|_1 = \lceil \frac{1}{2}n \rceil \text{ and } \|\vec{w}\|_1 = \lceil \frac{1}{2}n \rceil .\}$.

A self-avoiding walk $\omega = \langle \omega_0, \omega_1, \ldots, \omega_n \rangle$ *crosses* L_n^d if $\omega_j \in L_n^d$ for $0 \leq j \leq m$ and if ω_0 and ω_m are located at the (lexicographic) bottom and top antipodal corners of ∂L_n^d.

Consider the model in $\mathbb{L}^2$. Let $\varsigma_n(m)$ be the number of walks of length m crossing square L_n^2. The partition function of this model is

$$\varsigma_n(z) = \sum_{m=0}^{2\lfloor n(n+2)/2 \rfloor} \varsigma_n(m)\, z^m, \tag{9.83}$$

where z is conjugate to the steps in the walk. The shortest walk which crosses L_n^2 has length n_{min}, and the longest walk has length n_{max}, where

$$n_{min} = 2n \text{ and } n_{max} = 2 \left\lfloor \tfrac{1}{2} n(n+2) \right\rfloor. \tag{9.84}$$

In figure 9.16, an example of a walk which crosses L_6^2 is shown. Each walk which crosses the shaded squares can be chosen in two ways by reflecting them in the diagonal of the shaded squares. This shows that $\varsigma_2(8) = 2$. Thus, $\varsigma_6(48) \geq 2^4$. This argument can be generalised to obtain

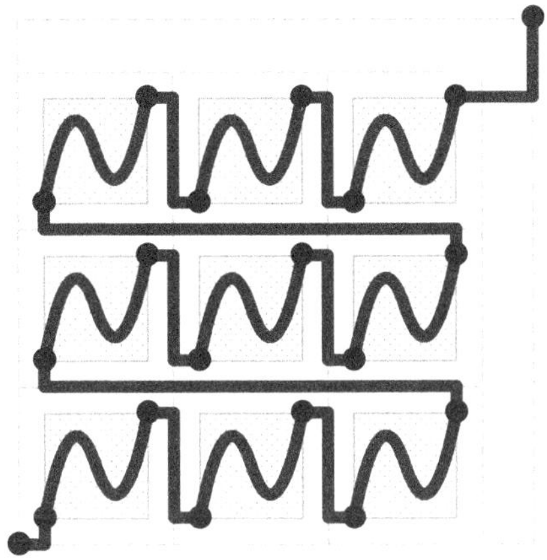

FIG. 9.17. Let $p = \lfloor \frac{n}{M+2} \rfloor$. Then L_n^2 contains p^2 smaller squares of size $M \times M$. There is a walk which crosses L_n^2.

$$\varsigma_{2^k-2}(2^k(2^k-2)) \geq \left(2\varsigma_{2^{k-1}-2}(2^{k-1}(2^{k-1}-2))\right)^4. \tag{9.85}$$

The result is the following lower bound:

$$\varsigma_{2^k-2}(2^k(2^k-2)) \geq 2^{4+4^2+\cdots+4^{k-1}} \geq 2^{4(4^{k-1}-1)/3}. \tag{9.86}$$

Thus, since $\varsigma_n(z) \leq \varsigma_{n+1}(z)$ for $z \geq 1$, and $\varsigma_n(z) \leq z^2 \varsigma_{n+1}(z)$ for $0 \leq z < 1$,

$$\liminf_{n\to\infty} \tfrac{1}{n} \log \varsigma_n(z) \geq \liminf_{k\to\infty} \tfrac{1}{2^k-2} \log(2^{4(4^{k-1}-1)/3} z^{2^k(2^k-2)}). \tag{9.87}$$

The right-hand side is infinite for all $z \geq 1$.

Theorem 9.43 *The following limits exist for $z \in [0, \infty]$:*

$$\mathcal{F}_\varsigma(z) = \lim_{n\to\infty} \tfrac{1}{n} \log \varsigma_n(z), \text{ and } \mathcal{H}_\varsigma(z) = \lim_{n\to\infty} \tfrac{1}{n^2} \log \varsigma_n(z).$$

Moreover, $\mathcal{F}_\varsigma(z) = \infty$ if $z \geq 1$, and if $\mathcal{F}_\varsigma(z) < \infty$ then $\mathcal{H}_\varsigma(z) = 0$, while $\log z \leq \mathcal{H}_\varsigma(z) \leq \log \mu_2 + \log z$ if $z \geq 1$.

Proof A walk crossing $L_{n_1}^2$ of length $m - m_1$ can be concatenated with a walk crossing $L_{n_2}^2$ of length m_1 (by identifying the lexicographic top endpoint of the first walk with the lexicographic bottom endpoint of the second). This gives a walk crossing $L_{n_1+n_2}^2$ of length m. That is,

$$\sum_{m_1=0}^{m} \varsigma_{n_1}(m - m_1)\varsigma_{n_2}(m_1) \leq \varsigma_{n_1+n_2}(m).$$

Multiplying by z^m and summing over m gives

$$\varsigma_{n_1}(z)\varsigma_{n_2}(z) \leq \varsigma_{n_1+n_2}(z).$$

By lemma A.1 in appendix A, $\mathcal{F}_\varsigma(z)$ exists. If $z \geq 1$, then $\mathcal{F}_\varsigma(z) = \infty$, as shown above.

If $z \leq 1$, then the second limit can be bounded using bridges. By prepending an edge in the $\vec{e}_1$ direction onto the first vertex of each walk, it follows that $\varsigma_n(m) \leq b_{m+1}$. Thus, $\varsigma_n(z) \leq \sum_{m=n_{min}}^{n_{max}} b_{m+1}\, z^m$. If $z \leq 1$, then this gives $\varsigma_n(z) \leq (n_{max} - n_{min}) b_{n_{max}+1}$, and, if $z > 1$, then $\varsigma_n(z) \leq (n_{max} - n_{min}) b_{n_{max}+1}\, z^{n_{max}}$. Since $n_{max} = O(n^2)$, this shows that

$$\limsup_{n\to\infty} \tfrac{1}{n^2} \log \varsigma_n(z) = \log \mathcal{H}_\varsigma(z) < \infty. \tag{9.88}$$

To see that the limit exists, consider L_n^2 in figure 9.17, which contains p^2 squares L_{M+2}^2, where $p = \left\lfloor \frac{n}{M+2} \right\rfloor$, and $n = p(M+2) + q$.

Each of the smaller L_M^2 (contained in the interior of L_{M+2}^2) can be crossed independently by a walk; label these squares by ij, where $1 \leq i, j \leq p$, and let the number of edges in the walk which crosses the ij-th square be n_{ij}. These walks can be made into a longer walk which crosses L_n^2 by adding $2p(p-1)(M+2) + 2q + 4$ edges, as shown in figure 9.17.

Since these walks cross each of the smaller squares independently, the number of walks in L_n^2 is at least $\prod_{i,j=1}^{p} \varsigma_M(n_{ij})$. Multiplying this with $z^{n_{11}+\cdots+n_{pp}}$ and summing over all the n_{ij} gives

$$(\varsigma_M(z))^{p^2} \leq (2M^2 - 2M)^{p^2} z^{-2p(p-1)(M+2)-2q-4} \varsigma_n(z), \tag{9.89}$$

since each L_M^2 sub-square contains at least $2M$ edges, and at most $2M^2$ edges.

From equation (9.88), for every $\epsilon > 0$, there exists an infinite set of integers $S(\epsilon)$ such that

$$\mathcal{H}_\varsigma(z) - \tfrac{\epsilon}{2} \leq \tfrac{1}{n^2} \log \varsigma_n(z) \leq \mathcal{H}_\varsigma(z), \tag{9.90}$$

whenever $n \in S(\epsilon)$.

From equation (9.89) and whenever $M \in S(\epsilon)$,

$$\begin{aligned}\liminf_{n\to\infty} \tfrac{1}{n^2} \log \varsigma_n(z) \geq{}& \frac{p^2 \log \varsigma_M(z)}{(p(M+2)+q)^2} - \frac{p^2 \log(2M(M-1))}{(p(M+2)+q)^2} \\ &+ \frac{(2p(p-1)(M+2)+2q+4)|\log z|}{(p(M+2)+q)^2}.\end{aligned}$$

Take $M \to \infty$ in $S(\epsilon)$ in the above and use equation (9.89) to obtain

$$\liminf_{n\to\infty} \tfrac{1}{n^2} \log \varsigma_n(z) \geq \mathcal{H}_\varsigma(z) - \tfrac{\epsilon}{2} \geq \limsup_{n\to\infty} \tfrac{1}{n^2} \log \varsigma_n(z) - \epsilon.$$

If $\epsilon \to 0^+$, then this establishes the existence of the limit.

If $z \geq 1$, then $\varsigma_n(z) \leq \sum_{m=n_{min}}^{n_{max}} c_m\, z^{n_{max}}$, where c_m is the number of walks of length m. Thus,

$$\log z \leq \lim_{n\to\infty} \tfrac{1}{n^2} \log \varsigma_n(z) \leq \log \mu_2 + \log z,$$

from equation (9.84). □

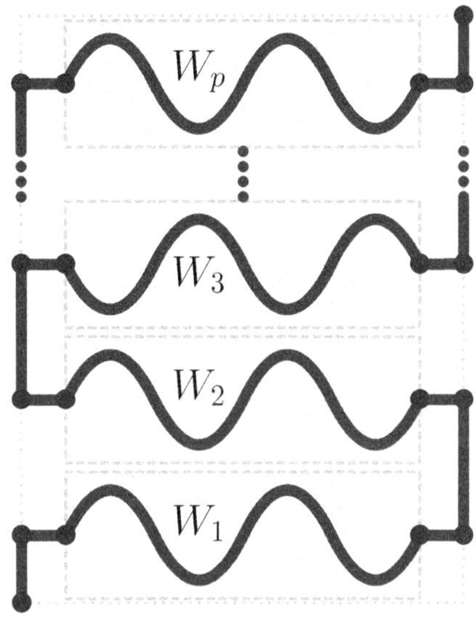

FIG. 9.18. Each shaded rectangle contains an unfolded walk and these walks can be combined into a single walk that crosses a rectangle of dimensions $[0, jM+1] \times [0, L]$, where $L = 2(p+1)m$.

9.6.2 *The dilute-dense critical point*

The construction in figure 9.16 implies that $\mathcal{F}_\varsigma(z) > 0$ if $z > 0$. It is also the case that $\sum_{n\geq 0} \varsigma_n(z) \leq \sum_{m\geq 0} c_m\, z^m = C(z)$, where $C(z)$ is the generating function of self-avoiding walks (see equation (1.56)). By theorem 1.1, $C(z)$ is finite if $z < \mu_2^{-1}$. In other words, there is a critical point z_c in the interval $[\mu_2^{-1}, 1]$. To see that $z_c = \mu_2^{-1}$, argue as in reference [395].

Define the number of bridges from $\vec{0}$ to the point (n_1, n_2) by $b_n(n_1, n_2)$. A walk which crosses L_n^2 and is not a bridge can be made into a bridge by prepending a single edge in the $\vec{e}_1$ direction to its bottom vertex. Thus,

$$\varsigma_n(m) \leq b_{m+1}(n+1, n) \leq b_{m+1}. \tag{9.91}$$

Notice that bridges are supermultiplicative: $b_{n_1} b_{n_2} \leq b_{n_1+n_2}$; therefore, by theorem A.1 in appendix A, for every $\epsilon > 0$, there exists an $N_0 \in \mathbb{N}$ such that

$$(\mu_2 - \epsilon)^n \leq b_n \leq \mu_2^n \text{ for all } n \geq N_0. \tag{9.92}$$

In the next lemma, it is seen that equation (9.92) is also valid if the second Cartesian coordinate of the endpoint of the bridge is fixed.

Lemma 9.44 *Let $\epsilon > 0$. Then there exist integers m and M, both even or both odd, such that*

$$b_m(M, 0) > (\mu_2 - \epsilon)^{2\lfloor m/2 \rfloor}.$$

Proof By equation (7.2), there exists an N_1 such that, for all $m \geq N_1$,

$$\tfrac{1}{(2m+1)^2} b_m > (\mu_2 - \epsilon)^m$$

There is a most popular endpoint for bridges of length m, say (k_1, k_2). Since there are at most $(2m+1)^2$ endpoints,

$$b_m(k_1, k_2) > (\mu_2 - \epsilon)^{m+1}.$$

By symmetry, $b_m(k_1, k_2) = b_m(k_1, \pm k_2)$, and so it may be assumed that $k_2 \geq 0$.

A bridge counted by $b_m(k_1, k_2)$ can be concatenated with a bridge counted by $b_m(k_1, -k_2)$ (identify the last vertex in the first with the first vertex in the second). This gives a bridge for which the last vertex has the coordinates $(2k_1, 0)$. Thus,

$$b_{2m}(2k_1, 0) \geq b_m(k_1, k_2) b_m(k_1, -k_2) \geq (b_m(k_1, k_2))^2 > (\mu_2 - \epsilon)^{2m}.$$

Thus, choose $M = 2k_1$ and replace $2m$ by m to obtain two even integers. Odd integers are obtained if the concatenation above is done by inserting an extra horizontal edge between the two walks. This gives a bridge for which the last vertex has the coordinates $(2k_1 + 1, 0)$; thus,

$$b_{2m+1}(2k_1 + 1, 0) \geq b_m(k_1, k_2) b_m(k_1, -k_2) \geq (b_m(k_1, k_2))^2 > (\mu_2 - \epsilon)^{2m}.$$

Put $M = 2k_1 + 1$ and replace $2m + 1$ by m to obtain two odd integers. □

The bridges in lemma 9.44 can be stacked vertically to create walks crossing a square (see in figure 9.18). This gives the following result.

Lemma 9.45 *Let $\epsilon > 0$ and choose m and M as in lemma 9.44. Let p be an odd number. Then the number of walks of length $n = 2mp + mpj + \frac{1}{2}(p+1)$ which cross the rectangle $[0, jM + 1] \times [0, 2mp]$ is at least $(\mu_2 - \epsilon)^{2jp\lfloor m/2 \rfloor}$.*

Proof Consider bridges of length jm counted by $b_{jm}(jM, 0)$, which are confined to the rectangle $[0, jM] \times (-m, m)$, and with span in the $\vec{e}_1$ direction equal to jM. Denote the number of these bridges by $b^*_{jm}(jM, 0)$.

Notice that $b^*_{jm}(jM, 0) \geq (b^*_m(M, 0))^j$ by concatenation.

Moreover, every walk counted by $b_m(M, 0)$ is in the rectangle $[0, M] \times (-m, m)$, so $b^*_m(M, 0) = b_m(M, 0)$. Thus,

$$b^*_{jm}(jM, 0) \geq (b_m(M, 0))^j \geq (\mu_2 - \epsilon)^{2j\lfloor m/2 \rfloor},$$

by the choices of m and M and from lemma 9.44.

Choose p such bridges $\{W_i\}$ and arrange them as in figure 9.18, with the first vertex of W_j placed at $(0, (2j-1)m)$. Add the extra edges as shown in figure 9.18 to join them into a single walk of length $2mp + mpj + \frac{1}{2}(p+1)$ and which crosses the rectangle $[0, jM + 1] \times [0, 2mp]$. The total number of such walks crossing the rectangle $[0, jM + 1] \times [0, 2mp]$ is surely larger than $(b^*_{jm}(jM, 0))^p$, since there are $b^*_{jm}(jM, 0)$ choices for each W_i. This completes the proof. □

A corollary of lemmas 9.44 and 9.45 is that the critical point of the dilute-dense transition is located at $z_c = \mu_2^{-1}$. This may be seen as follows. By lemma 9.45, the number of walks which cross $[0, jM+1] \times [0, 2mp]$ (with m and M as in lemma 9.44 and with p odd) is at least $(\mu_2 - \epsilon)^{2jp\lfloor m/2 \rfloor}$.

Choose $z > \mu_2^{-1}$ such that $z(\mu_2 - \epsilon) > 1$ and consider the partition function of walks which cross L^2_{jM+1}, namely, $\varsigma_{jM+1}(z)$. A class of walks contributing to this partition function has length $2mp + mjp + \frac{1}{2}(p+1) + 1$. The number of walks in this class is at least the number of walks of length $2mp + mjp + \frac{1}{2}(p+1)$ and which cross a rectangle of dimensions $[0, jM+1] \times [0, jM]$.

Proceed by choosing $2mp = jM$ in lemma 9.45 so that $p = \frac{jM}{2m}$ in the previous paragraph.

Choose m and M odd and let $j = 2mt$ to give an odd value for p. This implies that $p = Mt$ and that $mjp = 2m^2Mt^2$.

Therefore, the rectangle has dimensions $[0, jM+1] \times [0, jM]$, and the walks crossing it have length $2mMt + 2m^2Mt^2 + \frac{1}{2}(Mt+1)$. The number of these walks is at least $(\mu_2 - \epsilon)^{4mMt^2\lfloor m/2 \rfloor}$, and hence

$$\varsigma_{2mMt+1}(z) \geq (\mu_2 - \epsilon)^{4mMt^2\lfloor m/2 \rfloor} z^{2mMt + 2m^2Mt^2 + (Mt+1)/2 + 1}. \tag{9.93}$$

This implies the following.

$$\begin{aligned} &\lim_{n\to\infty} \tfrac{1}{n^2} \log \varsigma_n(z) \\ &\geq \liminf_{n\to\infty} \frac{\log\left((\mu_2 - \epsilon)^{4mMt^2\lfloor m/2 \rfloor} z^{2mMt + 2m^2Mt^2 + (Mt+1)/2 + 1} \right)}{(2mMt+1)^2} \\ &\geq \tfrac{1}{2M} \log((\mu_2 - \epsilon)z) > 0. \end{aligned} \tag{9.94}$$

For every $\epsilon > 0$, a z and a pair (m, M) can be found; thus, $\lim_{n\to\infty} \frac{1}{n^2} \log \varsigma_n(z) > 0$ if $z > \mu_2^{-1}$. This result proves the following theorem.

Theorem 9.46 *The free energy* $\mathcal{H}_\varsigma(z) = \lim_{n\to\infty} \frac{1}{n^2} \log \varsigma_n(z) > 0$ *if* $z > \mu_2^{-1}$. □

Corollary 9.47 *The free energy* $\mathcal{F}_\varsigma(a) = \lim_{n\to\infty} \frac{1}{n} \log \varsigma_n(z) = \infty$ *if* $z > \mu_2^{-1}$. *That is,* $z_c = \mu_2^{-1}$. □

It is also known that $\mathcal{F}_\varsigma(z_c) = 0$ and that this free energy is strictly negative if $z < z_c$; these results can be proven with the help of the mass of the self-avoiding walk [395].

9.6.3 *The dense phase*

The dense phase of walks was examined in references [160, 169]. The appropriate scaling form for the partition function should be given by equation (4.38) and, for the free energy, by equation (4.40). In these expressions σ is a surface exponent

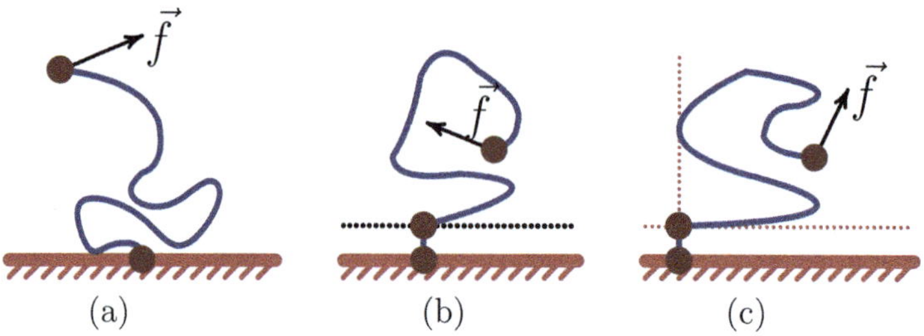

FIG. 9.19. (a) A pulled positive walk. (b) A pulled half-space walk. (c) A pulled tail.

---

dependent on the shape of the containing region (in the case here, a square, so it is expected that $\sigma = \frac{d-1}{d}$).

The exponent $\alpha_- - \alpha_t$ in equation (4.38) is usually replaced by $\gamma_D - 1$. This gives the scaling form $\varsigma_n \simeq n^{\gamma_D - 1} \mu_1^{n^\sigma} \mu_z^n$ (where the free energy is given by $\mathcal{F}_\varsigma(z) = \log \mu_z$). The exponent γ_D is not the usual entropic exponent and it is dominated by the surface term $\mu_1^{n^\sigma}$.

The ratio of the partition function of dense walks and dense polygons, (ς_n/ρ_n), was studied as well (where ρ_n is the partition function of polygons in a square). This gives the relation $[\varsigma_n/\rho_n] \simeq n^{\gamma_D}$. The exact value of the exponent γ_D is $\frac{19}{16}$ [160, 169]. This is in contrast with a model of walks collapsing to a Hamiltonian walk, where $\gamma_H = 1$ [20].

The metric exponent of the walk in the dense phase is $\nu = \frac{1}{2}$. The collapsed phase in a θ-collapse of a walk has the same metric exponent but the relation between these phases is uncertain. If R is the root mean square radius of gyration of dense walks, then $R \simeq \langle n \rangle^2$ if $z > z_c$ (where $\langle n \rangle$ is the mean number of edges in the walk). On the other hand, if $z < z_c$, then $R \simeq \langle n \rangle$ [84, 584]. At the critical point $z = z_c$, $R \simeq \langle n \rangle^{1/g}$, with $g = \frac{1}{\nu} = \frac{4}{3}$ and where ν is the metric exponent of a linear polymer in a good solvent. This is supported by a renormalisation group argument, which gave the estimate $g \approx 1.35$ [472].

9.7 Pulled self-avoiding walks

A positive walk from $\vec{0}$ in $\mathbb{L}_+^d$ pulled at its free end by a force is a model of a linear polymer attached at one endpoint to a *plate* $\partial \mathbb{L}_+^d$ and pulled by an externally applied force at the other end. This is illustrated in figure 9.19(a) and it is a *pulled positive walk*.

If the pulling force is in a direction away from the plate, then the model is in the *tensile* regime. In this regime the walk is elongated in the pulling direction. The forces are *pulling* forces and they *stretch* the polymer in the force direction.

If the force is directed towards the plate (and in particular towards $\vec{0}$, where the polymer is attached to the plate), then the polymer is in a *compressive* regime. The applied forces are *pushing* the polymer towards the plate.

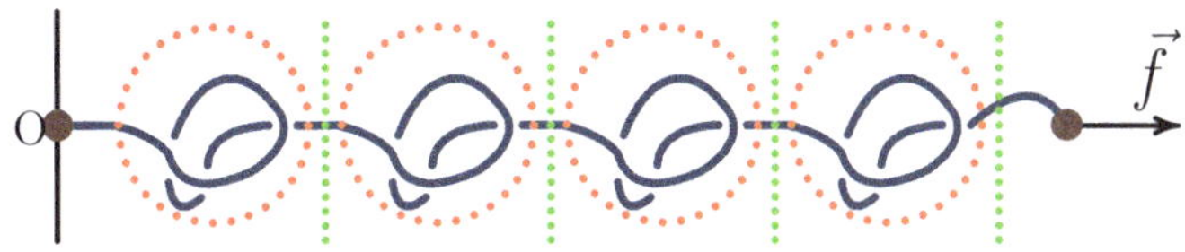

FIG. 9.20. The ballistic phase of a pulled polymer. According to the Pincus theory, the pulled polymer partitions into a set of independent coils called *Pincus balls* separated by *cut-planes.* The conformations of the polymer inside each Pincus ball are independent of the force and other Pincus balls. The size of the Pincus balls is dependent on the pulling force $\vec{f}$ and sets a length scale in addition to the length scale set by the size of the pulled polymer.

In the tensile regime the polymer goes through a phase transition from its *coiled (free) phase* to a *ballistic phase* characterised by the metric exponent $\nu = 1$ (in the coiled phase $\nu < 1$).

Models of pulled walks with a nearest-neighbour interaction were examined in reference [239]. The results indicate a transition to a ballistic phase as the force increases in magnitude. A model of a collapsing walk pulled from an adsorbing surface was investigated in reference [362]; this model also showed a transition to a ballistic phase for large pulling forces.

The application of a force to a polymer has been described using the Pincus theory [117, 465]. The Pincus theory identifies length scales in a walk stretched by a pulling force and examines weak and strong force regimes.

The strong force regime is ballistic [314, 315], and the polymer separates into *Pincus balls* which are partitioned by *cut-planes* normal to the direction of the force. Inside each Pincus ball the polymer explores its conformations independently from the conformations in other balls. The entanglement complexity of the polymer in each Pincus ball is large in the scaling limit, even for large applied forces [313]. The strong force stretching of the polymer into Pincus balls is illustrated in figure 9.20.

9.7.1 *The free energy of vertically pulled positive walks*

Let ω be a positive walk from $\vec{0}$ in $\mathbb{L}^d_+$ pulled by a force $\vec{f}$ from its endpoint; (see figure 9.19(a)). Suppose that $\omega = \langle \omega_0, \omega_1, \ldots, \omega_n \rangle$ and denote the d-th components of the vertices ω_j by $\omega_j(d)$. Since ω is a positive walk, $\omega_j(d) \geq 0$ for $j = 0, 1, \ldots, n$.

The number of positive walks from the origin and with the last vertex fixed at $\vec{v}$ is denoted by $c_n^+(\vec{v})$. If $\omega_n(d) = \vec{v}(d) = h$, then the *height* of the endpoint of ω is h, and the number of positive walks from the origin ending in a vertex at height h is $c_n^+(h)$. The partition function of this model is

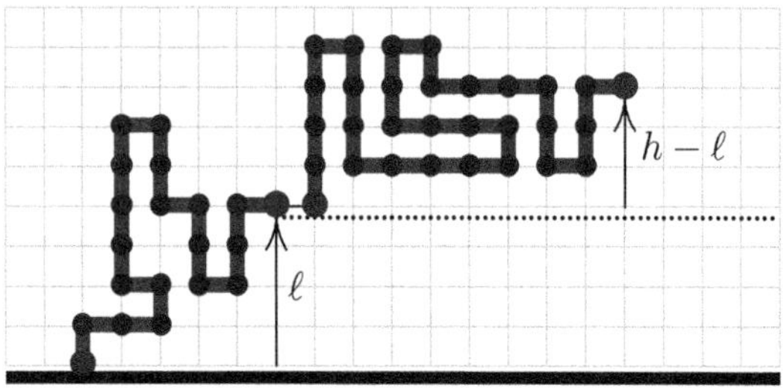

FIG. 9.21. Concatenating two tails by placing the second such that its bottom vertex has a height equal to the height of the final vertex of the first. The tails are concatenated by inserting a single edge from the top vertex of the first into the bottom vertex of the second. Since the endpoints of the tails are left-most or right-most in the $\vec{e}_1$ direction, there cannot be intersections between them if they are placed in this way. The concatenated walk is similarly a tail, and its final vertex has a height that is the sum of the heights of the two original tails.

$$T_n^+(y_d) = \sum_{h=0}^{n} c_n^+(h)\, {y_d}^h, \tag{9.95}$$

where y_d is conjugate to h.

Introduce *pulled half-space walks*, *pulled bridges* and *pulled tails* in order to analyse $T_n^+(y_d)$. Half-space walks are positive walks with $\omega_j(d) > 0$ for all $j = 1, 2, \ldots, n$; the half-space walk steps away from $\partial\mathbb{L}_+^d$ on its first step and never returns. A *pulled bridge* is a pulled half-space walk with its endpoint at the maximum height, that is, $\omega_n(d) \geq \omega_j(d) > 0$ for all $j = 0, 1, \ldots, n$. Notice that bridges, positive walks and half-space walks are oriented in the $\vec{e}_d$ direction.

The partition functions of pulled half-space walks and pulled bridges are $T_n^h(y_d)$ and $T_n^b(y_d)$, respectively. Notice that $T_n^+(y_d) \geq T_n^h(y_d) \geq T_n^b(y_d)$.

If a pulled half-space walk is doubly unfolded in the $\vec{e}_1$ direction (see figure (9.19)(c)), then it is a pulled *tail*. Denote the number of tails of length n and with endpoints at height h above $\partial\mathbb{L}_+^d$ by $c_n^t(h)$.

It follows mutatis mutandis from the unfolding constructions in section 7.15 that, if $\gamma_0 > \pi\sqrt{\frac{2}{3}}$, then there exist $\epsilon > 0$ and an $N_\epsilon \in \mathbb{N}$, such that $c_{n-1}^+(h) \leq e^{(\gamma_0-\epsilon)\sqrt{n}} c_n^t(h)$ for all $n \geq N_\epsilon$. (Notice that unfolding in the $\vec{e}_1$ direction does not change the height h of the last vertex in the positive walk.) This gives the following relations between the partition functions of pulled positive walks, pulled half-space walks and pulled tails. For all $n \geq N_\epsilon$,

$$T_n^+(y_d) \geq T_n^h(y_d) \geq T_n^t(y_d) \geq e^{-(\gamma_0-\epsilon)\sqrt{n}}\, T_{n-1}^+(y_d). \tag{9.96}$$

Two tails can be concatenated as shown in figure 9.21. This shows that

$$\sum_{\ell=0}^{h} c_n^t(\ell)\, c_m^t(h-\ell) \le c_{n+m+1}^t(h). \tag{9.97}$$

This relation shows that $c_{n-1}^t(h)$ satisfies the relation in equation (3.5) and so the partition function of pulled tails satisfies a supermultiplicative inequality similar to equation (3.4). By theorem 3.1, this gives the existence of a limiting free energy:

$$\mathcal{T}(y_d) = \lim_{n\to\infty} \tfrac{1}{n} \log T_n^t(y_d) = \lim_{n\to\infty} \tfrac{1}{n} \log T_n^h(y_d) = \lim_{n\to\infty} \tfrac{1}{n} \log T_n^+(y_d). \tag{9.98}$$

This follows from the inequalities (9.96).

Theorem 9.48 *The limiting free energy of pulled positive walks is a convex function of* $\log y_d$ *given by*

$$\mathcal{T}(y_d) = \lim_{n\to\infty} \tfrac{1}{n} \log T_n^+(y_d) = \lim_{n\to\infty} \tfrac{1}{n} \log T_n^h(y_d) = \lim_{n\to\infty} \tfrac{1}{n} \log T_n^t(y_d).$$

If $y_d \le 1$*, then* $\mathcal{T}(y_d) = \log \mu_d$*; if* $y_d > 1$*, then*

$$\max\{\log \mu_d, \log y_d\} \le \mathcal{T}(y_d) \le \log \mu_d + \log y_d.$$

There is a critical point $y_c \ge 1$ *such that, for all* $y_d > y_c$*,* $\mathcal{T}(y_d) > \log \mu_d$*.*

Proof If $y_d \le 1$, then $c_n^+ \ge T_n^+(y_d) \ge \ell_n^o$, where ℓ_n^o is the number of hoops of length n (see section 9.1.1). This shows that $\mathcal{T}(y_d) = \log \mu_d$ for $0 \le y_d \le 1$.

It only remains to prove the bounds on $\mathcal{T}(y_d)$ for $y_d > 1$.

Since $T_n^+(y_d) \ge \sum_{h=0}^{n} c_n^+(h)\, {y_d}^h > c_n^+$, $\mathcal{T}(y_d) \ge \log \mu_d$. In addition, $T_n^+(y_d) \ge {y_d}^n$. This shows that $\mathcal{T}(y_d) \ge \log y_d$.

On the other hand, if $y_d > 1$, then $T_n^t(y_d) \le \sum_{h=0}^{n} c_n^+(h)\, {y_d}^n = {y_d}^n c_n^+$. This shows that $\mathcal{T}(y_d) \le \log y_d + \log \mu_d$.

Moreover, $\mathcal{T}(y_d)$ is a convex function of $\log y_d$ by theorem (3.3). □

Theorem 9.48 does not define a free energy for pulled bridges. This is done separately in the next theorem.

Theorem 9.49 *The free energy* $\mathcal{T}(y_d) = \lim_{n\to\infty} \tfrac{1}{n} \log T_n^b(y_d)$.

Proof By concatenating pulled bridges similar to tails (see figure 9.21), it follows that the limit $\lim_{n\to\infty} \frac{1}{n} \log T_n^b(y_d)$ exists. Since $T_n^b(y_d) \le T_n^+(y_d)$, it is the case that $\lim_{n\to\infty} \frac{1}{n} \log T_n^b(y_d) \le \mathcal{T}(y_d)$, for all $y_d \ge 0$.

If $y_d \ge 1$, then half-space walks can be unfolded into bridges. This increases the height and the weight of the terms in the partition function. The arguments in section 7.15 show that, if $\gamma_0 > \pi\sqrt{\frac{2}{3}}$, then there exist $\epsilon > 0$ and an $N_\epsilon \in \mathbb{N}$ such that

$$T_n^h(y_d) \le e^{(\gamma_0-\epsilon)\sqrt{n}}\, T_n^b(y_d)$$

for all $n \ge N_\epsilon$. Thus, $\lim_{n\to\infty} \frac{1}{n} \log T_n^b(y_d) \ge \mathcal{T}(y_d)$, for all $y_d \ge 1$.

The case $y_d < 1$ remains:

$$\lim_{n\to\infty} \tfrac{1}{n} \log T_n^b(y_d) \leq \log \mu_d = \mathcal{T}(y_d).$$

To find a lower bound, consider walks in slabs $\mathcal{S}_w$ of width w.

Let $S_n^h([0w], a, b; w)$ be the partition function of half-space walks in a slab $\mathbb{S}_w$ of width w (see theorem 8.8). Then $\lim_{n\to\infty} \frac{1}{n} \log S_n^h([0w], a, b; w) = \mathcal{S}_w(a, b)$ (the limiting free energy of half-space walks from the origin, with a last endpoint in the top wall of $\mathbb{S}_w$, and adsorbing into the walls $\mathbb{S}_w$). By equation (8.20), $\log \mu_w = \mathcal{S}_w(1,1)$. Put $c_n^w = S_n^h([0w], 1, 1; w)$; then c_n^w is the number of walks in $\mathbb{S}_w$ starting in the bottom wall $B(\mathbb{S}_w)$ and ending in the top wall $T(\mathbb{S}_w)$ (and moreover, $c_n^w = \mu_w^{n+o(n)}$). Since the height of these walks is w, they can be turned into bridges of height $w+1$ by prepending a vertical edge; this gives $c_n^w \leq b_{n+1}(w+1)$. Multiply this by $y_d{}^{w+1}\, t^{n+1}$ to form the generating function of pulled bridges:

$$B(y_d, t) = \sum_{n=0}^{\infty} \sum_{w=0}^{\infty} b_n(w)\, y_d{}^w t^n = \sum_{n=0}^{\infty} T_n^b(y_w)\, t^n \geq t \sum_{n=0}^{\infty} c_n^w\, y_d{}^{w+1}\, t^n. \tag{9.99}$$

This shows that the radius of convergence of $B(y_d, t)$ is t_c given by

$$-\log t_c = \lim_{n\to\infty} \tfrac{1}{n} \log T_n^b(y_d).$$

The series $\sum_{n=0}^{\infty} c_n^w\, t^n$ on the right-hand side of equation (9.99) has radius of convergence $t = \mu_w^{-1}$ for each fixed w, given $y_d \in (0,1)$. This shows that $t_c \leq \mu_w^{-1}$ for every $w \in \mathbb{N}$. By lemma 8.18 and theorem 8.19, $\mu_{w-1} < \mu_w \to \mu_d$ as $w \to \infty$. This shows that $t_c \leq \inf_w\{\mu_w^{-1}\} = \mu_d^{-1}$. Thus

$$\lim_{n\to\infty} \tfrac{1}{n} \log T_n^b(y_d) \geq \log \mu_d.$$

Since $\mathcal{T}(y_d) = \log \mu_d$ if $y_d \leq 1$ by theorem (9.48), $\lim_{n\to\infty} \frac{1}{n} \log T_n^b(y_d) = \log \mu_d$ for $y_d < 1$. □

The free energy $\mathcal{T}(y_d)$ is convex in $\log y_d$ and it is differentiable almost everywhere. For almost every y_d the energy is given by

$$\mathcal{E}(y_d) = y_d \tfrac{d}{dy_d} \mathcal{F}^t(y_d). \tag{9.100}$$

Observe that $\mathcal{E}(y_d) \geq 0$ and that it is monotone non-decreasing in y_d and differentiable almost everywhere (see figure 9.22). Moreover, since $\mathcal{E}(y_d)$ is monotone, it has left- and right-limits everywhere.

There is a most popular height ℓ^* of the endpoint of tails in the partition function $T_n^t(y_d)$ such that $c_n^t(\ell^*)\, y_d{}^{\ell^*} \leq T_n^t(y_d) \leq (n+1)\, c_n^t(\ell^*)\, y^{\ell^*}$. It follows that the limit $\lim_{n\to\infty} \frac{1}{n} \log \left(c_n^t(\ell^*)\, y_d{}^{\ell^*}\right) = \mathcal{T}(y_d)$ exists. This shows that

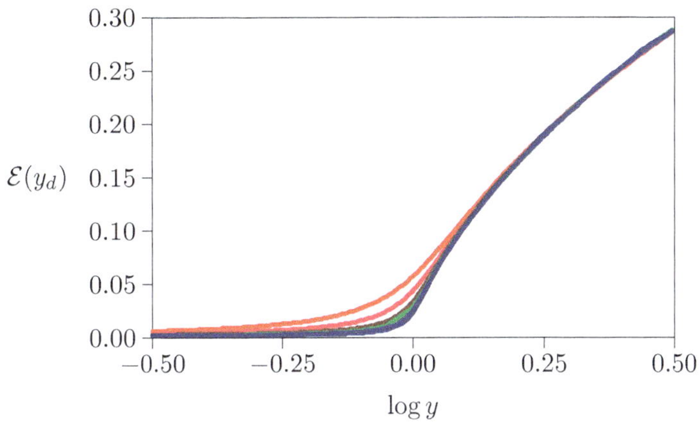

FIG. 9.22. The energy $\mathcal{E}(y_d)$ of pulled walks as a function of y for lengths 500, 1000, 2000, 3000, 4000 and 5000 in $d = 3$ dimensions [315].

$\lim_{n\to\infty} \frac{1}{n}\ell^* = \mathcal{E}(y_d)$ for almost every y_d. By equation (3.53), this is the energy density of the model defined by

$$\lim_{n\to\infty} \tfrac{1}{n}\ell^* = y_d \tfrac{d}{dy_d} \mathcal{T}(y_d) = \mathcal{E}^t(y_d) \text{ for almost all } y_d; \tag{9.101}$$

see also, for example, equation (3.56).

The bounds $c_n^t(\ell^*)\, {y_d}^{\ell^*} \leq T_n^t(y_d) \leq (n+1) c_n^t(\ell^*)\, {y_d}^{\ell^*}$ show that

$$\mathcal{T}(y_d) = \log \mathcal{P}^t(\epsilon^*) + \epsilon^* \log y_d \text{ for almost all } y_d, \tag{9.102}$$

where $\log \mathcal{P}^t(\epsilon)$ is the microcanonical density function or the Legendre transform of $\mathcal{T}(y_d)$ and $\epsilon^* = \mathcal{E}(y_d)$ by equation (9.101).

The lower bound $T_n^t(y_d) \geq c_n^t(\ell^*)\, y^{\ell^*}$ may be refined by noting that $c_n^t(\ell^*) \geq c_{n-\ell^*+1}^t(1)$; this is obtained by only considering tails of length $n - \ell^* + 1$ ending at height 1 and then adding a sequence of $\ell^* - 1$ vertical edges to the last vertex. Take logarithms, divide by n and let $n \to \infty$. In view of equation (9.101),

$$\mathcal{T}(y_d) \geq \big(1 - \mathcal{E}^t(y_d)\big) \log \mu_d + \mathcal{E}^t(y_d) \log y_d. \tag{9.103}$$

This gives the following lemma.

Lemma 9.50 *Let $f(x) = \mathcal{T}(e^x)$. Then f satisfies the differential inequality*

$$f(x) - \log \mu_d \geq f'(x)\,(x - \log \mu_d).$$

Proof Put $x = \log y_d$ in equations (9.103). Then $f'(x) = \mathcal{E}^t(y_d)$ almost everywhere, and the inequality follows as claimed. □

The differential inequality in lemma 9.50 may be integrated from z to x (provided that $z > \mu_d$). This gives the following theorem.

Theorem 9.51 *The free energy of vertically pulled walks satisfies the following inequality:*

$$\frac{\mathcal{T}(y_d) - \log \mu_d}{\log y_d - \log \mu_d} \leq \frac{\mathcal{T}(z) - \log \mu_d}{\log z - \log \mu_d}$$

whenever $y_d > z > \mu_d$.

Proof Let $z > \mu_d$, separate variables in the differential inequality in lemma 9.50 and integrate over (z, x):

$$\int_z^x \frac{f'(u)}{f(u) - \log \mu_d}\, du = \int_z^x \frac{du}{u - \log \mu_d}.$$

Integrate both sides and collect terms. Substitute $x = \log y_d$. This gives the claimed relation. □

If $z > 1$, then $\mathcal{T}(z) \leq \log \mu_d + \log z$ by theorem 9.48. Substitute this in theorem 9.51. This gives

$$\frac{\mathcal{T}(y_d) - \log \mu_d}{\log y_d - \log \mu_d} \leq \frac{\log z}{\log z - \log \mu_d}. \tag{9.104}$$

The corollary of this is that $\mathcal{T}(y_d)$ is asymptotic to $\log y_d$.

Corollary 9.52 *For every $\delta > 0$, there exists a z_δ such that, for all $y_d > z_\delta$,*

$$\log y_d \leq \mathcal{T}(y_d) \leq (1+\delta) \log y_d - \delta \log \mu_d,$$

where $\log y_d < \mathcal{T}(y_d)$ for all $y_d > \frac{\mu_d^2}{1+\mu_d}$.

Proof Increase z on the right-hand side of equation (9.104) until it is less than $1+\delta$. It remains to show that $\log y_d < \mathcal{T}(y_d)$ for all y_d.

Consider tails of length $n+1$ ending in a vertex at height $n - \lfloor \epsilon n \rfloor$ for some $\epsilon \in (0,1)$. These paths give $\lfloor \epsilon n \rfloor$ horizontal steps and may be chosen such that the horizontal steps (1) are in the $\vec{e}_1$ direction and (2) are at distinct heights. Then there are at least $\binom{n-\lfloor \epsilon n \rfloor}{\lfloor \epsilon n \rfloor}$ such tails, and this gives the lower bound

$$T^t_{n+1}(y_d) \geq \binom{n - \lfloor \epsilon n \rfloor}{\lfloor \epsilon n \rfloor} {y_d}^{n - \lfloor \epsilon n \rfloor}. \tag{9.105}$$

Take logarithms, divide by n and let $n \to \infty$. This shows

$$\mathcal{T}(y_d) \geq \log y_d + \log\left(\frac{(1-\epsilon)^{1-\epsilon} {y_d}^{-\epsilon}}{\epsilon^\epsilon (1-2\epsilon)^{1-2\epsilon}}\right). \tag{9.106}$$

The parameter ϵ may be chosen to maximise the lower bound on $\mathcal{F}^t(y_d)$. By equation (D.16) in appendix D this is realised when $\epsilon = \frac{\sqrt{1+4\overline{y}_d}-1}{2\sqrt{1+4\overline{y}_d}}$, where $\overline{y}_d = \frac{1}{y_d}$. The result is

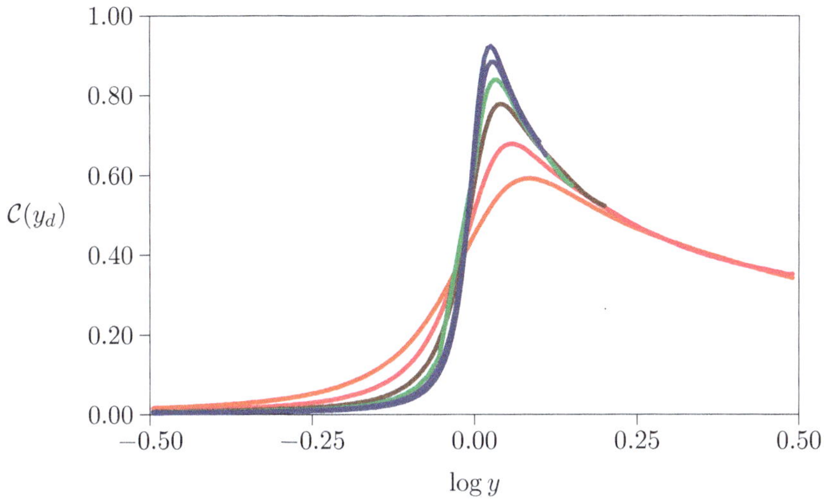

FIG. 9.23. The specific heat $\mathcal{C}(y_d)$ of pulled walks as a function of y for lengths 500, 3000, 4000 and 5000 in $d = 3$ dimensions [315].

$$\mathcal{T}(y_d) \geq \log\Big(y_d + \sqrt{{y_d}^2 + 4y_d}\Big) - \log 2 > \log y_d. \tag{9.107}$$

In other words, the lower bound in theorem 9.48 and corollary 9.52 is a strict lower bound when $y_d > \frac{\mu_d^2}{1+\mu_d}$. □

Since $\mathcal{T}(y_d) > \log \mu_d$ if $y_d > \frac{\mu_d^2}{1+\mu_d}$, an upper bound on the critical point y_c in theorem 9.48 is obtained.

Theorem 9.53 *There is a critical point y_c separating a free phase from a ballistic phase in the limiting free energy $\mathcal{T}(y_d)$ of (vertically) pulled positive walks.*

For $y_d \leq y_c$, $\mathcal{T}(y_d) = \log \mu_d$.

If $y_d > y_c$, then

$$\max\{\log y_d, \log \mu_d\} < \mathcal{T}(y_d) \leq \log y_d + \log \mu_d.$$

Moreover, $\mathcal{T}(y_d)$ is asymptotic to $\log y_d$ for large y_d, and $y_c \in \Big[1, \frac{\mu_d^2}{1+\mu_d}\Big]$. □

Introduce a pulling force f in the model by setting $y_d = e^f$. The last theorem states that there is a critical force $f_c \in \big[0, \log\big(\mu_d^2/(1+\mu_d)\big)\big]$ such that, for $f > f_c$, the model is ballistic.

The specific heat of pulled walks is plotted as a function of $\log y_d$ in figure 9.23 for different values of n. These curves pass (almost) through a single point when $\log y_d = 0$. This is the location of the critical point y_c in the model.

An argument due to N Beaton establishes that $y_c = 1$. To see this, consider the pulled bridge generating function $B(y_d, t)$ (see the proof of theorem 9.49). Define the pulled half-space walk generating function $H(y_d, t)$.

A self-avoiding walk $\omega = \langle \omega_0, \omega_1, \ldots, \omega_n \rangle$ in $\mathbb{L}^d$ has *height* $h = \omega_n(d) - \omega_0(d)$. Let $c_n(h)$ be the number of self-avoiding walks of height h and length n from $\vec{0}$ in $\mathbb{L}^d$. The concatenation in figure 1.3 shows that

$$\sum_{h_1=-(n+m)}^{n+m} c_n(h-h_1) c_m(h_1) \geq c_{n+m}(h). \tag{9.108}$$

Let $T_n^c(y_d)$ be the partition function of these *pulled* walks. By multiplying the last inequality by ${y_d}^h$ and summing over h, $T_n^c(y_d) T_m^c(y_d) \geq T_{n+m}^c(y_d)$. By theorem A.1 in appendix A, this shows that the limit $\lim_{n\to\infty} \frac{1}{n} \log T_n^c(y_d) = \log \kappa$ exists and that $T_n^c(y_d) \geq \kappa^n$. If $y_d \geq 1$, then $\kappa \geq \mu_d > 1$.

The generating function of pulled walks is defined by

$$C^t(y_d, t) = \sum_{n=0}^{\infty} T_n^c(y_d)\, t^n. \tag{9.109}$$

Since $T_n^c(y_d) \geq \kappa^n$, this shows that $C^t(y_d,t) \geq \frac{1}{1-\kappa t}$, with the result that $C^t(y_d, t_c) = \infty$, where $\log t_c = -\log \kappa = -\lim_{n\to\infty} \frac{1}{n} \log T_n^c(y_d)$. This shows the following.

Lemma 9.54 *The generating function $C^t(y_d,t)$ of pulled walks is divergent when $\log t = -\log \kappa = -\lim_{n\to\infty} \frac{1}{n} \log T_n^c(y_d)$.* □

Introduce the generating functions of pulled half-space walks, pulled bridges and pulled prime bridges, namely, $H^t(y_d,t)$, $B^t(y_d,t)$ and $Q^t(y_d,t)$, respectively. The Hammersley-Welsh decomposition of theorem 7.9 can be directly applied. In particular, it follows that the generating functions $H^t(y_d,t)$ and $B^t(y_d,t)$ are divergent when $\log t = -\log \kappa = -\lim_{n\to\infty} \frac{1}{n} \log T_n^c(y_d)$.

The prime decomposition of bridges (see equation (7.15)) is equally valid, so that

$$B^t(y_d, t) = \frac{1}{1 - Q^t(y_d,t)}. \tag{9.110}$$

This shows that $B^t(y_d,t) < \infty$ if and only if $Q^t(y_d,t) < 1$. This defines the critical value of t, namely $t_c(y_d)$, in the model.

Notice that, for any $y_d > 1$, $Q^t(1,t) < Q^t(y_d,t) < 1$ if $B^t(y_d,t) < \infty$. Since $Q^t(y_d,t)$ is a strictly increasing function of t for given y_d, this shows for the critical point that $t_c(y_d) < t_c(1) = \frac{1}{\mu_d}$ for any $y_d > 1$.

Since $t_c(y_d)$ is the radius of convergence of the generating function of pulled half-space walks, this shows that

$$\lim_{n\to\infty} \tfrac{1}{n} \log T_n^h(y_d) = \mathcal{T}(y_d) = -\log t_c(y_d) > \log \mu_d = \mathcal{T}(1), \tag{9.111}$$

by theorem 9.48. Since $\mathcal{T}(y_d) = \log \mu_d$ for $0 < y_d \leq 1$, it follows that it is strictly increasing for $y_d > 1$, in particular because it is convex in $\log y_d$.

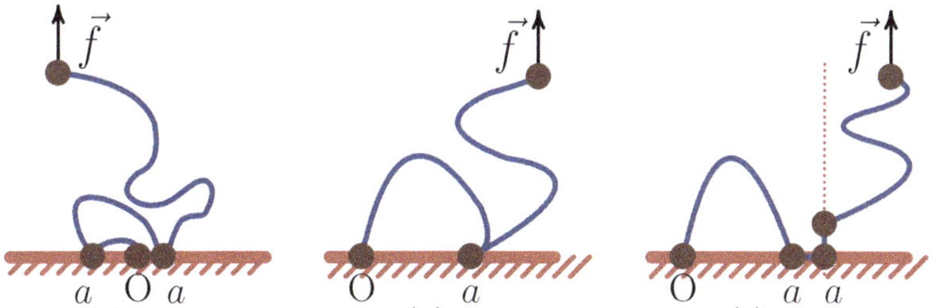

FIG. 9.24. Pulled adsorbing positive walks. (a) A pulled adsorbing walk has a last visit in the adsorbing plane. The basic structure is shown in (b); the walk is an adsorbing hoop followed by a pulled half-space walk. (c) A lower bound on the partition function is obtained by decomposing the walk into an adsorbing bridge and a tail in its last visit.

Theorem 9.55 *The limiting free energy of pulled positive walks, $\mathcal{T}(y_d)$, is strictly increasing when $y_d > 1$. Since $\mathcal{T}(y_d) = \log \mu_d$ for all $y_d \leq 1$, this shows that $y_c = 1$. For $y_d > 1$, the energy of the model (see equation (9.100)) is strictly positive, thus showing that the model is ballistic in this phase.* □

9.7.2 *Vertically pulled adsorbing walks*

Let the number of positive walks from $\vec{0}$ in $\mathbb{L}^d_+$ of length n, with v visits in $\partial\mathbb{L}^d_+$ and with endpoints at height h be $c^+_n(v,h)$. The partition function of a model of pulled adsorbing walks is

$$T^+_n(a, y_d) = \sum_{h=0}^{n} \sum_{v=0}^{n} c^+_n(v,h)\, a^v {y_d}^h, \tag{9.112}$$

where a is conjugate to visits, and y_d is conjugate to the height of the walk.

The basic decomposition of a pulled adsorbing walk is given in figure 9.24; it is an adsorbing hoop followed by a pulled half-space walk. A lower bound on $T^+_n(a, y_d)$ is obtained from figure 9.24(c), which has the structure of an adsorbing bridge followed by a tail.

Since the free energy of an adsorbing hoop is given in equation (9.5), and the free energy of a tail is given in equation (9.98), it follows that

$$\liminf_{n\to\infty} \tfrac{1}{n} \log T^+_n(a, y_d) \geq \max\{\mathcal{A}^+(a), \mathcal{T}(y_d)\}. \tag{9.113}$$

This follows because $T^+_n(a, y_d)$ is bounded from below by either an adsorbing bridge, or a pulled tail.

An upper bound is similarly obtained by considering figure 9.24(a). Decompose a pulled adsorbing positive walk into an adsorbing hoop followed by a pulled half-space walk at its last visit in $\partial\mathbb{L}^d_+$. By ignoring the intersections between the adsorbing hoop and the half-space walk, an upper bound on $T_n^+(a,y_d)$ is obtained from a convolution of $L_n^o(a)$ (the partition function of adsorbing hoops), and $T_n^h(y_d)$ (the partition function of pulled half-space walks). This gives

$$T_n^+(a,y_d) \leq \sum_{m=0}^{n} L_m^o(a) T_{n-m}^h(y_d). \tag{9.114}$$

Take logarithms, divide by n and take $n \to \infty$. It follows from equations (9.6) and (9.98) that

$$\limsup_{n\to\infty} \tfrac{1}{n} \log T_n^+(a,y_d) \leq \max\{\mathcal{A}^+(a), \mathcal{T}(y_d)\}. \tag{9.115}$$

This establishes the following theorem (see corollary 9.3 and theorem 9.55).

Theorem 9.56 *The limiting free energy of pulled adsorbing half-space walks is given by* $\mathcal{T}(a,y_d) = \lim_{n\to\infty} \frac{1}{n} \log T_n^+(a,y_d) = \max\{\mathcal{A}^+(a), \mathcal{T}(y_d)\}$.

In addition, $\mathcal{T}(a,y_d) = \log \mu_d$ *if* $a \leq 1$, *and* $y_d \leq 1$, *but* $\mathcal{T}(a,y_d) > \mu_d$ *if either* $a > a_c^+$ *or* $y > 1$. □

By the above, $\mathcal{T}(a,y_d) = \max\{\mathcal{A}^+(a), \mathcal{T}(y_d)\}$, where $\mathcal{T}(y_d)$ has the properties in theorems 9.53 with $y_c = 1$ (by theorem 9.55). Similarly, $\mathcal{A}^+(a)$ has the properties given in theorem 9.2 and corollary 9.3, and the critical point a_c^+. By corollary 9.17, $a_c^+ > 1$.

In addition, $\mathcal{T}(a,y_d) = \log \mu_d$ if $(a,y_d) \in [0,a_c^+) \times [0,1]$. If both $y_d \geq 1$ and $a < a_c^+$, then $\mathcal{T}(a,y_d) = \mathcal{T}(y_d)$; and if both $y_d \leq 1$ and $a \geq a_c^+$, then $\mathcal{T}(a,y_d) = \mathcal{A}^+(a)$. This shows that there are critical lines in the phase diagram of this model.

First, there is an adsorption transition at $a = a_c^+$ for $0 \leq y_d \leq 1$, and there is a ballistic transition at $y_d = 1$ for $0 \leq a \leq a_c^+$.

Finally, if both $y > 1$ and $a > a_c^+$, then there is a critical curve determined by $\mathcal{T}(y_d) = \mathcal{A}^+(a)$. Since $\mathcal{T}(y_d)$ is convex and strictly increasing for $y_d > 1$, its inverse function is concave and strictly increasing. This shows that the critical curve in the model is given by

$$y_c(a) = (\mathcal{T})^{-1} \circ \mathcal{A}^+(a), \quad \text{for } a > a_c^+. \tag{9.116}$$

The critical curve $y_c(a)$ is a monotonic function of $a > a_c^+$ with $y_c(a) > 1$. Since both $\mathcal{T}(y)$ and $\mathcal{A}^+(a)$ are finite functions, $y_c(a)$ is finite for every $a > a_c^+$.

By using the bounds in theorem 9.2 for $a > 1$, and theorem 9.48 for $y_d > 1$, it follows that $\mathcal{A}^+(a) > \mathcal{T}(y_d)$ if $y_d < \frac{\mu_{d-1}}{\mu_d} a$, and $\mathcal{A}^+(a) < \mathcal{T}(y_d)$ if $y_d > \mu_d a$. This shows that $y_c(a)$ is confined to a wedge in the ay_d-plane given by

$$\tfrac{\mu_{d-1}}{\mu_d} a \leq y_c(a) \leq \mu_d a. \tag{9.117}$$

Take logarithms of this, divide by $\log a$ and let $a \to \infty$ to see that

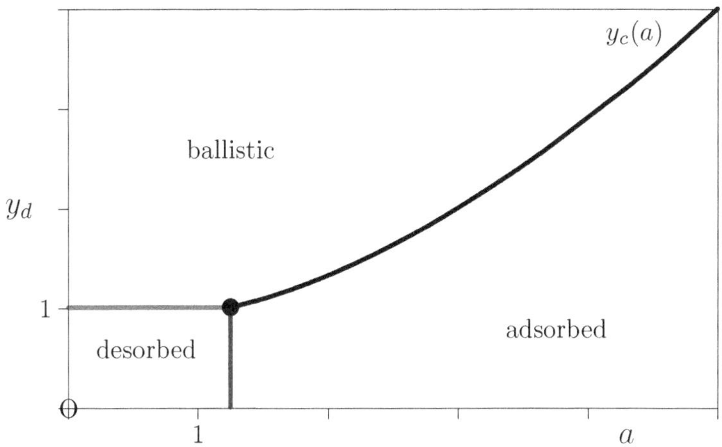

FIG. 9.25. The phase diagram of pulled adsorbing positive walks. For small values of a and y_d, the walk is in a desorbed and free phase with the usual self-avoiding walk statistics. There are critical lines separating this phase from an adsorbed and a ballistic phase. The phase boundary $y_c(a)$ is the locus of a line of first order transitions separating the adsorbed and ballistic phases; $\log y_c(a)$ is asymptotic to $\log a$.

$$\lim_{a\to\infty} \frac{\log y_c(a)}{\log a} = 1. \tag{9.118}$$

Let $\delta > 0$. Compare the bounds in theorems 9.7 and 9.52 to see that, for large enough values of a,

$$\mu_{d-1} a \leq (y_c(a))^{1+\delta}, \quad \text{and } y_c(a) \leq \mu_{d-1} a^{1+\delta}. \tag{9.119}$$

Since $y_c(a)$ approaches infinity with increasing a, for every $\delta > 0$, there is an a_δ such that these inequalities are true for all $a > a_\delta$. This, in particular, shows that, for any $\delta > 0$,

$$\liminf_{a\to\infty} \frac{(y_c(a))^{1+\delta}}{a} \geq \mu_{d-1}, \quad \text{and } \limsup_{a\to\infty} \frac{y_c(a)}{a^{1+\delta}} \leq \mu_{d-1}. \tag{9.120}$$

From these results it may be conjectured that $\lim_{a\to\infty} \frac{y_c(a)}{a} = \mu_{d-1}$.

The phase diagram is illustrated in figure 9.25. The transition between the ballistic and adsorbed phases is a first order transition. To see this, consider the order parameter $\mathcal{H}(y_d) = y_d \frac{\partial}{\partial y_d} \mathcal{F}^p(a, y_d)$, which is the expected height of the endpoint. For values of (a, y_d) in the desorbed or adsorbed phases, $\mathcal{H}(y_d) = 0$. However, in the ballistic phase $\mathcal{H}(y_d) > 0$, and by theorem 9.56, $\mathcal{H}(y_d)$ is independent of a in this phase. On approach to the critical curve $y_c(a)$ from the ballistic phase with fixed y_d, $\mathcal{H}(y_d) > 0$ is a positive constant, and it steps down to $\mathcal{H}(y_d) = 0$ into the adsorbed phase. Hence $\mathcal{H}(y_d)$ is discontinuous along $y_c(a)$, and this is the locus of a curve of first order transitions. This also shows that $y_c(a)$ is a transition where the force overcomes the adsorption of the walk and pulls it completely off the adsorbed plane into a ballistic phase.

10

ADSORBING WALKS IN THE HEXAGONAL LATTICE

The growth constant of self-avoiding walks in the hexagonal lattice was determined over thirty years ago by Nienhuis [436], and the location of the adsorption critical point was found by Batchelor and Yung [25] in 1995. These exact results (which are not rigorous) were proven more recently by Duminil-Copin and Smirnov [158] and Beaton *et al.* [28], respectively.

10.1 Walks and half-space walks in the hexagonal lattice

Let $\mathbb{H}$ be the two-dimensional hexagonal lattice oriented as shown in figure 10.1. The *half-hexagonal lattice* $\mathbb{H}_+$ is obtained by placing $\vec{0}$ in the midpoint of a vertically oriented edge in $\mathbb{H}$ and drawing a horizontal boundary $\partial\mathbb{H}_+$ through $\vec{0}$ to bisect a sequence of vertical edges (this is illustrated in figure 10.1).

A half-space walk in the hexagonal lattice starts in $\vec{0}$ (in the midpoint of an edge), steps into $\mathbb{H}_+$ and terminates in the midpoint of an edge (see figure 10.1).

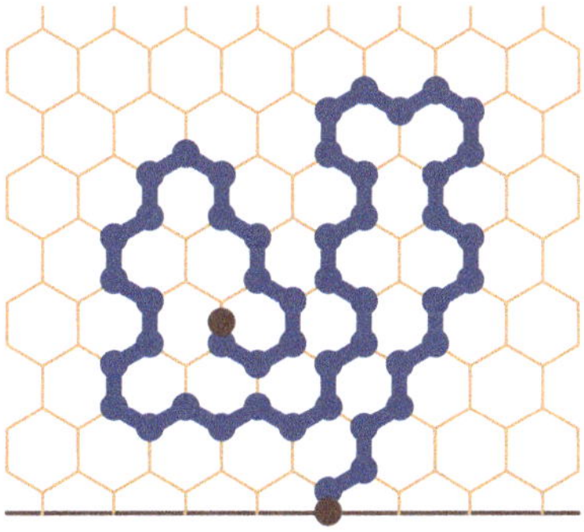

FIG. 10.1. A self-avoiding walk from the origin in the half-hexagonal lattice. The convention is that the origin is located at the midpoint of a vertical edge. The walk also terminates in the midpoint of an edge.

Denote the number of half-space walks from $\vec{0}$ of length n *steps* in $\mathbb{H}_+$ by $\mathfrak{h}_n$. Then $\mathfrak{h}_0 = 1$; $\mathfrak{h}_1 = 1$; $\mathfrak{h}_2 = 2$; $\mathfrak{h}_3 = 4$; and so on.

The number of self-avoiding walks of length n in $\mathbb{H}$ from $\vec{0}$ is denoted $\mathfrak{c}_n$ and is similarly defined. These walks terminate in the midpoint of an edge as well. Then $\mathfrak{c}_1 = 2$; $\mathfrak{c}_2 = 4$; $\mathfrak{c}_3 = 8$; and so on. Walks in $\mathbb{H}$ are in one-to-one correspondence with walks in the brick lattice. For example, the half-space walk in figure 10.1 may be represented by the walk in figure 10.2.

The Statistical Mechanics of Interacting Walks, Polygons, Animals and Vesicles, 2nd edition, E.J. Janse van Rensburg. © E.J. Janse van Rensburg. Published in 2015 by Oxford University Press.

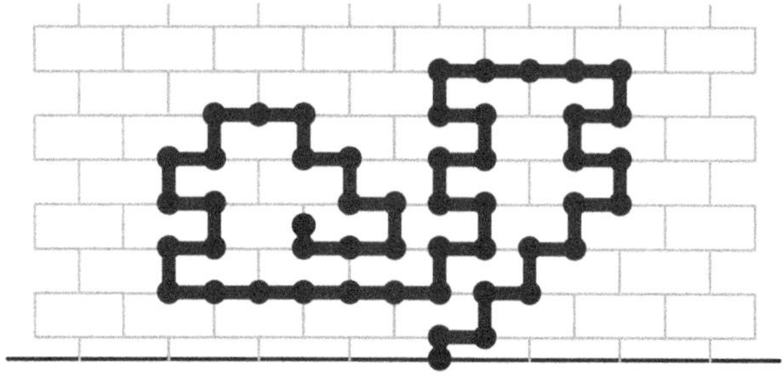

FIG. 10.2. A self-avoiding walk in the brick lattice with endpoints which are midpoints of edges. Since this walk is confined to the upper half-lattice, it is a positive walk. Walks in the brick lattice are in one-to-one correspondence with walks in $\mathbb{H}$. Since the brick lattice is obtained from the square lattice by deleting edges, the constructions for square lattice walks can be used in the brick and hexagonal lattices.

The brick lattice is obtained by deleting edges in $\mathbb{L}^2$ so that every walk in the brick lattice is also a square lattice walk. This gives unambiguous definitions to walks in the hexagonal lattice: For example, the definition of a bridge now extends to the hexagonal lattice (with minor modification to account for the half-edge at the beginning and end of walks). Moreover, the constructions on square lattice self-avoiding walks, amongst them the Hammersley-Welsh bridge decomposition described in section 7.1.2, can be directly applied to half-space walks in the hexagonal lattice.

The existence of a growth constant for self-avoiding walks in $\mathbb{H}$ is a corollary of theorem 1.1:

Theorem 10.1 *There exists a connective constant $\kappa_h = \log \mu_h$ defined by*

$$\lim_{n\to\infty} \tfrac{1}{n} \log \mathfrak{c}_n = \inf_{n>0} \tfrac{1}{n} \log \mathfrak{c}_n = \kappa_h.$$

The growth constant μ_h is bounded by $\sqrt{2} \leq \mu_h \leq 2$, and $\mathfrak{c}_n \geq \mu_h^n$. □

Determining the bounds on $\mathfrak{c}_n$ is not difficult. The growth constant was conjectured to have exact value $\mu_h = \sqrt{2+\sqrt{2}}$ in reference [436].

The number of bridges (from the origin) in the hexagonal lattice of length n steps is $\mathfrak{b}_n$. Notice that $\mathfrak{b}_n \leq \mathfrak{h}_n \leq \mathfrak{c}_n$, by inclusion. The bridge decomposition of lemma 7.1 and the bound of theorem 7.2 apply mutatis mutandis to the situation here, with the result that

$$\lim_{n\to\infty} \tfrac{1}{n} \log \mathfrak{c}_n = \lim_{n\to\infty} \tfrac{1}{n} \log \mathfrak{h}_n = \lim_{n\to\infty} \tfrac{1}{n} \log \mathfrak{b}_n = \log \mu_h. \tag{10.1}$$

Define the generating functions

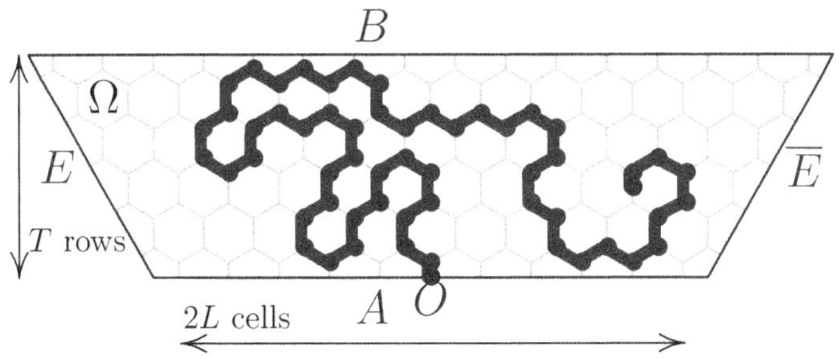

FIG. 10.3. A trapezoidal region Ω with width $2L$ cells and height T rows in the hexagonal lattice. The origin O is located in the midpoint of the bottom boundary. Walks in this region start in the bottom boundary in a midpoint of an edge and end in the midpoint of an edge.

$$\mathfrak{B}(t) = \sum_{n=0}^{\infty} \mathfrak{b}_n\, t^n;\ \mathfrak{H}(t) = \sum_{n=0}^{\infty} \mathfrak{h}_n\, t^n;\ \text{ and } \mathfrak{C}(t) = \sum_{n=0}^{\infty} \mathfrak{c}_n\, t^n \tag{10.2}$$

of bridges, half-space walks and walks, respectively, in $\mathbb{H}$. It follows that $\mathfrak{B}(t) \leq \mathfrak{H}(t) \leq \mathfrak{C}(t)$, and, by theorem 7.9, $t\,\mathfrak{C}(t) \leq (\mathfrak{H}(t))^2$, and $\log \mathfrak{H}(t) \leq \mathfrak{B}(t) - 1$. That is, $\mathfrak{C}(t) < \infty$ if and only if $\mathfrak{B}(t) < \infty$ (and $\mathfrak{H}(t) < \infty$).

Note that $\mathfrak{C}(t_c) = \infty$ by theorem 10.1 and that $t_c = \mu_h^{-1}$.

The *span* of a bridge (oriented vertically) is the vertical distance between its endpoints. Define $\mathfrak{B}_T(t)$ to be the generating function of bridges of span T. By theorem 7.9, it follows that

$$\mathfrak{H}(t) \leq \prod_{T=1}^{\infty} (1 + \mathfrak{B}_T(t)) < \infty \text{ if and only if } t < t_c. \tag{10.3}$$

10.1.1 *The generating functions in finite domains*

The hexagonal lattice $\mathbb{H}$ can be parameterised in the complex plane $\mathbb{C}$. Then, every vertex $\vec{v} \in \mathbb{H}$ is represented by a $q \in \mathbb{C}$. This allows the definition of a parafermionic operator related to the winding of walks [158].

Choose a finite simply connected domain Ω in the complex plane $\mathbb{C}$ (including a part of $\mathbb{H}$; see figure 10.3). The boundary of Ω is denoted $\partial\Omega$, and the interior of Ω is $V(\Omega) = \Omega \setminus \partial\Omega$.

The *winding* of a hexagonal lattice walk is the sum of the signed curvature along its length, starting from $\vec{0}$. This is also called the *rotation* of the walk. For example, the winding of an oriented polygon is $\pm 2\pi$, and the winding of a loop in $\mathbb{H}_+$ is $\pm\pi$ (see section 7.1.4).

Let ω_{ab} be a walk from $a \in \partial\Omega$ to a point $b \in \Omega$ (and recall that a and b are the midpoints of edges). Then the winding of ω_{ab} is denoted by $w_\omega(a,b)$. The *parafermionic operator* for (a,b) is defined by

FIG. 10.4. The walks (a) ω_1 and (b) ω_2 may be paired if they occupy the same vertices in the lattice and almost close a polygon but in opposite directions.

$$f_a(b) \equiv f(a,b,t,\sigma) = \sum_{\omega \in \Omega : a \to b} e^{-i\sigma w_\omega(a,b)} t^{\ell(\omega)}, \tag{10.4}$$

where $\ell(\omega)$ is the length of ω. Define

$$\tau_c^{-1} = \sqrt{2+\sqrt{2}}. \tag{10.5}$$

The parafermionic operator satisfies a symmetric identity.

Lemma 10.2 *If $t = \tau_c$, and if $\sigma = \frac{5}{8}$, then $f_a(b)$ satisfies the identity*

$$(p-v) f_a(p) + (q-v) f_a(q) + (r-v) f_a(r) = 0$$

for every $v \in V(\Omega)$, for $a \in \partial\Omega$, and where $\{p,q,r\}$ are the midpoints of the three edges incident with v.

Proof Let v and $\{p,q,r\}$ be as defined. Let the contribution of a walk ω to the summation in the identity be denoted by $C(\omega)$. The method of proof is to explicitly compute the contribution of pairs or triples of walks to the sum in equation (10.4).

If a walk ω ends on p next to v, then its contribution to the left-hand side of the identity is $(p-v)e^{-i\sigma w_\omega(a,p)} \tau_c^{\ell(\omega)}$.

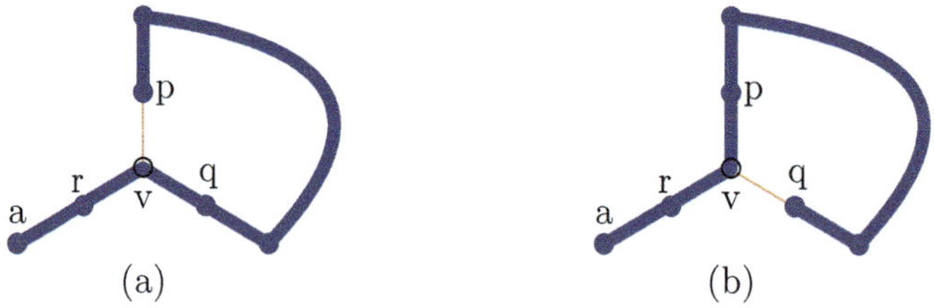

FIG. 10.5. A schematic representation of the walks in figure 10.4.

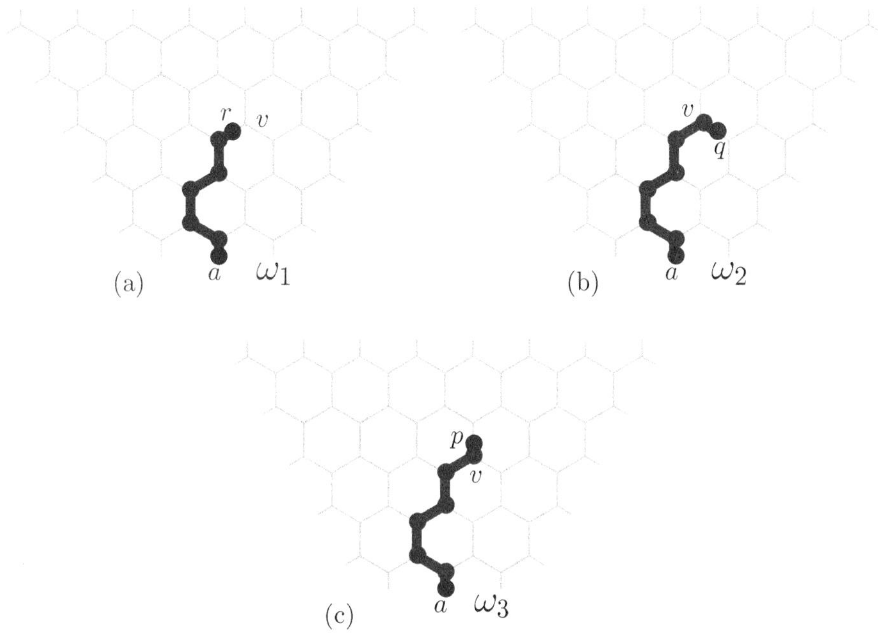

FIG. 10.6. An example of a triple set of walks from a and terminating in the midpoint of an edge incident with v.

Consider walks ending in p or q and passing through r and v. Let ω_1 be one such walk and suppose that it passes through the vertices in sequence $a \to r \to v \to q \to p$. Let ω_2 be the walk passing through the same vertices as ω_1 but in the sequence $a \to r \to v \to p \to q$. (That is, ω_2 is obtained by reversing the orientation of ω_1 between q and p and then moving a half-edge $\langle v \sim q \rangle$ to $\langle v \sim p \rangle$.) A pair (ω_1, ω_2) is illustrated in figure 10.4 and schematically represented in figure 10.5. The contribution of this pair is

$$C(\omega_1) + C(\omega_2) = (p - v)e^{-i\sigma w_{\omega_1}(a,p)}\tau_c^{\ell(\omega_1)} + (q - v)e^{-i\sigma w_{\omega_1}(a,q)}\tau_c^{\ell(\omega_2)}.$$

The points $\{p, q, r\}$ and v are points in the complex plane. It follows that $(p - v) = (r - v)e^{-2\pi i/3}$, and $(q - v) = (r - v)e^{2\pi i/3}$. Continuing the path to make a polygon of winding 2π or -2π gives $w_{\omega_1}(a,p) = w_{\omega_1}(a,r) - \frac{4}{3}\sigma\pi i$. Similarly, $w_{\omega_2}(a,q) = w_{\omega_1}(a,r) + \frac{4}{3}\sigma\pi i$ (note that $w_{\omega_1}(a,r) = w_{\omega_2}(a,r)$). Substituting these into the above gives

$$\mathcal{C}(\omega_1) + C(\omega_2) = (r - v)e^{-i\sigma w_{\omega_1}(a,r)}\tau_c^{\ell(\omega_1)}\left(e^{-2\pi i/3}e^{-4\sigma\pi i/3} + e^{2\pi i/3}e^{4\sigma\pi i/3}\right).$$

Putting $\sigma = \frac{5}{8}$ gives identically zero.

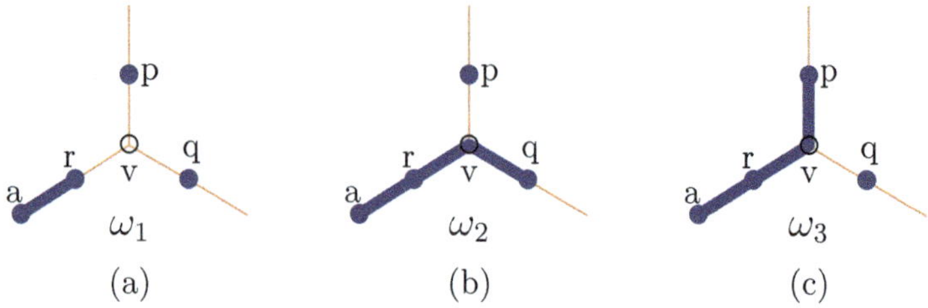

FIG. 10.7. A schematic illustration of three walks comprising a triple.

Consider next the case of walks which terminate in one of $\{p, q, r\}$ but do not (short of a half-step) complete a polygon rooted in v. These walks are grouped in triples $(\omega_1, \omega_2, \omega_3)$ (see figure 10.6 and the schematic illustration in figure 10.7). The walk ω_1 terminates in r (a half-step short of v, and it does not pass through v). The walk ω_2 passes through r and v and terminates in q (but avoids p). Similarly, the walk ω_3 passes through r and v and terminates in p (but avoids q). The walk ω_1 is one step shorter than ω_2 or ω_3. By using arguments similar to those used in the case illustrated in figure 10.4, the combined contribution of these walks is

$$\begin{aligned}&C(\omega_1) + C(\omega_2) + C(\omega_3)\\&\quad = (r-v)e^{-i\sigma w_{\omega_1}(a,r)}\tau_c^{\ell(\omega_1)}\left(1 + \tau_c e^{2\pi i/3}e^{\sigma\pi i/3} + \tau_c e^{-2\pi i/3}e^{-\sigma\pi i/3}\right).\end{aligned}$$

Substitute $\sigma = \frac{5}{8}$ and substitute $\tau_c^{-1} = \sqrt{2+\sqrt{2}}$ to see that this is 0. This completes the proof. □

A corollary of lemma 10.2 is that $f_a(b)$ satisfies a discrete (complex) divergence theorem in Ω and on $\partial\Omega$ so that the contour integral along $\partial\Omega$ vanishes. Suppose $\partial\Omega$ is a simple piecewise linear and closed curve bounding a simply connected domain Ω in $\mathbb{C}$ and that $\partial\Omega$ intersects lattice edges $\langle p(w) \sim p(v)\rangle$ in points $p \in \partial\Omega$ exactly once so that $p(v) \in \Omega$. Then

$$\sum_{p\in\partial\Omega} (p - p(v))\, f_a(p) = 0. \tag{10.6}$$

This is a discrete version of the divergence theorem in the complex plane and it suggests that the parafermionic operator is discrete holomorphic in the complex plane. Showing that it is holomorphic with a conformally invariant scaling limit would be sufficient to prove that the scaling limit is a Schramm-Loewner evolution [285, 506, 515].

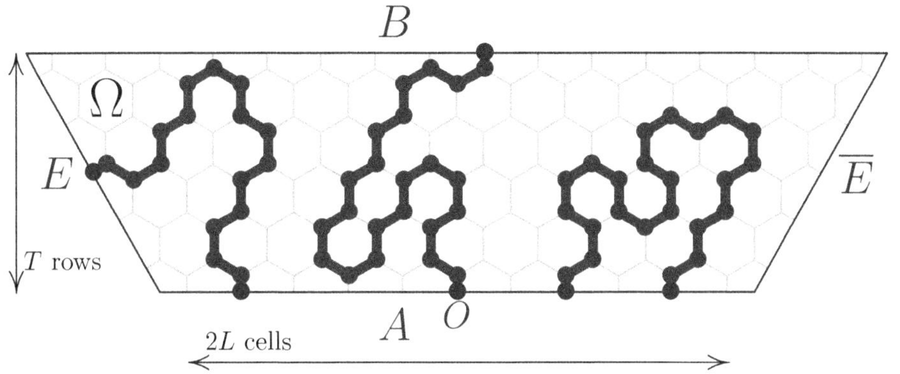

FIG. 10.8. Walks from the bottom boundary A of Ω. The generating function of walks from $\vec{0}$ in A to $A \setminus \{0\}$ is $A_\Omega^{T,L}(t)$. The generating function of walks from $\vec{0}$ in A to B is $B_\Omega^{T,L}(t)$, and the generating function of walks from the origin in A to $E \cup \overline{E}$ is $E_\Omega^{T,L}(t)$.

10.1.2 *The growth constant μ_h equals $\sqrt{2+\sqrt{2}}$*

The parafermionic operator and lemma (10.2) can be used to show that $\mu_h = \sqrt{2+\sqrt{2}}$. The proof is due to Duminil-Copin and Smirnov [158].

Let Ω be the trapezoidal region of $\mathbb{H}$ in figure 10.3. The point O is the midpoint of the bottom part A of $\partial\Omega$ and is also located at the origin in $\mathbb{C}$, with the real axis horizontally along line segment A.

The boundary $\partial\Omega$ consists of the two horizontal line segments A and B, with imaginary parts $\mathfrak{Im}(A) = 0$ and $\mathfrak{Im}(B)$ a height T rows in the hexagonal lattice above the real axis. The slanted line segments of $\partial\Omega$ are labelled by E and $\overline{E}$, so $\partial\Omega = A \cup E \cup B \cup \overline{E}$.

The length of A is $2L$ widths of a cell in $\mathbb{H}$.

There are several different walks between the boundaries of Ω, and these are illustrated in figure 10.8. Walks from boundary A may terminate in E or $\overline{E}$, or may cross Ω to terminate in B. Alternatively, a path may also return to A.

Define the generating functions

$$A_\Omega^{T,L}(t) = \sum_{\omega:0\to A\setminus\{0\}} t^{\ell(\omega)}; \; B_\Omega^{T,L}(t) = \sum_{\omega:0\to B} t^{\ell(\omega)}; \text{ and } E_\Omega^{T,L}(t) = \sum_{\omega:0\to E\cup\overline{E}} t^{\ell(\omega)},$$

where the summations are over walks confined to Ω. These are also called the generating functions of *arches*, *bridges* and *stretched walks*, respectively. Arches are walks that both start from and terminate in the boundary A. Bridges are walks from A to B, while stretched walks are walks from A to either of E or $\overline{E}$.

As the winding of a walk ending in B is 0, it follows from equation (10.4) that

$$B_\Omega^{T,L}(t) = \sum_{p\in B} f_O(p). \tag{10.7}$$

Similarly, the winding of a path ending in E is $\frac{\pi}{3}$. Multiplying by $-\sigma$ gives the angle $-\frac{5\pi}{24}$ in equation (10.4). The winding of a path ending in $\overline{E}$ is $-\frac{\pi}{3}$. After multiplication by $-\sigma$, this becomes $\frac{5\pi}{24}$. Notice that $\sum_{\omega:0\to E} t^{\ell(\omega)} = \sum_{\omega:0\to\overline{E}} t^{\ell(\omega)}$ by symmetry. Hence,

$$\begin{aligned}\sum_{p\in E} f_O(p) &= e^{-5\pi i/24} \sum_{\omega:0\to E} t^{\ell(\omega)} \\ &= \tfrac{1}{2} e^{-5\pi i/24} \sum_{\omega:0\to E\cup\overline{E}} t^{\ell(\omega)} = \tfrac{1}{2} e^{-5\pi i/24} E_\Omega^{T,L}(t). \end{aligned} \tag{10.8}$$

Similarly,

$$\sum_{p\in\overline{E}} f_O(p) = \tfrac{1}{2} e^{5\pi i/24} E_\Omega^{T,L}(t). \tag{10.9}$$

Next, consider walks from O returning to O. Each walk may be traversed in either of two directions, winding π in the one direction and $-\pi$ in the other. This shows that these cancel in pairs; thus, only the empty walk at O survives. Hence, $f_O(0) = \sum_{\omega:0\to 0} e^{-i\sigma w_\omega(0,0)} t^{\ell(\omega)} = 1$; therefore,

$$\sum_{p\in A} f_O(p) = 1 + \sum_{p\in A\setminus\{0\}} f_O(p). \tag{10.10}$$

It remains to compute the last summation. The walks ending towards the left of O in A have winding π, and those to the right have winding $-\pi$. Since $\sigma = \frac{5}{8}$,

$$\sum_{p\in A\setminus\{0\}} f_O(p) = \tfrac{1}{2}\left(e^{5\pi i/8} + e^{-5\pi i/8}\right) \sum_{p\in A\setminus\{0\}} t^{\ell(\omega)} = -\cos(\tfrac{3\pi}{8})\, A_\Omega^{T,L}(t). \tag{10.11}$$

This shows that

$$\sum_{p\in A} f_O(p) = 1 - \cos(\tfrac{3\pi}{8})\, A_\Omega^{T,L}(t). \tag{10.12}$$

Theorem 10.3 *Let Ω be the trapezoidal region in figure 10.3, with boundary $\partial\Omega$. If $t = \tau_c$, and if $\sigma = \frac{5}{8}$, then*

$$A_\Omega^{T,L}(\tau_c)\cos(\tfrac{3\pi}{8}) + B_\Omega^{T,L}(\tau_c) + E_\Omega^{T,L}(\tau_c)\cos(\tfrac{\pi}{4}) = 1.$$

Proof Proceed by applying the result in equation (10.6) to the trapezoidal region in figure 10.3. If $p \in A$, then $p - p(v) = -i$. Similarly, if $p \in B$, then

$p - p(v) = i$; for $p \in E$, $p - p(v) = ie^{2\pi i/3}$, and, if $p \in \overline{E}$, then $p - p(v) = ie^{-2\pi i/3}$. If $t = \tau_c$, then simplification gives

$$-\sum_{p \in A} f_a(p) + \sum_{p \in B} f_a(p) + e^{2\pi i/3} \sum_{p \in E} f_a(p) + e^{-2\pi i/3} \sum_{p \in \overline{E}} f_a(p) = 0. \quad (10.13)$$

Substitute $a = 0$ and use the results in equations (10.7), (10.8), (10.9) and (10.12). This completes the proof. □

The generating functions $A_\Omega^{T,L}(\tau_c)$, $B_\Omega^{T,L}(\tau_c)$ and $E_\Omega^{T,L}(\tau_c)$ are finite polynomials in τ_c. Moreover, $A_\Omega^{T,L}(\tau_c)$ and $B_\Omega^{T,L}(\tau_c)$ are positive and non-decreasing with L, while $E_\Omega^{T,L}(\tau_c)$ is non-negative and, by theorem 10.3, non-increasing. Thus, these generating functions are bounded and monotone in L, so the limits

$$\lim_{L \to \infty} A_\Omega^{T,L}(\tau_c) = \mathfrak{A}_T(\tau_c) \text{ and } \lim_{L \to \infty} B_\Omega^{T,L}(\tau_c) = \mathfrak{B}_T(\tau_c), \quad (10.14)$$

as well as the limit

$$\lim_{L \to \infty} E_\Omega^{T,L}(\tau_c) = \mathfrak{E}_T(\tau_c) \geq 0 \quad (10.15)$$

exist and are finite. Observe that $\mathfrak{B}_T(t)$ is the generating function of bridges of span T defined just before equation (10.3).

Taking $L \to \infty$ in theorem 10.3 gives the identity

$$\mathfrak{A}_T(\tau_c)\cos(\tfrac{3\pi}{8}) + \mathfrak{B}_T(\tau_c) + \mathfrak{E}_T(\tau_c)\cos(\tfrac{\pi}{4}) = 1. \quad (10.16)$$

A consequence of this relation is the following lemma.

Lemma 10.4 *The growth constant* $\mu_h \geq \sqrt{2 + \sqrt{2}}$.

Proof It is enough to show that the generating function $\mathfrak{C}(t)$ of half-space hexagonal lattice walks (see equation (10.2)) is divergent if $t = \tau_c$; this means that $\mu_h^{-1} = t_c \leq \tau_c$.

Consider equation (10.16) and note that $\mathfrak{E}_T(\tau_c) \geq 0$. If $\mathfrak{E}_T(\tau_c) > 0$, then

$$\mathfrak{C}_T(\tau_c) \geq \sum_{L=0}^{\infty} E_\Omega^{T,L}(\tau_c) \geq \sum_{L=0}^{\infty} \mathfrak{E}_T(\tau_c) = \infty,$$

in particular since $E_\Omega^{T,L}(\tau_c) \geq \mathfrak{E}_T(\tau_c)$. In this case, $\mathfrak{C}(\tau_c)$ is divergent.

On the other hand, suppose that $\mathfrak{E}_T(\tau_c) = 0$. Equation (10.16) simplifies to

$$\mathfrak{A}_T(\tau_c)\cos(\tfrac{3\pi}{8}) + \mathfrak{B}_T(\tau_c) = 1. \quad (10.17)$$

A loop ω which contributes to $\mathfrak{A}_{T+1}(\tau_c)$ but not to $\mathfrak{A}_T(\tau_c)$ has to visit a vertex v next to B in Ω of height $T+1$ cells. This loop may be partitioned into two bridges, both of which span Ω of width $T+1$ from A to B, by cutting it in v and adding two half-edges to joining the resulting walks to B. Thus,

$$\mathfrak{A}_{T+1}(\tau_c) - \mathfrak{A}_T(\tau_c) \leq \tau_c \left(\mathfrak{B}_{T+1}(\tau_c)\right)^2. \tag{10.18}$$

This gives the following claim.

CLAIM: The value of $T\mathfrak{B}_T(\tau_c) \geq \min\left\{\mathfrak{B}_1(\tau_c), \left(\tau_c \cos(\frac{3\pi}{8})\right)^{-1}\right\} > 0$.

PROOF OF CLAIM: To see this, combine equation (10.18) with equation (10.17) for T and for $T+1$. This gives

$$\begin{aligned}0 &= (\mathfrak{A}_{T+1}(\tau_c) - \mathfrak{A}_T(\tau_c)) \cos(\tfrac{3\pi}{8}) + (\mathfrak{B}_{T+1}(\tau_c) - \mathfrak{B}_T(\tau_c)) \\ &\leq \tau_c \cos(\tfrac{3\pi}{8}) \left(\mathfrak{B}_{T+1}(\tau_c)\right)^2 + (\mathfrak{B}_{T+1}(\tau_c) - \mathfrak{B}_T(\tau_c)).\end{aligned}$$

This may be written as

$$\tau_c \cos(\tfrac{3\pi}{8}) \left(\mathfrak{B}_{T+1}(\tau_c)\right)^2 + \mathfrak{B}_{T+1}(\tau_c) \geq \mathfrak{B}_T(\tau_c).$$

The proof of the claim is completed by induction using this inequality. △

Hence, by this claim,

$$\mathfrak{C}(\tau_c) \geq \sum_{T=0}^{\infty} \mathfrak{B}_T(\tau_c) \geq \sum_{T=0}^{\infty} \tfrac{1}{T} \min\left\{\mathfrak{B}_1(\tau_c), \left(\tau_c \cos(\tfrac{3\pi}{8})\right)^{-1}\right\} = \infty.$$

This completes the proof. □

In view of lemma 10.4, it only remains to show that $\mu_h \leq \sqrt{2+\sqrt{2}}$. By equation (10.3), it is enough to show that $\mathfrak{H}(t) < \infty$ if $t \leq \tau_c$.

By equation (10.16), $\mathfrak{B}_T(\tau_c) \leq 1$. Hence, if $\mathfrak{b}_{T,n}$ is the number of bridges of length n and span T, then, for $t \leq \tau_c$,

$$\begin{aligned}\mathfrak{B}_T(t) &= \sum_{n=0}^{\infty} \mathfrak{b}_{T,n}\, t^n = \sum_{n=0}^{\infty} \mathfrak{b}_{T,n}\, \tau_c^n \left(\tfrac{t}{\tau_c}\right)^n \\ &\leq \left(\tfrac{t}{\tau_c}\right)^T \sum_{n=0}^{\infty} \mathfrak{b}_{T,n}\, \tau_c^n = \left(\tfrac{t}{\tau_c}\right)^T \mathfrak{B}_T(\tau_c) \leq \left(\tfrac{t}{\tau_c}\right)^T,\end{aligned}$$

since the shortest bridge will have length T.

This, in particular, shows that $\mathfrak{H}(t) \leq \prod_{T=1}^{\infty} (1+\mathfrak{B}_T(t)) < \infty$ if $t < \tau_c$; therefore, $\mathfrak{C}(t)$ is finite if $t < \tau_c$. Thus, $\mu_h \leq \sqrt{2+\sqrt{2}}$.

Theorem 10.5. (Duminil-Copin and Smirnov [158]) *The growth constant of self-avoiding walks in the hexagonal lattice is* $\mu_h = \sqrt{2+\sqrt{2}}$. □

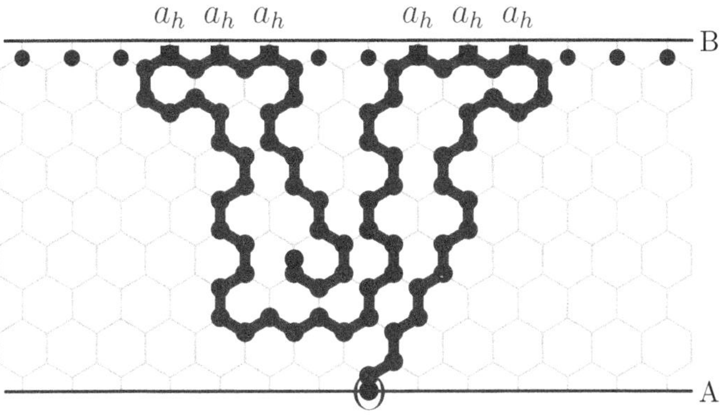

FIG. 10.9. A walk from $\vec{0}$ in a strip (or slit) in $\mathbb{H}$. The walk crosses over the strip and adsorbs on the opposite boundary with fugacity a_h.

10.2 Adsorption of walks in a slit in the hexagonal lattice

The model is presented in figure 10.9. A self-avoiding walk from $\vec{0}$ in a strip in $\mathbb{H}$ adsorbs on the opposite boundary (or *hard wall*) with fugacity a_h. As before, let $\mathbb{H}_+$ be the half-hexagonal lattice with $\vec{0}$ located in $\partial\mathbb{H}_+$.

Let Ω be defined as before (see figure 10.8) and let its boundary be $\partial\Omega = A \cup B \cup E \cup \overline{E}$. Let the length of A be $2L$. If $L \to \infty$, the Ω is a *horizontal strip or slit* of width T (where T is the vertical distance between A and B).

Suppose ω is a walk from $\vec{0}$ in Ω of length n. Then ω adsorbs onto B with activity a_h, as illustrated in figure 10.9. Taking first the limit $L \to \infty$ and then $n \to \infty$. Then the limit $T \to \infty$ gives an adsorbing walk in $\mathbb{H}_+$. This walk undergoes an adsorption transition at a critical value a_c^h of a_h.

If $L = \infty$, then this an adsorbing walk in a slit of width T in $\mathbb{H}$. Proving the existence of a thermodynamic limit in this model follows the same methods as in the proof of theorem 9.1. Taking $T \to \infty$ shows that a_c^h is the critical point where the walk adsorbs, as in theorem 9.2. In this case the limiting free energy of the model is similar to figure 9.4.

A conjecture by Batchelor and Yung [25] is that

$$a_c^h = 1 + \sqrt{2}. \tag{10.19}$$

This was proven by Beaton *et al.* [28]. In this section a brief outline of the method of proof is given.

Define the winding of the walk as in section 10.1 and denote the winding of a walk ω_{ab} from a point a to a point b by $w_\omega(a, b)$ (recall that points in Ω are parameterised by complex numbers). The parafermionic operator of this model in a *strip* Ω in $\mathbb{H}$ is

$$g_a(b) \equiv g(a, b, t, \sigma, a_h) = \sum_{\omega \in \Omega : a \to b} e^{-i\sigma w_\omega(a,b)} a_h^{v(\omega)} t^{\ell(\omega)}, \tag{10.20}$$

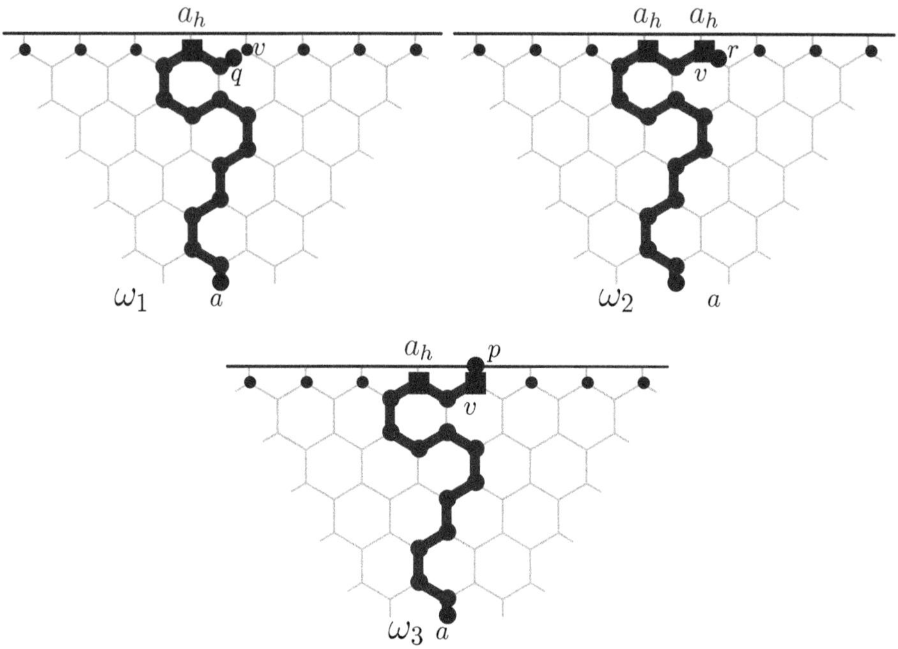

FIG. 10.10. Adsorbing walks ending in the midpoint of an edge next to a vertex v which is adjacent to the top adsorbing wall of the strip Ω. Such walks either end in q, r or p.

---

where $\ell(\omega)$ is the length of ω, and $v(\omega)$ is the number of visits of ω to vertices of $\mathbb{H}$ next to the top wall B of the strip Ω. This should be compared to equation (10.4).

Define vertices in Ω which are not next to the adsorbing wall B to be *bulk vertices.* Vertices next to B are weighted by a_h and are *surface vertices.*

The identity in lemma 10.2 applies to bulk vertices in Ω (the proof is identical to the proof of lemma 10.2) but with $f_a(b)$ replaced by $g_a(b)$ throughout.

Surface vertices satisfy a different identity, giving a modification of lemma 10.2. Define τ_c as before in equation (10.5).

Lemma 10.6 *If $t = \tau_c$, and $\sigma = \frac{5}{8}$, and $v \in \Omega$ is a bulk vertex, then the parafermionic operator satisfies*

$$S(v) = (p-v)\, g_a(p) + (q-v)\, g_a(q) + (r-v)\, g_a(r) = 0,$$

where $\{p, q, r\}$ are the midpoints of the three edges incident with v.

If, on the other hand, v is a vertex next to the adsorbing wall, with p directly above v and $\{p, q, r\}$ as the midpoints of the three edges incident with v and anti-clockwise about v, then

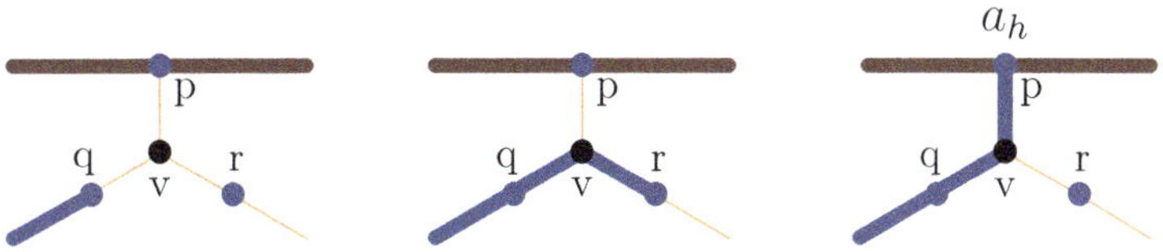

(a) (b) (c)

FIG. 10.11. (a) A walk with endpoint in q may be continued (b) either to r via v , or (c) to p in the boundary of Ω via v.

$$
\begin{aligned}
S(v) &= (p-v)g_a(p) + (q-v)g_a(q) + (r-v)g_a(r) \\
&= \frac{(q-v)(1-a_h)}{\tau_c a_h \lambda} \sum_{\omega:a\to q,p} e^{-i\sigma w_\omega(a,p)} a_h^{v(\omega)} \tau_c^{\ell(\omega)} \\
&\quad + \frac{(r-v)(1-a_h)\lambda}{\tau_c a_h} \sum_{\omega:a\to r,p} e^{-i\sigma w_\omega(a,p)} a_h^{v(\omega)} \tau_c^{\ell(\omega)},
\end{aligned} \tag{10.21}
$$

where $\lambda = e^{-i\sigma\pi/3}$, the sum over $\{\omega : a \to q, p\}$ is over all walks ending in p passing through q, and the sum over $\{\omega : a \to r, p\}$ is over all walks ending in p passing through r.

Proof If v is a bulk vertex, then the proof is the same as for lemma 10.2.

Suppose that v is a vertex next to the top wall B of Ω. Walks ending in a point which is a midpoint of an edge incident with v are illustrated in figure 10.10. Observe that there are no walks visiting p before v is visited.

Consider walks passing q and ending next to v. The endpoints of such paths are illustrated in figure 10.11, and there are three cases. Let the walk in figure 10.11 ending in v be ω. Computing the contribution of these three paths to the left-hand side of the identity gives

$$
\left(1 + \tau_c a_h \overline{\lambda} e^{2i\pi/3} + \tau_c a_h \lambda e^{-2i\pi/3}\right)(q-v)e^{-i\sigma(w_\omega(a,v)+\pi/3)} a_h^{v(\omega)-1} \tau_c^{\ell(\omega)-1},
$$

where $\overline{\lambda} = \frac{1}{\lambda}$. Since this vanishes when both $a_h = 1$ and $\sigma = \frac{5}{8}$, the leading factor in parenthesis must be equal to $1 - a_h$. This is the first term on the right-hand side of equation (10.21).

The second term is similarly determined by considering walks passing through r. This completes the proof. □

10.2.1 *The generating function*

As before, let Ω be the trapezoidal region in $\mathbb{H}$ (see figure 10.12), with bottom length $2L$ cells and height T rows. Let the boundary $\partial\Omega$ of Ω be $A \cup B \cup E \cup \overline{E}$.

Define the generating functions of walks from the origin in the middle vertex of the bottom wall and ending in A (but not in $\vec{0}$) by $A_\Omega^{T,L}$. This is given by

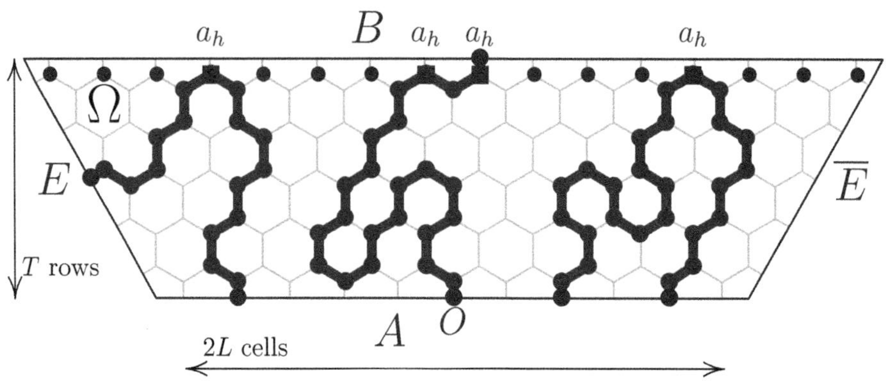

FIG. 10.12. Walks between the boundary components of Ω.

$$A_\Omega^{T,L}(a_h,t) = \sum_{\omega:0\to A\setminus\{0\}} a_h^{v(\omega)} t^{\ell(\omega)}, \tag{10.22}$$

where $v(\omega)$ is the number of visits of ω to vertices next to the adsorbing wall, and $\ell(\omega)$ is the length of ω.

Define the generating functions

$$B_\Omega^{T,L}(a_h,t) = \sum_{\omega:0\to B} a_h^{v(\omega)} t^{\ell(\omega)}, \tag{10.23}$$

$$C_\Omega^{T,L}(a_h,t) = \sum_{\omega:0\to E} a_h^{v(\omega)} t^{\ell(\omega)}, \tag{10.24}$$

$$\overline{C}_\Omega^{T,L}(a_h,t) = \sum_{\omega:0\to \overline{E}} a_h^{v(\omega)} t^{\ell(\omega)} \tag{10.25}$$

similarly.

Let $A_\Omega^{T,L}(a_h,t)$ be the generating function of *arches* or *hoops*, let $B_\Omega^{T,L}(a_h,t)$ be the generating function of (vertically oriented) *bridges* and let $C_\Omega^{T,L}(a_h,t)$ or $\overline{C}_\Omega^{T,L}(a_h,t)$ be the generating function of leftwards or rightwards *stretched walks*.

Proceed by generalising theorem 10.3 to the model here. Let $S(v)$ be given by the left-hand side of the identities in lemma 10.6.

Summing $S(v)$ over vertices in Ω gives 0 for all vertices not next to the adsorbing hard wall B.

For vertices v next to B, observe that $q-v=\frac{1}{2}e^{-5i\pi/6}$, and $r-v=\frac{1}{2}e^{-i\pi/6}$. Putting $S=\sum_{v\in\Omega} S(v)$ then gives for $p\in B$

$$2S = \frac{(1-a_h)e^{-5i\pi/6}}{\tau_c a_h \lambda} \sum_{\omega:0\to q,p} a_h^{v(\omega)} \tau_c^{\ell(\omega)} + \frac{(1-a_h)\lambda e^{-i\pi/6}}{\tau_c a_h} \sum_{\omega:0\to r,p} a_h^{v(\omega)} \tau_c^{\ell(\omega)}.$$

Reflecting walks $\omega : 0 \to r, p$ through the vertical line through the origin gives walks ω' ending in a vertex p' and passing through q'. Since $\ell(\omega) = \ell(\omega')$, and $v(\omega) = v(\omega')$, the two summations above can be combined after reflection to give

$$2S = \left(e^{-5i\pi/6}\overline{\lambda} + e^{-i\pi/6}\lambda\right) \frac{(1-a_h)}{\tau_c a_h} \sum_{\omega:0\to q,p,\, p\in B} a_h^{v(\omega)} \tau_c^{\ell(\omega)}.$$

Upon simplification and comparison to equation (10.23), this becomes

$$S = -i\,\frac{(1-a_h)}{4\tau_c^2 a_h}\, B_\Omega^{T,L}(a_h, \tau_c). \tag{10.26}$$

This provides the first step in proving the following theorem relating the generating functions in this model.

Theorem 10.7 *If* $a_h^* = 1 + \sqrt{2}$, *and* $\sigma = \frac{5}{8}$, *then*

$$\cos(\tfrac{3\pi}{8})\, A_\Omega^{T,L}(a_h, \tau_c) + \left(\frac{a_h^* - a_h}{a_h(a_h^* - 1)}\right) B_\Omega^{T,L}(a_h, \tau_c) + \cos(\tfrac{\pi}{4})\, C_\Omega^{T,L}(a_h, \tau_c) = 1.$$

Proof The proof is completed by computing S in equation (10.26) in terms of the generating functions $A_\Omega^{T,L}(a_h, \tau_c)$, $B_\Omega^{T,L}(a_h, \tau_c)$ and $C_\Omega^{T,L}(a_h, \tau_c)$.

Sum $g_0(b)$ in equation (10.20) for $b \in A$ in equation (10.12). This is a summation over adsorbing walks from $\vec{0}$ returning to A (these walks are loops or arches). Let $A = A^- \cup \{\vec{0}\} \cup A^+$, where A^- is the part of A to the left of $\vec{0}$, and A^+ is the part of A to the right of $\vec{0}$.

There is only one walk returning to $\vec{0}$, namely, the walk of length 0 and weight $a_h^0 \tau_a^0 = 1$.

Walks ending in A^+ have a total winding of minus π, and walks ending in A^- have total winding of π. Since $\lambda = e^{-i\sigma\pi/3}$, the winding of walks into A^+ produces the factor λ^3 in equation (10.20), while the winding of walks into A^- gives a factor $\overline{\lambda}^3$. Thus,

$$\sum_{b\in A} g_0(b) = 1 + \sum_{\omega\in\Omega:0\to A^-} \overline{\lambda}^3 a_h^{v(\omega)} \tau_c^{\ell(\omega)} + \sum_{\omega\in\Omega:0\to A^+} \lambda^3 a_h^{v(\omega)} \tau_c^{\ell(\omega)}.$$

This may be simplified to

$$\sum_{b\in A} g_0(b) = 1 + \sum_{\omega\in\Omega:0\to A^-} \left(\lambda^3 + \overline{\lambda}^3\right) a_h^{v(\omega)} \tau_c^{\ell(\omega)} = A_\Omega^{T,L,0}(\tau_c, a_h) - \cos(\tfrac{3\pi}{8})\, A_\Omega^{T,L}.$$

Similar calculations show that

$$\begin{aligned} e^{-i\pi/3} \sum_{b\in \overline{E}} g_0(b) + e^{i\pi/3} \sum_{b\in E} g_0(b) &= \left(e^{-i\pi/3}\lambda^2 + e^{i\pi/3}\overline{\lambda}^2\right) \sum_{\omega\in\Omega:0\to \overline{E}} a_h^{v(\omega)} \tau_c^{\ell(\omega)} \\ &= -\cos(\tfrac{\pi}{4})\, C_\Omega^{T,L}(a_h, \tau_c), \end{aligned}$$

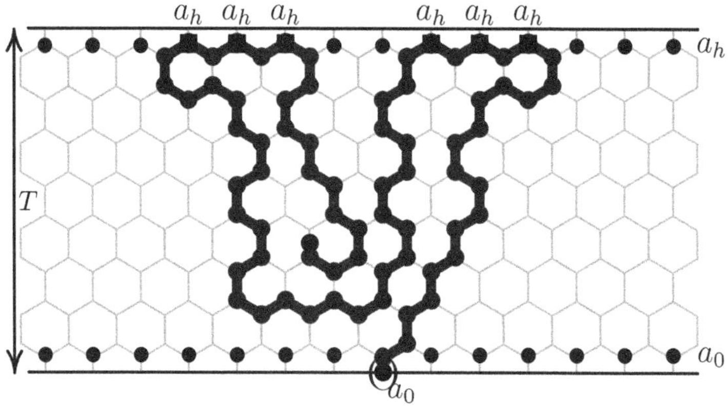

FIG. 10.13. A walk in Ω adsorbing on the bottom hard wall with fugacity a_0, and on the top hard wall with fugacity a_h.

and

$$\sum_{b \in B} g_0(b) = B_\Omega^{T,L}(a_h, \tau_c).$$

Determine S from the above as follows. Note that $S = \sum_{v \in \Omega} S(v)$, with $S(v)$ given in lemma (10.6). This may instead be considered a summation over the midpoints of all edges in Ω, including those midpoints on the boundary. If $\langle v_1 \sim v_2 \rangle \in \Omega$ is a bulk edge (that is, when its midpoint b is not in the boundary $\partial\Omega$), then there are contributions to S from each endpoint. These two terms have opposite signs, since $(b - v_1) = -(b - v_2)$, and they cancel.

This shows that S can be computed by considering edges in Ω with midpoints in the boundary $\partial\Omega$. If $\langle u_1 \sim u_2 \rangle$ is such an edge with midpoint p, and $u_2 \in \Omega$, then computing $p - u_j$ for edges on the boundary shows that

$$2S = -i \sum_{b \in A} g_0(b) + e^{-5i\pi/6} \sum_{b \in \overline{E}} g_0(b) + e^{-i\pi/6} \sum_{b \in E} g_0(b) + \sum_{b \in B} g_0(b).$$

As all the terms on the right-hand side have been determined above, comparison of this to equation (10.26) completes the proof. □

10.2.2 *Adsorbing walks in a slit*

Let $\mathbb{S}$ be the slit of width T in $\mathbb{H}$ obtained by taking $L \to \infty$ in Ω. Let ω be a self-avoiding walk from $\vec{0}$ in $\mathbb{S}$ (see figure 10.13). The walk ω is assumed to adsorb on both walls of $\mathbb{S}$. Visits next to wall A are weighted by a_0, and visits next to wall B are weighted by a_h.

The existence of a limiting free energy $\mathcal{H}_T(a_0, a_h)$ in this model may be shown using the same arguments as used in section 8.1. The free energy $\mathcal{H}_T(a_0, a_h)$ is convex in both $\log a_0$ and $\log a_h$ and is non-decreasing with a_0 and a_h. It is also independent of the location of the endpoints of ω. By symmetry,

$$\mathcal{H}_T(a_0, a_h) = \mathcal{H}_T(a_h, a_0). \tag{10.27}$$

Define the partition functions of adsorbing walks from $\vec{0}$ with endpoints in A by $H_n^A(a_0, a_h)$, and those with endpoints in B by $H_n^B(a_0, a_h)$. Let $H^C(a_0, a_h)$ be the partition function of walks not ending in either A or B. Then the following is similar to theorem 8.8.

Theorem 10.8 *There exists a limiting free energy for walks adsorbing onto the walls of a slit $\mathbb{S}$ of height T in $\mathbb{H}$ defined by*

$$\mathcal{H}_T(a_0, a_h) = \lim_{n\to\infty} \tfrac{1}{n} \log H_n^A(a_0, a_h).$$

Moreover, $\mathcal{H}_T(a_0, a_h) = \lim_{n\to\infty} \tfrac{1}{n} \log H_n^B(a_0, a_h) = \lim_{n\to\infty} \tfrac{1}{n} \log H_n^C(a_0, a_h)$.

The free energy $\mathcal{H}_T(a_0, a_h)$ is symmetric in its arguments and is a convex function of both $\log a_0$ and $\log a_h$. Moreover, $\mathcal{H}_T(a_0, 1) < \mathcal{H}_{T+1}(a_0, 1) < \mathcal{A}^+(a_0)$, where $\mathcal{A}^+(a_0)$ is the limiting free energy of adsorbing walks in $\mathbb{H}_+$ with $\partial\mathbb{H}_+ = A$. If $0 \le a_h \le 1$, then $\lim_{T\to\infty} H_T(a_0, a_h) = \mathcal{A}^+(a_0)$.

Additionally, $\mathcal{A}^+(a_0) = \log \mu_h$ if $0 \le a_0 \le 1$, where μ_h is the growth constant of self-avoiding walks in the hexagonal lattice. Since $\mathcal{A}^+(a_0) > \log a_0$, it follows that there exists a critical point $a_c^h \ge 1$ such that $\mathcal{A}^+(a_0) > \log \mu_h$ if $a_0 > a_c^h$. If $a_0 > a_c^h$, then the walk is adsorbed in $B(\mathbb{S})$.

Proof To see that $\mathcal{H}_T(a_0, 1) < \mathcal{H}_{T+1}(a_0, 1)$, a proof similar to that of lemma 8.20 may be used. The result, $\lim_{T\to\infty} H_T(a_0, a_h) = \mathcal{A}^+(a_0)$ whenever $0 \le a_h \le 1$, follows from the methods of theorem 9.18. That $\mathcal{A}^+(a_0) = \log \mu_h$ if $0 \le a_0 \le 1$ follows by the methods of theorem 9.2, which also shows that $\mathcal{A}^+(a_0) \ge \log a_0$. □

Properties of the adsorption critical point a_c^h in $\mathbb{H}_+$ may be described using the methods of section 9.1.

Define the generating functions $\mathfrak{A}_T(a_h, t)$, $\mathfrak{B}_T(a_h, t)$ and $\mathfrak{C}_T(a_h, t)$ of the partition functions $H_n^A(1, a_h)$, $H_n^B(1, a_h)$ and $H_n^C(1, a_h)$, respectively. By taking the width $L \to \infty$ in Ω (see figure 10.12),

$$\mathfrak{A}_T(a_h, t) = \lim_{L\to\infty} A_\Omega^{T,L}(a_h, t), \tag{10.28}$$

and similarly for $\mathfrak{B}_T(a_h, t)$ and $\mathfrak{C}_T(a_h, t)$.

The following is a consequence of the arguments leading to corollary 8.15.

Theorem 10.9 *The radius of convergence $t_T^h(a_h)$ in the t-plane of the generating functions $\mathfrak{A}_T(a_h, t)$, $\mathfrak{B}_T(a_h, t)$ and $\mathfrak{C}_T(a_h, t)$ is $t_T^h(a_h)$ given by*

$$-\log t_T^h(a_h) = \mathcal{H}_T(1, a_h).$$

Moreover, $-\log t_T^h(a_h)$ is strictly increasing to the limit $\mathcal{A}^+(a_h)$ as $T \to \infty$.

There exists a unique a_T such that $-\log t_T(a_T) = -\log \tau_c = \log \mu_h$ for each T. If $t = \tau_c$ (see equation (10.5)), then the generating functions $\mathfrak{A}_T(a_h, \tau_c)$, $\mathfrak{B}_T(a_h, \tau_c)$ and $\mathfrak{C}_T(a_h, \tau_c)$ have the radius of convergence a_T in the a_h-plane.

Proof That the radius of convergence of $\mathfrak{A}_T(a_h, t)$, $\mathfrak{B}_T(a_h, t)$ and $\mathfrak{C}_T(a_h, t)$ in the t-plane is $t_T^h(a_h)$ given by

$$-\log t_T^h(a_h) = \mathcal{H}_T(1, a_h)$$

follows from the methods leading to corollary 8.15.

It follows from theorem 10.8 that $-\log t_T^h(a_h) \nearrow \mathcal{A}^+(a_h)$ as $T \to \infty$.

There exists a unique a_T as claimed by the intermediate value theorem because $\mathcal{H}_T(1, a_h)$ is convex and continuous in $\log a_h$ and since $\mathcal{H}_T(1, a_h) < \log \mu_h$ if $a_h < 1$, and $\mathcal{H}(1, a_h) > \log \mu_h$ if a_h is large. Notice that, since $\mathcal{H}_T(1, a_h) \leq \mathcal{H}^+(1, a_h)$, $a_T \geq a_c^h$, by theorem 10.8.

This also implies that

$$t_T^h(a_h) < t_T^h(a_T) \Leftrightarrow a_h > a_T, \quad \text{and } t_T^h(a_h) > t_T^h(a_T) \Leftrightarrow a_h < a_T. \tag{10.29}$$

Consider the series $\mathfrak{A}_T(a_h, t)$. This is convergent if $t < t_T^h(a_h)$, and divergent if $t > t_T^h(a_h)$. By definition, $t_T^h(a_T) = \tau_c$ and so (by equation (10.29)), if $a_h < a_T$, then $t_T^h(a_h) > t_T^h(a_T) = \tau_c$. This shows that $\mathfrak{A}_T(a_h, \tau_c)$ is convergent if $a_h < a_T$. On the other hand, if $a_h > a_T$, then $t_T^h(a_h) < t_T^h(a_T) = \tau_c$. This shows that $\mathfrak{A}_T(a_h, \tau_c)$ is divergent if $a_h > a_T$. Hence, a_T is the radius of convergence of $\mathfrak{A}(a_h, \tau_c)$ in the a_h-plane.

It remains to prove that a_T decreases to a_c^h (the adsorption critical point for a walk adsorbing in the hard wall, defined in theorem 10.8).

Since $\log t_T(a_h) = -\mathcal{H}_T(1, a_h)$, it follows from theorem 10.8 that $t_T(a_h) > t_{T+1}(a_h) > e^{-\mathcal{A}^+(a_h)}$. This shows that $\langle a_T \rangle$ is a decreasing sequence and its limit $\overline{a}$ exists. Since $a_T > a_c^h$, it follows that $\overline{a} \geq a_c^h$.

On the other hand, $\overline{a} < a_T$, and, by equation (10.29), $t_T^h(\overline{a}) > t_T^h(a_T) = \tau_c$. This means that $\overline{a} \leq a_c^h$.

Hence, $\overline{a} = a_c^h$. □

10.2.3 *The critical point $a_c^h = 1 + \sqrt{2}$*

Consider the identity in theorem 10.7. Taking $L \to \infty$ *increases* $A_\Omega^{T,L}(\tau_c, a_h)$ and $B_\Omega^{T,L}(\tau, a_h)$ but *decreases* $C_\Omega^{T,L}(\tau_c, a_h)$. Hence, these limits exist as also seen in equation (10.28), and they are the generating functions $\mathfrak{A}_T(a_h, t)$, $\mathfrak{B}_T(a_h, t)$ and $\mathfrak{C}_T(a_h, t)$.

In the context here, follow the definitions in equations (10.22), (10.23) and (10.25) by calling $\mathfrak{A}_T(a_h, t)$ the generating function of *arches* or *loops*, $\mathfrak{B}_T(a_h, t)$ the generating function of (vertically oriented) *bridges* and $\mathfrak{C}_T(a_h, t)$ the generating function of leftwards or rightwards *stretched walks*.

Recall that $a_h^* = 1 + \sqrt{2}$ (see theorem 10.7). If $a_h < a_h^*$, then this gives

$$\cos(\tfrac{3\pi}{8})\mathfrak{A}_T(a_h, \tau_c) + \left(\frac{a_h^* - a_h}{a_h(a_h^* - 1)}\right)\mathfrak{B}_T(a_h, \tau_c) + \cos(\tfrac{\pi}{4})\mathfrak{C}_T(a_h, \tau_c) = 1, \tag{10.30}$$

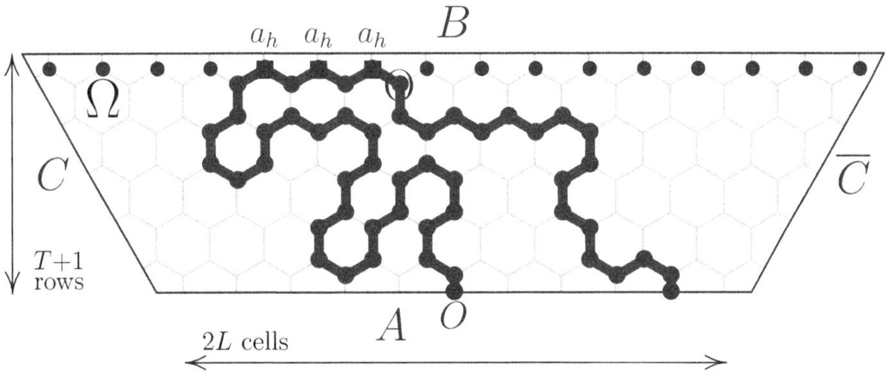

FIG. 10.14. An arch in Ω. This arch can be decomposed into two bridges by deleting the vertex marked by O and the two half-edges incident with it.

---

and it is significant that the coefficients are positive so that the generating functions are bounded. This shows that $a_h < a_T$ if $a_h < a_h^*$, by theorem (10.9). Hence, $a_h^* \leq a_T$ and thus $a_h^* \leq a_c^h$.

It remains to prove the opposite inequality. First observe the following lemma.

Lemma 10.10 *If $a_h < a_T$, then*

$$\cos(\tfrac{3\pi}{8})\mathfrak{A}_T(a_h, \tau_c) + \left(\frac{a_h^* - a_h}{a_h(a_h^* - 1)}\right)\mathfrak{B}_T(z, \tau_c) = 1.$$

Proof To see this, it may be shown that $\mathfrak{C}_T(a_h, \tau_c) = 0$.

Note that $C_\Omega^{T,L}(\tau_c, a_h)$ counts self-avoiding walks of length at least L. By theorem 10.9, the generating functions $\mathfrak{A}_T(a_h, \tau_c)$, $\mathfrak{B}_T(a_h, \tau_c)$ and $\mathfrak{C}_T(a_h, \tau_c)$ converge if $a_h < a_T$. The remainder, $\mathfrak{C}_T(a_h, \tau_c) - C_\Omega^{T,L}(\tau_c, a_h)$, includes all walks contributing to $C_\Omega^{T,L+1}(\tau_c, a_h)$ and so is an upper bound on $C_\Omega^{T,L+1}(\tau_c, a_h)$. But the remainder tends to 0 with increasing L, and hence $C_\Omega^{T,L+1}(\tau_c, a_h)$ tends to 0 as well. This shows that $\mathfrak{C}_T(a_h, \tau_c) = 0$. □

It only remains to take $T \to \infty$ to recover the model of adsorbing walks in a half-space. This will also produce $a_h^* \geq a_c^h$, establishing that the adsorption critical point in $\mathbb{H}_+$ is at $1 + \sqrt{2}$.

Consider arches contributing to $\mathfrak{A}_{T+1}(a_h, \tau_c)$. These arches have height of at most $T+1$, but also include arches of height at most T. Arches of height at most T do not have visits next to the top adsorbing wall and so these are not weighted by factors of a_h.

The difference $\mathfrak{A}_{T+1}(a_h, \tau_c) - \mathfrak{A}_T(a_h, \tau_c)$ is the generating function of arches of height of exactly $T+1$. One such arch is illustrated in figure 10.14. By deleting the vertex marked by O and the two half-edges incident with it, the arch is decomposed into two bridges: the first an adsorbing bridge of height $T+1$, and the second a bridge of height T. This proves that

$$\mathfrak{A}_{T+1}(a_h, \tau_c) - \mathfrak{A}_T(a_h, \tau_c) \leq \tau_c \mathfrak{B}_{T+1}(a_h, \tau_c)\mathfrak{B}_T(1, \tau_c). \tag{10.31}$$

Supposing that $a_h < a_{T+1}$ and then using lemma 10.10 to eliminate $\mathfrak{A}_T(a_h, \tau_c)$ from this inequality produces

$$0 \leq \frac{1}{\mathfrak{B}_{T+1}(a_h, \tau_c)} \leq \tau_c \cos(\tfrac{3\pi}{8}) + \left(\frac{a_h^* - a_h}{a_h(a_h^* - 1)}\right)\left(\frac{1}{\mathfrak{B}_T(1, \tau_c)}\right). \tag{10.32}$$

This is true for all $a_h < a_c^h$, since a T can be found such that $a_h < a_{T+1}$.

The last component in the proof is to show that $\lim_{T\to\infty} B_T(1, \tau_c) = 0$. This shows that the generating function of bridges with endpoints 'far' apart tends to 0 at the critical point, as those endpoints are forced ever further apart. The proof of this is a very lengthy probabilistic argument (see the appendix in reference [28]).

Theorem 10.11 *The limit* $\lim_{T\to\infty} \mathfrak{B}_T(1, \tau_c) = 0$. □

Suppose that $a_c^h > a_h^*$. Then there is an $a_h < a_c^h$ such that $a_h > a_h^*$. Thus, $a_h^* - a_h$ is negative in equation (10.32) and, by taking $T \to \infty$, the right-hand side becomes an arbitrarily large negative value, since $\mathfrak{B}_T(1, \tau_c) \to 0$. This is a contradiction, unless $a_c^h \leq a_h^*$.

This gives the following theorem.

Theorem 10.12 *The critical point* $a_c^h = 1 + \sqrt{2}$. □

11

INTERACTING MODELS OF ANIMALS, TREES AND NETWORKS

Self-interacting lattice animals and trees (see section 2.1) are models of branched polymer entropy. These may be considered in several different ensembles, namely, bond-animals or bond-trees (these are *weakly embedded animals or trees* (a tree is an acyclic animal)), or site-animals or site-trees (these are *strongly embedded animals or trees*), and may be counted by *size* (number of edges) or by *order* (number of vertices). In this chapter, interacting models of bond-animal and bond-trees, counted by size, are examined. Most of the results can also be proven for the other ensembles.

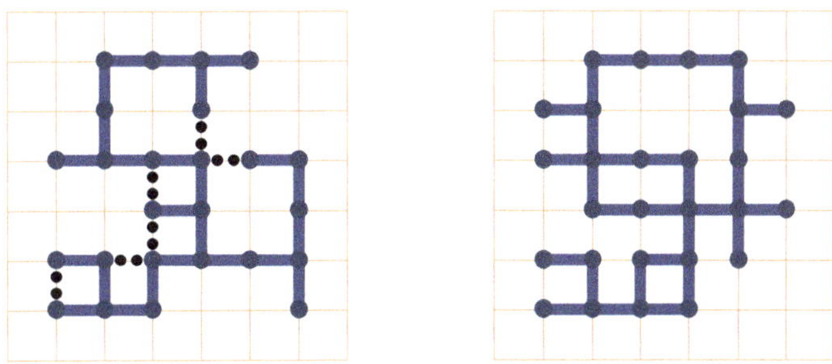

(a) (b)

FIG. 11.1. Models of bond-animals. (a) A collapsing animal with nearest-neighbour contacts between vertices. (b) A lattice animal with a cyclomatic index equal to 3.

The two animals or trees α_1 and α_2 in $\mathbb{L}^d$ are equivalent if $\alpha_1 = \alpha_2 + \vec{v}$ for some vector $\vec{v} \in \mathbb{R}^d$. The number of distinct bond-animals of size n is denoted by a_n, and the number of distinct bond-trees of size n is denoted by t_n.

Interacting models of animals are obtained by assigning an energy to each animal. For example, in figure 11.1(a) a *contact* in an animal α is an edge $\langle \vec{v} \sim \vec{u} \rangle \in \mathbb{L}^d$ such that both $\vec{v}, \vec{u} \in \alpha$ and $\langle \vec{v} \sim \vec{u} \rangle \notin \alpha$. This is a model of a *collapsing animal* or, if the animal is acyclic, of a *collapsing tree*. A *perimeter edge* of α is an edge $\langle \vec{v} \sim \vec{u} \rangle \in \mathbb{L}^d$ such that $\langle \vec{v} \sim \vec{u} \rangle \notin \alpha$ and at least one of $\vec{v} \in \alpha$ or $\vec{u} \in \alpha$.

In figure 11.1(b) a different model of a collapsing animal is shown; independent cycles in this animal may be weighted to define a model of an animal

The Statistical Mechanics of Interacting Walks, Polygons, Animals and Vesicles, 2nd edition, E.J. Janse van Rensburg. © E.J. Janse van Rensburg. Published in 2015 by Oxford University Press.

in the cycle-collapse ensemble. Both these are models of the branched polymer θ-transition.

11.1 The pattern theorem for interacting lattice animals

The basic quantity in models of interacting lattice animals is $a_n(k)$, the number of lattice animals of size n and energy k. By lemma 2.1, $a_n(k) \le a_n \le K^n$ for all $n \ge 0$, and finite K.

Assume that $a_n(k) = 0$ if $k > Ln$, where $L > 0$ is fixed. Assume that concatenation of animals (see figure 2.3) gives the supermultiplicative relation

$$a_n(k_1)\, a_m(k_2) \le \sum_{j=-q}^{q} a_{n+m+1}(k_1+k_2+j) \tag{11.1}$$

for some finite $q \ge 0$ independent of $\{n, m, k_1, k_2\}$ (this assumption can be relaxed to $q = o(n+m)$ in what follows). Then the model satisfies the properties described in section 3.3.1.3, provided that the model is *regular*. That is, there exists an m (independent of n and k) such that both

$$a_n(k) \le a_{n+m}(k), \quad \text{and } a_n(k) \le a_{n+m}(k+1). \tag{11.2}$$

Make the additional assumption that the energy of the model is *local*. That is, appending a single edge anywhere in the animal changes its energy by at most q. That is,

$$a_n(k) \le \sum_{j=-q}^{q} a_{n+1}(k+j). \tag{11.3}$$

Denote the partition function of the model by $A_n(z) = \sum_{k=0}^{Ln} a_n(k)\, z^n$. Multiply equation (11.1) by $z^{k_1+k_2}$ and sum the left-hand side over k_1 and k_2. Since $a_n(k) = 0$ if $k > Ln$, it follows that

$$\begin{aligned}
A_n(z)\, A_m(z) &\le \sum_{j=-q}^{q} \sum_{k_1=0}^{\infty} \sum_{k_2=0}^{\infty} a_{n+m+1}(k_1+k_2+j)\, z^{k_1+k_2} \\
&\le \sum_{j=-q}^{q} \sum_{k=0}^{\infty} (k+1)\, a_{n+m+1}(k+j)\, z^k \\
&\le \sum_{j=-q}^{q} z^{-q} \sum_{k=0}^{L(n+m+1)+q} (k+1)\, a_{n+m+1}(k+j)\, z^{k+j} \\
&\le (L(n+m+1)+q+1) \sum_{j=-q}^{q} z^{-q} A_{n+m+1}(z).
\end{aligned} \tag{11.4}$$

Define $\phi_n(z) = (Ln+q+1) \sum_{j=-q}^{q} z^{-q}$; then, $\log \phi_n(z) = o(n)$ if $q = o(n)$, and the partition function satisfies the supermultiplicative inequality

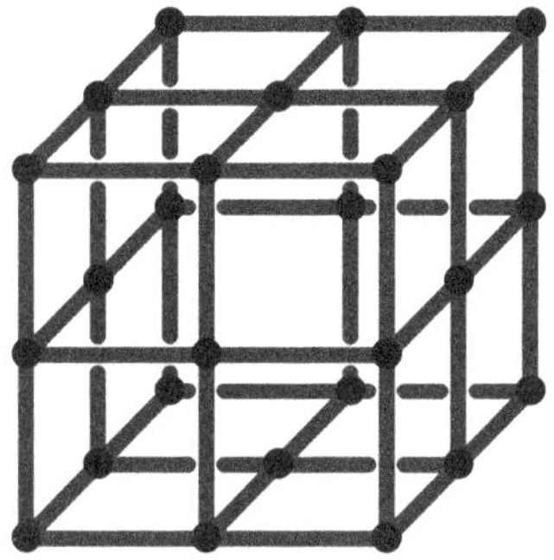

FIG. 11.2. A rectangular box-animal encompassing an empty interior.

$$A_n(z)\,A_m(z) \leq \phi_{n+m+1}(z)\,A_{n+m+1}(z). \tag{11.5}$$

By theorem A.4 (in appendix A) and theorem 3.3, there is a limiting free energy

$$\mathcal{A}(z) = \lim_{n\to\infty} \tfrac{1}{n} \log A_n(z) \tag{11.6}$$

which is a convex function of $\log z$ (and is the limit of a sequence of convex functions).

11.1.1 *Patterns in lattice animals*

A *pattern* P is a connected subgraph of an animal. Notice that a pattern is both the presence as well as the absence of edges. A pattern P is said to occur in an animal α if there is a vector $\vec{v}$ such that $P+\vec{v} \subseteq \alpha$. A pattern P occurs m times in α if there are m distinct vectors $\{\vec{v}_1, \vec{v}_2, \ldots, \vec{v}_m\}$ such that $P+\vec{v}_i \subset \alpha$.

If P occurs in α, and $Q \subseteq P$, then Q occurs as well, even if Q is not connected.

Let $\vec{a}, \vec{b} \in \mathbb{Z}^d$ be vertices and let $\vec{a}(j)$ denote the j-th Cartesian component of $\vec{a}$. Put $V = \{\vec{x} \in \mathbb{L}^d \mid \vec{a}(j) \leq \vec{x}(j) \leq \vec{b}(j) \text{ for } 1 \leq j \leq q\}$. The interior of V is $\mathring{V} = \{\vec{x} \in \mathbb{L}^d \mid \vec{a}(j) < \vec{x}(j) < \vec{b}(j) \text{ for } 1 \leq j \leq q\}$ and the boundary of V is $\partial V = V \setminus \mathring{V}$. Define the animal $B = \{\langle \vec{x} \sim \vec{y} \rangle \mid \vec{x}, \vec{y} \in \partial V\}$. Then B is a 1-complex which is a rectangular *box-animal* (see figure 11.2). The interior of B is empty.

Let α_1 and α_2 be lattice animals in $\mathbb{L}^d$ and suppose the top vertex of α_1 is $\vec{t}_1$, and the bottom vertex of α_2 is $\vec{b}_2$. Place α_2 in $\mathbb{L}^d$ such that $\vec{b}_2 = \vec{t}_1 + \vec{e}_1$. Then α_1 and α_2 can be concatenated (see figure 2.3) by adding $\langle \vec{t}_1 \sim \vec{b}_2 \rangle$ to obtain $\alpha_1 \circ \alpha_2$. A pattern P is *prime* if there do not exist animals $\{\alpha_1, \alpha_2\}$ such that $P \not\subseteq \alpha_1$, and $P \not\subseteq \alpha_2$, but $P \subseteq \alpha_1 \circ \alpha_2$. The box-animal in figure 11.2 is an example of a prime pattern. That is, P cannot be created by concatenating two animals.

Denote the number of animals of size n and energy k on which the pattern P occurs exactly m times as $a_n(k; [m, P])$. The partition function is defined by

$$A_n(z; [m, P]) = \sum_{k=0}^{\infty} a_n(k; [m, P])\, z^k. \tag{11.7}$$

Suppose that P is a *prime pattern* and concatenate two animals as illustrated in figure 2.3. This gives

$$\sum_{m_1=0}^{m} a_{n_1}(k_1;[m_1,P])\,a_{n_2}(k_2;[m-m_1,P]) \\ \leq \sum_{j=-q}^{q} a_{n_1+n_2+1}(k_1+k_2+j;[m,P]) \qquad (11.8)$$

by equation (11.1). Multiply the above by $z^{k_1+k_2}$ and sum over $\{k_1,k_2\}$. Putting $m_2 = m - m_1$ and then following the same steps as in equations (11.4) and (11.5) give

$$A_{n_1}(z;[m_1,P])\,A_{n_2}(z;[m_2,P]) \\ \leq \phi_{n_1+n_2+1}(z)\,A_{n_1+n_2+1}(z;[m_1+m_2,P]). \qquad (11.9)$$

As shown in section 3.3.1.3, it follows that the density function of prime patterns is given by

$$\log \mathcal{Q}_A(z,\epsilon) = \lim_{n\to\infty} \tfrac{1}{n} \log A_n(z;[\lfloor \epsilon n \rfloor, P]). \qquad (11.10)$$

By theorem 3.9, $\log \mathcal{Q}_A(z,\epsilon)$ is a concave function of ϵ.

Define the function $\widetilde{a}_n(k;\zeta) = \sum_{m=0}^{\infty} a_n(k;[m,P])\,\zeta^m$. Multiply equation (11.8) by ζ^m and sum over m. This gives

$$\widetilde{a}_{n_1}(k_1;\zeta)\widetilde{a}_{n_2}(k_2;\zeta) \leq \sum_{j=-q}^{q} \widetilde{a}_{n_1+n_2+1}(k_1+k_2+j;\zeta). \qquad (11.11)$$

Comparison to equations (11.1), (11.4), (11.5) and (11.6) gives existence of the limit

$$\begin{aligned} \alpha(z,\zeta) &= \lim_{n\to\infty} \tfrac{1}{n} \log \sum_{k=0}^{\infty} \widetilde{a}_n(k;\zeta)\,z^k \\ &= \lim_{n\to\infty} \tfrac{1}{n} \log \sum_{m=0}^{\infty} \left(\sum_{k=0}^{\infty} a_n(k;[m,P])\,z^k \right) \zeta^m \\ &= \lim_{n\to\infty} \tfrac{1}{n} \log \sum_{m=0}^{\infty} A_n(z;[m,P])\,\zeta^m. \end{aligned} \qquad (11.12)$$

As per section 3.3.1.4, it follows that

$$\alpha(z,\zeta) = \sup_{\epsilon}\{\log \mathcal{Q}_A(z,\epsilon) + \epsilon \log \zeta\}, \qquad (11.13)$$

and the joint density function may be defined

$$\mathcal{P}_A(\delta,\epsilon) = \inf_{z>0}\{\log \mathcal{Q}_A(z,\epsilon) - \delta \log z\}. \qquad (11.14)$$

11.1.2 *A pattern theorem for animals*

The limiting free energy of the model of interacting animals is given by $\mathcal{C}\!\!\!\alpha(z) \equiv \mathcal{C}\!\!\!\alpha(z,1)$ in equation (11.6) (and in this case the prime patterns P are not weighted by ζ). (The limiting free energy is bnoth regular and local; see equations (11.2) and (11.3).)

By equation (11.13), $\mathcal{C}\!\!\!\alpha(z) = \sup_\epsilon \log \mathcal{Q}_A(z,\epsilon)$ if $\zeta = 1$. This shows that $\mathcal{C}\!\!\!\alpha(z) \geq \log \mathcal{Q}_A(z,\epsilon)$. The pattern theorem is proven if it is shown that there exists an $\epsilon_c > 0$ such that $\mathcal{C}\!\!\!\alpha(z) > \log \mathcal{Q}_A(z,\epsilon)$ for all $\epsilon \in [0,\epsilon_c)$. This proves that the prime pattern P occurs with density at least ϵ_c in almost all, except exponentially few, animals.

Theorem 11.1 *Let P be a prime pattern and suppose that $\mathcal{Q}_A(z,\epsilon)$ is its density function. Then, for every (fixed) value of $z \geq 0$, there exists an $\epsilon_c > 0$ such that $\mathcal{Q}_A(z,0) < \mathcal{Q}_A(z,\epsilon_c)$.*

Proof Let P be a prime pattern and let α be an animal of size n, energy k and in which there are no occurrences of P.

Let B be the smallest box-animal which may contain P in its interior and suppose that the volume (number of vertices) in B and its interior is V. If P contains M edges, then $V \geq \frac{M}{2d}$.

The animal α contains at least $\lceil \frac{n}{d} \rceil$ vertices, so it can be intersected with at least $\lfloor \frac{n}{dV} \rfloor$ disjoint copies of B. Choose m such copies of B from $\lfloor \frac{n}{dV} \rfloor$ and delete all edges of α in the interior of each B but retain all the edges (and not just only those in α) of B. This leaves a connected animal which contains m copies of the box-animal B, all with empty interiors.

The prime pattern P can be inserted into each of the empty interiors of a chosen set of m box-animals B. Insert a copy of P into the interior of B and join P to B by adding extra edges (while care is taken not to violate the restriction on the presence and absence of edges in P).

Since P is a prime pattern, this creates exactly m disjoint copies of P, one in each of the m box-animals B. Since there are V vertices in each B, this may change the number of edges in α by at most mVd. Since the energy is local, this may change the energy by up to $mVdq$ (see equations (11.2) and (11.3)).

Since there are $a_n(k,[0,P])$ animals of size n, of energy k and with no occurrences of P, and α is covered by $\lfloor \frac{n}{dV} \rfloor$ box-animals of which m are selected, it follows that

$$\binom{\lfloor \frac{n}{dV} \rfloor}{m} a_n(k,[0,P]) \leq \sum_{l=-mVd}^{mVd} \left(\sum_{j=-mVdq}^{mVdq} a_{n+l}(k+j,[m,P]) \right).$$

Multiply this by z^k and sum over k:

$$\binom{\lfloor \frac{n}{dV} \rfloor}{m} A_n(z;[0,P]) \leq \sum_{l=-mVd}^{mVd} \left(\sum_{j=-mVdq}^{mVdq} z^{-j} A_{n+l}(z,[m,P]) \right).$$

This simplifies to

$$\binom{\lfloor \frac{n}{dV} \rfloor}{m} A_n(z;[0,P]) \leq \phi_{mdVq}(z) \sum_{l=-mVd}^{mVd} A_{n+l}(z,[m,P]),$$

where $\phi_k(z) = \sum_{j=-k}^{k} z^j$. Put $\overline{z} = \frac{1}{z}$ so that $\lim_{k\to\infty} \phi_k(z)^{1/k} = \max\{z, \overline{z}\}$.

Choose $m = \lfloor \epsilon n \rfloor$ (with $0 < \epsilon < \frac{1}{dV}$), take the power $\frac{1}{n}$ and take $n \to \infty$. Use the bound in $\mathcal{A}(z) \geq \mathcal{Q}_A(z,\epsilon)$ in equation (11.13) to see that

$$\left(\frac{\left(\frac{1}{dV}\right)^{1/dV}}{\epsilon^\epsilon \left(\frac{1}{dV} - \epsilon\right)^{1/dV-\epsilon}} \right) \mathcal{Q}_A(z,0) \leq (\max\{z,\overline{z}\})^{\epsilon dVq}\, e^{\epsilon V d \mathcal{A}(z)}\, \mathcal{Q}_A(z,\epsilon).$$

Put $\psi = (\max\{z,\overline{z}\})^{-dVq}\, e^{-Vd\mathcal{A}(z)}$. Then the above is the same as

$$\left(\frac{\left(\frac{1}{dV}\right)^{1/dV} \psi^\epsilon}{\epsilon^\epsilon (\frac{1}{dV} - \epsilon)^{1/dV-\epsilon}} \right) \mathcal{Q}_A(z,0) \leq \mathcal{Q}_A(z,\epsilon).$$

Choose $\epsilon = \frac{\psi}{(1+\psi)Vd} < \frac{1}{Vd}$. Then the above simplifies to

$$(1+\psi)^{1/Vd}\, \mathcal{Q}_A(z,0) \leq \mathcal{Q}_A(z,\epsilon).$$

Since $(1+\psi)^{1/Vd} > 1$, this completes the proof. □

The proof of theorem 11.1 can be slightly altered to show that $\mathcal{Q}_A(z,\epsilon)$ is strictly increasing for small values of ϵ. Notice that $\mathcal{Q}_A(z,\epsilon)$ is a concave function of ϵ, and so it is enough to show that there are two values of ϵ, say $0 < \epsilon_1 < \epsilon_2$, such that $\mathcal{Q}_A(z,0) < \mathcal{Q}_A(z,\epsilon_1) < \mathcal{Q}_A(z,\epsilon_2)$.

Theorem 11.2 *Let P be a prime pattern and suppose that $\mathcal{Q}_A(z,\epsilon)$ is its density function. There exists an $\epsilon_c > 0$ such that $\mathcal{Q}_A(z,\epsilon_c) = \sup_{\epsilon\geq 0} \mathcal{Q}_A(z,\epsilon)$, and $\mathcal{Q}_A(z,\epsilon) < \mathcal{Q}_A(z,\epsilon_c)$ whenever $\epsilon < \epsilon_c$. In other words, since $\log \mathcal{Q}_A(z,\epsilon)$ is concave in $[0,\epsilon_c]$, it is also strictly increasing in this interval.*

Proof In the proof of theorem 11.1 let α instead be an animal of size n and energy k and with the prime pattern P occurring $\lfloor \epsilon n \rfloor$ times in α.

Consider intersections of α with the box-animal B: at most $2^d \lfloor \epsilon n \rfloor$ of these box-animals intersect a copy of P; that is, there are at least $(\lfloor \frac{n}{Vd} \rfloor - 2^d \lfloor \epsilon n \rfloor)$ box-animals B disjoint with copies of P in α. Choose $\lfloor \delta n \rfloor$ of the B which are disjoint with occurrences of P. Delete all the edges of α in these B and place a new copy of P in each box. This shows that

$$\begin{aligned} &\binom{\lfloor \frac{n}{Vd} \rfloor - 2^d \lfloor \epsilon n \rfloor}{\lfloor \delta n \rfloor} a_n(k,[\lfloor \epsilon n \rfloor, P]) \\ &\leq \sum_{i=-\lfloor \delta n \rfloor Vd}^{\lfloor \delta n \rfloor Vd} \left(\sum_{j=-\lfloor \delta n \rfloor Vdq}^{\lfloor \delta n \rfloor Vdq} a_{n+i}(k+j;[\lfloor \epsilon n \rfloor + \lfloor \delta n \rfloor, P]) \right). \end{aligned}$$

Multiply this by z^k and sum over k. Take the power $\frac{1}{n}$ and let $n \to \infty$. Define ψ as in theorem 11.1. Simplification gives

$$\left(\frac{\left(\frac{1}{Vd} - 2^d \epsilon\right)^{1/Vd - 2^d \epsilon} \psi^\delta}{\delta^\delta \left(\frac{1}{Vd} - 2^d \epsilon - \delta\right)^{1/dV - 2^d \epsilon - \delta}} \right) \mathcal{Q}_A(z, \epsilon) \leq \mathcal{Q}_A(z, \epsilon + \delta).$$

The factor on the left-hand side is a maximum if $\delta = \delta^*$, with $\delta^* = \frac{\left(\frac{1}{Vd} - 2^d\epsilon\right)\psi}{1+\psi} > 0$. This gives

$$(1 + \psi)^{1/dV - 2^d \epsilon} \mathcal{Q}_A(z, \epsilon) \leq \mathcal{Q}_A(z, \epsilon + \delta^*).$$

If $2^d \epsilon < \frac{1}{dV}$, then such a δ^* can be found.

Since $\log \mathcal{Q}_A(z, \epsilon)$ is a concave function of ϵ, this shows that, for fixed z, the function $\mathcal{Q}_A(z, \epsilon)$ is strictly increasing in some interval $(0, \epsilon_c)$. □

Any subgraph of the hypercubic lattice can be inserted in the interior of a box-animal which is large enough. Thus, any finite subgraph can be a subset of a prime pattern, and theorem 11.2 generalises to any pattern P which can occur once in an animal. For example, the empty pattern (an absence of edges) can be inserted in the interior of a box-animal.

If $A_n(z; [m, P])$ is the partition function of animals of size n containing the pattern P exactly m times, let $A_n(z; [\leq m, P])$ be the partition function of animals of size n containing the pattern P *at most* m times.

Theorem 11.3 *Let P be any pattern which can occur once in a lattice animal. Then there exists an $\epsilon_0 > 0$ such that, for any $\epsilon \in [0, \epsilon_0)$, a $k_z > 0$, and an $N_0 > 0$ can be found such that*

$$A_n(z; [\lfloor \epsilon n \rfloor, P]) \leq A_n(z; [\leq \lfloor \epsilon n \rfloor, P]) < e^{-k_z n} A_n(z),$$

for all $n > N_0$. The conclusions of this theorem also apply to the equivalent models of bond-trees.

Proof Since $\mathcal{Q}_A(z, \epsilon) < \mathcal{Q}_A(z, \epsilon_c)$, whenever $\epsilon < \epsilon_c$, it follows from theorem 11.2 that

$$\limsup_{n \to \infty} \tfrac{1}{n} \log\left(\frac{A_n(z; [\lfloor \epsilon n \rfloor, P])}{A_n(z; [\lfloor \epsilon_c n \rfloor, P])} \right) = \log\left(\frac{\mathcal{Q}_A(z, \epsilon)}{\mathcal{Q}_A(z, \epsilon_c)} \right) = -2k_z < 0.$$

This defines a suitable value for k_z. Thus, there exists an $N_0 \in \mathbb{N}$ such that, if $n > N_0$, then $A_n(z; [\lfloor \epsilon n \rfloor, P]) \leq e^{-2k_z} A_n(z; [\lfloor \epsilon_c n \rfloor, P]) \leq e^{-2k_z} A_n(z)$.

Finally, notice that, if $\epsilon \in [0, \epsilon_c)$, then

$$\limsup_{n \to \infty} \tfrac{1}{n} \log A_n(z; [\leq \lfloor \epsilon n \rfloor, P]) \leq \sup\{\log \mathcal{Q}_A(\delta) \mid 0 \leq \delta \leq \epsilon\} = \log \mathcal{Q}_A(z, \epsilon),$$

since $\mathcal{Q}_A(z, \epsilon)$ is strictly increasing on $(0, \epsilon_c)$. Thus,

$$\limsup_{n \to \infty} \tfrac{1}{n} \log\left(\frac{A_n(z; [\leq \lfloor \epsilon n \rfloor, P])}{A_n(z; [\lfloor \epsilon_c n \rfloor, P])} \right) = \limsup_{n \to \infty} \tfrac{1}{n} \log\left(\frac{A_n(z; [\lfloor \epsilon n \rfloor, P])}{A_n(z; [\lfloor \epsilon_c n \rfloor, P])} \right).$$

This completes the proof for animals. The proof for bond-trees is mutatis mutandis but with $A_n(z)$ replaced with the partition function of bond-trees. □

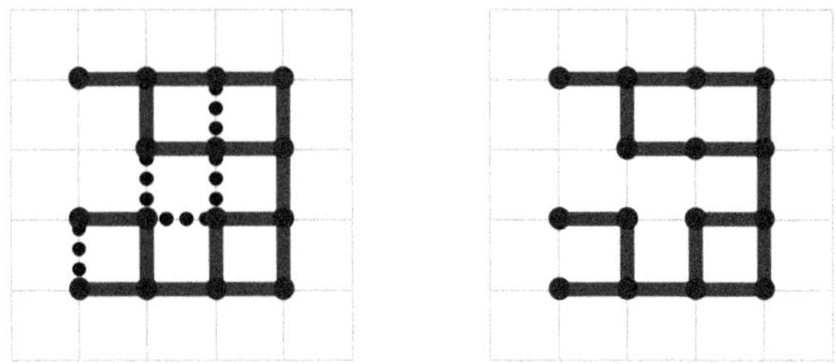

FIG. 11.3. (a) Animals in the contact model are counted by the number of nearest-neighbour contacts between vertices adjacent in the lattice. (b) Animals in the cycle model are counted by a cyclomatic index; this animal has two independent cycles.

A corollary of this is that the growth constant of lattice trees τ_d is strictly less than the growth constant of lattice animals λ_d: $\tau_d < \lambda_d$ (see theorem 2.2).

A pattern theorem for models of lattice animals (and lattice trees) was proven in reference [397]. The conclusions of theorem 11.3 apply with some additional work to models of site-animals and site-trees. It is also possible to generalise the above to models of bond-animals counted by vertices, or site-animals counted by edges. The proof for site-trees requires more care than the proof given above.

11.2 Self-interacting or collapsing lattice animals

Self-interacting branched polymers undergo a *collapse transition* similar to self-interacting linear polymers [117] (see section 9.4). The critical point of the collapse transition is the *branched polymer θ-point* (which may have a connection to bond percolation [226]).

There are two models of collapsing animals. The first is the *contact model*, where collapse is induced by nearest-neighbour contacts in animals (see figure 11.3(a)). The second is the *cycle model* in figure 11.3(b)), where collapse is induced by cycles. These models have been examined in numerous studies [198–201, 214, 216, 217].

In the contact model, animals are weighted by the number of *nearest-neighbour contacts* between vertices in the animal. This model is necessarily a bond-animal model, since neither site-animals nor site-trees can contain contacts. It is, however, possible to define site-animal and site-trees with *next-nearest-neighbour contacts*. In the cycle model the animals are weighted by the number of independent cycles.

Collapsing animals may be studied as site- or bond-animals, or site- or bond-trees, and counted by vertices or by edges, with respect to either nearest-neighbour contacts or cycles. This gives a large number of different models

[198, 199] which are in the same universality class. In this section the discussion will be limited to models of bond-animals counted by edges.

11.2.1 *The cycle model of lattice animals*

Let $a_n(c)$ be the number of bond-animals of size n edges and with c independent cycles in $\mathbb{L}^d$. The partition function of this model is $A_n(x_1) = \sum_{c\geq 0} a_n(c)\, x_1^c$.

Concatenating two animals as illustrated in figure 2.3 gives the supermultiplicative inequality $\sum_{c_1=0}^{c} a_n(c_1) a_m(c-c_1) \leq a_{n+m+1}(c)$. Multiply this by x_1^c and sum over c to see that

$$A_n(x_1) A_m(x_1) \leq A_{n+m+1}(x_1). \tag{11.15}$$

Since there exists a $K > 0$ such that $c_n(k) \leq K^n$ (see lemma 2.1) for all $n \geq 0$ and for finite K, this shows that the thermodynamic limit in the model is given for all $x_1 \geq 0$ by

$$\alpha_C(x_1) = \lim_{n\to\infty} \tfrac{1}{n} \log A_n(x_1). \tag{11.16}$$

By theorem 3.9 and as shown in section 3.3.1.3, the limit

$$\log \mathcal{Q}_C(\epsilon) = \lim_{n\to\infty} \tfrac{1}{n} \log a_n(\lfloor \epsilon n \rfloor) \tag{11.17}$$

exists and $\mathcal{Q}_C(\epsilon)$ is the density function of cycles. By theorem 3.9, $\log \mathcal{Q}_C(\epsilon)$ is a concave function of ϵ.

Lemma 11.4 *The density function $\mathcal{Q}_C(\epsilon) \leq \lambda_d$. The domain of $\mathcal{Q}_C(\epsilon)$ is $[0, 1-\frac{1}{d}]$.*

Proof Since $a_n(\lfloor \epsilon n \rfloor) \leq a_n$ it follows by theorem 2.2 that $\mathcal{Q}_C(\epsilon) \leq \lambda_d$. Clearly, $\mathcal{Q}_C(0) = \tau_d \geq 1$.

Let α be a bond-animal with n edges in $\mathbb{L}^d$. Notice that, if α is a tree, then it has zero cycles; therefore, $\mathcal{Q}_C(0) = \tau_d < \lambda_d$, by theorems 2.2 and 11.2.

The number of cycles in α can be maximised if α is chosen to be the 1-skeleton (or 1-complex) of a d-dimensional hypercube. The largest hypercube with n edges has at least $\lfloor \frac{n}{d} \rfloor$ vertices. Its side-length will be at least $\lfloor \lfloor \frac{n}{d} \rfloor^{1/d} \rfloor$ and it has at most $\lceil \frac{n}{d} \rceil$ vertices and a side-length of at most $\lceil \lceil \frac{n}{d} \rceil^{1/d} \rceil$. An animal is said to be *almost-hypercube* if it contains a hypercube of side-length $\lfloor \lfloor \frac{n}{d} \rfloor^{1/d} \rfloor$ and is itself contained in a hypercube of side-length $\lceil \lceil \frac{n}{d} \rceil^{1/d} \rceil$.

Let α be an almost-hypercube of size n and with v vertices. By Euler's relation, the number of cycles in α is $c = n - v + 1$. By hand shaking, $2n$ is equal to the sum of the degrees of vertices, which is at most $2dv$, so $v \geq \lfloor \lfloor \frac{n}{d} \rfloor^{1/d} \rfloor^d$.

Therefore, α has at most $n - \lfloor \lfloor \frac{n}{d} \rfloor^{1/d} \rfloor^d + 1$ independent cycles.

On the other hand, animals which are almost hypercubes have at most $\lceil \lceil \frac{n}{d} \rceil^{1/d} \rceil^d$ vertices and at least $n - \lceil \lceil \frac{n}{d} \rceil^{1/d} \rceil^d + 1$ cycles.

If $C(n)$ is the maximum number of cycles in α, then $n - \lceil \lceil \frac{n}{d} \rceil^{1/d} \rceil^d + 1 \leq C(n) \leq n - \lfloor \lfloor \frac{n}{d} \rfloor^{1/d} \rfloor^d + 1$. This shows that $\lim_{n\to\infty} \frac{1}{n} C(n) = 1 - \frac{1}{d}$.

That is, $\limsup_{n\to\infty} \frac{1}{n} \log a_n(C(n)) = \log \mathcal{Q}_C(1 - \frac{1}{d}) \geq 0$ is defined because $a_n(C(n)) > 0$ for an infinite sequence of values of n. □

Since $\log \mathcal{Q}_C(\epsilon)$ is concave, it is continuous on $(0, 1-\frac{1}{d})$ and differentiable almost everywhere in this interval.

Cycles can be removed from an animal by deleting edges [402].

Lemma 11.5 *If $C(n)$ is the maximum number of independent cycles in an animal of size n in the hypercubic lattice, then*

$$\binom{c+j}{j} a_{n+j}(c+j) \leq \binom{C(n)-c}{j} a_n(c).$$

Proof Let $\alpha_n(c)$ be an animal with n edges and c cycles.

An animal $\alpha_{n+j}(c+j)$ has at least one set of $c+j$ independent edges which may be deleted to give a tree. Pick one such set and delete j edges in $\binom{c+j}{j}$ ways. This gives an animal with n edges and c cycles.

However, more than one animal can give the same outcome in the above. An upper bound is needed on the possible number of outcomes.

If j edges are added to $\alpha_n(c)$, then there are at most $C(n)-c$ nearest neighbour contacts between vertices where an edge can be added to form a cycle (since there are at most $C(n)$ cycles). Choose j of these possible contacts; this shows that the number of different outcomes if j edges are deleted in $A_{n+j}(c+j)$ is at most $\binom{C(n)-c}{j}$. This completes the proof. □

Theorem 11.6 *The limit $\lim_{\epsilon\to 0^+} \mathcal{Q}_C(\epsilon) = \mathcal{Q}_C(0)$; that is, $\mathcal{Q}_C(\epsilon)$ is continuous on $[0, 1-\frac{1}{d})$. In addition, $\frac{d^+}{d\epsilon}\mathcal{Q}_C(\epsilon) \mid_{\epsilon=0^+} = \infty$.*

Proof By theorem 11.2, there is an $\epsilon_c > 0$ such that, if $\epsilon \in (0, \epsilon_c)$, then $\mathcal{Q}_C(0) < \mathcal{Q}_C(\epsilon)$. In lemma 11.5, put $c = 0$, and $j = \lfloor \epsilon n \rfloor$. Since $C(n) \leq n$, this shows that $a_{n+\lfloor \epsilon n\rfloor}(\lfloor \epsilon n \rfloor) \leq \binom{n}{\lfloor \epsilon n \rfloor} a_n(0)$. Take the power $\frac{1}{n}$ and let $n \to \infty$. This gives

$$\mathcal{Q}_C(0) \leq \mathcal{Q}_C^{1+\epsilon}\left(\tfrac{\epsilon}{1+\epsilon}\right) \leq \frac{\mathcal{Q}_C(0)}{\epsilon^\epsilon (1-\epsilon)^{1-\epsilon}},$$

provided that ϵ is small enough. Take $\epsilon \to 0^+$ to see that $\mathcal{Q}_C(\epsilon)$ is right-continuous at $\epsilon = 0$.

To see that the right-derivative at $\epsilon = 0$ is infinite, argue as follows. Let $t_n = a_n(0)$ be the number of trees of size n. Let $t_n[m, P]$ be the number of trees of size n in which the pattern P occurs m times independently and let $t_n[\geq m, P]$ be the number of trees in which the pattern P occurs at least m times. Define $t_n[\leq m, P]$ similarly.

Let P be the pattern consisting of three edges arranged in a ⊓ (a unit square with the bottom edge absent).

This pattern occurs with positive density in almost all trees, by theorem 11.3. If $\epsilon > 0$ is small enough, there exist a $k > 0$ and an N_0 such that

$$t_n[\geq \lfloor \epsilon n \rfloor, P] = t_n - t_n[< \lfloor \epsilon n \rfloor, P] \geq \left(1 - e^{-kn}\right) t_n$$

for all $n > N_0$.

Let $0 < \delta < \epsilon$. Choose $\lfloor \delta n \rfloor$ of the P's and add a fourth edge to each to complete a 4-cycle. This shows that

$$\binom{\lfloor \epsilon n \rfloor}{\lfloor \delta n \rfloor}(1 - e^{-kn}) t_n \leq a_{n+\lfloor \delta n \rfloor}(\lfloor \delta n \rfloor).$$

Take the $\frac{1}{n}$-th power and let $n \to \infty$. Since $a_n(0) = t_n$ and $\mathcal{Q}_C(\epsilon) \leq \lambda_d$, it follows by lemma 11.4 that

$$\frac{\epsilon^\epsilon \mathcal{Q}_C(0)}{\delta^\delta (\epsilon - \delta)^{\epsilon - \delta}} \leq \mathcal{Q}_C^{1+\delta}\left(\tfrac{\delta}{1+\delta}\right) \leq \lambda_d^\delta \mathcal{Q}_C\left(\tfrac{\delta}{1+\delta}\right).$$

In other words,

$$\frac{\mathcal{Q}_C\left(\frac{\delta}{1+\delta}\right) - \mathcal{Q}_C(0)}{\left(\frac{\delta}{1+\delta}\right)} \geq \frac{\mathcal{Q}_C(0)}{\left(\frac{\delta}{1+\delta}\right)}\left(\frac{\epsilon^\epsilon \lambda_d^{-\delta}}{\delta^\delta(\epsilon - \delta)^{\epsilon-\delta}} - 1\right).$$

The right-hand side diverges if $\delta \to 0^+$. This completes the proof. □

The above results generalise to models of site-animals counted by size (number of edges), so the cycle density function of collapsing site-animals counted by edges has the properties in theorem 11.6. These methods also apply to the case of collapse by nearest-neighbour contacts in bond-animals and bond-trees, counted either by number of edges or number of vertices.

Theorem 11.7 *The value of* $\mathcal{Q}_C(1 - \frac{1}{d}) = 1$, *and* $\mathcal{Q}_C(\epsilon)$ *is left-continuous at* $\epsilon = 1 - \frac{1}{d}$. *The left-derivative of* $\mathcal{Q}_C(\epsilon)$ *at* $\epsilon = 1 - \frac{1}{d}$ *is infinite. That is,* $\frac{d^-}{d\epsilon} \mathcal{Q}_C(\epsilon) \mid_{\epsilon = 1 - 1/d} = -\infty$.

Proof It is a corollary of lemma 11.4 that $\mathcal{Q}_C(\epsilon) \geq 1$.

To see that $\mathcal{Q}_C(\epsilon)$ is left-continuous at $\epsilon = 1 - \frac{1}{d}$, let $\epsilon \in \left(0, 1 - \frac{1}{d}\right)$ and let α be an animal with n edges and $c = C(n) - \lceil \epsilon n \rceil$ cycles. The number of vertices in α is $v = n - C(n) + \lceil \epsilon n \rceil + 1$.

A *perimeter edge* of α is a lattice edge with at least one endpoint in α (but is itself not in α). There are at most $b = 2dv - 2n = 2(d-1)n + 2d(\lceil \epsilon n \rceil - C(n)) + 2d$ perimeter edges in α.

Let $a_{n,v,b}$ be the number of animals with n edges, v vertices and b perimeter edges.

Consider a bond percolation process in $\mathbb{L}^d$. Let p be the density of open edges. Then the probability that the cluster at the origin has exactly n edges, $C(n) - \lceil \epsilon n \rceil$ cycles and b perimeter edges is

$$\sum_{v \geq 1} v\, a_{n,v,b}\, p^n (1-p)^b \leq 1.$$

Take the power $\frac{1}{n}$ and let $n \to \infty$. This shows that

$$\mathcal{Q}_C(1-\tfrac{1}{d}-\epsilon)\,p(1-p)^{2d\epsilon} \leq 1.$$

The left-hand side is a maximum if $p = \frac{1}{1+2d\epsilon}$, in which case it follows that

$$\mathcal{Q}_C(1-\tfrac{1}{d}-\epsilon) \leq \frac{(1+2d\epsilon)^{1+2d\epsilon}}{(2d\epsilon)^{2d\epsilon}}.$$

Taking $\epsilon \to 0^+$ shows that $\lim_{\epsilon\to 0^+} \mathcal{Q}_C(1-\frac{1}{d}-\epsilon) \leq 1$. This proves the left-continuity of $\mathcal{Q}_C(\epsilon)$ at $\epsilon = 1-\frac{1}{d}$, with $\mathcal{Q}_C(1-\frac{1}{d}) = 1$.

It remains to prove that the left-derivative of $\mathcal{Q}_C(\epsilon)$ at $\epsilon = 1-\frac{1}{d}$ is infinite.

In lemma 11.5 put $c+j = \lfloor \epsilon n \rfloor$ and $c = \lfloor \delta n \rfloor$. Take the $\frac{1}{n}$-th power and let $n \to \infty$. After simplification of the result,

$$\left(\frac{\epsilon^\epsilon(1-\frac{1}{d}-\epsilon)^{1-1/d-\epsilon}}{\delta^\delta(1-\frac{1}{d}-\delta)^{1-1/d-\delta}}\right) \mathcal{Q}_C^{1+\epsilon-\delta}\left(\tfrac{\epsilon}{1+\epsilon-\delta}\right) \leq \mathcal{Q}_C(\delta).$$

Since $1 \leq \mathcal{Q}_C(\epsilon) \leq \lambda_d$, and $0 < \delta < \epsilon$, this implies

$$\left(\frac{\epsilon^\epsilon(1-\frac{1}{d}-\epsilon)^{1-1/d-\epsilon}}{\delta^\delta(1-\frac{1}{d}-\delta)^{1-1/d-\delta}}\right) \mathcal{Q}_C\left(\tfrac{\epsilon}{1+\epsilon-\delta}\right) \leq \mathcal{Q}_C(\delta).$$

Take $\epsilon \to 1-\frac{1}{d}$ from below and put $\delta = 1-\frac{1}{d}-\kappa$. Then

$$\left(\frac{(1-\frac{1}{d})^{1-1/d}}{\kappa^\kappa(1-\frac{1}{d}-\kappa)^{1-1/d-\kappa}}\right) \mathcal{Q}_C\left(\tfrac{d+1}{d(1+\kappa)}\right) \leq \mathcal{Q}_C(1-\tfrac{1}{d}-\kappa).$$

This shows that

$$\left(1-\frac{(1-\frac{1}{d})^{1-1/d}}{\kappa^\kappa(1-\frac{1}{d}-\kappa)^{1-1/d-\kappa}}\right) \mathcal{Q}_C\left(\tfrac{d+1}{d(1+\kappa)}\right) \geq \mathcal{Q}_C\left(\tfrac{d+1}{d(1+\kappa)}\right) - \mathcal{Q}_C(1-\tfrac{1}{d}-\kappa).$$

Divide this by $\left(\frac{d+1}{d(1+\kappa)}\right) - \left(1-\frac{1}{d}-\kappa\right)$ and let $\kappa \to 0^+$. The right-hand side is the left-derivative at $1-\frac{1}{d}$, while the left-hand side diverges to $-\infty$. □

The proofs of theorems 11.6 and 11.7 are similar to proofs in references [402, 403]. The connection between percolation and collapsing animals was pointed out in reference [150].

The asymptotic behaviour of lattice trees is given by $t_n \sim n^{-\theta}\,\tau_d^n$ (see equation (2.9)), where θ is the branch polymer entropic exponent (see table 2.2). This suggest that $\lim_{n\to\infty} \frac{1}{\log n}\log(t_n\tau_d^{-n}) = -\theta$, but it is not known whether this limit exists. If it exists, then $a_n(c) \sim n^{-\theta_c}\tau_d^n$, and $\theta_c = \theta - c$ [526].

Theorem 11.8 *Suppose that the limit* $\theta = -\lim_{n\to\infty} \frac{1}{\log n}\log(t_n\tau_d^{-n})$ *exists. Then the limit*

$$\theta_c = -\lim_{n\to\infty} \frac{1}{\log n}\log(a_n(c)\,\tau_d^{-n})$$

exists for $c = 1, 2, \ldots$, *and, moreover,* $\theta_c = \theta - c$.

Proof By lemma 11.5,

$$a_{n+c}(c) \leq \binom{C(n)}{n} t_n.$$

Thus, $\limsup_{n\to\infty} \frac{1}{\log n}\left(\log(a_n(c)\tau_d^{-n})\right) \leq -\theta + c$.

Next, let P be the square-cap pattern $\sqcap$ and consider theorem 11.3 for lattice trees. Let $t_n[\geq \lfloor \epsilon n \rfloor, \sqcap]$ be the number of trees of size n and which contain the pattern $\sqcap$ at least $\lfloor \epsilon n \rfloor$ times. If $\epsilon > 0$ is small enough, then $\lim_{n\to\infty} (t_n[\geq \lfloor \epsilon n \rfloor, \sqcap])^{1/n} = \tau_d$. Choose $c < \lfloor \epsilon n \rfloor$ of the $\sqcap$'s and form cycles by adding an edge to each in the obvious way. This shows that

$$\binom{\lfloor \epsilon n \rfloor}{c} t_n[\geq \lfloor \epsilon n \rfloor, \sqcap] \leq a_{n+c}(c).$$

By theorem 11.3, there exist a $k > 0$ and an N_0 such that $t_n \geq t_n[\geq \lfloor \epsilon n \rfloor, \sqcap] \geq (1 - e^{-kn})t_n$. Thus, multiply the above by τ_d^{-n}, take logarithms and divide by $\log n$. Take the limit inferior of the right-hand side of the resulting inequality to see that $\liminf_{n\to\infty} \frac{1}{\log n}(\log(a_n(c)\tau_d^{-n})) \geq -\theta + c$. □

11.2.2 *Collapse in the cycle-contact model*

Let $a_n(c, k)$ be the number of bond-animals of size n edges, with c cycles and with k nearest-neighbour contacts (see figure 11.3) [198–200].

Let $\alpha = \{\langle \vec{x} \sim \vec{y} \rangle \mid 0 \leq x_i, y_i \leq L\}$ be a bond-animal in $\mathbb{L}^d$ which is a hypercube of side-length L with all possible edges present. The number of vertices in α is $v = (L+1)^d$, and the number of edges is n, where $dL^d \leq n \leq d(L+1)^d$. The number c of cycles in α is given by $c = n + 1 - v$, and so $c \leq 1 + d(L+1)^d - (L+1)^d$, and $c \geq 1 + dL^d - (L+1)^d$. Choose $\lfloor \delta n \rfloor$ edges in independent cycles and delete them. This leaves an animal with $n - \lfloor \delta n \rfloor$ edges, $c - \lfloor \delta n \rfloor$ cycles and $\lfloor \delta n \rfloor$ contacts. Divide c by the number of edges and take $n \to \infty$. This shows that if the density of contacts is δ, then the density of cycles is *at most* $\frac{d-1-\delta}{d-\delta}$.

This defines a region Δ over which the joint density function of animals in the contact-cycle model will be defined, as illustrated in figure 11.4.

Let α_1 be an animal of n_1 edges, $c - c_1$ cycles and $k - k_1$ contacts, and let α_2 be an animal of n_2 edges, c_1 cycles and k_1 contacts. Concatenating α_1 and α_2 as illustrated in figure 2.3 gives

$$\sum_{c_1, k_1} a_{n_1}(c_1, k_1) a_{n_2}(c - c_1, k - k_1) \leq a_{n_1+n_2+1}(c, k). \tag{11.18}$$

Multiplying equation (11.18) by x_2^k and summing over k (while putting $c_2 = c - c_1$) gives

$$\mathfrak{a}_{n_1}(c_1; x_2)\, \mathfrak{a}_{n_2}(c_2; x_2) \leq \mathfrak{a}_{n_1+n_2+1}(c_1 + c_2; x_2), \tag{11.19}$$

where $\mathfrak{a}_n(c; x_2) = \sum_{k\geq 0} a_n(c, k)\, {x_2}^k$ is the partition function of animals of size n with c cycles. By equation (3.47), it follows that

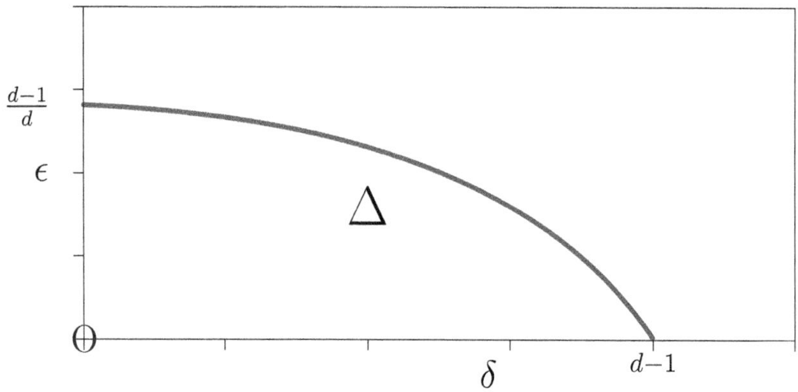

FIG. 11.4. The domain Δ of the joint density function of the contact-cycle bond-animal model has as boundaries the curve $\epsilon(d-\delta) = d-1-\delta$, and the horizontal and vertical axes.

$$\log \mathcal{Q}_C(\epsilon; x_2) = \lim_{n\to\infty} \tfrac{1}{n} \log \mathfrak{a}_n(\lfloor \epsilon n \rfloor; x_2) \tag{11.20}$$

exists and is the density function of cycles at given contact activity x_2.

A similar approach shows that that the limit

$$\log \mathcal{Q}_K(x_1, \delta) = \lim_{n\to\infty} \tfrac{1}{n} \log \mathfrak{a}_n(x_1, \lfloor \delta n \rfloor) \tag{11.21}$$

exists and is the density function of contacts at given cycle activity x_1. As shown in section 3.3.1.4, there is a joint density function in this model given by

$$\log \mathcal{Q}_{CK}(\epsilon, \delta) = \inf_{x_1>0} \{\log \mathcal{Q}_K(x_1; \delta) - \epsilon \log x_1\} = \inf_{x_2>0} \{\log \mathcal{Q}_C(\epsilon; x_2) - \delta \log x_2\}$$

(see equation (3.48)). Existence of $\mathcal{Q}_{CK}(\epsilon, \delta)$ implies the existence of the thermodynamic limit in the model; the limiting free energy is

$$\mathcal{A}(x_1, x_2) = \lim_{n\to\infty} \tfrac{1}{n} \log A_n(x_1, x_2). \tag{11.22}$$

In equation (11.18) put $k_1 = \lfloor \delta n_1 \rfloor$, and $k_2 = \lfloor \delta n_2 \rfloor$. This gives

$$\sum_{c_1} a_{n_1}(c_1, \lfloor \delta n_1 \rfloor) a_{n_2}(c - c_1; \lfloor \delta n_2 \rfloor) \leq a_{n_1+n_2+1}(c; \lfloor \delta n_1 \rfloor + \lfloor \delta n_2 \rfloor). \tag{11.23}$$

Multiply by x_1 and sum over c to obtain

$$\overline{a}_{n_1}(x_1, \lfloor \delta n_1 \rfloor) \overline{a}_{n_2}(x_1; \lfloor \delta n_2 \rfloor) \leq \overline{a}_{n_1+n_2+1}(x_1; \lfloor \delta n_1 \rfloor + \lfloor \delta n_2 \rfloor), \tag{11.24}$$

where $\overline{a}_n(x_1, k) = \sum_{c\geq 0} a_n(c, k)\, x_1^c$. As shown in section 3.3.1.4, there exists a density function

$$\alpha_K(x_1;\delta) = \lim_{n\to\infty} \tfrac{1}{n} \log \overline{a}_n(x_1, \lfloor \delta n \rfloor). \tag{11.25}$$

This gives the joint density function

$$\log \mathcal{Q}_{CK}(\epsilon,\delta) = \inf\{\alpha_K(x_1;\delta) - x_1 \log \epsilon\}. \tag{11.26}$$

By theorem (3.17), there is a sequence $\langle \epsilon_n \rangle$ such that $\epsilon_n = \lfloor \epsilon n \rfloor + o(n)$ and

$$\log \mathcal{Q}_{CK}(\epsilon,\delta) = \lim_{n\to\infty} \tfrac{1}{n} \log a_n(\epsilon_n, \lfloor \delta n \rfloor). \tag{11.27}$$

Notice that $\log \mathcal{Q}_{CK}(\epsilon,\delta)$ is concave in both ϵ and in δ.

11.2.2.1 *Properties of* $\mathcal{Q}_{CK}(\epsilon,\delta)$*:*

Lemma 11.9 *The function* $a_n(c,k)$ *satisfies the following inequalities:*

(1) $a_n(c,k) \leq a_{n+1}(c,k)$;

(2) $\binom{c+k_0}{k_0} a_n(c+k_0,k) \leq \binom{k+k_0}{k_0} a_{n-k_0}(c,k+k_0)$; *and*

(3) $\binom{k+c_0}{c_0} a_n(c,k+c_0) \leq \binom{c+c_0}{c_0} a_{n+c_0}(c+c_0,k)$.

Proof The first inequality is shown by appending a single edge to the (lexicographic) top vertex of the animal. The second inequality is seen by noting that there is at least one set of $c+k_0$ independent edges in an animal counted by $a_n(c+k_0,k)$ such that deleting these edges will leave a tree. Choose k_0 of these edges and remove them from the animal; this removes k_0 cycles and creates k_0 new contacts. There are a total of $k+k_0$ contacts, of which k_0 must be replaced with edges to find the original animal. The third inequality is shown using a similar method. □

Theorem 11.10 *The joint density function* $\mathcal{Q}_{CK}(\epsilon,\delta)$ *is continuous in* Δ *and on the boundaries of* Δ *if* $\delta = 0$ *or* $\epsilon = 0$.

Proof Let ϵ_n and $\lfloor \delta n \rfloor$ be as in equation (11.27) and put $\delta_n = \lfloor \delta n \rfloor$.

Since $\log \mathcal{Q}_{CK}(\delta,\epsilon)$ is concave, it is also continuous in the interior of Δ.

To show continuity if $\epsilon = 0$ in the second inequality in lemma 11.9, let $c = 0$, let $k_0 = \epsilon_n$ and let $k = \delta_n$. Use the first inequality of lemma 11.9 to see that

$$a_n(\epsilon_n,\delta_n) \leq \binom{\epsilon_n+\delta_n}{\epsilon_n} a_n(0,\epsilon_n+\delta_n).$$

Take logarithms on both sides, the power $\frac{1}{n}$ and then take $n \to \infty$. By the proof of theorem 3.16, $\limsup_{n\to\infty} \frac{1}{n} \log a_n(0,\epsilon_n+\delta_n) \leq \log \mathcal{Q}_{CK}(0,\epsilon+\delta)$ on the right-hand side. Thus, it follows that

$$\log \mathcal{Q}_{CK}(\epsilon,\delta) \leq \log \left[\frac{(\epsilon+\delta)^{\epsilon+\delta}}{\epsilon^\epsilon \delta^\delta} \right] + \log \mathcal{Q}_{CK}(0,\epsilon+\delta).$$

This shows that $\lim_{\epsilon\to 0^+} \mathcal{Q}_{CK}(\epsilon,\delta) \leq \mathcal{Q}_{CK}(0,\delta)$. Since $\mathcal{Q}_{CK}(\epsilon,\delta) \geq \mathcal{Q}_{CK}(0,\delta)$,

$$\lim_{\epsilon\to 0^+} \mathcal{Q}_{CK}(\epsilon,\delta) = \mathcal{Q}_{CK}(0,\delta).$$

Next, notice that $\mathcal{Q}_{CK}(0,\delta)$ is the density function of contacts in a model of bond-trees. Then, $\log \mathcal{Q}_{CK}(0,\delta)$ is concave and continuous. Thus, $\mathcal{Q}_{CK}(\epsilon,\delta)$ is a continuous function on that part of the boundary Δ with $\epsilon = 0$ and $\delta \in (0, d-1)$.

Continuity along the boundary of Δ with $\delta = 0$ is shown as follows. In the third equation in lemma 11.9 put $k = 0$, put $c = \epsilon_n$ and put $c_0 = \delta_n$. Taking logarithms, dividing by n and letting $n \to \infty$ give, as above,

$$\log \mathcal{Q}_{CK}(\epsilon,\delta) \leq \log\left(\frac{(\epsilon+\delta)^{\epsilon+\delta}}{\epsilon^\epsilon \delta^\delta}\right) + (1+\delta)\log \mathcal{Q}_{CK}\left(\tfrac{\epsilon+\delta}{1+\delta}, 0\right).$$

This shows that $\lim_{\delta\to 0^+} \mathcal{Q}_{CK}(\epsilon,\delta) \leq \mathcal{Q}_{CK}(\epsilon,0)$. Since $\mathcal{Q}_{CK}(\epsilon,\delta) \geq \mathcal{Q}_{CK}(\epsilon,0)$, this shows that

$$\lim_{\delta\to 0^+} \mathcal{Q}_{CK}(\epsilon,\delta) = \mathcal{Q}_{CK}(\epsilon,0).$$

Since $\mathcal{Q}_{CK}(\epsilon,0)$ is the cycle density function of a model of site-animals counted by size or number of edges, it follows that $\mathcal{Q}_{CK}(\epsilon,0)$ is continuous (and a concave function) of $\log \epsilon$ for $\epsilon \in \left(0, 1-\frac{1}{d}\right)$. In fact, by the same methods leading to theorem 11.6, $\mathcal{Q}_{CK}(\epsilon,0)$ is right-continuous at the point $\epsilon = 0$.

To see continuity at $(0,0)$, notice that

$$\log \mathcal{Q}_{CK}(0,0) \leq \log \mathcal{Q}_{CK}(\epsilon,\delta) \leq \log\left(\frac{(\epsilon+\delta)^{\epsilon+\delta}}{\epsilon^\epsilon \delta^\delta}\right) + (1+\delta)\log \mathcal{Q}_{CK}\left(\tfrac{\epsilon+\delta}{1+\delta}, 0\right).$$

Put $\delta = \delta(\epsilon) \geq 0$, a continuous function of ϵ such that $\delta(\epsilon) \to 0^+$ as $\epsilon \to 0^+$, and $\delta(\epsilon) > 0$ if $\epsilon > 0$. Take $\epsilon \to 0^+$ in the last set of inequalities; by the right-continuity of $\mathcal{Q}_{CK}(\epsilon,0)$ at $\epsilon = 0$, it follows that $\lim_{\epsilon\to 0^+} \log \mathcal{Q}_{CK}(\epsilon,\delta(\epsilon)) = \log \mathcal{Q}_{CK}(0,0)$. This shows that $\mathcal{Q}_{CK}(\epsilon,\delta)$ is continuous at the origin in Δ. □

The function $\mathcal{Q}_{CK}(\epsilon,\delta)$ interpolates between different ensembles of site- and bond-animals and site- and bond-trees, all counted with respect to size (number of bonds or edges). The number $\mathcal{Q}_{CK}(0,0)$ is the growth constant of site-trees counted by size, and $\sup_{\epsilon>0} \mathcal{Q}_{CK}(\epsilon,0)$ is the growth constant of site-animals counted by size.

The growth constant of bond-trees counted by size is $\tau_d = \sup_{\delta>0} \mathcal{Q}_{CK}(0,\delta)$. The growth constant of bond-animals counted by size is $\lambda_d = \sup_{\epsilon>0,\delta>0} \mathcal{Q}_{CK}(\epsilon,\delta)$ (see section (2.1)).

Continuity along the boundary $\epsilon = \frac{d-1-\delta}{d-\delta}$ of Δ is unknown. Bounds are stated in the next theorem.

Theorem 11.11 *For* $0 \leq \delta \leq d-1$,

$$\mathcal{Q}_{CK}\left(\tfrac{d-1-\delta}{d-\delta},\delta\right) \geq \frac{\left(\tfrac{d-1-\delta}{d-\delta}\right)^{(d-1-\delta)/(d-\delta)}}{\left(\tfrac{d-1-\delta}{d-\delta}-\delta\right)^{(d-1-\delta)/(d-\delta)-\delta}\delta^{\delta}}, \quad \textit{and}$$

$$\mathcal{Q}_{CK}\left(\tfrac{d-1-\delta}{d-\delta},\delta\right) \leq \frac{\left(\tfrac{2d}{d-\delta}-\delta-1\right)^{(2d/(d-\delta))-\delta-1}}{\left(\tfrac{2d}{d-\delta}-\delta-2\right)^{(2d/(d-\delta))-\delta-2}}.$$

Proof Let $C(n)$ be the largest number of cycles in an animal of size n edges. Consider animals of $C(n)-\delta_n$ cycles and ϵ_n contacts, where (δ_n,ϵ_n) is defined as in the proof of theorem 11.10.

Since there exists an animal with $C(n)$ cycles, delete δ_n edges, one per independent cycle, to show that there are at least $\binom{C(n)}{\delta_n}$ such animals. Take logarithms, divide by n and let $n\to\infty$ to find the lower bound (since $\frac{1}{n}C(n)\to 1-\frac{1}{d}$),

To find the upper bound, argue as follows. Let $a_{n,v,b}$ be the number of lattice animals with n edges, v vertices, $C(n)-\epsilon_n$ cycles, δ_n contacts and b perimeter edges.

Consider bond percolation with density p in the hypercubic lattice. The probability that the cluster at $\vec{0}$ has n edges is

$$\sum_{v\geq 0} v\, a_{n,v,b}\, p^n (1-p)^b \leq 1.$$

In an animal with v vertices and n edges, the number of cycles is $c=n-v+1$. Moreover, the size of the perimeter is $b = 2d\epsilon_n + 2d(n+1-C(n)) - 2n - \delta_n$. Taking the $\frac{1}{n}$-th power and taking $n\to\infty$ gives

$$\mathcal{Q}_{CK}\left(\tfrac{d-1-\delta}{d-\delta}-\epsilon,\delta\right)p(1-p)^{2d\epsilon-\delta-2+2d/(d-\delta)} \leq 1.$$

The left-hand side is a maximum if $p = \frac{1}{2d\epsilon-\delta-2+2d/(d-\delta)}$. If $\epsilon\to 0^+$, then the claimed upper bound is found. □

Right partial derivatives to ϵ and δ of $\mathcal{Q}_{CK}(\epsilon,\delta)$ are infinite along the $\delta=0$ boundary and the $\epsilon=0$ boundary of Δ.

Theorem 11.12 *The right partial derivatives of* $\mathcal{Q}_{CK}(\epsilon,\delta)$ *to* ϵ *(for fixed* $0\leq\delta<d-1$*) at* $\epsilon=0$ *and to* δ *(for fixed* $0\leq\epsilon<1-\frac{1}{d}$*) at* $\delta=0$ *are infinite.*

Proof Let $\mathfrak{a}_n(c;x_2;[\geq\lfloor\gamma n\rfloor,P_0])$ be the partition function of animals with c cycles, with contact activity x_2 and which contain the pattern P_0 in figure 11.5 at least $\lfloor\gamma n\rfloor$ times (see equation (11.20)).

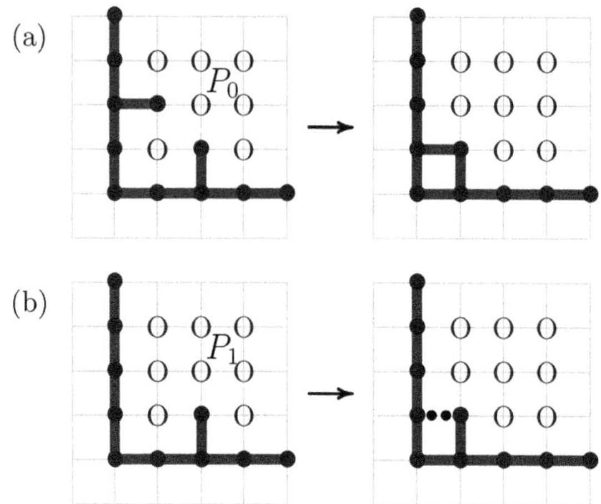

FIG. 11.5. (a) A cycle may be created in P_0 by moving two edges as shown. (b) A contact may be created in P_1 by moving a single edge as shown.

Let $\epsilon > 0$ be small and suppose ϵ_n is defined as in equation (11.27) (so that $\epsilon_n = \lfloor \epsilon n \rfloor + o(n)$). Put $c = 0$, choose ϵ_n of the patterns P_0 and change the pattern to create a cycle (see figure 11.5). This shows that

$$\binom{\lfloor \gamma n \rfloor}{\epsilon_n} \mathfrak{a}_n(0; x_2; [\geq \lfloor \gamma n \rfloor, P_0]) \leq \mathfrak{a}_n(\epsilon_n; x_2).$$

By theorem 11.3, there is a k_{x_2} such that

$$\mathfrak{a}_n(0; x_2; [\geq \lfloor \gamma n \rfloor, P_0]) \geq \left(1 - e^{-k_{x_2} n}\right) \mathfrak{a}_n(0; x_2)$$

for all sufficiently large n. Combine these inequalities, take the power $\frac{1}{n}$ and let $n \to \infty$. By equation (11.20), the result is

$$\left(\frac{\gamma^\gamma}{\epsilon^\epsilon (\gamma - \epsilon)^{\gamma - \epsilon}}\right) \mathcal{Q}_C(0; x_2) \leq \mathcal{Q}_C(\epsilon; x_2).$$

Multiply both sides by ${x_2}^\delta$ and take the infimum over $x_2 > 0$. This gives

$$\left(\frac{\gamma^\gamma}{\epsilon^\epsilon (\gamma - \epsilon)^{\gamma - \epsilon}}\right) \mathcal{Q}_{CK}(0, \delta) \leq \mathcal{Q}_{CK}(\epsilon, \delta).$$

Subtract $\mathcal{Q}_{CK}(0, \delta)$ from both sides and divide by ϵ:

$$\frac{1}{\epsilon}\left(\frac{\gamma^\gamma}{\epsilon^\epsilon (\gamma - \epsilon)^{\gamma - \epsilon}} - 1\right) \mathcal{Q}_{CK}(0, \delta) \leq \tfrac{1}{\epsilon}(\mathcal{Q}_{CK}(\epsilon, \delta) - \mathcal{Q}_{CK}(0, \delta)).$$

Take $\epsilon \to 0^+$ on both sides; then the left-hand side diverges, since $\mathcal{Q}_{CK}(0, \delta) > 0$ (this is the case because this is the growth constant of trees with a density of contacts; the number of such trees grows at least as fast as animals with the same density of cycles).

That the right-derivative to δ is infinite at $\delta = 0$ follows by a similar argument, where the pattern P_1 in figure 11.5 is used instead. □

11.2.2.2 *Collapse in the cycle-contact model:* Consider bond percolation in $\mathbb{L}^d$ at density p. The probability that $\vec{0}$ belongs to a cluster C with exactly n edges is

$$\mathrm{P}_r(|C|{=}n) = \sum_{k,c\geq 0} v\, a_n(c,k)\, p^n (1-p)^{s+k}, \qquad (11.28)$$

where $a_n(c,k)$ is the number of bond-animals with n edges, c cycles and k contacts, as well as $v = n - c + 1$ vertices and $s = 2dv - 2k - 2n$ perimeter edges.

Put $q = 1 - p$ and simplify $\mathrm{P}_r(|C|{=}n)$ to obtain

$$\mathrm{P}_r(|C|{=}n) = p^n q^{2n(d-1)+2d} \sum_{k,c\geq 0} (n+1-c)\, a_n(c,k)\, q^{-k-2dc}. \qquad (11.29)$$

Since $1 \leq (n+1-c) \leq (n+1)$, this shows that

$$\lim_{n\to\infty} \tfrac{1}{n} \log \mathrm{P}_r(|C|{=}n) = \mathcal{A}(q^{-2d}, q^{-1}) + \log p + 2(d-1)\log q, \qquad (11.30)$$

by equation (11.22). The left-hand side is equal to the percolation function $\zeta(p)$ (see theorems E.2 and E.13 in appendix E). That is,

$$\mathcal{A}(q^{-2d}, q^{-1}) = -\zeta(p) - \log p - 2(d-1)\log q. \qquad (11.31)$$

Since $\zeta(p)$ is a non-analytic function of p by theorem E.13, this shows that $\mathcal{A}(x_1, x_2)$ is a non-analytic function. Thus, $\mathcal{Q}_{CK}(\epsilon, \delta)$ is a non-analytic function. This gives the following theorem.

Theorem 11.13 *The limiting free energy of bond-animals in the cycle-contact ensemble is non-analytic at least at one point which corresponds to the critical percolation point:* $(x_1, x_2) = (q_c^{-2d}, q_c^{-1})$*, where* $q_c = 1 - p_c(d)$*, and* $p_c(d)$ *is the critical percolation density.* □

This shows that the limiting free energy of collapsing animals in the cycle-contact ensemble is non-analytic at a point corresponding to the critical bond percolation density. In fact, there is a curve in the phase diagram of the model along which animals are weighted as percolation clusters. This curve crosses the collapse critical curve in the bond percolation critical point (see figure 11.6).

Theorem 11.14 *There are non-analyticities in the limiting free energy of collapsing bond-animals at points other than the bond percolation critical point.*

Proof Let $0 \le b \le 1$, put $q = 1-p$ and define the function

$$Q_n(b,p) = p^n q^{2(d-1)n+2d} \sum_{k,c\geq 0} (n+1-c)\, a_n(c,k) \left(\tfrac{b}{q}\right)^k q^{-2dc}. \tag{11.32}$$

Then $Q_n(b,p) \le P_r(|C|{=}n)$ by equation (11.29). Thus,

$$\limsup_{n\to\infty} (Q_n(b,p))^{1/n} \le \lim_{n\to\infty} (P_r(|C|{=}n))^{1/n} = e^{-\zeta(p)} < 1, \text{ provided that } p < p_c(d),$$

by theorems E.2 and E.13. Since $2dv = 2n+2k+s$, it follows that

$$Q_n(b,p) = \sum_{k,c\geq 0} (n+1-c)\, a_n(c,k) \left(\tfrac{p}{b^2}\right)^n \left(\tfrac{q}{b}\right)^{s+k} b^{2dv}.$$

If $b \in (0,1]$, then $1 - \frac{p}{b^2} \le \frac{q}{b}$ for all $p \in [0,1]$. Thus,

$$Q_n(b,p) \ge \sum_{k,c\geq 0} (n+1-c)\, a_n(c,k) \left(\tfrac{p}{b^2}\right)^n \left(1-\tfrac{p}{b^2}\right)^{s+k} b^{2dv} = R_n(b,p).$$

The function $R_n(b,p)$ is the probability that the cluster C at $\vec{0}$ has size n in a combined edge-site percolation model where edges are open with probability $\frac{p}{b^2}$, and sites are occupied with probability b^{2d} (as long as $p \le b^2$ and where every vertex in the cluster is occupied). If b is close enough to 1, then p can be picked large enough that the open cluster at $\vec{0}$ is infinite with non-zero probability.

Thus, there are both $b_c < 1$ and $p_b < 1$ such that $\lim_{n\to\infty}(Q_n(b,p))^{1/n} = 1$, for all $b \ge b_c$ and for all $p \ge p_b$.

Comparing equation (11.22) to equation (11.32) shows that the free energy $\alpha\left(\frac{1}{q^{2d}}, \frac{b}{q}\right)$ is related to $\lim_{n\to\infty}(Q_n(b,p))^{1/n}$:

$$\lim_{n\to\infty} \tfrac{1}{n} \log Q_n(b,p) = \alpha\left(\tfrac{1}{q^{2d}}, \tfrac{b}{q}\right) + \log p + 2(d-1)\log q.$$

If $b > b_c$, then there is a critical point p_b such that $\lim_{n\to\infty}(Q_n(b,p))^{1/n}$ is singular at (b,p_b). This shows that the free energy is non-analytic at (b,p_b) for all $b > b_c$. □

11.2.2.3 *The θ-transition in lattice animals:* A schematic drawing of the phase diagram of collapsing lattice animals in the cycle-contact model is shown in figure 11.6. A critical curve (the θ-transition) separates a phase of free or *expanded* animals from a phase of compact or *collapsed* animals [312].

Animals weighted as percolation clusters are found along a curve in the diagram, and this curve intersects the critical curve in the critical percolation point [403]. At this critical point the animals are weighted as critical percolation clusters.

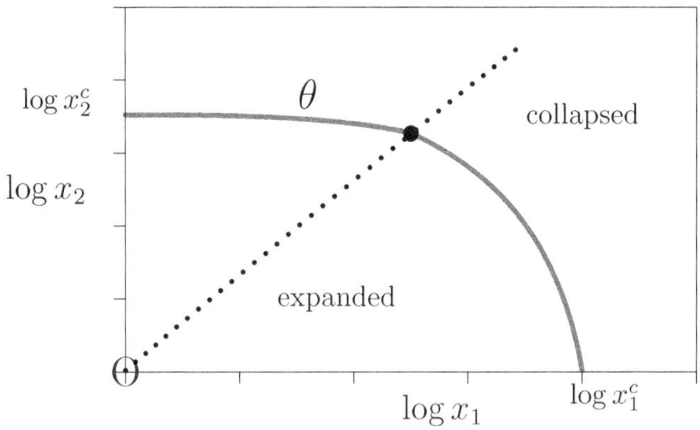

FIG. 11.6. The phase diagram of collapsing bond-animals in the contact-cycle ensemble. The θ-transitions occur along a critical curve separating expanded and collapsed phases. Bond-animals weighted as percolation clusters occur along the dotted curve. This crosses the θ-transition in the critical percolation point.

The θ-transition has been interpreted as a critical point in the one-component Potts model [226, 440, 575, 598, 599] (see reference [206] as well). This connection suggests a single collapsed phase, driven by either a contact or a cycle fugacity [510, 511, 537]. This single phase picture is also supported by extended Potts-model calculations on a Bethe lattice [279], and by a renormalisation group calculation in two dimensions [279].

The free energy $\mathcal{A}(1, x_2)$ (see equation (11.22)) is a model of *contact-collapse* from expanded animals to collapsed animals rich in nearest-neighbour contacts [147]. The critical point for contact-collapse (if $x_1 = 1$) has been estimated to be $x_2^c = 1.93(5)$ in $\mathbb{L}^2$ [295] (see reference [312] for additional results). The interpretation of the contact-collapse transition as a percolation critical point [279] was examined numerically in reference [295].

The scaling of branch polymers at the θ-point is described by its entropic exponent θ_θ (the θ-point value of the entropic exponent θ in equation (2.9)). Numerical estimates of θ_θ are listed in table 11.1 (the numerical data in this table were taken from the results in reference [334]). The entropic exponent θ_θ increases to its (exact) mean field value in $d = 6$ (the upper critical dimension of branched polymer collapse).

In $\mathbb{L}^2$ it was proposed that the branched polymer θ-point is in the Ising universality class with the crossover exponent $\phi_\theta = \frac{8}{15}$ [510, 511] (this gives the specific heat exponent $\alpha_\theta = \frac{1}{8}$). Monte Carlo simulations gave a range of results for this exponent, namely, $\phi_\theta = 0.406(26)$, and $\phi_\theta = 0.619(12)$ [537], while exact enumeration studies of contact-collapse in a model of lattice trees [201] gave $\phi_\theta = 0.60(3)$ [217].

TABLE 11.1. θ-Exponents for branched polymers

d	θ_θ	ϕ_θ	ν_θ	$2-\alpha_t$	$2-\alpha_u$
2	1.96(4)	$\frac{8}{15}$	$\frac{8}{15}$	$\frac{16}{15}$	2
3	2.13(2)	0.427(5)	0.396(7)	1.19(3)	2.8(2)
4	2.277(5)	0.469(1)	0.329(2)	1.316(8)	2.81(3)
5	2.4025(6)	0.49383(2)	0.2849(2)	1.425(1)	2.885(3)
6	$\frac{5}{2}$	$\frac{1}{2}$	$\frac{1}{4}$	$\frac{3}{2}$	3

The exact value of the metric exponent at the θ-transition in $\mathbb{L}^2$ is $\nu_\theta = \frac{8}{15}$ [110, 440, 535]. Numerical simulations gave $\nu_\theta = 0.555(5)$ in reference [312]. A renormalisation group calculation in reference [334] produced $\nu_\theta = 0.52(3)$. This should be compared with the (free) branched polymer value $\nu = 0.644(2)$ [308]. The remaining tricritical exponents in table 11.1 were computed from the tricritical scaling relations in equations (4.14) and (4.53).

In reference [110], the θ-transition of cycle-collapse in $\mathbb{L}^2$ is modelled by a tricritical zero-state Potts model. This approach gives a larger crossover exponent $\phi_\theta = \frac{2}{3}$, while $\nu_\theta = \frac{8}{15}$ (and numerically it is found that $\nu_\theta \approx 0.51$ [110]); see references [130, 368, 369] as well). A renormalisation group calculation in reference [334] gives $\nu_\theta = 0.52(3)$. In reference [558] a study of vesicles without holes as a model of cycle-collapse for animals in $\mathbb{L}^2$ gave the results $\phi_\theta = \frac{2}{3}$, and $\nu_\theta = \frac{1}{2}$.

Three-dimensional vesicles were argued in reference [534] to be in the same universality class as animals undergoing θ-transition by cycle-collapse; this gives the exponents $\phi_\theta = 1$ and $\nu_\theta = \frac{1}{2}$ if $d = 3$. However, renormalisation group calculations give values smaller than this: $\phi_\theta = 0.427(5)$, and $\nu_\theta = 0.396(7)$ [334]. In reference [368] the estimate $\phi_\theta = 0.521$ is claimed in $\mathbb{L}^3$.

11.3 Adsorbing lattice trees

Positive trees rooted at $\vec{0}$ in the half-lattice $\mathbb{L}^d_+$ were introduced in section 2.1.5 (see figure 2.4 and equation (2.20)). If α is a positive tree, and $\vec{v}$ is a vertex in α, then $\vec{v}$ is a *visit* if $\vec{v} \in \partial\mathbb{L}^d_+$. Positive trees weighed by visits are *adsorbing trees.*

A pair of vertices $\{\vec{v}, \vec{u}\}$ in α is a *contact* if $\langle \vec{v} \sim \vec{u} \rangle \notin \alpha$ but $\langle \vec{v} \sim \vec{u} \rangle \in \mathbb{L}^d$. Positive trees weighted by visits and contacts are *adsorbing* and *collapsing* trees.

Trees in the bulk lattice $\mathbb{L}^d$, rooted at $\vec{0}$, may also be weighted by visits (to $\partial\mathbb{L}^d_+$) and by contacts. These trees adsorb at a *defect plane* $\partial\mathbb{L}^d_+$ in $\mathbb{L}^d$.

Let $t^+_n(k, v)$ be the number of positive trees of size n edges, k contacts and v visits. Similarly, define $t_n(k, v)$ to be the number of trees rooted at $\vec{0}$ in $\mathbb{L}^d$, with k contacts and v visits to the defect plane $\partial\mathbb{L}^d_+$. Define the partition functions

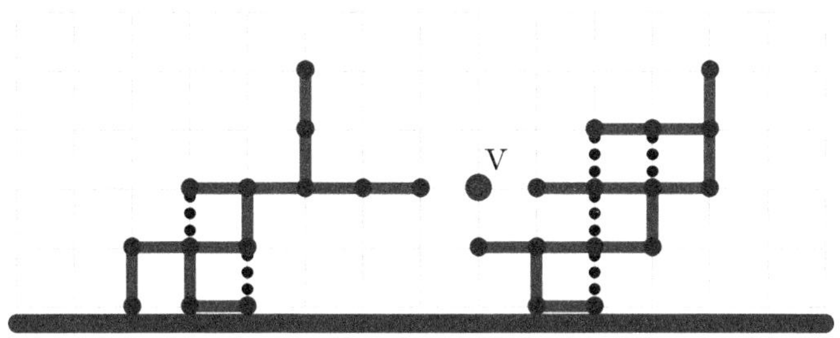

FIG. 11.7. Concatenation of two positive trees. The trees are placed so that there are paths of length 2 which can join them. By placing a new vertex v on one of these paths, the trees are concatenated by adding two edges. This may create up to $2(d-1)$ new contacts and at most one extra visit (if v is in $\partial\mathbb{L}^d_+$).

$$T^+_n(x,a) = \sum_{k,v\geq 0} \mathrm{t}^+_n(k,v)\, x^k a^v, \quad \text{and } T_n(x,a) = \sum_{k,v\geq 0} \mathrm{t}_n(k,v)\, x^k a^v. \qquad (11.33)$$

Theorem 11.15 *There exist limiting free energies $\mathcal{T}^+_v$ and $\mathcal{T}_v$ of adsorbing and collapsing trees given by*

$$\mathcal{T}^+_v(x,a) = \lim_{n\to\infty} \tfrac{1}{n} \log T^+_n(x,a), \quad \textit{and } \mathcal{T}_v(x,a) = \lim_{n\to\infty} \tfrac{1}{n} \log T_n(x,a).$$

The limiting free energies $\mathcal{T}^+_v(x,a)$ and $\mathcal{T}_v(x,a)$ are convex functions of $\log x$ *and* $\log a$ *and are monotone non-decreasing.*

Proof The proof is given for $\mathcal{T}^+_v(x,a)$; a similar proof works for $\mathcal{T}_v(x,a)$.

Let $t^+_n(k,v)$ be the number of (unrooted) positive trees of n edges, v visits and k contacts counted up to equivalence by translation parallel to $\partial\mathbb{L}^d_+$. Then $t^+_n(k,v) \leq \mathrm{t}^+_n(k,v) \leq v t^+_n(k,v) \leq (n+1) t^+_n(k,v)$, since a positive tree can be rooted in at most v visits to $\partial\mathbb{L}^d_+$ and $v \leq n+1$.

Let α_1 and α_2 be two positive trees with n_1 edges, v_1 visits and k_1 contacts, and n_2 edges, $v-v_1$ visits and $k-k_1$ contacts, respectively (see figure 6.8). Locate the lexicographic bottom and lexicographic top vertices of α_1 and α_2. Denote the bottom vertex of α_2 by $\vec{b}$ and the top vertex of α_1 by $\vec{t}$. Denote the i-th Cartesian component of a vector $\vec{c}$ by $\vec{c}(i)$.

Translate α_2 parallel to $\partial\mathbb{L}^d_+$ until $\vec{b}(i) = \vec{t}(i)$ for $2 \leq i \leq d-1$. Next, translate α_2 parallel to $\vec{e}_1$ until $\vec{b}(1) = \vec{t}(1)+2$. Finally, translate α_2 in the $-\vec{e}_1$ direction until there exists a path of length two steps (but no paths of length one step) between α_1 and α_2 (this must happen, because α_2 was translated such that $\vec{b}(i) = \vec{t}(i)$ for $2 \leq i \leq d-1$).

Let the endpoints of a path of length 2 joining α_1 and α_2 be $\vec{w}_1 \in \alpha_1$, and $\vec{w}_2 \in \alpha_2$. Suppose that the path is $\vec{w}_1 {\sim} \vec{u} {\sim} \vec{w}_2$. Concatenate α_1 and α_2 by adding the edges $\langle \vec{w}_1 \sim \vec{u} \rangle$, and $\langle \vec{u} \sim \vec{w}_2 \rangle$.

The new vertex $\vec{u}$ may be next-nearest-neighbour to at most $2(d-1)$ vertices. Thus, at most $2(d-1)$ new contacts may be created. If the new vertex is in $\partial\mathbb{L}^d_+$, then it is also a new visit.

The concatenated tree has n_1+n_2+2 edges, v or $v+1$ visits and $k+j$ contacts for some $(j \in \{0,1,\ldots,2(d-1)\})$.

Each distinct pair of trees will give a different outcome and can be uniquely recovered by finding that unique cut-vertex which cut a tree into two subtrees of size n_1 and n_2. There are $t^+_{n_2}(k_1,v_1)$ choices for α_1, and $t_{n_2}(k-k_1,v-v_1)$ choices for α_2. This shows that

$$\sum_{k_1=0}^{k}\sum_{v_1=0}^{v} t^+_{n_1}(k_1,v_1)t^+_{n_2}(k-k_1,v-v_1) \leq \sum_{i=0}^{1}\sum_{j=0}^{2(d-1)} t^+_{n_1+n_2+2}(k+j,v+i). \tag{11.34}$$

Multiply by $x^k a^v$ and sum over k and v. Define the partition function $Z^+_n(x,a) = \sum_{k,v\geq 0} t^+_n(k,v)\,x^k a^v$, and define the function $\phi(x) = x+1+\frac{1}{x}$. Then

$$Z^+_{n_1}(x,a)\,Z^+_{n_2}(x,a) \leq \phi^{2(d-1)}(x)\,\phi(a)\,Z^+_{n_1+n_2+2}(x,a).$$

The function $(\phi^{2(d-1)}(x)\,\phi(a))^{-1}\,Z^+_{n-1}(x,a)$ satisfies a supermultiplicative relation. By lemma 2.1, $t^+_n < K^n$ for some constant K and, since any tree contains at most $n+1$ visits and $d(n+1)$ nearest-neighbour contacts,

$$Z^+_n(x,a) \leq t^+_n\,(\max\{1,x\})^{d(n+1)}\,(\max\{1,a\})^{n+1}.$$

By theorem A.1 in appendix A, the limit

$$\mathcal{T}^+_v(x,a) = \lim_{n\to\infty}\tfrac{1}{n}\log\Big((\phi^{2(k-1)}(x)\phi(a))^{-1}\,Z^+_n(x,a)\Big) = \lim_{n\to\infty}\tfrac{1}{n}\log Z^+_n(x,a)$$

exists, is finite and is a convex function of $\log x$ and $\log a$. Since $T^+_n(x,a) \leq Z^+_n(x,a) \leq (n+1)T^+_n(x,a)$, this completes the proof. □

Notice that $\mathcal{T}^+_v(1,1) = \log \tau^+_d$, where τ^+_d is the growth constant of positive lattice trees. Let t_n be the number of lattice trees of size n edges (counted up to equivalence under translation). Each lattice tree can be placed in $\mathbb{L}^d_+$ with its root at $\vec{0}$. This shows that $t_n \leq t^+_n \leq \mathrm{t}^+_n \leq (n+1)t_n$, so $\tau^+_d = \tau_d$ (see theorem 2.3).

11.3.1 *The phase diagram*

The phase diagram of adsorbing and collapsing trees will include free, adsorbed and collapsed phases (see figure 11.8).

At small values of the parameters (x,a) the trees are desorbed and expanded in a *free phase*. Increasing the adsorption activity a gives a transition to an *adsorbed phase*, and increasing the contact activity x instead gives a θ-transition to a *contact-collapsed phase*. For large values of x and a, the model may be in a phase which is both *collapsed and adsorbed*, although this phase may not be present.

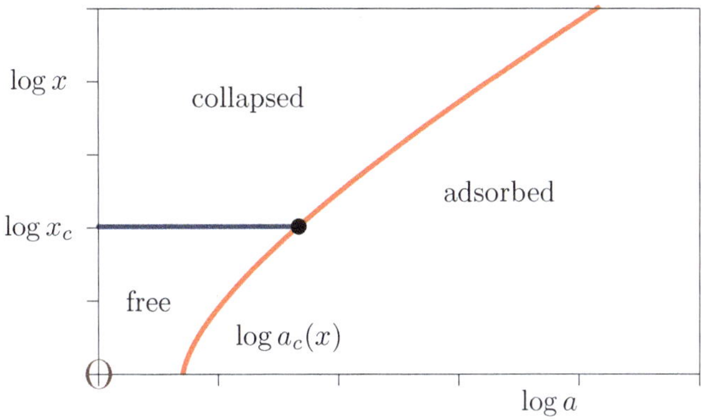

FIG. 11.8. The phase diagram of adsorbing and collapsing positive trees. A combined collapsed and adsorbed phase may be present at large values of x and a.

It will be shown that there exists an adsorption critical point $a_c^+(x) > 1$ for all $x \geq 0$ (see section 11.3.2) and that, if there is a collapsed phase in the model, then the critical point x_c for a collapse transition is independent of a.

Let α be a positive tree in $\mathbb{L}_+^d$ (rooted at $\vec{0}$). Construct the tree $\alpha_1 = (\alpha + \vec{e}_d) \cup \{\langle \vec{0} \sim \vec{e}_d \rangle\}$ by translating α one step in the $\vec{e}_d$ direction and then reconnecting it to $\vec{0}$. Then α_1 is a positive tree with exactly one visit (at $\vec{0}$) and $n+1$ edges. Since α can be recovered from α_1, this shows that $\mathrm{t}_n^+(k,v) \leq \mathrm{t}_{n+1}^+(k,1)$. In terms of partition functions this gives

$$T_n^+(x,a) \leq \frac{1}{1-a} \sum_{k \geq 0} \mathrm{t}_{n+1}^+(k,1)\, x^k \leq \frac{1}{a(1-a)} T_{n+1}^+(x,a) \text{ for } a \in (0,1). \quad (11.35)$$

Next, consider a model of collapsing trees. Let $t_n(k)$ be the number of unrooted trees of size n and with k contacts in $\mathbb{L}^d$. Let α_k be such a tree of size n edges and k contacts. Place α_k in $\mathbb{L}_+^d$ such that it can be rooted at $\vec{0}$ in $\partial \mathbb{L}_+^d$ by adding a single edge. This shows that $t_n(k) \leq \mathrm{t}_{n+1}^+(k,1)$.

On the other hand, suppose that α_ℓ is a rooted positive tree of size $n+1$, with k contacts and with exactly one visit to $\partial \mathbb{L}_+^d$. Deleting the root in α_ℓ gives an unrooted tree with k contacts. Since the edge with the root can be put back in at most $n+1$ possible ways, this shows that $\mathrm{t}_{n+1}^+(k,1) \leq (n+1) t_n(k)$. Together with the previous inequality, this gives $t_n(k) \leq \mathrm{t}_{n+1}^+(k,1) \leq (n+1) t_n(k)$.

If τ_1 is an (unrooted) tree of size n_1 and with $k - k_1$ nearest-neighbour contacts, and τ_2 is similarly a tree of size n_2 and with k_1 nearest-neighbour contacts, then τ_1 and τ_2 may be concatenated as illustrated in figure 2.3 to obtain the supermultiplicative relation $\sum_{k_1=0}^{k} t_{n_1}(k-k_1) t_{n_2}(k_1) \leq t_{n_1+n_2+1}(k)$. By theorem 3.1, this implies the existence of the free energy of collapsing trees:

$$\mathcal{T}_k(x) = \lim_{n\to\infty} \tfrac{1}{n} \log \sum_{k\geq 0} t_n(k)\, x^k. \tag{11.36}$$

Since it was shown above that $t_n(k) \leq \mathrm{t}^+_{n+1}(k,1) \leq (n+1) t_n(k)$, this gives

$$\mathcal{T}_k(x) = \lim_{n\to\infty} \tfrac{1}{n} \log \sum_{k\geq 0} \mathrm{t}^+_n(k,1)\, x^k. \tag{11.37}$$

By equation (11.35) and theorem 11.15, this shows that

$$\mathcal{T}^+_v(x,a) = \lim_{n\to\infty} \tfrac{1}{n} \log \sum_{k\geq 0} t_n(k)\, x^k = \mathcal{T}_k(x) \text{ for } a \in (0,1], \tag{11.38}$$

where the case $a = 1$ is dealt with in a similar way to the above.

Notice that $T^+_n(x,a) \geq \sum_k \mathrm{t}^+_n(k,n+1)\, x^k a^n$, since there are trees with $n+1$ visits. Since $\mathrm{t}^+_n(k,n+1)$ is the number of positive trees with k contacts and all its vertices in $\partial \mathbb{L}^d_+ \simeq \mathbb{L}^{d-1}$, this shows by theorem 11.15 that $\mathcal{T}^+_v(x,a) \geq \mathcal{T}^{(d-1)}_k(x) + \log a$ (where $\mathcal{T}_k(x)$ in $d-1$ dimensions is denoted $\mathcal{T}^{(d-1)}_k(x)$; if $d = 2$, then $\mathcal{T}^{(1)}_k(x) = 0$). These arguments, and similar arguments for $\mathcal{T}_v(x,a)$, give the following lemma.

Lemma 11.16 *If $x \geq 0$, then $\mathcal{T}^+_v(x,a) = \mathcal{T}_v(x,a) = \mathcal{T}_k(x)$ for all $a \in [0,1]$ and*

$$\max\left\{\mathcal{T}_k(x), \mathcal{T}^{(d-1)}_k(x) + \log a\right\} \leq \mathcal{T}^+_v(x,a) \leq \mathcal{T}_k(x) + \log a, \quad \textit{and}$$
$$\max\left\{\mathcal{T}_k(x), \mathcal{T}^{(d-1)}_k(x) + \log a\right\} \leq \mathcal{T}_v(x,a) \leq \mathcal{T}_k(x) + \log a,$$

for all $a > 1$.

Proof It only remains to prove the upper bound.

Since $\mathrm{t}^+_n(k,v)\, x^k a^v \leq \mathrm{t}^+_n(k,v)\, x^k a^{n+1}$ if $a > 1$, it follows that $T^+_n(x,a) \leq T^+_n(x,1)\, a^{n+1}$. Taking logarithms, dividing by n and letting $n \to \infty$ completes the proof for positive trees. A similar proof works for $\mathcal{T}_v(x,a)$. □

For every fixed $x \geq 0$, the free energy is independent of a for $a \in [0,1]$ and is a function of a for large $a > 0$. Define $a^+_c(x) = \inf\{a{>}0 \mid \mathcal{T}^+_c(x,a){>}\mathcal{T}_k(x)\}$.

Theorem 11.17 *For every fixed $x \geq 0$, there is a critical adsorption point $a^+_c(x) \geq 1$ in the free energy $\mathcal{T}^+_v(x,a)$. Moreover,*

$$\log a^+_c(x) \in \left[0, \log \tau_d - \tfrac{1}{2}\log \tau_{d-1}\right], \qquad \textit{if } x \leq 1;$$
$$\log a^+_c(x) \in \left[0, \log \tau_d - \log \tau_{d-1} + (d-1)\log x\right], \qquad \textit{if } x > 1.$$

Proof It only remains to prove the upper bounds on the intervals containing $\log a^+_c(x)$. By comparing the lower and upper bounds in lemma 11.16, it follows that $\log a^+_c(x) \leq \mathcal{T}_k(x) - \mathcal{T}^{(d-1)}_k(x)$.

If $x > 1$, then a tree has at most $(d-1)n$ contacts; therefore, $T_n(x,1) \leq t_n\, x^{(d-1)n}$. This gives $\mathcal{T}_k(x) \leq \log \tau_d + (d-1)\log x$. Also, $\mathcal{T}_k^{(d-1)}(x) \geq \log \tau_{d-1}$ if $x \geq 1$. This gives $\mathcal{T}_k(x) - \mathcal{T}_k^{(d-1)}(x) \leq \log \tau_d - \log \tau_{d-1} + (d-1)\log x$ if $x > 1$. This completes the case for $x > 1$.

If $x \leq 1$, then $\mathcal{T}(x) \leq \mathcal{T}(1) = \log \tau_d$.

Consider the partition function $T_n^{(d-1)}(x,1)$ of collapsing trees in $(d-1)$ dimensions. Notice that $T_n^{(d-1)}(x,1)$ includes all trees with zero contacts, so $T_n^{(d-1)}(x,1) \geq t_n^{(d-1)}(0)$. By subdividing the edges in each tree in $(d-1)$ dimensions, it is seen that $t_{2n}^{(d-1)}(0) \geq t_n^{(d-1)}$. Hence, $T_{2n}^{(d-1)}(x,1) \geq t_n^{(d-1)}$. Take logarithms, divide by $2n$ and let $n \to \infty$ to see that $\mathcal{T}_k^{(d-1)}(x) \geq \frac{1}{2}\log \tau_{d-1}$. Thus, $\mathcal{T}_k(x) - \mathcal{T}_k^{(d-1)}(x) \leq \log \tau_d - \frac{1}{2}\log \tau_{d-1}$.

This completes the proof. □

Theorem 11.17 establishes the existence of an adsorption critical point $a_c^+(x) \geq 1$ for each $x \geq 0$ (see figure 11.8). Since the free energy $\mathcal{T}_v^+(x,a)$ is non-decreasing, it follows that $a_c(x)$ is a non-decreasing function of x and so is continuous and differentiable almost everywhere.

The presence of a collapse critical point in the model is not known rigorously. However, assuming that $\mathcal{T}_k(x)$ has a critical point at x_c implies that $\mathcal{T}_v^+(x,a)$ has a line of collapse transitions at $x = x_c$ for all $0 \leq a < a_c(x_c)$.

Theorem 11.18 *Suppose that the free energy of a model of contact-collapsing trees $\mathcal{T}_k(x)$ is singular at $x = x_c$. Then $\mathcal{T}_v^+(x,a)$ is singular at $x = x_c$ for every $0 \leq a \leq a_c^+(x_c)$.*

Proof Choose $\epsilon > 0$ and let $a_d < \inf\{a_c^+(x) \mid x \in [x_c - \epsilon, x_c + \epsilon]\}$. By lemma 11.16, $\mathcal{T}_v^+(x,a) = \mathcal{T}_k(x)$ for all $x \in [x_c - \epsilon, x_c + \epsilon]$, and $0 \leq a \leq a_d$. This shows that $\mathcal{T}_v^+(x,a)$ is singular at $x = x_c$ for every $0 \leq a \leq a_d$.

If $\epsilon \to 0^+$, then a_d may be taken to $a_c^+(x_c)$ (since $a_c^+(x)$ is continuous). This completes the proof. □

Similar arguments can be used to prove the same results for rooted trees adsorbing at a defect plane (with partition function $T_n(x,a)$; see equation (11.33)).

Theorem 11.19 *For every $x \geq 0$, there is a critical adsorption point $a_c(x) \geq 1$ in $\mathcal{T}_v(x,a)$. Moreover,*

$$\log a_c(x) \in \left[0, \log \tau_d - \tfrac{1}{2}\log \tau_{d-1}\right], \qquad \text{if } x \leq 1;$$
$$\log a_c(x) \in \left[0, \log \tau_d - \log \tau_{d-1} + (d-1)\log x\right], \qquad \text{if } x > 1.$$

In addition, $a_c(x) \leq a_c^+(x)$ for all $x \geq 0$.

If $a < a_c(x)$ and if $\mathcal{T}_v^+(x,a)$ is a singular function of x at $x = x_c$, then $\mathcal{T}_v(x,a)$ is a singular function of x at $x = x_c$, and $a < a_c(x)$.

Proof The proof is similar to the proof of theorem 11.17.

Notice that $\mathcal{T}_v^+(x,a) \leq \mathcal{T}_v(x,a)$ for all (x,a). By lemma 11.16, $\mathcal{T}_v^+(x,a) = \mathcal{T}_v(x,a)$ for $x \leq 1$; this implies that $x_c(a) \leq x_c^+(a)$ and also shows that, if

TABLE 11.2. Adsorbing branched polymer crossover exponent

	$d=2$	$d=3$	$d=4$	$d=5$
Lam *et al.* [370]	0.716	--	--	--
Orlandini *et al.* [447]	0.70(6)	--	--	--
Hsu *et al.* [278]	$\frac{1}{2}$	--	--	--
de Queiroz [119]	0.505(15)	--	--	--
Vujic [570]	0.503(3)	--	--	--
You *et al.* [604]	0.50(3)	--	--	--
Hsu *et al.* [284]	0.480(4)	0.50(1)	0.50(2)	0.51(3)

$a < a_c(x)$, then $\mathcal{T}_v(x,a)$ is a singular function of x at $x = x_c$ (if $\mathcal{T}_v^+(x,a)$ is a singular function of x at $x = x_c$). □

The critical line $x = x_c$ for $a \in (0, a_c^+(x))$ in figure 11.8 separating the free and collapsed phases is a line of θ-transitions to a compact phase. The crossover and other critical exponents are the θ-exponents for branch polymer collapse (see table 11.1).

Adsorption transitions are seen along the critical curve $a_c^+(x)$. For $x < x_c$, the branched polymer adsorption transition is seen. The crossover exponent for branched polymer adsorption is $\phi = \frac{1}{2}$ in $d = 2$ dimensions [119, 333]. Numerical estimates for ϕ gave somewhat inconsistent results initially but more extensive simulations settled down on values consistent with $\phi = \frac{1}{2}$ (see table 11.2). This crossover exponent is thought to be *superuniversal* (has the same value in all dimensions) [278, 333].

Estimates of the critical adsorption points are listed in table 11.3. If $x = x_c$, and $a \in [0, a_c^+(x_c))$, then the model is critical with respect to the collapse transition and is at the θ-point. Increasing a to its critical value (when $x = x_c$) takes the model to criticality with respect to adsorption at the θ-point; this is the *special point.* The adsorption transition exponents at the special point have special point values; for example, the crossover exponent for adsorption at the special point is denoted ϕ_s. It has been argued that $\phi_s = \frac{1}{3}$ in two dimensions in references [510, 511].

The entropic exponent θ_1 for positive trees (see equation (2.22)) has the special point value θ_s given by

$$\theta_s = \frac{d-3}{d-2}(\theta - 1) + (2 - \phi_\theta) \tag{11.39}$$

(see references [333, 558]). Taking $d \to 2$ in this gives $\theta_s = 1 - \phi_\theta$ (see table 11.4 for numerical estimates of θ_s).

TABLE 11.3. Adsorbing tree and animal critical points

	$d=2$	$d=3$	$d=4$	$d=5$
Bond-trees:				
De'Bell *et al.* [123]	2.865(5)	– –	– –	– –
You *et al.* [604]	2.247(68)	– –	– –	– –
Site-animals:				
Hsu *et al.* [284]	2.2778(8)	1.4747(6)	1.2674(6)	1.1786(5)

11.3.2 *The location of the adsorption transition*

Multiplying equation (11.34) by a^v and summing over v gives

$$Z_{n_1}^+(k_1;a)\,Z_{n_2}^+(k_2;a) \leq \phi(a) \sum_{j=0}^{2(d-1)} Z_{n_1+n_2+2}^+(k_1+k_2+j;a) \tag{11.40}$$

where $Z_n^+(k;a) = \sum_{v\geq 0} t_n^+(k,v)\,a^v$. By theorem (3.16), the limit

$$\mathcal{Q}_v^+(\epsilon;a) = \lim_{n\to\infty} \left(Z_n^+(\lfloor \epsilon n \rfloor + \sigma_n;a)\right)^{1/n} = \lim_{n\to\infty} \left(T_n^+(\lfloor \epsilon n \rfloor + \sigma_n;a)\right)^{1/n} \tag{11.41}$$

exists for some function $\sigma_n = o(n)$ (and where $T_n^+(k;a) = \sum_{v\geq 0} \mathfrak{t}_n^+(k,v)\,a^v$).

Similarly, if $Z_n^+(x;v) = \sum_{k\geq 0} t_n^+(k,v)\,x^k$, then

$$\mathcal{Q}_v^+(x;\delta) = \lim_{n\to\infty} \left(Z_n^+(x;\lfloor \delta n \rfloor + \sigma_n')\right)^{1/n} = \lim_{n\to\infty} \left(T_n^+(x;\lfloor \delta n \rfloor + \sigma_n')\right)^{1/n} \tag{11.42}$$

for some function $\sigma_n' = o(n)$ (and where $T_n^+(x;v) = \sum_{k\geq 0} \mathfrak{t}_n^+(k,v)\,x^k$). As shown in section 3.3.1.4, there is a joint density function $\mathcal{Q}_v^+(\epsilon,\delta)$.

A similar approach establishes the existence of density functions $\mathcal{Q}_v(\epsilon;a)$ and $\mathcal{Q}_v(x;\delta)$ for a collapsing tree at a defect plane (see equation (11.33) and theorem 11.15).

TABLE 11.4. Adsorbing tree special point exponent θ_s

	$d=2$	$d=3$	$d=4$	$d=5$
Hsu *et al.* [284]	0.870(9)	1.476(7)	1.91(1)	2.18(4)
Eqn (11.39)	$\frac{101}{120}$	1.573(5)	1.948(2)	2.2204(2)

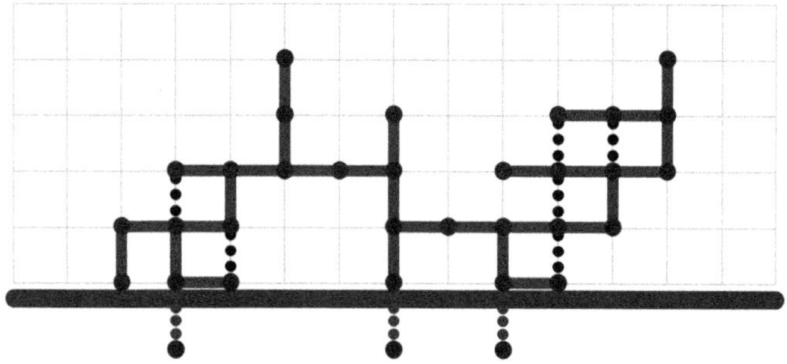

FIG. 11.9. A positive tree adsorbing in $\partial \mathbb{L}^d_+$. Edges can be appended into the negative half-space as shown.

As shown in section 3.3.2.5 and equation (3.62), the critical adsorption points (for fixed x) are located at

$$\log a_c^+(x) = -D^+ \log \mathcal{Q}_v^+(x;0), \quad \text{and } \log a_c(x) = -D^+ \log \mathcal{Q}_v(x;0), \qquad (11.43)$$

where $D^+ f(0)$ is the right-derivative of $f(x)$ and $x = 0$.

Notice that $\log \mathcal{Q}_v^+(x;0) = \log \mathcal{Q}_v(x;0) = \mathcal{T}_k(x)$ (the free energy of collapsing trees) (see lemma 11.16). Since $\log \mathcal{Q}_k(x;\delta)$ and $\log \mathcal{Q}_k^+(x;\delta)$ are concave functions of δ, and $a_c^+(x) \geq a_c(x) \geq 0$, the above implies that $\log \mathcal{Q}_k(x;\delta)$ and $\log \mathcal{Q}_k^+(x;\delta)$ are monotone non-increasing functions of δ and that they are right-continuous at $\delta = 0$. Subtract $\log a_c(x)$ from $\log a_c^+(x)$ and take the limit $\delta \to 0^+$. This gives the following lemma.

Lemma 11.20 *For every finite value of $x \geq 0$,*

$$\log a_c^+(x) - \log a_c(x) = e^{-\mathcal{T}_k(x)} \lim_{\delta \to 0^+} \tfrac{1}{\delta}\big(\mathcal{Q}_v(x;\delta) - \mathcal{Q}_v^+(x;\delta)\big). \qquad \square$$

Let α be a positive tree with size n edges, k nearest-neighbour contacts and $\lfloor \delta n \rfloor$ visits (see figure 11.9). By appending edges to α into the negative half-lattice $\mathbb{L}^d_- = \mathbb{L}^d \setminus \mathbb{L}^d_+$ a tree at the defect plane $\partial \mathbb{L}^d_+$ is obtained. Such edges can be added without creating nearest-neighbour contacts, (for example, by appending edges only onto visits with even or with odd parity). There are at least $\frac{1}{2}\lfloor \delta n \rfloor$ visits which can be chosen, so choose $\lfloor \kappa n \rfloor$ visits and append edges in each α. Since there are $t_n^+(k, \lfloor \delta n \rfloor)$ choices for α (α is not rooted), this shows that

$$\binom{\frac{1}{2}\lfloor \delta n \rfloor}{\lfloor \kappa n \rfloor} t_n^+(k, \lfloor \delta n \rfloor) \leq t_{n+\lfloor \kappa n \rfloor}(k, \lfloor \delta n \rfloor). \qquad (11.44)$$

Multiply this by x^k and sum over k. Take the power $\frac{1}{n}$ and let $n \to \infty$. Then

$$\frac{(\delta/2)^{\delta/2}}{\kappa^\kappa(\delta/2-\kappa)^{\delta/2-\kappa}}\,\mathcal{Q}_v^+(x;\delta) \le \left(\mathcal{Q}_v\left(x;\tfrac{\delta}{1+\kappa}\right)\right)^{1+\kappa} \le e^{\kappa\mathcal{T}_k(x)}\,\mathcal{Q}_v\left(x;\tfrac{\delta}{1+\kappa}\right), \quad (11.45)$$

since $\log \mathcal{Q}_v(x;\delta) \le \mathcal{T}_k(x)$ for all $\delta \in (0,1)$. Rearrange the above into

$$\left(\frac{(\delta/2)^{\delta/2}e^{-\kappa\mathcal{T}_k(x)}}{\kappa^\kappa(\delta/2-\kappa)^{\delta/2-\kappa}}\right)\mathcal{Q}_v^+(x;\delta) \le \mathcal{Q}_v\left(x;\tfrac{\delta}{1+\kappa}\right). \quad (11.46)$$

The left-hand side is a maximum if $2\kappa = \frac{\delta}{1+e^{\mathcal{T}_k(x)}}$. This gives

$$\left(1+e^{-\mathcal{T}_k(x)}\right)^{\delta/2}\mathcal{Q}_v^+(x;\delta) \le \mathcal{Q}_v\left(x;\frac{2\delta(1+e^{\mathcal{T}_k(x)})}{2(1+e^{\mathcal{T}_k(x)})+\delta}\right). \quad (11.47)$$

The corollary of this is the following theorem.

Theorem 11.21 *The difference* $\log a_c^+(x) - \log a_c(x) \ge \frac{1}{2}\log(1+e^{-\mathcal{T}_k(x)}) > 0$ *for finite* $x > 0$.

Proof Since $a_c(x) \ge 0$ the right-derivative of $\mathcal{Q}_v(x;\delta)$ to δ is not positive.

There are two cases to consider.

Assume that the right-derivative to δ at $\delta = 0$ is $D^+ \log \mathcal{Q}_v(x;0) = 0$. Then $\log \mathcal{Q}_v(x;\delta) = \log \mathcal{Q}_v(x;0) + f(\delta)$ (where $\frac{1}{\delta}f(\delta) \to 0$ as $\delta \to 0^+$). Thus,

$$\frac{\mathcal{Q}_v\left(x;\frac{2\delta(1+e^{\mathcal{T}_k(x)})}{2(1+e^{\mathcal{T}_k(x)})+\delta}\right)}{\mathcal{Q}_v(x;\delta)} = 1 + f(\delta).$$

Substitute this into equation (11.47) and simplify to see that

$$\mathcal{Q}_v(\delta) \ge \left(1+e^{-\mathcal{T}_k(x)}\right)^{\delta/2}\mathcal{Q}_v^+(\delta) + f(\delta).$$

Substitute this into the right-hand side of lemma 11.20 and take $\delta \to 0^+$. This gives $\log a_c^+(x) - \log a_c(x) \ge \frac{1}{2}\log(1+e^{-\mathcal{T}_k(x)})$ as required.

Assume that the right-derivative to δ at $\delta = 0$ is negative: $D^+ \log \mathcal{Q}_v(x;0) < 0$. For a given $\delta > 0$, there is an $\eta > 0$ such that

$$\mathcal{Q}_v(x;0) + \delta(1+\eta)D^+\mathcal{Q}_v(x;0) \le \mathcal{Q}_v(x;\delta) \le \mathcal{Q}_v(x;0) + \delta(1-\eta)D^+\mathcal{Q}_v(x;0).$$

Notice that, if $\delta \to 0^+$, then $\eta \to 0^+$ may be taken.

Use the above to bound $\mathcal{Q}_v\left(x;\frac{2\delta(1+e^{\mathcal{T}_k(x)})}{2(1+e^{\mathcal{T}_k(x)})+\delta}\right)$ from above and $\mathcal{Q}_v(x;\delta)$ from below. Expand and collect terms to shows that there is a function η_2 of x such that

$$\frac{\mathcal{Q}_v\left(x;\frac{2\delta(1+e^{\mathcal{T}_k(x)})}{2(1+e^{\mathcal{T}_k(x)})+\delta}\right)}{\mathcal{Q}_v(x;\delta)} \le 1 - 2\eta\delta\,D^+\log\mathcal{Q}_v(x;0) + \eta_2 f(\delta)$$

for some function $f(\delta) > 0$, where $f(\delta) \to 0$ as $\delta \to 0^+$. Substitute this in the right-hand side in lemma 11.20. Take the limit $\delta \to 0^+$. This shows that

$$\log a_c^+(x) - \log a_c(x) \geq \tfrac{1}{2}\log(1+e^{-\mathcal{T}_k(x)}) + 2\eta D^+ \log \mathcal{Q}_v(x;0).$$

Finally, take $\eta \to 0^+$ to finish the proof. □

Since $a_c(x) \geq 1$, a corollary of theorem 11.21 is the following.

Corollary 11.22 *For every* $x \geq 0$, $a_c^+(x) \geq \sqrt{1+e^{-\mathcal{T}_k(x)}} > 1$. □

If $x = 1$, then the lower bound on $a_c^+(1)$ becomes $a_c^+(1) \geq \sqrt{1+\tau_d^{-1}}$.

11.4 Adsorbing percolation clusters

A bond percolation cluster α is a bond-animal composed of *open bonds* in $\mathbb{L}^d$ (see appendix E). An edge $\langle \vec{v} \sim \vec{u} \rangle$ in $\mathbb{L}^d$ is a *perimeter edge* of α if $\langle \vec{v} \sim \vec{u} \rangle \not\in \alpha$ but at least one of $\vec{v} \in \alpha$ or $\vec{u} \in \alpha$. That is, the perimeter of α is the set of closed bonds in $\mathbb{L}^d$ incident with α. A cluster α_1 is equivalent to a cluster α_2 if $\alpha_1 = \alpha_2 + \vec{v}$ for a vector $\vec{v} \in \mathbb{Z}^d$.

Bond percolation clusters are defined similarly in the half-lattice $\mathbb{L}_+^d$ (see equation (2.20)). A bond percolation cluster α_+ in $\mathbb{L}_+^d$ is a bond-animal in $\mathbb{L}_+^d$ of open edges. The edge $\langle \vec{v} \sim \vec{u} \rangle \in \mathbb{L}_+^d$ is a *perimeter edge* if $\langle \vec{v} \sim \vec{u} \rangle \not\in \alpha$ but at least one of $\vec{v} \in \alpha$ or $\vec{u} \in \alpha$. A vertex $\vec{u} \in \alpha$ is a *visit* if $\vec{u} \in \partial\mathbb{L}_+^d$. A cluster $\alpha_1 \in \mathbb{L}_+^d$ is equivalent to a cluster $\alpha_2 \in \mathbb{L}_+^d$ if $\alpha_1 = \alpha_2 + \vec{v}$ for a vector $\vec{v} \in \mathbb{Z}^d$ which is parallel to $\partial\mathbb{L}_+^d$.

Let $a_n(\rho)$ be the number of bond-animals of size n and perimeter ρ in $\mathbb{L}^d$ and let $a_n^+(\rho, v)$ be the number of bond-animals of size n in $\mathbb{L}_+^d$, with perimeter of size ρ with v visits to $\partial\mathbb{L}_+^d$.

Denote the number of bond-animals rooted at $\vec{0}$ in $\mathbb{L}^d$ of size n and perimeter ρ by $\mathfrak{a}_n(\rho)$. Let the number of bond-animals in $\mathbb{L}_+^d$ of size n (rooted at $\vec{0}$), with perimeter ρ and with v visits to $\partial\mathbb{L}_+^d$ be $\mathfrak{a}_n^+(\rho, v)$. Then $na_n^+(\rho, v) \geq \mathfrak{a}_n^+(\rho, v) \geq a_n^+(\rho, v)$ since a cluster can be rooted at any visit in $\partial\mathbb{L}_+^d$.

Suppose that the density of open edges in $\mathbb{L}^d$ and in $\mathbb{L}_+^d$ is p and that visits in $\partial\mathbb{L}_+^d$ are weighted by a parameter a. The partition function of the model is

$$A_n^+(p,a) = \sum_{\rho, v \geq 0} \mathfrak{a}_n^+(\rho, v)\, q^{\rho - v} a^v, \tag{11.48}$$

where $q = 1 - p$. Define the partition function of unrooted clusters as well by

$$C_n^+(p,a) = \sum_{\rho, v \geq 0} a_n^+(\rho, v)\, q^{\rho - v} a^v. \tag{11.49}$$

These are models of adsorbing bond percolation clusters [120, 121]. It follows that $aC_n^+(p,q) \geq A_n^+(p,a) \geq C_n^+(p,a)$.

Concatenating two clusters similar to the concatenation of trees in figure 11.7 gives (similar to equation (11.34))

$$C^+_{n_1}(p,a)C^+_{n_2}(p,a) \le \phi^{2(d-1)}(q)\phi(\tfrac{a}{q})C^+_{n_1+n_2+2}(p,a), \tag{11.50}$$

where $\phi(x) = x+1+\frac{1}{x}$. The number of clusters has an exponential upper bound by lemma 2.1. Thus, by theorem A.1 in appendix A,

$$\mathcal{C}^+(p,a) = \lim_{n\to\infty} \tfrac{1}{n}\log C^+_n(p,a) = \lim_{n\to\infty} \tfrac{1}{n}\log A^+_n(p,a). \tag{11.51}$$

Notice that $\mathcal{C}^+(p,0) = \mathcal{C}(p)$ is the limiting free energy of percolation clusters.

There is an adsorption transition in this model for every value of the percolation probability $p \in (0,1)$.

Theorem 11.23 *The partition function $\mathcal{C}^+(p,a)$ is a non-analytic function of a for every $p \in (0,1)$. Moreover, if the critical curve is defined by*

$$a^+_c(p) = \inf\{a \mid \mathcal{C}^+(p,a) > \mathcal{C}^+(p,0)\},$$

then $a^+_c(p) \ge q$.

Proof Let $a \ge 0$ and consider clusters with exactly one visit in $\partial\mathbb{L}^d_+$ to see the following:

$$A^+_n(p,a) \ge a\sum_{\rho\ge 0} a^+_n(\rho,1)\,q^{\rho-1} \ge a\sum_{\rho\ge 0} a_{n-1}(\rho-2(d-1))\,q^{\rho-1}. \tag{11.52}$$

Take logarithms, divide by n and let $n\to\infty$. This shows that $\mathcal{C}^+(p,a) \ge \mathcal{C}(p)$.

Translate an animal one step vertically away from $\partial\mathbb{L}^d_+$ to see that $a^+_n(\rho,v) \le a_n(\rho+v)$. Multiply by $q^\rho a^v$ and sum over ρ and v. If $a<q$, then

$$A^+_n(p,a) \le \sum_{\rho,v\ge 0} a_n(\rho+v)\,q^\rho a^v \le \frac{q}{q-a}\sum_{\rho\ge 0} a_n(\rho)\,q^\rho. \tag{11.53}$$

Take logarithms, divide by n and let $n\to\infty$ to obtain $\mathcal{C}^+(p,a) \le \mathcal{C}(p)$ (provided that $a<q$).

This shows that $\mathcal{C}^+(p,a) = \mathcal{C}(p)$ for all $0<a<q$.

On the other hand, $A^+_n(p,a) \ge q^{2(d-1)n+2}a^{n+1}$ (consider the straight line animal of length $n+1$ in $\partial\mathbb{L}^d_+$ with perimeter $\rho = (2d-3)n+2$). This shows that $\mathcal{C}^+(p,a)$ is a strictly increasing function of a for large a.

That is, for each $p \in (0,1)$, there is a critical value of $a \ge q$, denoted $a^+_c(p)$, where adsorption of the cluster occurs. □

The probability that the cluster C at the origin has size n in ordinary bond percolation is

$$P_r(|C|=n) = \sum_{\rho\ge 0} \mathfrak{a}_n(\rho)\,p^n q^\rho, \tag{11.54}$$

where $\mathfrak{a}_n(\rho)$ is the number of bond-animals of size n and perimeter ρ rooted at $\vec{0}$. Since $a_n(\rho) \le \mathfrak{a}_n(\rho) \le (n+1)a_n(\rho)$, it follows that

$$\zeta(p) = -\lim_{n\to\infty} \tfrac{1}{n} \log P_r(|C|=n) = -\lim_{n\to\infty} \tfrac{1}{n} \log \sum_{\rho\ge 0} a_n(\rho)\, p^n q^\rho. \tag{11.55}$$

By equations (11.52) and (11.53),

$$a \sum_{\rho\ge 0} a_{n-1}(\rho - 2(d-1))\, q^{\rho-1} \le A_n^+(p,a) \le \frac{q}{q-a} \sum_{\rho\ge 0} a_n(\rho)\, q^\rho. \tag{11.56}$$

Multiply this by p^n, take logarithms, divide by n and let $n \to \infty$ to see from equation (11.55) that

$$\mathcal{C}^+(p,a) = -\log p - \zeta(p) \text{ if } a < q. \tag{11.57}$$

By theorem E.13 in appendix E, $\zeta(p) > 0$ if $p < p_c(d)$, where $p_c(d)$ is the critical density for percolation in $\mathbb{Z}^d$, and $\zeta(p) = 0$ if $p > p_c(d)$. This proves that $\mathcal{C}^+(p,a) = \mathcal{C}(p)$ is non-analytic at $p = p_c(d)$, provided that $a < 1 - p_c(d)$.

These observations give the following theorem.

Theorem 11.24 *Let $p \in (0,1)$. Then $\mathcal{C}^+(p,a)$ is a non-analytic function of a at $a = a_c^+(p) \ge q$. It is also a non-analytic function of p at $p = p_c(d)$, provided that $a < a_c^+(p_c(d))$; this corresponds to bulk percolation in the model.*

If $a_c^+(p)$ is a continuous function of p at $p = p_c(d)$, then $\mathcal{C}^+(p,a)$ is an analytic function of p for $p \in [p_c(d) - \epsilon, p_c(d) + \epsilon] \setminus \{p_c(d)\}$ (for $a < a_c^+(p_c(d))$).

Proof It only remains to prove that $\mathcal{C}^+(p,a)$ is a non-analytic function of p at $p = p_c(d)$ for all $a < a_c^+(p_c(d))$ (if $a_c^+(p)$ is a continuous function of p at $p = p_c(d)$). Let $\epsilon > 0$ and choose a_d such that

$$a_d < \inf\{a_c^+(p) \mid p \in [p_c(d) - \epsilon, p_c(d) + \epsilon]\}.$$

The partition function $\mathcal{C}^+(p,a)$ is an analytic function of p for all $p \in [p_c(d) - \epsilon, p_c(d) + \epsilon] \setminus \{p_c(d)\}$, for $a < a_d$. By taking $\epsilon \to 0^+$, a_d can be chosen arbitrarily close to $a_c^+(p_c)$. □

The point $(p_c(d), a_c^+(p_c(d)))$ is the meeting point of at least three critical curves in the phase diagram. At least three phases meet here, namely, *desorbed subcritical percolation*, *desorbed supercritical percolation* and *adsorbed subcritical percolation*.

Increasing (p,a) so that $p \in [0,1]$ while $a < a_c^+(p)$ takes the model from desorbed subcritical percolation to desorbed supercritical percolation at the critical point $p_c(d)$. If instead p is increased in $[0,1]$ with $a > a_c^+(p)$, then the model is in an adsorbed state. It is not known whether there is a bulk percolation critical point where the adsorbed cluster percolates into bulk.

11.5 Embeddings of abstract graphs

An abstract *graph* $G = (V, E)$ is a set of vertices V joined in pairs by edges in E. To distinguish (graph-theoretic) edges from lattice edges (unit length line segments joining lattice vertices), edges in a graph will be called *branches.*

Let G be a graph and suppose it has a vertex set $V = \{v_i\}$ and an edge set or branches $E = \{\langle v_i \sim v_j \rangle\}$. The *degree* of a vertex $v \in V$ is the number of branches incident with it. This is denoted by $\deg v$. The maximal degree in a graph G will be denoted by $\operatorname{maxdeg} G$. A *path* in G is a sequence of branches $P = \langle \langle v_0 \sim v_1 \rangle, \langle v_1 \sim v_2 \rangle, \langle v_2 \sim v_3 \rangle, \ldots, \langle v_{n-1} \sim v_n \rangle \rangle$. The *endpoints* or *end-vertices* of P are v_0 and v_n. P is also denoted by $v_0 \leftrightsquigarrow v_n$.

A graph G is *connected* if there is a path with endpoints $\{v, w\}$ in G for each pair of vertices $v, w \in V$. A connected graph is also called a *network.*

A *loop* is a branch of the form $\langle v \sim v \rangle$. A *double edge* contains two branches in E with the same endpoints: (for example, $\langle v \sim w \rangle, \langle v \sim w \rangle$; thus, strictly speaking, E is a multiset).

A graph or a network is *simple* if it contains no loops and no double edges.

Let $\mathcal{G}$ be the set of connected simple graphs with no vertices of degree equal to 2, but place the *circle graph* in $\mathcal{G}$ (a circle graph has one vertex v and one branch $\langle v \sim v \rangle$; it is one vertex and one loop).

Two graphs are said to be identical if they are isomorphic (that is, there is a bijection between vertices which preserves adjacency). Two graphs are *homeomorphic* if both can be obtained from the same graph by inserting vertices of degree 2 in the interior of branches. Such graphs are said to have the same *homeomorphism-type.*

Let $g_n(G)$ be number of lattice animals in $\mathbb{L}^d$ of size n edges (counted up to equivalence under translation in $\mathbb{L}^d$) which are (as graphs) homeomorphic to a given graph $G \in \mathcal{G}$. If there is an $n > 0$ such that $g_n(G) > 0$, then G is *embeddable.* Any lattice animal which is homeomorphic to G is said to be an *embedding* of G. An embedding of a graph G as an animal A is also said to be a *realisation* of G or a *placement* of G in $\mathbb{L}^d$.

Not all graphs are embeddable in a given lattice. A non-planar graph may be embeddable in $\mathbb{L}^3$ but is by definition not embeddable in $\mathbb{L}^2$. A graph with a vertex of degree larger than $2d$ cannot be embedded in $\mathbb{L}^d$.

The embedded graph is itself an abstract graph consisting of lattice vertices and lattice edges, but it may also contain vertices of degree 2 (which are the *internal vertices* of the branches).

Examples of embedded graphs are self-avoiding walks, which are embeddings of the line graph $L = (\{v, w\}, \{\langle v \sim w \rangle\})$, and lattice knots, which are embeddings of the circle graph $S = (\{v\}, \{\langle v \sim v \rangle\})$. Notice that embeddings of graphs into the lattice $\mathbb{L}^d$ are also placements of the graphs in the ambient space $\mathbb{R}^d$, so such placements may be topologically distinct. For example, lattice knots of different knot types are topologically distinct placements of S in $\mathbb{R}^3$ (see section 1.7.1).

Examples of embedded graphs are given in references [590] (figure eights), trees with restricting branching [523], lattice animals with fixed cyclomatic index [252], and brushes and combs [385] (see section 2.2).

In an embedding of a graph the branches are mapped to self-avoiding walks. If each branch is mapped to a self-avoiding walk of length L, then the embedding is *monodispersed* or *uniform*. If branches are mapped to walks of differing lengths, then the embedding is *polydispersed* (see for example references [102, 109, 215, 519, 520, 588, 589]).

11.5.1 *Embedded graphs*

Let $\mathcal{G}_N$ be the subset of $\mathcal{G}$ with graphs G with $\operatorname{maxdeg} G \leq N$. A 4-star graph $s_4 \in \mathcal{G}_4$ has $\operatorname{maxdeg} s_4 = 4$ and is embeddable in $\mathbb{L}^2$ because it is planar (see figure 2.5). Any planar graph in $\mathcal{G}_4$ is embeddable in $\mathbb{L}^2$.

In lemma 11.25 it is shown that any graph in $\mathcal{G}_6$ is embeddable in $\mathbb{L}^3$ [524]. Key concepts in the proof of the lemma are the *top plane* and the *top line*.

If $\vec{v} \in \mathbb{L}^d$, then denote its i-th Cartesian coordinate by $\vec{v}(i)$. The *top plane* $T_p(\alpha)$ of an embedding α of a graph G in $\mathbb{L}^d$ is defined by

$$T_p(\alpha) = \big\{\vec{v} \in \mathbb{L}^d \mid \vec{v}(d) = \max\{\vec{u}(d) \mid \vec{u} \in \alpha\}\big\}. \tag{11.58}$$

The *top line* is

$$L_p(\alpha) = \{\vec{v} \in T_p(\alpha) \mid \vec{v}(d-1) = \max\{\vec{u}(d-1) \mid \vec{u} \in \alpha \cap T_p(\alpha)\}\}. \tag{11.59}$$

If $\vec{v} \in \alpha$ has $\deg \vec{v} = 1$, then it is an *end-vertex*, and, if $\deg \vec{v} > 2$, then it is a *branching point*.

Lemma 11.25 *Let $G \in \mathcal{G}$ be a graph. Then there is an embedding of G in $\mathbb{L}^3$ if and only if $G \in \mathcal{G}_6$. Moreover, there is also an embedding α of G in $\mathbb{L}^3$ such that every end-vertex is in $L_p(\alpha)$ and every branch not incident on an end-vertex has exactly one edge in $L_p(A)$.*

Proof Suppose that α is an embedding of the graph $G \in \mathcal{G}$ in $\mathbb{L}^3$.

Then $\operatorname{maxdeg} \alpha \leq 6$; therefore, $G \in \mathcal{G}_6$, since α is homeomorphic to G.

Conversely, suppose that $G \in \mathcal{G}_6$. The proof is a double induction. First it is proven that trees are embeddable as claimed; then, there will be an induction on the number of cycles in G.

To see that trees are embeddable as claimed, induction will be done on the number of branching points in trees. Let τ be a tree and suppose that it has b branching points.

If $b = 1$, then τ is homeomorphic to a star graph with six or fewer arms. Such graphs can be embedded directly by construction, and, moreover, by extending the arms to $L_p(\tau)$ in an obvious way, an embedding with the required properties can be obtained.

Suppose that the lemma is true for all trees with b or fewer branching points. Select a tree τ_0 from $\mathcal{G}_6$ with $b+1$ branching points. Without loss of generality,

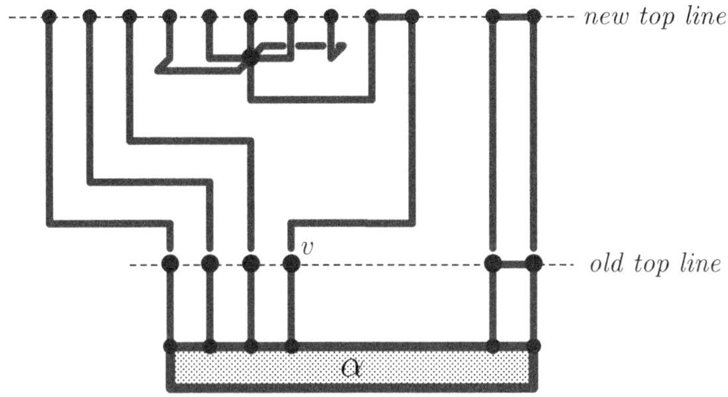

FIG. 11.10. A graph α with b branching points, with each branch visiting its top plane T_p such that all end-vertices are in its top line L_p, and every branch not incident with an end-vertex has exactly one edge in L_p. A new branch point can be appended to α by adding a six-armed star to any of its end-vertices in L_p. By extending the existing branches, the new graph can be changed such that it is uniform, all its branches have edges in T_p and all its end-vertices are in L_p.

---

it may be supposed that every branching point in this tree has degree equal to 6 (if not, add branches until all branching points have degrees equal to 6, embed the tree, and then remove these branches to obtain an embedding of the original tree).

The tree τ_0 has $4b+2$ end-vertices, and so there is at least one branching point v which has five branches which end in an end-vertex each. Delete these five branches to obtain a tree τ_1 which is homeomorphic to a tree graph with b branching points and is embeddable in the required way (see figure 11.10).

By the induction hypothesis, τ_1 has the required properties and has an end-vertex $\vec{v}$ in the top line $L_p(\tau_1)$. Append a six-armed star to the end-vertex $\vec{v}$ as shown in figure 11.10 and extend the rest of the branches in the embedding so that all end-vertices are again in the top line.

This finishes the proof for embeddings of trees.

If a graph $G \in \mathcal{G}_6$ is not a tree, then cut cycles in G until it is a tree τ and label the newly created end-vertices with the same labels so that the original cycles can be recovered. Embed τ in the lattice with all the end-vertices in the top line and join the end-vertices with the same labels by adding edges in the obvious way. The edges and end-vertices in the top line and top plane can again be extended until an embedding with the required properties is obtained. □

11.5.2 *Knotted embeddings of graphs*

Let $G \in \mathcal{G}_6$ be a graph and suppose that α is an embedding of G in $\mathbb{L}^3$. Such an embedding of G as a network induces an injection $k : G \to \mathbb{R}^3$ representing G.

The injection k is a *knotted graph*; this notion is abused by calling the image of k in $\mathbb{R}^3$ a knotted graph and denoting it k. The image k is also a 1-complex in $\mathbb{R}^3$, and it is piecewise linear if the map k is piecewise linear.

Generally, embeddings of G in $\mathbb{L}^3$ as networks induce piecewise linear embeddings of k in $\mathbb{R}^3$.

The two knotted embeddings k_1 and k_2 of G are equivalent if there exists an orientation preserving homeomorphism $H : (\mathbb{R}^3, k_1) \to (\mathbb{R}^3, k_2)$. The equivalence class of embeddings is denoted by K, and k is a representative of K.

The collection of all equivalence classes of embeddings k is denoted $\mathcal{K}(G)$. If G is the circle graph, then its embeddings are lattice polygons and there is an equivalence class of planar embeddings which defines the *unknot*. If G is a general graph, then it may be more difficult to define a class of embeddings as being the unknot.

A projection $P : G \to \mathbb{R}^3 \to \mathbb{R}^2$ of an embedded graph G into a plane in $\mathbb{R}^3$ is *regular* if the following conditions are met:

- each point in the projection is the image of at most two points in the embedding,
- each double point in the image is the transverse intersection of two projected arcs (this is a *crossing*), and
- the projected image of each branching point or end-vertex is the image of exactly one point (the branch point or end-vertex) in the embedding.

Crossings of arcs are signed by the convention in figure 1.9.

The *crossing number* $C(K)$ of an equivalence class K is the minimal number of crossings over all regular projections of all representatives of that equivalence class. The *minimal crossing number* of $\mathcal{K}(G)$ is

$$C_m(G) = \min_{K \in \mathcal{K}(G)} C(K). \tag{11.60}$$

The *unknot* is an equivalence class of embeddings K of G which has $C(K) = C_m(G)$. Generally, the unknot may not be unique (although it is unique if G is the circle graph).

If G is a planar graph, then the unknot is the equivalence class of planar embeddings of G in $\mathbb{R}^3$ (and these are equivalent [524] in $\mathbb{R}^3 \cup \{\infty\}$).

If a graph G has an equivalence class K of embeddings k with minimal crossing number $C(K) > C_m(G)$, then K is a *knotted embedding* of G.

Some graphs are intrinsically complex with respect to embeddings in 3-space. For example, every embedding of the complete graph K_{4k+3} is chiral (there are an even number of non-equivalent unknotted representatives of embeddings of K_{4k+3}), and every embedding of K_7 has an odd number of Hamiltonian cycles which are knotted circles; the unknotted embedding of K_7 will always contain knotted cycles).

11.5.3 *A pattern theorem for embeddings of graphs*

Let $G \in \mathcal{G}_6$; then there is an embedding of G in $\mathbb{L}^3$, by lemma 11.25. Let $g_n(G)$ be the number of embeddings of G in $\mathbb{L}^3$ and with n edges, counted modulo equivalence under translations in the lattice.

Let α be any embedding of G with m edges and let ρ be a lattice polygon of length $n-m$ edges in $\mathbb{L}^3$. Then ρ can be concatenated onto ω in the following way. Let $\vec{t}$ be the lexicographic top vertex of α. If $\deg \vec{t} \geq 2$, then there is an edge in α incidence with $\vec{t}$ perpendicular to $\vec{e}_1$. The polygon ρ can be concatenated onto this edge in the usual way (see figure 1.5). If $\deg \vec{t} = 1$, then ρ is concatenated onto α as follows: translate ρ so that its bottom vertex $\vec{b} = \vec{t} + \vec{e}_1$. Delete an edge in ρ incident with $\vec{b}$ and concatenate ρ onto $\vec{t}$ by adding an edge $\langle \vec{t} \sim \vec{b} \rangle$.

Since there are $g_n(G)$ choices for α and at least $\frac{1}{2} p_{n-m}$ choices for ρ (the factor $\frac{1}{2}$ arises because $\vec{b}$ (the bottom edge of ρ) must be parallel to $\vec{t}$ (the top edge of α)),

$$g_m(G) p_{n-m} \leq 2 g_n(G). \tag{11.61}$$

Take logarithms, divide by n and let $n \to \infty$ to see that

$$\log \mu_3 \leq \liminf_{n \to \infty} \tfrac{1}{n} \log g_n(G) \tag{11.62}$$

by theorem 7.7. This gives the following theorem.

Theorem 11.26 *Let $g_n(G)$ be the number of distinct embeddings of G in $\mathbb{L}^3$. Then, for any $G \in \mathcal{G}_6$, $\lim_{n \to \infty} \frac{1}{n} \log g_n(G) = \log \mu_3$.*

Proof Let α be any embedding of $G \in \mathcal{G}_6$ with n edges. Suppose that G has b branches labelled $\{1, 2, \ldots, b\}$ and suppose branch i contains n_i edges. Suppose also that G has v branching points and end-vertices.

Branch i has at most c_{n_i} conformations, and it can be placed into at most one of $\binom{v}{2}$ possible ways into G to reconstruct α.

This shows that

$$g_n(G) \leq \binom{\binom{v}{2}}{b} \sum_{\{n_i\}} \prod_{i=1}^{b} c_{n_i} \delta(n - \sum n_i).$$

However, $c_n = \mu_3^{n+o(n)}$. Take logarithms, divide by n and let $n \to \infty$. Then at least one of the n_i in the above goes to infinity, showing that $\limsup_{n \to \infty} \frac{1}{n} \log g_n(G) \leq \log \mu_3$. Comparison to equation (11.62) completes the proof. □

The argument in the proof of theorem 11.26 gives a pattern theorem for embeddings of graphs in $\mathbb{L}^3$. Let $g_n[\leq \lfloor \epsilon n \rfloor, P \mid G]$ be the number of embeddings of G with n edges which contain the Kesten pattern P at most $\lfloor \epsilon n \rfloor$ times.

Then each branch of length n_i of the animal α in the proof of theorem 11.26 has at most $c_{n_i}[\leq\lfloor \epsilon n_i \rfloor, P]$ conformations. That is,

$$g_n[\leq\lfloor \epsilon n \rfloor, P \mid G] \leq \binom{\binom{v}{2}}{b} \sum_{\{n_i\}} \prod_{i=1}^{b} c_{n_i}[\leq\lfloor \epsilon n_i \rfloor, P] \, \delta(n - \sum n_i). \tag{11.63}$$

By theorem 7.13, $\limsup_{n\to\infty} \frac{1}{n} \log c_n[\leq\lfloor \epsilon n \rfloor P] < \log \mu_3$. Thus, the pattern theorem for embeddings of $G \in \mathcal{G}_6$ is the following.

Theorem 11.27 *Suppose that $G \in \mathcal{G}_6$ and let P be a Kesten pattern. If $g_n[\leq mP \mid G]$ is the number of embeddings of G with n edges containing P at most m times, then $\limsup_{n\to\infty} \frac{1}{n} \log g_n[\leq\lfloor \epsilon n \rfloor P \mid G] < \log \mu_3$.* □

A *cut-edge* in a graph is a single branch whose deletion will also disconnect the graph. A graph without any cut-edges is 2-edge-connected.

Notice that an embedding of a 2-edge-connected graph is a 2-edge-connected animal. A 2-edge-connected graph cannot have any end-vertices. This gives a Frisch-Wasserman-Delbruck result [126, 212] for embedded networks.

Theorem 11.28 *Let $G \in \mathcal{G}_6$ be 2-edge-connected and suppose that P is a Kesten-pattern. Then, for sufficiently large n, all but exponentially few embeddings of G of size n in $\mathbb{L}^3$ are knotted. Moreover, there is an $\epsilon > 0$ such that all but exponentially few embeddings contain at least $\lfloor \epsilon n \rfloor$ distinct copies of P.*

Proof Let $C_m(G)$ be the minimal crossing number of G. Put $q = \lceil 1 + \frac{1}{3} C_m(G) \rceil$. Let T denote the tight trefoil arc in figure 1.11 and let T^q be obtained by concatenating q copies of T into a string in the obvious way.

By theorem 11.27,

$$\limsup_{n\to\infty} \tfrac{1}{n} \log g_n[\leq\lfloor \epsilon n \rfloor, T^q \mid G] < \log \mu_3.$$

Since G is 2-edge-connected, each copy of T^q must be in a cycle, and this cycle is a (topological) knot of minimal crossing number at least $3q$. Take q large enough so that $3q > C_m(G)$, in which case the embeddings of G are knotted.

This completes the proof. □

See references [521, 522, 524] for more about the Frisch-Wasserman-Delbruck conjecture and knotted lattice embeddings of graphs.

11.6 Uniform networks

Let $\mathcal{G}$ be the set of abstract simple connect graphs with no vertices of degree equal to 2 (but add the circle graph consisting of one vertex $\vec{v}$ and one edge $\langle \vec{v} \sim \vec{v} \rangle$ to $\mathcal{G}$). Let $\mathcal{G}_N$ be the subset of $\mathcal{G}$ such that, if $G \in \mathcal{G}_N$, then $\operatorname{maxdeg} G \leq N$.

Realisations of graphs $G \in \mathcal{G}$ in $\mathbb{L}^d$ are, as before, lattice animals with the same *homeomorphism type* as G (see section 11.5). Branches in G are realised as

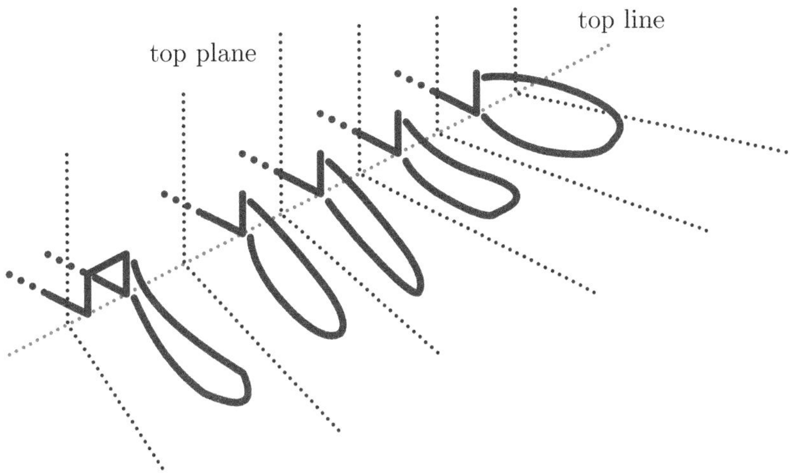

FIG. 11.11. An embedding of a uniform network exists such that each end-vertex is in the top line, and every branch without an end-vertex has one edge in the top line. By concatenating polygons in narrow fg-wedges as shown, with edges in the top plane, and then deleting the edges in the polygons which are in the top plane, a larger animal is obtained. Moreover, these polygons can expand in their own fg-wedges as the limit is taken.

---

branches (self-avoiding walks) in the lattice animal, joined together at branching points which corresponds to vertices in G. The realisation of a graph G as a lattice animal with the same homeomorphism type is, as before, a *network*.

A network is uniform if all its branches have the same length. A uniform network is *monodispersed* (see section 2.2).

Let $G \in \mathcal{G}_6$ be an abstract graph. A cycle in G is even if it is the union of an even number of branches in G (if it is not even, then it is odd). By lemma 11.25, any graph in $\mathcal{G}_6$ has an embedding α in $\mathbb{L}^3$ such that every end-vertex is in the top line L_p and every branch without an end-vertex has one edge in L_p. By concatenating walks of the shape ⁊⊔⊓⊔⊓∟ on end-vertices of branches in L_p, and polygons of shape ▭ on the edges of branches in L_p, α may be changed into a uniform embedding υ of G with the same property that every branch either has its end-vertex in the top line or has an edge in the top line.

Since $\mathbb{L}^3$ is bipartite, it can be shown that, if n is even, then there exist uniform embeddings υ of size n of G. If n is odd, then there exist uniform embeddings υ of G for n large enough if and only if G has no odd (length) cycles.

Let $u_n(f)$ be the number of uniform embeddings of G with exactly n edges per branch and a total of f branches. Then $u_n(f) > 0$ for all n large enough (and possibly only for even n).

If $\vec{v}$ is a vertex, or $\langle \vec{v} \sim \vec{w} \rangle$ is an edge, in the top plane $T_p(v)$ of v, then $\vec{0}$ may be chosen at $\vec{v}$, and the $\vec{e}_1$ direction may be set up perpendicular to $T_p(v)$. By subdividing v if necessary, enough space may be created about each vertex and edge in $T_p(v)$ so that a wedge $W \equiv W_{fg}$ may be set up at each vertex and edge in $T_p(v)$ such that the wedges W are mutually disjoint and have the properties of the wedge in theorem 8.25.

Polygons confined to the wedges W can be concatenated on the edges and vertices in the top plane of v as illustrated in figure 11.11. The resulting animal is uniform and is ambient isotopic to v.

Suppose that v has f branches and that the wedges are $\{W_1, W_2, \ldots, W_f\}$. This shows

$$u_n(f) \geq \prod_{j=1}^{f} p_{n-w_j}(W_j), \tag{11.64}$$

where $p_n(W_j)$ is the number of polygons of length n rooted at the origin of each wedge W_j and where w_j is the number of edges in the j-th branch of v. Take logarithms, divide by n and let $n \to \infty$. Then

$$\liminf_{n\to\infty} \tfrac{1}{n} \log u_n(f) \geq f \log \mu_3. \tag{11.65}$$

This result gives the following theorem.

Theorem 11.29 *The limit* $\lim_{n\to\infty} \frac{1}{n} \log u_n(f) = f \log \mu_3$.

Proof If v is a uniform network of size nf edges and with f branches, then v can be decomposed into f self-avoiding walks by cutting the branches from v.

The f branches can be put together in at most $2^f f!$ ways (by selecting one of two endpoints in each and then arranging them in $f!$ ways). Since each branch has c_n conformations, this shows that $u_n(f) \leq (2^f f!) c_n^f$. Take logarithms, divide by n and let $n \to \infty$ to complete the proof □

Theorem 11.29 is also applicable to (non-uniform) embeddings of animals. Let $v_n(G)$ be the number of embeddings of a graph G of f branches and with a total of n edges. Clearly, every uniform embedding of G is also an embedding; therefore, $u_n(f) \leq v_{fn}(G)$. By concatenating polygons onto the top edge of embeddings of G, it follows that $u_n(f) \leq v_{fn+2k}(G)$ for every $k \in \mathbb{N}$. Take logarithms, divide by $N = fn + 2k$ and take $N \to \infty$ with $k \in \{0, 1, \ldots, M\}$ ($M > f$ fixed) to find the limit inferior on the right-hand side. This shows that $\log \mu_d \leq \liminf_{n\to\infty} \frac{1}{n} \log v_n(G)$.

On the other hand, dissect every embedding of G into f branches, each of which is a self-avoiding walk. This shows that $v_n(G) \leq s^f f! \sum_{\{n_i\}} \prod_{i=1}^{f} c_{n_i}$, where the sum over all $\{n_i\}$ is such that $\sum_i n_i = n$. Since $c_n = \mu_d^{n+o(n)}$, it follows that $\limsup_{n\to\infty} \frac{1}{n} \log v_n(G) \leq \mu_d$. This gives the following theorem [109, 527, 589].

Theorem 11.30 *The limit* $\lim_{n\to\infty} \frac{1}{n} \log v_n(G) = \log \mu_3$. □

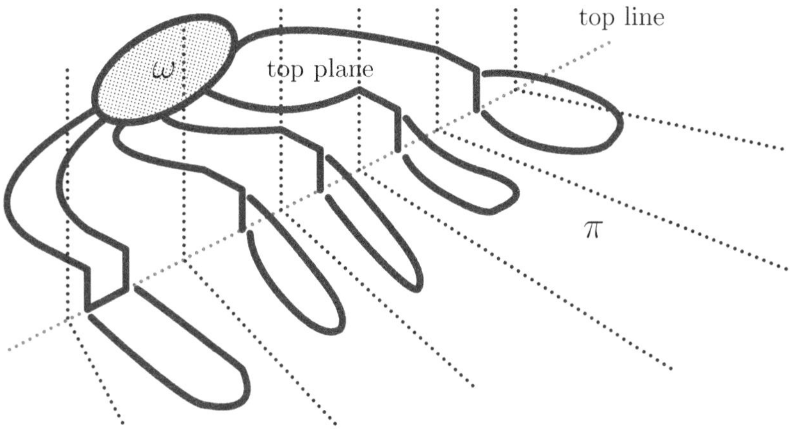

FIG. 11.12. A slight change in figure 11.11 can be made to find an embedding of a uniform network adsorbing into the plane π. The top line is in π such that each end-vertex is in the top line, and each branch without an end-vertex has one edge in the top line. Adsorbing polygons in narrow fg-wedges can be concatenated onto the animal as shown.

11.6.1 *Adsorbing uniform networks*

Theorem 11.29 can be generalised to adsorbing uniform animals. Slightly modify the construction in figure 11.11 to figure 11.12. The discussion is limited to $d = 3$ dimensions but can be generalised to $d \geq 3$.

Let $\pi = \partial \mathbb{L}^3_+$ be the adsorbing plane in the positive half-lattice $\mathbb{L}^3_+$ (see equation (2.20)). A graph $G \in \mathcal{G}_6$ can be embedded disjoint with π. Choose a top plane normal to π such that it intersects π in the top line (see figure 11.12).

By lemma 11.25, there is an embedding υ of G such that each branch which is in a cycle has one edge in the top line, and each endpoint is in the top line.

Positive polygons are concatenated onto the vertices and edges in the top plane as illustrated and then confined to wedges W_j chosen such that each satisfies the conditions of theorem 8.25.

Suppose υ has m edges and f branches $\langle b_1, b_2, \ldots, b_f \rangle$, each of size m_i.

Further, suppose that the polygon in W_j is concatenated on the j-branch of υ, has size n_j and makes a total of v_j visits to the plane π, such that $m_j + n_j = n$. Then the embedded animal is a uniform embedding of G with each branch of length n. Denote the number of such uniform animals by $u^+_n(G, v)$, counted such that each animal has at least one vertex in the adsorbing plane π. This shows that

$$u^+_n(G, v) \geq \prod_{j=1}^{f} p^+_{n_j}(W_j, v_j)\, \delta(v - \sum v_i) \prod_{k=1}^{f} \delta(n - m_j - n_j). \qquad (11.66)$$

Multiply this by a^v and sum over v (while keeping $v_j > 0$ for each j) to obtain a lower bound on the partition function for adsorbing uniform networks:

$$U_n^+(G,a) = \sum_{v\geq 1} u_n^+(G,v)\, a^v \geq \prod_{j=1}^{f} \sum_{v_j>0} p_{n_j}^+(v_j)\, a^{v_j} \prod_{k=1}^{f} \delta(n - m_j - n_j). \quad (11.67)$$

Take logarithms, divide by n and let $n \to \infty$. Then each $n_j \to \infty$, since m_j is fixed. By theorem 9.21, this gives

$$\liminf_{n\to\infty} \tfrac{1}{n} \log U_n^+(G,a) \geq f \mathcal{A}_p^+(a), \quad (11.68)$$

where $\mathcal{A}_p^+(a)$ is the limiting free energy of adsorbing polygons in $\mathbb{L}_+^3$ (see theorem 9.25).

To find an upper bound on $u_n^+(G,v)$, decompose v into f branches. The branches can again be put together in $2^f f!$ ways (see the proof of theorem 11.29). However, each branch may have endpoints at heights h_1 and h_2 which could be a distance fn from π. Moreover, if the branch is disjoint with π, then it may have at most hc_n conformations, where h is the height of one of its endpoints above π. Since h can take up to fn values, this shows that the number of conformations of each branch is at most fnc_n if it has zero visits.

If a branch has $v > 0$ visits to π, then it has at most $\sum_k c_k^+ c_{n-k}^+(v-1)$ conformations (cut it into two positive walks in its first visit).

Put $b_n(v) = fnc_n$ if $v = 0$, and $b_n(v) = \sum_k c_k^+ c_{n-k}^+(v-1)$ if $v > 0$. This gives

$$u_n^+(G,v) \leq 2^f f! \prod_{j=1}^{f} \sum_{v_j\geq 1} b_n(v_j)\, \delta(v - \sum_j v_j). \quad (11.69)$$

Multiply by a^v and sum over v. This gives

$$U_n^+(G,a) \leq 2^f f! \Big(\sum_{v\geq 1} b_n(v)\, a^v\Big)^f. \quad (11.70)$$

Take logarithms, divide by n and let $n \to \infty$ to see that

$$\limsup_{n\to\infty} \tfrac{1}{n} \log U_n^+(G,a) \leq f \limsup_{n\to\infty} \tfrac{1}{n} \log \sum_{v\geq 1} b_n(v)\, a^v. \quad (11.71)$$

Notice that $c_k^+ = \mu_3^{n+o(n)}$ and that $\sum_v c_n^+(v)\, a^v = \big(\mathcal{A}_p^+(a)\big)^{n+o(n)}$ by theorem 9.23. Thus, since $\mathcal{A}_p^+(a) \geq \mu_d$, it follows that $\limsup_{n\to\infty} \frac{1}{n} \log \sum_{v\geq 1} b_n(v)\, a^v = \mathcal{A}_p^+(a)$. Comparison of the above to equation (11.68) gives

$$\lim_{n\to\infty} \tfrac{1}{n} \log U_n(G,a) = f \mathcal{A}_p^+(a). \quad (11.72)$$

Since $\mathcal{A}^+(a) = \mathcal{A}_p^+(a)$ in $\mathbb{L}_+^3$, where $\mathcal{A}^+(a)$ is the limiting free energy of adsorbing positive walks, the following theorem is obtained.

Theorem 11.31 *Let $u_n^+(G,v)$ be the number of uniform embeddings of the graph G in $\mathbb{L}_+^3$, with f branches of length n and with a total of v visits in the adsorbing plane and let $U_n^+(G,a) = \sum_v u_n^+(G,v)\,a^v$ be the partition function of this model. Then the limiting free energy of the adsorbing uniform graph is*

$$\lim_{n\to\infty} \tfrac{1}{n} \log U_n^+(G,a) = f\mathcal{A}^+(a),$$

where $\mathcal{A}^+(a)$ is the limiting free energy of adsorbing positive walks. □

More generally, a similar theorem can be proven for non-uniform adsorbing embeddings of G. Below is an outline.

Let $v_n^+(G,v)$ be the number of embeddings of a graph G in $\mathbb{L}_+^3$ and with v visits to the adsorbing plane. Clearly, $u_n^+(G,v) \leq v_{fn}^+(G,v) \leq v_{fn+2k}^+(G,v)$ for $k \in \mathbb{N}$, where the last inequality follows by concatenating polygons of length $2k$ onto the top edge of the lattice animal. Let $k \in \{0,1,\ldots,M\}$, where $M > f$. Multiply by a^v and sum over v to see that $U_n^+(G,a) \leq \sum_v v_{fn+2k}^+(G,v)\,a^v$. Take logarithms, divide by $N = fn+2k$ and take $N \to \infty$. This gives $\mathcal{A}^+(a) \leq \liminf_{n\to\infty} \frac{1}{n} \log \sum_v v_n^+(G,v)\,a^v$.

On the other hand, animals can be decomposed into f walks in $\mathbb{L}_+^d$ of total length n and with at least v and at most $v+2f$ visits. Each of these walks either has or does not have visits to π. If a walk of length n_i does not have visits, then it has at most nc_{n_i} conformations, since its one endpoint (say the lowest endpoint) may have at most n different heights above the adsorbing line.

If a walk of length n_i has v visits, then it can be decomposed into two positive subwalks by cutting it in its first visit. This shows that the branch has at least $\sum_m c_m^+ c_{n_i-m}^+(v)$ conformations.

Notice that the f branches can be put together in at most $2^f f!$ ways.

Define $b_n(v)$ as above; then $\sum_v v_n^+(G,v)\,a^v \leq 2^f f! \prod_{i=1}^f \left(\sum_{v_i} b_{n_i}(v_i)\,a^{v_i}\right)$. Take logarithms, divide by n and let $n \to \infty$. Then at least one of the $n_i \to \infty$, and this shows that $\limsup_{n\to\infty} \frac{1}{n} \log \sum_v v_n^+(G,v)\,a^v \leq \mathcal{A}^+(z)$. This establishes the following theorem:

Theorem 11.32 *Let $v_n^+(G,v)$ be the number of embeddings of size n edges of the graph G in $\mathbb{L}_+^3$, with f branches and with a total of v visits in the adsorbing plane.*

Let $\sum_v v_n^+(G,v)\,a^v$ be the partition function of this model. Then the limiting free energy of the adsorbing uniform graph is

$$\lim_{n\to\infty} \tfrac{1}{n} \log \sum_v v_n^+(G,v)\,a^v = \mathcal{A}^+(a),$$

where $\mathcal{A}^+(a)$ is the limiting free energy of adsorbing positive walks. □

The above generalises to higher dimensions in the half-lattice $\mathbb{L}_+^d$ for $d \geq 3$. The situation is different in two dimensions (in $\mathbb{L}_+^2$), since a polygon of length n in $\mathbb{L}_+^2$ has a maximum of $\frac{n}{2}$ visits; while a walk of length n may have n visits; cycles and branches with endpoints interact differently with π in $\mathbb{L}_+^2$.

Existence of the limiting free energy for adsorbing uniform networks in two dimensions has not been shown for general graphs (but is known for walks and polygons) [519]. Uniform networks in a slit geometry were considered in reference [529], where it is shown that the growth constant is dependent on the width of the slit. Similar results exist for slab geometry [102] (for some models of uniform animals).

12

INTERACTING MODELS OF VESICLES AND SURFACES

Lattice surfaces in $\mathbb{L}^d$ are the unions of *plaquettes*, which are closed geometric unit area squares with corners which are vertices in $\mathbb{L}^d$.

Surfaces are defined as follows. Let p and q be two plaquettes in $\mathbb{L}^d$. Then p and q are adjacent if their intersection is a lattice edge: $p \cap q = e$, where e is an edge in $\mathbb{L}^d$. The plaquettes p and q are said to be *incident* on e.

A set of plaquettes $S = \{p_1, p_2, \ldots, p_n\}$ is *connected* if their union is a connected subset of $\mathbb{R}^d$. If S is connected, then it is a lattice surface if $p_i \cap p_j$ is either an edge or the empty set (for $1 \leq i, j \leq n$), and if at most two plaquettes are incident on any edge.

The boundary of a lattice surface S is denoted by ∂S, and it is the union of all edges incident with exactly one plaquette. If $\partial S = \emptyset$, then S is a *closed surface.* A closed surface is a piecewise linear embedding of a 2-manifold in $\mathbb{L}^d$ (and by inclusion, in $\mathbb{R}^d$). If S is a topological sphere, then its embedding is a *lattice sphere.*

In this chapter, two models of lattice vesicles and surfaces are examined. In section 12.1 the Fisher-Guttmann-Whittington square lattice vesicle [195] is examined. This model is an undirected version of the models of directed vesicles examined in chapter 6.

Closed orientable lattice surfaces in $\mathbb{L}^3$ are models of vesicles if they are weighted by volume (and surface area). Such a model is the model of *crumpling surfaces*, presented in section 12.2. There are numerous studies of self-avoiding surfaces in the literature; see, for example, references [47, 525, 543, 591].

12.1 Square lattice vesicles

A *lattice vesicle* in $\mathbb{L}^2$ is a lattice polygon weighted by its area [26, 192, 195, 379] (see reference [581] for vesicles in $\mathbb{L}^3$). The partition function of this model is

$$P_n(q) = \sum_{m=0}^{n^2/4} p_n(m)\, q^m, \tag{12.1}$$

where $p_n(m)$ is the number of square lattice polygons of length n enclosing an area of m plaquettes or unit squares: $d_m = \sum_n p_n(m)$ is the number of square lattice disks of area m [329]. The area generating variable is q. Concatenation of two polygons (see figure 1.5) gives

$$\sum_{m_1=0}^{m} p_{n_1}(m_1) p_{n_2}(m - m_1) \leq p_{n_1+n_2}(m + 1). \tag{12.2}$$

The Statistical Mechanics of Interacting Walks, Polygons, Animals and Vesicles, 2nd edition, E.J. Janse van Rensburg. © E.J. Janse van Rensburg. Published in 2015 by Oxford University Press.

Multiply by q^m and sum over m to obtain

$$P_{n_1}(q)P_{n_2}(q) \leq q^{-1} P_{n_1+n_2}(q). \tag{12.3}$$

By theorem A.1 in appendix A, the limiting free energy

$$\mathcal{V}(q) = \lim_{n\to\infty} \tfrac{1}{n} \log P_n(q) \tag{12.4}$$

exists, where $P_n(q) \leq q^{-1} e^{n\mathcal{V}(q)}$. Since $P_n(q) \leq p_n$ if $q \leq 1$, this also shows that that $\mathcal{V}(q) \leq \log \mu_2 < \infty$ if $q \in (0,1]$. Observe that $\mathcal{V}(1) = \log \mu_2$.

There are square-shaped vesicles of perimeter $4k$ and area k^2. This shows that $\mathcal{V}(q) = \infty$ if $q > 1$.

Theorem 12.1 *The limit* $\mathcal{V}(q) = \lim_{n\to\infty} \frac{1}{n} \log P_n(q)$ *exists. If* $q \leq 1$, *then* $\mathcal{V}(q) \leq \log \mu_2$, *and, if* $q > 1$, *then* $\mathcal{V}(q) = \infty$. *If* $q > 1$, *then*

$$\lim_{n\to\infty} \tfrac{1}{n^2} \log P_n(q) = \tfrac{1}{16} \log q, \quad \textit{and} \quad \lim_{n\to\infty} \tfrac{d}{d\log q}\big(\tfrac{1}{n^2} \log P_n(q)\big) = \tfrac{1}{16}.$$

Proof It only remains to prove existence of the last limits.

If $n = 4k$, then the area of a square-shaped polygon is k^2, and, if n is not a multiple of 4, then the maximum area is $\frac{1}{16}(n^2-4)$. Thus, $\liminf_{n\to\infty} \frac{1}{n^2} \log P_n(q) \geq \frac{1}{16} \log q$. If $q > 1$, then $P_n(q) \leq p_n\, q^{n^2/16}$. Thus, $\limsup_{n\to\infty} \frac{1}{n^2} \log P_n(q) \leq \frac{1}{16} \log q$. This proves existence of the limit $\lim_{n\to\infty} \frac{1}{n^2} \log P_n(q) = \frac{1}{16} \log q$.

By theorem 3.3, $\frac{1}{n^2} \log P_n(q)$ is a convex function of $\log q$ for every n. The sequence $\langle \frac{1}{n^2} \log P_n(q) \rangle$ is a sequence of convex functions convergent to a convex limit. By theorem B.9 (in appendix B), the sequence of derivatives converges to the derivative of the limit. This shows that

$$\lim_{n\to\infty} \tfrac{d}{d\log q} \tfrac{1}{n^2} \log P_n(q) = \tfrac{d}{d\log q} \lim_{n\to\infty} \tfrac{1}{n^2} \log P_n(q) = \tfrac{d}{d\log q} \tfrac{1}{16} \log q = \tfrac{1}{16}.$$

This completes the proof. □

The limiting free energy $\mathcal{V}(q)$ is continuous in $(0,1)$ since it is convex and finite on $[0,1]$. It is right-continuous at $q = 0$ because $\log P_n(q)$ is both increasing and convex. It is also left-continuous at $q = 1$.

Theorem 12.2 *The limiting free energy* $\mathcal{V}(q)$ *is continuous on* $(0,1)$, *right-continuous at* $q = 0$ *and left-continuous at* $q = 1$. *It is finite on* $[0,1]$, *and* $\mathcal{V}(q) = \infty$ *if* $q > 1$.

Proof It only remains to show left-continuity at $q = 1$. Let both $\epsilon > 0$ and $\delta > 0$ be arbitrary and small. Since $\frac{1}{n} \log P_n(q) \to \mathcal{V}(q)$ and, by equation (12.4), there exists an integer N_0 such that for all $n > N_0$,

$$\begin{aligned} \mathcal{V}(1) - \tfrac{1}{n} \log P_n(1) &< \tfrac{1}{3}\epsilon, \\ \tfrac{1}{n} \log P_n(1) - \tfrac{1}{n} \log P_n(1-\delta) &< \tfrac{1}{3}\epsilon, \\ \tfrac{1}{n} \log P_n(1-\delta) - \mathcal{V}(1-\delta) &\leq -\tfrac{1}{n} \log(1-\delta). \end{aligned}$$

Add the last three inequalities to obtain

$$0 \leq \mathcal{V}(1) - \mathcal{V}(1-\delta) \leq \tfrac{2}{3}\epsilon - \tfrac{1}{n}\log(1-\delta) \leq \epsilon$$

for n large enough. This shows that $\mathcal{V}(q)$ is left-continuous at $q = 1$. □

12.1.1 *The phase diagram of square lattice vesicles:*

The generating function of the model is

$$V(q,t) = \sum_{n=0}^{\infty} P_n(q)\, t^n. \tag{12.5}$$

The radius of convergence of $V(q,t)$ is $t_c(q) = e^{-\mathcal{V}(q)}$, so $t_c(q) = 0$ if both $q > 1$ and $t_c(1) = \frac{1}{\mu_2}$. Note that $V(1,t)$ is the square lattice polygon generating function and that $V(1,t)$ is convergent if $t < \frac{1}{\mu_2}$, and divergent if $t > \frac{1}{\mu_2}$, by theorem 7.7.

Theorem 12.3 *The radius of convergence of $V(q,t)$ is $t_c(q)$, where*

$$t_c(q) \begin{cases} \sim \dfrac{1}{\sqrt{q}}, & \textit{if } q < 1; \\ = \dfrac{1}{\mu_2}, & \textit{if } q = 1; \\ = 0, & \textit{if } q > 1. \end{cases}$$

Moreover, $t_c(q)$ is a continuous function q for all $q \neq 1$.

Proof The continuity properties follow by theorem 12.2.

The minimal area of a polygon of length n is $\frac{1}{2}(n-2)$. Thus, $P_n(q) \leq p_n q^{(n-2)/2}$; therefore, $t_c(q) \geq \frac{1}{\mu_2\sqrt{q}}$. Concatenating two polygons of lengths n and m and of minimal areas (see figure 1.5) gives

$$p_n(\tfrac{1}{2}(n-2))p_m(\tfrac{1}{2}(m-2)) \leq p_{n+m}(\tfrac{1}{2}(n+m-2)).$$

This shows that the limit $\lim_{n\to\infty} \frac{1}{n}\log p_n\left(\frac{1}{2}(n-2)\right) = \log\overline{\mu} \leq \log\mu_2$ exists and is finite and bigger than 0. But $P_n(q) \geq p_n\left(\frac{1}{2}(n-2)\right) q^{(n-2)/2}$. This shows that $t_c(q) \leq \frac{1}{\overline{\mu}\sqrt{q}}$. □

Inverting $t_c(q)$ gives the phase boundary $q_c(t)$ of the vesicle. By theorem 12.3,

$$q_c(t) \begin{cases} = 1, & \text{if } t \leq \frac{1}{\mu_2}; \\ \sim \dfrac{1}{t^2}, & \text{if } t > \frac{1}{\mu_2}. \end{cases} \tag{12.6}$$

The phase diagram is illustrated in figure 12.1.

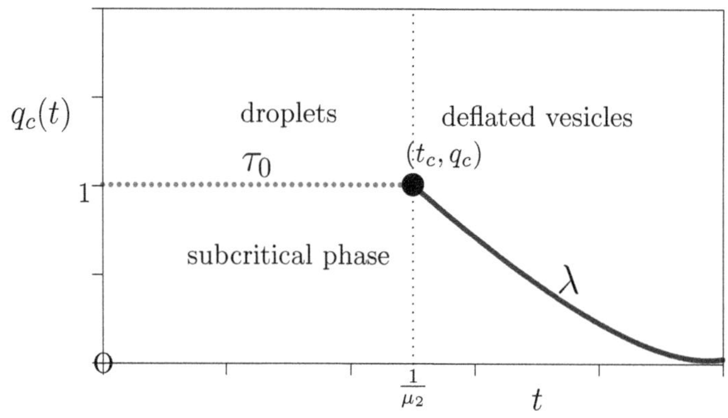

FIG. 12.1. The phase diagram of square lattice vesicles. The phase boundary is a critical line τ_0 of essential singularities meeting a λ-isotherm of simpler singularities in the tricritical point (t_c, q_c).

As $V(1,t)$ is the polygon generating function, it is finite if $t \in [0, \frac{1}{\mu_2})$. Derivatives of $V(q,t)$ with respect to q are given by

$$\frac{\partial^k}{\partial q^k} V(q,t) \leq \sum_{n=0}^{\infty} n^{2k} P_n(q)\, t^n, \tag{12.7}$$

since the maximum area of any polygon of length n is less that n^2. If both $q = 1$ and $t < \frac{1}{\mu_2}$, then this is finite for all finite $k \in \mathbb{N}$. This shows that that the phase boundary $q = 1$ with $0 \leq t < \frac{1}{\mu_2}$ is a line of essential singularities in the generating function. This is the τ_0-line shown in figure 12.1. The transition along the τ_0-line is a first order transition (or condensation) to inflated vesicles or *droplets* [7, 289].

The singularities along the critical curve $q_c(t)$ for $t > \frac{1}{\mu_2}$ are due to simpler singularities in the generating function at $q < 1$. This is the λ-isotherm, and along it the area of polygons are minimal (since $q < 1$). The transition is to a phase of deflated polygons (in the branched polymer universality class).

The tricritical point is located at the junction of the τ_0-line and the λ-isotherm. Approaching this point along the τ_0-line with $q = 1$ gives polygon exponents. For example, by equations (1.55), (1.59) and (4.11) and table 1.4, $2 - \alpha_u = 2 - \frac{1}{2} = \frac{3}{2}$.

The density of area is the Legendre transform

$$\log \mathcal{W}(\epsilon) = \inf_{q>0} \{\mathcal{V}(q) - \epsilon \log q\} \quad \text{for } \epsilon \in \left[\tfrac{1}{2}, \infty\right). \tag{12.8}$$

It follows from equation (3.62) that $\frac{d^+}{d\epsilon} \log \mathcal{W}(\epsilon) \mid_{\epsilon=1/2} = 0$, since $q_c = 1$. The function $\mathcal{W}(\epsilon)$ is concave and non-increasing for $\epsilon \in [\frac{1}{2}, \infty)$. It may be shown

TABLE 12.1. Square lattice vesicle tricritical exponents

ϕ	ν_t	$2-\alpha_t$	$2-\alpha_u$	$2-\alpha_+$	ν_+
$\frac{2}{3}$	$\frac{1}{2}$	1	$\frac{3}{2}$	0	$\frac{5}{8}$

that that $\mathcal{W}\left(\frac{1}{2}\right) < \mu_2$ (by using the pattern theorem for polygons in $\mathbb{L}^2$); this shows that $\mathcal{W}(\epsilon) < \mu_2$ for $\epsilon \in [\frac{1}{2}, \infty)$.

A variation on the above is the model of polygons in $\mathbb{L}^2$ of length n enclosing area of size $\lfloor n^\epsilon \rfloor$ (for $\epsilon \in [0,2]$). For large but finite n, $1 \leq p_n(\lfloor n^\epsilon \rfloor) \leq p_n$. Thus, if $\epsilon \in (0,2)$, then

$$0 \leq \liminf_{n\to\infty} \tfrac{1}{n} \log p_n\left(\lfloor n^\epsilon \rfloor\right) \leq \limsup_{n\to\infty} \tfrac{1}{n} \log p_n\left(\lfloor n^\epsilon \rfloor\right) \leq \log \mu_2. \tag{12.9}$$

If $\epsilon = 2$, then $\lim_{n\to\infty} \frac{1}{n} \log p_n\left(\lfloor n^2 \rfloor\right) = -\infty$.

If $\epsilon = 0$, then $\lim_{n\to\infty} \frac{1}{n} \log p_n(1) = -\infty$.

Let $\epsilon_{max}(n)$ be that least value of ϵ which maximises $p_n(\lfloor n^\epsilon \rfloor)$. Put $\varepsilon = \limsup_{n\to\infty} \frac{1}{n}\epsilon_{max}$; then $\varepsilon \leq 2$, and the mean area enclosed in a polygon is roughly proportional to n^ε. Numerical studies suggest that $\varepsilon \approx \frac{3}{2}$ [86, 180].

12.1.2 *Tricritical scaling in two-dimensional lattice vesicles*

Along the λ-isotherm in figure 12.1 it is expected that $V(q_c(t),t) \simeq (q_c(t)-q)^{2-\alpha_+}$ (see equation (4.9)). Since $q_c(t) < 1$, this gives a transition to a *deflated vesicle* or *branched polymer* phase. The value of $2-\alpha_+$ can be guessed from the known branched polymer exponents. Since the generating function of lattice trees is $T(t) \simeq |\log(\tau_2 x)|^{\theta-1}$, where θ is the (branched polymer) entropic exponent (see table 2.5), it follows that $2-\alpha_+ = \theta - 1$. The exact value of θ is 1 (see reference [460] and also references [191, 379, 556, 558]). Thus, $2-\alpha_+ = 0$.

Numerical simulations give the value $\nu = 0.65(4)$ for the metric exponent along the λ-isotherm [379]. This is consistent with the branched polymer value (in table 2.5) and with the Flory value $\frac{5}{8}$ in equation (2.14).

The critical point (t_c, q_c) is the lattice polygon critical point, and in $d=2$ the metric exponent has the exact value $\nu = \frac{3}{4}$ [165], from which the specific heat exponent α can be computed by equation (1.55). This gives $2-\alpha = \frac{3}{2}$. The crossover exponent is computed using equation (4.46): $\phi = \frac{2}{3}$. By equation (4.14), $2-\alpha_t = 1$.

The remaining tricritical exponents can be computed from the results above. By equation (4.53), $\nu_t = \frac{1}{2}$ (this is the metric exponent of inflated vesicles), and, by equation (4.54), $y_t = \frac{4}{3}$.

The arguments before and after equation (12.9) show that the mean area A_n of a lattice polygon of length n should scale as $A_n \sim n^\epsilon$, where $\epsilon \approx \frac{3}{2}$.

On the other hand, the mean area for critical vesicles (when $t = \frac{1}{\mu_2}$) may be computed by taking the derivative $\langle A \rangle = \frac{\partial}{\partial \log q} \log V(q,t) \mid_{q=1}$. This shows (by

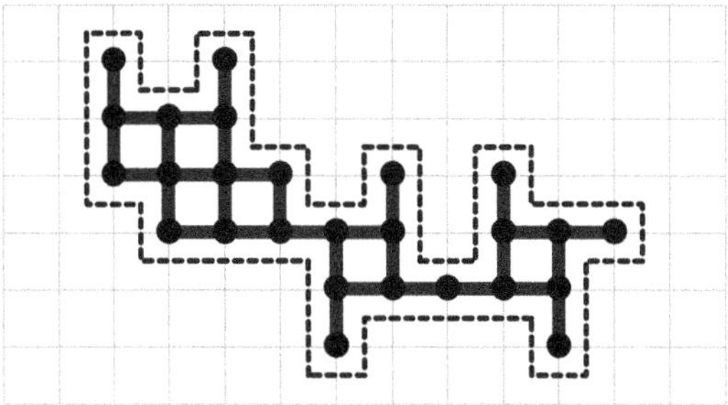

FIG. 12.2. A lattice disk and its dual animal.

equation (4.16)) that $\langle A \rangle \sim (1-q)^{2-\alpha_u-\Delta}$, where Δ is the gap exponent. Taking $q \to 1^-$ and noting that the polygon generating function $V(1,t)$ is convergent at the tricritical point shows that $2-\alpha_u = \Delta = \frac{1}{\phi}$ and hence $\Delta = \frac{3}{2}$.

Assuming that the main contribution to $\langle A \rangle$ comes from the subset of polygons with area proportional to n^δ gives $\langle A \rangle \sim (1-q)^{2-\alpha_u-\delta}$ (as the tricritical point is approached). This gives $\delta = \Delta = \epsilon = \frac{3}{2}$, as expected. The tricritical exponents of vesicles are listed in table 12.1.

12.1.3 *The perimeter of square lattice vesicles*

The above relates the area of a square lattice vesicle to its perimeter. Consider the converse of this. What is the relation of the perimeter of a polygon counted by area (a lattice disk) to its area?

The area of a lattice disk is paved by unit squares in $\mathbb{L}^2$ called *plaquettes.* The *dual animal* of a disk is obtained by placing vertices at the centres of the plaquettes in its interior (see figure 12.2). Such animals are site-animals in the dual lattice.

Perimeter edges in a dual animal are themselves dual to edges in the perimeter of the disk (which is a lattice polygon). Thus, the perimeter of a dual animal has the same size as the length of the lattice polygon bounding the disk. This shows that, if a dual animal has v vertices, s edges and c cycles, then its perimeter has size

$$\varpi = 2v + 2 - 2c = 2s + 4 - 4c. \tag{12.10}$$

This animal corresponds to a polygon of length ϖ and area v. Every edge in the dual animal is either a cut-edge or is in a 4-cycle, so that a fundamental set of independent cycles can be chosen which only contains 4-cycles.

By equation (12.10), the perimeter of a disk is a maximum if the cyclomatic index of its dual animal is a minimum. Thus, consider dual animals counted by size and cyclomatic index. Let $A_v(c)$ be the number of site-animals dual to disks with v vertices and c 4-cycles. Concatenating two such animals can be done as illustrated in figure 1.5. This shows that

$$A_{v_1}(c_1)A_{v_2}(c_2) \leq A_{v_1+v_2}(c_1+c_2). \tag{12.11}$$

By theorem 3.9, there is a density function for such animals given by

$$\log \mathcal{S}_C(\epsilon) = \lim_{v\to\infty} \tfrac{1}{v} \log A_v(\lfloor \epsilon v \rfloor); \tag{12.12}$$

$\log \mathcal{S}_C(\epsilon)$ is concave on $[0,1)$ and continuous in $(0,1)$.

Each dual animal is a site-animal, and each site-animal in the square lattice has at least $v-1$ and at most $2v$ edges. That is, if $a_n(c)$ is the number of bond-animals of n edges and cyclomatic index c, then

$$A_v(c) \leq \sum_{n=v-1}^{2v} a_n(c) \leq (v+1)a_{2v}(c), \tag{12.13}$$

since $a_n(c) \leq a_{n+1}(c)$. A consequence is the following theorem.

Theorem 12.4 *If* $\epsilon = 0$ *then* $\mathcal{S}_C(0) \geq \sqrt{\mu_2}$. *Moreover,* $\lim_{\epsilon\to 1^-} \mathcal{S}_C(\epsilon) \leq 1$.

Proof By theorem 11.7, the density function of cycles in bond-animals has the value $\mathcal{Q}_C(\frac{1}{2}) = 1$.

Putting $c = \lfloor \epsilon v \rfloor$, taking logarithms, dividing by v and then taking $v \to \infty$ in equation (12.13) give $\log \mathcal{S}_C(\epsilon) \leq 2\log \mathcal{Q}_C(\frac{\epsilon}{2})$. Taking $\epsilon \to 1^-$ shows that $\lim_{\epsilon\to 1^-} \log \mathcal{S}_C(\epsilon) \leq 1$.

Each self-avoiding walk of step-length 2 in the square lattice is a site-animal of cyclomatic index 0 dual to a disk. Thus, $A_{2v+1}(0) \geq c_v$, where c_v is the number of self-avoiding walks of length v edges. Take logarithms, divide by $2v+1$ and let $v \to \infty$ to see that $\log \mathcal{S}_C(0) \geq \frac{1}{2}\log \mu_2$. □

Corollary 12.5 *There exists an* $\epsilon_0 \in [0,1)$ *such that* $\mathcal{S}_C(\epsilon_0) = \sup_{\epsilon\in[0,1]}\{\mathcal{S}_C(\epsilon)\} > \mathcal{S}_C(1^-)$. *Moreover,* $\mathcal{S}_C(\epsilon_0) > 1$. □

The *mean perimeter length per plaquette* of a disk with v plaquettes is

$$\left\langle \frac{\varpi}{v} \right\rangle = \frac{\sum_{c\geq 0}(2(v-c)+2)A_v(c)}{v\sum_{c\geq 0} A_v(c)}. \tag{12.14}$$

Both the numerator and denominator grows exponentially fast with v; in fact, at an exponential rate given by $A_v(\lfloor \epsilon_0 v \rfloor) = (\mathcal{S}_C(\epsilon_0))^{v+o(v)}$. This shows that, as $v \to \infty$, the mean perimeter length per plaquette approaches a non-zero constant since $\epsilon_0 < 1$.

Theorem 12.6 *The limit* $\liminf_{v\to\infty} \left\langle \frac{\varpi}{v} \right\rangle = 2(1-\epsilon_0) > 0$. □

That is, the mean perimeter of a disk is proportional to its area. This is in contrast with the numerical data on square lattice polygons, where the mean area grows as the $\frac{3}{2}$-power of the perimeter length.

FIG. 12.3. A lattice sphere of area 78 and with curvature 156.

12.2 Crumpling self-avoiding surfaces

Lattice spheres are realisations of spheres as lattice surfaces in $\mathbb{L}^d$. The area of a lattice sphere is the number of plaquettes it contains. If $d = 3$, then the volume of a lattice sphere is the size of the volume in its interior (the number of unit lattice cubes in its interior).

Let S_1 and S_2 be two lattice spheres in $\mathbb{L}^d$. Then $S_1 \equiv S_2$ if $S_1 = S_2 + \vec{v}$ for some $\vec{v} \in \mathbb{L}^d$. Let s_n be the number of distinct lattice spheres with area n plaquettes. Then $s_6 = \binom{d}{3}$ in $\mathbb{L}^d$, so that $s_6 = 1$ in $\mathbb{L}^3$.

The *bottom plaquette* p_b and *top plaquette* p_t of a closed lattice surface are found by a lexicographic ordering of plaquettes by the coordinates of their midpoints. It is necessarily the case that p_b and p_t are normal to $\vec{e}_1$. Thus, the bottom and top plaquettes have at most $\binom{d-1}{2}$ orientations.

Let p_1 be the top plaquette of a lattice sphere S_1, and p_2 be the bottom plaquette of a lattice sphere S_2. If p_1 is parallel to p_2, then S_1 and S_2 may be placed in $\mathbb{L}^3$ such that $p_1 = p_2$. This gives a concatenated surface (obtained by deleting p_1 and p_2). There are s_{n_1} choices for S_1, and $s_{n_2}\binom{q-1}{2}^{-1}$ choices of S_2 so that its bottom plaquette p_2 has the same orientation as the top plaquette p_1 of S_1. Next, S_2 is placed in $\mathbb{L}^d$ so that its bottom plaquette is on the top plaquette of S_1. By deleting the top and bottom plaquettes, a lattice sphere of area $n_1 + n_2 - 2$ is obtained. This shows that

$$s_{n_1} s_{n_2} \leq \binom{d-1}{2} s_{n_1+n_2-2}. \tag{12.15}$$

Let $\vec{n}_p$ be the normal unit vector to a plaquette p. Two plaquettes p and q are *perpendicular* if $\vec{n}_p \cdot \vec{n}_q = 0$. If $p \cap q = e$, where e is an edge in a surface S, and $\vec{n}_p \cdot \vec{n}_q = 0$, then the edge e is a *fold*. The total number of folds in S is its *curvature*.

Let $s_n(f)$ be the number of lattice spheres of area n and with f folds. By concatenating two lattice spheres as above, at most eight folds can be created or destroyed. This shows that

$$\sum_{g=0}^{f} s_{n_1}(f-g)s_{n_2}(g) \leq \binom{d-1}{2} \sum_{i=-8}^{8} s_{n_1+n_2-2}(f+i). \tag{12.16}$$

The number of surfaces with f folds can be bounded from above and below as shown in the next two lemmas.

Lemma 12.7 *The functions $s_n(f)$ and s_n have upper bounds*

$$s_n(f) \leq \sum_{k=0}^{f} \binom{n}{k} (2d-4)^k \quad \textit{and } s_n \leq d(2d-3)^{n-1}.$$

Proof Let S be a lattice sphere in $\mathbb{L}^d$. Label the bottom plaquette of S with 1, and order the four edges incident with the bottom plaquette lexicographically by their midpoints. The plaquette incident with the lexicographic least edge gets label 2, the plaquette incident with the next least edge gets label 3 and so on.

If j plaquettes have been labelled recursively in this way, let p be the least label so that p is incident with an unlabelled plaquette. Order the edges of p incident with unlabelled plaquettes lexicographically by their midpoints. Then label the plaquette incident with the bottom edge of p, with $j+1$.

This gives a canonical labelling for the plaquettes in S.

The surface S can also be constructed as follows. Place a plaquette in the cubic lattice in one of d possible orientations and label this plaquette with 1 (this will be the bottom plaquette of S). Order the edges of this plaquette lexicographically by their midpoints and append plaquette 2 in one of $2d-3$ possible ways to the lexicographic least edge of plaquette 1. Continue by adding plaquettes to the remaining edges of plaquette 1 by adding plaquettes 3, 4 and 5 in lexicographic order.

Once j plaquettes have been added, let i be the smallest label such that plaquette i has an edge not paired with another plaquette. Order the unpaired edges of i lexicographically and append plaquettes $j+1$, $j+2$, ...

Repeat this process until n plaquettes have been labelled. At each step in the construction there is a unique edge to which to append the next plaquette, which may be added in one of $2d-3$ possible orientations. Thus, the maximum number of surfaces which may be constructed in this way is $d(2d-3)^{n-1}$. This shows that $s_n \leq 3^n$ in $\mathbb{L}^3$ [171, 225], and $s_n \leq d(2d-3)^{n-1}$ in $\mathbb{L}^d$, generally.

A bound on $s_n(f)$ can be obtained using the arguments above as follows. Consider the construction of a surface S with f folds; at the j-th step, let i be the smallest label with an unpaired edge. Let $j+1$ be the label of the next plaquette to be added. There is a choice between (1) adding the plaquette with label $j+1$ so that a fold is created (in one of $2d-4$ ways), or (2) adding this plaquette with the same orientation as the plaquette with label i so that no fold is created.

Plaquettes are added a total of n times, and the choice to create a fold is available a maximum of f times. If choice (1) is assumed to occur k times, then this can be done in $\binom{n}{k}(2d-4)^k$ ways. A surface with f folds may be created by choosing $k \leq f$ from the n plaquettes for creating a fold. This shows that $s_n(f) \leq \sum_{k=0}^{f} \binom{n}{k}(2d-4)^k$. □

Since lemma 12.7 gives an exponential bound on s_n, it follows from the supermultiplicative inequality in equation (12.15) and from theorem A.1 in appendix A that

$$\lim_{n\to\infty} \tfrac{1}{n} \log s_n = \log \eta_d. \tag{12.17}$$

This defines the growth constant η_d of lattice spheres. Numerical simulations in references [228–230] show that $\log \eta_3 \approx 0.55$ in $\mathbb{L}^3$.

Multiply equation (12.16) by ς^f and sum over f. Define the partition function $S_n(\varsigma) = \sum_{f=0}^{2n} s_n(f)\, \varsigma^f$; then

$$S_{n_1}(\varsigma)\, S_{n_2}(\varsigma) \leq \binom{d-1}{2} \phi(\varsigma)\, S_{n_1+n_2-2}(\varsigma), \tag{12.18}$$

where $\phi(\varsigma) = \sum_{i=-8}^{8} \varsigma^i$. By theorem A.1 and lemma 12.7, the limiting free energy of *crumpling surfaces* is

$$\mathcal{S}(\varsigma) = \lim_{n\to\infty} \tfrac{1}{n} \log \sum_{f\geq 0} s_n(f)\, \varsigma^f. \tag{12.19}$$

The partition function $\mathcal{S}(\varsigma)$ is a convex function of $\log \varsigma$.

12.2.1 *The density of folds in crumpling surfaces in* $\mathbb{L}^3$

By corollary 3.14, the limits

$$\begin{aligned} \log \mathcal{W}_s(\leq\epsilon) &= \lim_{n\to\infty} \tfrac{1}{n} \log s_n(\leq\lfloor \epsilon n \rfloor), \quad \text{and} \\ \log \mathcal{W}_s(\geq\epsilon) &= \lim_{n\to\infty} \tfrac{1}{n} \log s_n(\geq\lfloor \epsilon n \rfloor) \end{aligned} \tag{12.20}$$

exist. By theorem 3.16, the density of folds is

$$\mathcal{W}_s(\epsilon) = \min\{\mathcal{W}_s(\leq\epsilon), \mathcal{W}_s(\geq\epsilon)\} \text{ for } \epsilon \in [0,2]. \tag{12.21}$$

Notice that $\eta_3 = \sup_\epsilon \{\mathcal{W}_s(\leq\epsilon), \mathcal{W}_s(\geq\epsilon)\}$.

Some properties of $\mathcal{W}(\epsilon)$ will be a consequence of the following lemma.

Lemma 12.8 *For any $n \geq 1$ and for every $\epsilon \in (0,2]$, there exists a finite positive integer $m_0(\epsilon)$ such that, for every fixed $m > m_0(\epsilon)$,*

$$s_{6nm^2-2(n-1)}\left(\leq\lceil \epsilon(6nm^2 - 2(n-1))\rceil\right) \geq 4^{n-1}.$$

For every positive $\epsilon < \frac{3}{2}$,

$$s_{6n^2+4\lfloor \epsilon n^2 \rfloor}\left(\lceil \leq 12(n+\epsilon n^2)\rceil\right) \geq \binom{6\lfloor (n-2)^2/4 \rfloor}{\lfloor \epsilon n^2 \rfloor}.$$

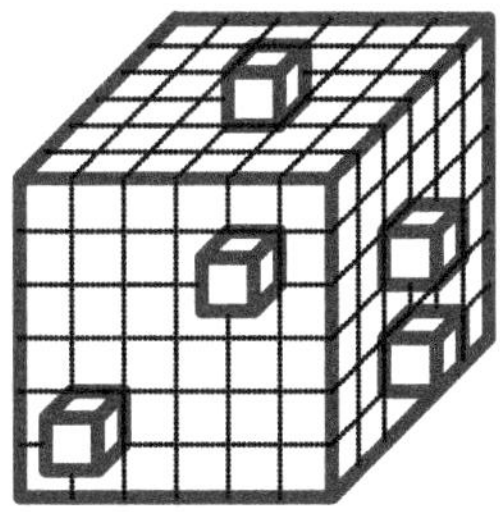

FIG. 12.4. An m-cube with 1-cubes fused to its outer surface.

Proof The first inequality is obtained by the construction of a family of lattice spheres of area $N = 6nm^2 - 2(n-1)$ and with fewer than ϵN folds.

Define an m-cube to be a lattice sphere of area $6m^2$ with the geometry of a cube of side-length m in $\mathbb{L}^3$. The number of folds in an m-cube is $12m$.

A top plaquette t^{++} and a bottom plaquette b^{++} of an m-cube can be found by a lexicographic ordering of the plaquettes by their midpoints with respect to the directions $(\vec{e}_1, \vec{e}_2, \vec{e}_3)$.

Similarly, if the lexicographic ordering is done with respect to the directions $(\vec{e}_1, -\vec{e}_2, -\vec{e}_3)$ instead, then the top and bottom plaquettes, t^{--} and b^{--}, respectively, are found. Continuing in this way, the plaquettes t^{+-}, b^{+-}, t^{-+} and b^{-+} are found by ordering with respect to $(\vec{e}_1, \vec{e}_2, -\vec{e}_3)$ and $(\vec{e}_1, -\vec{e}_2, \vec{e}_3)$. The plaquettes $\{t^{++}, t^{+-}, t^{-+}, t^{--}\}$ and $\{b^{++}, b^{+-}, b^{-+}, b^{--}\}$ are distinct if $m > 1$.

Two m-cubes can be concatenated (if $m > 1$) by identifying t^{++} on the first with b^{++} on the second, or by identifying t^{--} on the first with b^{--} on the second. Alternatively, t^{+-} may be identified with b^{+-}, or t^{-+} may be identified with b^{-+}.

If n such m-cubes are concatenated in a row, then there are 4^{n-1} possible distinct outcomes. Each concatenation deletes two plaquettes from the m-cubes but the number of folds remains unchanged. The total area is $6nm^2 - 2(n-1)$, and the total number of folds is $12nm$.

Thus, $s_{6nm^2-2(n-1)}(12nm) \geq 4^{n-1}$. Increase m (if necessary) until $\epsilon > \frac{12m}{6m^2-2}$. Since $\frac{12m}{6m^2-2} \geq \frac{12nm}{6nm^2-2(n-1)}$ for any $n \geq 1$, this value of m is sufficient (put $m_0(\epsilon) = \lceil \epsilon^{-1} + \sqrt{\epsilon^{-2} + 1/3}\,\rceil$).

This completes the proof of the first inequality.

The second inequality is seen by considering figure 12.4. A 1-cube can be 'fused' onto the outside of an n-cube by identifying a plaquette on the 1-cube with a plaquette in the n-cube and then deleting the plaquette.

Perform this construction by selecting l plaquettes disjoint with folds and with each other in the n-cube. This selection can be done in at least $\binom{6\lfloor (n-2)^2/4 \rfloor}{l}$ ways. By counting the number of folds and plaquettes, the resulting surfaces have degree of folding $12n + 12l$, and area $6n^2 + 4l$.

Thus, $s_{6n^2+4l}(\leq 12(n+l)) \geq \binom{6\lfloor (n-2)^2/4 \rfloor}{l}$. Put $l = \lfloor \epsilon n^2 \rfloor$ to complete the proof. □

Theorem 12.9 *The function* $\log \mathcal{W}_s(\epsilon)$ *is a concave function of* ϵ*, and it is continuous in* $[0,2)$*. Moreover,* $\lim_{\epsilon\to 0^+} \mathcal{W}_s(\epsilon) = \mathcal{W}_s(0) = 1$.

Proof By theorem 3.17, $\log \mathcal{W}_s(\epsilon)$ is a concave function of ϵ and is continuous in $(0,2)$. Define $\mathcal{W}_s(0) = 1$.

Since $\log \mathcal{W}_s(\epsilon)$ is concave, it has right-limits everywhere in its essential domain; so, it remains to show that $\lim_{\epsilon\to 0^+} \mathcal{W}_s(\epsilon) = 1$. For every $\epsilon > 0$, there exists a fixed $m > 0$ so that $s_n(\leq \lceil \epsilon n \rceil) \geq 4^{(n-2)/(6m^2-2)} > 1$ for infinitely many values of n, by lemma 12.8. This shows $\lim_{\epsilon\to 0^+} \mathcal{W}_s(\epsilon) \geq 1$.

By lemma 12.7, $s_n(\leq \lceil \epsilon n \rceil) \leq \sum_{f \leq \lceil \epsilon n \rceil} \sum_{k \leq f} \binom{n}{k} 2^k$. For small ϵ, the sum over k can be bounded from above by $(f+1)\binom{n}{f} 2^f$, since the maximum is obtained by putting $k = f$.

Thus, $s_n(\leq \lceil \epsilon n \rceil) \leq \sum_{f \leq \lceil \epsilon n \rceil} (f+1)\binom{n}{f} 2^f \leq (\lceil \epsilon n \rceil + 1) \sum_{f \leq \lceil \epsilon n \rceil} \binom{n}{f} 2^f$. Take the power $\frac{1}{n}$, and the limit superior as $n \to \infty$.

This shows that $\limsup_{n\to\infty} [s_n(\leq \lceil \epsilon n \rceil)]^{1/n} \leq \frac{2^\epsilon}{\epsilon^\epsilon (1-\epsilon)^{1-\epsilon}}$ by lemma D.3 in appendix D, provided that $0 \leq \epsilon \leq \frac{2}{3}$. Hence, it follows that

$$\mathcal{W}_s(\leq \epsilon) \leq \frac{2^\epsilon}{\epsilon^\epsilon (1-\epsilon)^{1-\epsilon}}. \tag{12.22}$$

Taking $\epsilon \to 0^+$ gives the result $\lim_{\epsilon\to 0^+} \mathcal{W}_s(\leq \epsilon) \leq 1$. Since $\mathcal{W}_s(\epsilon) \leq \mathcal{W}_s(\leq \epsilon)$, this completes the proof. □

Theorem 12.10 *The right derivative* $\frac{d^+}{d\epsilon} \mathcal{W}_s(\epsilon) \mid_{\epsilon=0^+} = \infty$.

Proof By lemma 12.8,

$$\lim_{m\to\infty} \tfrac{1}{6m^2} \log s_{6m^2+4\lfloor \epsilon m^2 \rfloor} (\leq \lceil 12(m+\epsilon m^2) \rceil) \geq \lim_{m\to\infty} \tfrac{1}{6m^2} \log \binom{6\lfloor (m-2)^2/4 \rfloor}{\lfloor \epsilon m^2 \rfloor}.$$

If $\epsilon < \frac{3}{2}$, then these limits can be evaluated to obtain

$$\mathcal{W}_s^2(2\epsilon) \geq \left(\mathcal{W}_s\left(\tfrac{2\epsilon}{1+2\epsilon/3}\right)\right)^{1+2\epsilon/3} \geq \frac{(\frac{1}{4})^{1/4}}{(\frac{1}{6}\epsilon)^{\epsilon/6} (\frac{1}{4} - \frac{1}{6}\epsilon)^{1/4-\epsilon/6}}.$$

By equation (12.22), this becomes

$$\left(\frac{27}{\epsilon^\epsilon (3-\epsilon)^{3-\epsilon}}\right)^{1/24} \leq \mathcal{W}_s(\epsilon) \leq \frac{2^\epsilon}{\epsilon^\epsilon (1-\epsilon)^{1-\epsilon}}, \tag{12.23}$$

where the lower bound is valid if $\epsilon < 2$, and the upper bound is valid if $\epsilon < \frac{1}{2}$.

By the squeeze theorem for limits,

$$\frac{d^+}{d\epsilon} \mathcal{W}_s(\epsilon) \Big|_{\epsilon=0^+} = \infty.$$

This completes the proof. □

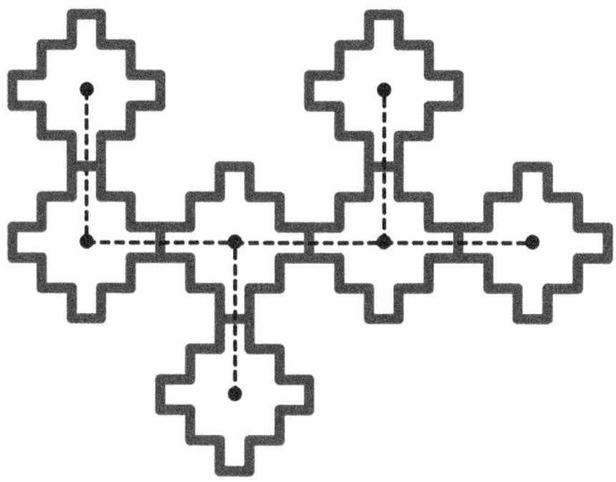

FIG. 12.5. A cross-section of a site-tree generated by gluing together the surfaces in figure 12.3. The vertices are separated by five steps.

The density function $\mathcal{W}_s(\epsilon)$ approaches 1 with infinite slope as $\epsilon \to 0^+$. This result is a limited pattern theorem for lattice spheres, since it implies a natural density of folds.

12.2.2 *The free energy of crumpling self-avoiding surfaces*

The limiting free energy $\mathcal{S}(\zeta)$ of crumpling lattice spheres can be bounded from above and below by using the combinatorial properties of $s_n(f)$.

If $\zeta \geq 1$, then the partition function of the model is $S_n(\zeta) = \sum_{f\geq 0} s_n(f)\,\zeta^f \leq s_n \sum_{l=0}^{2n} \zeta^f \leq 2n\, s_n\, \zeta^{2n}$. By taking logarithms, dividing by n and letting $n \to \infty$, the following theorem is obtained.

Theorem 12.11 *If $\zeta = 1$, then $\mathcal{S}(1) = \log \eta_3$. Moreover, $\mathcal{S}(\zeta) \leq \log \eta_3 + 2\log \zeta$, for $\zeta \geq 1$.* □

A lower bound can be constructed on $\mathcal{S}(\zeta)$ by considering figure 12.3. If the width of this lattice sphere is p, then it has area $3(p^2+1)$, and has $6(p^2+1)$ folds. These lattice spheres can be concatenated in tree-like conformations (see figure 12.5) in the way illustrated in figure 12.3. The number of different ways this can be done is related to the number of site-trees.

Let the midpoint (barycentre) of each lattice sphere in figure 12.3 be a vertex in a site-tree and let two vertices be adjacent if the corresponding lattice spheres are joined.

The length of the edge between vertices in the site-tree is p, and the number of different ways in which q lattice spheres can be joined into a surface is T_q, the number of (unrooted) site-trees in $\mathbb{L}^3$ and with q vertices.

Each time two lattice spheres are concatenated, two plaquettes are lost, as well as eight folds. Thus, in a surface built from q of the lattice spheres, as in figure 12.5, the area is $n = 3(p^2+1)q - 2(q-1) = (3p^2+1)q + 2$, and the number of folds is $f = 6(p^2+1)q - 8(q-1) = (6p^2-2)q+8$. Hence, $s_{(3p^2+1)q+2}(\zeta) \geq T_q\,\zeta^{(6p^2-2)q+8}$. Fix p and let $q \to \infty$. Then $n \to \infty$, and so

$$\mathcal{S}(\zeta) \geq \lim_{q\to\infty} \left(\left(\frac{1}{(3p^2+1)q+2} \right) \log T_q + \left(\frac{(6p^2-2)q+8}{(3p^2+1)q+2} \right) \log \zeta \right). \tag{12.24}$$

The number of site-trees in $\mathbb{L}^3$ is known to be $T_q = \Lambda_s^{q+o(q)}$ (this result follows directly from the concatenation of site-trees [354]), where $\Lambda_s \geq 3$ is the growth constant of site-trees.

Consequently, from equation (12.24),

$$\mathcal{S}(\zeta) \geq \frac{\log \Lambda_s}{3p^2+1} + \frac{(6p^2-2)\log\zeta}{3p^2+1} \tag{12.25}$$

for odd $p \geq 1$, provided that $\zeta \geq 1$. If $p=1$ in equation (12.25), then $\mathcal{S}(\zeta) \geq \frac{1}{4}\log 3 + \log\zeta$, and, if $p \to \infty$ instead, then $\mathcal{S}(\zeta) \geq 2\log\zeta$.

No improvement is gained by using other values of p as all the lines defined by $y = \frac{1}{3p^2+1}(\log 3 + (6p^2-2)\log\zeta)$ intersect in the point $(\frac{1}{4}\log 3, \frac{1}{2}\log 3)$. This gives the following lower bounds.

Theorem 12.12 *The limiting free energy*

$$\mathcal{S}(\zeta) \geq \begin{cases} \frac{1}{4}\log 3 + \log\zeta, & \textit{if } 0 \leq \log\zeta \leq \frac{1}{4}\log 3; \\ 2\log\zeta, & \textit{if } \log\zeta > \frac{1}{4}\log 3. \end{cases} \qquad \square$$

If $\zeta \leq 1$, then bounds may be obtained from the generating function $S(\zeta,q) = \sum_{n\geq 0} S_n(\zeta)\,q^n$ (where q is conjugate to area). The radius of convergence of $S(\zeta,q)$ in the q-plane is $q_c(\zeta)$, so $\mathcal{S}(\zeta) = -\log q_c(\zeta)$ (where $S_n(\zeta)$ is the partition function).

Theorem 12.13 *For every $\zeta > 0$, the radius of convergence is bounded by $q_c(\zeta) < 1$. If $\zeta \in [0,1]$, then $q_c(\zeta) \leq \min\{1, \eta_3^{-1}\zeta^{-2}\}$.*

Proof To see that $q_c(\zeta) \leq 1$, note that, if $\zeta \leq 1$, then the generating function is bound from below by $S(\zeta,q) \geq \sum_{n\geq 0} s_n(\leq\lfloor\epsilon n\rfloor)\,\zeta^{\lfloor\epsilon n\rfloor} q^n$. For small ϵ, $s_n(\leq\lfloor\epsilon n\rfloor) = (\mathcal{W}_s(\leq\epsilon))^{n+o(n)}$, by equation (12.20). Hence, $q_c(\zeta) \leq 1/(\mathcal{W}_s(\leq\epsilon)\,\zeta^{\epsilon})$. Take $\epsilon \to 0^+$ to see that $q_c(\zeta) \leq 1$ by theorem 12.9.

Next, notice that $S(\zeta,q) \geq \sum_{n\geq 0} s_n\,\zeta^{2n} q^n$ if $\zeta \in [0,1]$. This shows that $q_c(\zeta) \leq \eta_3^{-1}\zeta^{-2}$ by equation (12.17). $\square$

Theorem 12.14 *For $\zeta \in [0,1]$, $q_c(\zeta) \leq e^{-\zeta^{24}/8e}$.*

Proof The generatig function $S(\zeta,q) \geq \sum_{n\geq 0} s_n(\leq\lfloor\epsilon n\rfloor)\,\zeta^{\lfloor\epsilon n\rfloor} q^n$, and $s_n(\leq\lfloor\epsilon n\rfloor) = (\mathcal{W}_s(\leq\epsilon))^{n+o(n)}$ for small $\epsilon < 2$, by equation (12.20). Thus, the bound in equation (12.23) gives

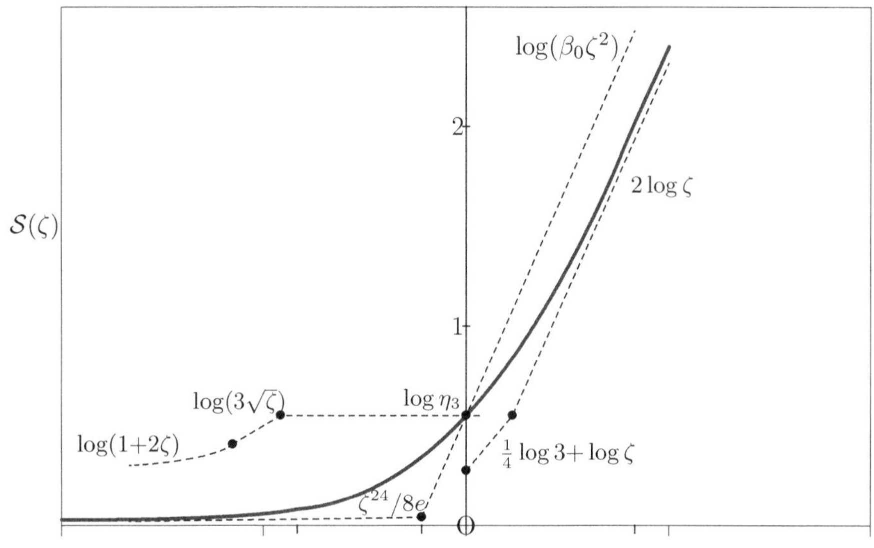

FIG. 12.6. Bounds on the limiting free energy of crumpling surfaces.

$$S(\zeta, q) \geq \sum_{n=0}^{\infty} \left[\frac{27}{\epsilon^{\epsilon}(3-\epsilon)^{(3-\epsilon)}} \right]^{n/24+o(n)} \zeta^{\epsilon n} q^n.$$

The factor in the square brackets is bounded from below by $\frac{3}{\epsilon}$. Thus, $q_c(\zeta) \leq (\epsilon/(3\zeta^{24}))^{\epsilon/24} < 1$ if $\epsilon < 3\zeta^{24}$. Minimising this bound shows that $\log q_c(\zeta) \leq -\zeta^{24}/8e$. □

The two bounds in theorems 12.13 and 12.14 give the combined lower bound

$$\mathcal{S}(z) \geq \begin{cases} \log\eta_3 + 2\log\zeta, & \text{if } -0.275\ldots \leq \log\zeta \leq 0; \\ \frac{1}{8e}\zeta^{24}, & \text{if } \log\zeta \leq -0.275\ldots. \end{cases} \tag{12.26}$$

A lower bound on $q_c(\zeta)$ is found by writing $S(\zeta, q)$ as the sum of two terms which will be bounded from above: by lemma 12.7,

$$\begin{aligned} S(\zeta, q) &= \sum_{n=0}^{\infty} \sum_{f=0}^{2n} s_n(f)\, \zeta^f q^n \\ &= \sum_{n=0}^{\infty} \sum_{f \leq \lfloor \epsilon n \rfloor} s_n(f)\, \zeta^f q^n + \sum_{n=0}^{\infty} \sum_{f > \lfloor \epsilon n \rfloor} s_n(f)\, \zeta^f q^n \\ &\leq \sum_{n=0}^{\infty} \sum_{f \leq \lfloor \epsilon n \rfloor} \sum_{k=0}^{f} \binom{n}{k} 2^k\, \zeta^f q^n + \sum_{n=0}^{\infty} \sum_{f > \lfloor \epsilon n \rfloor} s_n(f)\, \zeta^f q^n. \end{aligned} \tag{12.27}$$

Since $s_n \leq 3^n$, the last term is bounded for $\epsilon = \frac{1}{2}$ by

$$\sum_{n=0}^{\infty} \sum_{f > \lfloor n/2 \rfloor} s_n(f)\, \zeta^f q^n \leq \sum_{n=0}^{\infty} (2n - \lfloor \tfrac{n}{2} \rfloor)\, 3^n \zeta^{n/2} q^n. \tag{12.28}$$

It remains to bound the first term. This is finite if $q < \frac{1}{3\sqrt{\zeta}}$. By equations (D.19) and (D.5) in appendix D,

$$\begin{aligned}
&\sum_{n=0}^{\infty} \sum_{f=0}^{\lfloor n/2 \rfloor} \left(\sum_{k=0}^{f} \binom{n}{k} 2^k \right) \zeta^f\, q^n \leq \sum_{n=0}^{\infty} \sum_{f=0}^{\lfloor n/2 \rfloor} \left(\frac{2^{f/n}}{(f/n)^{f/n}(1-f/n)^{(1-f/n)}} \right)^n \zeta^f\, q^n \\
&= \sum_{n=0}^{\infty} \sum_{f=0}^{\lfloor n/2 \rfloor} \left(\frac{n^n}{f^f (n-f)^{(n-f)}} \right) (2\zeta)^f\, q^n \leq \sum_{n=0}^{\infty} \frac{72\sqrt{2\pi n}}{110} \left(\sum_{f=0}^{\lfloor n/2 \rfloor} \binom{n}{f} (2\zeta)^l \right) q^n \\
&\leq \frac{72\sqrt{2\pi}}{110} \sum_{n=0}^{\infty} \sqrt{n}\,(1+2\zeta)^n\, q^n.
\end{aligned}$$

This is finite if $q < \frac{1}{1+2\zeta}$. These arguments give the following theorem.

Theorem 12.15 *The radius of convergence*

$$q_c(\zeta) \geq \begin{cases} \frac{1}{3}\sqrt{\zeta}, & \text{if } \zeta \in [\frac{1}{4}, 1]; \\ \frac{1}{1+2\zeta}, & \text{if } \zeta \in [0, \frac{1}{4}]. \end{cases}$$ □

Lower bounds on $q_c(\zeta)$ give upper bounds on $\mathcal{S}(\zeta)$. The bounds derived above on $\mathcal{S}(\zeta)$ are illustrated in figure 12.6.

The limiting free energy $\mathcal{S}(\zeta)$ is asymptotic to $S_0 + 2\log\zeta$, where S_0 is the limiting entropy in the model. It can be shown that $S_0 > 0$ using an argument based on figure 12.5: the entropy of surfaces is at least that of site-trees in a sub-lattice of $\mathbb{L}^3$. This shows in particular that $\mathcal{W}_s(2) > 1$, and, since $\mathcal{S}(\zeta) = \sup_\epsilon\{\log \mathcal{W}_s(\epsilon) + \epsilon \log\zeta\}$, it follows that $S_0 = \log \mathcal{W}_2(2) > 0$. Since $\eta_3 \geq \mathcal{W}_s(\epsilon)$, it follows that $S_0 \lessapprox 0.55$.

Numerical simulations [446] suggest that $q_c(\zeta)$ approaches very close to 1 for modest values of ζ. For example, $q_c(0.635) = 0.987(10)$, while the upper bound in theorem 12.14 is $q_c(0.635) \leq 0.99999915\ldots$. This upper bound is due to lattice spheres similar to the example in figure 12.4, with isolated folds exploring large smooth surface areas. In this regime the surfaces have an inflated or disk-like appearance [446], which is realised along a τ_0-curve in the phase diagram.

For larger values of $\zeta \leq 1$, the bound on $q_c(\zeta)$ is given by theorem 12.13. This bound is due to *tree-like* surfaces such as those in figure 12.5; this suggests a branch polymer phase. The transition to this phase should be along a λ-isotherm in the phase diagram.

TABLE 12.2. Crumpling vesicle tricritical exponents

ϕ	ν_t	$2-\alpha_t$	$2-\alpha_u$	$2-\alpha_+$	ν_+
0.6908(23)	0.4825(25) [446]	0.78(3)	1.17(5)	$\frac{1}{2}$	$\frac{1}{2}$

Along the λ-isotherm the limiting theory is in a branched polymers phase. Along this curve, $2-\alpha_+ = \theta - 1 = \frac{1}{2}$, since $\theta = \frac{3}{2}$ for branched polymers in three dimensions (see equations (4.33), (2.9) and (2.11)).

The τ_0-curve of transitions to a smooth phase meets the λ-isotherm at a tricritical point in a model of tricritical branched polymers. Numerical simulations in reference [446] give the following tricritical exponents: the tricritical entropic exponent is $\theta_t = 1.78(3)$, while the metric exponent is $\nu_t = 0.4825(25)$. From these results the rest of the tricritical exponents can be computed. This gives $2-\alpha_t = \theta_t - 1 = 0.78(3)$, while the crossover exponent can be computed from the hyperscaling relation $3\nu = \frac{1}{\phi}$, giving $\phi = 0.6908(23)$; so, $2-\alpha_u = 1.17(5)$.

Assuming the Flory values for branched polymer exponents gives $\nu_t = \frac{1}{2}$ so that $\phi = \frac{2}{3}$. It follows that the specific heat exponent is $\alpha = \frac{1}{2}$, by equation (4.53) that $2-\alpha_t = \frac{3}{2}$ and by equation (4.14) that $2-\alpha_u = 1$.

13

MONTE CARLO METHODS FOR THE SELF-AVOIDING WALK

A Monte Carlo algorithm [419] is a method for sampling state space. The algorithm normally consists of two parts. The first part is a method for sampling states of the model, and the second is a (probabilistic) rule for implementing the sampling method. For example, a method for sampling states in the system may be implemented using the Metropolis algorithm [277, 418].

There are numerous different algorithms for the self-avoiding walk and related models. The oldest is the Rosenbluth method [260, 496]. This *static* or *kinetic growth* algorithm samples self-avoiding walks from a non-uniform distribution by growing them recursively and independently. This algorithm serves as a template for a number of more sophisticated algorithms, including PERM [14, 235, 281, 470] and GARM [489].

In contrast, dynamic Monte Carlo algorithms for the self-avoiding walk are implementations of elementary moves with a Metropolis style rule. Such algorithms sample along (correlated) Markov chains in state space. The first dynamic algorithms were implemented in the 1960s (see for example the Verdier-Stockmayer algorithm [560], and reference [400] for analysis of algorithms in this class).

Good dynamic Monte Carlo algorithms (the BFACF algorithm [11, 12, 35], the Beretti-Sokal algorithm [37] and the pivot algorithm [401]) were invented in the 1980s. Implementation of these algorithms using umbrella sampling techniques [549], multiple Markov chain sampling techniques [223, 547], and advanced data structures to improve computational complexity [105, 346] greatly expanded efficiency and applicability.

Dynamic Monte Carlo algorithms are *canonical* if they are implemented to sample walks of (given) fixed length, and *grand canonical* if they sample from distributions over state space of walks of arbitrary length.

Monte Carlo algorithms for models of self-avoiding walks are built on sets of elementary moves by implementing them using recipes such as the Metropolis algorithm [418] or kinetic growth style sampling [496]. In this way, a large and diverse set of algorithms for self-avoiding walks, including the Rosenbluth algorithm [260, 496], and the PERM [235, 281] and flatPERM algorithms [14, 470], have been discovered. These algorithms have been generalised to the scanning method [411] and GARM-style or GAS-style algorithms [489].

The Statistical Mechanics of Interacting Walks, Polygons, Animals and Vesicles, 2nd edition, E.J. Janse van Rensburg. © E.J. Janse van Rensburg. Published in 2015 by Oxford University Press.

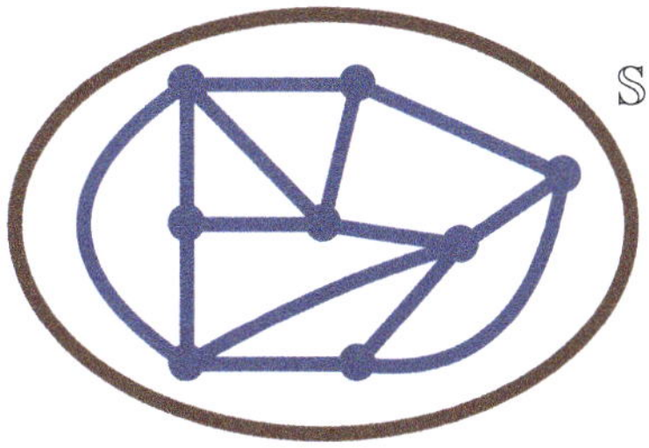

FIG. 13.1. States in state space $\mathbb{S}$ joined by a set of reversible elementary moves are represented by vertices in a graph. If the graph is connected, then the set of elementary moves is irreducible on $\mathbb{S}$.

Algorithms which sample along correlated Markov chains in state space include the BFACF algorithms [12, 35], the Beretti-Sokal algorithm [37] and the pivot algorithm [367, 401] (see for example reference [300]).

13.1 Dynamic Markov chain Monte Carlo algorithms

Let $\mathbb{S}$ be the (conformational) state space of a self-avoiding walk. If $\sigma \in \mathbb{S}$, then an *elementary move* is a sampling rule applied to update σ to a state $\tau \in \mathbb{S}$. The elementary move creates a *linkage* $\langle \sigma \rightsquigarrow \tau \rangle$ if it updates σ to τ. If $\langle \sigma \rightsquigarrow \tau \rangle$ is a linkage if and only if $\langle \tau \rightsquigarrow \sigma \rangle$ is a linkage, then the elementary move is *reversible* (and denoted by $\langle \sigma \sim \tau \rangle$).

The linkages of a reversible elementary move on $\mathbb{S}$ create a graph on $\mathbb{S}$ (see figure 13.1). If there is a sequence of elementary moves from any state $\sigma \in \mathbb{S}$ to any other state $\tau \in \mathbb{S}$, then the elementary move is said to be *irreducible* on $\mathbb{S}$.

Dynamic Monte Carlo algorithms sample along Markov chains in $\mathbb{S}$. The basic implementation is as follows: suppose that $\langle \sigma_1, \sigma_2, \ldots, \sigma_N \rangle$ is a realisation of a Markov chain in $\mathbb{S}$ under an implementation of a set of elementary moves using a sampling rule. The next state is obtained by applying an elementary move to update σ_N and to generate σ_{N+1}. Thus, the Markov chain is a locally biased and correlated random walk over linkages in $\mathbb{S}$, generated using a probabilistic rule (of which the Metropolis algorithm is one implementation).

13.1.1 *The implementation of dynamic Markov chain Monte Carlo algorithms*

The implementation of a sampling rule to sample along linkages $\langle \sigma \sim \tau \rangle$ defines a transition matrix $P_{\sigma\tau}$ such that $P_{\sigma\tau} = \mathrm{P}_r(\sigma \to \tau)$ is the probability of sampling state τ if σ is the current state.

Note that $P_{\sigma\tau}$ is the transition matrix of a homogeneous Markov chain if the elementary moves are independent of time. Moreover,

$$\sum_\tau P_{\sigma\tau} = \sum_\tau \mathrm{P}_r(\sigma \to \tau) = 1, \tag{13.1}$$

since the chain has a next state with probability 1.

This shows that P is a probability matrix with largest eigenvalue 1. The algorithm is said to be *reversible* if there is a distribution π over state space such that

$$\pi(\sigma) P_{\sigma\tau} = \pi(\tau) P_{\tau\sigma}. \tag{13.2}$$

This is also a *condition of detailed balance.* If π is the uniform distribution, then the Markov chain is said to be *symmetric.*

Summing equation (13.2) over σ gives $\sum_\sigma \pi(\sigma) P_{\sigma\tau} = \pi(\tau)$ so that π is the eigenvector of $P_{\sigma\tau}$ associated with its largest eigenvalue. The distribution π over state space is the *stationary distribution* of the algorithm, provided that $P_{\sigma\tau}$ is reversible with respect to π. The algorithm is *irreducible* if there is a finite value of n such that $[P^n]_{\sigma\tau} > 0$. That is, for any two given states $\sigma, \tau \in \mathbb{S}$, the algorithm may realise a Markov chain such that σ and τ are states in the chain.

An irreducible Monte Carlo algorithm is *aperiodic* if, for any two given states σ and τ, there is an N such that $[P^n]_{\sigma\tau} > 0$ for all $n \geq N$. An algorithm which is not aperiodic may still be irreducible. A dynamic Monte Carlo algorithm which is both aperiodic and irreducible is *ergodic.*

An ergodic and reversible Markov chain Monte Carlo algorithm has a unique stationary distribution π (this is a consequence of the Perron-Frobenius theorem). This also shows that π is the *stationary limit distribution* of the algorithm because $\rho(\sigma)[P^n]_{\sigma\tau} \to \pi_\tau$ as $n \to \infty$ for any initial distribution ρ over $\mathbb{S}$.

This shows that, if $\langle \sigma_1, \sigma_2, \ldots, \sigma_N \rangle$ is a realisation of a Markov chain by an ergodic and reversible algorithm sampling from $\mathbb{S}$, then the algorithm samples states $\sigma_j \in \mathbb{S}$ asymptotically from the stationary distribution π (along the Markov chain), irrespective of the choice of an initial state σ_1. Thus, if $\mathcal{O}$ is an observable on the states σ_i, then a natural estimator for the expected value $\langle \mathcal{O} \rangle$ (with respect to π) is the average

$$[\mathcal{O}]_N^{est} = \frac{1}{N} \sum_{i=1}^{N} \mathcal{O}(\sigma_i) \to \langle \mathcal{O} \rangle \text{ as } N \to \infty. \tag{13.3}$$

The convergence of $[\mathcal{O}]_N^{est}$ to $\langle \mathcal{O} \rangle$ as $N \to \infty$ is a consequence of the ergodic theorem (see for example reference [224]). By the central limit theorem, the distribution of $[\mathcal{O}]_N^{est}$ is asymptotically normal about $\langle \mathcal{O} \rangle$. The states σ_i are correlated, and care must be taken in the analysis of variance of observables measured along the realised Markov chain.

13.1.2 *The Metropolis-Hastings algorithm*

A model of interacting clusters (such as collapsing or adsorbing walks) introduces a Boltzmann distribution over $\mathbb{S}$; this distribution is obtained from the partition function

$$Z_n(\beta) = \sum_{|\sigma|=n} e^{\beta E(\sigma)} = \sum_m q_n(m)\, z^m, \tag{13.4}$$

where the first sum is over all clusters σ of size $|\sigma| = n$ with energy $E(\sigma)$ in $\mathbb{S}$, and where $q_n(m)$ is the number of clusters of *size* n and *energy* m. In the above, $z = e^{\beta}$, and $\beta = \frac{1}{k_B T}$, where T is temperature, and k_B is Boltzmann's constant. The Boltzmann distribution is given by

$$P_\beta(\sigma) = \frac{e^{\beta E(\sigma)}}{Z_n(\beta)}. \tag{13.5}$$

A Monte Carlo algorithm sampling along a symmetric Markov chain in $\mathbb{S}$ can be used to sample from the distribution P_β by implementing it with the Metropolis algorithm [277, 418]. States proposed by the underlying elementary moves are accepted by a Metropolis rejection criterion such that the algorithm realises a Markov chain with the stationary distribution P_β.

More generally, if π is a distribution over $\mathbb{S}$, then an algorithm sampling along a symmetric Markov chain can be implemented to sample asymptotically from π by using the Metropolis-Hastings algorithm [277].

The Metropolis-Hastings algorithm Let π be a distribution over $\mathbb{S}$. Suppose that algorithm A samples along an ergodic and symmetric Markov chain in $\mathbb{S}$. Then algorithm A may be modified to sample along a reversible Markov chain with stationary distribution π:

(1) let σ_1 be an arbitrary initial state;

(2) suppose that $\langle \sigma_1, \sigma_2, ..., \sigma_N \rangle$ is a realisation of a Markov chain in S;

(3) select σ_{N+1} by applying algorithm A to σ_N to find a possible next state τ. The state σ_{N+1} is determined the following rule. Compute the ratio

$$q = \frac{\pi(\tau)}{\pi(\sigma_N)}.$$

Put $\sigma_{N+1} = \tau$ with probability $\min\{q, 1\}$, and the default is the rejection of τ by putting $\sigma_{N+1} = \sigma_N$. □

The *rejection technique* central to the Metropolis-Hastings algorithm implies that any implementation with an underlying set of irreducible elementary moves samples along aperiodic Markov chains in $\mathbb{S}$. Hence, the Metropolis algorithm is ergodic if the elementary moves are irreducible: it samples along a Markov chain from a unique distribution (which must be π) asymptotically. Normally, π is a Boltzmann distribution.

The canonical expectation value of an observable $\mathcal{O}$ is defined by

$$\langle \mathcal{O} \rangle = \sum_\sigma \mathcal{O}(\sigma)\pi(\sigma). \tag{13.6}$$

The estimator of this is given by equation (13.3).

By noting that

$$\langle \mathcal{O} \rangle_\rho = \sum_\sigma \mathcal{O}(\sigma)\rho(\sigma) = \frac{\sum_\sigma \mathcal{O}(\sigma)(\frac{\rho(\sigma)}{\pi(\sigma)})\pi(\sigma)}{\sum_\sigma (\frac{\rho(\sigma)}{\pi(\sigma)})\pi(\sigma)} = \frac{\left\langle \mathcal{O}(\sigma)(\frac{\rho(\sigma)}{\pi(\sigma)})\right\rangle}{\left\langle \frac{\rho(\sigma)}{\pi(\sigma)} \right\rangle}, \tag{13.7}$$

expectation values may be estimated with respect to a (different) distribution ρ over state space S by taking ratios of the canonical estimators. In particular, by using equation (13.3), the estimator for $\langle \mathcal{O} \rangle_\rho$ is given by the ratio estimator

$$[\mathcal{O}]^{est}_{N,\rho} = \frac{\left[\mathcal{O}(\frac{\rho}{\sigma})\right]^{est}_N}{\left[\frac{\rho}{\sigma}\right]^{est}_N} = \frac{\frac{1}{N}\sum_{j=1}^N \mathcal{O}(\phi_j)(\frac{\rho(\phi_j)}{\pi(\phi_j)})}{\frac{1}{N}\sum_{j=1}^N \frac{\rho(\phi_j)}{\pi(\phi_j)}}. \tag{13.8}$$

This is the principle of *importance sampling*: by sampling from a distribution π, expectation values of observables with respect to another distribution ρ over $\mathbb{S}$ can be estimated. As long as ρ and π have sufficient overlap, this will give good results. The utility of this approach is that a Monte Carlo simulation from the distribution π may have higher mobility than sampling from ρ.

13.1.3 *Initialisation bias*

The time series of an observable $\mathcal{O}$ measured along a (correlated) Markov chain realised by a dynamic Monte Carlo algorithm is dependent on the first (chosen) state which initialised the chain. This is of significance if the initial state is slowly mixing, leading to *initialisation bias* in the Markov chain. Since the dynamics of Monte Carlo algorithms are generally not well understood, it is not possible to know a priori if a chosen initial state will be slowly mixing.

Initialisation bias may be reduced by discarding the (say) first K states along the Markov chain and then recording the next M states, for a simulation of length $K+M$ of which only the last M states are kept.

Dynamic Monte Carlo algorithms are asymptotically homogeneous; any initialisation bias will decay as the initial state mixes until it disappears in statistical noise. The rate of decay of initialisation bias is exponential at a rate of τ_{exp}, which is the *exponential autocorrelation time* of the algorithm (see section 13.1.6.1 and equation (13.23)). It should suffice to discard the first $K \gg \tau_{exp}$ states to avoid results skewed by the choice of an initial state.

13.1.4 *Umbrella sampling*

Markov chains realised at large values of β by a Monte Carlo algorithm sampling from the Boltzmann distribution $P_\beta(\sigma)$ (see equation (13.5)) may have poor mobility. This makes it difficult to obtain statistically sound averages of observables along the chain. This phenomenon is the *quasi-ergodicity* problem (the Markov chain spends long periods sampling from small regions of state space, giving poor numerical results).

Umbrella sampling [549] is a generalisation of the principle of importance sampling and is a solution for simulations beset by quasi-ergodicity problems.

Its underlying mechanism is the deliberate broadening of the stationary distribution π of the algorithm to be a composite and wide stationary distribution called the *umbrella distribution*. The umbrella distribution has large overlaps with a sequence of Boltzmann distributions, some which may have poor mobility, and others with good mobility. Sampling from the umbrella distribution realises Markov chains which sample more freely from large regions in state space with improved mobility.

Let ρ be a distribution over $\mathbb{S}$ and suppose that the Metropolis algorithm has poor mobility when sampling from ρ. Let π be a distribution over $\mathbb{S}$ overlapping ρ and suppose that the Metropolis algorithm is very mobile when it samples from π. By using importance sampling (see equation (13.8)), estimators of observables over ρ may be estimated by sampling from ρ.

In some cases, ρ and π do not overlap and importance sampling fails. However, there is considerable flexibility in choosing π (but ρ is fixed). Exploit the freedom in choosing π by replacing it with a sequence of distributions $\langle \phi_1, \phi_2, \ldots, \phi_M \rangle$, where $\rho = \phi_M$. The distributions ϕ_j are chosen such that ϕ_j and ϕ_{j+1} have large overlap and that the Metropolis algorithm has good mobility on ϕ_1 (and poor mobility over $\phi_M \equiv \rho$).

The distributions ϕ_j are combined in a single *umbrella distribution* Φ by

$$\Phi(\sigma) = \sum_{j=1}^{M} w_j \phi_j(\sigma), \tag{13.9}$$

where the w_j are normalising weights. These weights are chosen so that the individual distributions ϕ_j have more or less the same overlap with Φ: That is, $\sum_\sigma \Phi(\sigma)\phi_j(\sigma) \approx \frac{1}{M}$ for all $1 \leq j \leq M$. This can be done by using 'training runs' where the average $\langle \phi_j \rangle_\Phi$ is computed.

In most applications the component distributions ϕ_j are Boltzmann distributions over state space, so the umbrella distribution is given by

$$\Phi(\sigma) = \sum_{j=1}^{M} w_j e^{\beta_j E(\sigma)}, \tag{13.10}$$

where $e^{\beta_j E(\sigma)}$ is the summand in equation (13.4). In this case the sequence of component distributions is parameterised by $\langle \beta_j \rangle$ [547]. Notice that the weights w_j are related to the normalising partition function in equation (13.5). The sequence of β_j includes values of β where the Markov chain is mobile, as well as values of β where the Markov chain has low mobility.

It remains to choose the weights w_j such that each of the (not normalised) component distributions contributes approximately the same amount to Φ. That is, the sampling should be approximately uniform in the parameter β over the entire range $[\beta_1, \beta_M]$ in parameter space. This is achieved by noting that

$$\Pi = \sum_\sigma \Phi(\sigma) = \sum_\sigma \sum_{j=1}^M w_j e^{\beta_j E(\sigma)} = \sum_{j=1}^M w_j \sum_\sigma e^{\beta_j E(\sigma)} = \sum_{j=1}^M w_j Z_n(\beta_j), \tag{13.11}$$

by equation (13.4). Since Π should be equal to 1, choose

$$w_j Z_n(\beta_j) = \tfrac{1}{M}, \text{ for } j = 1, 2, \ldots, M. \tag{13.12}$$

Since $F_n(\beta_j) = \log Z_n(\beta_j)$ is the extensive free energy, this shows that a convenient choice for the weights is

$$w_j = \tfrac{1}{M} e^{-F_n(\beta_j)} \approx \tfrac{1}{M} e^{-n\mathcal{F}(\beta_j)}, \tag{13.13}$$

where $\mathcal{F}(\beta)$ is the limiting (intensive) free energy of the model.

This may implemented by first doing simulations to estimate $\mathcal{F}(\beta)$ over the range of β. In practice it is found that even rough estimates for $\mathcal{F}(\beta)$ may be used to obtain a good umbrella.

13.1.5 *Multiple Markov chain Monte Carlo*

Multiple Markov chain Monte Carlo improves mobility by swapping chains realised from distributions with poor mobility with chains realised from distributions with high mobility [223].

Let ϕ_β be a distribution over $\mathbb{S}$ parameterised by β and assume that algorithm A is an ergodic algorithm sampling along reversible Markov chains from ϕ_β. Assume that the mobility of algorithm A is low at β. Suppose as well that algorithm A has high mobility when $\beta = \beta_1$. The basic idea of multiple Markov chain Monte Carlo is to select a sequence of parameter values $\langle \beta_1, \beta_2, \ldots, \beta_M \equiv \beta \rangle$ interpolating between β_1 and β.

Thus, realise M Markov chains by algorithm A, one at each value of β_j, in parallel for a specified number of iterations. Let chain number j be realised from distribution ϕ_{β_j}. Suppose that σ_j is the final state in chain j.

A pair of states (σ_j, σ_{j+1}) is selected from an adjacent pair of chains j and $j+1$. An attempt to swap these states between the chains is made. That is, an attempt is made to exchange states σ_j in chain j with state σ_{j+1} in chain $j+1$, and vice versa. Accept this proposed swap of states σ_j and σ_{j+1} with probability

$$r(j, j+1) = \min\left\{1, \frac{\phi_{\beta_{j+1}}(\sigma_j)\phi_{\beta_j}(\sigma_{j+1})}{\phi_{\beta_j}(\sigma_j)\phi_{\beta_{j+1}}(\sigma_{j+1})}\right\}. \tag{13.14}$$

The swapping of states may be repeated between other pairs of chains before sampling by algorithm A resumes, only to be later interrupted again for another attempt at swapping states between chains.

The set of Markov chains together with swapping is itself a Markov chain which is the *composite Markov chain.*

If each Markov chain is ergodic, then the composite chain is also ergodic, and its unique stationary distribution is the product of the stationary distributions of the separate chains.

To see this, notice that if $r(j, j+1) < 1$, then the probability of swapping chains $(j, j+1)$ is $\frac{1}{M-1} r(j, j+1)$, so that the product distribution is invariant under the swap:

$$\frac{r(j,j+1)}{M-1} \prod_{k=1}^{M} \phi_{\beta_k}(\sigma_k) = \frac{1}{M-1} \left(\frac{\phi_{\beta_{j+1}}(\sigma_j)\phi_{\beta_j}(\sigma_{j+1})}{\phi_{\beta_j}(\sigma_j)\phi_{\beta_{j+1}}(\sigma_{j+1})} \right) \prod_{k=1}^{M} \phi_{\beta_k}(\sigma_k) \tag{13.15}$$
$$= \frac{1}{M-1} \left(\prod_{k \neq j, j+1} \phi_{\beta_k}(\sigma_k) \right) \times \left(\phi_{\beta_{j+1}}(\sigma_j)\phi_{\beta_j}(\sigma_{j+1}) \right).$$

A similar argument demonstrates this for the case that $r(j, j+1) = 1$.

Each successful swap between chains j and $j+1$ corresponds to a large change of the state in each chain. This reduces correlations along the chains. States in chains with low mobility are swapped up to distributions with high mobility before being swapped back again. This alleviates quasi-ergodicity problems.

The implementation of multiple Markov chain Monte Carlo requires that a suitable set of distributions ϕ_β be introduced. Adjacent distributions should have significant overlap to ensure that swapping will occur frequently. There should also not be an unnecessarily large number of parallel chains, because this makes sampling inefficient by applying too much computer resources to chains with high mobility. Observe that swapping two states is computationally inexpensive and so can be attempted frequently. For examples of the implementation of multiple Markov chain Monte Carlo simulations, see references [321, 546, 547].

13.1.6 *Analysis of variance*

The statistical properties of an observable $\mathcal{O}$ measured along a Markov chain realised by a dynamic Monte Carlo algorithm may be estimated by repeating the simulation, independently, M times. The average of $\mathcal{O}$ is computed independently for each of the simulations to give a collection of independent measurements $[\mathcal{O}]_j^{est}$ for $j = 1, 2, \ldots, M$.

By the central limit theorem, the distribution of $[\mathcal{O}]_j^{est}$ is asymptotically normal and unbiased about the mean $\langle \mathcal{O} \rangle$. Thus,

$$[\mathcal{O}]^{est} = \frac{1}{M} \sum_{j=1}^{M} [\mathcal{O}]_j^{est} \tag{13.16}$$

is an estimate of $\langle \mathcal{O} \rangle$. A statistical confidence interval on $[\mathcal{O}]^{est}$ is obtained by calculating

$$\sigma^2[\mathcal{O}]^{est} = \frac{1}{M-1} \left([\mathcal{O}^2]^{est} - \left([\mathcal{O}]^{est} \right)^2 \right). \tag{13.17}$$

An estimated 68% confidence interval about $[\mathcal{O}]^{est}$ is $\sigma[\mathcal{O}]^{est} = \sqrt{\sigma^2[\mathcal{O}]^{est}}$. However, results are more commonly stated with an estimated 95% confidence interval (given by $2\sigma[\mathcal{O}]^{est}$).

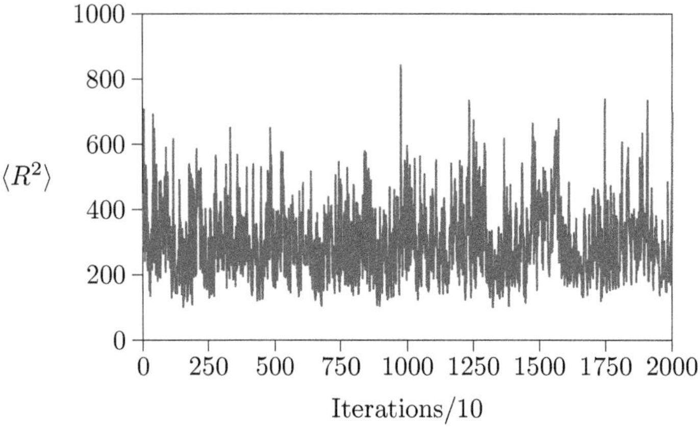

FIG. 13.2. A times series of an observable measured along a Markov chain realised by a dynamic Monte Carlo algorithm. In this example the square radius of gyration of a self-avoiding walk of length 200 in $\mathbb{L}^2$ is sampled by the pivot algorithm.

13.1.6.1 *Integrated autocorrelation time analysis:* Suppose the time series of correlated states $\langle \sigma_1, \sigma_2, \ldots, \sigma_N \rangle$ is realised by a dynamic Markov chain Monte Carlo algorithm and let $\langle \mathcal{O}(\sigma_1), \mathcal{O}(\sigma_2), \ldots, \mathcal{O}(\sigma_N) \rangle$ be a (correlated) time series of observable $\mathcal{O}$ along the chain (see for example figure 13.2). The average of $\mathcal{O}(\sigma_i)$ is given by

$$[\mathcal{O}]_N^{est} = \tfrac{1}{N} \sum_{i=1}^{N} \mathcal{O}(\sigma_i). \tag{13.18}$$

The autocorrelation function of the time series for $\mathcal{O}$ is given by

$$S_{\mathcal{O}}(k) = \langle \mathcal{O}(\sigma_i)\mathcal{O}(\sigma_{i+k}) \rangle - \langle \mathcal{O} \rangle^2 . \tag{13.19}$$

Assuming that $N \gg k$, this may be estimated by

$$S_{\mathcal{O}}(k) \approx \tfrac{1}{N-k} \sum_{i=1}^{N-k} \Big(\mathcal{O}(\sigma_i)\mathcal{O}(\sigma_{i+k}) - ([\mathcal{O}]_N^{est})^2 \Big). \tag{13.20}$$

Since the time series is asymptotically homogeneous, it follows that $S_{\mathcal{O}}(k)$ is independent of i on the right-hand side of equation (13.19). With increasing k, the autocorrelation function $S_{\mathcal{O}}(k)$ decays to leading order at an exponential rate given by $\tau_{exp,\mathcal{O}}$:

$$S_{\mathcal{O}}(k) \simeq C_{\mathcal{O}} e^{-k/\tau_{exp,\mathcal{O}}}, \tag{13.21}$$

where $\tau_{exp,\mathcal{O}}$ is the *exponential autocorrelation time* associated with the observable $\mathcal{O}$. A formal definition for the exponential autocorrelation time associated with the observable $\mathcal{O}$ is

$$\tau_{exp,\mathcal{O}}^{-1} = -\limsup_{k\to\infty} \tfrac{1}{k} \log S_{\mathcal{O}}(k). \tag{13.22}$$

The exponential autocorrelation time of the underlying Monte Carlo algorithm is the supremum of this over all observables:

$$\tau_{exp} = \sup_{\mathcal{O}} \tau_{exp,\mathcal{O}}, \tag{13.23}$$

which is the time scale of the longest autocorrelations in the simulation.

In general, it is not possible to compute the exponential autocorrelation time. Instead, the integrated autocorrelation time $\tau_{int,\mathcal{O}}$ is used. Choose $M \approx m\tau_{exp,\mathcal{O}}$, where m is sufficiently large for autocorrelations to have decayed into the statistical noise of the time series. If $|k| > M$, then $S_{\mathcal{O}}(k)$ should be less than statistical noise in the data. Thus, make the estimate

$$2\tau_{exp,\mathcal{O}} \approx \sum_{k=-\infty}^{\infty} S_{\mathcal{O}}(k) \gtrsim 2\tau_{int,\mathcal{O}} = \sum_{k=-M}^{M} S_{\mathcal{O}}(k) \approx 1 + 2\sum_{k=1}^{M} \frac{S_{\mathcal{O}}(k)}{S_{\mathcal{O}}(0)} \tag{13.24}$$

for M large. This gives the *integrated autocorrelation* time $\tau_{int,\mathcal{O}}$. For a window of size M, it is

$$\tau_{int,\mathcal{O}}^{est} = \tfrac{1}{2} + \sum_{k=1}^{M} \frac{S_{\mathcal{O}}(k)}{S_{\mathcal{O}}(0)}, \tag{13.25}$$

where $S_{\mathcal{O}}(k)$ is as given in equation (13.20). This is a good approximation when $M \gtrsim m\tau_{int,\mathcal{O}}$, with m in the range 20–50. The underlying assumption in the definition of $\tau_{int,\mathcal{O}}$ is that the decay of the autocorrelation function is purely exponential. This is generally not the case, as there may be more than one exponentially decaying mode in the time series, each one with its own characteristic autocorrelation time. If the times series of an observable is correlated, then $\tau_{int,\mathcal{O}} > \frac{1}{2}$.

Assuming that, for an observable $\mathcal{O}$, its associated integrated autocorrelation time is the dominant rate of decay, estimate the variance of $\mathcal{O}$ by

$$\begin{aligned} \sigma^2 [\mathcal{O}]_N^{est} &= \left[(\mathcal{O} - [\mathcal{O}]_N^{est})^2\right]_N^{est} \\ &= \tfrac{1}{N^2}\Big[\sum_{i,j=1}^{N} (\mathcal{O}(\sigma_i) - [\mathcal{O}]_N^{est})(\mathcal{O}(\sigma_j) - [\mathcal{O}]_N^{est})\Big]_N^{est} \\ &= \tfrac{1}{N^2} \sum_{i,j=1}^{N} S_{\mathcal{O}}(|i-j|) \\ &\approx \tfrac{2}{N} S_{\mathcal{O}}(0) \Big(\tfrac{1}{2} + \sum_{k=1}^{\infty} \tfrac{S_{\mathcal{O}}(k)}{S_{\mathcal{O}}(0)}\Big) \\ &= \tfrac{2}{N} \tau_{\mathrm{int},\mathcal{O}} S_{\mathcal{O}}(0). \end{aligned} \tag{13.26}$$

The analysis of variance by estimating autocorrelation times of times series is fraught with practical difficulties which may lead to unreliable results (which

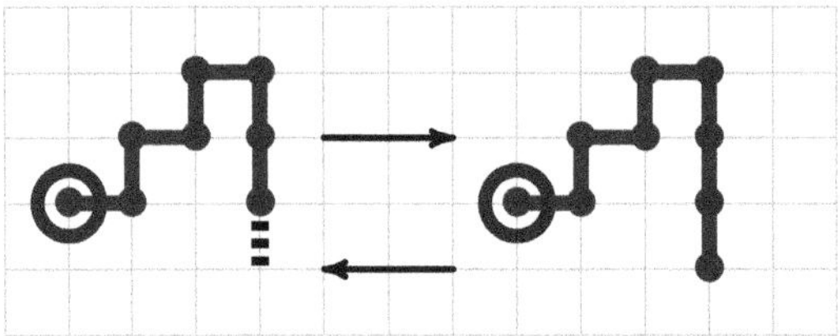

FIG. 13.3. The elementary moves of the Beretti-Sokal algorithm. A positive move appends an edge to the endpoint of the walk, and a negative move deletes the last edge in the walk. Each positive move is uniquely reversible by a negative move.

---

are seemingly acceptable). The only cure for such problems is a sufficiently long time series (that is, an effective algorithm and long computer simulations). Good results are obtained if the time series has a length of at least $1000\tau_{\mathrm{int},\mathcal{O}}$.

13.1.6.2 *Block data analysis:* A *block data analysis* is as follows. The time series $\langle\sigma_1, \sigma_2, \ldots, \sigma_N\rangle$ is partitioned into $\lfloor\frac{N}{M}\rfloor$ *blocks*, each of length M (and where N is large compared to M). The estimator for $\langle\mathcal{O}\rangle$ is given by equation (13.18). For the k-th block,

$$[\mathcal{O}]^{est}_{M,k} = \tfrac{1}{M} \sum_{i=(k-1)M+1}^{kM} \mathcal{O}(\sigma_i). \tag{13.27}$$

If $M \gg \tau_{int,M}$, then these blocks may be assumed to be statistically independent and unbiased estimators of the expected value $\langle\mathcal{O}\rangle$. They are asymptotically normally distributed about the mean, so equations (13.16) and (13.17) may be used to determine a statistical confidence interval. Since there are $\lfloor\frac{N}{M}\rfloor$ blocks of data which are treated as independent time series, this gives the estimate

$$\sigma^2[\mathcal{O}]^{est} = \tfrac{1}{\lfloor N/M\rfloor - 1}\left(\left[[\mathcal{O}^2]^{est}_{M,k}\right]^{ave} - \left([\mathcal{O}]^{est}_N\right)^2\right) \tag{13.28}$$

for the variance (where $[[\mathcal{O}^2]^{est}_{M,k}]^{ave}$ is the average of $\left[\mathcal{O}^2\right]^{est}_{M,k}$ over the $\lfloor N/M\rfloor$ blocks).

In practical applications it is enough to determine $\sigma\,[\mathcal{O}]^{est}$ as a function of M. This increases with M before levelling off for sufficiently large M (large compared to $\tau^{est}_{int,M}$).

13.2 The Beretti-Sokal algorithm

The Beretti-Sokal algorithm implements an elementary move at the endpoint of a self-avoiding walk in $\mathbb{L}^d$ from $\vec{0}$ (see figure 13.3) to sample from the Boltzmann distribution on the state space of walks [37]. The *positive elementary move* is implemented by appending an edge to the endpoint of the walk, and the *negative elementary move* reverses this by deleting the last edge in the walk.

Let σ be the current state of length $|\sigma|$ and suppose that there are a_+ different edges which may be appended to σ to extend its length by one step, while respecting self-avoidance.

If σ has length larger than 0, then there is always a last edge in σ which may be deleted. This is the negative elementary move.

Implement the algorithm as follows. Perform a particular positive elementary move with probability P_+, and a negative elementary move with probability P_-. Since $a_+ \leq 2d-1$, a constraint on the probabilities is that $(2d-1)P_+ + P_- \leq 1$. Introduce the parameter $t > 0$ and choose $P_+ = tP_-$ so that the algorithm samples asymptotically from the Boltzmann distribution

$$P_t(\sigma) = \frac{t^{|\sigma|}}{Z(t)} \tag{13.29}$$

over the lengths $|\sigma|$ of walks σ from $\vec{0}$. The normalising factor $Z(t) = \sum_{n=0}^{\infty} c_n\, t^n$ is the self-avoiding walk generating function. The mean length of the walk is controlled by t.

Let $P_+ = \frac{t}{N}$, and $P_- = \frac{1}{N}$, where N is a normalising function. Optimising these choices subject to the constraint that $(2d-1)P_+ + P_- \leq 1$ gives

$$P_+ = \frac{t}{1+(2d-1)t}, \quad \text{and } P_- = \frac{1}{1+(2d-1)t}. \tag{13.30}$$

With these choices, the transition probabilities from state σ to state τ are given by

$$P_r(\sigma \to \tau) = \begin{cases} \dfrac{t}{1{+}(2d-1)t}, & \text{for positive elementary moves;} \\ \dfrac{1}{1{+}(2d-1)t}, & \text{for negative elementary moves.} \end{cases} \tag{13.31}$$

By equation (13.31), it follows that

$$t^{|\sigma|}\, P_r(\sigma \to \tau) = t^{|\tau|}\, P_r(\tau \to \sigma). \tag{13.32}$$

This is the *condition of detailed balance* for the algorithm. By equation (13.2), it follows that the Beretti-Sokal algorithm samples along a Markov chain asymptotically from the stationary Boltzmann distribution in equation (13.29).

The Boltzmann distribution is normalisable only if $\sum_n c_n\, t^n < \infty$, or, in other words, if $0 < t < \lim_{n\to\infty} c_n^{-1/n} = \mu_d^{-1}$ (see theorem 1.1).

In this algorithm there is a trade-off between efficiency and sampling longer walks. If t is fixed close to μ_d^{-1}, then the algorithm samples walks with longer average length along a time series with longer autocorrelations.

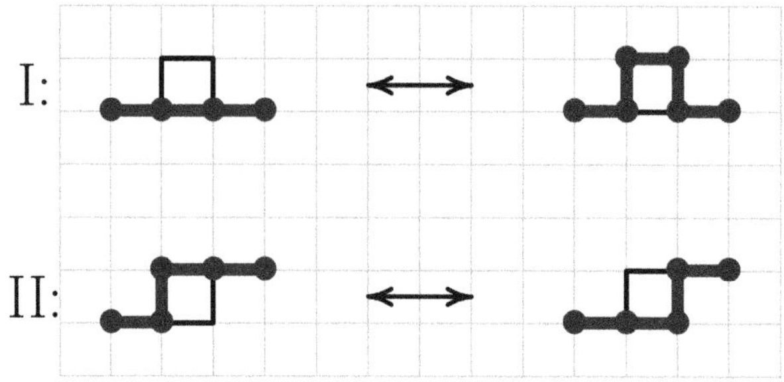

FIG. 13.4. The elementary moves of the BFACF algorithm.

The mean length of walks is given by

$$\langle n \rangle_t = \frac{\sum_n n c_n t^n}{\sum_n c_n t^n} \simeq \frac{\gamma}{-\log(\mu_d t)}, \tag{13.33}$$

assuming that $c_n \simeq A n^{\gamma-1} \mu_d^n$ (so that $\langle n \rangle_t \to \infty$ as t approaches μ_d^{-1}.

It is known that the autocorrelation time τ of the algorithm is estimated by $O(\langle n \rangle^2) \lesssim \tau \lesssim O(\langle n \rangle^{1+\gamma})$ [37, 517, 518] (in a simulation where the average length of the walks is $\langle n \rangle$). An application of the Beretti-Sokal algorithm to an interacting model was done in reference [435].

13.3 The BFACF algorithm

The BFACF algorithm is an implementation of BFACF elementary moves (see figure 13.4) [12, 35] to sample self-avoiding walks with fixed endpoints or to sample lattice polygons in $\mathbb{L}^d$.

Elementary moves of type I are implemented by replacing one edge by three edges in a ⊔-conformation or reversing this by deleting three edges and replacing it by a single edge. In one direction this increases the length of a walk by 2 (a *positive elementary move*), while in the other direction it decreases the length of the walk by 2 (a *negative elementary move*). Observe that the positive elementary move is immediately reversible by a negative elementary move, and vice versa.

Elementary moves of type II are performed at vertices where two edges meet at a right angle. These edges join the diagonally opposite edges of a unit square (plaquette) in the lattice and may be replaced by exchanging them as shown in figure 13.4. These moves do not change the length of the walk; they are *neutral elementary moves.*

The irreducibility properties of BFACF elementary moves depends on the implementation of the algorithm (that is, walks with fixed endpoints sampled or polygons sampled, etc.). Proofs of the irreducibility of the elementary moves are lengthy, so the following theorems are stated without proofs.

Theorem 13.1 *Let $\mathbb{S}(\vec{x})$ be the state space of self-avoiding walks from the origin to $\vec{x}$ in $\mathbb{L}^2$. Then the BFACF elementary moves are irreducible on $\mathbb{S}(\vec{x})$.* □

The proof of theorem 13.1 may found in reference [399]; see theorem 9.7.2 therein.

In $\mathbb{L}^3$ the situation is more complex. It is known that the BFACF elementary moves are irreducible on $\mathbb{S}(\vec{x})$, provided that $\|\vec{x}\|_\infty > 1$ [293].

Theorem 13.2 *Let $\mathbb{S}(\vec{x})$ be the state space of self-avoiding walks from the origin to $\vec{x}$ in $\mathbb{L}^3$. Then the BFACF elementary moves are irreducible on $\mathbb{S}(\vec{x})$, provided that $\|x\|_\infty > 1$.* □

If $\|x\|_\infty = 1$ in theorem 13.2, then the BFACF elementary moves are not irreducible on $\mathbb{S}(\vec{x})$ in $\mathbb{L}^3$. This follows because walks (viewed as piecewise continuous embeddings of a curve C in $\mathbb{R}^3$) have fixed knot types when their endpoints are joined by a line segment of length at most $\sqrt{3}$. BFACF elementary moves, which are ambient isotopies on $(C, \mathbb{R}^3)$, cannot change these knot types.

The issue of knotting and the irreducibility of BFACF elementary moves were analysed in reference [330] in the case of (unrooted) lattice polygons.

Theorem 13.3 *Let $\mathbb{S}_K$ be the state space of unrooted lattice polygons in $\mathbb{L}^3$ of fixed knot type K. Then the BFACF elementary moves are irreducible on $\mathbb{S}_K$.* □

The generalisation of theorem 13.3 to BFACF-style algorithms in the face-centred and body-centred cubic lattices was done in reference [324].

BFACF elementary moves can be implemented using the Metropolis-Hastings algorithm. Let σ be the current state (walk or polygon) of length $|\sigma|$. Choose an edge uniformly along σ, and, again uniformly, a direction (one of $2(d-1)$) normal to the chosen edge. Translating the chosen edge in this direction will result in either a positive move (accept this with probability t^2), a neutral move (accept this with probability 1) or a negative move (accept it with probability 1). This produces a state σ'. If σ' is self-intersecting, then reject it and find the next state ϕ by putting $\phi = \sigma$. Otherwise, put $\phi = \sigma'$.

The probability of obtaining ϕ is

$$P_r(\sigma \to \phi) = \begin{cases} \dfrac{t^2}{2(d-1)|\sigma|}, & \text{for a positive move;} \\ \dfrac{1}{2(d-1)|\sigma|}, & \text{for a neutral move;} \\ \dfrac{1}{2(d-1)|\sigma|}, & \text{for a negative move.} \end{cases} \tag{13.34}$$

This shows that

$$|\sigma| t^{|\sigma|} P_r(\sigma \to \phi) = |\phi| t^{|\phi|} P_r(\phi \to \sigma). \tag{13.35}$$

Comparison to equation (13.2) gives the stationary distribution of the algorithm:

$$D(t) = \sum_{\sigma \in S} |\sigma|\, t^{|\sigma|}. \tag{13.36}$$

This is a stretched Boltzmann distribution over the state space $\mathbb{S}$.

The standard implementation of the BFACF algorithm differs from the Metropolis implementation in that the transition probabilities are optimised. Let P_+ be the probability of attempting a positive elementary move, let P_0 be the probability of attempting a neutral elementary move and let P_- be the probability of attempting a negative elementary move. Then $P_+ + P_0 + P_- \leq 1$.

As before, let σ be the current state. To perform an elementary move, choose an edge e uniformly in σ and execute a (particular) positive elementary move with probability P_+, a (particular) neutral elementary move with probability P_0 or a (particular) negative elementary move with probability P_-. This generates the next state ϕ.

The implementation uses *rejection*. If no elementary move is executed (this may occur because it may be the case that $P_+ + P_0 + P_- < 1$) then the next state is by default $\phi = \sigma$.

With this implementation, the algorithm is initialised by a state σ_1 and it samples along an aperiodic and irreducible Markov chain in the state space of walks or polygons (subject to the conditions in theorems 13.1, 13.2 and 13.3 above).

It remains to specify the probabilities P_+, P_- and P_0. Since the algorithm will sample from a Boltzmann distribution on state space, require that $P_+ = t^2 P_-$ (in addition to $P_+ + P_0 + P_- \leq 1$). Consider all the local conformations of an edge e and its neighbouring edges in a walk or a polygon to determine the values of P_+, P_0 and P_-. This gives the following.

- All $2(d-1)$ directions normal to e give positive elementary moves. Thus, $2(d-1)P_+ \leq 1$.
- There are $2d-3$ possible positive elementary moves and just one neutral elementary move available. Hence, $(2d-3)P_+ + P_0 \leq 1$.
- There are $2d-4$ possible positive elementary moves and two neutral elementary moves available. Hence, $(2d-4)P_+ + 2P_0 \leq 1$.
- Finally, there are $2d-3$ possible positive elementary moves and one negative elementary move available. Hence, $(2d-3)P_+ + P_- \leq 1$.

The second condition is redundant. Put $P_+ = t^2 P_-$ in the last condition to see that $(2d-3)t^2 P_+ + P_- \leq 1$. Maximise P_- by putting this equal to 1. Then solve for P_+ and put $P_0 = \frac{1}{2}(P_+ + P_-)$ to obtain

$$P_+ = \frac{t^2}{1+(2d-3)t^2}; \; P_- = \frac{1}{1+(2d-3)t^2}; \text{ and } P_0 = \frac{\frac{1}{2}(1+t^2)}{1+(2d-3)t^2}. \tag{13.37}$$

If the state ϕ follows the state σ in a Markov chain realised by the algorithm, then the transition matrix of the algorithm is given by

$$P_r(\sigma \to \phi) = \begin{cases} \frac{1}{|\sigma|} \frac{t^2}{1+(2d-3)t^2}, & \text{if } |\phi| = |\sigma| + 2; \\ \frac{1}{2|\sigma|} \frac{1+t^2}{1+(2d-3)t^2}, & \text{if } |\phi| = |\sigma|; \\ \frac{1}{|\sigma|} \frac{1}{1+(2d-3)t^2}, & \text{if } |\phi| = |\sigma| - 2. \end{cases} \tag{13.38}$$

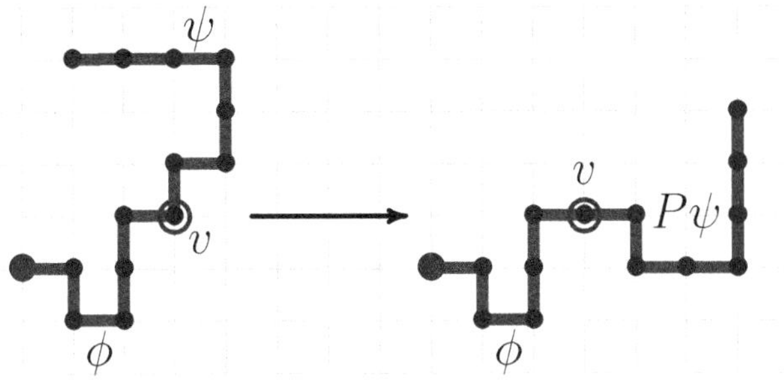

FIG. 13.5. An example of a pivot elementary move. By rotating the subwalk ψ starting at the vertex v through an angle 90° clockwise, the subwalk $P\psi$ on the right is obtained. This changes the conformation of the walk $\phi\psi \to \phi P\psi$. In general, p can be any element of the symmetry group of the underlying lattice.

The condition of detailed balance is given by equation (13.35); the algorithm samples asymptotically from the stretched Boltzmann distribution in equation (13.36). The mean of an observable $\mathcal{O}$ over $D(t)$ is given by

$$\ll\!\mathcal{O}\!\gg_t = \tfrac{1}{D(t)} \sum_{\sigma \in S} |\sigma|\, \mathcal{O}(\sigma)\, t^{|\sigma|}. \tag{13.39}$$

Averages of $\mathcal{O}$ computed along Markov chains realised by the algorithm will converge to $\ll\!\mathcal{O}\!\gg_t$, provided that $D(t)$ is convergent. Estimates of canonical expectation are given by using the ratio estimate

$$\langle \mathcal{O} \rangle_t = \frac{\ll\!\mathcal{O}/|\sigma|\!\gg_t}{\ll\!1/|\sigma|\!\gg_t}. \tag{13.40}$$

This ratio estimator convergences to $\langle \mathcal{O} \rangle$ in probability because the algorithm is ergodic over the state space $\mathbb{S}$.

In references [93, 399] it is shown that the integrated autocorrelation time diverges since $\langle |\sigma| \rangle_t$ diverges and $\tau_{int} > C \langle |\sigma| \rangle_t^{4\nu}$ (where ν is the metric exponent of the model, and C is a constant).

Generalisations of the BFACF algorithm can be found in the literature [92], and a variant of it has also been used to sample from the state space of lattice ribbons [445].

13.4 The pivot algorithm

The pivot algorithm was introduced by Lal [367] and popularised by Madras and Sokal [401]. The algorithm uses global neutral elementary moves to sample in the state space of self-avoiding walks of fixed length (the canonical ensemble).

The basic elementary move of the pivot algorithm is illustrated in figure 13.5. Choose a vertex $\vec{v}$ uniformly along a walk ω. Cut ω into the two subwalks

ϕ and ψ at $\vec{v}$ so that $\omega = \phi\psi$. With $\vec{v}$ as the origin, operate on the shorter of the subwalks, say ψ, with an element of the symmetry group of the underlying lattice, say P (for $\mathbb{L}^d$, there are $d!2^d$ choices for P, including the identity). It is sufficient to choose P uniformly but other implementations are also possible, Rotate and reflect ψ with P to obtain $P\psi$. Put $\omega' = \phi P\psi$. If ω' is self-avoiding, then it is the next state; otherwise, ω' is rejected, and ω is read again as the next state.

The following theorem is from reference [401].

Theorem 13.4 *Let S_n be the state space of unrooted self-avoiding walks of length n in $\mathbb{L}^d$. If a set A of pivot elementary moves includes all reflections through coordinate (hyper)-planes and all plane rotations by $\pm\frac{\pi}{2}$ (in planes with basis which is two coordinate directions), then A is irreducible when applied to S_n.*

Proof The basic idea is to reflect and rotate subwalks in a walk ω through the walls of the smallest rectangular box containing it.

Let $\omega = \langle \omega_0, \omega_1, \ldots, \omega_n \rangle$ be a walk and denote the k-coordinate of ω_i by $\omega_i(k)$. Define the *span* in the $\vec{e}_k$ direction by $M_k(\omega) = \max_{i,j} |\omega_i(k) - \omega_j(k)|$. The *diameter* of the smallest rectangular box containing ω is

$$D(\omega) = \sum_{k=1}^{d} M_k(\omega).$$

In addition, if $\omega_i = \frac{1}{2}(\omega_{i-1} + \omega_{i+1})$, then ω has an *straight internal angle* at vertex ω_i. Let $A(\omega)$ be the number of straight internal angles in ω.

Define the smallest rectangular box $B(\omega)$ containing ω and with faces parallel to coordinate surfaces.

The theorem is proven by the following claim.

CLAIM: By using reflections through coordinate (hyper)-planes and plane rotations by $\pm\frac{\pi}{2}$, ω can be updated to ω' such that $A(\omega) + B(\omega) \leq A(\omega') + B(\omega')$.
PROOF OF CLAIM: There are two cases to consider.
Case 1: Suppose there is a face of $B(\omega)$ which does not contain an endpoint of ω. Without loss of generality, suppose that this is the face K given by $x_1 = \max_i\{\omega_i(1)\} = N_1$ and which is normal to the $\vec{e}_1$ direction.

Let ω_k be the last vertex of ω in K; then $0 < k < n$.

Define the walk ω' by defining the vertices ${\omega_i}' = \omega_i$ if $i \leq k$, and

$${\omega_i}'(j) = \begin{cases} 2M_1 - \omega_i(1), & \text{if } j = 1; \\ \omega_i(j), & \text{if } j > 1, \end{cases} \qquad \text{if } i > k.$$

That is, ω' is obtained by reflecting vertices of ω though the face K of $B(\omega)$, starting at ω_k. Observe that $A(\omega) = A(\omega')$, since no right angles are created or destroyed by the reflection.

Consider $B(\omega)$ next. Denote the subwalk of ω from ω_i to ω_j by $\omega[i,j]$. Then the span of ω' in the $\vec{e}_1$ direction is

$$M_1(\omega') = M_1(\omega[1,k]) + M_1(\omega[k,n]),$$

since $\omega[1,k]$ and $\omega[k,n]$ are on either side of the hyperplane containing K. Since neither ω_0 or ω_n are in K, either $M_1(\omega[1,k]) = M_1(\omega) > 0$, and $M_1(\omega[k,n]) > 0$, or $M_1(\omega[1,k]) > 0$, and $M_1(\omega[k,n]) = M_1(\omega) > 0$. This shows that $M_1(\omega') > M_1(\omega)$.

Since the other coordinates of ω in the $\vec{e}_\ell$ directions for $\ell = 2,3,\ldots,d$ remain unchanged, $M_i(\omega') = M_i(\omega)$ for $i = 2,3,\ldots,d$.

This shows that $B(\omega) < B(\omega')$ and hence

$$A(\omega) + B(\omega) < A(\omega') + B(\omega').$$

Case 2: Suppose that there are no faces of $B(\omega)$ which do not contain an endpoint of ω. Then the endpoints ω_0 and ω_n are on diagonally opposite corners of $B(\omega)$.

Let ω_k be the last vertex of ω, where ω has a right angle. Then $0 < k < n$, and $\omega[k,n]$ is a straight line segment, which we suppose is in the $\vec{e}_2$ direction without loss of generality.

Rotate $\omega[k,n]$ outside of $B(\omega)$ to straighten out the right angle at ω_k and to obtain the walk ω'. Notice that the subwalk $\omega'[k-1,n]$ is a straight line and suppose without loss of generality that it is in the $\vec{e}_1$ direction. In addition, $M_1(\omega) = M_1(\omega'[0,k])$. Thus, it follows that

$$M_1(\omega') = M_1(\omega) + n - k.$$

Comparing the spans in the other directions shows that

$$M_2(\omega') = M_2(\omega'[0,k]) \geq M_2(\omega) - (n-k),$$

where $M_j(\omega') = M_j(\omega)$ for $j > 2$. Hence, $B(\omega') \geq B(\omega)$.

One right angle was lost, so $A(\omega) = A(\omega') - 1$.

This shows that

$$A(\omega) + B(\omega) < A(\omega') + B(\omega')$$

in this case as well. $\triangle$

This completes the proof, since recursive application of the above will maximise $A(\omega) + B(\omega)$ (at which point ω is a straight line segment). Observe that this straight line segment may be rotated in any other coordinate direction by application of rotations about the origin. $\square$

The implementation of the elementary move is by rejection (of proposed self-intersecting states). This makes the algorithm aperiodic. Reversibility of the pivot algorithm is demonstrated as follows in the case that pivots are selected uniformly from the symmetry group of the hypercubic lattice, The probability

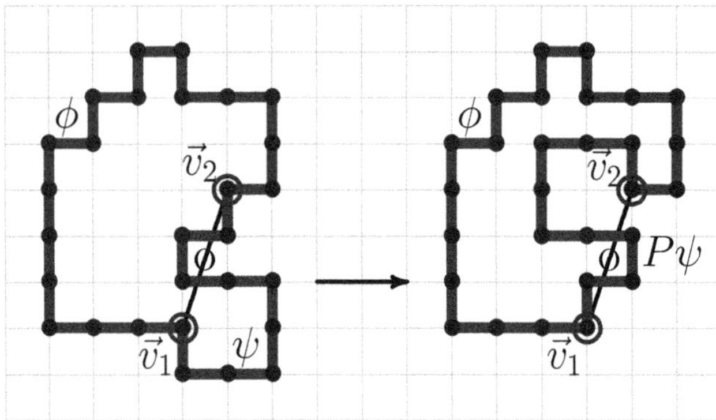

FIG. 13.6. An elementary move of the pivot algorithm on a square lattice polygon $\omega = \phi\psi$. In this illustration the subwalk ψ is reflected through the centre-of-mass $\frac{1}{2}(\vec{v}_1 + \vec{v}_2)$ of its endpoints $(\vec{v}_1, \vec{v}_2)$. This gives a new polygon $\phi P\psi$, where P is an inversion move.

---

of obtaining a state ψ from a state ϕ is $\mathrm{P}_r(\phi \to \psi) = \frac{1}{n+1}\frac{2^{-d}}{d!}$, since the pivot can be chosen uniformly from one of $n+1$ vertices and there are $d!2^d$ choices for the elementary move. Thus, $\mathrm{P}_r(\phi \to \psi) = \mathrm{P}_r(\psi \to \phi)$. This shows that the algorithm is ergodic and samples along a Markov chain from the uniform distribution.

Fast implementations of the pivot algorithm using specialised data structures and algorithms can be found in references [105, 346]. Proofs that the pivot algorithm is irreducible in the half-lattice $\mathbb{L}^d_+$ (see equation (2.20)) and in a lattice with a cut-plane can be found in reference [347].

The pivot algorithm can be implemented with *dimerisation* as follows. Let ω_1 and ω_2 be two walks obtained independently (for example, by sampling states far apart along a time series realised by the pivot algorithm). Then $\omega_1\omega_2$ dimerises if the concatenated conformation $\omega_1\omega_2$ of the walks is itself a self-avoiding walk. The dimerisation of walks gives rise to the *dimerisation* algorithm [5].

Dimerisation may be implemented as follows. Generate a set S_0 of N_0 *independent* self-avoiding walks of length n, using, for example, the pivot algorithm.

The dimerisation of walks in S_0 is a recursive process. Suppose that a set of walks S_m is obtained by a recursive dimerisation of walks in S_{m-1} (where $m = 1$ initially). Then the walks in S_m have length $2^m n$; suppose there are N_m walks in S_m.

Construct the walks in S_{m+1} as follows. Choose a pair of walks in S_m and concatenate them. If they dimerise (that is, if the resulting walk is self-avoiding), then place the dimerised walk in S_{m+1}. Otherwise, reject the pair of walks (and do not choose them again). Continue by choosing a new pair until all the walks, except perhaps one in S_m, have been selected. This gives a set of N_{m+1} walks in S_{m+1} and of length $2^{m+1}n$. Notice that $N_{m+1} \leq \frac{1}{2}N_m$. Stop the algorithm if

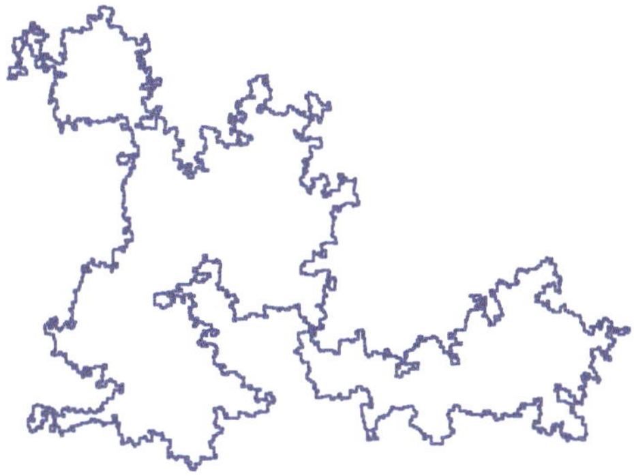

FIG. 13.7. The pivot algorithm can be used to sample very long walks and polygons. Above is a polygon of length 10 000 steps sampled from a time series of the pivot algorithm for square lattice polygons.

N_{m+1} is too small for calculating meaningful averages; otherwise, increment m and repeat the algorithm.

If the algorithm terminates with set S_m, then N_m independent walks of length $2^m n$ have been sampled. Notice that walks in the set S_{m-1} are not independent of walks in S_m. The probability that a pair of walks dimerise when they are concatenated may be approximated as follows. Two walks in $\mathbb{L}^d$ from $\vec{0}$ avoid one another with probability $P_n = \frac{c_{2n}}{c_n^2}$. Substitute $c_n \approx An^{\gamma-1}\mu_d^n$ into this to obtain $P_n \simeq Cn^{1-\gamma}$ for some constant C. Since $\gamma > 1$ in low dimensions, this argument shows that walks dimerise with decreasing probability as n increases.

Thus, the attrition of walks from set S_0 to S_1 is given by approximately $|S_1| \approx \frac{1}{2}|S_0|n^{1-\gamma}$. In $\mathbb{L}^3$ this shows that $|S_1| \approx \frac{1}{2}|S_0|n^{-1/6}$ (see reference [5]). In general, dimerisation is an effective tool for enhancing the sampling of walks obtained by other means.

The pivot algorithm may also be generalised to lattice polygons. Choose two vertices $V = \{\vec{v}_1, \vec{v}_2\}$ in a polygon ω (see figure 13.6). These vertices cut the polygon into the two subwalks ϕ and ψ. Without loss of generality, let ψ be the shorter of the two subwalks.

Choose $\vec{0}$ at the centre of mass $\frac{1}{2}(\vec{v}_1 + \vec{v}_2)$ of ψ and operate on ψ with an element P of the symmetry group of $\mathbb{L}^d$, chosen such to leave the locations of $\vec{v}_1$ and $\vec{v}_2$ unchanged or to interchange them (there is always at least one non-trivial operation available, namely the reflection of ψ through $\frac{1}{2}(\vec{v}_1 + \vec{v}_2)$; this is called an *inversion* of ψ).

If there is more than one possible operation available, then choose one uniformly from those available. This gives a new conformation for ψ, which is denoted $P\psi$. If $\omega' = \phi P\psi$ is a polygon (that is, if ϕ and $P\psi$ are self-avoiding), then this is a successful pivot move on ω. This algorithm can be used to sample very long polygons efficiently (see figures 13.7 and 13.8).

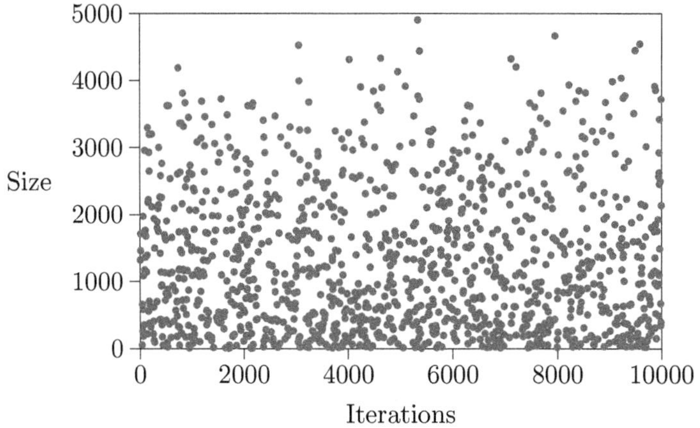

FIG. 13.8. A scatter plot of the size of a pivot against the iterations for square lattice polygons of length 10 000. The maximum size of a subwalk involved in a pivot is 5 000. These data show that, while the pivots are concentrated to smaller lengths, there is still a sizeable fraction occurring at lengths larger than 2500 (or one quarter of the length of the polygon). Similar statistics are known for the pivot algorithm for self-avoiding walks.

Theorem 13.5 *Let $\mathbb{P}_n$ be the state space of unrooted lattice polygons of length n in $\mathbb{L}^3$. If a set A of pivot elementary moves includes inversions and rotations by $\pm\frac{\pi}{2}$ about coordinate axes of $\mathbb{L}^3$, then A is irreducible when applied to $\mathbb{P}_n$.*

Proof Let $\omega = \langle \omega_0, \omega_1, \ldots, \omega_n \rangle$ be an (unrooted) polygon in $\mathbb{L}^3$, where $\omega_k = \omega_{k+Nn}$ for all $N \in \mathbb{Z}$. A projection of ω in the $\vec{e}_i\vec{e}_j$-coordinate plane is a closed path (with self-intersections). Denote this projected path by $P_{ij}\omega$.

Consider $P_{12}\omega$ and let its convex hull be $C_{12}\omega$. The convex hull $C_{12}\omega$ is a geometric polygon, consisting of sides which are straight line segments, and each side contains the projected images of at least two distinct vertices of ω.

Let the area of $C_{12}\omega$ be denoted by $A_{12}(\omega)$ and denote its boundary by $\partial C_{12}\omega$. Let the number of self-intersections of the path $P_{12}\omega$ be $M_{12}(\omega)$ (this is the number of pairs of vertices in ω which projects to the some (lattice) vertex in $P_{12}\omega$).

Denote the subwalk of ω from the vertex with label i to the vertex with label j by $\omega[i, j]$.

CLAIM: By doing pivot operations on the polygon ω, its conformation can be changed in a finite number of steps so that $P_{12}\omega = \partial C_{12}\omega$.
PROOF OF CLAIM: Suppose that there is a vertex ω_k which projects to the interior of $C_{12}\omega$. There is a first ℓ (cyclically after k) such that $P_{12}\omega_\ell \in \partial C_{12}\omega$. Without loss of generality, let ω_ℓ be in side 1 of $C_{12}\omega$. Then there is a last j preceding k such that $P_{12}\omega_\ell$ is in side 1 of $C_{12}\omega$.

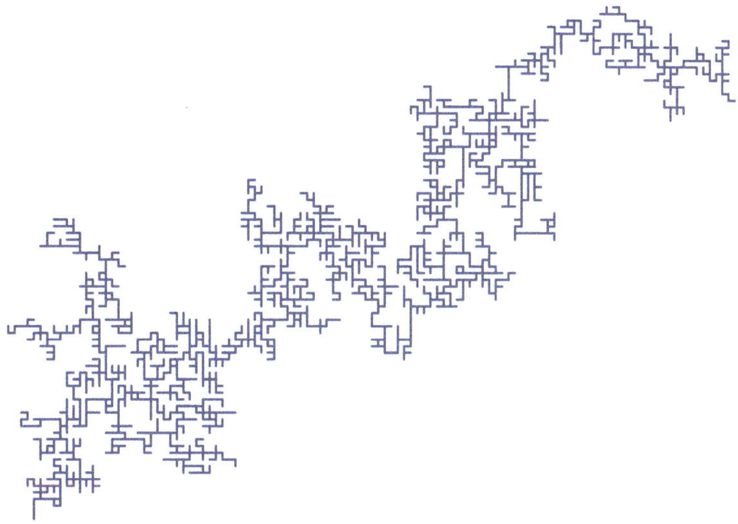

FIG. 13.9. A lattice animal of size 1600 edges sampled by a cut-and-paste algorithm; see references [307, 308] for a detailed implementation of the algorithm.

The subwalk $\omega[j,\ell]$ has projected endpoints in side 1 of $C_{12}\omega$, and ω_k is in the interior of $C_{12}\omega$. Do an inversion of the subwalk $\omega[j,\ell]$ through the centre of mass of its endpoints to obtain a new polygon ω'.

If $P_{12}\omega[j,\ell]$ is disjoint with the rest of the projected polygon, then this strictly increases $A_{12}(\omega)$, while $M_{12}(\omega)$ remains unchanged. If $P_{12}\omega[j,\ell]$ is not disjoint with the rest of the projected polygon then this strictly decreases $M_{12}(\omega)$.

Since $A_{12}(\omega)$ is bound from above, a recursive application of inversions through the sides of $C_{12}\omega$ will eventually give a polygon ω such that $P_{12}\omega = \partial C_{12}\omega$. $\triangle$

Notice that, if $P_{12}\omega = \partial C_{12}\omega$, then $P_{12}\omega$ is a rectangle in the $\vec{e}_1\vec{e}_2$-plane; in this case, all vertices of ω projects to $\partial C_{12}\omega$.

If there are sequences of edges which form a $\sqcup$-shaped sequence of line segments, then these can be rotated outside $C_{12}\omega$ to increase the size of $C_{12}(\omega)$. If this gives a polygon which does not project into $\partial C_{12}(\omega)$, then the claim above can be used to make it do so again.

Recursive application of the above gives a polygon ω such that $P_{ij}(\omega) \in \partial C_{ij}(\omega)$ for every pair (i,j). Then the diagonally opposite corners of the smallest rectangular box containing ω must be occupied. It is a simple exercise to check that such polygons can be made planar by using inversions and rotations of $\pm\frac{\pi}{2}$ about coordinate axes and by judicious choices of pivot points. $\square$

A corollary of the above is the following. It was first proven in reference [399].

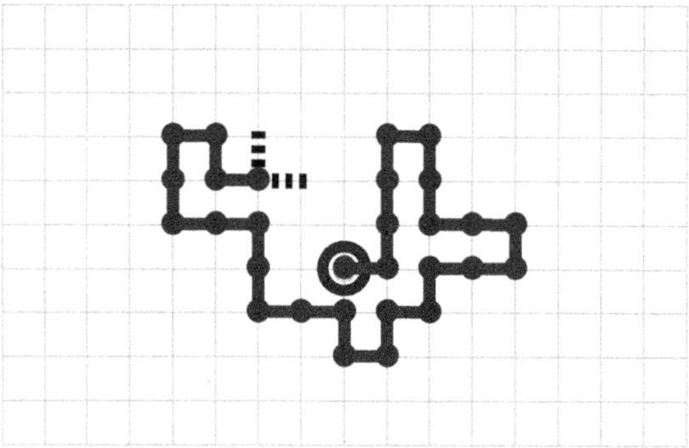

FIG. 13.10. The Rosenbluth method grows self-avoiding walks from the origin by adding steps (or edges) to the endpoint. In this example there are two possible ways to extend the walk by adding a step. One of the two steps is chosen uniformly as the next edge.

Corollary 13.6 *Let $\mathbb{P}_n$ be the state space of unrooted lattice polygons of length n in $\mathbb{L}^2$. If a set A of pivot elementary moves includes inversions and reflections through coordinate axes of $\mathbb{L}^2$, then A is irreducible when applied to $\mathbb{P}_n$.*

Pivot moves can be generalised in several possible ways. For example, cut a walk or a polygon into N pieces of almost equal length, do a pivot on each and then reassemble them in random order into a walk or a polygon. Such moves are irreducible on walks or polygons because they include the standard implementation of the pivot algorithm as a special case. These algorithms are *cut-and-paste* algorithms, and they have also been used to sample lattice trees and animals [307, 308] (see figure 13.9).

13.5 The Rosenbluth method and the PERM algorithm

13.5.1 *The Rosenbluth method*

The Rosenbluth algorithm [260, 496] is implemented by kinetically growing walks from $\vec{0}$ in $\mathbb{L}^d$ by adding steps to the endpoint. This is illustrated in figure 13.10, where a partially grown walk is shown.

The Rosenbluth method is simple to implement, works quite well (especially in higher dimensions), and remains a useful method. In $\mathbb{L}^2$ it suffers from attrition of walks, limiting effective sampling to walks of length about seventy steps.

Let ω be a walk from the origin with endpoint ω_n. The set $A_+(\omega)$ of edges which may be appended to ω_n to extend ω by one step is its *positive atmosphere*, which is of size $a_+(\omega) = |A_+(\omega)|$. For example, the walk in figure 13.10 has $a_+(\omega) = 2$. If $a_+(\omega) = 0$ for a given walk, then it cannot be grown. In this case the walk is *trapped.*

Let $\mathbb{S}$ be the state space of self-avoiding walks from $\vec{0}$ and denote the empty walk (with one vertex at the origin) by $\emptyset$. Create a graph G with $\mathbb{S}$ as vertex set, and a linkage between $\omega, \upsilon \in \mathbb{S}$ if υ can be obtained from ω by adding an edge to the endpoint of ω. Then G is a connected graph, since there is a (directed) path from $\emptyset$ to any walk ω.

The implementation of the Rosenbluth algorithm is as follows. Choose $\emptyset$ as the starting state and grow walks down the linkages in G by appending edges one at a time. This samples down a sequence of walks $\langle \emptyset, \phi_1, \phi_2, \ldots, \phi_n \rangle$, where ϕ_{j+1} is obtained by adding an edge from the positive atmosphere of ϕ_j. The weight of ϕ_n is recursively defined by

$$W(\phi_n) = a_+(\phi_{n-1}) W(\phi_{n-1}) = \prod_{j=0}^{n-1} a_+(\phi_j). \tag{13.41}$$

In $\mathbb{L}^d$, $a_+(\emptyset) = 2d$, and the sequence has the final state $\omega = \phi_n$ with weight $W(\omega)$.

If a walk is trapped and cannot be grown, then it is assigned weight 0 and discarded. A new walk is instead started from $\vec{0}$. Notice that a discarded walk is present in the set of sampled walks with weight 0, so that sample averages (see equations (13.44) and (13.45)) of observables are computed by dividing by the *number of started walks.*

Since the probability of selecting the $(m+1)$-th step in the Rosenbluth algorithm is $\frac{1}{a_+(\phi_m)}$, the probability of sampling a walk ω is

$$P_r(\omega) = \prod_{m=0}^{n-1} \frac{1}{a_+(\phi_m)} = \frac{1}{W(\omega)}. \tag{13.42}$$

By repeatedly sampling along sequences of walks, a sample of N started walks $\{\omega_j\}$ with associated weights $\{W_j\}$ is obtained. The walks are independently sampled, but not from the uniform distribution on the state space of walk of length n.

The Rosenbluth algorithm (as implemented above) is the prototype of kinetic growth algorithms for self-avoiding walks. Such algorithms sample walks independently but not uniformly. For example, a walk σ is sampled with probability $P_r(\sigma)$ and has an assigned *weight* $W(\sigma) = \frac{1}{P_r(\sigma)}$.

The expected weight of walks of length n is

$$\langle W \rangle_n = \sum_\sigma P_r(\sigma) W(\sigma) = \sum_\sigma 1 = c_n, \tag{13.43}$$

since the summation is over all walks of length n.

In other words, by estimating $\langle W \rangle_n$, an approximate count of the number of self-avoiding walks of length n is obtained. This is the *Rosenbluth counting theorem*, and this shows that this algorithm is in particular an *approximate enumeration algorithm* [299].

Suppose that an implementation of the Rosenbluth algorithm produced a sequence of independent walks $\langle \sigma_1, \sigma_2, \ldots, \sigma_M \rangle$ such that walk σ_i was sampled with probability p_i. The weight of σ_i is $W_i(\sigma_i) = \frac{1}{p_i}$. The *sample average* of the weights is

$$[W]_N^{sample} = \tfrac{1}{N} \sum_{i=1}^{N} W_i(\sigma_i) \to \langle W \rangle_n = c_n \tag{13.44}$$

by the strong law of large numbers. This follows in particular from equation (13.43) if the algorithm is irreducible on state space (any walk of length n can be grown with positive probability).

An estimator $[\mathcal{O}]_N^{est}$ for the (canonical) average of an observable $\mathcal{O}$ over the uniform distribution may be computed using weighted averages. It follows that it is a ratio estimator:

$$[\mathcal{O}]_N^{est} = \frac{\sum_{i=1}^{N} W_i(\sigma_i)\mathcal{O}(\sigma_i)}{\sum_{i=1}^{N} W_i(\sigma_i)} = \frac{[W\mathcal{O}]_N^{sample}}{[W]_N^{sample}}. \tag{13.45}$$

By the strong law of large numbers, $[\mathcal{O}]_N^{est} \to \langle \mathcal{O} \rangle$ (the canonical expectation value of $\mathcal{O}$ over all walks of length n) as $N \to \infty$; this is a ratio estimate of $\langle \mathcal{O} \rangle$.

The analysis of variance in these algorithms is frequently done using the block data analysis of section 13.1.6.2. Repeated measurements of $[\mathcal{O}]_M^{est}$ gives a collection of (say) N estimates $\{[\mathcal{O}]_M^{est}(j)\}_{j=1}^N$ which are assumed to be asymptotically normally distributed about $\langle \mathcal{O} \rangle$. These *block averages* are statistically independent. Thus, the average of the block averages is given by

$$[\mathcal{O}]_*^{est} = \tfrac{1}{N} \sum_{j=1}^{N} [\mathcal{O}]_N^{est}(j). \tag{13.46}$$

This is an average over $N \times M$ walks in N blocks each containing M walks. Since the block averages are independent of one another, the confidence interval of the above is given by

$$\sigma^2 [\mathcal{O}]_*^{est} = \frac{1}{N-1} \left(\left[\mathcal{O}^2\right]_*^{est} - ([\mathcal{O}]_*^{est})^2 \right), \tag{13.47}$$

similar to the result in equation (13.17), but with the block averages given by the ratio estimator in equation (13.45).

High quality data can be obtained using this approach by carefully balancing the choices for N and M; M should be large enough that the block averages are tightly squeezed about the average, and N should be large to make $\sigma^2 [\mathcal{O}]_*^{est}$ small.

It is generally the case that the weights W_i disperse over many orders of magnitude for walks of increasing length (see for example figure 13.11). This problem, together with attrition due to trapped walks (figure 13.12), restricts the applicability of these algorithms, but methods have been developed for variance reduction [211]. These were eventually incorporated in the PERM and GARM algorithms [235, 489].

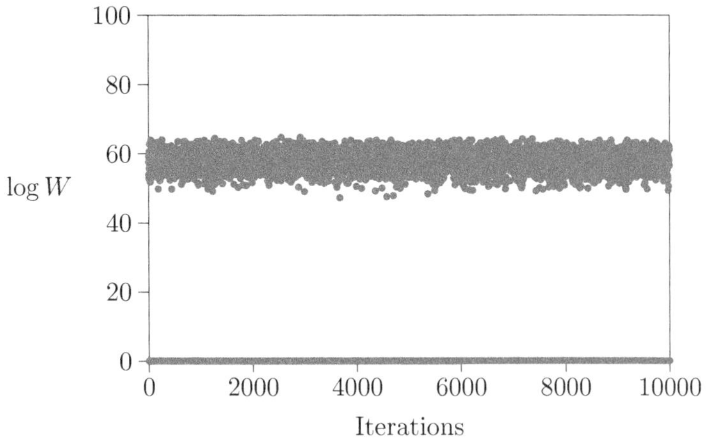

FIG. 13.11. The logarithm of the weight of walks of length sixty in two dimensions sampled by the Rosenbluth method. Computing the variance of $\log W$ only for non-zero weights about its mean gives about 2.72, so the spread of the data is in a 95% confidence band of width 10.88. The weights are scattered over about five orders of magnitude, roughly between $10^{22.2}$ at the low end and $10^{27.5}$ at the high end. This means that walks with larger weights may contribute 10^5 times more to the sample averages in equations (13.44) and (13.45). The data points on the horizontal axis are from walks which were trapped and then discarded with zero weight. The fraction of such discarded states increases quickly with the length of the walk in low dimensions. This, together with the dispersion of weights, is the main reason why the Rosenbluth algorithm cannot be used to sample walks longer than about seventy or eighty steps, effectively.

13.5.2 *The Rosenbluth method for interacting models*

The Rosenbluth method may be used to sample walks in a state space $\mathbb{S}$ of an interacting model from a Boltzmann distribution

$$P_\beta(\omega) = \frac{e^{\beta E(\omega)}}{\sum_{\omega\in\mathbb{S}} e^{\beta E(\omega)}}, \tag{13.48}$$

where $E(\omega)$ is the energy of the walk ω, and $\beta = \frac{1}{k_B T}$ is the inverse temperature. The denominator is the partition function of the model.

The algorithm is implemented as follows. Let $\phi \equiv \phi_j$ be obtained from $\psi \equiv \phi_{j-1}$ in a sequence of states grown by the algorithm.

Let $\Delta E(\phi, \psi) = E(\psi) - E(\phi)$ be the change in energy in the transition. The probability of selecting ψ if ϕ is the current state is

$$\mathrm{P}_r(\phi \to \psi) = \frac{e^{\beta\Delta E(\phi,\psi)}}{\sum_\psi e^{\beta\Delta E(\phi,\psi)}}, \tag{13.49}$$

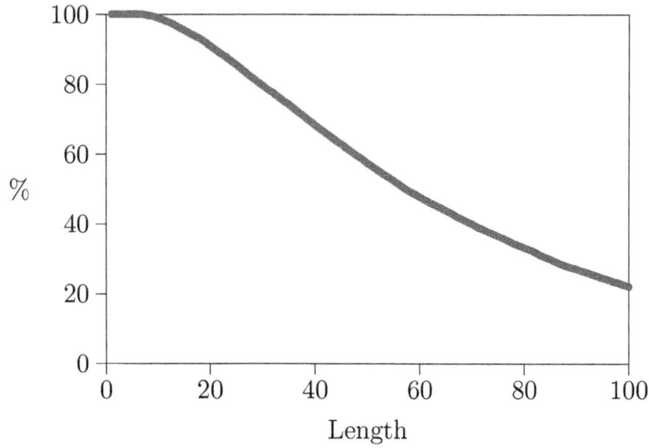

FIG. 13.12. The attrition of walks started by Rosenbluth sampling as a function of length. In this example, walks were grown to a maximum length of 100 in $\mathbb{L}^2$. The number of walks surviving at each length is given as a percentage of the number of started walks. At length 100 only about 20% of started walks survive, and their weights are dispersed over many orders of magnitude, thus making the algorithm inefficient.

---

where the summation in the denominator is over all walks ψ such that (ϕ, ψ) is a linkage (ψ can be reached from ϕ by appending an edge to ϕ).

The probability of sampling a sequence of walks $\langle \phi_0, \phi_1, \ldots, \phi_n \rangle$ with final walk of length n and energy $E(\phi_n)$ is

$$\mathrm{P}_r(\phi_n) = \prod_{j=1}^{n} P(\phi_{j-1} \to \phi_j) = \frac{e^{\beta(E(\phi_n - \phi_0)}}{\prod_{j=0}^{n-1} \sum'_{\psi_j} e^{\beta \Delta E(\phi_j, \psi_j)}}, \tag{13.50}$$

where the primed summation is over all walks ψ_j where (ϕ_j, ψ_j) is a linkage.

This gives the weight of a walk $\omega = \phi_n$, which was realised by sampling along the sequence $\langle \phi_0, \phi_1, \ldots, \phi_n \rangle$ as

$$W(\omega) = \prod_{j=0}^{n-1} \sum_{\psi_j}{}' e^{\beta \Delta E(\phi_j, \psi_j)}. \tag{13.51}$$

The average of the weight $W(\omega)$ where $\omega = \phi_n$ over walks of length n is

$$\begin{aligned} \langle W \rangle_n &= \sum_{\omega} \mathrm{P}_r(\omega) W(\omega) = \sum_{\phi_n} e^{\beta(E(\phi_n - \phi_0))} \\ &= e^{-\beta E(\phi_0)} \sum_{\phi_n} e^{\beta E(\phi_n)} = e^{-\beta E(\phi_0)} Z_n(\beta), \end{aligned} \tag{13.52}$$

where $Z_n(\beta) = \sum_\phi e^{\beta E(\phi)}$ is the partition function of the model (where $n = |\phi|$).

Since ϕ_0 is the starting walk of zero length, most models have $E(\phi_0) = 0$. This shows that the average of the weights of walks is $\langle W \rangle_n = Z_n(\beta)$, the partition function of the model, as expected.

Averages of observables are computed with respect to the Boltzmann distribution P_β using exactly the same approach as in equation (13.45). A confidence interval can be estimated using a blocked data analysis.

13.5.3 *Microcanonical sampling with the Rosenbluth method*

Let $c_n(m)$ be the number of self-avoiding walks of length n and with energy m. Rosenbluth sampling may be used to estimate $c_n(m)$ as follows. Let $\omega \equiv \phi_n$ be the final state in a sequence $\langle \phi_0, \phi_1, \ldots, \phi_n \rangle$ realised by the algorithm.

Similar to equation (13.41), the weight of ω is defined by

$$W(\omega) = \prod_{j=0}^{n-1} a_+(\phi_j). \tag{13.53}$$

The probability of sampling ω is $\Pr(\omega) = \frac{1}{W(\omega)}$, and the average of the weights for walks of energy m and length n is

$$\langle W \rangle_{n,m} = \sum_{\omega} P(\omega) W(\omega) = c_n(m), \tag{13.54}$$

since $W(\omega) = 0$ if ω does not have energy m; hence, there are exactly $c_n(m)$ non-zero terms (all equal to 1) left in the summation over ω.

13.5.4 *The scanning method: variance reduction*

The scanning method [411] is an algorithm for reducing the dispersion of Rosenbluth weights. The basic idea is as follows: compute Rosenbluth weights by scanning walks into the future to find the correct probability for choosing a possible step. If it is possible to scan walks ahead to length n, then it is in principle possible to determine the probabilities for selecting a given step in such a way that the algorithm samples walks of length n from the uniform distribution.

Suppose that a walk of length m was grown along a sequence of walks (denoted by $\langle \phi_0, \phi_1, \ldots, \phi_m \rangle$) by appending one step at a time to the endpoint.

The positive endpoint atmosphere $a_+(\phi_m)$ of the current walk is the number of open edges available for the next step. Selecting the step uniformly from those available gives Rosenbluth sampling. More generally, there is a 'correct' probability for selecting the next state σ by selecting one of the $a_+(\phi_m)$ available edges, and this probability depends on all the previous states $\langle \phi_0, \phi_1, \ldots, \phi_m \rangle$.

Denote this conditional probability by

$$p_m(\sigma \mid \phi_0, \phi_1, \ldots, \phi_m) = \Pr\left(\sigma \text{ is the next state given } \langle \phi_0, \phi_1, \ldots, \phi_m \rangle\right).$$

The scanning method is implemented by exactly computing the conditional probability $p_m(\sigma \mid \phi_0, \phi_1, \ldots, \phi_m)$ at each step, further conditioned on the fact that the final walk must have length n. Thus, the probability of the $(m+1)$-th step is computed by taking into account all the partially grown walks of length $n-m$ and which will complete the current walk of length m into a walk of length n.

The number of partial walks of length $n-m$ and which can complete the sequence $\langle \phi_0, \phi_1, \ldots, \phi_m \rangle$ into a walk of length n, given that the m-th step is σ, is denoted by $N_n(\sigma \mid \phi_0, \phi_1, \ldots, \phi_m)$. For each possible $(m+1)$-th step to σ, this number will vary, and it is determined by an exact enumeration procedure.

That is, define the (conditional) transition probability for the $(m+1)$-th step to the state σ to be given by

$$p_m(\sigma \mid \phi_0, \phi_1, \ldots, \phi_m) = \frac{N_n(\sigma \mid \phi_0, \phi_1, \ldots, \phi_m)}{\sum_\tau N_n(\tau \mid \phi_0, \phi_1, \ldots, \phi_m)}, \tag{13.55}$$

and the $(m+1)$-th step to σ is determined by stepping with probability $p_m(\sigma \mid \phi_0, \phi_1, \ldots, \phi_m)$.

Once a walk of length n is constructed, then a new walk is started, and eventually a collection of independent walks will be generated. These will be a sample from the uniform distribution over walks of length n. To see this, note that the probability of obtaining a given sequence $\langle \phi_0, \phi_1, \ldots, \phi_n \rangle$ with final state $\omega = \phi_n$ is given by

$$\mathrm{P}_r(\omega) = \frac{1}{2d} \prod_{m=1}^{n+1} p_m(\phi_{m+1} \mid \phi_0, \phi_1, \ldots, \phi_m). \tag{13.56}$$

Since the algorithm enumerates all walks up to length n at each step of the algorithm, each walk is generated with the same probability, and so $\mathrm{P}_r(\omega) = c_n^{-1}$.

The implementation above is computationally expensive, since the probability for each step must be determined by scanning all possible paths to length n. This situation becomes better if approximations instead of exact counts are used but the resulting algorithm does not sample from the uniform distribution.

The modification is as follows. Implement the algorithm as above but do not scan ahead $n-m$ steps at each iteration. Instead, scan ahead a maximum of b steps (more precisely, scan ahead $\min\{b, n-m\}$ steps at each iteration)

The parameter b should be chosen small enough to make the algorithm fast, and large enough to give a distribution over the walks, which are approximately uniform.

In this implementation, the number of possible ways ϕ_m can be extended by adding $\min\{b, n-m\}$ steps is denoted $N_n(\sigma, b \mid \phi_0, \phi_1, \ldots, \phi_m)$. Thus, the probability of the next state σ is

$$p_m(\sigma, b \mid \phi_0, \phi_1, \ldots, \phi_m) = \frac{N_n(\sigma, b \mid \phi_0, \phi_1, \ldots, \phi_m)}{\sum_\tau N_n(\tau, b \mid \phi_0, \phi_1, \ldots, \phi_m)}, \tag{13.57}$$

and it is a function of b. The probability of a given walk ω is

$$\mathrm{P}_r(\omega) = \frac{1}{2d} \prod_{m=1}^{n+1} p_m(\phi_{m+1}, b \mid \phi_0, \phi_1, \ldots, \phi_m). \tag{13.58}$$

Since not every given walk can be completed to a walk of length n, the result is that the probabilities $\mathrm{P}_r(\omega)$ are not normalised over walks of length n. This shows that $\sum_\omega \mathrm{P}_r(\omega) < 1$, and this must be taken into account.

That is, averages can be computed by assigning a weight $W(\omega) = \frac{1}{\mathrm{P}_r(\omega)}$ to each walk; then the average of an observable $\mathcal{O}$ is approximated by

$$[\mathcal{O}]_{N,b}^{est} = \frac{\sum_{i=1}^N W(\omega)\mathcal{O}(\omega_i)}{\sum_{i=1}^N W(\omega)} \tag{13.59}$$

if N walks $\{\omega_i\}_{i=1}^N$ were sampled, with walk ω_i obtained with probability $\mathrm{P}_r(\omega_i)$.

This approximation depends on b and $\mathrm{P}_r(\omega)$. Increasing b improves the quality the estimates but increases the CPU-time used by the algorithm. Taking $b = 1$ uncovers Rosenbluth sampling.

If b is taken equal to $n - m$ at the m-th step, then the original scanning algorithm is uncovered and $\mathrm{P}_r(\omega)$ is the uniform distribution. This reduces the variance of the estimates and is the essential variance reduction part of the scanning method.

Generally, the larger b, the smaller the variance in estimates $[\mathcal{O}]_{N,b}^{est}$.

In implementations of the scanning method, b is usually increased to be at most equal to 10, and then the approximations are extrapolated to infinite b. This procedure generally gives good results [412, 416].

13.5.5 *The PERM algorithm*

PERM is a variance reduction method for the Rosenbluth method and works by the systematic enrichment of walks of large weights in the sample and the systematic pruning of walks of low weight in the sample. PERM (an acronym for 'pruned and enriched Rosenbluth method') was introduced in references [235, 571] (see reference [211] as well).

Suppose that S is a collection of walks of length n grown by Rosenbluth sampling. The weight of a walk $\omega \in S$ is given by $W(\omega)$ in equation (13.41).

Introduce a cut-off T_n on $W(\omega)$ for walks of length n. If $W(\omega) > T_n$, then enrich ω in S by adding a copy of ω to S and by reducing $W(\omega)$ by a factor of 2. Then there are two copies of ω in S, each with a weight of $\frac{1}{2}W(\omega)$.

Continue growing each copy of ω *independently* by using the Rosenbluth algorithm. While this *enrichment* does not change the sample average at n, continued growth of the walks from this value of n will result in the copies growing along different trajectories with reduced weights in state space; their enrichment has the effect of reducing the dispersion of the weights by reducing large weights in a systematic way.

Walks with weights which become too small are dealt with by pruning them (removing them from S). This is implemented systematically by introducing a lower threshold t_n at length n on $W(\omega)$. If $W(\omega) < t_n$, then the walk is *pruned* and discarded with probability $1 - \frac{1}{q}$ where q is a parameter of the algorithm. If the walk is not pruned (with probability $\frac{1}{q}$), then its weight is increased by a factor of q.

Enriching a walk in S does not disturb the sample average of the weights. If N is the number of started walks, then the sample average of the weights is given by

$$[W]_N^{sample} = \frac{1}{N} \sum_{\phi \in S} W(\phi). \tag{13.60}$$

This is so because enriching a walk by two copies and reducing the weight by a factor of 2 leaves the summation unchanged. A similar argument shows that pruning has no effect on the sample average (if it is repeatedly computed and averaged). Hence, the same arguments leading to equation (13.44) applies here, and $[W]_N^{sample}$ must converge to $\langle W \rangle_n = c_n$.

Thresholds dependent on the running averages

$$t_N = c[W]_N^{sample}, \quad \text{and } T_N = C\,[W]_N^{sample} \tag{13.61}$$

may be used, where $\frac{C}{c} = K$. Choosing $K \approx 10$ reduces the dispersion of the weights of the walks to about one order of magnitude.

13.5.6 *Flat histogram PERM*

The dispersion of weights in PERM may be further reduced by taking the thresholds in its implementation to be equal: $t_N = T_N$.

This may be implemented as follows [470]. Let $[W]_N^{sample}$ be the running average of the weights of walks of length n after N walks in a sample S have been grown. If the N-th walk ϕ has PERM weight $W(\phi)$, then compute the ratio

$$r = \frac{W(\phi)}{[W]_N^{sample}}, \tag{13.62}$$

and observe that the weight $W(\phi)$ is also included in the calculation of $[W]_N^{sample}$.

If $r > 1$, then ϕ has a larger than expected weight and it should be enriched. This is done by computing $c = \lfloor r \rfloor$ with probability $p = \lceil r \rceil - r$, and with default $c = \lceil r \rceil$. The enrichment is done by placing c copies of the walk in the sample S, each with reduced weight $\frac{1}{c} W_m$.

If $r < 1$, on the other hand, then ϕ has weight smaller than expected and it can be pruned with probability r. If it is kept in S (not pruned), then its weight is increased by the factor $\frac{1}{r}$.

In flatPERM simulations the running average $[W]_N^{sample}$ of the weight is initially poor but improves quickly. The number of walks sampled at each length $m \leq n$ is roughly a constant, giving a flat histogram when the number of walks seen at each length $m \leq n$ is plotted against n. In figure 13.13 the attrition of walks is plotted against length n for both the Rosenbluth method and for

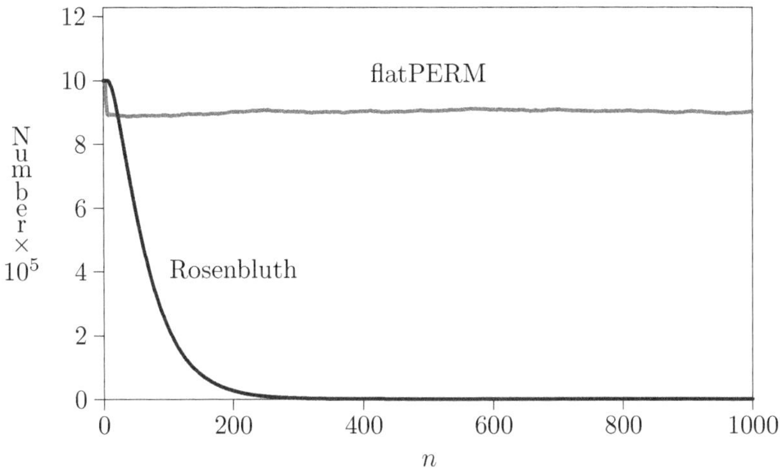

FIG. 13.13. Attrition of started walks in the Rosenbluth and flatPERM algorithms in $\mathbb{L}^2$ for walks up to length 1000. A million walks were started in each algorithm. In the Rosenbluth algorithm more than 80% of the walks were trapped and discarded by $n = 150$. In flatPERM sampling, there is an initial attrition of about 10% of the walks, but then the number stabilises to almost a constant for lengths up to 1000. The nearly constant number of completed walks after an initial decrease for small values of n shows that flatPERM is highly effective in self-tuning to produce a flat histogram over the lengths of completed walks.

flatPERM. While attrition is severe in the Rosenbluth method, a flat histogram is seen for flatPERM, with virtually no attrition when walks grow successfully beyond a few steps in length. The variance reduction in flatPERM gives algorithms which converge efficiently to a steady state with weights confined in a narrow range, as for example shown in figure 13.14.

The generalisation of PERM and flatPERM to interacting models in the canonical and microcanonical ensembles proceeds in the same way as for the Rosenbluth method.

13.6 The GARM algorithm

The GARM algorithm [323, 489] is based on very general implementations of elementary moves, which are collectively called *atmospheric moves*. The GARM algorithm is based on a generalisation of equation (13.43), giving rise to the *GARM counting theorem* [489].

Suppose that a set of elementary moves applied to a lattice cluster (such as a walk, polygon, tree or animal) gives rise to a collection A of potentially successful updates of the cluster. The collection A is the *atmosphere* of the cluster, and each

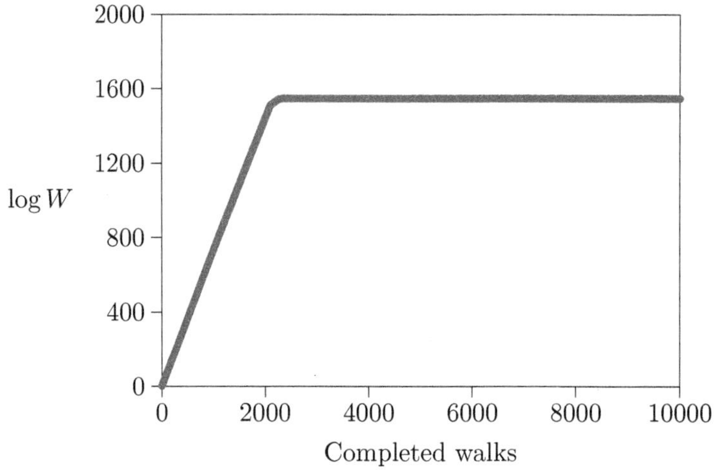

FIG. 13.14. The logarithm of the weight of walks of length 1000 in $\mathbb{L}^3$ and sampled by flatPERM, plotted against the number of completed walks. The weights increases quickly to a constant by about iteration number 2000. Thereafter, the flat histogram sampling maintains a constant value for the weights.

---

atmospheric (elementary) move is a linkage between the cluster and its updated version if the move is executed successfully.

The elementary moves seen in the algorithms above include endpoint elementary moves, BFACF moves, pivots and cut-and-paste moves, and collectively these moves may be implemented using GARM or GAS dynamics. The elementary moves in an application may be *local*, in which case they update a cluster by changing a limited number of edges. Otherwise the elementary moves are *global* (making a chance in a part of a cluster comparable in size to the entire cluster (for example, pivot elementary moves).

13.6.1 *Atmospheric elementary moves*

The most commonly used atmospheric moves are the *endpoint elementary moves* [489], *BFACF elementary moves* [323] and the *generalised atmospheric elementary moves* [489]. These are discussed in turn below.

An atmospheric move which increases the size (or length) of a lattice cluster is a *positive atmospheric move*. If a move leaves the size of the cluster unchanged, then it is a *neutral atmospheric move*, and, if it reduces the size of the cluster, then it is a *negative atmospheric move*.

13.6.1.1 *Endpoint atmospheric moves for walks:* The implementation of the Rosenbluth method relied on extending a walk by appending edges (new steps) at its endpoint. This is illustrated in figure 13.10. The collection of perimeter edges which may be appended to a walk ω gives its *positive atmosphere*, denoted

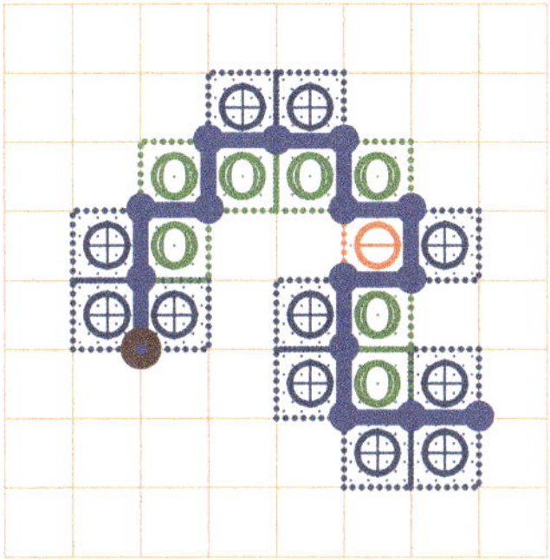

FIG. 13.15. The BFACF atmosphere of a self-avoiding walk. Places where a positive BFACF elementary move can be done are denoted by $+$. Similarly, places where a neutral move can be done are denoted by 0, and the one place where a negative move can be done is indicated by $-$. These labelled plaquettes constitute the positive, neutral and negative atmospheres of the walk, respectively.

---

by $a_+(\omega)$. For the walk in figure 13.10, $a_+(\omega) = 2$, and these moves are *positive endpoint atmospheric moves.*

The reverse of a positive endpoint atmospheric move is a deletion of the last edge in ω. This shortens the walk by one step and defines a *negative atmosphere* for ω. Notice that $a_-(\omega) = 1$ if ω has length more than 0, and that a negative atmospheric move is a reversal of a positive atmospheric move.

An endpoint atmosphere may also be defined for lattice trees. The positive endpoint atmosphere is the collection of perimeter edges with one endpoint in the tree, and the negative endpoint atmosphere is the collection of leaves in the tree.

13.6.1.2 *BFACF atmospheres for walks and polygons:* Positions along a walk or polygon where a BFACF elementary move may be performed are indicated in figure 13.15. Each elementary move is indicated by a plaquette incident with the lattice cluster, and these are *atmospheric plaquettes* labelled by the type of move they represent (positive (+), neutral (0) or negative (−)).

The set of plaquettes where positive moves can be performed is the positive atmosphere. Similarly, the sets of plaquettes where negative or neutral moves can be performed, are negative and neutral atmospheres, respectively.

Observe that each execution of a positive BFACF atmospheric move is immediately reversible by a negative BFACF atmospheric move in precisely the same location where the positive move was done.

Unlike the situation for endpoint atmospheres for walks, the positive BFACF atmosphere on its own does not induce an elementary moves irreducible on the set of polygons or walks with fixed endpoints. By theorems 13.1, 13.2 and 13.3, the full set of positive, neutral and negative moves is necessary for some models

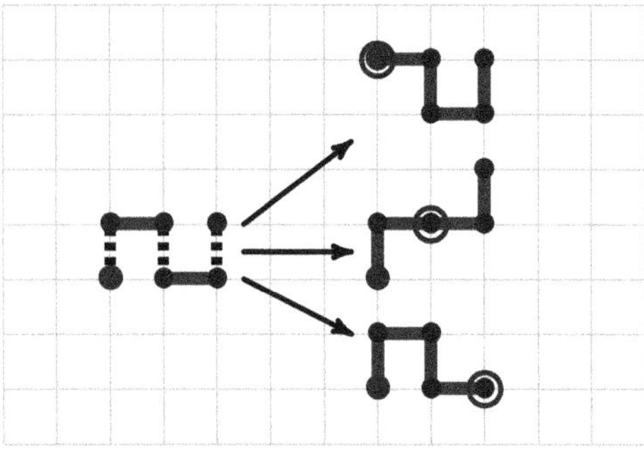

FIG. 13.16. The contraction of edges in a walk. There are three edges in the walk on the left; these may be contracted to give a new self-avoiding walk, displayed on the right.

of polygons and walks with fixed endpoints in $\mathbb{L}^2$ and $\mathbb{L}^3$. (In $\mathbb{L}^3$ the elementary moves are irreducible on polygons of fixed knot types.)

13.6.1.3 *Generalised atmospheric elementary moves:* Let $\widehat{e}$ be an edge in a self-avoiding walk $\omega = \phi\widehat{e}\psi$. Then $\widehat{e}$ is said to be *contracted* if its length is taken to zero, giving the cluster $\phi(\psi - \widehat{e})$, which may or may not be a new walk (see figure 13.16).

The above contraction may be reversed by the *insertion* of an edge. Let $\vec{v}$ be a vertex in ω, and let $\widehat{e}$ be a lattice edge incident with $\vec{v}$. Cut ω into two subwalks ϕ and ψ in $\vec{v}$ so that $\omega = \phi\psi$. Then $\psi_e = \psi + \widehat{e}$ is the walk obtained by translating ψ one step in the direction of $\widehat{e}$.

A new walk $\omega' = \phi\widehat{e}\psi_e$ may be constructed from ϕ and ψ_e, provided that ω' is self-avoiding. The walk ω' is obtained by the *insertion* of $\widehat{e}$ in ω at vertex $\vec{v}$.

Observe that the contraction of $\widehat{e}$ in ω' recovers ω.

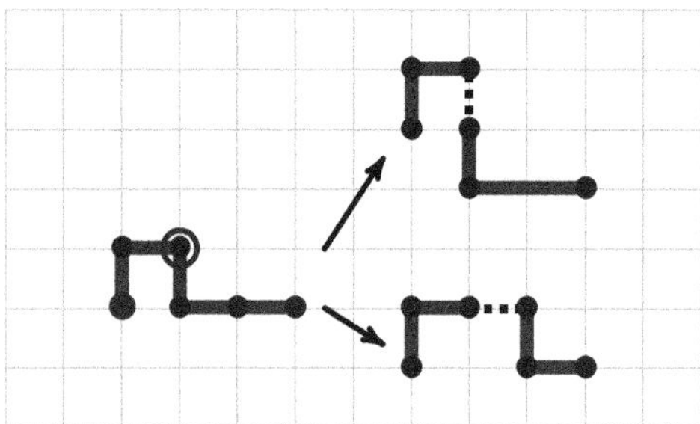

FIG. 13.17. An insertion move in a self-avoiding walk. The marked vertex is replaced by an edge as shown. In each case a new self-avoiding walk is obtained.

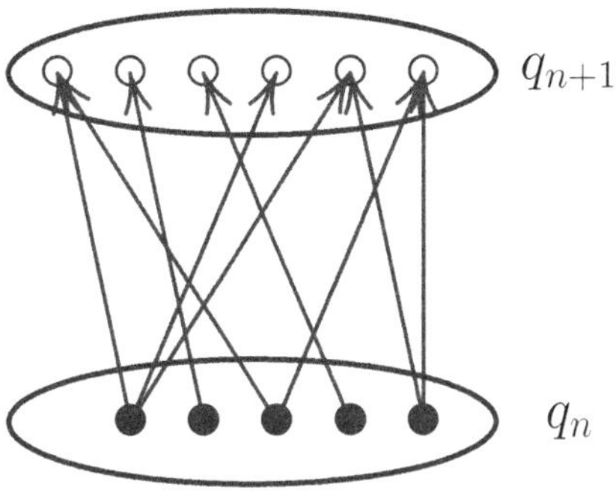

FIG. 13.18. Atmospheric elementary moves set up linkages between clusters of sizes n and $n+1$ (or $n+2$). These linkages are denoted schematically by arrows above. An arrow from a cluster of size n (•) to a cluster of size $n+1$ (∘) is a positive atmospheric move (and can be reversed by a negative atmospheric move).

The insertion of an edge at a vertex in a walk is illustrated in figure 13.17. There are two possible choices for the insertion of an edge: either vertically down or horizontally to the right. In each case a new conformation is created. The reverse of an insertion is a contraction, which can be used to recover the original walk.

The set of vertices and edges which may be inserted in a walk ω (to recover a walk of increased length) composes the *positive generalised atmosphere* of ω. The size of the positive generalised atmosphere of ω is denoted by $a_+(\omega)$. If ω is the walk on the left in figure 13.17, then $a_+(\omega) = 14$.

Similarly, the set of edges which may be contracted in a walk ω to recover a shorter walk composes the *negative generalised atmosphere* of ω. The size of the negative generalised atmosphere of ω is denoted by $a_-(\omega)$. If ω is the walk on the left in figure 13.17, then $a_-(\omega) = 4$.

13.6.2 *Growth constants and atmospheric elementary moves*

The implementation of an elementary move on a cluster ω turns it into a new cluster ω'. This creates a *linkage* (ω, ω') between the two clusters.

The linkages due to elementary moves may be represented graphically, as in figure 13.18. Represent clusters of size n as set of vertices and suppose that a positive elementary move may be used to update these clusters to size $n+1$. The linkage between a cluster ω of size n and a cluster ω' of size $n+1$ is represented by an arc (ω, ω').

Let S_n be the set of all clusters of size n and let $q_n = |S_n|$ be the number of clusters of size n. Then the number of linkages from S_n to S_{n+1} (this is the number of arcs in figure 13.18) is given by $\sum_{\omega:|\omega|=n} a_+(\omega)$.

On the other hand, consider negative atmospheric moves to recover ω from ω' in a linkage (ω, ω'). If every positive atmospheric move is immediately reversible by a negative atmospheric move, then this shows that the number of arcs in figure 13.18) is given by $\sum_{\omega:|\omega|=n+1} a_-(\omega)$.

Hence, the number of arcs from S_n to S_{n+1} in figure 13.18 is

$$\#\{\text{Number of arcs}\} = \sum_{\omega:|\omega|=n} a_+(\omega) = \sum_{\omega:|\omega|=n+1} a_-(\omega). \tag{13.63}$$

Since $\sum_{\omega:|\omega|=n} a_+(\omega) = q_n \langle a_+ \rangle_n$, where $\langle a_+ \rangle_n$ is the average of the size of the positive atmosphere of walks of length n, and $\sum_{\omega:|\omega|=n+1} a_-(\omega) = q_{n+1} \langle a_- \rangle_{n+1}$, where $\langle a_- \rangle_{n+1}$ is the average of the positive atmosphere of walks of length $n+1$, it follows that

$$\frac{\langle a_+ \rangle_n}{\langle a_- \rangle_{n+1}} = \frac{q_{n+1}}{q_n}. \tag{13.64}$$

This result is quite general and not dependent on the types of elementary moves; it simply requires that there are positive elementary moves linking clusters of length n to clusters of length $n+1$, and that these moves can be reversed uniquely to become negative atmospheric moves.

A similar result is valid for BFACF atmospheric moves when applied to walks or polygons. An argument similar to the argument leading to equation (13.64) gives

$$\frac{\langle a_+ \rangle_n}{\langle a_- \rangle_{n+2}} = \frac{c_{n+2}(\vec{x})}{c_n(\vec{x})} \tag{13.65}$$

for walks ω from the origin to the lattice site $\vec{x}$. In the above $c_n(\vec{x})$ may be replaced by lattice polygons.

The limit ratio theorem 7.17 then shows that $\frac{c_{n+2}(\vec{x})}{c_n(\vec{x})} \to \mu_d^2$ for self-avoiding walks. That is, the ratio of average atmospheres should approach μ_d^2 as $n \to \infty$, so atmospheric statistics may be used to estimate μ_d [487].

Similar arguments apply to lattice trees and animals and to knotted polygons, with suitable interpretation of the above to those models.

13.6.3 *GARM sampling and the GARM counting theorem*

Consider a model of lattice clusters, and suppose that there are q_n cluster of size n. Suppose that a set of atmospheric elementary moves, containing positive moves and neutral moves, is implemented on the lattice clusters. In addition, suppose that every positive atmospheric move is immediately reversible by a corresponding negative move.

Using a starting state ϕ_0, a sequence of states $\Phi = \langle \phi_0, \phi_1, \phi_2, \ldots, \phi_m \rangle$ is grown using positive moves and neutral moves. Each transition $\phi_j \to \phi_{j+1}$ is a *linkage* (ϕ_j, ϕ_{j+1}), where ϕ_{j+1} is obtained from ϕ_j by the application of a neutral or positive atmospheric move.

Each of the states ϕ_j has a positive atmosphere, a neutral atmosphere and a negative atmosphere, and these have sizes (number of possible moves) $a_+(\phi_j)$, $a_0(\phi_j)$ and $a_-(\phi_j)$, respectively.

Each transition $\phi_j \to \phi_{j+1}$ is performed by selecting uniformly from the positive moves and neutral moves available, so the probability of realising ϕ_{j+1} from ϕ_j is

$$\Pr_r(\phi_j \to \phi_{j+1}) = \Pr_r(\phi_{j+1} \mid \phi_j) = \frac{1}{a_+(\phi_j) + a_0(\phi_j)}. \tag{13.66}$$

Hence, the probability of realising the sequence Φ if the starting state is ϕ_0 is

$$\Pr_r(\Phi \mid \phi_0) = \prod_{j=0}^{|\Phi|-1} \Pr_r(\phi_{j+1} \mid \phi_j) = \prod_{j=0}^{|\Phi|-1} \frac{1}{a_+(\phi_j) + a_0(\phi_j)}. \tag{13.67}$$

Assign a weight $W(\Phi)$ to each sequence by

$$W(\Phi \mid \phi_0) = \prod_{j=1}^{|\Phi|} \frac{a_+(\phi_{j-1}) + a_0(\phi_{j-1})}{a_-(\phi_j) + a_0(\phi_j)}. \tag{13.68}$$

The average value of the weight over all sequences Φ with final state ϕ_n is

$$\langle W(\phi_m)\rangle = \sum_{\Phi\to\phi_m} \Pr_r(\Phi \mid \phi_0) W(\Phi \mid \phi_0) = \sum_{\Phi\to\phi_m} \prod_{j=1}^{|\Phi|} \frac{1}{a_-(\phi_j) + a_0(\phi_j)}, \tag{13.69}$$

where the summation over $\Phi \to \phi_m$ is over all sequences ending in state ϕ_m.

The right-hand side of equation (13.69) may be interpreted as the probability that a sequence of states starting in state ϕ_m terminates in state ϕ_0 if neutral moves and/or negative moves are used. This observation gives rise to the GARM counting theorem.

Theorem 13.7 (GARM counting theorem) *Suppose that positive atmospheric moves and neutral atmospheric moves may be implemented on a model of lattice clusters and that this implementation is irreducible (that is, every state ω can be grown along a finite sequence of positive moves and neutral moves from a unique starting state ϕ_0). Then*

$$\langle W\rangle_n = q_n,$$

where $\langle W\rangle_n$ is the average weight of sequences ending in a state of size n, and q_n is the number of distinct clusters of size n.

Proof If the starting state ϕ_0 is unique, then the sequence of negative moves and neutral moves in equation (13.69) must terminate in ϕ_0 with probability 1. This shows that $\langle W(\phi_m)\rangle = 1$ for every state of ϕ_m.

The average weight of all sequences ending in a state ϕ_m of length n is $\langle W\rangle_n = \sum_{|\phi|=n} \langle W(\phi)\rangle = \sum_{|\phi|=n} 1 = q_n$. This completes the proof. □

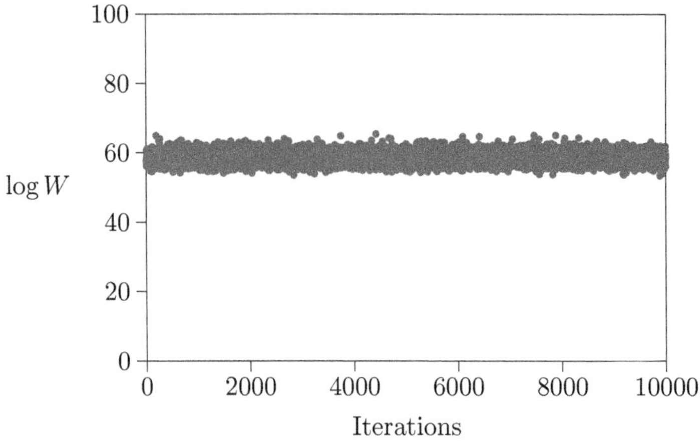

FIG. 13.19. The logarithm of the weight of walks of length 1000 from the origin in $\mathbb{L}^2$ and sampled by GARM (implemented with the generalised atmospheric elementary moves in figure 13.17 and with the negative atmospheric elementary move shown in figure 13.16). The variance of $\log W$ for non-zero weights is about 1.48 and the spread of the data is in a 95% confidence band of width 5.92. The GARM weights are scattered over 2.5 orders of magnitude (this is far better than the spread seen in Rosenbluth weights in figure 13.11).

The average GARM weights are computed in the same way as for Rosenbluth sampling. The sample average of weights is

$$[W]_N^{sample} = \tfrac{1}{N}\sum_{i=1}^{N} W_i(\sigma_i) \to \langle W \rangle_n = q_n \tag{13.70}$$

by the strong law of large numbers.

An estimator for the (canonical) average of the observable $\mathcal{O}$ over the uniform distribution may be computed in a similar way using weighted averages. It follows that

$$[\mathcal{O}]_N^{est} = \frac{\sum_{i=1}^{N} W_i(\sigma_i)\mathcal{O}(\sigma_i)}{\sum_{i=1}^{N} W_i(\sigma_i)} = \frac{[W\mathcal{O}]_N^{sample}}{[W]_N^{sample}}. \tag{13.71}$$

This is exactly the same expression used in Rosenbluth sampling (see equation (13.45)).

The dispersion of measured average weights $\langle W \rangle_n$ in a GARM simulation with more general atmospheric elementary moves is far smaller than in the Rosenbluth algorithm (which is GARM with an endpoint elementary move). In figure 13.19 a scatter plot of $\log W(\phi)$ is plotted against iterations for sequences which terminate in walks of length $n = 1000$. The weights scatter about a very narrow band of about one order of magnitude. This should be compared to the five orders of magnitude seen in figure 13.11 for walks of length $n = 60$. In addition to this, this implementation of GARM suffered no attrition due to trapped conformations (as illustrated for the Rosenbluth method in figure 13.13).

GARM may be implemented with pruning and enrichment in exactly the same way that these variance reduction methods were implemented in the PERM algorithm by tracking the weights of a given sequence.

13.6.4 *Flat histogram GARM*

A flat histogram implementation of GARM using pruning and enrichment moves proceeds similarly to the implementation of flatPERM. This gives rise to the *flatGARM algorithm* [489].

Suppose that the sequence $\Phi = \langle \phi_0, \phi_1, \ldots, \phi_j, \ldots \rangle$ is realised by flatGARM sampling, and that the state ϕ_j has weight $W(\phi_j)$. The implementation proceeds by computing the parameter

$$r = \frac{W(\phi_j)}{\langle W(n(\phi_j)) \rangle} \tag{13.72}$$

at step j, where $\langle W(n(\phi_j)) \rangle$ is the average of the weights of states of length (number of edges) $n(\phi_j)$ seen before in the simulation. The GARM weights are given by equation (13.68), so

$$W(\phi_j|\phi_0) = \prod_{i=1}^{j} \frac{a_+(\phi_{i-1}) + a_0(\phi_{i-1})}{a_-(\phi_i) + a_0(\phi_i)}. \tag{13.73}$$

To determine the next state ϕ_{j+1} requires that an elementary move is selected and that r is computed. Observe that the average in the denominator of r *includes* the statistics of ϕ_{j+1}, which are as yet unknown. That is, it is necessary to compute $a_-(\phi_{j+1})$ and $a_0(\phi_{j+1})$ in order to determine the weight $W(\phi_{j+1}|\phi_0)$ to include it in the average $\langle W(n(\phi_j)) \rangle$ and to determine r.

Since $a_-(\phi_{j+1})$ and $a_0(\phi_{j+1})$ cannot be known before ϕ_{j+1} is known, it becomes necessary to approximate the atmospheric statistics of ϕ_{j+1} and then to compute an estimate for r. A method giving good practical results is to guess that $a_-(\phi_{j+1}) = a_-(\phi_j)$, and $a_0(\phi_{j+1}) = a_0(\phi_j)$, if a neutral move will be performed, or $a_-(\phi_{j+1}) = a_-(\phi_j) + 1$, and $a_0(\phi_{j+1}) = a_0(\phi_j)$, if a positive move will be performed. These estimates for generalised atmospheres work well in actual simulations.

Once the parameter r has been computed, pruning and enrichment is implemented as follows: if $r < 1$, then retain the current sequence with probability r and update the weight $W(s_j) \to \frac{1}{r} W(s_j)$. Otherwise, prune the sequence (with probability $1 - r$). If $r > 1$, then enrich the sequence: compute $c = \lceil r \rceil$ with probability $r - \lfloor r \rfloor$, and $c = \lfloor r \rfloor$ otherwise. Make c copies of the sequence, each with weight $\frac{1}{c} W(\phi_j)$, and continue growing from each, recursively pruning and enriching at each step.

This implementation of flatGARM introduces some initial attrition of started sequences due to pruning but the parameter r is designed such that a flat histogram is obtained; the pruned sequences are eventually replaced by enriched sequences. A major advantage over PERM and flatPERM is that correlations are suppressed in the enrichment process. While endpoint atmospheric moves in PERM or flatPERM leave the walk unchanged up to the enrichment point, generalised atmospheric moves quickly update edges even along the early part of the walk, so correlations between enriched copies soon become statistically insignificance.

The implementation of GARM and flatGARM sampling from the Boltzmann distribution, or in the microcanonical ensemble, proceeds in the same way as for implementations of PERM.

13.7 The GAS algorithm

GARM implements generalised elementary moves as illustrated in figures 13.16 and 13.17 in a systematic way to obtain an algorithm which works in more or less the same way as the Rosenbluth algorithm. The mix of elementary moves in a GARM simulation is quite arbitrary; pivot or cut-and-paste moves may be used as neutral elementary moves with (say) positive elementary BFACF moves and negative elementary BFACF moves to construct a working algorithm.

However, GARM is limited to an implementation using only positive elementary moves and neutral elementary moves. The atmospheric statistic $a_-(\phi)$ of negative moves inverting the positive moves must be computed in order to determine the weight of a growing sequence (as in equation (13.68)) but these moves are not implemented.

The introduction of negative elementary moves in a GARM-style algorithm leads to a new class of algorithms with different dynamical properties. These are the GAS algorithms [323] .

13.7.1 *Generalised atmospheric sampling*

Suppose that ϕ is a lattice cluster (for example a walk or a tree) together with a set of reversible elementary moves with the atmospheric statistics $a_+(\phi)$, $a_0(\phi)$ and $a_-(\phi)$ of positive, neutral and negative atmospheres, respectively. Assume that positive elementary moves increase the size of ϕ by one unit, neutral elementary moves leave it unchanged, and negative elementary moves decrease the size of ϕ by one unit.

Suppose that any two lattice clusters ϕ and ψ communicate along at least one sequence of elementary moves (that is, the set of elementary moves is irreducible on the state space $\mathbb{S}$). The elementary moves define linkages in $\mathbb{S}$ as illustrated in figure 13.1.

More generally, suppose that a transition of state $\phi \in \mathbb{S}$ to $\psi \in \mathbb{S}$ under the action of an elementary move induces a map $f : \mathbb{S} \to \mathbb{S}$ such that $(\phi, \psi) \in f$ if ϕ can be updated to ψ by an elementary move. Then f induces a digraph on $\mathbb{S}$

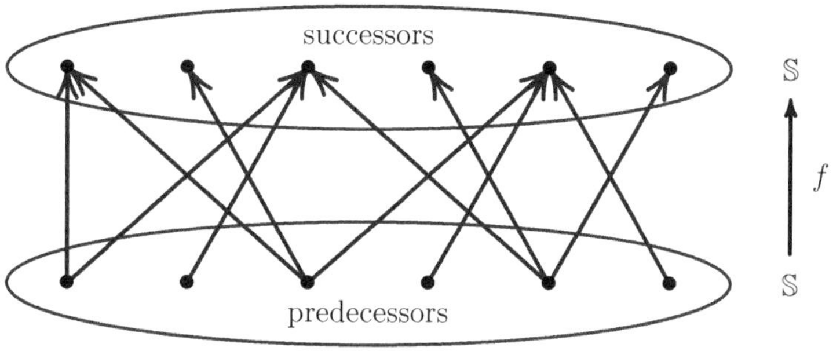

FIG. 13.20. Generalised atmospheric moves in the GAS algorithm define a map $f : \mathbb{S} \to \mathbb{S}$ which maps predecessors in $\mathbb{S}$ to successors in $\mathbb{S}$. The map may be represented by a digraph illustrated above. Each arc from a predecessor state to a successor state corresponds to an elementary move.

---

and with arcs from $\phi \to \psi$ if there is an elementary move taking ϕ to ψ; ϕ is the *predecessor state*, and ψ is the *successor state*. This is illustrated in figure 13.20.

Since all elementary moves are assumed to be reversible (positive moves are reversible by negative moves, and neutral moves are reversible by neutral moves), the indegree of any successor vertex ϕ is equal to the outdegree of its corresponding predecessor vertex ϕ.

States are sampled in state space $\mathbb{S}$ along sequences $\Phi = \langle \phi_0, \phi_1, \ldots, \phi_L \rangle$ of arbitrary length, starting in a seed state ϕ_0. The transition from $\phi_j \to \phi_{j+1}$ is through an elementary move selected from a distribution over those available. While it is not necessary, the algorithm will always select uniformly from the set of available elementary moves (and thus uniformly from the set of successors of ϕ_j).

The *level* of a state $\phi_j \in \Phi$ is its position in Φ. That is, ϕ_0 has level 0, and ϕ_j has level j. The algorithm samples ϕ_j by sampling in level j. The number of levels or length of Φ is $|\Phi| = L + 1$.

Generally, Φ is an element in the product space $\mathbb{S}^L = \bigotimes_{i=0}^{L} \mathbb{S} = \mathbb{S} \otimes \mathbb{S} \otimes \mathbb{S} \otimes \cdots \otimes \mathbb{S}$. It remains to show that the algorithm samples along a Markov chain in $\mathbb{S}$ from a distribution over sequences $\Phi \in \mathbb{S}^L$.

The state space $\mathbb{S}$ is infinite, containing lattice clusters (such as walks) of arbitrary size. Thus, the algorithm may sample along sequences containing clusters of arbitrary (large) size.

A simple way to correct this is to introduce a cut-off on the size of states, say n_{max}. In that case, the state space $\mathbb{S}$ is a finite space of clusters of sizes up to n_{max}. The product space $\mathbb{S}^L$ is finite, and it is possible to determine a distribution over $\mathbb{S}^L$. The cut-off on $\mathbb{S}$ is a sufficient but not necessary condition; in practice, it greatly simplifies simulations.

The distribution over $\mathbb{S}^L$ is normalised by introducing weights to sequences Φ. Suppose that state ϕ_j at level j along Φ has been realised. Introduce the parameter β (possibly dependent on the length of state ϕ_j) and perform the next elementary move with probabilities

$$\Pr(\text{positive atmos. move}) = \frac{\beta a_+(\phi_j)}{a_-(\phi_j) + a_0(\phi_j) + \beta a_+(\phi_j)}, \tag{13.74}$$

$$\Pr(\text{neutral atmos. move}) = \frac{a_0(\phi_j)}{a_-(\phi_j) + a_0(\phi_j) + \beta a_+(\phi_j)}, \text{ and} \tag{13.75}$$

$$\Pr(\text{negative atmos. move}) = \frac{a_-(\phi_j)}{a_-(\phi_j) + a_0(\phi_j) + \beta a_+(\phi_j)}, \tag{13.76}$$

which are normalised to sum up to unity. These are implemented by first selecting the type of elementary move (positive, neutral or negative) and then choosing uniformly from those moves available.

Define $P_+(\Phi)$ to be the number of positive atmospheric moves performed along a sequence Φ of length $|\Phi|$. Similarly, denote by $P_-(\Phi)$ the number of negative atmospheric moves performed along a sequence Φ of length $|\Phi|$. Then $P_+(\Phi) - P_-(\Phi) + |\phi_0|$ is the size (number of edges) in the final state of the sequence ϕ.

The probability of realising Φ is clearly given by

$$\Pr(\Phi) = \left(\prod_{j=0}^{|\Phi|-1} \frac{1}{a_-(\phi_j) + a_0(\phi_j) + \beta a_+(\phi_j)} \right) \beta^{P_+(\Phi)}. \tag{13.77}$$

The *weight* of a sequence Φ is determined as follows. Define ℓ_j to be the length (number of edges) of state ϕ_j (which is a cluster of size at most n_{max} in S). Define the function $\sigma(\phi_j, \phi_{j+1})$ along Φ by

$$\sigma(\phi_j, \phi_{j+1}) = \begin{cases} -1, & \text{if } \phi_{j+1} \text{ follows } \phi_j \text{ through a positive move;} \\ +1, & \text{if } \phi_{j+1} \text{ follows } \phi_j \text{ through a negative move;} \\ 0, & \text{if } \phi_{j+1} \text{ follows } \phi_j \text{ through a neutral move.} \end{cases} \tag{13.78}$$

Then the weight of Φ is

$$W(\Phi) = \left(\prod_{j=0}^{|\Phi|-1} \frac{a_-(\phi_j) + a_0(\phi_j) + \beta a_+(\phi_j)}{a_-(\phi_{j+1}) + a_0(\phi_{j+1}) + \beta a_+(\phi_{j+1})} \right) \prod_{j=0}^{|\Phi|-1} \beta^{\sigma(\phi_j, \phi_{j+1})}. \tag{13.79}$$

Notice that $\sum_j \sigma(\phi_j, \phi_{j+1}) = P_-(\Phi) - P_+(\Phi)$, since each positive move gives a negative sign to the sum, and each negative move gives a positive sign.

Thus, the weight of Φ telescopes and may be written as

$$W(\Phi) = \left(\frac{a_-(\phi_0) + a_0(\phi_0) + \beta a_+(\phi_0)}{a_-(\phi_L) + a_0(\phi_L) + \beta a_+(\phi_L)} \right) \beta^{P_-(\Phi) - P_+(\Phi)}, \tag{13.80}$$

where $L = |\Phi| - 1$, and ϕ_L is the last state in Φ.

Let $\tau = \phi_L$ be a fixed state in $\mathbb{S}$. Then the average weight over all sequences of length $L+1$ from ϕ_0 to τ is given by

$$\langle W(\Phi) \rangle_\tau = \sum_{\Phi:\phi_0 \to \tau} \mathrm{P}_r(\Phi)\, W(\Phi), \tag{13.81}$$

where the summation over $\Phi : \phi_0 \to \tau$ is over all sequences starting from the initial state ϕ_0 and ending in state τ. This simplifies to

$$\langle W(\Phi) \rangle_\tau = \sum_{\Phi:\phi_0 \to \tau} \left(\prod_{j=0}^{|\Phi|-1} \frac{1}{a_-(\phi_{j+1}) + a_0(\phi_{j+1}) + \beta a_+(\phi_{j+1})} \right) \beta^{P_-(\Phi)}. \tag{13.82}$$

Let Ψ be the reverse sequence of states seen in Φ. That is, this sequence of states is $\Psi = \langle \psi_0, \psi_1, \ldots, \psi_{L-1}, \psi_L \rangle = \langle \phi_L, \phi_{L-1}, \ldots, \phi_1, \phi_0 \rangle$, where $\psi_j = \phi_{L-j}$, and $\tau \equiv \phi_L = \psi_0$.

Every positive elementary move along Φ becomes a negative elementary move along Ψ, and vice versa, while neutral elementary moves along Φ are neutral elementary moves along Ψ. Thus, $P_+(\Psi) = P_-(\Phi)$, and $P_-(\Psi) = P_+(\Phi)$.

In terms of Ψ, equation (13.82) may be written as

$$\langle W(\Phi) \rangle_\tau = \sum_{\Psi:\tau \to \phi_0} \left(\prod_{j=0}^{|\Psi|-1} \frac{1}{a_-(\psi_{L-j-1}) + a_0(\psi_{L-j-1}) + \beta a_+(\psi_{L-j-1})} \right) \beta^{P_+(\Psi)},$$

and this becomes

$$\langle W(\Phi) \rangle_\tau = \sum_{\Psi:\tau \to \phi_0} \left(\prod_{j=0}^{|\Psi|-1} \frac{1}{a_-(\psi_j) + a_0(\psi_j) + \beta a_+(\psi_j)} \right) \beta^{P_+(\Psi)}.$$

Comparison to equation (13.77) shows that the summand in the above is the probability that $\Psi : \tau \to \phi_0$ is a sequence of length $L+1$, starting in τ, terminating in ϕ_0 and composed of a sequence of elementary moves from states ψ_j to states ψ_{j+1} (for $j = 0, 1, 2, \ldots, |\Psi| - 1$), and with transition probabilities

$$\mathrm{P}_r(\text{positive atmos. move}) = \frac{\beta a_+(\psi_j)}{a_-(\psi_j) + a_0(\psi_j) + \beta a_+(\psi_j)}, \tag{13.83}$$

$$\mathrm{P}_r(\text{neutral atmos. move}) = \frac{a_0(\psi_j)}{a_-(\psi_j) + a_0(\psi_j) + \beta a_+(\psi_j)}, \quad \text{and} \tag{13.84}$$

$$\mathrm{P}_r(\text{negative atmos. move}) = \frac{a_-(\psi_j)}{a_-(\psi_j) + a_0(\psi_j) + \beta a_+(\psi_j)}. \tag{13.85}$$

Hence, the average weight is

$$\langle W(\Phi) \rangle_\tau = \sum_{\Psi:\tau \to \phi_0} \mathrm{P}_r(\Psi \to \phi_0 \mid \text{starting state } \tau \text{ and } |\Psi| = L+1), \tag{13.86}$$

where $\mathrm{P}_r(\Psi \to \phi_0 \mid \tau)$ is the conditional probability that Ψ ends in state ϕ_0 if it started in state τ. Summing over sequences Ψ starting in τ gives the conditional probability that a GAS sequence Ψ will terminate in state ϕ_0 if it was conditioned to have started in state τ:

$$\langle W(\Phi)\rangle_\tau = \mathrm{P}_r(\phi_0 \mid \text{starting state } \tau \text{ and } |\Psi| = L+1)\,. \tag{13.87}$$

Here the right-hand side is the probability that the sequence Ψ terminates in state ϕ_0 if the sequence was conditioned to start in state τ and with length $L+1$. This is also the probability of Ψ hitting state ϕ_0 at time L if it was conditioned to start in state τ.

Suppose that the elementary moves are irreducible on $\mathbb{S}$ and that β is such that P_r(positive atmos. move) < 1 for all possible transitions in realisations of Ψ. Define the size of a state ψ_j by $n(\psi_j)$. Then $\langle n(\psi_j)\rangle$ is a biased random walk on $\mathbb{N}_0$, the non-negative integers. For small values of β, this walk is recurrent and aperiodic, which means that it will end in the state ϕ_0 with a positive probability which is asymptotically independent of τ and L. This shows that $\langle W(\Phi)\rangle_\tau > 0$ and is asymptotically independent of τ and L if β is small [530]. In fact,

$$\lim_{L\to\infty} \langle W(\Phi)\rangle_\tau = \mathrm{P}_r(\phi_0) > 0, \tag{13.88}$$

where $\mathrm{P}_r(\phi_0)$ is the asymptotic probability that the chain hits state ϕ_0.

Summing over starting states τ of length n and taking $L \to \infty$ gives

$$\lim_{L\to\infty} \sum_{|\tau|=n} \langle W(\Phi)\rangle_\tau = q_n \mathrm{P}_r(\phi_0) > 0, \tag{13.89}$$

where q_n is the number of lattice clusters of size n. Hence, it follows that

$$\frac{\lim_{L\to\infty} \sum_{|\tau|=n} \langle W(\Phi)\rangle_\tau}{\lim_{L\to\infty} \sum_{|\tau|=m} \langle W(\Phi)\rangle_\tau} = \frac{q_n}{q_m}. \tag{13.90}$$

Notice that $\sum_{|\tau|=n} \langle W(\Phi)\rangle_\tau = \langle W(\Phi)\rangle_n$ is the average weight of states of size $|\tau| = n$ along Φ , so the above becomes

$$\frac{\lim_{L\to\infty} \langle W(\Phi)\rangle_n}{\lim_{L\to\infty} \langle W(\Phi)\rangle_m} = \frac{q_n}{q_m}. \tag{13.91}$$

Hence, by choosing L large and estimating the averages of the weights of states of sizes n and m along sequence Φ, the ratio $\frac{q_n}{q_m}$ is approximated by

$$\frac{q_n}{q_m} \approx \frac{[W(\Phi)]_n^{est}}{[W(\Phi)]_m^{est}}. \tag{13.92}$$

If q_m is known by other means, then it may be used to estimate q_n.

The above is valid if the algorithm samples along recurrent ergodic sequences in $\mathbb{S}$. For this to be the case, it would be sufficient if P_r(positive atmos. move) $<$ P_r(negative atmos. move) for any state $\phi \in \mathbb{S}$ in equations (13.83) and (13.85).

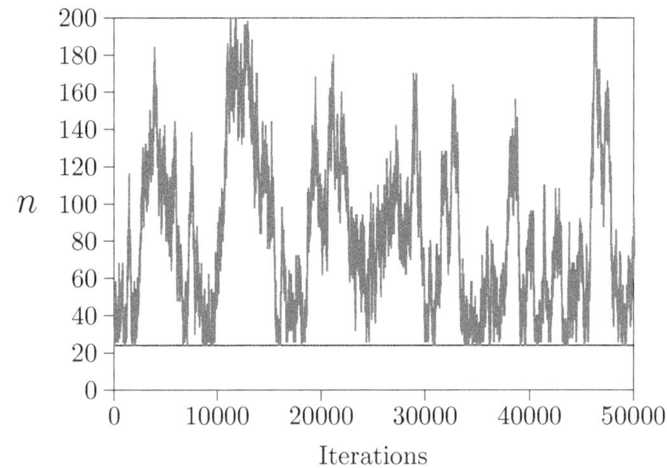

FIG. 13.21. Time series realised by a GAS simulation of the length of knotted polygons of fixed knot type 3_1 (the trefoil knot) using BFACF elementary moves. The large fluctuations in n quickly explore state space, sampling polygons efficiently at large n. In this simulation the maximum length of polygons was $n = 200$, while the minimal length of polygon of knot type 3_1 is $n = 24$ [143].

This would be the case if $\beta a_+(\psi) < a_-(\psi)$ for all ψ. This gives a very strong condition on β, since states with $a_-(\psi) = 0$ can be found frequently in applications.

In practical applications it has been found that the condition

$$\beta \lesssim \frac{[a_-(\psi)]_n^{est}}{[a_+(\psi)]_n^{est}} \tag{13.93}$$

gives good results, where $[a_-(\psi)]_n^{est}$ is the estimated average size of the negative atmospheres of states of size n, and $[a_+(\psi)]_n^{est}$ is the estimated average size of the positive atmospheres of states of size n. The larger β is, the more effectively the sequence explores $\mathbb{S}$, with better numerical results.

A time series realised by a GAS implementation of BFACF elementary moves on lattice knots of type 3_1 is shown in figure 13.21; the length of the polygons are plotted against iterations. It is clear that the length fluctuates quickly as GAS explores state space; this reduces autocorrelations along the time series of observables and improves the efficiency of the algorithm.

13.7.2 *Flat histogram generalised atmospheric sampling*

GAS weights over sequences Φ are given by equations (13.79) and (13.80). These weights do not disperse as much as Rosenbluth or GARM weights because the power of β increases strictly with length, and the prefactor is a simple ratio only dependent on the first and last states in the sequence. Thus, there is no reason to implement enrichment or pruning as variance reduction techniques.

The state space $\mathbb{S}$ in a GAS simulation contains clusters of all sizes, indexed by n. Let $\mathbb{S}_n$ be the set of clusters of size n. Then $|\mathbb{S}_n| = q_n$, and $\mathbb{S} = \cup_{n=0}^{n_{max}} \mathbb{S}_n$ is the disjoint union of the $\mathbb{S}_n$.

The execution of positive elementary moves and negative elementary moves along a realisation of a GAS sequence Φ is a random walk on $\mathbb{N}$ in the size of clusters. An implementation of a GAS algorithm such that states in $\mathbb{S}$ are sampled uniformly in length is a flat histogram implementation called *flatGAS*.

Define the probabilities for atmospheric moves on a state ϕ similarly to those in equations (13.74), (13.75) and (13.76) by $Q_+(\phi) = \mathrm{P}_r(\text{positive atmos. move})$, $Q_0(\phi) = \mathrm{P}_r(\text{neutral atmos. move})$ and $Q_-(\phi) = \mathrm{P}_r(\text{negative atmos. move})$.

A flatGAS algorithm may be implemented by choosing $Q_+(\phi)$ approximately equal to $Q_-(\phi)$, on average, for all states in each of the sets $\mathbb{S}_n$. That is, assume that the algorithm may be implemented such that $Q_+(\phi) \approx Q_-(\phi)$. The only degree of freedom available for achieving this is the parameter β, so replace β by β_n, such that β_n is a function of ϕ, and in this particular case, a function of the length or size n of ϕ. Requiring that $\langle Q_+(\phi)\rangle_n = \langle Q_-(\phi)\rangle_n$, where the average is over $\mathbb{S}_n$, gives

$$\beta_n \langle a_+(\phi)\rangle_n = \langle a_-(\phi)\rangle_n. \tag{13.94}$$

Assuming that states in $\mathbb{S}_n$ are uniformly populated by the algorithm, this would imply that, on average, a positive elementary move on a state $\phi \in \mathbb{S}_n$ is selected with the about same frequency as a negative elementary move on a state $\phi \in \mathbb{S}_n$. That is, the sequence of states Φ realised by the algorithm is a random walk stepping between sets in the collection of sets $\{\mathbb{S}_n\}$, and visiting each of the sets $\mathbb{S}_n$ about the same number of times if Φ is very long (see figure 13.21). This gives a flat histogram on the size of the clusters (see figure 13.22).

The modification of the GAS algorithm by replacing β with β_ℓ also modifies the weights in equation (13.79) to

$$W(\Phi) = \left(\frac{a_-(\phi_0) + a_0(\phi_0) + \beta_{\ell_0} a_+(\phi_0)}{a_-(\phi_L) + a_0(\phi_L) + \beta_{\ell_L} a_+(\phi_L)}\right) \prod_{j=0}^{|\Phi|-1} \beta_{\ell_j}^{\sigma(\phi_j,\phi_{j+1})}, \tag{13.95}$$

where ℓ_j is the size of state ϕ_j. The transition probabilities of positive, neutral and negative elementary moves from state ϕ_j to ϕ_{j+1} along the sequence Φ are similarly modified to become

$$\mathrm{P}_r(\text{positive atmos. move}) = \frac{\beta_{\ell_j} a_+(\phi_j)}{a_-(\phi_j)+a_0(\phi_j)+\beta_{\ell_j} a_+(\phi_j)}, \tag{13.96}$$

$$\mathrm{P}_r(\text{neutral atmos. move}) = \frac{a_0(\phi_j)}{a_-(\phi_j)+a_0(\phi_j)+\beta_{\ell_j} a_+(\phi_j)}, \quad \text{and} \tag{13.97}$$

$$\mathrm{P}_r(\text{negative atmos. move}) = \frac{a_-(\phi_j)}{a_-(\phi_j)+a_0(\phi_j)+\beta_{\ell_j} a_+(\phi_j)}, \tag{13.98}$$

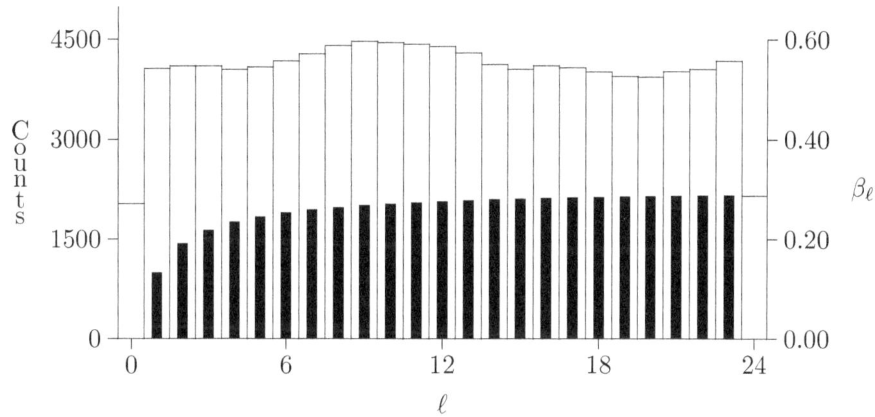

FIG. 13.22. Flat histogram sampling using the flatGAS algorithm implemented with endpoint elementary moves on self-avoiding walks from $\vec{0}$ in $\mathbb{L}^3$. The solid bars are values of β_ℓ as a function of ℓ on the right-hand side scale (see equation (13.99)). The cut-off $n_{max} = 24$ is implemented by putting $\beta_{24} = 0$. Open bars are the counts of walks of length ℓ observed in the GAS sequence. The histogram is reasonably flat over the entire range of ℓ.

---

respectively, and these are now explicitly dependent on the length or size of state ϕ_j because of the parameter β_{ℓ_j}.

The implementation of flatGAS is similar to GAS. Sequences are started in an initial state ϕ_0 and grown for L iterations. A total of N sequences generated in $\mathbb{S}$ with a cut-off n_{max} on the maximum size of a cluster are grown, each independent of the others. If L is large enough, then the data from N sequences are independent measurements of observables on the state space $\mathbb{S}$, which may be analysed to determine averages and confidence intervals.

A flat histogram obtained by flatGAS sampling is shown in figure 13.22. This is an implementation using endpoint elementary moves on walks from $\vec{0}$ in $\mathbb{L}^3$ with a cut-off at $n_{max} = 24$ (this may be implemented by putting $\beta_{24} = 0$).

The values of the parameters β_ℓ were computed using the ratio

$$\beta_\ell = \frac{[a_-(\phi)]_n^{est}}{[a_+(\phi)]_n^{est}}, \tag{13.99}$$

where the estimated averages on the right-hand side are running estimates of the average atmosphere of walks of length n; these approximate the averages in equation (13.94).

TABLE 13.1. Approximate enumeration of walks in $\mathbb{L}^3$

n	c_n	GAS $\langle W_n \rangle$	flatGAS $\langle W_n \rangle$
0	1	1	1.00040
1	6	6.00013	6.00254
2	30	30.0131	29.9971
3	150	150.202	149.95
4	726	727.451	725.929
5	3534	3541.02	3535.28
6	16926	16958.1	16944.4
7	81390	81540.4	81497.5
8	387966	388757	388447
9	1853886	1.85830×10^6	1.85568×10^6
10	8809878	8.83188×10^6	8.81789×10^6
11	41934150	4.20464×10^7	4.19746×10^7
12	198842742	1.99482×10^8	1.99077×10^8
13	943974510	9.47294×10^8	9.45217×10^8
14	4468911678	4.48663×10^9	4.47232×10^9
15	21175146054	2.12653×10^{10}	2.11767×10^{10}
20	49917327838734	5.00652×10^{13}	4.98981×10^{13}
25	116618841700433358	1.16875×10^{17}	1.16462×10^{17}
30	270569905525454674614	2.71271×10^{20}	2.70128×10^{20}

States were binned along the flatGAS sequence Φ of length $L = 10^5$ by length $\ell \in [0, 24]$. Notice that the shortest and largest states were sampled about half as frequently as the other states.

In general, GAS and flatGAS are approximate enumeration algorithms which determine ratios of the numbers of clusters of different sizes by equation (13.92). In table 13.1 the estimates of c_n for walks in $\mathbb{L}^3$ are given using both a GAS and a flatGAS simulation. In these simulations GAS was implemented using endpoint atmospheres. The implementation of GAS used $\beta = 0.212$ and realised 100 sequences of length 10^7 with a cut-off $n_{max} = 71$. The flatGAS simulation quickly settled down into flat histogram sampling. The observed errors show that the GAS and flatGAS data are scattered about the exact enumeration data (obtained from [108]). Overall, these relatively short runs produced estimates that deviate by less than 1% from the exact data. More accurate results can be obtained by increasing either the length of the sequences, or by increasing the number of started sequences.

An implementation of GAS or flatGAS on polygons in $\mathbb{L}^3$ with BFACF elementary moves may be used to sample lattice knots (see theorem 13.3). This was done in references [322, 324]. In table 13.2 approximate enumeration results of unknotted polygons and lattice knots of knot types 3_1 and 4_1 are displayed.

TABLE 13.2. Approximate enumeration of lattice knots with GAS

n	$p_n(0_1)$	$p_n(3_1)$	$p_n(4_1)$
4	3		
6	22		
8	$(2.0699 \pm 0.0024) \times 10^2$		
10	$(2.4143 \pm 0.0047) \times 10^3$		
12	$(3.1863 \pm 0.0081) \times 10^4$		
14	$(4.541 \pm 0.014) \times 10^5$		
16	$(6.856 \pm 0.023) \times 10^6$		
18	$(1.084 \pm 0.0041) \times 10^8$		
20	$(1.774 \pm 0.0032) \times 10^9$		
22	$(2.987 \pm 0.015) \times 10^{10}$		
24	$(5.156 \pm 0.027) \times 10^{11}$	3328	
26	$(9.053 \pm 0.051) \times 10^{12}$	$(2.8105 \pm 0.0063) \times 105$	
28	$(1.618 \pm 0.010) \times 10^{14}$	$(1.4339 \pm 0.0053) \times 10^7$	
30	$(2.931 \pm 0.020) \times 10^{15}$	$(5.769 \pm 0.028) \times 10^8$	3648
32	$(5.380 \pm 0.039) \times 10^{16}$	$(2.005 \pm 0.011) \times 10^{10}$	$(5.414 \pm 0.017) \times 10^5$
34	$(9.980 \pm 0.076) \times 10^{17}$	$(6.342 \pm 0.039) \times 10^{11}$	$(4.341 \pm 0.025) \times 10^7$
36	$(1.871 \pm 0.015) \times 10^{19}$	$(1.862 \pm 0.012) \times 10^{13}$	$(2.495 \pm 0.018) \times 10^9$
38	$(3.537 \pm 0.030) \times 10^{20}$	$(5.204 \pm 0.038) \times 10^{14}$	$(1.164 \pm 0.010) \times 10^{11}$

An implementation of GAS with endpoint atmospheres is a generalisation of Beretti-Sokal dynamics [37]. This implementation is called GABS in reference [323], and its flat histogram implementation flatGABS is an efficient approximate enumeration tool for self-avoiding walks, primarily because the implementation performs elementary moves in $O(1)$ CPU-time.

Microcanonical implementation of GAS and flatGAS proceeds similarly to microcanonical implementations of Rosenbluth and GARM sampling. States of length n and energy E are binned and their average weights are determined over sequences. The ratios of average weights give estimates of counts in exactly the same way as for canonical simulations via equation (13.92).

APPENDIX A

SUBADDITIVITY

A.1 The basic subadditive theorem

A sequence $\langle a_n \rangle$ is said to be subadditive if for n, m in a subset of $\mathbb{N}$ one has

$$a_n + a_m \geq a_{n+m}. \tag{A.1}$$

Fekete's lemma is a fundamental results on subadditive functions and shows the existence of the limit $\lim_{n\to\infty} \frac{1}{n} a_n$ (see, for example, [280]).

Theorem A.1 (Fekete's lemma) *Suppose that $M \subseteq \mathbb{N}$ is closed under addition and let $a_n : M \to \mathbb{Z}$ be a subadditive function on M. Then*

$$\lim_{n\to\infty} \tfrac{1}{n} a_n = \inf_{n\geq 1} \tfrac{1}{n} a_n = \upsilon,$$

where $\upsilon \in [-\infty, \infty)$ and where the limit is taken in M. If $a_n \geq An$ for a finite A, then υ is finite and so $a_n \geq n\upsilon$.

Proof Fix $k \in M$ and let $n \in M$. Choose $A_k = \max\{a_l \mid l \leq k, \text{ and } l \in M\}$. Put $j = \lfloor \frac{n}{k} \rfloor$; then $n = jk + r$ for some $0 \leq r \leq k$ and, since M is closed under addition, $r \in M$, and $n \in M$.

Repeated application of equation (A.1) gives $a_n = a_{jk+r} \leq j a_k + A_k$. Hence,

$$\tfrac{1}{n} a_n \leq \tfrac{j}{n} a_k + \tfrac{1}{n} A_k.$$

Take the limit superior on the left-hand side as $n \to \infty$ in M for fixed k. Then $j \to \infty$ in $\mathbb{N}$ and so

$$\limsup_{n\to\infty} \tfrac{1}{n} a_n \leq \tfrac{1}{k} a_k.$$

Taking the infimum of the right-hand side establishes the existence of the limit. Since $\frac{1}{k} a_k$ is finite for every $k \in M$, the limit is not $+\infty$ but could be equal to $-\infty$.

If $a_n \geq An$, and A is finite, then $\inf_{n\geq 1} \frac{1}{n} a_n = \upsilon \geq A > -\infty$. □

A.2 The Wilker and Whittington generalisation of Fekete's lemma

Fekete's lemma can be generalised [593].

Theorem A.2 *Let $a_n : \mathbb{N} \to \mathbb{Z}$ be a sequence and suppose that $f : \mathbb{N} \to \mathbb{Z}$ is a function with the property that $\lim_{n\to\infty} \frac{1}{n} f_n = 1$.*

If $\langle a_n \rangle$ satisfies the generalised subadditivity condition

$$a_n + a_m \geq a_{n+f_m}$$

for all $n, m > N_0$, where $N_0 \in \mathbb{N}$ is finite, then

$$\lim_{n\to\infty} \tfrac{1}{n} a_n = \upsilon$$

exists, and $a_n \geq \upsilon f_n$ for $n > N_0$.

Proof Fix $n \geq m > N_0$ and let p be the largest integer such that $n - m \geq (p-1)f_m$. Then there is a function r_m, with $0 \leq r_m \leq f_m$, such that

$$n - m = (p-1)f_m + r_m.$$

Repeated application of the generalised subadditive inequality gives

$$a_n = a_{(p-1)f_m+m+r_m} \leq (p-1)a_m + a_{m+r_m}.$$

Divide this by n to obtain

$$\tfrac{1}{n} a_n \leq \frac{(p-1)a_m}{(p-1)f_m + m + r_m} + \frac{a_{m+r_m}}{(p-1)f_m + m + r_M}.$$

For fixed m, take the limit superior of the left-hand side as $n \to \infty$. Then $p \to \infty$ on the right-hand side; the result is that

$$\upsilon = \limsup_{n\to\infty} \tfrac{1}{n} a_n \leq \frac{a_m}{f_m} = \frac{a_m}{m} \times \frac{m}{f_m}.$$

Take the limit inferior as $m \to \infty$ on the right-hand side. This proves the existence of the limit. The claimed inequality follows directly from the above. □

There is a generalisation of theorem A.2, which may be useful in some cases:

Theorem A.3 *Suppose that $a_n : \mathbb{N} \to \mathbb{Z}$, and $\lim_{n\to\infty} \frac{1}{n}(a_n - a_{n-1}) = 0$.*
Suppose $f_{n,m}$ is non-negative, and a_n satisfies the generalised subadditivity relation

$$a_n + a_m \geq a_{n+f_{n,m}}.$$

Assume that $\inf_{n\geq 0} f_{n,m} = \psi_m$, and $\lim_{m\to\infty} \frac{1}{m}\psi_m = 1$. If $\sup_{n,m\geq 0} \frac{1}{m} f_{n,m} = \phi$ is finite, then

$$\lim_{n\to\infty} \tfrac{1}{n} a_n = \nu$$

exists, and, moreover, $a_n \geq \psi_n \nu$.

Proof Fix $m > 0$ and recursively define m_p by

$$m_0 = m, \quad \text{and } m_p = m_{p-1} + f_{m_{p-1},m_0} \quad \text{for } p = 1, 2, \ldots .$$

Then $m_p \geq m_{p-1} + \psi_{m_0}$ and, recursively, $m_p \geq p\psi_{m_0}$.

By the hypothesis for $p = 1$ in the above, $a_{m_1+f_{m_0,m_0}} \leq a_{m_0} + a_{m_0} = 2a_{m_0}$ and, by recursively applying the subadditive relation,

$$a_{m_p} = a_{m_{p-1}+f_{m_{p-1},m_0}} \leq a_{m_{p-1}} + a_{m_0} \leq \cdots \leq p a_{m_0}.$$

This shows that

$$\frac{1}{m_p} a_{m_p} \leq \frac{p}{m_p} a_{m_0} \leq \frac{1}{\psi_{m_0}} a_{m_0},$$

since $m_p \geq p\psi_{m_0}$. Taking the limit superior as $p \to \infty$ on the left-hand side,

$$\limsup_{p\to\infty} \frac{1}{m_p} a_{m_p} \leq \frac{1}{\psi_{m_0}} a_{m_0}. \tag{A.2}$$

Since $m_p = m_{p-1} + f_{m_{p-1},m_0}$, one has $m_p \leq m_{p-1} + \phi m_0$, by definition of ϕ.

Define $a_{n+1} - a_n = u(n)$; then $u(n) = o(n)$.

Let $n > m_0$ be the smallest n such that $m_p \leq n$; say, $n = m_p + r$. Then $r \leq \phi m_0$, and $\frac{1}{n} m_p \to 1$ as $n \to \infty$. It follows that

$$\begin{aligned} a_n = a_{m_p+r} &= a_{m_p} + \sum_{j=m_p}^{n-1} u(j) \\ &\leq a_{m_p} + r \max_{m_p \leq j \leq n} |u(j)| \leq a_{m_p} + \phi m_0 \max_{m_p \leq j \leq n} |u(j)|. \end{aligned}$$

Divide this by n and note that $u(j) = o(j)$. Take the limit superior as $n \to \infty$ on the left-hand side. Then $p \to \infty$ on the right-hand side, showing that

$$\limsup_{n\to\infty} \frac{1}{n} a_n \leq \limsup_{p\to\infty} \frac{1}{m_p} a_{m_p},$$

since $r \leq \phi m_0$. Compare this to equation (A.2) to see that

$$\limsup_{n\to\infty} \frac{1}{n} a_n \leq \frac{1}{\psi_{m_0}} a_{m_0}.$$

Take the limit inferior as $m_0 \to \infty$ on the right-hand side. This completes the proof, since $\frac{1}{m}\psi_m \to 1$. □

A.3 The generalisation by JM Hammersley

The following theorem is due to Hammersley [259].

Theorem A.4 *Let $a_n : \mathbb{N} \to \mathbb{Z}$ be a sequence.*

Let $n, m \geq M$ for some $M \in \mathbb{N}$ and suppose there exists a real-valued sequence $\langle g_n \rangle$ such that a_n satisfies the subadditive inequality

$$a_n + a_m + g_{n+m} \geq a_{n+m}, \quad \text{for all } n, m \geq M.$$

If there exists an $N_0 \in \mathbb{N}$ such that g_n is non-decreasing for all $n \geq N_0$, then

$$\sum_{n=N_0}^{\infty} \frac{1}{n(n+1)} g_n < \infty \quad \Leftrightarrow \quad \lim_{n\to\infty} \tfrac{1}{n} a_n = \upsilon \quad \text{exists.}$$

Moreover, for $n \geq N_0$,

$$\tfrac{1}{n} a_n \geq \lim_{m\to\infty} \tfrac{1}{m} a_m + \tfrac{1}{n} g_n - 4 \sum_{m=2n}^{\infty} \frac{1}{m(m+1)} g_m.$$

Proof The proof is rather lengthy and is presented as a series of claims:

CLAIM: If $\sum_{n=N_0}^{\infty} \frac{1}{n(n+1)} g_n = \infty$, then the limit $\lim_{n\to\infty} \frac{1}{n} a_n$ does not exist.
PROOF OF CLAIM: Define the functions

$$h_n = \begin{cases} g_n - g_{N_0} & \text{if } n \geq N_0; \\ 0 & \text{otherwise,} \end{cases} \quad \text{and} \quad \phi_n = \left\lceil n \sin \sum_{m=1}^{n-1} \left(\frac{h_m}{m(m+1)} \right) \right\rceil.$$

Then ϕ_n oscillates to infinity with n, and so $\lim_{n\to\infty} \frac{1}{n}\phi_n$ does not exist.

However, ϕ_n satisfies the generalised subadditivity relation. To see this, note that $|\sin x - \sin(x - \delta)| = |\delta \cos(x - \theta\delta)| \leq |\delta|$ by the mean value theorem.

Let $n, m \in \mathbb{N}$, put $N = n + m$, put $p = \frac{n}{N}$, and put $q = \frac{m}{N}$, so that $p + q = 1$.

Since $p\lceil a \rceil \leq \lceil p(a+1) \rceil$, and $\lceil a - b \rceil \geq \lceil a \rceil - \lceil b \rceil$, it follows that

$$\begin{aligned}
&p\phi_N - \phi_{pN} \\
&= p\left\lceil N \sin\left(\sum_{j=1}^{N-1} \frac{h_j}{j(j+1)} \right) \right\rceil - \left\lceil pN \sin\left(\sum_{j=1}^{N-1} \frac{h_j}{j(j+1)} - \sum_{j=pN}^{N-1} \frac{h_j}{j(j+1)} \right) \right\rceil \\
&\leq \left\lceil pN \sin\left(\sum_{j=1}^{N-1} \frac{h_j}{j(j+1)} + p \right) \right\rceil - \left\lceil pN \sin\left(\sum_{j=1}^{N-1} \frac{h_j}{j(j+1)} - \sum_{j=pN}^{N-1} \frac{h_j}{j(j+1)} \right) \right\rceil \\
&\leq \left\lceil pN \sin\left(\sum_{j=pN}^{N-1} \frac{h_j}{j(j+1)} + p \right) \right\rceil.
\end{aligned}$$

This may be simplified further as follows:

$$
\begin{aligned}
&p\phi_N - \phi_{pN} \\
&\leq \left\lceil pNh_N\left(\sum_{j=pN}^{N-1}\frac{1}{j(j+1)}+p\right)\right\rceil, \quad \text{since } h_n \text{ is non-decreasing;} \\
&\leq \left\lceil pNh_N\left(\tfrac{1}{pN}-\tfrac{1}{N}\right)+p\right\rceil, \quad \text{since } \sum_{j=pN}^{N-1}\left(\frac{1}{j(j+1)}\right)=\frac{1}{pN}-\frac{1}{N}; \\
&= \lceil(1-p)h_N+p\rceil \leq \lceil(1-p)h_N\rceil+1 \leq qh_N+2.
\end{aligned}
$$

Next, interchange p and q to obtain $q\phi_N - \phi_{qN} \leq ph_N + 2$. Add these inequalities to obtain

$$\phi_{n+m} - \phi_n - \phi_m \leq h_{n+m} + 4.$$

Hence, since $h_n = g_n - g_{N_0}$, it follows that $\phi_n + 4 - g_{N_0}$ satisfies the generalised subadditive inequality for large enough n but the limit does not exist.

This completes the proof of the claim. △

The converse of the above claim is:

CLAIM: If $\sum_{n=N_0}^{\infty}\left|\frac{1}{n(n+1)}g_n\right| < \infty$, then the limit $\lim_{n\to\infty}\frac{1}{n}a_n$ exists.

PROOF OF CLAIM: Put $M_0 = \max\{N_0, M\}$ and let $\phi_n = a_n + g_{N_0}$.

Then $\phi_{n+m} \leq \phi_n + \phi_m + h_{n+m}$ for all $n \geq M_0$, for all $m \geq M_0$ and for all $h_n \geq 0$, and non-decreasing if $n \geq M_0$.

Let $p > q > 2M_0 > 0$ be integers and define

$$M(p,q) = \max\{\phi(q), \phi(q+1), \ldots, \phi(p)\}.$$

Hence, $\phi(n) \leq M(p,q)$ if $q \leq n \leq p$.

Let r and s be natural numbers, with $r > M_0$, and $s > 5r$ and define $s = nr + z$, where $3r \leq z < 5r$, and $n = 2^{n_1} + 2^{n_2} + \cdots + 2^{n_k}$, with natural numbers $n_1 > n_2 > \ldots > n_k \geq 0$. By the generalised subadditive inequality and the bound on ϕ_n,

$$\phi_s \leq \phi_{nr} + \phi_z + h_{nr+z} \leq \phi_{nr} + M(3r,4r) + h_{4nr},$$

since $5r \leq s \leq nr + z < (n+4)r$ (and this implies $n \geq 2$, and $(n+4)r \leq 4nr$).

Since h_n is non-decreasing, and $\frac{1}{x} = \sum_{l=x}^{\infty}\frac{1}{l(l+1)}$, it follows that

$$h_{4nr} = 4nr\,h_{4nr}\sum_{l=4nr}^{\infty}\frac{1}{l(l+1)} \leq 4nr\sum_{l=4nr}^{\infty}\frac{h_l}{l(l+1)}.$$

Define $\phi_n \equiv \phi(n)$ for convenience and then repeatedly apply the generalised subadditivity relation to obtain

$$\begin{aligned}
\phi_{nr} &= \phi((2^{n_1} + 2^{n_2} + \cdots + 2^{n_k})r) \\
&\leq \phi(2^{n_1}r) + \phi((2^{n_2} + \cdots + 2^{n_k})r) + h((2^{n_1} + 2^{n_2} + \cdots + 2^{n_k})r) \\
&\leq \phi(2^{n_1}r) + \phi((2^{n_2} + \cdots + 2^{n_k})r) + h\big(2^{n_1+1}r\big) \\
&\leq \cdots \\
&\leq \sum_{j=1}^{k} \big(\phi(2^{n_j}r) + h\left(2^{n_j+1}r\right)\big).
\end{aligned} \tag{A.3}$$

The generalised subadditive relation implies that $\phi(2^\nu r) \leq 2\phi(2^{\nu-1}r) + h(2^\nu r)$. Multiply this by $2^{n_j-\nu}$ to obtain

$$\phi(2^\nu r)2^{n_j-\nu} \leq 2 \cdot 2^{n_j-\nu}\phi(2^{\nu-1}r) + 2^{n_j-\nu}h(2^\nu r).$$

Sum this last equation over $\nu = 1, 2, \ldots, n_j$; then

$$\begin{aligned}
\sum_{\nu=1}^{n_j} 2^{n_j-\nu}\phi(2^\nu r) &\leq \sum_{\nu=1}^{n_j} 2^{n_j-\nu}\left(2\phi(2^{\nu-1}r) + h(2^\nu r)\right) \\
&= \sum_{\nu=0}^{n_j-1} 2^{n_j-\nu}\phi(2^{\nu-1}r) + \sum_{\nu=1}^{n_j} 2^{n_j-\nu}h(2^\nu r).
\end{aligned}$$

The result is that

$$\phi(2^{n_j}r) \leq 2^{n_j}\phi(r) + \sum_{\nu=1}^{n_j} 2^{n_j-\nu}h(2^\nu r),$$

and this is trivially valid if $n_j = 0$, and so is valid for all $n_j \geq 0$. Substitute this now into equation (A.3) above. Then

$$\begin{aligned}
\phi(nr) &\leq \sum_{j=1}^{k}\left(2^{n_j}\phi(r) + h(2^{n_j+1}) + \sum_{\nu=1}^{n_j} 2^{n_j-\nu}h(2^\nu r)\right) \\
&\leq \sum_{j=1}^{k} 2^{n_j}\phi(r) + \sum_{m=0}^{n_1}\left(h(2^{m+1}r) + \sum_{\nu=1}^{m} 2^{m-\nu}h(2^\nu r)\right) \\
&\leq n\phi(r) + \sum_{m=0}^{n_1+1} 2^{n_1+1-m}h(2^m r) \leq n\phi(r) + 2n\sum_{m=1}^{n_1+1} 2^{-m}h(2^m r).
\end{aligned}$$

In addition,

$$\begin{aligned}\sum_{m=1}^{n_1+1} 2^{-m} h(2^m r) &= 2r \sum_{m=1}^{n_1+1} h(2^m r) \left(\sum_{l=2^m r}^{(2^{m+1}r-1)} \frac{1}{l(l+1)} \right) \\ &\leq 2r \sum_{m=1}^{n_1+1} \left(\sum_{l=2^m r}^{(2^{m+1}r-1)} \frac{h(l)}{l(l+1)} \right) \\ &= 2r \sum_{l=2r}^{(2^{n_1+2}r-1)} \frac{h(l)}{l(l+1)} \leq 2r \sum_{l=2r}^{4nr-1} \frac{h(l)}{l(l+1)}.\end{aligned}$$

Taken together, this shows that

$$\phi_s \leq n\phi_r + 4nr \sum_{l=2r}^{\infty} \frac{h_l}{l(l+1)} + M(3r, 4r).$$

Divide this by $s = nr + z$ and let $n \to \infty$ with r and z fixed. Then take the limit superior of the left-hand side as $s \to \infty$:

$$\limsup_{s\to\infty} \tfrac{1}{s}\phi_s \leq \tfrac{1}{r}\phi_r + 4 \sum_{l=2r}^{\infty} \frac{1}{l(l+1)} h_l.$$

Taking the limit inferior as $r \to \infty$ on the right-hand side establishes existence of the limit, since the series is absolutely convergent. Since the series is convergent, the limit is also finite.

Finally, $a_n = \phi_r - g(N_0)$, and so

$$\lim_{n\to\infty} \tfrac{1}{n} a_n \leq \tfrac{1}{r} a_r + 4 \sum_{l=2r}^{\infty} \frac{1}{l(l+1)} h_l - \tfrac{1}{r} g(N_0).$$

Since $g(n)$ is non-decreasing, replace $g(N_0)$ by $g(r)$ in the above to obtain the claimed bound. □

A.4 A ratio limit theorem by H Kesten

The following ratio limit theorem is a key ingredient in the proof of ratio limits such as in equations (1.3) and (1.8). The theorem works if one replaces $n+2$ by $n+1$ everywhere (or by relabelling the functions such that $2n$ is replaced by n).

Theorem A.5 *Let $a_n : \mathbb{N} \to \mathbb{R}$ be a sequence of positive numbers ($a_n > 0$) and define the ratio $\phi_n = \frac{a_{n+2}}{a_n}$. Assume that*

- $\lim_{n\to\infty} a_n^{1/n} = \mu$;
- $\liminf_{n\to\infty} \phi_n > 0$;
- *there exists an $A > 0$ such that $\phi_n \phi_{n+2} \geq \phi_n^2 - \frac{1}{n} A$ for all sufficiently large values of n.*

Then the limit

$$\lim_{n\to\infty} \phi_n = \mu^2 \text{ exists.}$$

Proof The fact that $\liminf_{n\to\infty} \phi_n > 0$, and the existence of an $A > 0$ such that $\phi_n\phi_{n+2} \geq \phi_n^2 - \frac{1}{n}A$ for all sufficiently large n, imply that there exists a $B > 0$ such that

$$\phi_{n+2} \geq \phi_n - \tfrac{1}{n}B \tag{A.4}$$

for all sufficiently large n. Define $\sigma_n = \phi_n - \mu^2$.

CLAIM: The limit superior $\limsup_{n\to\infty} \sigma_n \leq 0$.

PROOF OF CLAIM: Assume that $\limsup_{n\to\infty} \sigma_n > 0$. Then there exists an infinite sequence $\langle n_j \rangle$ in $\mathbb{N}$ such that $\lim_{j\to\infty} \sigma_{n_j} = \epsilon > 0$. Define, for each $j \in \mathbb{N}$, $M_j = \left\lfloor \frac{1}{2B} n_j \sigma_{n_j} \right\rfloor$. It follows that $M_j \to \infty$ as $j \to \infty$.

For sufficiently large j, and for every $0 \leq k < M_j$, one has

$$\phi_{n_j+2k} \geq \phi_{n_j} - \tfrac{k}{n_j} B \geq \mu^2 + \sigma_{n_j} - \tfrac{1}{n_j} M_j B \geq \mu^2 + \tfrac{1}{2}\sigma_{n_j}.$$

Hence,

$$\prod_{k=0}^{M_j-1} \phi_{n_j+2k} = \left(\tfrac{1}{a_{n_j}}\right) a_{n_j+2M_j} \geq \left(\mu^2 + \tfrac{1}{2}\sigma_{n_j}\right)^{M_j}.$$

Take the power $\frac{1}{M_j}$ and then take $j \to \infty$. This shows that $\mu^2 \geq \mu^2 + \frac{1}{2}\epsilon^2$. This is a contradiction. △

CLAIM: The limit inferior $\liminf_{n\to\infty} \sigma_n \geq 0$.

PROOF OF CLAIM: The proof is similar to that of the last claim.

Suppose that $\liminf_{n\to\infty} \sigma_n < 0$. Since σ_n is bounded below, there exist an $\epsilon > 0$ and a sequence $\langle n_j \rangle$ such that $\sigma_{n_j} < 0$ and $\lim_{j\to\infty} \sigma_{n_j} = -\epsilon$.

We have already shown that $\phi_{n+2} \geq \phi_n - \frac{1}{n}B$, and one may assume without loss of generality that $B \geq \mu^2$ (by scaling the ϕ_n, if necessary).

Define, for each $j \in \mathbb{N}$, $N_j = \left\lfloor \frac{1}{4B} n_j \left|\sigma_{n_j}\right| \right\rfloor$. Since $-\mu^2 < \sigma_{n_j} < 0$, $N_j \leq \frac{1}{4B} n_j \mu^2 \leq \frac{1}{4} n_j$. For every $0 < k \leq N_j$ and for j large enough, it follows from equation (A.4) that

$$\phi_{n_j-2k} \leq \phi_{n_j} + \left(\frac{k}{n_j - 2k}\right) B \leq \mu^2 - \left|\sigma_{n_j}\right| + \frac{n_j \left|\sigma_{n_j}\right|}{4(n_j - \frac{1}{2}n_j)} = \mu^2 - \tfrac{1}{2}\left|\sigma_{n_j}\right|.$$

Hence,

$$\prod_{k=1}^{N_j} \phi_{n_j-2k} = \frac{a_{n_j}}{a_{n_j} - 2N_j} \leq \left(\mu^2 - \tfrac{1}{2}\left|\sigma_{n_j}\right|\right)^{M_j}.$$

Take the power $\frac{1}{N_j}$ and then take $j \to \infty$. This gives the contradiction $\mu^2 \leq \mu^2 - \frac{1}{2}\epsilon$. This completes the proof of the claim. △

By the two claims, the theorem is proven. □

APPENDIX B

CONVEX FUNCTIONS

Free energies $\mathcal{F}$ of models of interacting lattice clusters are convex functions (as shown in theorem 3.3). This has certain consequences; namely, that $\mathcal{F}$ is differentiable almost everywhere (if it is finite), and finite size approximations to the free energy of an infinite system may be shown to converge to the free energy almost everywhere. The convexity of $\mathcal{F}$ is related to the stability of the thermodynamic equilibrium of a system.

If f is a convex function, then $-f$ is a concave function, and with appropriate reinterpretation, all results for convex functions apply to concave functions. There are numerous classical and more recent references on convex functions; see, for example, the book by GH Hardy, JE Littlewood and G Polya [274], or references [43, 497], for more details.

B.1 Convex functions and the midpoint condition

An extended real-valued function $f : \mathbb{R} \to \mathbb{R} \cup \{\infty\}$ is *convex* on $I = [a, b]$ if, for each $0 \leq \lambda \leq 1$, and $a \leq x < y \leq b$,

$$\lambda f(x) + (1-\lambda) f(y) \geq f(\lambda x + (1-\lambda) y) \,. \tag{B.1}$$

This reduces to the *midpoint condition*

$$f(x) + f(y) \geq 2 f\big(\tfrac{1}{2}(x+y)\big) \tag{B.2}$$

if $\lambda = \frac{1}{2}$.

Convexity may also be defined in terms of the epigraph of f. This is the subset in $\mathbb{R} \times \mathbb{R}$ defined by

$$\operatorname{epi} f = \{(x, y) \mid y \geq f(x)\}. \tag{B.3}$$

The function f is *closed* if $\operatorname{epi} f$ is a closed set and is convex if and only if $\operatorname{epi} f$ is a convex set in $\mathbb{R}^2$, as may be seen immediately from the definition of convexity in equation (B.1). Observe that, if f is a lower semicontinuous extended real-valued function, then f is necessarily closed. (A function f is lower semicontinuous at a point x_0 if $\liminf_{x \to x_0} f(x) \geq f(x_0)$).

Denote the Lebesgue (outer)-measure by μ. In the next theorem convex functions are shown to be continuous and thus Lebesgue measurable.

Theorem B.1 *Suppose that f is convex on a closed interval I. If f is bounded in I, then f is continuous on I, except perhaps at the endpoints of I.*

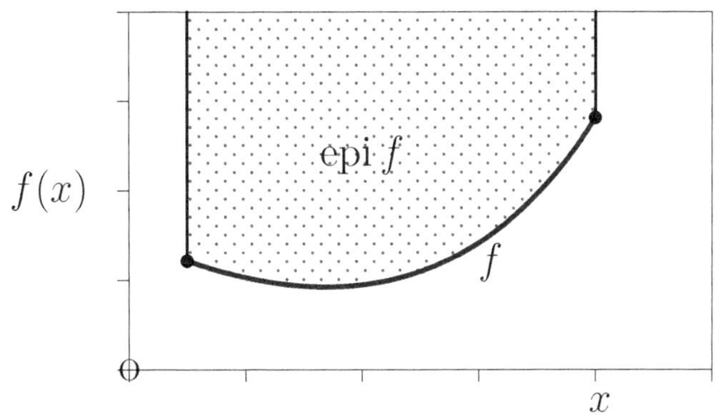

FIG. B.1. The epigraph of a function f is the shaded region with boundary f and vertical lines as indicated. The function f is convex if and only if epi$f$ is a convex set in the plane.

Proof Put $I = [a, b]$. Then f is bounded in (a, b), say $|f| < C$ in (a, b).

Let $x \in (a, b)$. Choose $n > m$ in $\mathbb{N}$, and $\delta > 0$ a small real number, such that $x + n\delta \in (a, b)$. Then

$$f(x + m\delta) = f\big(\tfrac{1}{n}(m(x + n\delta) + (n - m)x)\big) \leq \tfrac{m}{n} f(x + n\delta) + \tfrac{n-m}{n} f(x).$$

This can be rearranged into

$$\tfrac{1}{n}(f(x + n\delta) - f(x)) \geq \tfrac{1}{m}(f(x + m\delta) - f(x)),$$

and, if $\delta \to -\delta$, then

$$\tfrac{1}{m}(f(x) - f(x - m\delta)) \geq \tfrac{1}{n}(f(x) - f(x - n\delta)).$$

Put $m = 1$ and note that $f(x + \delta) + f(x - \delta) \geq 2f(x)$, and $|f| < C$;

$$\tfrac{1}{n}(C - f(x)) \geq f(x + \delta) - f(x) \geq f(x) - f(x - \delta) \geq \tfrac{1}{n}(f(x) - C).$$

Next, take $\delta \to 0$ and $n \to \infty$ such that $x \pm n\delta \in (c, d)$, and $n\delta \to 0$. Then, by the squeeze theorem for limits, f is left- and right-continuous at x. $\square$

Theorem B.2 *Suppose that f satisfies the midpoint condition on a closed interval I. If f is bounded above in some open interval $J \subset I$, then f is convex and bounded in J, and $f(x) = +\infty$ in $I \setminus J$.*

Proof The proof proceeds by generalising the midpoint condition through proving the following two claims.

CLAIM: If $\lambda_r \in \mathbb{Q}$, and for $x, y \in I$, then

$$\lambda_r f(x) + (1 - \lambda_r) f(y) \geq f(\lambda_r x + (1 - \lambda_r) y).$$

PROOF OF CLAIM: Put $m = 2^n$ and consider m points $\{x_i\}_{i=1}^{m}$ in I.

Then repeated application of the midpoint condition shows that

$$f(x_1) + f(x_2) + \cdots + f(x_m) \geq m f\big(\tfrac{1}{m}(x_1 + x_2 + \cdots + x_m)\big) \qquad \text{(B.4)}$$

If this relation is true for $m-1$, then it will be true for all $m \in \mathbb{N}$.

Consider $m-1$ points $\{x_i\}_{i=1}^{m-1}$ and put $x_m = \frac{1}{m-1}(x_1 + x_2 + \cdots + x_{m-1})$. Then $mx_m = (m-1)x_m + x_m = (x_1 + x_2 + \cdots + x_m)$ and, by equation (B.4), it follows that

$$m f(x_m) = m f\big(\tfrac{1}{m}(x_1 + x_2 + \cdots + x_m)\big) \leq f(x_1) + f(x_2) + \cdots + f(x_m)$$

In other words, subtracting $f(x_m)$ from both sides gives

$$f(x_1) + f(x_2) + \cdots + f(x_{m-1}) \geq (m-1) f\Big(\tfrac{1}{m-1}(x_1 + x_2 + \cdots + x_{m-1})\Big).$$

This shows that equation (B.4) is true for any $m \in \mathbb{N}$.

If $p+q = m$, and $x = x_1 = x_2 = \cdots = x_p$, while $y = x_{p+1} = x_{p+2} = \cdots = x_m$, then a consequence of equation (B.4) is that

$$\lambda_p f(x) + (1 - \lambda_p) f(y) \geq f(\lambda_p x + (1-\lambda_p) y), \qquad \text{(B.5)}$$

for any rational number $\lambda_p = \frac{p}{m}$. This completes the proof of the claim. $\triangle$

If f is continuous, then a limit $\lambda_p \to \lambda$ can be taken through rational numbers, and this will complete the proof. If f is not necessarily continuous, then prove the following claim.

Claim: It may be assumed that f is bounded on $J \subseteq I$ and infinite on $I \setminus J$.
Proof of claim: Put $I = [a,b]$, put $J = (c,d)$, and assume there is a $y \in I \setminus J$ such that $|f(y)| < \infty$. Without loss of generality, suppose that $a < y < c$.

Let $x \in (y,c)$ and choose integers $p > q$ such that $\chi = y + \frac{p}{q}(x-y) \in (c,d)$.

By equation (B.5),

$$f(x) = f(\tfrac{1}{p}(q\chi + (p-q)y)) \leq \tfrac{q}{p} f(x) + \tfrac{p-q}{p} f(y) < \infty.$$

Thus, $f(x) < \infty$, and put $c = y$ to find $f(x)$ bounded in an enlarged interval (c,d).

A similar argument shows that (c,d) can be grown by increasing d if there are points y with $d < y < b$, with $|f(y)| < \infty$.

By theorem B.1, f is continuous in J since it is bounded on J. $\triangle$

By theorem B.1 and the last claim, f is continuous in J. □

B.2 Derivatives of convex functions

Finite convex functions on open intervals are continuous and hence measurable.

Theorem B.3 *Suppose that f is a finite and convex function on an open interval I. Then f has finite and non-decreasing left- and right-derivatives in I. Moreover, if $D^- f = \frac{d^-}{dx} f$ is the left-derivative of f, and $D^+ f = \frac{d^+}{dx} f$ is the right-derivative of f, then $D^- f \leq D^+ f$.*

Proof Put $I = (a, b)$ and choose $x \in I$. Let $n \in \mathbb{N}$ and let $h > 0$ small, such that $x + (\frac{n+1}{n})h \in I$. Then define $q(x, h) = f(x+h) - f(x)$. It follows that

$$q(x+\tfrac{h}{n}, h) - q(x, h) = \sum_{i=0}^{n-1} \left(f(x+\tfrac{h}{n}(i+2)) - 2f(x+\tfrac{h}{n}(i+1)) + f(x+\tfrac{h}{n}i)\right) \geq 0,$$

by convexity of f. This shows that $q(x+\frac{h}{n}, h) \geq q(x, h)$. Repeat the above to obtain a sequence of inequalities:

$$q(x, h) \leq q(x+\tfrac{h}{n}, h) \leq \cdots \leq q(x+\tfrac{mh}{n}, h) \quad \text{for some } m \in \mathbb{N}.$$

Choose $\frac{m}{n} \in \mathbb{Q}$ such that $\frac{m}{n} \to \frac{\delta}{n}$ for some $\delta > 0$. By continuity of f, $q(x, h) \leq q(x+\delta, h)$. Thus $q(x, h)$ is a non-decreasing function of x.

It remains to construct the derivatives of f.

Claim: The right-derivative of f exists and is finite in I.
Proof of claim: It follows from the above that

$$q(x, \tfrac{h}{n}) \leq q(x+\tfrac{h}{n}, \tfrac{h}{n}) \leq \cdots \leq q(x+(n-1)\tfrac{h}{n}, \tfrac{h}{n}).$$

Thus, for $0 < m < n$,

$$\frac{1}{m}\sum_{i=0}^{m-1} q(x+\tfrac{ih}{n}, \tfrac{h}{n}) \leq \frac{1}{n}\sum_{i=0}^{n-1} q(x+\tfrac{ih}{n}, \tfrac{h}{n}).$$

These sums telescope to

$$\tfrac{n}{mh}\left(f(x+\tfrac{1}{n}mh) - f(x)\right) \leq \tfrac{1}{h}(f(x+h) - f(x))$$

if $q(x, h) = f(x+h) - f(x)$ is substituted. Let $0 < h' < h$ and suppose that $\frac{m}{n}$ is a sequence of rational numbers which converges to $\frac{1}{h}h'$. Taking $n \to \infty$ and assuming that $x < y$ give

$$\tfrac{1}{h'}(f(x+h') - f(x)) \leq \tfrac{1}{h}(f(x+h) - f(x)) \leq \tfrac{1}{h}(f(y+h) - f(y)) \qquad \text{(B.6)}$$

by continuity of f and since $q(x, h)$ is non-decreasing with x. This shows that $\frac{1}{h}(f(x+h) - f(x))$ is a non-decreasing function of h. Thus, $\lim_{h \to 0^+} \frac{1}{h}(f(x+h) - f(x))$ exists in the extended real numbers, for every $x \in (a, b)$.

Taking $h' \to 0$ in equation (B.6) with y and h fixed shows that $D^+f < \infty$. Since $x < y$, taking $h \to 0^+$ in equation (B.6) also shows that $D^+f(x)$ is non-decreasing with x.

An argument similar to the above gives, for $0 > h' > h$, and $x > y$,

$$\tfrac{1}{h'}(f(x+h') - f(x)) \geq \tfrac{1}{h}(f(x+h) - f(x)) \geq \tfrac{1}{h}(f(y+h) - f(y)).$$

A similar argument to the above proves the existence of the left-derivative D^-f, and shows that it is finite and non-decreasing.

Finally, since $q(x,h)$ is non-decreasing with x, and for $h > 0$,

$$-\tfrac{1}{h}(f(x-h) - f(x)) = \tfrac{1}{h}(f(x) - f(x-h)) \le \tfrac{1}{h}(f(x+h) - f(x)).$$

Taking $h \to 0^+$ shows that $D^- f \le D^+ f$, as required. This proves the claim. △

This completes the proof. □

Thus, a finite convex function on an open interval has left- and right-derivatives everywhere. These derivatives are non-decreasing and, hence, measurable.

Finite convex functions are differentiable almost everywhere. This follows from a standard theorem in measure theory, namely, that a non-decreasing real-valued function on a closed interval is differentiable almost everywhere in the interval; see for example reference [497].

The proof that a convex function is differentiable almost everywhere uses *Vitali converings.* Let $I = [a, b]$ be an interval. The *length* of I is $l(I) = b - a$. A set E is *covered (in the sense of Vitali)* if there is an (infinite) collection of intervals $\mathcal{I} = \{I_i\}$ such that $E \subset \bigcup_i I_i$ and, for every $x \in E$, and $\epsilon > 0$, there is an $I \in \mathcal{I}$ such that $x \in I$ and $l(I) < \epsilon$; see for example reference [497].

The critical property about Vitali's coverings is given by Vitali's theorem.

Theorem B.4 (Vitali's theorem) *Let E be a finite measure subset of the real line and let $\mathcal{I}$ be a collection of intervals which cover E in the sense of Vitali. Given an $\epsilon > 0$, there is a finite and disjoint collection of intervals in $\mathcal{I}$, say $\langle I_1, I_2, \ldots, I_N \rangle$, such that*

$$\mu^*\left[E \setminus \cup_{i=1}^N I_n\right] < \epsilon,$$

where μ^* is Lebesgue outer measure. □

To show that finite convex functions are differentiable almost everywhere, it is only necessary to show that non-decreasing functions are differentiable almost everywhere. To see this, suppose that $x < y < z$ and that f is a convex function. Put $\lambda = \frac{y-z}{x-z}$. Then $0 < \lambda < 1$, and

$$\lambda f(x) + (1-\lambda) f(z) \ge f(y).$$

If $f(x) \le f(y)$, then replacing $f(x)$ by $f(y)$ in the last inequality gives $f(y) \le f(z)$.

In other words, if f is non-decreasing in an interval containing a point x, then it remains non-decreasing for larger values of x. A similar argument shows that, if f is non-increasing on an interval containing x, then it is non-increasing for all smaller values of x.

Hence, a convex function is either non-increasing or non-decreasing, or first non-increasing and then non-decreasing.

Theorem B.5 *Let f be a non-decreasing real-valued function on $[a, b]$. Then f is differentiable almost everywhere.*

Proof Define the following (Dini)-derivatives of f:

$$\begin{aligned}
D^+f(x) &= \limsup_{h\to 0^+} \tfrac{1}{h}(f(x+h)-f(x))\,; \\
D^-f(x) &= \limsup_{h\to 0^+} \tfrac{1}{h}(f(x)-f(x-h))\,; \\
D_+f(x) &= \liminf_{h\to 0^+} \tfrac{1}{h}(f(x+h)-f(x))\,; \text{ and} \\
D_-f(x) &= \liminf_{h\to 0^+} \tfrac{1}{h}(f(x)-f(x-h))\,.
\end{aligned}$$

Since f is non-decreasing, $D^+f(x) \geq D_-f(x)$, and similar relations are valid between the other derivatives.

By theorem B.3, $D^+f(x) = D_+f(x)$, and $D^-f(x) = D_-f(x)$.

For each pair of rational numbers (u, v), define the set

$$E_{u,v} = \{x | D^+f(x) > u > v > D_-f(x)\}\,.$$

Then $F = \bigcup_{u,v\in\mathbb{Q}} E_{u,v}$ is the set of points where $D^+f(x) > D_-f(x)$.

CLAIM: The set F is a null set: $\mu F = 0$

PROOF OF CLAIM: It is shown that $\mu E_{u,v} = 0$ for all pairs $(u, v) \in \mathbb{Q}^2$.

Put $\mu E_{u,v} = s$ and let $\epsilon > 0$. There is an open set $U \supseteq E_{u,v}$ such that $\mu U < s+\epsilon$. By the definition of $D_-f(x)$, there is an h_x such that $I_x = [x-h_x, x] \subset U$, and $f(x) - f(x-h_x) < vh_x$, for each $x \in U$.

The set of intervals $[x - h_x, x]$ is a Vitali covering, and there is a finite disjoint set $\{I_1, I_2, \ldots, I_N\}$ which almost covers $E_{u,v}$: if I_j^o is the interior of interval I_j, then $\left(\bigcup I_j^o\right) \cap U > s-\epsilon$. Put $A = \left(\bigcup I_j^o\right) \subseteq U$. Then $\mu A \leq \mu U \leq s+\epsilon$.

Let $I_j = [x_j - h_j, x_j]$ and put $\ell(I_j) = h_j$, the length of the j-th interval. Sum over the intervals:

$$\sum_{i=1}^{N} (f(x_i) - f(x_i - h)) < v \sum_{i=1}^{N} h_i < v\mu A < v(s+\epsilon). \tag{B.7}$$

Each $y \in A$ is the left endpoint of $J_y = (y, y+k) \in I_n$ for some value of n and for small enough k (k can be found such that $f(y+k) - f(y) > uk$, by the definition of $D^+f(y)$).

The collection of intervals $\{J_y\}$ is a Vitali covering of A, and there is a finite disjoint collection $\{J_1, J_2, \ldots, J_m\}$ almost covering A: $(\bigcup J_j) \cap A > s - 2\epsilon$.

Define $J_j = (y_j + k_j, y_j)$ so that $\ell(J_j) = k_j$. By the definition of $D^+f(y)$,

$$\sum_{i=1}^{m} [f(y_i + k_i) - f(y_i)] > u \sum_{i=1}^{m} k_i > u(s - 2\epsilon). \tag{B.8}$$

Each of the J_j is contained in some interval I_i. Sum over those $J_j \subset I_i$ and index the J_j by j_i to obtain

$$\sum_{j_i=1} (f(y_{j_i}+k_{j_i}) - f(y_{j_i})) \leq f(x_i) - f(x_i - h_i),$$

since f is increasing. Thus, by summing this over all i, one obtains

$$\sum_{j=1}^{m} (f(y_j + k_j) - f(y_j)) \leq \sum_{i=1}^{n} (f(x_i) - f(x_i - h_i)).$$

This shows by equations (B.7) and (B.8) that $v(s+\epsilon) \geq u(s-2\epsilon)$.

Take $\epsilon \to 0^+$ to find that $vs \geq us$. Since $u > v$, this implies that $s = 0$, with the result that $\mu E_{u,v} = 0$. This completes the proof of the claim. $\triangle$

By this claim, $\mu F = 0$, and $D^+ f(x) = D_- f(x)$, for almost every x. This shows that f is differentiable almost everywhere (since it is real valued and finite). This completes the proof. □

Corollary B.6 *If f is a real-valued non-decreasing function on $[a,b]$, then f' is measurable.*

Proof By theorem B.5, $g = f'$ is defined almost everywhere in $[a,b]$, except on a null set A (a zero-measure set).

Define $g_n(x) = n\big(f(x+\frac{1}{n}) - f(x)\big)$ for $x \in [a,b] \backslash A$ and put $f(x) = f(a)$ if $x \leq a$, and put $f(x) = f(b)$ if $x \geq b$. Then $g_n(x) \to g(x)$ pointwise for almost all x, so that g is measurable. Since $g = f'$ almost everywhere in $[a,b]$, this shows that f' is measurable in $[a,b]$. □

Since a convex function is either non-increasing, non-decreasing or first non-increasing and then non-decreasing, it follows from theorem B.5 and corollary B.6 that, if f is convex on $[a,b]$, then f is differentiable almost everywhere in (a,b), and f' is measurable in $[a,b]$. Since the Lebesgue measure is σ-finite, these properties extend naturally to all of $\mathbb{R}$.

By theorem B.3, if f is a convex function, then f' is non-decreasing, and so f' is continuous and differentiable almost everywhere.

Theorem B.7 *If f is a real-valued convex function, then f is differentiable everywhere, except on a countable set of points.*

Proof Since f is convex, f' is non-decreasing.

Thus, f' is continuous almost everywhere.

To see that the number of points where f' is not continuous is countable, consider first the case that f is convex and real valued (and so finite) on $[a,b]$. Put $A = f(b) - f(a)$ and define the jump function

$$J(x) = \lim_{h\to 0^+} f'(x+h) - \lim_{h\to 0^+} f'(x-h).$$

If $J(x) = 0$, then f' is continuous at x. If $J(x) > 0$, then f' is said to be *discontinuous* at x.

The *gap* at x is the interval $I_x = (f'(x^-), f'(x^+))$, where the left- and right-limits are $f'(x^\pm) = \lim_{h\to 0^+} f'(x\pm h)$. It follows that $J(x) = \text{length}\,(I_x)$, the length of this interval.

Since f' is non-decreasing, the intervals I_x compose a disjoint family.

Put N_k equal to the number of intervals with $J(x) \geq 2^{-k}$. Then $2^{-k}N_k \leq A$, or $N_k \leq 2^k A < \infty$. Thus, N_k is finite for each k. Since the number of points where f' is discontinuous is less than $\bigcup_k N_k$ the number of I_x is countable. Hence, the number of points where f' is discontinuous is countable. This completes the proof. □

Thus, finite real-valued convex functions are differentiable except on a countable sets of points.

B.3 Convergence

If $\langle f_n \rangle$ is a sequence of convex functions, then, by Fatou's lemma, $\liminf f_n$ is measurable. If, in addition, $f_n \to f$ pointwise almost everywhere (or in measure), then f is measurable. By Lusin's theorem, there is a continuous function g such that $f = g$, except in a set of small Lebesgue measure.

More generally, if $\langle f_n \rangle$ is a Cauchy in measure sequence of functions on a finite measure set, then there exists a measurable f such that $f_n \to f$ in measure on the finite measure set. By the σ-finiteness of the Lebesgue measure, these results extend to all of $\mathbb{R}$.

Lemma B.8 *Suppose that $\langle f_n \rangle$ is a sequence of convex functions converging pointwise to a limit f almost everywhere. Then f is convex.*

Proof Without loss of generality the ae condition may be ignored.

Suppose first that f_n and f are defined on the closed interval $[a, b]$.

Suppose that f is not convex on $[a, b]$. That is, there exist points $c, d \in [a, b]$ such that $f(c) + f(d) < 2f(\frac{1}{2}(c+d))$.

Put $A = 2f(\frac{1}{2}(c+d)) - f(c) - f(d) > 0$. Since $f_n(x)$ is convex,

$$f_n(c) + f_n(d) \geq 2f_n(\tfrac{1}{2}(c+d)).$$

For every $\epsilon > 0$, there exist N_c, N_d and N in $\mathbb{N}$ such that

(1) for all $n > N_c$, $|f_n(c) - f(c)| < \epsilon$;
(2) for all $n > N_d$, $|f_n(d) - f(d)| < \epsilon$; and
(3) for all $n > N$, $\left|f_n(\frac{1}{2}(c+d)) - f(\frac{1}{2}(c+d))\right| < \epsilon$.

However,

$$\begin{aligned} |f_n(c) - f(c)| + |f_n(d) - f(d)| &\geq |f_n(c) + f_n(d) - f(c) - f(d)| \\ &\geq 2\left(f_n(\tfrac{1}{2}(c+d)) - f(\tfrac{1}{2}(c+d))\right) + A. \end{aligned}$$

Take $n > \max\{N_c, N_d, N\}$. Then, by (1), (2) and (3) above, $2\epsilon \geq A - 2\epsilon$, or $A \leq 4\epsilon$. This shows that $A \leq 0$, since $\epsilon > 0$ is arbitrary. This is a contradiction, and hence f is convex on $[a, b]$. Since the Lebesgue measure is σ-finite, this extends to $\mathbb{R}$. □

By lemma B.8, the limit of a sequence of convex functions is itself convex. The limit is differentiable almost everywhere. It is also the limit of the sequence of derivatives almost everywhere:

Theorem B.9 *Suppose that $\langle f_n \rangle$ is a sequence of convex functions converging pointwise to a limit f almost everywhere. Then f is convex. Moreover, the sequence of derivatives $\langle f_n' \rangle$ converges to f' almost everywhere.*

Proof Without loss of generality the ae condition may be ignored. Put $D^+ \equiv \frac{d^+}{dx}$ and $D^- \equiv \frac{d^-}{dx}$.

By lemma B.8, f is convex and differentiable almost everywhere. Without loss of generality, assume that f and f_n are differentiable everywhere.

By theorem B.3, f_n and f have non-decreasing left- and right-derivatives everywhere.

Next, note by convexity of f_n that, for fixed $x < y$, and $\lambda \in (0,1)$,

$$\begin{aligned} f_n(x + \lambda(y - x)) &= f_n(\lambda y + (1 - \lambda)x) \\ &\le \lambda f_n(y) + (1 - \lambda) f_n(x) = f_n(x) + \lambda (f_n(y) - f_n(x)) \end{aligned}$$

Rearrange this to see that

$$\frac{f_n(x + \lambda(y - x)) - f_n(x)}{\lambda(y - x)} \le \frac{f_n(y) - f_n(x)}{y - x}.$$

Take $\lambda \to 0^+$ on the left-hand side and put $y = x + h$. This gives

$$D^+ f_n(x) \le \tfrac{1}{h}(f_n(x + h) - f_n(x)). \tag{B.9}$$

Choose an $\epsilon > 0$. Fix $x \in \mathbb{R}$. Then $f_n(x) \to f(x)$, and there is an $N_0 \in \mathbb{N}$ such that for all $n \ge N_0$,

$$|f_n(x) - f(x)| \le \epsilon. \tag{B.10}$$

Suppose that $n \ge N_0$. By equations (B.9) and (B.10),

$$D^+ f_n(x) \le \tfrac{1}{h}(f_n(x + h) - f_n(x)) \le \tfrac{1}{h}(f_n(x + h) - f(x) + \epsilon).$$

Take the limit superior of the left-hand side. This gives

$$\limsup_{n \to \infty} D^+ f_n(x) \le \tfrac{1}{h}(f(x + h) - f(x) + \epsilon).$$

Take $\epsilon \to 0^+$ and then $h \to 0^+$ to find that $\limsup_{n\to\infty} D^+ f_n(x) \le D^+ f(x)$.

A similar argument shows that $\liminf_{n\to\infty} D^- f_n(x) \ge D^- f(x)$.

Since f is differentiable almost everywhere by theorem B.5, $\lim_{n\to\infty} \frac{d}{dx} f_n(x) = f'(x)$ for almost all x, and thus $f_n' \to f'$ pointwise almost everywhere. This completes the proof. □

B.4 The Legendre transform

Suppose that f is an extended real-valued function and that $f > -\infty$. The essential domain of f is defined by the set

$$\operatorname{dom} f = \{x \in \mathbb{R} \bigm| f(x) < \infty\}. \tag{B.11}$$

If $f < \infty$, then $\operatorname{dom} f = \mathbb{R}$.

The Legendre transform of an extended real-valued function f of one variable is usually defined by the supremum

$$f^*(p) = \sup\{px - f(x) \mid \text{for } x \in \operatorname{dom} f\}. \tag{B.12}$$

The transform f^* is necessarily convex. To see this, consider

$$f^*(p) + f^*(q) \geq px + f(x) + qx + f(x) = (p+q)x + 2f(x) \tag{B.13}$$

for any $x \in \operatorname{dom} f$. Take the supremum on the right-hand side to see that $f^*(p) + f^*(q) \geq 2f(\frac{1}{2}(p+q))$.

If f is itself convex, finite and closed, then it can be recovered from its Legendre transform, since $f^{**} = f$, as will be seen below.

The above definition generalised to real-valued functions on $\mathbb{R}^n$ is the Legendre-Fenchel transformation [550].

Instead of using equation (B.12) as a definition, consider the alternative definition, where f^* is defined such that

$$p = \tfrac{d}{dx} f(x) \Leftrightarrow x = \tfrac{d}{dp} f^*(p). \tag{B.14}$$

Integrating the left-hand side with respect to x for fixed p shows that $px = f(x) + C_p$, and the right-hand side with respect to p for fixed x shows $px = f^*(p) + C'_x$. This shows that

$$px = f(x) + f^*(p), \tag{B.15}$$

a result which follows essentially by inverting the derivatives of f and f^* above.

This result is valid if f and f^* are differentiable almost everywhere, in which case the last equality is true almost everywhere. More generally, definition (B.12) is used and is defined for a more general class of (measurable) functions f.

In the case that f is convex, then it is differentiable almost everywhere, and its derivative is monotone and continuous almost everywhere. This shows that $p = \frac{d}{dx} f(x)$ has a solution $x^* = f^*(p)$ almost everywhere, which is a convex function in p.

The above can be implemented by considering the *subdifferential* of f instead of the derivative. The subdifferential of f at x is the set $\partial f(x)$ defined by

$$\partial f(x) = \{p \mid f(y) \geq f(x) + p(y-x), \text{ for all } y \in \mathbb{R}\}. \tag{B.16}$$

This defines a map $x \to \partial f(x)$, which is not a bijection but which may be inverted at p by finding an x_p such that $p \in \partial f(x_p)$. The function x_p is not necessarily unique.

This in particular shows that $f(y) - py \geq f(x_p) - px_p$ and so one may choose $x_p \in \operatorname{dom} f$ to maximise the right-hand side. This gives the definition for f^* in equation (B.12) by inverting the subdifferential of f and generalising the arguments following equation (B.14).

Lemma B.10 *The Legendre transform f^* is necessarily closed and convex.*

Proof Define $g_x(p) = px - f(x)$. Then $f^*(p)$ is the supremum of $g_x(p)$ over $x \in \mathbb{R}$. This shows that $\operatorname{epi} f^* = \bigcap_x \operatorname{epi} g_x$. The epigraph of g_x is a closed and convex set for each x. Thus, $\operatorname{epi} f^*$ is closed and convex, since it is the intersection of closed and convex sets. Hence, f^* is convex. □

Since f^* is convex, it is continuous and also differentiable ae. By theorem B.7, it is in fact differentiable everywhere except on a countable set of points in its essential domain.

Since $f > -\infty$ and f^* is convex, it follows that $f^* > -\infty$. Thus, the Legendre transform of f^* (or the *biconjugate* of f) is defined by

$$f^{**}(x) = \sup_p \{px - f^*(p) \mid \text{for } p \in \operatorname{dom} f^*\}. \tag{B.17}$$

In addition, f^{**} is convex and therefore continuous and also differentiable almost everywhere, by lemma B.10.

Theorem B.11 *For a measurable function f,*

(a) *f^{**} is convex, and $f^{**} \leq f$;*
(b) *if $f^{**} = f$, then f is closed and convex; and*
(c) *if f is closed and convex, then $f^{**} = f$ on the essential domain of f.*

Proof The function f^{**} is convex, and $\operatorname{epi} f^*$ is closed, by lemma B.10. For any $x \in \mathbb{R}$, and $p \in \operatorname{dom} f^*$, $f^*(p) \leq px - f(x)$; therefore,

$$px - f^*(p) \leq px - (px - f(x)) = f(x).$$

Take the supremum on the left-hand side for $p \in \operatorname{dom} f^*$ to see that $f^{**} \leq f$.

If $f^{**} = f$, then, by lemma B.10, f is closed and convex.

Fix $x \in \operatorname{dom} f$ and assume f is closed and convex. Then $f(x) < \infty$, and $\partial f(x) \neq \emptyset$. Let $p \in \partial f(x)$; then it follows from equation (B.16) that

$$f(y) \geq f(x) + p(y - x) \quad \text{for all } y \in \mathbb{R}.$$

Choose y such that $f^*(p) = py - f(y)$ (since f is closed, there exists such a y). Then $f^*(p) = py - f(y) \leq px - f(x)$, and this shows that $f(x) \leq px - f^*(p) \leq f^{**}(x)$. □

Hence, if f is finite and convex, then $f^{**} = f$ if f is closed. If f is a convex function on $\mathbb{R}$, then it is lower semicontinuous and so closed; thus $f^{**} = f$.

APPENDIX C

KESTEN'S PATTERN THEOREM

C.1 Patterns

A *pattern* is a finite subwalk in a self-avoiding walk.

A *pattern theorem* proves that there exists a class of patterns (called *proper patterns*) which will occur with positive density on all, except exponentially few, walks of length n, if n is large.

The definition of a pattern may be relaxed to include *events*, for example, the event that a walk fills a d-dimensional cube of side-length L and with its centre in a vertex of the walk.

In this appendix, Kesten's proof of a pattern theorem [348] for the self-avoiding walk is presented, using the approach of the presentation in reference [399].

Definition C.1 *A pattern P with vertices $(p_0, p_1, \ldots, p_N)$ is said to occur at the j-th vertex of a walk $\omega = (\omega_0, \omega_1, \ldots, \omega_n)$ if there is a vector $\vec{v} \in \mathbb{Z}^d$ such that $\omega_{j+\ell} = p_\ell + \vec{v}$ for $\ell = 0, 1, \ldots, N$. The pattern P occurs k times if there are k distinct vectors $\vec{v}_i \in \mathbb{Z}^d$ such $\omega_{j+\ell} = p_\ell + \vec{v}$ for $\ell = 0, 1, \ldots, N$ and for $i = 1, 2, \ldots, k$.* □

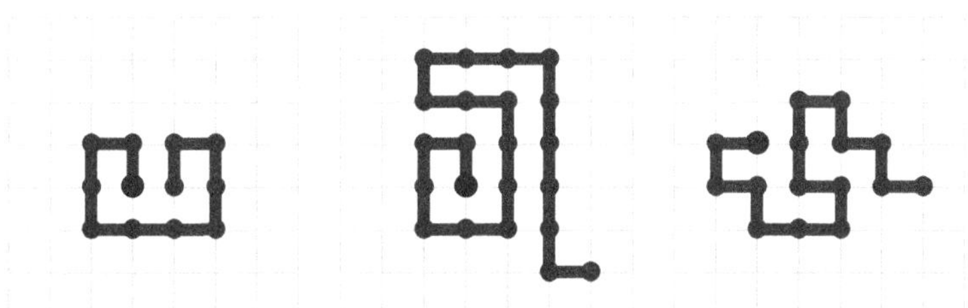

(a) (b) (c)

FIG. C.1. (a) An improper pattern. (b) A front pattern. (c) A proper internal pattern.

Not every pattern can occur in a given set of walks. For example, the pattern in figure C.1(a) cannot occur in walks of length larger than eleven steps. The pattern in figure C.1(b) can occur at most once at the beginning of a walk, and a second time at the end of a walk (with steps reversed); this is also called a *front pattern* or an *end-pattern*. The pattern in figure C.1(c) can occur an arbitrary number of times in a walk.

A pattern P is a *proper pattern* if, for every $k \in \mathbb{N}$, there is a walk on which P occurs at least k times. The following theorem characterises the class of proper patterns [261]. *The bellman's theorem* is named after a character (the bellman) in the poem *The Hunting of the Snark (An Agony in 8 Fits)* by Lewis Carroll , because the bellman proclaims that 'What I tell you three times is true' [98].

Similarly, the bellman's theorem states that, if a pattern can occur (at least) three times in a walk, then it is a proper pattern.

Theorem C.2 (The bellman's theorem) *If a pattern P occurs three times on a walk in $\mathbb{Z}^d$, then P is a proper pattern.*

Proof Denote the first three occurrences of P along a walk ω by P_1, P_2 and P_3, with P_i occurring before P_{i+1}.

Each P_i is contained in a *smallest* rectangular box B_i with faces parallel to lattice planes in $\mathbb{Z}^d$. Since the P_i are translations of the same pattern, the three boxes are translations of each other. Observe that no two of the B_i can be the same box.

Since $B_1 \neq B_2$, there is a vertex $\vec{v} \in P_1$ which is outside B_2. This vertex cannot be in $P_1 \cap P_2$, since it lies wholly outside B_2, and thus v precedes all vertices in P_2. Hence, there is a class of vertices in ω which precede P_2 and lie outside B_2. This implies that there is a last vertex v_1 outside B_2.

Similarly, there is a first vertex v_2 outside B_2 since $B_2 \neq B_3$.

Let ω_2 denote the part of ω from v_1 to v_2. Then P_2 occurs on ω_2, and only the first and last vertices of ω_2 lie outside B_2. Thus, there exists a rectangular box B_4 containing ω_2 such that the first and last vertices of ω_2 lie on the faces of B_4, and these are the only vertices of ω_2 in the faces of B_4.

If v_1 and v_2 do not lie on opposite faces of B_4, then this can be arranged by adding some extra steps to the faces of B_4 without disturbing the inside of B_4. Thus, one may assume that the vertices v_1 and v_2 lie on opposite faces of B_4.

Finally, add an extra edge to v_2 to a vertex outside B_4. This walk is itself a pattern S and is also a walk containing P at least once. The repetition $SSS\cdots S = S^N$ is a walk which contains the pattern P at least N times, for each $N \in \mathbb{N}$. This proves that P is a proper pattern. □

A *cube* is a subset

$$Q = \{\vec{x} \in \mathbb{Z}^d \mid \text{for } a_i \leq \vec{x}(i) \leq a_i + b \text{ and for all } i = 1, 2, \ldots, d\}, \qquad \text{(C.1)}$$

where $\vec{x}(i)$ is the i-th Cartesian coordinate of $\vec{x}$, and the a_i and $b > 0$ are integers. A cube has 2^d corners. If Q is a cube, then define $\overline{Q}$ as the cube which is two steps larger in all directions (that is, $a_i - 2 \leq \vec{x}(i) \leq a_i + 2 + b$ in the above). Define $\partial Q = \overline{Q} \setminus Q$.

A cube Q has radius r if $b = 2r$, and it is centred at the origin if $a_i = -r$ for each $i = 1, 2, \ldots, d$. If $\vec{v} = (v_1, v_2, \ldots, v_d)$ is a vertex in $\mathbb{Z}^d$ and if $a_i = v_i - r$, then Q is a cube of radius r centred at $\vec{v}$.

If $\omega = (\omega_0, \omega_1, \ldots, \omega_n)$ is a self-avoiding walk, then denote the k-th Cartesian coordinate of the vertex ω_i by $\omega_i(k)$. The walk ω is a *bridge* if $\omega_0(1) < \omega_i(1) \leq$

$\omega_n(1)$, for $i = 1, 2, \ldots, n$. A bridge is a *prime bridge* if there does not exist any $j \leq n$ such that $\omega_0(1) < \omega_i(1) < \omega_j(1)$, for $i = 1, 2, \ldots, j-1$. In other words, a prime bridge cannot be decomposed into two bridges by cutting it in a vertex.

A consequence of the bellman's theorem is the following definition of proper patterns. The proof is straightforward.

Theorem C.3 (Proper patterns) *Let P be a pattern. The following are equivalent.*

(a) There is a walk ω such that P occurs three times in ω.

(b) There exists a cube Q and a walk ω such that

  (1) P occurs at least once in ω,

  (2) $\omega \in Q$, and

  (3) the endpoints of ω are opposite corners of Q.

(c) There is a bridge ω such that P occurs on ω at least once.

(d) There is a prime bridge ω such that P occurs on ω at least once.

(e) P is a proper pattern. □

Suppose that Q is a cube, that P is a proper pattern, and ω is a walk. Suppose that $P = (p_0, p_1, \ldots, p_N)$, that p_0 and p_N are two corners of Q and that $\omega = (\omega_0, \omega_1, \ldots, \omega_n)$. If there exists a vector $\vec{v} \in \mathbb{Z}^d$ such that

$$\begin{aligned} \omega_{j+i} &= p_i + \vec{v}, \quad \text{for every } i = 0, 1, \ldots, N; \\ \omega_i &\notin Q + \vec{v}, \quad \text{for every } i < j, \text{ and } i > j + N, \end{aligned}$$

then (P, Q) is said to *occur* at vertex j in ω. Observe that the intersection of ω and $Q + \vec{v}$ is exactly the set of vertices of $P + \vec{v}$.

The proof of the following lemma is by induction for (a) and by direct construction for (b).

Lemma C.4 *Suppose that Q is a cube in $\mathbb{Z}^d$ and that P is a pattern in Q.*

(a) There exists a walk ω contained in Q and with endpoints on two corners of Q, such that ω visits every point in Q.

(b) Suppose $P = (p_0, p_1, \ldots, p_N)$ such that p_0 and p_N are corners of Q. Let $x, y \in \partial Q$ be two distinct outer points of ∂Q. Then there exists a self-avoiding walk $\omega = (\omega_0, \omega_1, \ldots, \omega_n)$ such that

  (1) ω is contained in $\overline{Q}$;

  (2) $\omega_0 = x$ and $\omega_n = y$;

  (3) there exists j such that $p_i = \omega_{j+i}$ for $i = 0, 1, \ldots, N$; and

  (4) $\omega_i \in \partial Q$ if $i < j$, or $i > j + N$.

  In other words, (P, Q) occurs at the j-th vertex of ω. □

Define $c_n[\leq k, (P, Q)]$ to be the number of self-avoiding walks of length n from $\vec{0}$ such that (P, Q) occurs at no more than k different vertices. Similarly, $c_n[\leq k, P]$ is the number of self-avoiding walks of length n from the origin such that the pattern P occurs at no more than k different vertices. (The function $c_n[k,(P,Q)]$ is the number of self-avoiding walks of length n from $\vec{0}$ such that (P,Q) occurs at exactly k different vertices, and so on.)

To examine the occurrence of patterns, both $c_n[\leq k,(P,Q)]$ and $c_n[\leq k, P]$ will be examined. In fact, the occurrence of the pair (P,Q) is less likely, as it requires that the walk avoids the cube Q except by passing through it in a subwalk which is exactly an occurrence of P.

A key ingredient in the proof of the pattern theorem is the following theorem.

Theorem C.5 *Let (P,Q) be a proper pattern and cube pair as in lemma C.4. If there exists an $\epsilon > 0$ such that*

$$\limsup_{n\to\infty} \left(c_n[\leq \epsilon n,(P,Q)]\right)^{1/n} < \mu_d, \text{ then } \limsup_{n\to\infty} \left(c_n[\leq \epsilon n, P]\right)^{1/n} < \mu_d.$$

Proof Let P be a proper pattern. Choose a pattern $P' \supset P$ such that (P',Q) is a proper pattern-cube pair as in C.3.

Consider walks of length n on which (P',Q) occurs no more than k times. On these walks P may occur any number of times, at least once for each occurrence of (P',Q), and then any number of other occurrences where P, but not Q, occurs.

Requiring that P cannot occur more than k times restricts the occurrences of (P',Q) to at most k times but also prevents walks on which P occurs more than k times while (P',Q) occurs less than k.

This shows that

$$c_n[\leq k, P] \leq c_n[\leq k,(P',Q)],$$

and hence, since P' is a proper pattern, the first inequality of the theorem must imply the second. This completes the proof. □

Thus, it is only necessary to show that $\limsup_{n\to\infty} \left(c_n[\leq \epsilon n,(P,Q)]\right)^{1/n} < \mu_d$ for sufficiently small $\epsilon > 0$ in order to prove the pattern theorem.

C.2 Proving Kesten's pattern theorem

Suppose ω is a self-avoiding walk of length n with vertices $(\omega_0, \omega_1, \ldots, \omega_n)$.

Define $Q(j)$ to be the cube of radius r centred at the j-th step ω_j of a walk ω. Then $\overline{Q}(j)$ and $\partial Q(j)$ are defined similarly to the definitions above.

The event E^* occurs at the vertex ω_j if $Q(j)$ is completely covered by ω (that is, there is an i for each vertex $v \in Q(j)$ such that $v = \omega_i$).

For every $k \geq 1$, E_k occurs at the vertex ω_j if at least k points of $\overline{Q}(j)$ are covered by ω. In addition, $\widetilde{E}_k$ occurs at ω_j if either E^* or E_k or both occur there.

Let E denote any of E^*, E_k or $\widetilde{E}_k$; $E(m)$ occurs at ω_j if E occurs at vertex ω_j of the subwalk with vertices $(\omega_{j-m}, \omega_{j-m+1}, \ldots, \omega_j, \omega_{j+1}, \ldots, \omega_{j+m})$. If $j < m$, then this subwalk has first vertex ω_0, and, if $j > n-m$, then this subwalk has last vertex ω_n.

The next two lemmas lie at the centre of the proof of the pattern theorem.

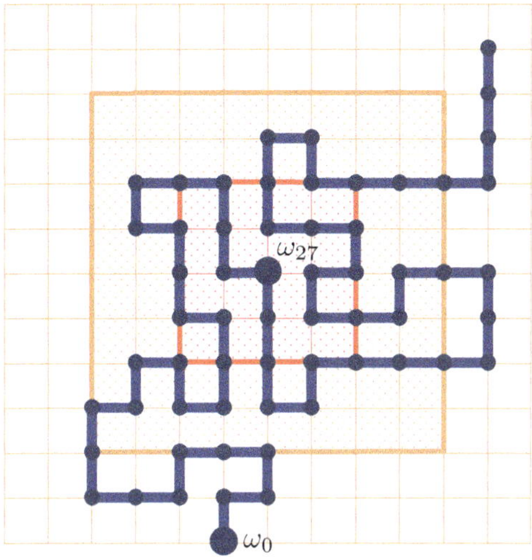

FIG. C.2. A cube $Q(j)$ centred at $j = \omega_{27}$ with radius $r = 2$. The event E^* occurs since every vertex of $Q(j)$ is occupied, and E_{47} occurs because forty–seven points of $\overline{Q}(j)$ are covered by the walk.

The first lemma states that, if E occurs on almost all walks, then $E(m)$ must occur on almost all walks (and that it occurs with density on almost all walks in a way which will be made precise).

The second lemma states that almost all walks fill cubes of given fixed side-length $2r$. This shows that some event (filling a cube) occurs with density on almost all walks.

In other words, if a walk is likely to fill a cubical volume in the lattice, then it is likely to do so often and each time in a bounded number of steps with high probability.

Lemma C.6 *Let E be any of E^*, E_k or $\widetilde{E}_k$. Then if*

$$\liminf_{n\to\infty} (c_n[0, E])^{1/n} < \mu_d,$$

then there exist an $\epsilon > 0$ and an integer m such that

$$\limsup_{n\to\infty} (c_n[\leq\epsilon n, E(m)])^{1/n} < \mu_d.$$

Proof The idea of the proof is to determine an upper bound on $c_n[\leq k, E(m)]$ by concatenating at most k subwalks which may contain occurrences of $E(m)$ with $M - k$ subwalks of the same length which cannot contain any occurrences of $E(m)$. This will give an upper bound which will be good enough to prove the bound on the limit superior.

Observe that the number of walks of length n in which E never occurs is the same as the number of walks of length n in which $E(n)$ never occurs: $c_n[0,E] = c_n[0,E(n)]$.

Thus, by the hypothesis, there exist an m and an $\epsilon > 0$ such that

$$c_m[0,E(m)] < (\mu_d(1-\epsilon))^m, \quad \text{and } c_m < (\mu_d(1+\epsilon))^m.$$

Let ω be an n-step self-avoiding walk with vertices (ω_j) and let $M = \lfloor \frac{n}{m} \rfloor$. If $E(m)$ occurs at most k times in ω, then $E(m)$ occurs in at most k of the M subwalks with vertices $(\omega_{(i-1)m}, \omega_{(i-1)m+1}, \ldots, \omega_{im})$ for $i = 1, 2, \ldots, M$.

In other words, since the number of walks with no occurrence of $E(m)$ and c_m are bounded as above, the result is the bound

$$\begin{aligned} c_n[\leq k, E(m)] &\leq \sum_{j=0}^{k} \binom{M}{j} (c_m)^j (c_m[0,E(m)])^{M-j} c_{n-Mm} \\ &\leq \mu^{Mm} c_{n-Mm} \sum_{j=0}^{k} \binom{M}{j} (1+\epsilon)^{jm} (1-\epsilon)^{Mm-jm}. \end{aligned} \tag{C.2}$$

Claim: There exist a $\rho > 0$ and a $t < 1$ such that $(c_n[\leq \rho M, E(m)])^{1/M} < t\mu^m$ for sufficiently large values of $M = \lfloor \frac{n}{m} \rfloor$.
Proof of claim: If ρ is a small positive number, then

$$\sum_{j=0}^{\rho M} \binom{M}{j} (1+\epsilon)^{jm} (1-\epsilon)^{Mm-jm} \leq (\rho M + 1) \binom{M}{\rho M} \left(\frac{1+\epsilon}{1-\epsilon} \right)^{\rho M m} (1-\epsilon)^{Mm}.$$

As $M \to \infty$ the right-hand converges to

$$\frac{(1-\epsilon)^m}{\rho^\rho (1-\rho)^{1-\rho}} \left(\frac{1+\epsilon}{1-\epsilon} \right)^{\rho M m}.$$

This is strictly less than 1 for $0 < \rho < \rho_1$, for some small positive ρ_1. Using this result in equation (C.2) proves the claim. △

For fixed values of m, the claim above implies that

$$\left(c_n \left[\tfrac{\rho M n}{n}, E(m) \right] \right)^{1/n} \leq (c_n[\rho M, E(m)])^{1/Mm} < t^{1/m} \mu_d < \mu_d,$$

and this finishes the proof. □

Lemma C.7 *The limit inferior* $\liminf_{n\to\infty} (c_n[0,E^*])^{1/n} < \mu_d$.

Proof The proof is by contradiction: assume that

$$\liminf_{n\to\infty} (c_n[0,E^*])^{1/n} = \mu_d.$$

A consequence of this assumption is the following.

CLAIM: There exists a K, with $r+3 \leq K \leq (2r+5)^d$, such that

$$\liminf_{n\to\infty} \left(c_n[0, \widetilde{E}_K]\right)^{1/n} < \mu_d, \text{ and } \liminf_{n\to\infty} \left(c_n[0, \widetilde{E}_{K+1}]\right)^{1/n} = \mu_d.$$

PROOF OF CLAIM: If E^* does not occur, then $E_{(2d+5)^d}$ cannot occur. Thus,

$$c_n[0, E^*] \leq c_n[0, \widetilde{E}_{(2d+5)^d}] \leq c_n.$$

Since $c_n[0, \widetilde{E}_k]$ is a non-decreasing function of k, this shows that

$$\lim_{n\to\infty} \left(c_n[0, \widetilde{E}_{(2d+5)^d}]\right)^{1/n} = \mu_d.$$

On the other hand, note that $c_n[0, \widetilde{E}_{r+3}] = 0$ for $n \geq r+2$. This completes the proof of the claim. △

By lemma C.6, there exist an $\epsilon > 0$ and an integer m such that

$$\limsup_{n\to\infty} \left(c_n[\leq \epsilon n, \widetilde{E}_K(m)]\right)^{1/n} < \mu_d.$$

Consider next the set of walks of length n where $\widetilde{E}_{K+1}$ never occurs but $\widetilde{E}_K(m)$ occurs at least ϵn times. More precisely, define the set of walks

$$T_n = \{\omega \mid |\omega| = n, \, E_K(m) \text{ occurs at least } \epsilon n \text{ times}, \, \widetilde{E}_{K+1} \text{ never occurs}\}.$$

Replacing $E_K(m)$ by $\widetilde{E}_K(m)$ does not change anything, since the fact that $\widetilde{E}_{K+1}$ does not occur implies that $E^*(m)$ cannot occur.

Now the number of walks in T_n is at least $c_n[0, \widetilde{E}_{K+1}] - c_n[\leq \epsilon n, \widetilde{E}_K(m)]$, and, by the claim above, it follows that

$$\lim_{n\to\infty} |T_n|^{1/n} = \mu_d.$$

In other words, this shows that there is a number K such that it is not rare to find cubes with exactly K points occupied but that there are no cubes with $K+1$ points occupied.

A contradiction will be obtained by showing that a class of walks exponentially larger than T_n but about of the same length can be constructed from the walks in T_n.

Kesten triples: Proceed by constructing Kesten triples. Suppose that ω is a walk of length n such that $\widetilde{E}_{K+1}$ never occurs but $E_K(m)$ occurs at the j_ℓ-th vertices for $\ell = 1, 2, \ldots, s$ (amongst all the occurrences of $E_K(m)$).

In addition, suppose that $j_1 > m$, that $j_s < n-m$, and that $j_{\ell+1} - m > j_\ell + m$ for $\ell = 1, 2, \ldots, s-1$. And finally, suppose that $\overline{Q}(j_\ell)$ are pairwise disjoint for $\ell = 1, 2, \ldots, s$.

Define

$$\sigma_\ell = \min\{i \mid \omega_i \in \overline{Q}(j_\ell)\}, \quad \text{and } \tau_\ell = \max\{i \mid \omega_i \in \overline{Q}(j_\ell)\}.$$

Since $E_K(m)$ occurs at the j_ℓ-th vertex, and E_{K+1} does not occur here, exactly K points of $\overline{Q}(j_\ell)$ must be occupied by vertices of ω, and those vertices must lie in the set $\{\omega_{j_\ell-m}, \omega_{j_\ell-m+1}, \ldots, \omega_{j_\ell+m}\}$.

Thus, $j_\ell - m \leq \sigma_\ell < j_\ell < \tau_\ell \leq j_\ell + m$, for $\ell = 1, 2, \ldots, s$.

Each subwalk $(\omega_{\sigma_\ell}, \ldots, \omega_{\tau_\ell})$ may be replaced by a walk which stays inside $\overline{Q}(j_\ell)$ (see lemma C.4) and completely covers all vertices of $Q(j_\ell)$. Since there are no overlaps between the cubes, this can be done independently in each cube. The result is a self-avoiding walk on which E^* occurs at least s times and with length at $n' < n + s(2r+5)^d$.

A Kesten triple (ω, ψ, J) consists of a self-avoiding walk $\omega \in T_n$, a choice of the set $J = \{j_1, j_2, \ldots, j_s\}$ (a subset of $N = \{1, 2, \ldots, n\}$ and where $E_K(m)$ occurs at each $j_\ell \in J$, and $s = \lfloor \delta n \rfloor$), and the self-avoiding walks ψ, which can be obtained by the construction in the last paragraph from ω. The value of δ will be determined later.

Counting Kesten triples: First determine a lower bound on Kesten triples.

Each cube $\overline{Q}(j)$ intersects $V = (4r+9)^d$ cubes of radius $r+2$ (because $\overline{Q}(j)$ intersects a cube of radius $r+2$ centred at x if the distance between x and ω_j is $\|\omega_j - x\|_\infty \leq 2(r+2)$; where $\|\cdot\|_\infty$ is the max-norm).

The number of Kesten triples is at least the cardinality of T_n times the number of choices for J for each $\omega \in T_n$.

Each $\omega \in T_n$ contains at least ϵn occurrences of $E_K(m)$; hence, there are at least u vertices ω_j with $j = h_1, h_2, \ldots, h_u$, with $h_i < h_{i+1}$ and where

- $u = \left\lfloor \frac{\epsilon n}{2(m+1)V} \right\rfloor - 2$;
- $E_K(m)$ occurs at the h_ℓ-th vertex of ω for $\ell = 1, 2, \ldots, u$;
- $0 < h_1 - m < h_u + m < n$, and $h_\ell + m < h_{\ell+1} - m$, for $\ell = 1, 2, \ldots, u-1$;
- and the cubes $\overline{Q}(h_\ell)$ for $\ell = 1, 2, \ldots, u$ are pairwise disjoint.

A possible choice for J is any subset of $\{h_1, h_2, \ldots, h_u\}$ of at least $\lfloor \delta n \rfloor$ elements. Put $\rho = \frac{\epsilon}{2(m+1)V}$; then the number of triples is at least

$$|\text{Kesten triples}| \geq |T_n| \binom{\lfloor \rho n \rfloor - 2}{\lfloor \delta n \rfloor}.$$

Next, derive an upper bound on the number of Kesten triples.

Note that E^* occurs at least $|J| = \lfloor \delta n \rfloor$ times on a walk ψ obtained from $\omega \in T_n$ by the construction, but E^* never occurs in ω.

On the other hand, making changes in ω to find a ψ may create extra copies of E^* in the cubes intersecting $\overline{Q}(j_\ell)$ but no more than $V|J|$ in total (since V cubes of radius $r+2$ intersects $\overline{Q}(j_\ell)$.

Thus, given a ψ, there are at most $\binom{V\lfloor \delta n \rfloor}{\lfloor \delta n \rfloor}$ possibilities for the locations of the cubes $\overline{Q}(j_\ell)$ for $\ell = 1, 2, \ldots, |J|$.

Given ψ and the locations of these $|J|$ cubes, each $\overline{Q}(j_\ell)$ determines a subwalk of ψ which replaces a piece of ω, of length at most $2m$. Thus, the total number of walks which were replaced by E^* in each cube is at most $C = \sum_{i=0}^{2m} c_i$, and this occurs in at most $\lfloor \delta n \rfloor$ disjoint cubes, if we know ψ and the locations of the cubes.

Next, if we know ω and the locations of the cubes, then J is uniquely determined.

Finally, the length of ψ is at most n', which we saw above is at most $n + s(2r+5)^d$, where $s = \lfloor \delta n \rfloor$ is the number of times E^* occurs.

This shows that

$$|\text{Kesten triples}| \leq \binom{V\lfloor \delta n \rfloor}{\lfloor \delta n \rfloor} C^{\lfloor \delta n \rfloor} \sum_{i=0}^{n+\lfloor \delta n \rfloor (2r+5)^d} c_i,$$

where the first factor on the right-hand side is a bound on the number of ways of selecting the locations of the cubes $\overline{Q}(j_\ell)$, the second factor is a bound on the number of choices for walks ω, and the last factor (the summation) is an upper bound on the number of walks ψ.

Since $\binom{V\lfloor \delta n \rfloor}{\lfloor \delta n \rfloor} \leq 2^{V\lfloor \delta n \rfloor}$, this gives the upper bound

$$|\text{Kesten triples}| \leq 2^{V\lfloor \delta n \rfloor} C^{\lfloor \delta n \rfloor} \sum_{i=0}^{n+\lfloor \delta n \rfloor (2r+5)^d} c_i.$$

Finally, compare the lower and upper bounds on the number of Kesten triples. Take the power $\frac{1}{n}$ and take $n \to \infty$. This shows that

$$\frac{\mu_d \rho^\rho}{\delta^\delta (\rho - \delta)^{\rho-\delta}} \leq \left(2^V C\right)^\delta \mu^{1+\delta(2r+5)^d}.$$

Simplify this further by putting $y = 2^V C \mu^{(2r+5)^d}$, and $t = \frac{\delta}{\rho}$. Then

$$\left(y^t t^t (1-t)^{1-t}\right)^\rho \geq 1.$$

It is now sufficient to show this is a contradiction if the left-hand side is less than 1 for small values of $t > 0$. But this must be the case since, if $f(t) = y^t t^t (1-t)^{1-t}$, then $\lim_{t\to 0^+} f(t) = 1$, and $\lim_{t \to 0^+} f'(t) = -\infty$.

This completes the proof. □

C.3 Kesten's pattern theorem

Lemmas C.6 and C.7 provide the basic results for proving Kesten's pattern theorem; the proof is completed by proving the first inequality in theorem C.5.

Theorem C.8 *Let (P,Q) be a proper pattern and cube pair as in lemma C.4. Then there exists an $\epsilon > 0$ such that*

$$\limsup_{n\to\infty} \left(c_n[\leq \epsilon n, (P,Q)]\right)^{1/n} < \mu_d.$$

Thus, by theorem C.5, there exists an $\epsilon > 0$ such that

$$\limsup_{n\to\infty} \left(c_n[\leq \epsilon n, P]\right)^{1/n} < \mu_d.$$

Proof Assume that the theorem is not true; it then follows that, for every $\epsilon > 0$ and for every proper pattern and cube pair (P,Q),

$$\limsup_{n\to\infty} \left(c_n[\leq \epsilon n, (P,Q)]\right)^{1/n} = \mu_d.$$

Assume that Q is the cube with radius r and centred at the origin.

The event E^{**} occurs at the j-th vertex of a walk ω if the cube $\overline{Q}(j)$ is completely covered by ω.

By lemmas C.6 and C.7, there exist an $\epsilon_1 > 0$ and an m_1 such that

$$\limsup_{n\to\infty} \left(c_n[\leq \epsilon_1 n, E^{**}(m_1)]\right)^{1/n} < \mu_d.$$

Choose $\epsilon > 0$, small and not specified, and let H_n denote the set of walks of length n and on which (P,Q) occurs at most ϵn times and on which $E^{**}(m_1)$ occurs at least $\epsilon_1 n$ times. Then

$$|H_n| \geq c_n[\leq \epsilon n, (P,Q)] - c_n[\leq \epsilon_1 n, E^{**}(m_1)],$$

so, by the hypothesis of the theorem,

$$\lim_{n\to\infty} H_n^{1/n} = \mu_d.$$

Construct a Kesten triple (ω, ψ, J) as follows: let ω be a walk in H_n and let $J = \{j_1, j_2, \ldots, j_s\}$ such that $E^{**}(m_1)$ occurs at j_ℓ, where $j_1 > m_1$, where $j_s < n - m_1$ and where $j_{\ell+1} - m_1 > j_\ell + m_1$, for $\ell = 1, 2, \ldots, s-1$. Let $s = \lfloor \delta n \rfloor$ for some small $\delta > 0$, to be specified later. The walk ψ in the Kesten triple is obtained by replacing occurrences of $E^{**}(m_1)$ by (P,Q) (since $E^{**}(m_1)$ fills $\overline{Q}(j_\ell)$, these cubes are necessarily disjoint).

Finding bounds on the number of Kesten triples will complete the proof. Use the approach in the proof of lemma C.7 to find the lower bound

$$|\text{Kesten triples}| \geq |H_n| \binom{\lfloor \rho n \rfloor - 2}{\lfloor \delta n \rfloor},$$

where $\rho = \frac{\epsilon_1}{2(m_1+1)V}$ (and where $V = (4r+9)^d$).

An upper bound is found by using the approach in the proof of lemma C.7. Notice that ψ has at most $\lfloor \epsilon n \rfloor + 2m_1 V \lfloor \delta n \rfloor$ occurrences of (P, Q) (since there are at most ϵn occurrences of (P, Q), and changing $E^{**}(m_1)$ to a (P, Q) may create several occurrences of (P, Q) by adding (P, Q) or by vacating other cubes intersecting $\overline{Q}(j)$).

Finally, ψ has length at most n. Taken together, it follows that

$$|\text{Kesten triples}| \leq 2^{\lfloor \epsilon n \rfloor + 2m_1 V \lfloor \delta n \rfloor} Z_1^{\lfloor \delta n \rfloor} \sum_{i=0}^{n} c_i,$$

where $Z_1 = \sum_{i=0}^{2m_1} c_i$.

Put $\delta = \epsilon$, take the power $\frac{1}{n}$ and then $n \to \infty$. This gives

$$\frac{\mu_d \rho^\rho}{\delta^\delta (\rho - \delta)^{\rho - \delta}} \leq 2^{\delta + 2m_1 V \delta} Z_1^\delta \mu_d.$$

Similar to the arguments at the end of the proof of lemma C.7, this gives a contradiction for sufficiently small $\delta > 0$. □

APPENDIX D

ASYMPTOTIC APPROXIMATIONS

The Stirling approximation for the factorial [187] is given by

$$n! = \sqrt{2\pi n}\, n^n e^{-n} \left(1 + \frac{1}{12n} + \frac{1}{288n^2} - \frac{139}{51840n^3} + O\left(\frac{1}{n^4}\right)\right). \tag{D.1}$$

Lower and upper bounds on the factorial are

$$\sqrt{2\pi n}\, n^n e^{-n} e^{1/(12n+1)} < n! < \sqrt{2\pi n}\, n^n e^{-n} e^{1/(12n)} \tag{D.2}$$

(see references [187, 494]), while the approximation

$$n! \simeq \sqrt{\left(2n + \tfrac{1}{2}\right)\pi}\, n^n e^{-n} \tag{D.3}$$

is from a numerical point of view very accurate.

Stirling's approximation can also be stated as

$$\frac{11\, e^k}{12\sqrt{2\pi k}} \leq \frac{k^k}{k!} \leq \frac{11\, e^k}{10\sqrt{2\pi k}}, \tag{D.4}$$

from which it follows, for example,

$$\frac{n^n}{\ell^\ell (n-\ell)^{n-\ell}} \leq \frac{72\sqrt{2\pi n}}{110} \binom{n}{\ell} \quad \text{for } \ell, n \in \mathbb{N}_0, \text{ and } 0 \leq \ell \leq n. \tag{D.5}$$

More on Stirling's approximation may be found in reference [592].

Catalan numbers are defined by

$$\mathcal{C}_n = \frac{1}{n+1}\binom{2n}{n}. \tag{D.6}$$

By putting both $n \to 2n$ and $\ell = n$ in equation (D.5) and keeping more terms in the Stirling approximation, one finds that

$$\left|\sqrt{\pi n^3}\, 4^{-n} \mathcal{C}_n - \left(1 - \tfrac{9}{8n} + \tfrac{145}{128n^2}\right)\right| < \tfrac{1}{n^3} M \tag{D.7}$$

for some fixed $M > 0$; see lemma 4.4 in reference [488].

D.1 Approximation of the binomial coefficient

The binomial coefficient can be approximated using approximations for the factorial. A more interesting approach is to consider random walks in $\mathbb{Z}$.

Consider a random walk from 0 in $\mathbb{Z}$ giving left or right steps. Assign a generating variable t with each right step so that $\bar{t} \equiv \frac{1}{t}$ represents a left step.

The generating function of walks of length n is $(t+\bar{t})^n$, and the coefficient of t^k is the number of walks ending at $k \in \mathbb{Z}$ from the origin. Thus, by the binomial theorem, the number of walks ending at k is $\binom{n}{(n+k)/2}$, since the walk ends at k if it gives $\frac{1}{2}(n+k)$ right steps (and $\frac{1}{2}(n-k)$ left steps).

Assume that $t \in (0,1)$ and suppose that the walk steps to the right with probability $P_+ = \frac{t}{t+\bar{t}}$ and to the left with probability $P_- = \frac{\bar{t}}{t+\bar{t}}$. Let X_k be the final position of the walk after k steps. Then $X_0 = 0$.

Note that X_k is a random variable and that $X_n = k$ for $k \in [-n, n]$ with probability

$$\mathrm{P_r}(X_n{=}k) = \begin{cases} \binom{n}{\frac{1}{2}(n+k)} \frac{t^k}{(t+\bar{t})^n}, & \text{if } \operatorname{mod}(n+k,2)=0; \\ 0, & \text{otherwise.} \end{cases} \tag{D.8}$$

The value of $\mathrm{P_r}(X_n{=}k)$ is 0 if n and k do not have the same parity (are not both even, or not both odd).

Asymptotically, n and k have the same parity with probability $\frac{1}{2}$, and the distribution of X_n is given by

$$\mathbb{E}[\mathrm{P_r}(X_n{=}k)] = \tfrac{1}{2}\binom{n}{\frac{1}{2}(n+k)} \frac{t^k}{(t+\bar{t})^n}\,(1+o(1))\,. \tag{D.9}$$

By the central limit theorem, the random variable X_n is asymptotically normally distributed with mean μn and variance $\sigma^2 n$, where

$$\mu = \langle X_1 \rangle = \frac{t-\bar{t}}{t+\bar{t}}, \quad \text{and } \sigma^2 = \langle X_1^2\rangle - \langle X_1\rangle^2 = \frac{t+\bar{t}}{t+\bar{t}} - \left(\frac{t-\bar{t}}{t+\bar{t}}\right)^2 = \frac{4}{(t+\bar{t})^2}.$$

Thus, it follows that

$$\mathbb{E}\left[\mathrm{P_r}(X_n{=}k)\right] = \frac{e^{-(k-\mu n)^2/2n\sigma^2}}{\sqrt{2\pi n\sigma^2}}\,(1+o(1))\,. \tag{D.10}$$

This is maximal when $k = \mu n$, with μ given above. In particular, if $k = \lfloor an \rfloor$, then one may solve for t to obtain

$$t_a = \pm\sqrt{\frac{n+\lfloor an\rfloor}{n-\lfloor an\rfloor}}\,. \tag{D.11}$$

Since $t \in (0,1)$, select the positive root t_a^+. In this event, it follows that, if $t = t_a^+$, then $\mu n = \lfloor an \rfloor + O(\sqrt{n})$. For this choice of k and with $t = t_a^+$, one may solve for the binomial coefficient in equation (D.9):

$$\binom{n}{\frac{1}{2}(n+\lfloor an \rfloor)} = \left[2\mathbb{E}[\mathrm{P}_r(X_n = \lfloor an \rfloor)] \, \frac{(t+\bar{t})^n}{t^{\lfloor an \rfloor}}\right]_{t=t_a^+} (1+o(1)) \,. \tag{D.12}$$

Substitute $t = t_a^+$ into equation (D.12) to obtain the following approximation for the binomial:

Theorem D.1 *The binomial coefficient is approximated by*

$$\binom{n}{\frac{1}{2}(n+\lfloor an \rfloor)} = \frac{2^{n+1}(1-\frac{1}{n}\lfloor an \rfloor)^{\lfloor an \rfloor}(1+o(1))}{\sqrt{2\pi n}\left(\sqrt{1-\frac{1}{n^2}\lfloor an \rfloor^2}\right)^{n+\lfloor an \rfloor+1}}. \qquad \square$$

Next, replace $\lfloor an \rfloor = 2\lfloor bn \rfloor - n$ for $b \in [0,1]$ and simplify to obtain

$$\binom{n}{\lfloor bn \rfloor} = \frac{1+o(1)}{\sqrt{2\pi n}\left(\frac{1}{n}\lfloor bn \rfloor\right)^{\lfloor bn \rfloor+1/2}\left(1-\frac{1}{n}\lfloor bn \rfloor\right)^{n-\lfloor bn \rfloor+1/2}}. \tag{D.13}$$

For large values of n, the above can be approximated by

$$\binom{n}{\lfloor bn \rfloor} = \frac{1+o(1)}{\sqrt{2\pi n}\, b^{bn+1/2}(1-b)^{(1-b)n+1/2}}. \tag{D.14}$$

Taking $n \to \infty$ gives corollary D.2.

Corollary D.2 *The rate of growth of binomial coefficients is given by*

$$B(b) = \lim_{n\to\infty} \binom{n}{\lfloor bn \rfloor}^{1/n} = \frac{1}{b^b(1-b)^{1-b}},$$

for $b \in [0,1]$. The function $B(b)$ has infinite right- and left-derivatives at $b=0$, and $b=1$, respectively. Moreover, $B(0) = B(1) = 1$, and $B(\frac{1}{2}) = 2 \geq B(b)$. $\square$

One may replace $\lfloor an \rfloor$ or $\lfloor bn \rfloor$ by $\lceil an \rceil$ or $\lceil bn \rceil$ with few, if any changes, in corollary D.2.

More generally, for $a, X > 0$, and $b \in [0,a]$, one may consider the function

$$B(a,b) = \frac{a^a X^b}{b^b(a-b)^{a-b}}, \tag{D.15}$$

which has a maximum for fixed $a > 0$ given by

$$\max_{b\in(0,a)} B(a,b) = (1+x)^a, \text{ realised when } b = \frac{ax}{1+x}. \tag{D.16}$$

Similar to $B(b)$ in corollary D.2, $B(a,b)$ has infinite right- and left-derivatives at $b=0$, and $b=a$, for fixed $a, x > 0$.

The function for $x > 0$, and $a \in (0,1)$, defined by

$$C(a) = \frac{(1-a)^{1-a}\,x^a}{a^a(1-2a)^{1-2a}}, \tag{D.17}$$

similarly has the maximum

$$\max_{a\in(0,1)} C(a) = \tfrac{1}{2}\left(1+\sqrt{1+4x}\right), \text{ realised when } a = \frac{\sqrt{1+4x}-1}{2\sqrt{1+4x}} \tag{D.18}$$

for fixed $a > 0$. The function $C(a)$ has infinite left- and right-derivatives at $a = 0$, and $a = 1$, respectively.

Lemma D.3 *Let $0 < q < \infty$ and suppose $\gamma \in [0, \frac{q}{q+1}]$. Then*

$$\limsup_{n\to\infty} \left(\sum_{i=0}^{\lfloor \gamma n\rfloor} \binom{n}{i} q^i\right)^{1/n} \le \frac{q^\gamma}{\gamma^\gamma(1-\gamma)^{1-\gamma}}.$$

Proof Let $p = \frac{q}{q+1}$ so that $0 \le \gamma \le p < 1$. Then

$$\sum_{i=0}^{\lfloor \gamma n\rfloor} \binom{n}{i} q^i = \sum_{i=0}^{\lfloor \gamma n\rfloor} \binom{n}{i} p^i(1-p)^{n-i}.$$

Consider p to be the probability that a biased tossed coin will land heads. Let η be the number of heads in n tosses of the coin. Then the probability that there are at most $\lfloor \gamma n\rfloor$ heads in n tosses is

$$\mathrm{P_r}(\eta \le \lfloor \gamma n\rfloor) = \sum_{i=0}^{\lfloor \gamma n\rfloor} \binom{n}{i} p^i(1-p)^{n-i}.$$

If ζ is a random variable equal to 1 if $\eta \le \lfloor \gamma n\rfloor$, and 0 otherwise, then the expectation of ζ is $\mathbb{E}\zeta = \mathrm{P_r}(\eta \le \lfloor \gamma n\rfloor$. By Jensen's inequality,

$$\mathbb{E}\zeta \le \mathbb{E}\left(e^{t(\lfloor \gamma n\rfloor-\eta)}\right) = e^{t\lfloor \gamma n\rfloor}\,\mathbb{E}\left(e^{-t\eta}\right) = e^{t\lfloor \gamma n\rfloor}\left(pe^{-t}+(1-p)\right).$$

The right-hand side is a minimum when $\lfloor \gamma n\rfloor(1-p) = p(n-\lfloor \gamma n\rfloor)e^{-t}$. Thus, substituting for t gives

$$\sum_{i=0}^{\lfloor \gamma n\rfloor} \binom{n}{i} p^i(1-p)^{n-i} \le \left(\frac{p}{\frac{1}{n}\lfloor \gamma n\rfloor}\right)^{\lfloor \gamma n\rfloor} \left(\frac{1-p}{1-\frac{1}{n}\lfloor \gamma n\rfloor}\right)^{n-\lfloor \gamma n\rfloor}. \tag{D.19}$$

Substitute $p = \frac{q}{q+1}$, take the power $\frac{1}{n}$ and take $n \to \infty$.

This completes the proof. □

D.2 Approximation of trinomial coefficients

Trinomial coefficients are defined by

$$(t+1+\bar{t})^n = \sum_{j=-n}^{n} \binom{n}{j}_2 t^j. \tag{D.20}$$

The term 'trinomial coefficient' is used as in reference [9]; that is, 'trinomial coefficient' is used synonymously with 'central trinomial coefficient'.

Trinomial coefficients are given in terms of binomial coefficients by

$$\binom{n}{m}_2 = \binom{n}{-m}_2 = \sum_{k=0}^{\lfloor (n-m)/2 \rfloor} \frac{n!}{k!(m+k)!(n-m-2k)!}. \tag{D.21}$$

The properties of a random walker in $\mathbb{Z}$ can be used to find approximations to trinomial coefficients. Let the position of the walker be x_n and suppose it gives right or left unit length steps in $\mathbb{Z}$, or a neutral step where it stays put at x_n, so that $x_{n+1} = x_n \pm 1$, or $x_{n+1} = x_n$.

Let $x_0 = 0$ and suppose the random walker gives right, neutral and left steps with probabilities P_+, P_0 and P_-, respectively. Suppose $t \in (0,1)$, let $\bar{t} = \frac{1}{t}$ and let

$$P_+ = \frac{t}{t+1+\bar{t}}; \quad P_0 = \frac{1}{t+1+\bar{t}}; \quad \text{and } P_- = \frac{\bar{t}}{t+1+\bar{t}}.$$

The probability that $x_n = k$ for $k \in [-n, n]$ is given by

$$\mathrm{P_r}\,(x_n = k) = \binom{n}{k}_2 \frac{t^k}{(t+1+\bar{t})^n}, \tag{D.22}$$

since $x_n = k$ if the walk gives $k+p$ right steps and p left steps, and since the number of walks ending at k is exactly given by the trinomial coefficient.

The central limit theorem states that the random variable x_n is asymptotically normal with mean μn and variance $\sigma^2 n$, where

$$\mu = \langle x_1 \rangle = \frac{t-\bar{t}}{t+1+\bar{t}}, \tag{D.23}$$

and

$$\sigma^2 = \langle x_1^2 \rangle - \langle x_1 \rangle^2 = \frac{t+\bar{t}}{t+1+\bar{t}} - \left(\frac{t-\bar{t}}{t+1+\bar{t}} \right)^2 = \frac{t+4+\bar{t}}{(t+1+\bar{t})^2}. \tag{D.24}$$

Thus, it follows that

$$P(x_n = k) = \frac{e^{-(k-\mu n)^2/2n\sigma^2}}{\sqrt{2\pi n \sigma^2}} (1+o(1)). \tag{D.25}$$

Substitute $k = \lfloor \alpha n \rfloor$, substitute $\mu = \frac{t-\bar{t}}{t+1+\bar{t}}$ and solve for t:

$$t_\alpha^\pm = \frac{\lfloor \alpha n \rfloor \pm \sqrt{4n^2 - 3\lfloor \alpha n \rfloor^2}}{2(n - \lfloor \alpha n \rfloor)}. \tag{D.26}$$

Since $t \in (0,1)$, select t_α^+ so that $\mu n = \lfloor \alpha n \rfloor + O(\sqrt{n})$ and solve for the trinomial coefficient in equation (D.22). If $k = \lfloor \alpha n \rfloor$, and $t = t_\alpha^+$, then this gives

$$\binom{n}{\lfloor \alpha n \rfloor}_2 = \left(P(x_n = \lfloor \alpha n \rfloor) \left(\frac{(t_\alpha^+ + 1 + \overline{t_\alpha^+})^n}{(t_\alpha^+)^{\lfloor \alpha n \rfloor}} \right) \right) (1 + o(1)). \tag{D.27}$$

Simplification gives theorem D.4.

Theorem D.4 *The trinomial coefficient is approximated by*

$$\binom{n}{\lfloor \alpha n \rfloor}_2 = \left(\frac{2^{\lfloor \alpha n \rfloor}}{\sqrt{2\pi n}} \right) \frac{A_\alpha^{n+1/2} (1 + o(1))}{B_\alpha^{n - \lfloor \alpha n \rfloor + 1/2} C_\alpha^{1/4} D_\alpha^{n + \lfloor \alpha n \rfloor + 1/2}},$$

where

$$A_\alpha = 4 - \tfrac{3}{n} \lfloor \alpha n \rfloor + \sqrt{4 - \tfrac{3}{n^2} \lfloor \alpha n \rfloor^2}; \qquad B_\alpha = 1 - \tfrac{1}{n} \lfloor \alpha n \rfloor;$$

$$C_\alpha = 4 - \tfrac{3}{n^2} \lfloor \alpha n \rfloor^2; \quad \textit{and } D_\alpha = \tfrac{1}{n} \lfloor \alpha n \rfloor + \sqrt{4 - \tfrac{3}{n^2} \lfloor \alpha n \rfloor^2}. \qquad \square$$

For large values of n, one may use corollary D.5.

Corollary D.5 *The trinomial coefficient*

$$\binom{n}{\lfloor \alpha n \rfloor}_2 = \left(\frac{2^{\alpha n}}{\sqrt{2\pi n} (1 - \alpha)^{(1-\alpha)n + 1/2}} \right) \frac{(4 - 3\alpha + \sqrt{\beta})^{n+1/2} (1 + o(1))}{\beta^{1/4} (\alpha + \sqrt{\beta})^{(1+\alpha)n + 1/2}},$$

where $\beta = 4 - 3\alpha^2$. □

The rate of exponential growth of the trinomial coefficient is given by corollary D.6.

Corollary D.6 *The limit*

$$\lim_{n \to \infty} \binom{n}{\lfloor \alpha n \rfloor}_2^{1/n} = \frac{2^\alpha \left(4 - 3\alpha + \sqrt{4 - 3\alpha^2}\right)}{(1 - \alpha)^{1-\alpha} \left(\alpha + \sqrt{4 - 3\alpha^2}\right)^{1+\alpha}}$$

$$= \frac{\left(2 - \alpha^2 - \alpha\sqrt{4 - 3\alpha^2}\right)^{\alpha/2} \left(13 - 9\alpha^2 + (7 - 3\alpha^2)\sqrt{4 - 3\alpha^2}\right)^{1/3}}{2^{\alpha/2} (1 - \alpha)(1 + \alpha)^{1+\alpha}}. \qquad \square$$

The rate of growth of the generating function of trinomial coefficients is defined by

$$F(z) = \lim_{n \to \infty} \left(\sum_{j=0}^{n} \binom{n}{j}_2 z^j \right)^{1/n}. \tag{D.28}$$

Given an n, there is a value of j, say $j = j_n$, which maximises the summand.

Assuming that $j_n = \lfloor \epsilon_m n \rfloor + o(n)$, where $\epsilon_m \in (0,1)$ is a function of z, it follows that

$$F(z) = \lim_{n\to\infty} \left(\binom{n}{j_n}_2 z^{j_n} \right)^{1/n}. \tag{D.29}$$

By replacing j_n in the above by $\lfloor \epsilon n \rfloor + o(n)$ and then using equation (D.6) when taking $n \to \infty$ in equation (D.29), it follows that

$$F(z) = \sup_{0<\epsilon<1} \left(\frac{z^\epsilon 2^\epsilon \left(4 - 3\epsilon + \sqrt{4-3\epsilon^2}\right)}{(1-\epsilon)^{1-\epsilon} \left(\epsilon + \sqrt{4-3\epsilon^2}\right)^{1+\epsilon}} \right). \tag{D.30}$$

By differentiating and solving for ϵ, the supremum is realised when

$$\epsilon = \epsilon_m = \frac{z - \overline{z}}{z + 1 + \overline{z}} = \mu \quad \text{in equation (D.23).} \tag{D.31}$$

It follows that $\epsilon_m \in (0,1)$ if $z > 1$, and so the supremum is achieved when $\epsilon_m = 0$ when $z \leq 1$ at ϵ_m. These results show that, after simplification,

Theorem D.7 *The function $F(z)$ is given by*

$$F(z) = \begin{cases} z + 1 + \overline{z}, & \text{if } z \in (1, \infty); \\ 3, & \text{if } z \in [0,1], \end{cases} \tag{D.32}$$

where $\overline{z} = \frac{1}{z}$. □

D.3 The Euler-Maclaurin formula

The Euler-Maclaurin formula (see for example [273]) is a useful method for approximating series by integrals. The formula is useful because it gives precise remainders, allowing estimates of the error in the approximation.

Theorem D.8 (Euler-Maclaurin) *If $f \in C^{2m}[0,N]$ for some $m, N \in \mathbb{N}$, and both $N > 0$ and $m \geq 1$, then*

$$\sum_{n=0}^{N} f(n) = \int_0^N f(x)\,dx + \tfrac{1}{2}(f(0) + f(N)) + \sum_{n=1}^{m} \frac{B_{2n}}{(2n)!} \left(f^{(2n-1)}(N) - f^{(2n-1)}(0) \right) + R_m$$

where the remainder term is given by

$$R_m = \int_0^N \left[\frac{-B_n(x - \lfloor x \rfloor)}{(2m)!} \right] f^{(2m)}(x)\,dx.$$

The function $B_n(x)$ is the n-th Bernoulli polynomial; $B_n = B_n(0)$, and $B_n(x - \lfloor x \rfloor)$ is the n-th periodic Bernoulli polynomial. □

In some applications the remainder term R_m approaches 0 with $m \to \infty$ (derivatives of f exist to all orders), and one may able to show that the series is convergent.

The remainder term can be bounded by

$$|R_m| \le \frac{2\zeta(2m)}{(2\pi)^{2m}} \int_0^N \left| f^{(2m)}(x) \right| dx.$$

The *Bernoulli functions* are defined by the recurrence $\frac{d}{dx}B_n(x) = nB_{n-1}(x)$, with $B_0(x) = 1$, and $\int_0^1 B_n(x)dx = 0$, for $n \ge 1$. The first few are given by $B_1(x) = x - \frac{1}{2}$; $B_2(x) = x^2 - x + \frac{1}{6}$; $B_3(x) = x^3 - \frac{3}{2}x^2 + \frac{1}{2}x$; and $B_4(x) = x^4 - 2x^3 + x^2 - \frac{1}{30}$.

The *Bernoulli numbers* B_n are defined by the exponential generating function $\frac{t}{e^t-1} = \sum_{m=0}^{\infty} \frac{B_m}{m!} t^m$. Notably, $B_{2n+1} = 0$ if $n \ge 1$ (that is $B_3 = B_5 = B_7 = \cdots = 0$), and the first few non-zero Bernoullui numbers are $B_1 = -\frac{1}{2}$; $B_2 = \frac{1}{6}$; $B_4 = -\frac{1}{30}$; and $B_6 = \frac{1}{42}$.

The Stirling approximation for $n!$ is obtained as follows. Consider the integral

$$\int_0^{\infty} \frac{1}{(z+x)^2}\,dx = \frac{1}{z}.$$

Applying the Euler-Maclaurin formula to the integral on the left-hand side gives

$$\frac{1}{z} = \frac{1}{2z^2} + \sum_{n=1}^{\infty} \frac{1}{(z+n)^2} - \sum_{n=1}^{\infty} \frac{B_n}{z^{2n+1}}.$$

The infinite series in the above is a polygamma function and

$$\sum_{n=1}^{\infty} \frac{1}{(z+n)^2} = \frac{d^2}{dz^2} \log z! = \frac{1}{z} - \frac{1}{2z^2} + \sum_{n=1}^{\infty} \frac{B_{2n}}{z^{2n+1}}.$$

Take the anti-derivative of this twice to obtain

$$\log z! = C + \left(z + \tfrac{1}{2}\right) \log z - z + \sum_{n=1}^{\infty} \frac{B_{2n}}{2n\,(2n-1)\,z^{2n-1}}.$$

The constant C is computed from the Legendre duplication formula

$$2^{2z+1} z!\,(z + \tfrac{1}{2})! = \sqrt{\pi}\,(2z+1)!.$$

This gives $C = \frac{1}{2}\log(2\pi)$. Finally,

$$\log z! = \tfrac{1}{2}\log(2\pi) + \left(z + \tfrac{1}{2}\right) \log z - z + \sum_{n=1}^{\infty} \frac{B_{2n}}{2n(2n-1)z^{2n-1}}. \tag{D.33}$$

This is an asymptotic series approximating $\log z!$. By truncating the infinite series, the error in the approximation is less than the magnitude of the first omitted term. If the series is discarded in its entirety, then the Stirling approximation in equation (D.1) is found.

D.4 Saddle point approximations of the integral

Suppose that f is an analytic complex function of the complex variable $z = x + iy$. Then $f = u + iv$, and $\{u, v\}$ satisfies the Cauchy-Riemann equations. Consequently, it may be shown that u satisfies the Laplace equation:

$$\frac{\partial^2}{\partial x^2} u + \frac{\partial^2}{\partial y^2} u = 0. \tag{D.34}$$

At any stationary point of f, if $\frac{\partial^2}{\partial x^2} u > 0$, then $\frac{\partial^2}{\partial y^2} u < 0$. That is, if f is real valued, then its stationary points are necessarily saddle points.

The function f may be approximated in the vicinity of a saddle point z_0 by

$$f(x) \approx f(z_0) + \tfrac{1}{2} f''(z_0) \left(z - z_0\right)^2. \tag{D.35}$$

This approximation of f near its saddle points allows the approximation of integrals of the form

$$\mathcal{I} = \int_a^b g(z) e^{N f(z)} \, dz, \tag{D.36}$$

where N is large, and $g(z)$ is slowly varying.

Insert the approximation of f in the above and replace $g(z) \approx g(z_0)$. Deform the contour of integration from a to b into a contour C in the complex plane passing through z_0. This gives

$$\mathcal{I} \simeq g(z_0) e^{N f(z_0)} \int_C e^{\frac{N}{2} f''(z_0)(z - z_0)^2} dz. \tag{D.37}$$

Put $z - z_0 = r e^{i\phi}$ and let $f''(z_0) = |f''(z_0)| \, e^{i\theta}$. The integration will be along r, with ϕ fixed at a convenient value, namely, $\theta + 2\phi = \pi$. The integration is approximated by integrating from $r = -\infty$ to $r = \infty$. These substitutions simplify the integral above into a Gaussian integral:

$$\mathcal{I} \simeq g(z_0) e^{N f(z_0)} e^{i\phi} \int_{-\infty}^{\infty} e^{-\frac{N}{2} |f''(z_0)| r^2} dr. \tag{D.38}$$

Evaluating the integral gives the saddle point approximation

$$\mathcal{I} \simeq \frac{\sqrt{2\pi}\, g(z_0) e^{N f(z_0)} e^{i\phi}}{\sqrt{N \, |f''(z_0)|}}. \tag{D.39}$$

The phase ϕ is determined by solving for $\phi = \frac{\pi - \theta}{2}$.

It is possible to refine the above approximations by keeping more terms in the expansion (D.35). This approximation also has the implicit assumption that the only significant contribution to the integral along the contour C is from the vicinity of the saddle point (this means that $u(z_0) \gg u(z)$ if $|z_0 - z|$ is 'large'. This may not be necessarily the case. For more on this method, which is related to steepest descent methods, see, for example, reference [13].

D.5 Asymptotic formulae for the q-factorial and related functions

The q-deformations of special functions are obtained from an arithmetic based on *q-integers*, and defined by

$$[n]_q = \frac{1-q^n}{1-q} = 1 + q + q^2 + \cdots + q^{n-1} \tag{D.40}$$

where $q \in \mathbb{C}$. These q-integers retain many of the properties of the usual integers. A q-integer is sometimes called a *q-bracket* or a *basic integer*.

The *q-Pochhammer function* is a finite product formally defined by

$$\begin{aligned}(t;q)_n &= \prod_{k=0}^{n-1}(1-tq^k) \\ &= (1-t)(1-tq)(1-tq^2)\cdots(1-tq^{n-1}) = \frac{(t;q)_\infty}{(tq^n;q)_\infty},\end{aligned} \tag{D.41}$$

$(t;q)_\infty$ is an infinite product.

The q-Pochhammer function $(t;q)_\infty$ is an analytic function inside the unit disk in the q-plane. Euler's function is defined by $(q;q)_\infty$. Observe that $(q)_\infty = (q;q)_\infty$ and

$$\begin{aligned}(q)_n &= \prod_{k=1}^{n}(1-q^k) = \prod_{k=1}^{\infty}\frac{1-q^k}{1-q^{n+k}} \\ &= \frac{(q;q)_\infty}{(q^{n+1},q)_\infty} = (1-q)(1-q^2)\cdots(1-q^n).\end{aligned} \tag{D.42}$$

Put $n=\infty$ in the above to find an infinite product representation of $(q)_\infty$.

A partition identity (see for example reference [8]) shows that

$$e(t,q) = \sum_{k=0}^{\infty}\frac{t^k}{(q)_k} = \frac{1}{(t;q)_\infty}. \tag{D.43}$$

Note that the usual factorial in the definition of the exponential is replaced by $(q)_k$. For this reason, $e(t,q)$ is sometimes called the *q-exponential*; however, see equation (D.46) below.

The natural definition of the q-analogue of the factorial is in terms of q-integers:

$$[n]_q! = \frac{(q;q)_n}{(1-q)^n} = \frac{(q;q)_\infty(1-q)^{-n}}{(q^{n+1},q)_\infty}, \tag{D.44}$$

and these are also called *basic factorials* or *Gaussian factorials.* This gives rise to the q-analogue of the gamma function or the *q-gamma function*:

$$\Gamma_q(z) = \frac{(q;q)_\infty(1-q)^{1-z}}{(q^z;q)_\infty} \tag{D.45}$$

which becomes $\Gamma_q(k+1) = [k]_q!$ if $k \in \mathbb{N}$. Observe that, if $q \to 1^-$ in each of the above, then the usual integers, factorials and gamma functions are recovered.

The natural definition of the *q-exponential* is

$$e_q^t = \sum_{k=0}^{\infty} \frac{t^k}{[k]!} = \frac{1}{((1-q)t;q)_\infty}, \tag{D.46}$$

or, alternatively,

$$E_q^t = \sum_{k=0}^{\infty} \frac{q^{\binom{k}{2}} t^k}{[k]!} = (-(1-q)t;q)_\infty. \tag{D.47}$$

Observe that $E_q^t = e_{1/q}^t = 1/e_q^{-t}$.

These definitions of the exponential give rise to q-analogues of the trigonometric functions sine and cosine. There are also q-analogues of special functions. For example, q-deformed Bessel functions are defined by

$$J(p,q,t) = \sum_{m=0}^{\infty} \frac{q^{\binom{m}{2}}(-t)^m}{(p;q)_m\,(q;q)_m}. \tag{D.48}$$

More on q-calculus can be found, for example, in reference [345].

D.5.1 *The q-binomial coefficient*

The above definitions extend to the *q-binomial coefficient*, defined by

$$\begin{bmatrix} n \\ k \end{bmatrix} = \frac{(q)_n}{(q)_k (q)_{n-k}}, \tag{D.49}$$

and, if $n \in \mathbb{N}$, and $k \in \mathbb{N}$, then this becomes

$$\begin{bmatrix} n \\ k \end{bmatrix} = \frac{[n]_q!}{[k]_q!\,[n-k]_q!}. \tag{D.50}$$

Observe that $[n]_q = \begin{bmatrix} n \\ 1 \end{bmatrix}$ and that $\begin{bmatrix} n \\ k \end{bmatrix}$ satisfies the following q-binomial identities:

$$\begin{bmatrix} n \\ k \end{bmatrix} = q^k \begin{bmatrix} n-1 \\ k \end{bmatrix} + \begin{bmatrix} n-1 \\ k-1 \end{bmatrix}, \quad \text{and} \quad \begin{bmatrix} n \\ k \end{bmatrix} = \begin{bmatrix} n-1 \\ k \end{bmatrix} + q^{n-r} \begin{bmatrix} n-1 \\ k-1 \end{bmatrix}. \tag{D.51}$$

Lemma D.9 is not obvious.

Lemma D.9 *If $n \in \mathbb{N}_0$, and $k \in \mathbb{N}_0$, then the Gaussian binomial coefficient is a polynomial of degree $k(n-k)$ with non-negative integer coefficients. In other words,*

$$\begin{bmatrix} n \\ k \end{bmatrix} = \sum_{\alpha=0}^{k(n-k)} N_{nki} q^\alpha,$$

where $N_{nk\alpha} \in \mathbb{N}_0$.

Proof Observe that $\begin{bmatrix} n \\ 0 \end{bmatrix} = 1$ and is a polynomial of degree 0.

The proof proceeds by induction.

Suppose that $\begin{bmatrix} m \\ j \end{bmatrix}$ is a polynomial with coefficients in $\mathbb{N}_0$ and with degree $j(m-j)$, where $m \leq n$, and $j \leq k$. By equation (D.51),

$$\begin{bmatrix} n+1 \\ k \end{bmatrix} = q^k \begin{bmatrix} n \\ k \end{bmatrix} + \begin{bmatrix} n \\ k-1 \end{bmatrix} = \sum_{\alpha=0}^{k(n-k)} N_{nk\alpha} q^{\alpha+k} + \sum_{\alpha=0}^{(k-1)(n-k+1)} N_{n(k-1)\alpha} q^{\alpha}.$$

The largest power of q in the above is $k + k(n-k) = k(n+1-k)$ as required, and the coefficients of q^α on the right-hand side are sums over integers in $\mathbb{N}_0$.

Similarly, again by equation (D.51),

$$\begin{bmatrix} n \\ k+1 \end{bmatrix} = \begin{bmatrix} n-1 \\ k \end{bmatrix} + q^{k+1} \begin{bmatrix} n-1 \\ k-1 \end{bmatrix}.$$

Hence, if $\begin{bmatrix} n-1 \\ k+1 \end{bmatrix}$ is a polynomial in q of degree $(k+1)((n-1)-(k+1))$ and with coefficients in $\mathbb{N}_0$, then $\begin{bmatrix} n \\ k+1 \end{bmatrix}$ is a polynomial in q of degree $(k+1)(n-(k+1))$ and with coefficients in $\mathbb{N}_0$. Notice that $\begin{bmatrix} k+1 \\ k+1 \end{bmatrix} = 1$. □

The q-binomial theorem is usually stated as

$$\sum_{n=0}^{\infty} \frac{(t;q)_n}{(q;q)_n} z^n = \frac{(tz;q)_\infty}{(z;q)_\infty} \tag{D.52}$$

but it is also encountered as

$$(tq;q)_n = \sum_{k=0}^{n} \begin{bmatrix} n \\ k \end{bmatrix} q^{\binom{n+1}{2}} t^k. \tag{D.53}$$

The proof of the last equality proceeds by the counting of partitions.

The q-binomial theorem gives rise to the *Jacobi triple product* identity

$$(-tq;q)_\infty (-\bar{t},q)_\infty (q;q)_\infty = \sum_{n=-\infty}^{\infty} q^{\binom{n+1}{2}} t^n, \tag{D.54}$$

where $\bar{t} = \frac{1}{t}$.

The identity

$$\frac{1}{(t;q)_n} = \sum_{k=0}^{\infty} \begin{bmatrix} n-1+k \\ k \end{bmatrix} t^k \tag{D.55}$$

is also useful [8]. By the q-binomial theorem in equation (D.53), it follows that

$$\left(\sum_{k=0}^{\infty} \begin{bmatrix} n-1+k \\ k \end{bmatrix} (tq)^k \right)^{-1} = \sum_{k=0}^{n} \begin{bmatrix} n \\ k \end{bmatrix} q^{\binom{n+1}{2}} t^k. \tag{D.56}$$

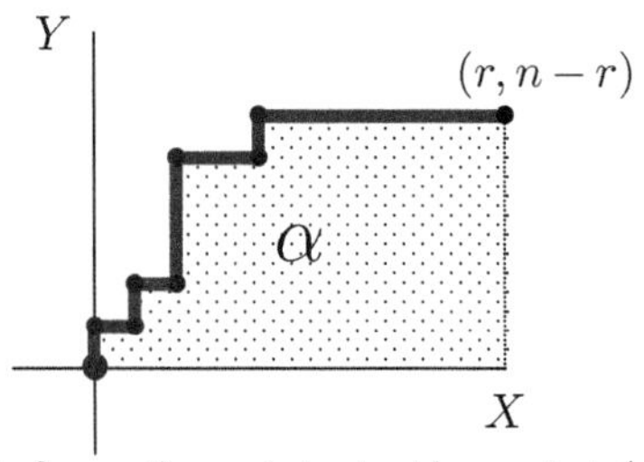

FIG. D.1. A staircase walk from the origin to the point $(r, n-r)$ overstepping an area α.

D.5.2 *Directed paths and the q-binomial coefficient*

There is a close connection between Gaussian binomial coefficients and D.9 (see reference [467]). Consider the square lattice $\mathbb{L}^2$ and let $\mathbb{L}^2_+$ be the positive half-lattice with the boundary $\partial\mathbb{L}^2_+$ as the x-axis. In figure D.1, directed paths from the origin in $\mathbb{L}^2_+$ give only north and east steps. The path *oversteps* area α in the first quadrant of $\mathbb{L}^2$ underneath the path and to the left of the endpoint of the path.

Then the number of such paths of length n is related to N_{nki} and Gaussian binomial coefficients (see lemma D.9) by the following lemma.

Lemma D.10 *The number of square lattice directed paths in $\mathbb{L}^2$, of length n, from the origin to the point $(r, n-r)$, giving only north and east steps and which overstep an area of size α in the first quadrant of $\mathbb{L}^2$ is $N_{nr\alpha}$.*

Proof This is true for any α if $n = 1$, and $0 \leq r \leq 1$. Suppose that the lemma is true for any $m \leq n$.

Consider $m = n+1$. By equation (D.51),

$$\sum_{\alpha=0}^{r(n+1-r)} N_{(n+1)r\alpha} q^\alpha = \sum_{\alpha=0}^{r(n-r)} N_{nr\alpha} q^\alpha + \sum_{\alpha=0}^{(r-1)(n-r+1)} N_{n(r-1)\alpha} q^{\alpha+n+1-r}.$$

If coefficients of q^α are compared, then

$$N_{(n+1)r\alpha} = N_{nr\alpha} + N_{n(r-1)(\alpha+r-n-1)}.$$

This result is interpreted in figure D.2. By the induction hypothesis, the term $N_{nr\alpha}$ counts the number of walks of length n, from $(0,0)$ to $(r, n-r)$, and overstepping area α.

Appending a vertical step to paths turn these into walks of length $n+1$ which contribute towards $N_{(n+1)r\alpha}$

In a similar fashion, $N_{n(r-1)(\alpha+r-n-1)}$ counts the number of walks from the origin to $(r-1, n-r+1)$ and which enclose the area $\alpha + r - n - 1$. If a horizontal edge is appended from $(r-1, n-r+1)$ to $(r, n-r+1)$, then it encloses area α.

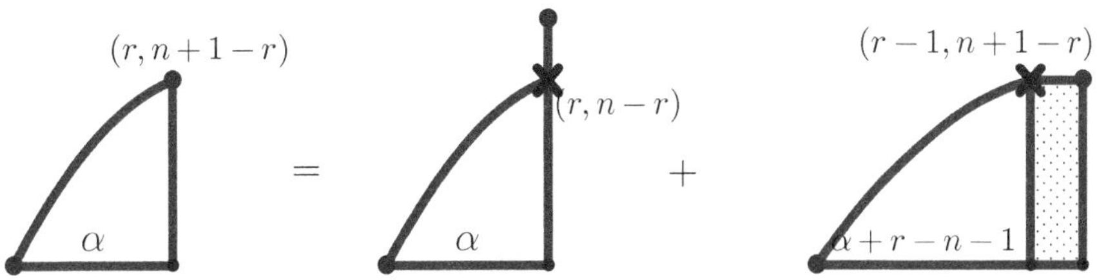

FIG. D.2. Any staircase walk of length $n+1$ from the origin to $(r, n+1-r)$ and which oversteps area α has either a north step or an east step as its last step. If the last step was north, then it can be deleted to find $N_{nr\alpha}$. If it was to the east, then deleting the last step also removes the last column, and this gives $N_{n(r-1)(\alpha+r-n-1)}$.

Thus, $N_{(n+1)r\alpha}$ counts the number of walks from $(0,0)$ to $(r, n-r+1)$ and which overstep area α. □

Putting $q=1$ gives the corollary

$$\binom{n}{r} = \sum_{\alpha=0}^{r(n-r)} N_{nr\alpha}. \tag{D.57}$$

Figure D.1 is also a Ferrer's diagram of a partition. Thus, let the number of partitions of $\alpha \in \mathbb{N}$ into r parts with largest part k be denoted $\rho_{k,r}(\alpha)$. Then $\rho_{k,r}(\alpha) = N_{(k+r)r\alpha}$.

The generating function of $\rho_{k,r}(\alpha)$ for fixed k and r is

$$\sum_{\alpha=0}^{kr} \rho_{k,r}(\alpha) q^\alpha = \begin{bmatrix} k+r \\ r \end{bmatrix}. \tag{D.58}$$

See reference [531] for a different proof of this result.

The generating function of partitions follows from this as

$$V_p(t,q) = \sum_{k=1}^{\infty}\sum_{r=1}^{\infty} \begin{bmatrix} k+r \\ r \end{bmatrix} t^{2(k+r)} = \sum_{k=1}^{\infty}\sum_{r=1}^{k-1} \begin{bmatrix} k \\ r \end{bmatrix} t^{2k}, \tag{D.59}$$

where t is the generating variable conjugate to perimeter length.

D.5.3 *Theta functions, and partial theta functions*

The Jacobi θ-function is defined by

$$\theta(t,q) = \sum_{n=-\infty}^{\infty} t^n q^{n^2}; \tag{D.60}$$

$\phi(q) = \theta(1, q)$ is the Ramanujan θ-function. The Jacobi triple product identity may be cast as

$$(q^2; q^2)_\infty(-tq; q^2)_\infty(-\bar{t}q; q^2)_\infty = \sum_{n=-\infty}^{\infty} t^n q^{n^2}. \tag{D.61}$$

By expanding the second and third products on the left-hand side, this becomes

$$\begin{aligned}\sum_{n=-\infty}^{\infty} t^n q^{n^2} &= (q^2; q^2)_\infty \prod_{k=1}^{\infty} \left(1 + (t + \bar{t})q^{2k-1} + q^{4k-2}\right) \\ &= (q^2; q^2)_\infty \prod_{k=1}^{\infty} \left(1 + 2\cos(2\pi z)q^{2k-1} + q^{4k-2}\right),\end{aligned} \tag{D.62}$$

where $t = e^{2\pi i z}$. Putting $t = 1$ gives

$$\sum_{n=-\infty}^{\infty} q^{n^2} = (q^2; q^2)_\infty(-q; q^2)_\infty^2 = \frac{(q)_\infty}{(-q; q)_\infty}. \tag{D.63}$$

This last identity may also be obtained from the Ramanujan θ-function, which is defined as

$$f(a, b) = \sum_{n=-\infty}^{\infty} a^{\frac{n(n+1)}{2}} b^{\frac{n(n-1)}{2}}. \tag{D.64}$$

The Jacobi triple product becomes

$$(-a; ab)_\infty(-b; ab)_\infty(ab; ab)_\infty = \sum_{n=-\infty}^{\infty} a^{\frac{n(n+1)}{2}} b^{\frac{n(n-1)}{2}}. \tag{D.65}$$

By choosing $a = b = q$ in the above, Ramanujan's ϕ-function $\phi(q) = f(q, q)$ is recovered:

$$\phi(q) = \sum_{n=-\infty}^{\infty} q^{n^2} = (q^2; q^2)_\infty(-q; q^2)_\infty^2 = \frac{(-q; q^2)_\infty(q^2; q^2)_\infty}{(-q^2; q^2)_\infty(q; q^2)_\infty}. \tag{D.66}$$

Other identities may be recovered; for example, if $a = q$, and $b = q^3$, then

$$\sum_{n=-\infty}^{\infty} q^{\binom{n+1}{2}} = \sum_{n=-\infty}^{\infty} q^{\frac{n(n+1)}{2}} = \frac{(q^2; q^2)_\infty}{(q; q^2)_\infty}, \tag{D.67}$$

and, if $a = -q$, and $b = -q^2$, then

$$\sum_{n=-\infty}^{\infty} (-1)^n q^{\frac{n(3n-1)}{2}} = (q; q)_\infty. \tag{D.68}$$

This identity is the famous *pentagonal number theorem* [184].

The summation in the series definitions of θ-functions given above is over all of $\mathbb{Z}$. It may in some instances be necessary to have to have the summation only over $\mathbb{N}$. This restriction gives rise to *partial θ-functions.* An example is encountered in equation (D.63), which may be stated as

$$\sum_{n=0}^{\infty} q^{n^2} = \tfrac{1}{2}(1+\phi(q)) = \tfrac{1}{2}\left(1+(q^2;q^2)_\infty(-q;q^2)_\infty^2\right). \tag{D.69}$$

This is related to Gauss's formula (see for example [8]):

$$\sum_{n=0}^{\infty} q^{\binom{n+1}{2}} = \sum_{n=0}^{\infty} q^{\frac{n(n+1)}{2}} = \frac{(q^2;q^2)_\infty}{(q;q^2)_\infty}. \tag{D.70}$$

More generally, a partial θ-function has the form

$$\theta_p(t,q) = \sum_{n=0}^{\infty}(-1)^n t^n q^{\binom{n}{2}}. \tag{D.71}$$

Some identities for partial θ-functions are known. For example, in reference [10] it is shown that

$$\theta_p(t,q) = (q)_\infty (t)_\infty \sum_{n=0}^{\infty} \frac{q^n}{(t)_n (q)_n} = (t)_\infty \sum_{n=0}^{\infty} \frac{t^n q^{n^2}}{(t)_n (q)_n}. \tag{D.72}$$

D.5.4 *Asymptotics for $(q;q)_\infty$*

A good starting point is the identity in equation (D.68). The left-hand side converges quickly if $|q| \ll 1$, but, when q is close to 1, another approach is necessary.

Theorem D.11 (Poisson's sum formula) *Suppose that f is continuous for all $t \in \mathbb{R}$, and $\sum_{k=-\infty}^{\infty} f(t+k)$ is uniformly convergent for $t \in [0,1]$. Then*

$$\sum_{n=-\infty}^{\infty} f(t+n) = \sum_{n=-\infty}^{\infty} e^{-2\pi i n t} \int_{-\infty}^{\infty} f(x) e^{2\pi i n x}\, dx,$$

for all values of t where the right-hand side converges.

Proof The function

$$h(t) = \sum_{k=-\infty}^{\infty} f(t+k)$$

is continuous and periodic with period $T = 1$. In a Fourier expansion of h the coefficients would be

$$a_n = \int_0^1 h(s) e^{2\pi i n s}\, ds.$$

Moreover, by Fejér's theorem, $\sum_{n=-\infty}^{\infty} a_n e^{-2\pi i n t}$ is Cesàro summable to $h(t)$ for all t (since h is continuous and periodic).

Hence, by replacing h in the last integral with the uniformly convergent series over f, one obtains

$$\sum_{k=-\infty}^{\infty} \int_0^1 f(s+k)e^{2\pi ins}\,ds = \sum_{k=-\infty}^{\infty} \int_k^{k+1} f(s)e^{2\pi ins}\,ds = \int_{-\infty}^{\infty} f(s)e^{2\pi ins}\,ds.$$

The right-hand side evaluates to a_n, and therefore the Fourier series for $h(t)$ is as claimed above. □

Corollary D.12 (The θ-transformation formula) *For all values of $t, s \in \mathbb{C}$, such that $\Re e(s) > 0$,*

$$\sum_{n=-\infty}^{\infty} e^{-\pi s\,(n+t)^2} = \frac{1}{\sqrt{s}} \sum_{n=-\infty}^{\infty} e^{-\frac{\pi n^2}{s}+2\pi i n t}.$$

Proof Put $f(t) = e^{-\pi s t^2}$ in theorem D.11. Then $\sum_{n=-\infty}^{\infty} f(t+n)$ converges uniformly for $t \in [0,1]$, and $\Re e(s) > 0$. The Poisson sum formula gives

$$\sum_{n=-\infty}^{\infty} e^{-\pi s(t+n)^2} = \sum_{n=-\infty}^{\infty} e^{-2\pi i n s} \int_{-\infty}^{\infty} e^{-\pi s x^2+2\pi i n x}\,dx, \tag{D.73}$$

where the integral above gives the Fourier coefficients. It remains to evaluate these coefficients.

Assume that $s > 0$ and substitute $y = \sqrt{s}x$:

$$\int_{-\infty}^{\infty} e^{-\pi s x^2+2\pi i n x}\,dx = \frac{e^{-\frac{\pi n^2}{s}}}{\sqrt{s}} \int_{-\infty}^{\infty} e^{-\pi\left(y-\frac{in}{\sqrt{s}}\right)^2}\,dy = \frac{e^{-\frac{\pi n^2}{s}}}{\sqrt{s}} \int_{-\infty-\frac{in}{\sqrt{s}}}^{\infty-\frac{in}{\sqrt{s}}} e^{-\pi y^2}\,dy.$$

The contour integral on the right evaluates as

$$\int_{-\infty-\frac{in}{\sqrt{s}}}^{\infty-\frac{in}{\sqrt{s}}} e^{-\pi y^2}\,dy = \int_{-\infty}^{\infty} e^{-\pi y^2}\,dy = 1,$$

using Cauchy's theorem. This completes the evaluation of the Fourier coefficients in equation (D.73).

Substitution of the result and replacing $n \to -n$ in the summation on the right-hand side of equation (D.73) completes the proof. □

Corollary D.12 gives a suitable expansion for $(q;q)_\infty$; see, for example, reference [359] for an expansive treatment.

Theorem D.13 *For Euler's function,*

$$(q;q)_\infty = \left(\frac{r}{q}\right)^{\frac{1}{24}} \sqrt{\frac{-2\pi}{\log q}} \sum_{n=-\infty}^{\infty} \left(r^{n(6n+1)} - r^{(3n+1)(2n+1)}\right),$$

where $r = e^{\frac{4\pi^2}{\log q}}$. For $q \in (0,1)$ the right-hand side converges quickly.

Proof Put $t = \frac{1}{6} + \frac{\pi}{3i\log q}$, put $q^3 = e^{-2\pi s}$ and define r by $(\log r)(\log q) = 4\pi^2$. Substitute these variables in the left-hand side of corollary D.12 and then simplify. This shows that

$$\sum_{n=-\infty}^{\infty} e^{-\pi s(n+t)^2} = \left(\frac{q}{r}\right)^{1/24} e^{-i\pi/6} \sum_{n=-\infty}^{\infty} (-1)^n q^{3n^2/2+n/2}.$$

Comparison to equation (D.68) gives

$$(q;q)_\infty = \left(\frac{r}{q}\right)^{1/24} \frac{e^{i\pi/6}}{\sqrt{s}} \sum_{n=-\infty}^{\infty} e^{-\pi n^2/s+2\pi int}.$$

It remains to simplify the right-hand side.

Substitute t and s on the right-hand side and simplify:

$$\sum_{n=-\infty}^{\infty} e^{-\pi n^2/2+2\pi int} = \sum_{n=-\infty}^{\infty} r^{n^2/6+n/6} e^{2\pi in/6}.$$

The sum on the right-hand side is absolutely convergent if $|r| < 1$. Collecting terms,

$$(q;q)_\infty = \left(\frac{r}{q}\right)^{1/24} \frac{e^{-i\pi/6}}{\sqrt{s}} \sum_{n=-\infty}^{\infty} r^{n^2/6+n/6} e^{2\pi in/6}.$$

The right-hand side is evaluated by summing along n in residue classes modulo 6: replace $n \to 6n + \ell$ and sum over ℓ from -3 to 2.

$$(q;q)_\infty = \left(\frac{r}{q}\right)^{1/24} \frac{e^{i\pi/6}}{\sqrt{s}} \sum_{n=-\infty}^{\infty} \sum_{\ell=-3}^{2} r^{(6n+\ell)^2/6+(6n+\ell)/6} e^{2\pi i(n+\ell/6)}.$$

Comparing the terms in the sum over ℓ shows that the terms for $\ell = -2$ and $\ell = 1$ cancel. The terms for both $\ell = -3$ and $\ell = 2$ are equal under the exchange of $n \to -n$, up to phase factors. This is similarly the case for $\ell = -1$ and $\ell = 0$. This simplifies the expressions above to

$$(q;q)_\infty = \left(\frac{r}{q}\right)^{1/24} \frac{2}{\sqrt{s}} \sum_{n=-\infty}^{\infty} \Big(r^{(3n+1)(2n+1)} \cos\big(\tfrac{2\pi}{6}(6n+\tfrac{5}{2})\big) + r^{n(n+6)} \cos\big(\tfrac{2\pi}{6}(6n+\tfrac{1}{2})\big)\Big).$$

Evaluate the cosines and replace $s = \frac{-3\log q}{2\pi}$ to complete the proof. □

Taking logarithms of the result in theorem D.13 and keeping the $n = 0$ term gives the following asymptotic formula for $(q;q)_\infty$ as $q \to 1^-$:

Theorem D.14 *The function* $\log(q;q)_\infty = \frac{\pi^2}{6\log q} + \frac{1}{2}\log\Big(\frac{2\pi}{-\log q}\Big) + O(\log q)$. □

This gives the approximation

$$(q;q)_\infty = \sqrt{\frac{-2\pi}{\log q}}\, e^{-\frac{\pi^2}{6|\log q|}+O(\log(q))} \quad \text{as } q \to 1^-. \tag{D.74}$$

For additional results, see, for example, the paper by Moak [425].

D.5.5 *Asymptotics for* $(t;q)_\infty$

First examine the limit $\lim_{n\to\infty}(t;q)_n$ for $q \in (0,1)$ and assume that $0 < tq^k < 1$ for all $k \geq 1$.

Theorem D.15 *If* $q < 1$, *and* $0 < tq^k < 1$ *for all* $k \geq 1$, *then* $0 < (t;q)_\infty \leq 1$.

Proof Surely, $0 \leq (t;q)_n \leq 1$ for all $n \geq 0$, and $(t;q)_n$ is monotone decreasing, so the limit $\lim_{n\to\infty}(t;q)_n$ exists and is less than or equal to 1.

It remains to show that the limit is not 0.

For any positive $p < 1$, $\log(1-p) \geq \frac{-p}{1-p}$. Moreover, for any small $\epsilon > 0$, there exists a $N_0 \in \mathbb{N}$ such that $0 < 1-\epsilon < 1-tq^k$ if $k > N_0$, since $q < 1$ and since $t \neq 0$. Hence, $\log(1-tq^k) \geq \frac{-tq^k}{1-\epsilon}$ if $k > N_0$.

Thus, assume that $n > N_0 + 1$; then

$$\begin{aligned}
(t;q)_n &= \exp\left(\sum_{k=0}^{n-1} \log(1-tq^k)\right) \\
&\geq \exp\left(\sum_{k=0}^{N_0} \log(1-tq^k) + \frac{-t}{1-\epsilon}\sum_{i=N_0+1}^{n-1} q^k\right), \quad \text{if } n > N_0+1; \\
&= \exp\left(\sum_{k=0}^{N_0} \log(1-tq^k) + \left(\frac{-tq}{1-\epsilon}\right)\left(\frac{1-q^{n-N_0-1}}{1-q}\right)\right).
\end{aligned}$$

Take $n \to \infty$ to obtain

$$\lim_{n\to\infty}(t;q)_n \geq \exp\left(\sum_{k=0}^{N_0} \log(1-tq^k) + \frac{-tq}{(1-\epsilon)(1-q)}\right)$$

which is greater than 0. □

Asymptotic approximations for the q-Pochhammer function $(t;q)_n$ can be determined using the Euler-Maclaurin theorem. This can be determined directly or indirectly by first determining an approximation for the q-gamma function (see, for example, references [425] and [468]).

Define $\mathcal{L}i_2$ to be the dilogarithm:

$$\mathcal{L}i_2 = \sum_{k=1}^{\infty} \frac{z^k}{k^2} = \int_z^0 \tfrac{1}{t}\log(1-t)\,dt. \tag{D.75}$$

Theorem D.16 *If $0 < t < 1$, and $0 < q < 1$, then*

$$\log(t;q)_n = \tfrac{1}{\log q}\left(\mathcal{L}i_2(t) - \mathcal{L}i_2(tq^{n-1})\right) + \tfrac{1}{2}\left(\log(1-t) - \log(1-tq^{n-1})\right) + R_1,$$

where

$$|R_1| \le \tfrac{4}{3}|t|^2|\log q|\left|\frac{1-q^{n-1}}{(1-t)(1-tq^{n-1})}\right|.$$

For fixed $t < 1$, this term vanishes as $q \nearrow 1^-$.

Proof The Euler-Maclaurin formula gives

$$\log(t;q)_n = \int_0^{n-1} \log(1-tq^x)\,dx + \tfrac{1}{2}\left(\log(1-t) + \log(1-tq^{n-1})\right) + R_1.$$

where

$$R_1 = \tfrac{1}{2}\int_0^{n-1} (B_2 - B_2(x-\lfloor x\rfloor))\left(\tfrac{d^2}{dx^2}\log(1-tq^x)\right)dx.$$

The integral over $\log(1-tq^x)$ can be done by expanding first the logarithm and then integrating term by term; the result is a sequence of terms giving rise to dilogarithms and logarithms.

It remains only to examine the remainder. Observe that

$$|B_2 - B_2(x-\lfloor x\rfloor)| = \left|-\tfrac{1}{3}(x-\lfloor x\rfloor) - (x-\lfloor x\rfloor)^2\right| \le \tfrac{4}{3} \text{ if } x \in [0, n-1].$$

Next, consider the derivative in R_1 above and compute it to see that

$$\begin{aligned}
|R_1| &\le \tfrac{1}{2}\int_0^{n-1} \left|(B_2 - B_2(x-\lfloor x\rfloor))\left(\tfrac{tq^x(\log q)^2}{(1-tq^x)^2}\right)\right| dx \\
&\le \tfrac{2}{3}\int_0^{n-1} \left|\frac{tq^x(\log q)^2}{(1-tq^x)^2}\right| dx = \tfrac{2}{3}t^2|\log q|\int_{tq^{n-1}}^{t} \frac{dz}{(1-z)^2} \\
&= \tfrac{2}{3}t^2|\log q|\left|\frac{1-q^N}{(1-t)(1-tq^N)}\right|.
\end{aligned}$$

This completes the proof. □

Taking $n \to \infty$ in theorem D.16 gives the following corollary:

Corollary D.17 *If $\epsilon > 0$, and $t \in [0, 1-\epsilon]$, and if $q \in (0,1)$, then*

$$\log(t;q)_\infty = \tfrac{1}{\log q}(\mathcal{L}i_2(t) - \tfrac{\pi^2}{6}) + \tfrac{1}{2}\log(1-t) + R_1,$$

or if $\mathcal{L}i_2(t) + \mathcal{L}i_2(1-t) = \frac{\pi^2}{6} - \log(t)\log(1-t)$ is used, then

$$\log(t;q)_\infty = -\tfrac{1}{\log q}\left(\mathcal{L}i_2(1-t) - \log(t)\log(1-t)\right) + \tfrac{1}{2}\log(1-t) + R_1,$$

where R_1 approaches 0 uniformly as $q \nearrow 1^-$ for any $t \in [0, 1-\epsilon]$. □

By this corollary,

$$(t;q)_\infty = \sqrt{1-t}\; e^{-\frac{1}{|\log q|}\left(\mathcal{L}i_2(t)-\frac{\pi^2}{6}\right)+R_1}, \tag{D.76}$$

where $R_1 \to 0$ uniformly as $q \to 1^-$. For example, if $t = q^n < q < 1$ for $n \in \mathbb{N}$, then

$$(q^n;q)_\infty = \sqrt{1-q^n}\; e^{-\frac{1}{|\log q|}\left(\mathcal{L}i_2(q^n)-\frac{\pi^2}{6}\right)+R_1}. \tag{D.77}$$

D.6 Asymptotics from the generating function

If the generating function $g(t)$ of a function q_n on $\mathbb{N}$, defined by

$$g(t) = \sum_{n=0}^{\infty} q_n t^n, \tag{D.78}$$

is known, then q_n may in principle be found by using Darboux's theorem. Darboux's theorem has the following formulation (see, for example, reference [443]).

Theorem D.18 (Darboux's theorem) *Let $\psi(t)$ be an analytic function with Laurent expansion*

$$\psi(t) = \sum_{n=-\infty}^{\infty} a_n t^n,$$

in the annulus $0 < |t| < r$.

Let $\chi(t)$ be a function which is analytic in $0 < |t| < r$ and with Laurent expansion given by

$$\chi(t) = \sum_{n=-\infty}^{\infty} b_n t^n,$$

in $0 < |t| < r$.

Suppose that the difference of the m-th derivatives of $\psi(t)$ and $\chi(t)$ has a finite number of singularities at $t = t_j$ such that

$$\psi^{(m)}(t) - \chi^{(m)}(t) = O([t-t_j]^{\sigma_j - 1}),$$

for some positive constants σ_j, as $t \to t_j$. Then

$$a_n = b_n + o(r^{-n} n^{-m}) \quad \text{as } n \to \infty.$$

□

D.7 Convergence of continued fractions

Finite fractions are denoted by

$$S_n = \mathop{K}_{k=1}^{n} \frac{a_k}{b_k} = \cfrac{a_1}{b_1 + \cfrac{a_2}{b_2 + \cfrac{a_3}{b_3 + \cfrac{\cdots}{b_n}}}}. \tag{D.79}$$

(For example, $S_1 = \frac{a_1}{b_1}$, and $S_2 = \frac{a_1}{b_1 + \frac{a_2}{b_2}}$).

Taking $n \to \infty$ gives the infinite continued fraction S_∞. The n-th approximation to S_∞ is $S_n(s)$, obtained by truncating the fraction at n and putting $b_n = c$:

$$S_n(c) = \mathop{K}_{k=1}^{n} \frac{a_k}{b_k} = \cfrac{a_1}{b_1 + \cfrac{a_2}{b_2 + \cfrac{a_3}{b_3 + \cfrac{\cdots}{c}}}}. \tag{D.80}$$

Useful theorems on the convergence of continued fractions are the following.

Theorem D.19 (Pringsheim [474]) *Let $\langle a_n \rangle$ and $\langle b_n \rangle$ be sequences in $\mathbb{R}$ such that $|b_n| > |a_n| + 1$. Then the continued fraction $K_{k=1}^{\infty} \frac{a_k}{b_k}$ converges absolutely to $f \in \mathbb{R}$ with $0 < |f| < 1$.* □

Theorem D.20 (Worpitzky [572]) *Let $\langle a_n \rangle$ be a sequence in $\mathbb{C}$ and let K be a continued fraction given by*

$$K = \mathop{K}_{k=1}^{\infty} \frac{a_k}{1} = \cfrac{a_1}{1 + \cfrac{a_2}{1 + \cfrac{a_3}{1 + \cfrac{a_4}{1 + \cdots}}}}.$$

If $0 < |a_n| \leq \frac{1}{4}$ for all $n \geq 2$, then K converges absolutely and $0 < |K| \leq \frac{1}{2}$. Moreover, if, in addition, $S_n(b) = K_{k=1}^{n} \frac{a_k}{1}$, where $a_n = b$ and $|b| \leq \frac{1}{2}$ is the n-th approximant of K, then $0 < |S_n(b)| \leq \frac{1}{2}$. □

Theorem D.21 (Van Vleck [554]) *If $r_n > 0$, then a condition that is necessary for the continued fraction*

$$K = \mathop{K}_{k=1}^{\infty} \frac{a_k}{1} = \cfrac{1}{1 + \cfrac{r_2 e^{i\theta_2}}{1 + \cfrac{r_3(1-r_2)e^{i\theta_3}}{1 + \cfrac{r_4(1-r_3)e^{i\theta_4}}{1 + \cdots}}}}$$

to converge for all values of θ_n *is that the series*

$$1 + \sum_{j=2}^{\infty} \prod_{\ell=2}^{j} \frac{r_\ell}{1 - r_\ell}$$

is convergent. This condition is also sufficient if $r_n < 1$ *for all values of* n. □

APPENDIX E

PERCOLATION IN $\mathbb{Z}^d$

Percolation occurs when a cluster of infinite size develops at a critical density of open edges or vertices in a lattice. A comprehensive discussion of percolation phenomena can be found in the book by Grimmett [240]; the older review articles by Essam [181] and Stauffer [533] are also good introductions.

E.1 Edge percolation

Consider the hypercubic lattice $\mathbb{L}^d$ and let $\mathbb{E}^d$ be the collection of unit length edges in the hypercubic lattice. An edge in $\mathbb{E}^d$ is *open* with probability $p \in [0, 1]$ and *closed* with complementary probability $q = 1 - p$.

Let μ_b be the Bernoulli measure on the set $\{0, 1\}$, and define the *sample space* $\mathbb{P} = \prod_{e\in\mathbb{E}^d}\{0, 1\}$. If $\tau \in \mathbb{P}$, then $\mu_b(\tau(e){=}1) = p$ and $\mu_b(\tau(e){=}0) = q$. This gives a product measure $P_p = \prod_{e\in\mathbb{E}^d}\mu_b$ with density p on $\mathbb{P}$. The probability space $(\mathbb{P}, \mathbb{F}, P_p)$ is a measure space with σ-algebra $\mathbb{F}$ of subsets of $\mathbb{P}$ endowed with the (probability) measure P_p. The parameter p is the *density* of the process.

The random graph of all open edges (with their end-vertices) in $\mathbb{E}^d$ consists of connected components which are open.

The open cluster containing a vertex $\vec{x} \in \mathbb{L}^d$ is denoted $C(\vec{x})$, and, since P_p is translational invariant, the distribution of $C(\vec{x})$ is independent of $\vec{x}$. The study of percolation is normally concerned with open clusters containing the origin $\vec{0}$ and which are denoted by $C(\vec{0}) \equiv C$.

The *size* of C is the number of edges in C and is denoted $|C|$. The *order* of C is the number of vertices it contains and is denoted by $\|C\|$.

E.1.1 *The size of the cluster at the origin*

The *percolation probability* $\theta(p)$ is the probability that $|C|{=}\infty$.

The probability $\theta(p)$ is best defined in terms of the measure by P_p as follows. Let $P_p(|C|{=}n)$ be the probability that the cluster at the origin has size n at density p. Then $P_p(|C|{=}\infty)$ is the probability that the cluster at the origin is infinite in size. Hence,

$$\theta_d(p) = P_p(|C|{=}\infty) = 1 - \sum_{n=0}^{\infty} P_p(|C|{=}n). \tag{E.1}$$

The *fundamental theorem of percolation* states that in two or higher dimensions there exists a critical probability $p_c(d) \in (0, 1)$ such that $\theta_d(p) = 0$ if $p < p_c(d)$, and $\theta_d(p) > 0$ if $p > p_c(d)$ [80].

The probability $p_c(d)$ is the *critical percolation probability* or the *critical density*, defined by

$$p_c(d) = \sup\{p \mid \theta_d(p) = 0\}. \tag{E.2}$$

Observe that in one dimension it is trivial that $p_c(1) = 1$; however, percolation is interesting because $0 < p_c(d) < 1$ if $d \geq 2$.

Since $\mathbb{L}^d$ is a subgraph of $\mathbb{L}^{d+1}$, the existence of an infinite cluster in $\mathbb{L}^d$ implies an infinite cluster in $\mathbb{L}^{d+1}$. Thus,

$$p_c(d+1) \leq p_c(d) \;\text{ for } d \geq 1. \tag{E.3}$$

The mean size of an open cluster is defined by $\chi_d(p) = \mathbb{E}_p|C|$, where $\mathbb{E}_p$ is the expectation at density p. This is explicitly given by

$$\chi_d(p) = \infty \cdot P_p(|C|{=}\infty) + \sum_{n=0}^{\infty} n P_p(|C|{=}n). \tag{E.4}$$

In other words, $\chi_d(p) = \infty$ if $p > p_c(d)$, and $\chi_d(p) < \infty$ if $p < p_c(d)$.

Theorem E.1 (The fundamental theorem of percolation [80]) *If $d \geq 2$, then there exists a critical percolation probability $p_c(d) \in (0, 1)$.*

Proof By equation (E.3), it is sufficient to show that $p_c(2) < 1$, and $p_c(d) > 0$.

CLAIM: The probability $p_c(2) < 1$.

PROOF OF CLAIM: The underlying approach is a Peierls argument [462], which exploits the self-duality of $\mathbb{L}^2$. Choose the origin in $\mathbb{L}^2$ and set up the standard x- and y-axes. Let the planar dual graph of $\mathbb{L}^2$ be denoted $(\mathbb{L}^2)^*$. Since $\mathbb{L}^2$ is self-dual, its edges are in one-to-one correspondence with the *dual edges* in $(\mathbb{L}^2)^*$.

Suppose that the cluster at the origin C is finite. The *perimeter edges* of C are closed edges with at least one endpoint in C. *Solvent edges* are closed edges with exactly one endpoint in C. The union of dual edges of the solvent edges is a lattice polygon in $(\mathbb{L}^d)^*$ and which contains C in its interior.

Define $p_n(C)$ as the number of polygons of length n in $(\mathbb{L}^d)^*$ and with C in their interior. Each polygon counted by $p_n(C)$ passes through the x-axis and must contain a point of the form $(k + \frac{1}{2}, \frac{1}{2})$, $|k| \leq n$.

Choose such a point as a root in the polygon. Let $p_n(r)$ be the number of rooted lattice polygons in $\mathbb{L}^2$ of length n and let p_n be the number of polygons of length n (counted modulo translation). Then it follows that $p_n(C) \leq 2n p_n(r)$, since k has at most $2n$ values, and $p_n(r) \leq n p_n$, since the root has at most n positions. This shows that $p_n(C) \leq 2n^2 p_n(r)$.

The probability that a polygon A in $\mathbb{L}^2$ contains the cluster C at the origin if the process has density p is at most

$$1 - \theta_2(p) \leq \sum_A \mathrm{P}_r\,(A \text{ is closed}) \leq \sum_{n=1}^{\infty} p_n(C)\, q^n \leq \sum_{n=1}^{\infty} 2n^2 p_n\, q^n,$$

where $q = 1 - p$. This is finite if $q < \frac{1}{\mu_2} < 1$, by theorem 1.4, and the right-hand side approaches 0 if $q \to 0^+$.

In other words, there is a $\epsilon < 1$ such that $\sum_{n=1}^{\infty} p_n(C)\, q^n < \frac{1}{2}$ if $1-\epsilon > q$, or $p > \epsilon > 0$. This shows that $\theta_2(p) \geq \frac{1}{2}$ if $p \leq \epsilon < 1$, and so there is a critical probability $p_c(2) \leq \epsilon < 1$. $\triangle$

CLAIM: The percolation critical point $p_c(d) > 0$.
PROOF OF CLAIM: Let m_n be the number of paths of length n open edges from the origin. Each path is open with probability p^n, and the expected number of open paths of length n is $\mathbb{E}_p m_n = c_n\, p^n$, where c_n is the number of self-avoiding walks of length n from the origin.

If the origin is in an infinite open cluster, then there are open paths of all lengths from the origin. This shows that $\theta_d(p) \leq \mathbb{E}_p m_n = c_n p^n$.

However, $c_n = \mu_d^{n+o(n)}$, by theorem 1.1; thus, $\theta_d(p) \leq p^n c_n \to 0$ if $p < \frac{1}{\mu_d}$. Hence, $p_c(d) \geq \frac{1}{\mu_d} > 0$ for all d. $\triangle$

This completes the proof. □

E.2 The decay of the percolation cluster

In section E.4 it will be shown that $P_p(|C|{=}n) \geq e^{\eta n^{(d-1)/d}}$ for a finite $\eta > 0$ and for all $n > 0$ if $p > p_c(d)$ (see theorem E.12). Since $P_p(|C|{=}n) \leq 1$, it follows that

$$\lim_{n\to\infty} \tfrac{1}{n} \log P_p(|C|{=}n) = 0 \;\text{ for all } p > p_c(d). \tag{E.5}$$

This limit also exists for $p \leq p_c(d)$, and it can be shown to be strictly negative if $p < p_c(d)$. The existence of the limit can be shown using an argument in reference [364].

Theorem E.2 *There exists a function $\zeta(p) \geq 0$ such that*

$$\zeta(p) = -\lim_{n\to\infty} \tfrac{1}{n} \log P_p(|C|{=}n).$$

Moreover, $\zeta(p)$ is finite if $p \in (0,1]$.

Proof It is shown that $P_p(|C|{=}n)$ satisfies a supermultiplicative relation.

The probability that a cluster A at the origin has exactly n edges is $\pi_n = P_p(|C|{=}n)$. The probability that A also has a lexicographic most vertex at the origin is at least $\frac{1}{2n}\pi_n$, since A has at most $2n$ vertices.

Similarly, the probability that a cluster B containing the vertex $v = (1,0,\ldots)$ has size m is π_m by translational invariance. The probability that such a cluster has its lexicographic least vertex at v and has size m is at least $\frac{1}{2m}\pi_m$.

Clusters A and B may be concatenated into the single cluster $A{\star}B$ by adding the edge between the origin and v (see figure 2.3). The probability of $A{\star}B$ is $\frac{p}{(1-p)^2} P_p(A)P_p(B)$, since the perimeter edge of A and B between the origin and v had been closed, but is now open. Hence, $P_p(A{\star}B) = \frac{p}{(1-p)^2} P_p(A)P_p(B)$.

Summing over all clusters A with a lexicographic most vertex at the origin gives $\sum_A P_p(A) = P_p(|A|{=}n$, top vertex at the origin$) \geq \frac{1}{2n}\pi_n$. Similarly, summing over all clusters B with a lexicographic least vertex at the v and of size m gives $\sum_B P_p(B) \geq \frac{1}{2m}\pi_m$.

This shows that

$$\sum_A \sum_B P_p(A \star B) \geq \frac{p}{(1-p)^2}(\tfrac{1}{2n}\pi_n)(\tfrac{1}{2m}\pi_m).$$

By translational invariance,

$$\sum_A \sum_B P_p(A \star B) \leq P_p(|C| = n+m, \text{ top vertex at the origin}) \leq \tfrac{d}{n+m}\pi_{n+m},$$

since each cluster of size n edges has at least $\frac{n}{d}$ vertices. This shows that

$$p(1-p)^2(\tfrac{1}{2n}\pi_n)(\tfrac{1}{2m}\pi_m) \leq \tfrac{d}{n+m}\pi_{n+m}. \tag{E.6}$$

By lemma (A.1), the limit $\zeta(p) = -\lim_{n\to\infty} \frac{1}{n}\log \pi_n$ exists.

Consider a cluster which is a straight walk of length n. By determining its perimeter, it follows that $p^n q^{2(d-1)n+2d} \leq \pi_n \leq 1$. Thus, $\zeta(p)$ is finite if $p \in (0,1]$. This completes the proof. □

E.3 Exponential decay of the subcritical cluster

To show that $\zeta(p) > 0$ if $p < p_c(d)$ requires some work: the proof below is due to Aizenman and Newman [3, 240]. There are several ingredients in this proof, most notably, (1) events, (2) the BK inequality, (3) some results from graph theory, and (4) a shift of focus from edges to vertices in clusters. Consider these in turn.

E.3.1 *Ingredients*

(1) and (2): Let a and b be two states sampled independently from the probability space $(\mathbb{P}, \mathbb{F}, P_p)$. Then $a(e) = 0$ if the edge e is closed, and $a(e) = 1$ otherwise. This is similarly the case for b. If two states a and b are related by $a(e) \leq b(e)$ on all subsets in $\mathbb{F}$, then $a \leq b$.

An *event* A is a collection of edges. The event A *occurs* if every edge in A is open; it occurs in a state a if for all edges $e \in A$, $a(e) = 1$ and so $\mathbb{1}_A(a) = 1$. An event is *increasing* if $\mathbb{1}_A(a) \leq \mathbb{1}_A(b)$ for any pair of states $a \leq b$.

The function P_p is a probability measure on the space of events. The probability of an event A is its measure $P_p(A) = \mathbb{E}_p \mathbb{1}_A$. Observe that $P_{p_1}(A) \leq P_{p_2}(A)$ if $p_1 \leq p_2$, and A is an increasing event. By the FKG inequality [207], increasing events are positively correlated:

$$P_p(AB) \geq P_p(A)P_p(B), \tag{E.7}$$

where AB is the event that both A and B occur.

In some cases events A and B will be disjoint, denoted by $A \circ B$ (for example, A and B could be two open paths which are edge-disjoint). Then the BK inequality [552] states that

$$P_p(A \circ B) \le P_p(A) P_p(B). \tag{E.8}$$

(3): A *skeleton* is a tree graph with all vertices of degree 1 or 3, and with all vertices labelled in the set $M = \{1, 2, 3, \ldots, m\}$. The labelled vertices of degree 1 are *end-vertices*, and the unlabelled vertices of degree 3 are *internal vertices.*

If S is a skeleton, then denote its set of end-vertices by $D(S)$, and its internal vertices by $I(S)$. If $|D(S)| = m$, then $|I(S)| = m - 2$, and there are $2m - 3$ edges in S. The number of skeletons with $|D(S)| = m$ is denoted w_m.

The function w_m can be determined recursively: select one of the $2m - 3$ edges, say e, and add an edge $\langle w \sim u \rangle$, with w subdividing e, and u becoming a new end-vertex. Assign to u the label $m + 1$. This will generate $2m - 3$ new skeletons with $m + 1$ end-vertices from each skeleton with m end-vertices. Thus, $w_{m+1} = (2m - 3) w_m$ and, since $w_3 = 1$, it follows that

$$w_{m+1} = \frac{(2m-3)!}{2^{m-1}(m-1)!}. \tag{E.9}$$

In addition to the above, the following lemma will be useful.

Lemma E.3 *Let $G = (V, E)$ be a finite connected graph with vertex set V. Suppose that $W = \{v_i\} \subset V$. Then there is a vertex $v_i \in W$ such that $G - v_i$ has one component which contains all the vertices in $W \setminus \{v_i\}$.*

Proof Let T be a spanning tree of G with root u. Define $d(u, v_i)$ as the path length from u to v_i. There is a finite longest path, so choose i and u such that $d(u, v_i)$ is maximal. Then u is an end-vertex of T, and $T - v_i$ has a component which includes all the vertices in the set $W \setminus \{v_i\}$. Consequently, there is an i such that $G - v_i$ has one component which contains the vertices in $W \setminus \{v_i\}$, as claimed. □

(4): It is necessary to shift focus from $P_p(|C| = n)$ to $P_p(\|C\| = v)$, the probability that the cluster at the origin has v vertices.

The probability $P_p(\|C\| = v)$ is related to lattice animals in the following way: define $a_v(k, c)$ to be the number of lattice animals with n edges, v vertices, cyclomatic index c, and k nearest-neighbour *contacts* (these are lattice edges with both endpoints in the cluster but which are not in the cluster themselves). It follows that

$$P_p(\|C\| = v) = \sum_{k,c} v\, a_v(k, c)\, p^n q^\rho, \tag{E.10}$$

where $\rho = 2dv - 2n - k$ is the number of perimeter edges of the animal, and one may verify that $v = n + 1 - c$.

Lemma E.4 *The function $\zeta(p) = 0$ if and only if $\lim_{v \to \infty} \frac{1}{v} \log P_p(\|C\| = v) = 0$.*

Proof Exploit the relation with animals in equation (E.10): animals of size n edges have $\lfloor \frac{n}{d} \rfloor \le v \le n+1$.

Hence, $P_p(|C|=n) \le \sum_{v=\lfloor n/d \rfloor}^{n+1} P_p(\|C\|=v)$. Choose v_M, a value of v which maximises $P_p(\|C\|=v)$ in the right-hand summation.

Then $v_M \in \{\lfloor \frac{n}{d} \rfloor, \lfloor \frac{n}{d} \rfloor + 1, \dots, n+1\}$.

This shows that $P_p(|C|=n) \le (n+1)P_p(\|C\|=v_M)$. Take logarithms on both sides, divide by n and take $n \to \infty$. By theorem E.2,

$$\lim_{n\to\infty} \tfrac{1}{n} \log P_p(|C|=n) \le \alpha_m \limsup_{v\to\infty} \tfrac{1}{v} \log P_p(\|C\|=v),$$

where $\alpha_m = \limsup_{n\to\infty} \frac{1}{n} v_M \le 1$.

On the other hand, if an animal has v vertices, c cycles and k contacts, then it has at least $v-1$ edges and at most $(d-1)v$ edges.

Thus, $P_p(\|C\|=v) \le \sum_{n=v-1}^{(d-1)v} P_p(|C|=n)$.

Define n_M as that value of n in $\{v-1, v, \dots, (d-1)v\}$ which maximises $P_p(|C|=n)$. Then $P_p(\|C\|=v) \le (d-1)v P_p(|C|=n_M)$. Take logarithms and divide both sides by v. Take the limit superior on the left-hand side as $v \to \infty$ and define $\alpha_M = \limsup_{v\to\infty} \frac{1}{v} n_M$. Then, again by theorem E.2,

$$\limsup_{v\to\infty} \tfrac{1}{v} \log P_p(\|C\|=v) \le \alpha_M \lim_{n\to\infty} \tfrac{1}{n} \log P_p(|C|=n).$$

It only remains to show that the limit $\lim_{v\to\infty} \frac{1}{v} \log P_p(\|C\|=v) = 0$ exists. This limit exists by an argument exploiting supermultiplicativity in the same way as in the proof of theorem E.2. □

E.3.2 *Connectivity functions*

With the above concepts in place, one may prove that $\zeta(p) = 0$ if $p > p_c(d)$, and $\zeta(p) > 0$ if $p < p_c(d)$. The proof is lengthy and relies on the properties of *connectivity functions* in the percolation process.

Let $\mathbb{1}_{v_0 \leftrightarrow v_1}$ be the indicator function of the event that the vertices v_0 and v_1 belong to the same open cluster. Then the order of the open cluster which contains the vertex v_0 is $\|C(v_0)\| = \sum_{v_1} \mathbb{1}_{v_0 \leftrightarrow v_1}$, and the pair-connectivity function is defined by

$$\tau_p(v_0, v_1) = \mathbb{E}_p \mathbb{1}_{v_0 \leftrightarrow v_1}. \tag{E.11}$$

Clearly, the mean order of the cluster containing v_0 is

$$\chi'(p) = \sum_{v_1} \tau_p(v_0, v_1) = \mathbb{E}_p \|C(v_0)\| . \tag{E.12}$$

Then it follows that $\chi'(p) = \mathbb{E}_p \|C(v_0)\| \le \mathbb{E}_p \|C\| = \chi_d(p)$.

Thus, since $\chi_d(p) < \infty$ if $p < p_c(d)$, it follows that

$$\chi'(p) < \infty \quad \text{if } p < p_c(d). \tag{E.13}$$

More generally, one may define the connectivity function of the set of vertices $\{v_0, v_1, \ldots, v_m\}$ by

$$\begin{aligned}\tau_p(v_0, v_1, \ldots, v_m) &= P_p\left(\{v_i\} \text{ belongs to the same open cluster}\right) \\ &= \mathbb{E}_p(\mathbb{1}_{v_0 \leftrightarrow v_1} \mathbb{1}_{v_0 \leftrightarrow v_2} \cdots \mathbb{1}_{v_0 \leftrightarrow v_m})) \end{aligned} \tag{E.14}$$

Thus, by using transitivity and the FKG inequality,

$$\begin{aligned}\mathbb{E}_p \|C(v_0)\|^m &= \sum_{v_1, v_2, \ldots, v_m} (\mathbb{E}_p \mathbb{1}_{v_0 \leftrightarrow v_1})(\mathbb{E}_p \mathbb{1}_{v_0 \leftrightarrow v_2}) \cdots (\mathbb{E}_p \mathbb{1}_{v_0 \leftrightarrow v_m}) \\ &\leq \sum_{v_1, v_2, \ldots, v_m} \mathbb{E}_p(\mathbb{1}_{v_0 \leftrightarrow v_1} \mathbb{1}_{v_0 \leftrightarrow v_2} \cdots \mathbb{1}_{v_0 \leftrightarrow v_m}) \\ &= \sum_{v_1, v_2, \ldots, v_m} \tau_p(v_0, v_1, \ldots, v_m). \end{aligned} \tag{E.15}$$

Consider next the 3-point connectivity function $\tau_p(v_0, v_1, v_2)$.

Let $W = \{v_0, v_1, v_2\}$ and let C be the cluster of open edges containing the vertices in W. By lemma E.3, there is a shortest path α of open edges in C between two of the vertices in W (say v_0 and v_1) which avoids the third vertex (v_2). Moreover, there is a shortest path from v_2 which meets α in a vertex v. (In the cases that $v_0 = v_1 = v_2$, or $v_0 = v_1$, one may that find $v = v_0$ in the above). The paths from v to v_i are edge-disjoint by this construction. That is, if the event of three edge-disjoint paths from v to v_i are denoted by $\{v \leftrightarrow v_0\} \circ \{v \leftrightarrow v_1\} \circ \{v \leftrightarrow v_2\}$, then it follows that

$$\begin{aligned}\tau_p(v_0, v_1, v_2) &\leq \sum_v P_p(\{v \leftrightarrow v_0\} \circ \{v \leftrightarrow v_1\} \circ \{v \leftrightarrow v_2\}) \\ &\leq \sum_v \tau_p(v, v_0)\, \tau_p(v, v_1)\, \tau_p(v, v_2), \end{aligned} \tag{E.16}$$

where $\{v \leftrightarrow v_i\}$ is the event that there is an open path from v to v_i, and where the BK inequality was used (equation (E.15)).

Sum the above over v_1 and v_2 and use equation (E.12) and equation (E.15) with $m = 2$. This shows that

$$\mathbb{E}_p \|C\|^2 \leq (\chi'(p))^3. \tag{E.17}$$

This is a *skeleton-inequality*, and it will be generalised to find upper bounds on $\mathbb{E}_p \|C\|^m$.

A generalisation of equation (E.17) is needed. Consider the connectivity function $\tau_p(v_0, v_1, \ldots, v_m)$. A cluster contributing to this function will be mapped to a skeleton S by a map $\Phi_{\mathbf{v}}$ (where $\mathbf{v} = (v_0, v_1, \ldots, v_m)$ are the vertices incident with open edges and will be mapped to the labelled end-vertices of the skeleton).

In particular, define $\Phi_{\mathbf{v}}$ such that (1) the end-vertex of S with label i is mapped to the edge v_i: $\Phi(i) = v_i$, and (2) each of the $2m-1$ edges of S corresponds to edge-disjoint open paths joining the edges v_i in the cluster.

Lemma E.5 *The mapping* $\Phi_{\mathbf{v}}$ *exists.*

Proof Do induction on the number of end-vertices (of which there are $m+1$ in the skeleton).

If $m=2$, then S has three end-vertices and one internal vertex; this case has been considered above. Suppose that there exists a $\Phi_{\mathbf{v}}$ if there are m end-vertices in S. By lemma E.3, there is a label j such that the edges with endpoints in the set $\{v_0, \dots, v_{j-1}, v_{j+1}, \dots, v_m\}$ are in the same connected component of the cluster if v_j is removed. Without loss of generality, suppose that $j=m$.

If all the v_i are distinct from v_m, then first use the induction hypothesis to construct edge-disjoint open paths and a skeleton S' for $\{v_0, v_1, \dots, v_{m-1}\}$. Then add the vertex v_m and a path π which meets the skeleton in a first edge, as was done for the case that $m=2$.

On the other hand, if v_m is equal to another edge, say v_0, then the path π will consist of only a single vertex, and the outcome is still the same.

The internal vertices of S can be labelled $\{m+1, m+2, \dots, 2m-1\}$, and edges of S can then be indicated by $\langle i \sim j \rangle$. Under the mapping such an edge becomes a path $\Phi_{\mathbf{v}}(i) \leftrightsquigarrow \Phi_{\mathbf{v}}(j)$ in the cluster. □

Consider all the possible skeletons and possible mappings $\Phi_{\mathbf{v}}$ that can be defined in this way. These give an upper bound on the connectivity function:

$$\tau_p(v_0, \dots, v_m) \leq \sum_S \sum_{\Phi_{\mathbf{v}}} P_p(\exists \text{ edge-disjoint paths } \{\Phi_{\mathbf{v}}(i) \leftrightsquigarrow \Phi_{\mathbf{v}}(j)\},\ (i,j) \in S).$$

By the BK inequality, this becomes

$$\tau_p(v_0, v_1, \dots, v_m) \leq \sum_S \sum_{\Phi_{\mathbf{v}}} \prod_{(i,j) \in S} \tau_p(\Phi_{\mathbf{v}}(i), \Phi_{\mathbf{v}}(j)). \tag{E.18}$$

In other words, if equation (E.15) is used and the sums are executed over the external edges labelled $\{\Phi_{\mathbf{v}}(1), \Phi_{\mathbf{v}}(2), \dots, \Phi_{\mathbf{v}}(m)\}$ in the above, then

$$\mathbb{E}_p \|C\|^m \leq \sum_S \sum_{\Phi_{\mathbf{v}}} \prod_{\substack{(i,j) \in S \\ i,j>m}} \tau_p(\Phi_{\mathbf{v}}(i), \Phi_{\mathbf{v}}(j)) (\chi'(p))^m. \tag{E.19}$$

The remaining summations over the internal vertices due to the mappings can next be executed to obtain

$$\mathbb{E}_p \|C\|^m \leq \sum_S (\chi'(p))^{2m-1} = \frac{(2m-3)!}{2^{m-1}(m-1)!} (\chi'(p))^{2m-1}, \tag{E.20}$$

where the sum over S is done by using equation (E.9).

E.3.2.1 *Final steps in the proof that $\zeta(p) > 0$ if $p < p_c(d)$:* Proceed by computing the expectation

$$\begin{aligned}\mathbb{E}_p\Big(\|C\|\, e^{t\|C\|}\Big) &= \sum_{m=0}^{\infty} \frac{t^m}{m!}\, \mathbb{E}_p(\|C\|^{m+1}) \\ &\le \chi'(p)\left(1 + \sum_{m=1}^{\infty}\left(\frac{t^m}{m!}\right)\frac{(2m-3)!}{2^{m-1}(m-1)!}(\chi'(p))^{2m}\right) \\ &= \frac{\chi'(p)}{\sqrt{1-2t[\chi'(p)]^2}} \end{aligned} \tag{E.21}$$

By Markov's inequality,

$$P_p(\|C\| \ge v) = P_p(\|C\|\, e^{t\|C\|} \ge v e^{tv}) \le \frac{1}{v e^{tv}}\, \mathbb{E}_p\Big(\|C\|\, e^{t\|C\|}\Big), \quad \text{if } t \ge 0.$$

Thus,

$$P_p(\|C\| \ge v) \le \left(\frac{\chi'(p)}{v e^{tv}}\right) \frac{1}{\sqrt{1-2t(\chi'(p))^2}}. \tag{E.22}$$

Choose $t = \frac{1}{2(\chi'(p))^2} - \frac{1}{2v}$ if $v > (\chi'(p))^2$. Then $t > 0$, and the above simplifies to $P_p(\|C\| \ge v) \le \frac{1}{\sqrt{v}} e^{(1-v/(\chi'(p))^2)/2}$. In other words,

$$P_p(\|C\| = v) \le P_p(\|C\| \ge v) \le \frac{1}{\sqrt{v}} e^{(1-v/(\chi'(p))^2)/2}. \tag{E.23}$$

It follows that $\zeta(p) > 0$ if $p < p_c(d)$, since, by equation E.13, $\chi'(p) < \infty$ if $p < p_c(d)$:

$$\lim_{v\to\infty} \tfrac{1}{v} \log P_p(\|C\| = v) \le -\frac{1}{2(\chi'(p))^2} < 0. \tag{E.24}$$

By lemma E.4, it follows that

Theorem E.6 *There exists a finite function $\zeta(p) \ge 0$ such that*

$$\zeta(p) = -\lim_{n\to\infty} \tfrac{1}{n} \log P_p(|C| = n),$$

and, moreover, $\zeta(p) = 0$ if $p > p_c$, and $\zeta(p) > 0$ if $p < p_c(d)$. □

Moreover, by noting that a cluster of order v vertices has size n with $v-1 \le n \le (d-1)v$, one may cast equation (E.23) in the following form:

$$P_p(|C| \ge n) \le P_p(|C| \ge v-1) \le \sqrt{\frac{d-1}{n}}\, e^{(1-n/(d-1)(\chi'(p))^2)/2}. \tag{E.25}$$

This gives the following theorem.

Theorem E.7 *Suppose that $0 < p < p_c(d)$. Then there exists a $\lambda_p > 0$ such that*

$$P_p(|C|{=}n) \leq P_p(|C|{\geq}n) \leq e^{-\lambda_p n}$$

for all $n > 0$. □

In fact, by equation (E.23), one has

$$P_p(|C|{\geq}n) \leq \sqrt{\tfrac{d-1}{n}}\, e^{(1-n/(d-1)(\chi'(p))^2)/2} \leq 2e^{-n/(\chi'(p))^2)/2} \tag{E.26}$$

for $n > (\chi'(p))^2$.

E.4 Subexponential decay of the supercritical cluster

The cluster C at the origin has exponential decay in the subcritical phase, as shown in theorem E.7. This decay changes at the critical point, and the limit in theorem E.6 is 0 instead. Thus, a different bound on $P_p(|C|{=}n)$ is expected; it decays slower than exponential to zero with n.

Choose $m \in \mathbb{Z}$ and let B_m be a box of side-lengths $2m+1$ centred at the origin in $\mathbb{L}^d$. Denote edges in the boundary of B_m by ∂B_m and denote by $|B_m|$ the number of lattice edges in $|B_m|$.

Let $p_c(d) < p < 1$ and let R_m be the number of vertices in B_m which are in the infinite open cluster. Then the probability that R_m exceeds $\frac{1}{2}\theta(p)\,|B(m)|$ is at least $\frac{1}{2}\theta(p)$. To see this, follow arguments given in reference [240].

Lemma E.8 *The probability $P_p(R_m{\geq}\frac{1}{2}\theta(p)\,|B_m|) \geq \frac{1}{2}\theta(p)$.*

Proof It is the case that $R_m \leq |B_m|$. Hence, the expected value of R_m is bounded above by

$$\begin{aligned}\mathbb{E}_p R_m &\leq |B_m|\,P_p(R_m{\geq}\tfrac{1}{2}\theta(p)\,|B_m|) + \tfrac{1}{2}\theta(p)\,|B_m|\,P_p(R_m{<}\tfrac{1}{2}\theta(p)\,|B_m|)\\ &\leq |B_m|\,P_p(R_m{\geq}\tfrac{1}{2}\theta(p)\,|B_m|) + \tfrac{1}{2}\theta(p)\,|B_m|\,.\end{aligned}$$

However, $\mathbb{E}_p R_m = \theta(p)\,|B_m|$ on the left-hand side. Substitute this to complete the proof. □

Let A_m be the event that all the edges joining pairs of vertices in ∂B_m are open *and* all edges joining vertices in ∂B_m to ∂B_{m+1} are closed. By counting edges in ∂B_m and between ∂B_m and ∂B_{m+1}, there is a function π_p such that $0 < \pi_p < 1$ and

$$P_p(A_m) \geq \pi_p^{|\partial B_m|} \quad \text{for all } m \geq 1. \tag{E.27}$$

Let F_m be all the vertices in B_m which are joined by an open path to vertices in ∂B_m.

It follows that $F_m \geq R_m$; hence, by lemma E.8,

$$P_p(\tfrac{1}{2}\theta(p)\,|B_m| \leq F_m \leq |B_m|) \geq \tfrac{1}{2}\theta(p). \tag{E.28}$$

Notice that F_m is independent of all edges between vertices in the interior of B_m and between edges from a vertex in ∂B_{m-1} to a vertex in ∂B_m. Thus, F_m is independent of A_m, so that in particular, by equations (E.27) and (E.28),

$$P_p(A_m, \tfrac{1}{2}\theta(p)\,|B_m| \leq F_m \leq |B_m|) \geq \tfrac{1}{2}\theta(p)\,\pi_p^{|\partial B_m|}. \tag{E.29}$$

If the left-hand side occurs, then the vertex $(m, m, \ldots, m)$ is in a finite cluster of order at most $|B_m|$ and at least $\frac{1}{2}\theta(p)\,|B_m|$ vertices. By translational invariance, this shows that

$$P_p(\tfrac{1}{2}\theta(p)\,|B_m| \leq |C| \leq |B_m|) \geq \tfrac{1}{2}\theta(p)\,\pi_p^{|\partial B_m|}, \tag{E.30}$$

where C is the cluster at the origin. Since $|\partial B_m| = O(n^{(d-1)/d})$, and $\pi = \alpha p + \cdots$ for some constant $\alpha > 0$, the above shows that

$$P_p(n \leq |C| < \infty) \geq e^{-\eta n^{(d-1)/d}} \tag{E.31}$$

for some $\infty > \eta > 0$ and for all n. This gives the following lemma.

Lemma E.9 *Let $p_c(d) < p < 1$. Then the probability that the cluster at the origin has size n and is finite is bounded below as*

$$P_p(n \leq |C| < \infty) \geq e^{-\eta n^{(d-1)/d}}$$

for some finite $\eta > 0$ and for all $n > 0$. □

E.4.1 *Decay of the supercritical cluster*

It remains to tighten lemma E.9. Following the proof by Aizenman, Delyon and Souillard [2], the key lemma is the following.

Lemma E.10 *Let $p_c(d) < p < 1$ and let $\delta = 2^{d+3}(\theta(p))^{-2}$. Then there exists an $\eta(p) > 0$ and a sequence $\langle r(i)\rangle$ (with $r(i) \in \mathbb{N}$, $r(1) = 1$ and $2 \leq \frac{r(i+1)}{r(i)} \leq \delta$) such that*

$$P_p(|C| = r(i)) \geq r(i)\, e^{-\eta [r(i)]^{(d-1)/d}}.$$

Proof For each $m \geq 1$, there is a $k(m) \in \mathbb{N}$ such that, by equation (E.30),

$$P_p(|C| = k(m)) \geq \frac{1}{2|B_m|}\,\theta(p)\,\pi_p^{|\partial B_m|}, \tag{E.32}$$

where

$$\tfrac{1}{2}\theta(p)\,|B_m| \leq k(m) \leq |B_m| = (2m+1)^d.$$

Hence, since $|B_m| = (2m+1)^d$ and $|\partial B_m| \leq 2d(2m+1)^{d-1}$, it follows from equation (E.32) above that there exists an $\nu_p > 0$ such that

$$P_p(|C| = k(m)) \geq \frac{\theta(p)}{2|B_m|}\, e^{-\nu_p k(m)^{(d-1)/d}},$$

since one may choose ν_p such that, for all m, one has $|\partial B_m| \leq \nu_p k(m)^{(d-1)/d}$.

For large values of $k(m)$, the right-hand side becomes small due to the exponential term; hence, one may choose η_p such that

$$P_p(|C|=k(m)) \geq k(m)e^{-\eta_p k(m)^{(d-1)/d}}, \quad \text{and } P_p(|C|=1) \geq e^{-\eta_p}.$$

Next, the sequence $\langle r(i)\rangle$ must be constructed. This is done inductively. Choose $r(1) = 1$.

Suppose $r(i) = k(j)$ is chosen. Let ℓ be the smallest integer such that $|B_j| \leq \frac{1}{4}\theta(p)\,|B_\ell|$. Put $r(i+1) = k(\ell)$.

It remains to show that $\langle r(i)\rangle$ has the desired properties.

Now $k(\ell) \geq \frac{1}{2}\theta(p)\,|B_\ell| \geq 2\,|B_j| \geq 2k(j)$, since $\frac{1}{2}\theta(p)\,|B_m| \leq k(m) \leq |B_m|$. Hence, $2r(i) \leq r(i+1)$, since $r(i) = k(j)$, and $r(i+1) = k(\ell)$.

On the other hand, it is also the case that ℓ is the smallest integer such that $(2j+1)^d \leq \frac{1}{4}\theta(p)(2\ell+1)^d$. Hence, $(2j+1)^d > \frac{1}{4}\theta(p)(2\ell-1)^d$. This may be put into

$$(2\ell+1)^d \leq \left((2j+1)(\tfrac{4}{\theta(p)})^{1/d}+2\right)^d \leq \frac{2^{d+2}}{\theta(p)}(2j+1)^d.$$

Thus, since $k(\ell) \leq |B_\ell|$, it follows that

$$r(i+1) = k(\ell) \leq |B_\ell| \leq \frac{2^{d+2}}{\theta(p)}\,|B_j| \leq \frac{2^{d+3}}{\theta(p)^2}\,r(i)$$

because $|B_j| \geq \frac{2k(j)}{\theta(p)}$, and $k(j) = r(i)$.

This completes the proof. □

The sequence $\langle r(i)\rangle$ constructed in lemma E.10 may be used to express any $n \in \mathbb{N}$ as

$$n = \sum_{i=1}^{N} w_i r(i), \tag{E.33}$$

where $N = \max\{i|r(i) \leq n\}$ and the w_i are weights such that $0 \leq w_i \leq \delta$, where $\delta = \frac{2^{d+3}}{\theta(p)^2}$ (as in lemma E.10).

Showing this is a necessary component to strengthen lemma E.9 into its desired form.

Lemma E.11 *Define δ, the sequence $\langle r(i)\rangle$, and η as in lemma E.10. Then any $n \in \mathbb{N}$ where $n > 0$ can be written in the form $n = \sum_{i=1}^{N} w_i r(i)$, where $N = \max\{i|r(i) \leq n\}$, and the w_i are weights such that $0 \leq w_i \leq \delta$. Moreover,*

$$\sum_{i=1}^{N} w_i r(i)^{(d-1)/d} \leq 4\delta n^{(d-1)/d}.$$

Proof If $n = 1$, then take $N = w_1 = 1$. The result follows because $r(1) = 1$.

Suppose that the above is true for all $n \leq k$. Suppose that $n = k+1$. Put $N = \max\{i | r(i) \leq k+1\}$.

Then, by the induction hypothesis, $k+1-r(N) \leq k$, so

$$k+1 = r(N) + (k+1-r(N)) = r(N) + \sum_{i=1}^{N} w_i r(i),$$

for some weights w_i satisfying $0 \leq w_i \leq \delta$.

It must be verified that $1 + w_N \leq \delta$.

CLAIM: The sum $1 + w_N \leq \delta$.
PROOF OF CLAIM: By the definition of N, $r(N) \leq k+1 < r(N+1)$. Also, by construction of $r(i)$, $r(N+1) \leq \delta r(N)$. This gives

$$(1 + w_N) r(N) \leq k+1 < r(N+1) \leq \delta r(N).$$

This completes the proof of the claim. △

It remains to show that

$$\sum_{i=1}^{N} w_i r(i)^{(d-1)/d} \leq 4\delta n^{(d-1)/d}.$$

To see this, note that $w_i \leq \delta$ and, by lemma E.9, that $2 \leq \frac{r(i+1)}{r(i)} \leq \delta$; thus,

$$r(i) \leq \tfrac{1}{2} r(i+1) \leq 2^{i-N} r(N) \leq 2^{i-N} n$$

for $1 \leq i \leq N$. Hence,

$$\begin{aligned}
\sum_{i=1}^{N} w_i r(i)^{(d-1)/d} &\leq \delta n^{(d-1)/d} \sum_{i=1}^{N} 2^{(i-N)(d-1)/d} \\
&\leq \delta n^{(d-1)/d} \sum_{i=1}^{\infty} 2^{-i/2} \quad \text{since } d \geq 2 \\
&\leq 4\delta n^{(d-1)/d}.
\end{aligned}$$

This completes the proof. □

It only remains to strengthen lemma (E.9).

Theorem E.12 *Let $p_c(d) < p < 1$. Then the probability that the cluster at the origin has size n and is finite is bounded below as*

$$P_p(|C|=n) \geq e^{-\eta n^{(d-1)/d}},$$

for some finite $\eta > 0$ and for all $n > 0$. □

Proof The proof uses the sequence $\langle r(i)\rangle$ defined in lemma E.10 and with properties as shown in lemma E.11 to expand the result in lemma E.10 to all $n \in \mathbb{N}$.

Let $n > 0$. Then there are weights $w_i \in [0, \delta]$ such that $n = \sum_{i=1}^{N} w_i r(i)$.

By equation (E.6) in the proof of theorem E.2,

$$\tfrac{1}{r+s} P_p(|C|{=}(r+s)) \geq pq^{-2} \tfrac{1}{r} P_p(|C|{=}r) \tfrac{1}{s} P_r(|C|{=}s),$$

for $r > 0$, and $s > 0$, and for all $p \in (0,1)$, and $q \in (0,1)$. Applying this relation recursively for $n = \sum_{i=1}^{N} w_i r(i)$ gives the inequality

$$\tfrac{1}{n} P_p(|C|{=}n) \geq \prod_{i=1}^{N} \left(\frac{pq^{-2}}{r(i)} P_p(|C|{=}r(i)) \right)^{w_i}. \tag{E.34}$$

Now, $w_i \leq \delta$, and $r(i) > 2^{i-1}$, by lemmas E.10 and E.11. This shows that

$$N = \max\{i | r(i) {\leq} n\} \leq 1 + \log_2 n,$$

and

$$w = \sum_{i=1}^{N} w_i \leq \delta(1 + \log_2 n).$$

By lemma E.10,

$$P_p(|C|{=}r(i)) \geq r(i) e^{-\eta [r(i)]^{(d-1)/d}} \tag{E.35}$$

and, by lemma E.11,

$$\sum_{i=1}^{N} w_i r(i)^{(d-1)/d} \leq 4\delta n^{(d-1)/d}. \tag{E.36}$$

Define $x = \min\{1, pq^{-2}\}$ in equation (E.34) above. This gives

$$\begin{aligned}
\tfrac{1}{n} P_p(|C|{=}n) &\geq x^{\delta(1+\log_2 n)} \prod_{i=1}^{N} \left(e^{-\eta (r(i))^{(d-1)/d}} \right)^{w_i} && \text{by equation (E.35);} \\
&= x^{\delta(1+\log_2 n)} e^{-\eta \sum_{i=1}^{N} w_i (r(i))^{(d-1)/d}} \\
&\geq x^{\delta(1+\log_2 n)} e^{-4\eta\delta n^{(d-1)/d}} && \text{by equation (E.36).}
\end{aligned}$$

A suitable redefinition of η completes the proof. □

Thus, theorems E.7 and E.12 can be taken together:

Theorem E.13 *There exist a function $\zeta(p) = -\lim_{n\to\infty} \frac{1}{n} \log P_p(|C|{=}n)$ and a $p_c(d) \in (0,1)$ such that $\zeta(p) = 0$ if $p > p_c(d)$, and $\zeta(p) > 0$ if $p < p_c(d)$.* □

REFERENCES

[1] Adler, J, Meir, Y, Harris, AB, Aharony, A, and Duarté, JAMS (1988). Series study of random animals in general dimensions. *Phys Rev B*, **38**, 4941–4954.

[2] Aizenman, M, Delyon, F, and Souillard, B (1980). Lower bounds on the cluster size distribution. *J Stat Phys*, **23**, 267–280.

[3] Aizenman, M and Newman, CM (1984). Tree graph inequalities and critical behavior in percolation models. *J Stat Phys*, **36**, 107–143.

[4] Akcoglu, MA and Krengel, U (1981). Ergodic theorems for superadditive processes. *J Reine Angew Math*, **323**, 106–127.

[5] Alexandrowicz, Z (1969). Monte Carlo of chains with excluded volume: a way to evade sample attrition. *J Chem Phys*, **51**, 561–565.

[6] Alvarez, J and Janse van Rensburg, EJ (2008). Directed paths in a layered environment. *J Phys A: Math Theo*, **41**(46), 465003.

[7] Andreev, AF (1964). Singularity of thermodynamic quantities at a first order phase transition point (singularity of thermodynamic potential for liquid at boiling point). *Sov Phys-JETP*, **18**, 1415–1416.

[8] Andrews, GE (1976). *The Theory of Partitions*, Volume 2 of *Encycl Math Appl*. Addison-Wesley.

[9] Andrews, GE (1990). "Exemplum memorabile inductionis fallacis" and q-trinomial coefficients. *J Amer Math Soc*, **3**, 653–669.

[10] Andrews, GE and Warnaar, SO (2007). The product of partial theta functions. *Adv Appl Math*, **39**, 116–120.

[11] Aragao de Carvalho, C and Caracciolo, S (1983). A new Monte Carlo approach to the critical properties of self-avoiding random walks. *J de Phys*, **44**, 323–331.

[12] Aragao de Carvalho, C, Caracciolo, S, and Fröhlich, J (1983). Polymers and $g\phi^4$-theory in four dimensions. *Nucl Phys B*, **215**, 209–248.

[13] Arfken, GB and Weber, HJ (1995). *Mathematical Methods for Physicists*. Academic Press.

[14] Bachmann, M and Janke, W (2003). Multicanonical chain-growth algorithm. *Phys Rev Lett*, **91**, 208105.

[15] Baiesi, M and Orlandini, E (2012). Universal properties of knotted polymer rings. *Phys Rev E*, **86**, 031805.

[16] Baiesi, M, Orlandini, E, and Stella, AL (2001). Peculiar scaling of self-avoiding walk contacts. *Phys Rev Lett*, **87**, 70602.

[17] Baiesi, M, Orlandini, E, and Stella, AL (2007). Ranking knots of random, globular polymer rings. *Phys Rev Lett*, **99**, 58301.

[18] Baiesi, M, Orlandini, E, and Stella, AL (2010). The entropic cost to tie a knot. *J Stat Mech: Theo Expr*, **2010**, P06012.

[19] Banderier, C and Flajolet, P (2002). Basic analytic combinatorics of directed lattice paths. *Theo Comp Sci*, **281**, 37–80.

[20] Barber, MN (1970). Asymptotic results for self-avoiding walks on a manhattan lattice. *Physica*, **48**, 237–241.

[21] Barber, MN (1973). Finite size scaling. In *Phase Transitions and Critical Phenomena* (ed. C Domb and JL Lebowitz), Volume 8, pp. 145–266. Academic Press.

[22] Barber, MN (1973). Scaling relations for critical exponents of surface properties of magnets. *Phys Rev B*, **8**, 407–409.

[23] Barber, MN, Guttmann, AJ, Middlemiss, KM, Torrie, GM, and Whittington, SG (1978). Some tests of scaling theory for a self-avoiding walk attached to a surface. *J Phys A: Math Gen*, **11**, 1833–1842.

[24] Bastolla, U and Grassberger, P (1997). Phase transitions of single semistiff polymer chains. *J Stat Phys*, **89**, 1061–1078.

[25] Batchelor, MT and Yung, CM (1995). Exact results for the adsorption of a flexible self-avoiding polymer chain in two dimensions. *Phys Rev Lett*, **74**, 2026–2029.

[26] Baumgärtner, A (1992). Inflated vesicles: a lattice model. *Physica A*, **190**, 63–74.

[27] Baxter, RJ (1980). Hard hexagons: exact solution. *J Phys A: Math Gen*, **13**, L61–L70.

[28] Beaton, NR, Bousquet-Mélou, M, de Gier, J, Duminil-Copin, H, and Guttmann, AJ (2014). The critical fugacity for surface adsorption of self-avoiding walks on the honeycomb lattice is $1+\sqrt{2}$. *Commun Math Phys*, **326**, 727–754.

[29] Beaton, NR, Guttmann, AJ, and Jensen, I (2011). A numerical adaptation of self-avoiding walk identities from the honeycomb to other 2d lattices. *J Phys A: Math Theo*, **45**, 035201.

[30] Beaton, NR, Guttmann, AJ, and Jensen, I (2012). Two-dimensional self-avoiding walks and polymer adsorption: critical fugacity estimates. *J Phys A: Math Theo*, **45**, 055208.

[31] Bender, EA (1974). Convex n-ominoes. *Disc Math*, **8**, 219–226.

[32] Bennett-Wood, D, Brak, R, Guttmann, AJ, Owczarek, AL, and Prellberg, T (1994). Low-temperature two dimensional polymer partition function scaling: series analysis results. *J Phys A: Math Gen*, **27**, L1–L8.

[33] Bennett-Wood, D, Enting, IG, Gaunt, DS, Guttmann, AJ, Leask, JL, Owczarek, AL, and Whittington, SG (1997). Exact enumeration study of free energies of interacting polygons and walks in two dimensions. *J Phys A: Math Gen*, **31**, 4725–4741.

[34] Bennett-Wood, D and Owczarek, AL (1996). Exact enumeration results for self-avoiding walks on the honeycomb lattice attached to a surface. *J Phys A: Math Gen*, **29**, 4755–4768.

[35] Berg, B and Foerster, D (1981). Random paths and random surfaces on a digital computer. *Phys Lett B*, **106**, 323–326.

[36] Berndt, BC (1991). *Ramanujan's Notebooks: Part III.* Springer-Verlag.

[37] Berretti, A and Sokal, AD (1985). New Monte Carlo method for the self-avoiding walk. *J Stat Phys*, **40**, 483–531.

[38] Bétréma, J and Penaud, J-G (1993). Modèles avec particules dures, animaux dirigés et séries en variables partiellement commutatives. arXiv: math.co/0106210.

[39] Binder, K, Domb, C, and Green, MS (1983). Critical behaviour at surfaces. In *Phase Transitions and Critical Phenomena* (ed. C Domb and JL Lebowitz), Volume 8, pp. 1–144. Academic Press.

[40] Binder, K and Landau, DP (1990). Critical phenomena at surfaces. *Physica A*, **163**, 17–30.

[41] Biskup, M and Den Hollander, F (1999). A heteropolymer near a linear interface. *Ann Appl Prob*, **9**, 668–687.

[42] Blote, HWJ and Hilhorst, HJ (1984). Spiralling self-avoiding walks: an exact solution. *J Phys A: Math Gen*, **17**, L111–115.

[43] Boas, RP and Boas, HP (1996). *A primer of real functions* (4 edn), Volume 13 of *Math Assoc Amer.* Cambridge University Press.

[44] Bonin, J, Shapiro, L, and Simion, R (1993). Some q-analogues of the Schröder numbers arising from combinatorial statistics on lattice paths. *J Stat Plann Inf*, **34**, 35–55.

[45] Borgs, C, Chayes, J, van der Hofstad, R, and Slade, G (1999). Mean-field lattice trees. *Ann Comb*, **3**, 205–221.

[46] Borisov, OV, Boulakh, AB, and Zhulina, EB (2003). Annealing polyelectrolytes at charged interfaces. *EuroPhys J E*, **12**, 543–551.

[47] Bouchard, E and Bouchard, J-P (1989). Self-avoiding surfaces at interfaces. *J de Phys*, **50**, 829–841.

[48] Bousquet-Mélou, M (1992). Convex polyominoes and heaps of segments. *J Phys A: Math Gen*, **25**, 1925–1934.

[49] Bousquet-Mélou, M (1994). Codage des polyominos convexes et équations pour l'énumération suivant l'aire. *Disc Appl Math*, **48**, 21–43.

[50] Bousquet-Melou, M (1994). Polyominoes and polygons. *Contemp Math*, **178**, 55–55.

[51] Bousquet-Mélou, M (1996). A method for the enumeration of various classes of column-convex polygons. *Disc Math*, **154**, 1–25.

[52] Bousquet-Mélou, M (2001). Walks on the slit plane: other approaches. *Adv Appl Math*, **27**, 243–288.

[53] Bousquet-Mélou, M (2006). Three osculating walkers. *J Phys: Conf Ser*, **42**, 35–47.

[54] Bousquet-Mélou, M and Brak, R (2009). Exactly solved models. In *Polygons, Polyominoes and Polycubes*, pp. 43–78. Springer.

[55] Bousquet-Mélou, M and Guttmann, AJ (1997). Enumeration of three-dimensional convex polygons. *Ann Comb*, **1**, 27–53.

[56] Bousquet-Mélou, M and Petkovšek, M (2000). Linear recurrences with constant coefficients: the multivariate case. *Disc Math*, **225**, 51–75.

[57] Bousquet-Mélou, M and Petkovšek, M (2003). Walks confined in a quadrant are not always *d*-finite. *Theo Comp Sci*, **307**, 257–276.

[58] Bousquet-Mélou, M and Rechnitzer, A (2002). Lattice animals and heaps of dimers. *Disc Math*, **258**, 235–274.

[59] Bousquet-Mélou, M and Schaeffer, G (2002). Walks on the slit plane. *Prob Theo Rel Fields*, **124**, 305–344.

[60] Bousquet-Mélou, M and Viennot, XG (1992). Empilements de segments et q-énumération de polyominos convexes dirigés. *J Comb Theo, Ser A*, **60**, 196–224.

[61] Bovier, A, Fröhlich, J, and Glaus, U (1984). Mathematical aspects of the physics of disordered systems. In *Critical Phenomena, Random Systems, Gauge Theories (Les Houghes, Session XLIII)* (ed. K Osterwalder and R Stora), pp. 725–893. Elsevier.

[62] Brak, R and Essam, JW (2004). Asymmetric exclusion model and weighted lattice paths. *J Phys A: Math Gen*, **37**, 4183–4217.

[63] Brak, R, Essam, J, Osborn, J, Owczarek, AL, and Rechnitzer, A (2006). Lattice paths and the constant term. *J Phys: Conf Ser*, **42**, 47–58.

[64] Brak, R, Essam, JW, and Owczarek, AL (1998). New results for directed vesicles and chains near an attractive wall. *J Stat Phys*, **93**, 155–192.

[65] Brak, R, Essam, JW, and Owczarek, AL (1999). Partial difference equation method for lattice path problems. *Ann Comb*, **3**, 265–275.

[66] Brak, R and Guttmann, AJ (1990). Exact solution of the staircase and row-convex polygon perimeter and area generating function. *J Phys A: Math Gen*, **23**, 4581–4588.

[67] Brak, R, Guttmann, AJ, and Whittington, SG (1991). A collapse transition in a directed walk model. *J Math Chem*, **8**, 255–267.

[68] Brak, R, Guttmann, AJ, and Whittington, SG (1992). A collapse transition in a directed walk model. *J Phys A: Math Gen*, **25**, 2437–2446.

[69] Brak, R, Iliev, GK, Rechnitzer, A, and Whittington, SG (2007). Motzkin path models of long chain polymers in slits. *J Phys A: Math Theo*, **40**, 4415–4437.

[70] Brak, R and Owczarek, AL (1995). On the analyticity properties of scaling functions in models of polymer collapse. *J Phys A: Math Gen*, **28**, 4709–4725.

[71] Brak, R, Owczarek, AL, and Prellberg, T (1993). A scaling theory of the collapse transition in geometric cluster models of polymers and vesicles. *J Phys A: Math Gen*, **28**, 4709–4725.

[72] Brak, R, Owczarek, AL, and Prellberg, T (1994). Exact scaling behaviour of partially convex vesicles. *J Stat Phys*, **76**, 1101–1128.

[73] Brak, R, Owczarek, AL, and Rechnitzer, A (2009). Exact solutions of lattice polymer models. *J Math Chem*, **45**, 39–57.

[74] Brak, R, Owczarek, AL, Rechnitzer, A, and Whittington, SG (2005). A directed walk model of a long chain polymer in a slit with attractive walls. *J Phys A: Math Gen*, **38**, 4309–4325.

[75] Branden, P, Claesson, A, and Steingrimsson, E (2002). Catalan continued fractions and increasing subsequences in permutations. *Disc Math*, **258**, 275–287.

[76] Bratko, D, Chakraborty, AK, and Shakhnovich, E (1998). Adsorption of random copolymers on disordered surfaces. *Comp Theor Polymer Sci*, **8**, 113–126.

[77] Bray, AJ and Moore, MA (1977). Surface critical exponents in terms of bulk exponents. *Phys Rev Lett*, **38**, 1046–1048.

[78] Brezin, E, Le Guillou, JC, and Zinn-Justin, J (1973). Approach to scaling in renormalized perturbation theory. *Phys Rev D*, **8**, 2418–2430.

[79] Briggs, GB and Alexander, JW (1927). *On types of knotted curves.* Princeton University Press.

[80] Broadbent, SR and Hammersley, JM (1957). Percolation processes I. Crystals and mazes. *Proc Camb Phil Soc*, **53**, 629–641.

[81] Brydges, DC and Imbrie, JZ (2003). Branched polymers and dimensional reduction. *Ann Math*, **158**, 1019–1039.

[82] Burde, G and Zieschang, H (2003). *Knots*, Volume 5. de Gruyter.

[83] Burkhardt, TW, Eisenriegler, E, and Guim, I (1989). Conformal theory of energy correlations in the semi-infinite two-dimensional $O(n)$ model. *Nucl Phys B*, **316**, 559–572.

[84] Burkhardt, TW and Guim, I (1991). Self-avoiding walks that cross a square. *J Phys A: Math Gen*, **24**, L1221–L1228.

[85] Bustingorry, S, Jagla, EA, and Lorenzana, J (2003). Solid-solid volume collapse transitions are zeroth order. *arXiv: cond-mat/0307134*.

[86] Camacho, CJ and Fisher, Michael E (1990). Tunable fractal shapes in self-avoiding polygons and planar vesicles. *Phys Rev Lett*, **65**, 9–12.

[87] Caracciolo, S, Causo, MS, and Pelissetto, A (1998). High-precision determination of the critical exponent γ for self-avoiding walks. *Phys Rev E*, **57**, R1215–R1218.

[88] Caracciolo, S, Gherardi, M, Papinutto, M, and Pelissetto, A (2011). Geometrical properties of two-dimensional interacting self-avoiding walks at the θ-point. *J Phys A: Math Theo*, **44**, 115004.

[89] Caracciolo, S and Glaus, U (1985). A new Monte Carlo simulation for two models of self-avoiding lattice trees in two dimensions. *J Stat Phys*, **41**, 95–114.

[90] Caracciolo, S, Guttmann, AJ, Jensen, I, Pelissetto, A, Rogers, AN, and Sokal, AD (2005). Correction-to-scaling exponents for two-dimensional self-avoiding walks. *J Stat Phys*, **120**, 1037–1100.

[91] Caracciolo, S, Pelissetto, A, and Sokal, AD (1990). Monte Carlo test of a hyperscaling relation for the two-dimensional self-avoiding walk ii. *J Phys A: Math Gen*, **23**, 4509–4518.

[92] Caracciolo, S, Pelissetto, A, and Sokal, AD (1990). Nonlocal Monte Carlo algorithm for self-avoiding walks with fixed endpoints. *J Stat Phys*, **60**, 1–53.

[93] Caracciolo, S, Pelissetto, A, and Sokal, AD (1990). Universal distance ratios for two-dimensional self-avoiding walks: corrected conformal-invariance predictions. *J Phys A: Math Gen*, **23**, L969–L974.

[94] Caracciolo, S and Sokal, AD (1987). Monte Carlo test of a hyperscaling relation for the two-dimensional self-avoiding walk. *J Phys A: Math Gen*, **20**, 2569–2576.

[95] Cardy, JL (1983). Conformal invariance. In *Phase Transitions and Critical Phenomena* (ed. C Domb and JL Lebowitz), Volume 11, pp. 55–126. Academic Press.

[96] Cardy, JL and Guttmann, AJ (1993). Universal amplitude combinations for self-avoiding walks, polygons and trails. *J Phys A: Math Gen*, **26**, 2485–2494.

[97] Cardy, JL and Saleur, H (1989). Universal distance ratios for two-dimensional polymers. *J Phys A: Math Gen*, **22**, L601–L604.

[98] Carroll, L (1876). *The Hunting of the Snark (An Agony in 8 Fits).* Macmillan.

[99] Carvalho, MCTP and Privman, V (1988). Directed walk models of polymers at interfaces. *J Phys A: Math Gen*, **21**, L1033–L1037.

[100] Chang, I and Meirovitch, H (1993). Collapse transition of self-avoiding walks on a square lattice in the bulk and near a linear wall: the universality classes of the θ- and θ'-points. *Phys Rev E*, **48**, 3656–3660.

[101] Chayes, JT and Chayes, L (1986). Ornstein-Zernike behavior for self-avoiding walks at all noncritical temperatures. *Comm Math Phys*, **105**, 221–238.

[102] Chee, M-N and Whittington, SG (1987). The growth constant of uniform star polymers in a slab geometry. *J Phys A: Math Gen*, **20**, 4915–4921.

[103] Cherayil, BJ, Douglas, JF, and Freed, KF (1985). Effect of residual interactions on polymer properties near the θ-point. *J Chem Phys*, **83**, 5293–5309.

[104] Cherayil, BJ, Douglas, JF, and Freed, KF (1985). Non-Gaussian corrections at the θ-point: comparison between theory, experiment, and numerical simulations. *Macromol*, **18**, 821–823.

[105] Clisby, N (2010). Accurate estimate of the critical exponent ν for self-avoiding walks via a fast implementation of the pivot algorithm. *Phys Rev Lett*, **104**, 055702.

[106] Clisby, N (2013). Calculation of the connective constant for self-avoiding walks via the pivot algorithm. *J Phys A: Math Theo*, **46**, 245001.

[107] Clisby, N and Jensen, I (2012). A new transfer-matrix algorithm for exact enumerations: self-avoiding polygons on the square lattice. *J Phys A: Math Theo*, **45**, 115202.

[108] Clisby, N, Liang, R, and Slade, G (2007). Self-avoiding walk enumeration via the lace expansion. *J Phys A: Math Theo*, **40**, 10973.

[109] Colby, SA, Gaunt, DS, Torrie, GM, and Whittington, SG (1987). Branched polymers attached in a wedge geometry. *J Phys A: Math Gen*, **20**, L515–L520.

[110] Coniglio, Antonio (1983). Potts model formulation of branched polymers in a solvent. *J Phys A: Math Gen*, **16**, L187–L191.

[111] Conway, AR and Guttmann, AJ (1996). Square lattice self-avoiding walks and corrections to scaling. *Phys Rev Lett*, **77**, 5284–5287.

[112] Cozzarelli, NR (1992). Evolution of DNA topology: implications for its biological roles. *Proc Symp Appl Math*, **45**, 1–16.

[113] Daoud, M and de Gennes, P-G (1977). Statistics of macromolecular solutions trapped in small pores. *J de Phys*, **38**, 85–93.

[114] Dayantis, J and Palierne, JF (1996). Correction of scaling exponents for self-avoiding walks in two dimensions. *Macromol Theo Simul*, **5**, 1143–1150.

[115] de Gennes, P-G (1972). Exponents for the excluded volume problem as derived by the wilson method. *Phys Lett A*, **38**, 339–340.

[116] de Gennes, P-G (1975). Collapse of a polymer chain in poor solvents. *J de Phys Lett*, **36**, 55–57.

[117] de Gennes, P-G (1979). *Scaling Concepts in Polymer Physics.* Cornell.

[118] de Gennes, P-G (1984). Tight knots. *Macromol*, **17**, 703–704.

[119] de Queiroz, SLA (1995). Surface crossover exponent for branched polymers in two dimensions. *J Phys A: Math Gen*, **28**, 6315–6321.

[120] De'Bell, K and Essam, JW (1981). Mean field theory of percolation with application to surface effects. *J Phys A: Math Gen*, **14**, 1993–2008.

[121] De'Bell, K and Lookman, T (1993). Surface phase transitions in polymer systems. *Rev Mod Phys*, **65**, 87–113.

[122] De'Bell, K, Lookman, T, and Whittington, SG (1990). Analysis of exact enumeration data for self-avoiding walks attached to a surface. *Phys Rev A*, **41**, 682–688.

[123] De'Bell, K, Lookman, T, Zhao, D et al. (1991). Exact result for trees attached to a surface and estimates of the critical Boltzmann factor for surface adsorption. *Phys Rev A*, **44**, 1390–1392.

[124] Deguchi, T and Tsurusaki, K (1993). A new algorithm for numerical calculation of link invariants. *Phys Lett A*, **174**, 29–37.

[125] Deguchi, T and Tsurusaki, K (1997). Universality of random knotting. *Phys Rev E*, **55**, 6245–6248.

[126] Delbrück, M (1962). Knotting problems in biology. *Proc Symp Appl Math*, **14**, 55–63.

[127] Delest, M-P and Viennot, XG (1984). Algebraic languages and polyominoes enumeration. *Theo Comp Sci*, **34**, 169–206.

[128] Derrida, B (1981). Phenomenological renormalisation of the self-avoiding walk in two dimensions. *J Phys A: Math Gen*, **14**, L5–L10.

[129] Derrida, B and De Seze, L (1982). Application of the phenomenological renormalization to percolation and lattice animals in two dimensions. *J de Phys*, **43**, 475–483.

[130] Derrida, B and Herrmann, HJ (1983). Collapse of branched polymers. *J de Phys*, **44**, 1365–1376.

[131] Derrida, B and Stauffer, D (1985). Corrections to scaling and phenomenological renormalization for 2-dimensional percolation and lattice animal problems. *J de Phys*, **46**, 1623–1630.

[132] Des Cloizeaux, J (1974). Lagrangian theory for a self-avoiding random chain. *Phys Rev A*, **10**, 1665–1669.

[133] Des Cloizeaux, J (1980). A method for determining by direct renormalization the properties of long polymers in solutions. *J de Phys Lett*, **41**, 151–155.
[134] Des Cloizeaux, J (1980). Short range correlation between elements of a long polymer in a good solvent. *J de Phys*, **41**, 223–238.
[135] Des Cloizeaux, J (1981). Polymers in solutions: principles and applications of a direct renormalization method. *J de Phys*, **42**, 635–652.
[136] Des Cloizeaux, J and Jannink, G (1990). *Polymers in Solution: their Modelling and Structure.* Clarendon.
[137] Dewdney, AK (1985). A computer microscope zooms in for a look at the most complex object in mathematics. *Sci Amer*, **253**, 16–24.
[138] Dewdney, AK (1989). Computer recreations. *Sci Amer*, **260**, 122–125.
[139] Dhar, D (1982). Equivalence of the two-dimensional directed-site animal problem to Baxter's hard-square lattice-gas model. *Phys Rev Lett*, **49**, 959–962.
[140] Dhar, D (1983). Exact solution of a directed-site animals-enumeration problem in three dimensions. *Phys Rev Lett*, **51**, 853–856.
[141] Dhar, D (1987). The collapse of directed animals. *J Phys A: Math Gen*, **20**, L847–L850.
[142] Di Francesco, P, Saleur, H, and Zuber, JB (1987). Relations between the Coulomb gas picture and conformal invariance of two-dimensional critical models. *J Stat Phys*, **49**, 57–79.
[143] Diao, Y (1993). Minimal knotted polygons on the cubic lattice. *J Knot Theo Ram*, **2**, 413–425.
[144] Diao, Y, Ernst, C, and Janse van Rensburg, EJ (1999). Thicknesses of knots. *Math Proc Camb Phil Soc*, **126**, 293–310.
[145] Diao, Y, Nardo, JC, and Sun, Y (2001). Global knotting in equilateral random polygons. *J Knot Theo Ram*, **10**, 597–607.
[146] Diao, Y, Pippenger, N, and Summers, DW (1994). On random knots. *J Knot Theo Ram*, **3**, 419–430.
[147] Dickman, R and Schieve, WC (1986). Collapse transition and asymptotic scaling behavior of lattice animals: low-temperature expansion. *J Stat Phys*, **44**, 465–489.
[148] DiMarzio, EA and Rubin, RJ (1971). Adsorption of a chain polymer between two plates. *J Chem Phys*, **55**, 4318–4336.
[149] Doi, M and Edwards, SF (1988). *The Theory of Polymer Dynamics*, Volume 73. Oxford University Press.
[150] Domb, C (1976). Lattice animals and percolation. *J Phys A: Math Gen*, **9**, L141–L148.
[151] Domocos, Virgil (1996). A combinatorial method for the enumeration of column-convex polyominoes. *Disc Math*, **152**, 115–123.
[152] Dotsenko, VS and Fateev, VA (1984). Conformal algebra and multipoint correlation functions in $2d$ statistical models. *Nucl Phys B*, **240**, 312–348.
[153] Douglas, JF, Cherayil, BJ, and Freed, KF (1985). Polymers in two dimensions: renormalization group study using three-parameter model. *Macromol*, **18**, 2455–2463.
[154] Douglas, JF and Freed, KF (1985). Polymer contraction below the θ-point: arenormalization group description. *Macromol*, **18**, 2445–2454.
[155] Doye, J, Sear, R, and Frenkel, D (1998). The effect of chain stiffness on the phase behaviour of isolated homopolymers. *J Chem Phys*, **108**, 2134–2142.
[156] Duarte, JAMS (1986). On Monte Carlo generation of lattice trees. *J Phys A: Math Gen*, **19**, 1979–1985.
[157] Duminil-Copin, H and Hammond, A (2014). Self-avoiding walk is sub-ballistic. *Commun Math Phys*, **324**, 401–423.
[158] Duminil-Copin, H and Smirnov, S (2012). The connective constant of the honeycomb lattice equals $\sqrt{2+\sqrt{2}}$. *Ann Math*, **175**, 1653–1665.
[159] Duplantier, B (1982). Lagrangian tricritical theory of polymer chains in solution near the θ-point. *J de Phys*, **43**, 991–1019.
[160] Duplantier, B (1986). Exact critical exponents for two-dimensional dense polymers. *J Phys A: Math Gen*, **19**, L1009–L1014.

[161] Duplantier, B (1986). Polymer network of fixed topology: renormalization, exact critical exponent γ in two dimensions, and $d = 4-\epsilon$. *Phys Rev Lett*, **57**, 941–944.

[162] Duplantier, B (1986). Tricritical polymer chains in or below three dimensions. *EuroPhys Lett*, **1**, 491–498.

[163] Duplantier, B (1987). Geometry of polymer chains near the θ-point and dimensional regularization. *J Chem Phys*, **86**, 4233–4244.

[164] Duplantier, B (1989). Fractals in two dimensions and conformal invariance. *Physica D*, **38**, 71–87.

[165] Duplantier, B (1989). Renormalization and conformal invariance for polymers. In *7th International Summer School on Fundamental Problems in Statistical Mechanics*, SACLAY-SPHT-T-89-162, pp. 171–223.

[166] Duplantier, Bertrand (1989). Statistical mechanics of polymer networks of any topology. *J Stat Phys*, **54**, 581–680.

[167] Duplantier, B (1989). Two-dimensional fractal geometry, critical phenomena and conformal invariance. *Phys Rep*, **184**, 229–257.

[168] Duplantier, B and Saleur, H (1986). Exact surface and wedge exponents for polymers in two dimensions. *Phys Rev Lett*, **57**, 3179–3182.

[169] Duplantier, B and Saleur, H (1987). Exact critical properties of two-dimensional dense self-avoiding walks. *Nucl Phys B*, **290**, 291–326.

[170] Duplantier, B and Saleur, H (1987). Exact tricritical exponents for polymers at the θ-point in two dimensions. *Phys Rev Lett*, **59**, 539–542.

[171] Durhuus, B, Fröhlich, J, and Jonsson, T (1983). Self-avoiding and planar random surfaces on the lattice. *Nucl Phys B*, **225**, 185–203.

[172] Durrett, Richard (1984). Oriented percolation in two dimensions. *Ann Prob*, 999–1040.

[173] Edwards, SF (1965). The statistical mechanics of polymers with excluded volume. *Proc Phys Soc*, **85**, 613–624.

[174] Edwards, SF (1967). Statistical mechanics with topological constraints: I. *Proc Phys Soc*, **91**, 513–519.

[175] Edwards, SF (1968). Statistical mechanics with topological constraints: II. *J Phys A: Gen Phys*, **1**, 15–28.

[176] Edwards, SF and Singh, P (1979). Size of a polymer molecule in solution. Part 1. Excluded volume problem. *J Chem Soc, Farad Trans 2*, **75**, 1001–1019.

[177] Enting, IG (1980). Generating functions for enumerating self-avoiding rings on the square lattice. *J Phys A: Math Gen*, **13**, 3713–3722.

[178] Enting, IG and Guttmann, AJ (1985). Self-avoiding polygons on the square, L and Manhattan lattices. *J Phys A: Math Gen*, **18**, 1007–1018.

[179] Enting, IG and Guttmann, AJ (1989). Polygons on the honeycomb lattice. *J Phys A: Math Gen*, **22**, 1371–1384.

[180] Enting, IG and Guttmann, AJ (1990). On the area of square lattice polygons. *J Stat Phys*, **58**, 475–484.

[181] Essam, JW (1980). Percolation theory. *Rep Prog Phys*, **43**, 833–912.

[182] Essam, JW, De'Bell, K, Adler, J, and Bhatti, FM (1986). Analysis of extended series for bond percolation on the directed square lattice. *Phys Rev B*, **33**, 1982–1986.

[183] Essam, JW, Guttmann, AJ, and De'Bell, K (1988). On two-dimensional directed percolation. *J Phys A: Math Gen*, **21**, 3815–3832.

[184] Euler, L (1783). Evolutio producti infiniti $(1-x)(1-x^2)(1-x^3)(1-x^4)(1-x^5)$ etc in seriem simplicem. In *Acta Academiae Scientarum Imperialis Petropolitinae 1780,* pp. 47–55, *Opera Omnia, Series Prima,* **3**, pp. 472–479.

[185] Family, F (1983). Cluster renormalisation study of site lattice animals in two and three dimensions. *J Phys A: Math Gen*, **16**, L97–L103.

[186] Fayolle, G, Iasnogorodski, R, and Malyšev, VA (1999). *Random Walks in the Quarter-plane: Algebraic Methods, Boundary Value Problems and Applications*, Volume 40. Springer.

[187] Feller, W (1968). *An Introduction to Probability Theory and its Applications* (3 edn), Volume 1. Wiley.

[188] Feynman, RP, Hibbs, AR, and Styer, DF (2010). *Quantum Mechanics and Path Integrals: Emended Edition*. Dover.

[189] Finsy, R, Janssens, M, and Bellemans, A (1975). Internal transition in an infinitely long polymer chain. *J Phys A: Math Gen*, **8**, L106–L109.

[190] Fisher, ME (1966). Shape of a self-avoiding walk or polymer chain. *J Chem Phys*, **44**, 616–622.

[191] Fisher, ME (1978). Yang-Lee edge singularity and ϕ^3-field theory. *Phys Rev Lett*, **40**, 1610–1613.

[192] Fisher, ME (1989). Fractal and nonfractal shapes in two-dimensional vesicles. *Physica D*, **38**, 112–118.

[193] Fisher, ME and Aharony, A (1974). Scaling function for two-point correlations I expansion near four dimensions. *Phys Rev B*, **10**, 2818–2833.

[194] Fisher, ME and Gaunt, DS (1964). Ising model and self-avoiding walks on hypercubical lattices. *Phys Rev A*, **133**, 224–239.

[195] Fisher, ME, Guttmann, AJ, and Whittington, SG (1991). Two-dimensional lattice vesicles and polygons. *J Phys A: Math Gen*, **24**, 3095–3106.

[196] Fisher, ME and Sykes, MF (1959). Excluded-volume problem and the Ising model of ferromagnetism. *Phys Rev*, **114**, 45–58.

[197] Flajolet, P and Sedgewick, R (2001). *Analytic Combinatorics: Functional Equations, Rational and Algebraic Functions*. INRIA Rap Resch No 4103.

[198] Flesia, S and Gaunt, DS (1992). Lattice animal contact models of a collapsing branched polymer. *J Phys A: Math Gen*, **25**, 2127–2137.

[199] Flesia, S, Gaunt, DS, Soteros, CE, and Whittington, SG (1992). Models for collapse in trees and *c*-animals. *J Phys A: Math Gen*, **25**, 3515–3521.

[200] Flesia, S, Gaunt, DS, Soteros, CE, and Whittington, SG (1993). Collapse transition in animals and vesicles. *J Phys A: Math Gen*, **26**, L993–L997.

[201] Flesia, S, Gaunt, DS, Soteros, CE, and Whittington, SG (1994). Statistics of collapsing lattice animals. *J Phys A: Math Gen*, **27**, 5831–5846.

[202] Flory, PJ (1949). The configuration of real polymer chains. *J Chem Phys*, **17**, 303–309.

[203] Flory, PJ (1956). Statistical thermodynamics of semi-flexible chain molecules. *Proc Royal Soc London Ser A Math Phys Sci*, **234**, 60–73.

[204] Flory, PJ (1969). *Statistical Mechanics of Chain Molecules*. Interscience.

[205] Forgacs, G, Privman, V, and Frisch, HL (1989). Adsorption–desorption transition of polymer chains interacting with surfaces. *J Chem Phys*, **90**, 3339–3345.

[206] Fortuin, CM and Kasteleyn, PW (1972). On the random-cluster model: I Introduction and relation to other models. *Physica*, **57**, 536–564.

[207] Fortuin, CM, Kasteleyn, PW, and Ginibre, J (1971). Correlation inequalities on some partially ordered sets. *Comm Math Phys*, **22**, 89–103.

[208] Foster, DP (1990). Exact evaluation of the collapse phase boundary for two-dimensional directed polymers. *J Phys A: Math Gen*, **23**, L1135–L1138.

[209] Foster, DP, Orlandini, E, and Tesi, MC (1992). Surface critical exponents for models of polymer collapse and adsorption: the universality of the θ and θ' points. *J Phys A: Math Gen*, **25**, L1211–L1217.

[210] Foster, DP and Yeomans, J (1991). Competition between self-attraction and adsorption in directed self-avoiding polymers. *Physica A*, **177**, 443–452.

[211] Fraser, SJ and Winnik, MA (1979). Variance reduction for lattice walks grown with markov chain sampling. *J Chem Phys*, **70**, 575–581.

[212] Frisch, HL and Wasserman, E (1961). Chemical topology. *J Amer Chem Soc*, **83**, 3789–3795.

[213] Fuller, FB (1971). The writhing number of a space curve. *Proc Nat Acad Sci*, **68**, 815–819.

[214] Gaunt, DS (1980). The critical dimension for lattice animals. *J Phys A: Math Gen*, **13**, L97–L101.

[215] Gaunt, DS and Colby, SA (1990). Branched polymers in a wedge geometry in three dimensions. *J Stat Phys*, **58**, 539–551.

[216] Gaunt, DS and Flesia, S (1990). Modelling the collapse transition in branched polymers. *Physica A*, **168**, 602–608.

[217] Gaunt, DS and Flesia, S (1991). The collapse transition for lattice animals. *J Phys A: Math Gen*, **24**, 3655–3670.

[218] Gaunt, DS, Lipson, JEG, Whittington, SG, and Wilkinson, MK (1986). Lattice models of branched polymers: uniform combs in two dimensions. *J Phys A: Math Gen*, **19**, L811–L816.

[219] Gaunt, DS and Peard, PJ (2000). $1/d$-Expansions for the free energy of weakly embedded site animal models of branched polymers. *J Phys A: Math Gen*, **42**, 7515–7540.

[220] Gaunt, DS, Peard, PJ, Soteros, CE, and Whittington, SG (1994). Relationships between growth constants for animals and trees (lattice theory). *J Phys A: Math Gen*, **27**, 7343–7352.

[221] Gaunt, DS and Ruskin, H (2001). Bond percolation processes in d dimensions. *J Phys A: Math Gen*, **11**, 1369–1380.

[222] Gaunt, DS, Sykes, MF, Torrie, GM, and Whittington, SG (1982). Universality in branched polymers on d-dimensional hypercubic lattices. *J Phys A: Math Gen*, **15**, 3209–3217.

[223] Geyer, CJ and Thompson, EA (1995). Annealing Markov chain Monte Carlo with applications to ancestral inference. *J Amer Stat Assoc*, **90**, 909–920.

[224] Gilks, WR, Richardson, S, and Spiegelhalter, DJ (1996). *Markov Chain Monte Carlo in Practice*, Volume 2. Chapman & Hall/CRC.

[225] Ginibre, Jean, Grossmann, Alexander, and Ruelle, David (1966). Condensation of lattice gases. *Comm Math Phys*, **3**, 187–193.

[226] Giri, M, Stephan, MJ, and Grest, GS (1977). Spin models and cluster distribution for bond and site percolation models. *Phys Rev B*, **16**, 4971–4977.

[227] Glaus, U (1985). Monte Carlo test of dimensional reduction for branched polymers in three dimensions. *J Phys A: Math Gen*, **18**, L609–L616.

[228] Glaus, U (1986). Monte Carlo simulation of self-avoiding surfaces in three dimensions. *Phys Rev Lett*, **56**, 1996–1999.

[229] Glaus, U (1988). Monte Carlo study of self-avoiding surfaces. *J Stat Phys*, **50**, 1141–1166.

[230] Glaus, U and Einstein, TL (1987). On the universality class of planar self-avoiding surfaces with fixed boundary. *J Phys A: Math Gen*, **20**, L105–L111.

[231] Gouyou-Beauchamps, D (1998). Enumeration of remarkable families of polyominoes. *Rapport de Recherche – Insitut National de Recherche en Informatique et en Automatique*, 3–8.

[232] Gouyou-Beauchamps, D and Viennot, XG (1988). Equivalence of the two-dimensional directed animal problem to a one-dimensional path problem. *Adv Appl Math*, **9**, 334–357.

[233] Grassberger, P (1993). Monte Carlo simulation of 3d self-avoiding walks. *J Phys A: Math Gen*, **26**, 2769–2776.

[234] Grassberger, P (1994). Nonuniform star polymers in two dimensions. *J Phys A: Math Gen*, **27**, L721–L725.

[235] Grassberger, P (1997). Pruned-enriched Rosenbluth method: Simulations of θ polymers of chain length up to 1000000. *Phys Rev E*, **56**, 3682–3693.

[236] Grassberger, P (2004). Simulations of grafted polymers in a good solvent. *J Phys A: Math Gen*, **38**, 323–331.

[237] Grassberger, P and Hegger, R (1995). Simulations of θ-polymers in 2 dimensions. *J de Phys I*, **5**, 597–606.

[238] Grassberger, P and Hegger, R (1995). Simulations of three-dimensional θ-polymers. *J Chem Phys*, **102**, 6881–6899.

[239] Grassberger, P and Hsu, H-P (2002). Stretched polymers in a poor solvent. *Phys Rev E*, **65**, 031807.

[240] Grimmett, G (1999). *Percolation* (2 edn), Volume 321. Springer Verlag.

[241] Grosberg, AY, Feigel, A, and Rabin, Y (1996). Flory-type theory of a knotted ring polymer. *Phys Rev E*, **54**, 6618–6622.

[242] Grosberg, A, Izrailev, S, and Nechaev, S (1994). Phase transition in a heteropolymer chain at a selective interface. *Phys Rev E*, **50**, 1912–1921.
[243] Guida, R and Zinn-Justin, J (1998). Critical exponents of the n-vector model. *J Phys A: Math Gen*, **31**, 8103–8122.
[244] Guttmann, AJ (1983). Bounds on connective constants for self-avoiding walks. *J Phys A: Math Gen*, **16**, 2233–2238.
[245] Guttmann, AJ (1987). On the critical behaviour of self-avoiding walks. *J Phys A: Math Gen*, **20**, 1839–1854.
[246] Guttmann, AJ (1989). On the critical behaviour of self-avoiding walks ii. *J Phys A: Math Gen*, **22**, 2807–2813.
[247] Guttmann, AJ and Conway, AR (2001). Square lattice self-avoiding walks and polygons. *Ann Comb*, **5**, 319–345.
[248] Guttmann, AJ and Enting, IG (1988). The size and number of rings on the square lattice. *J Phys A: Math Gen*, **21**, L165–L172.
[249] Guttmann, AJ and Gaunt, DS (1978). On the asymptotic number of lattice animals in bond and site percolation. *J Phys A: Math Gen*, **11**, 949–953.
[250] Guttmann, AJ and Hirschhorn, M (1984). Comment on the number of spiral self-avoiding walks. *J Phys A: Math Gen*, **17**, 3613–3614.
[251] Guttmann, AJ, Jensen, I, Wong, LH, and Enting, IG (2000). Punctured polygons and polyominoes on the square lattice. *J Phys A: Math Gen*, **33**, 1735–1764.
[252] Guttmann, AJ and Whittington, SG (1978). Two-dimensional lattice embeddings of connected graphs of cyclomatic index two. *J Phys A: Math Gen*, **11**, 721–729.
[253] Guttmann, AJ and Wormald, NC (1984). On the number of spiral self-avoiding walks. *J Phys A: Math Gen*, **17**, L271–L274.
[254] Hakim, V and Nadal, JP (1983). Exact results for 2d directed animals on a strip of finite width. *J Phys A: Math Gen*, **16**, L213–L218.
[255] Hammersley, JM (1957). Percolation processes II. The connective constant. *Proc Camb Phil Soc*, **53**, 642–645.
[256] Hammersley, JM (1960). Limiting properties of numbers of self-avoiding walks. *Phys Rev*, **118**, 656–656.
[257] Hammersley, JM (1961). The number of polygons on a lattice. *Proc Camb Phil Soc*, **57**, 516–523.
[258] Hammersley, JM (1961). On the rate of convergence to the connective constant of the hypercubical lattice. *Quart J Math*, **12**, 250–256.
[259] Hammersley, JM (1962). Generalisation of the fundamental theorem on subadditive functions. *Math Proc Camb Phil Soc*, **58**, 235–238.
[260] Hammersley, JM and Morton, KW (1954). Poor man's Monte Carlo. *J Roy Stat Soc Ser B (Meth)*, **16**, 23–38.
[261] Hammersley, JM and SG, Whittington (1985). Self-avoiding walks in wedges. *J Phys A: Math Gen*, **18**, 101–111.
[262] Hammersley, JM, Torrie, GM, and Whittington, SG (1982). Self-avoiding walks interacting with a surface. *J Phys A: Math Gen*, **15**, 539–571.
[263] Hammersley, JM and Welsh, DJA (1962). Further results on the rate of convergence to the connective constant of the hypercubical lattice. *Quart J Math*, **13**, 108–110.
[264] Hara, T and Slade, G (1990). On the upper critical dimension of lattice trees and lattice animals. *J Stat Phys*, **59**, 1469–1510.
[265] Hara, T and Slade, G (1991). Critical behaviour of self-avoiding walk in five or more dimensions. *Bull (New Ser) of Amer Math Soc*, **25**, 417–423.
[266] Hara, T and Slade, G (1992). The lace expansion for self-avoiding walk in five or more dimensions. *Rev Math Phys*, **4**, 235–327.
[267] Hara, T and Slade, G (1992). The number and size of branched polymers in high dimensions. *J Stat Phys*, **67**, 1009–1038.
[268] Hara, T and Slade, G (1992). Self-avoiding walk in five or more dimensions I. The critical behaviour. *Comm Math Phys*, **147**, 101–136.

[269] Hara, T and Slade, G (1994). Mean-field behaviour and the lace expansion. *NATO ASI Ser C Math Phys Sci – Adv Study Inst*, **420**, 87–122.

[270] Hara, T and Slade, G (1995). The self-avoiding-walk and percolation critical points in high dimensions. *Comb, Prob & Comp*, **4**, 197–215.

[271] Hara, T, Slade, G, and Sokal, AD (1993). New lower bounds on the self-avoiding-walk connective constant. *J Stat Phys*, **72**, 479–517.

[272] Hardy, GH (1940). *Ramanujan: Twelve Lectures on Subjects Suggested by his Life and Work*. Chelsea.

[273] Hardy, GH (1991). *Divergent Series* (2 edn). Chelsea.

[274] Hardy, GH, Littlewood, JE, and Pólya, G (1952). *Inequalities* (2 edn). Cambridge University Press.

[275] Hardy, GH and Ramanujan, S (1917). Asymptotic formulæ for the distribution of integers of various types. *Proc London Math Soc*, **16**, 112–132.

[276] Hardy, GH and Ramanujan, S (1918). Asymptotic formulæ in combinatory analysis. *Proc London Math Soc*, **2**, 75–115.

[277] Hastings, WK (1970). Monte Carlo sampling methods using markov chains and their applications. *Biometrika*, **57**, 97–109.

[278] Hegger, R and Grassberger, P (1994). Chain polymers near an adsorbing surface. *J Phys A: Math Gen*, **27**, 4069–4081.

[279] Henkel, Malte and Seno, Flavio (1996). Phase diagram of branched polymer collapse. *Phys Rev E*, **53**, 3662–3672.

[280] Hille, E and Phillips, RS (1957). *Functional Analysis and Semi-Groups*, Volume 31. AMS.

[281] Hsu, H-P and Grassberger, P (2011). A review of Monte Carlo simulations of polymers with perm. *J Stat Phys*, **144**, 597–637.

[282] Hsu, H-P, Nadler, W, and Grassberger, P (2004). Scaling of star polymers with 1-80 arms. *Macromol*, **37**, 4658–4663.

[283] Hsu, H-P, Nadler, W, and Grassberger, P (2005). Simulations of lattice animals and trees. *J Phys A: Math Gen*, **38**, 775–806.

[284] Hsu, H-P, Nadler, W, and Grassberger, P (2005). Statistics of lattice animals. *Comp Phys Comm*, **169**, 114–116.

[285] Ikhlef, Y and Cardy, J (2009). Discretely holomorphic parafermions and integrable loop models. *J Phys A: Math Theo*, **42**, 102001.

[286] Iliev, GK and Janse van Rensburg, EJ (2012). Directed path models of adsorbing and pulled copolymers. *J Stat Mech: Theo Expr*, **2012**, P01019.

[287] Imbrie, JZ (2003). Dimensional reduction and crossover to mean-field behaviour for branched polymers. In *Ann Henri Poincaré*, Volume 4, pp. 445–458. Birkäuser-Verlag.

[288] Isaacson, J and Lubensky, TC (1980). Flory exponents for generalized polymer problems. *J de Phys Lett*, **41**, 469–471.

[289] Isakov, SN (1984). Nonanalytic features of the first order phase transition in the Ising model. *Comm Math Phys*, **95**, 427–443.

[290] Ishinabe, T (1987). Examination of the θ-point from exact enumeration of self-avoiding walks ii. *J Phys A: Math Gen*, **20**, 6435–6453.

[291] James, EW and Whittington, SG (2002). Extent of self-averaging in the statistical mechanics of finite random copolymers. *J Phys A: Math Gen*, **35**, 3203–3211.

[292] Jane van Rensburg, EJ (2004). Inflating square and rectangular lattice vesicles. *J Phys A: Math Gen*, **37**, 3903–3932.

[293] Janse van Rensburg, EJ (1992). Ergodicity of the bfacf algorithm in three dimensions. *J Phys A: Math Gen*, **25**, 1031–1042.

[294] Janse van Rensburg, EJ (1992). On the number of trees in $\mathbb{Z}^d$. *J Phys A: Math Gen*, **25**, 3529–3547.

[295] Janse van Rensburg, EJ (2000). The cluster structure in collapsing animals. *J Phys A: Math Gen*, **33**, 3653–3667.

[296] Janse van Rensburg, EJ (2002). The probability of knotting in lattice polygons. *Contemp Math*, **304**, 125–136.

[297] Janse van Rensburg, EJ (2005). Square lattice directed paths adsorbing on the line $y = qx$. *J Stat Mech: Theo Expr*, **2005**, P09010.
[298] Janse van Rensburg, EJ (2006). Forces in Motzkin paths in a wedge. *J Phys A: Math Gen*, **39**, 1581–1608.
[299] Janse van Rensburg, EJ (2009). Approximate enumeration of self-avoiding walks. In *Algorithmic Probability and Combinatorics* (ed. ME Lladser, RS MAier, M Mishna and A Rechnitzer), pp. 127–152. American Mathematical Society.
[300] Janse van Rensburg, EJ (2009). Monte Carlo methods for the self-avoiding walk. *J Phys A: Math Theo*, **42**, 323001.
[301] Janse van Rensburg, EJ (2009). Thoughts on lattice knot statistics. *J Math Chem*, **45**, 7–38.
[302] Janse van Rensburg, EJ (2010). Adsorbing Motzkin paths. *J Phys A: Math Theo*, **43**, 485006.
[303] Janse van Rensburg, EJ (2010). The adsorption transition in directed paths. *J Stat Mech: Theo Expr*, **2010**, P08030.
[304] Janse van Rensburg, EJ (2010). Pulled directed lattice paths. *J Phys A: Math Theo*, **43**, 215001.
[305] Janse van Rensburg, EJ (2010). Pulled Motzkin paths. *J Phys A: Math Theo*, **43**, 335001.
[306] Janse van Rensburg, EJ and Ma, J (2006). Plane partition vesicles. *J Phys A: Math Gen*, **39**, 11171.
[307] Janse van Rensburg, EJ and Madras, N (1992). A nonlocal Monte Carlo algorithm for lattice trees. *J Phys A: Math Gen*, **25**, 303–333.
[308] Janse van Rensburg, EJ and Madras, N (1997). Metropolis Monte Carlo simulation of lattice animals. *J Phys A: Math Gen*, **30**, 8035–8066.
[309] Janse van Rensburg, EJ, Orlandini, E, Owczarek, AL, Rechnitzer, A, and Whittington, SG (2005). Self-avoiding walks in a slab with attractive walls. *J Phys A: Math Gen*, **38**, L823–L828.
[310] Janse van Rensburg, EJ, Orlandini, E, Sumners, DW, Tesi, MC, and Whittington, SG (1997). The writhe of knots in the cubic lattice. *J Knot Theo Ram*, **6**, 31–44.
[311] Janse van Rensburg, EJ, Orlandini, E, Sumners, DW, Tesi, MC, and Whittington, SG (1999). The writhe of a self-avoiding polygon. *J Phys A: Math Gen*, **26**, L981–L986.
[312] Janse van Rensburg, EJ, Orlandini, E, and Tesi, MC (1999). Collapsing animals. *J Phys A: Math Gen*, **32**, 1567–1584.
[313] Janse van Rensburg, EJ, Orlandini, E, Tesi, MC, and Whittington, SG (2008). Knot probability of polygons subjected to a force: a Monte Carlo study. *J Phys A: Math Theo*, **41**, 025003.
[314] Janse van Rensburg, EJ, Orlandini, E, Tesi, MC, and Whittington, SG (2008). Knotting in stretched polygons. *J Phys A: Math Theo*, **41**, 015003.
[315] Janse van Rensburg, EJ, Orlandini, E, Tesi, MC, and Whittington, SG (2009). Thermodynamics and entanglements of walks under stress. *J Stat Mech: Theo Expr*, **2009**, P07014.
[316] Janse van Rensburg, EJ, Orlandini, E, and Whittington, SG (2006). Self-avoiding walks in a slab: rigorous results. *J Phys A: Math Gen*, **39**, 13869–13902.
[317] Janse van Rensburg, EJ and Prellberg, T (2013). The pressure exerted by adsorbing directed lattice paths and staircase polygons. *J Phys A: Math Gen*, **46**, 115202.
[318] Janse van Rensburg, EJ and Promislow, SD (1995). Minimal knots in the cubic lattice. *J Knot Theo Ram*, **4**, 115–130.
[319] Janse van Rensburg, EJ and Rechnitzer, A (2002). Exchange symmetries in Motzkin path and bargraph models of copolymer adsorption. *Elect J Comb*, **9**, 1–24.
[320] Janse van Rensburg, EJ and Rechnitzer, A (2003). High precision canonical Monte Carlo determination of the growth constant of square lattice trees. *Phys Rev E*, **67**, 036116.
[321] Janse van Rensburg, EJ and Rechnitzer, A (2004). Multiple markov chain Monte Carlo study of adsorbing self-avoiding walks in two and in three dimensions. *J Phys A: Math Gen*, **37**, 6875–6898.

[322] Janse van Rensburg, EJ and Rechnitzer, A (2008). Atmospheres of polygons and knotted polygons. *J Phys A: Math Theo*, **41**, 105002.
[323] Janse van Rensburg, EJ and Rechnitzer, A (2009). Generalized atmospheric sampling of self-avoiding walks. *J Phys A: Math Theo*, **42**, 335001.
[324] Janse van Rensburg, EJ and Rechnitzer, A (2011). BFACF-style algorithms for polygons in the body-centered and face-centered cubic lattices. *J Phys A: Math Theo*, **44**, 165001.
[325] Janse van Rensburg, EJ and Rechnitzer, A (2011). On the universality of knot probability ratios. *J Phys A: Math Theo*, **44**, 162002.
[326] Janse van Rensburg, EJ and Rechnitzer, A (2012). The compressibility of minimal lattice knots. *J Stat Mech: Theo Expr*, **2012**, P05003.
[327] Janse van Rensburg, EJ, Sumners, DW, Wasserman, E, and Whittington, SG (1999). Entanglement complexity of self-avoiding walks. *J Phys A: Math Gen*, **25**, 6557–6566.
[328] Janse van Rensburg, EJ and Whittington, SG (1990). The knot probability in lattice polygons. *J Phys A: Math Gen*, **23**, 3574–3590.
[329] Janse van Rensburg, EJ and Whittington, SG (1990). Punctured discs on the square lattice. *J Phys A: Math Gen*, **23**, 1287–1294.
[330] Janse van Rensburg, EJ and Whittington, SG (1991). The BFACF algorithm and knotted polygons. *J Phys A: Math Gen*, **24**, 5553–5567.
[331] Janse van Rensburg, EJ and Whittington, SG (1991). The dimensions of knotted polygons. *J Phys A: Math Gen*, **24**, 3935–3948.
[332] Janse van Rensburg, EJ and Whittington, SG (1999). The knot probability in lattice polygons. *J Phys A: Math Gen*, **23**, 3573–3590.
[333] Janssen, H-K and Lyssy, A (1994). Surface adsorption of branched polymers: mapping onto the Yang-Lee edge singularity and exact results for three dimensions. *Phys Rev E*, **50**, 3784–3796.
[334] Janssen, H-K and Stenull, O (2011). Collapse transition of randomly branched polymers: renormalized field theory. *Phys Rev E*, **83**, 051126.
[335] Jensen, I (2001). Enumerations of lattice animals and trees. *J Stat Phys*, **102**, 865–881.
[336] Jensen, I (2003). Counting polyominoes: a parallel implementation for cluster counting. In *ICCS 2003* (ed. PMA Sloot, D Abramson, AV Bogdanov, YE Gorbachev, JJ Dongarra and AY Zomaya), pp. 203–212. Springer.
[337] Jensen, I (2003). A parallel algorithm for the enumeration of self-avoiding polygons on the square lattice. *J Phys A: Math Gen*, **36**, 5731–5746.
[338] Jensen, I (2004). Enumeration of self-avoiding walks on the square lattice. *J Phys A: Math Gen*, **37**, 5503–5524.
[339] Jensen, I (2004). Self-avoiding walks and polygons on the triangular lattice. *J Stat Mech: Theo Expr*, **2004**, P10008.
[340] Jensen, I and Guttmann, AJ (1999). Self-avoiding polygons on the square lattice. *J Phys A: Math Gen*, **32**, 4867–4876.
[341] Jensen, I and Guttmann, AJ (2000). Statistics of lattice animals (polyominoes) and polygons. *J Phys A: Math Gen*, **33**, L257–L263.
[342] Joyce, GS (1984). An exact formula for the number of spiral self-avoiding walks. *J Phys A: Math Gen*, **17**, L463–467.
[343] Jungreis, D (1994). Gaussian random polygons are globally knotted. *J Knot Theo Ram*, **3**, 455–464.
[344] Kac, VG (1979). Contravariant form for infinite-dimensional lie algebras and superalgebras. In *Group Theoretical Methods in Physics* (ed. W Beigblöck, A Bohm and E Takasugi), Volume 94 of *Lect Notes Phys*, pp. 441–445. Springer.
[345] Kac, VG and Cheung, P (2002). *Quantum Calculus*. Springer-Verlag.
[346] Kennedy, T (2002). A faster implementation of the pivot algorithm for self-avoiding walks. *J Stat Phys*, **106**, 407–429.
[347] Kennedy, T (2004). Conformal invariance and stochastic loewner evolution predictions for the 2d self-avoiding walk-Monte Carlo tests. *J Stat Phys*, **114**, 51–78.
[348] Kesten, H (1963). On the number of self-avoiding walks. *J Math Phys*, **4**, 960–969.
[349] Kesten, H (1964). On the number of self-avoiding walks ii. *J Math Phys*, **5**, 1128–1137.

[350] Kholodenko, AL and Vilgis, TA (1998). Some geometrical and topological problems in polymer physics. *Phys Rep*, **298**, 251–370.

[351] Kholodenko, AL and Vilgis, TA (1999). Path integral calculation of the writhe for circular semiflexible polymers. *J Phys A: Math Gen*, **29**, 939–948.

[352] Kim, D (1988). The number of convex polyominos with given perimeter. *Disc Math*, **70**, 47–51.

[353] Klarner, DA (1965). Some results concerning polyominoes. *Fibonacci Quart*, **3**, 9–20.

[354] Klarner, DA (1967). Cell growth problems. *Canad J Math*, **19**, 851–863.

[355] Klarner, DA and Rivest, RL (1973). A procedure for improving the upper bound for the number of n-ominoes. *Canad J Math*, **25**, 585–602.

[356] Klarner, DA and Rivest, RL (1974). Asymptotic bounds for the number of convex n-ominoes. *Disc Math*, **8**, 31–40.

[357] Klein, DJ (1981). Rigorous results for branched polymer models with excluded volume. *J Chem Phys*, **75**, 5186–5189.

[358] Klushin, LI, Polotsky, AA, Hsu, H-P, Markelov, DA, Binder, K, and Skvortsov, AM (2013). Adsorption of a single polymer chain on a surface: Effects of the potential range. *Phys Rev E*, **87**, 022604.

[359] Knopp, MI (2008). *Modular Functions in Analytic Number Theory.* American Mathematical Society.

[360] Koniaris, K and Muthukumar, M (1991). Knottedness in ring polymers. *Phys Rev Lett*, **66**, 2211–2214.

[361] Koshy, T (2009). Lobb's generalization of Catalan's parenthesization problem. *College Math J*, **40**, 99–107.

[362] Krawczyk, J, Owczarek, AL, Prellberg, T, and Rechnitzer, A (2005). Pulling absorbing and collapsing polymers from a surface. *J Stat Mech: Theo Expr*, **2005**, P05008.

[363] Kühn, R (1996). Equilibrium ensemble approach to disordered systems I. General theory, exact results. *Z Phys B Cond Mat*, **100**, 231–242.

[364] Kunz, H and Souillard, B (1978). Essential singularity in the percolation model. *Phys Review Lett*, **40**, 133–135.

[365] Kurtze, DA and Fisher, ME (1979). Yang-Lee edge singularities at high temperatures. *Phys Rev B*, **20**, 2785–2796.

[366] Lacher, RC and Sumners, DW (1991). Data structures and algorithms for computation of topological invariants of entanglements: link, twist and writhe. In *Computer Simulation of Polymers* (ed. RJ Roe), pp. 365–373. Prentice-Hall.

[367] Lal, M (1969). Monte Carlo computer simulation of chain molecules I. *Mol Phys*, **17**, 57–64.

[368] Lam, PM (1987). Specific heat and collapse transition of branched polymers. *Phys Rev B*, **36**, 6988–6992.

[369] Lam, PM (1989). Collapse of percolation clusters – a transfer matrix study. *J Stat Phys*, **54**, 1081–1089.

[370] Lam, PM and Binder, K (1988). Adsorption of branched polymers at surfaces: scaling and Monte Carlo analysis. *J Phys A: Math Gen*, **21**, L405–L409.

[371] Lawrie, ID and Sarlbach, S (1984). Tricriticality. In *Phase Transitions and Critical Phenomena* (ed. C Domb and JL Lebowitz), Volume 9, pp. 65–161. Academic Press.

[372] Lay, I (1958). *Introduction to Combinatorial Analysis.* Princeton University Press.

[373] Le Guillou, JC and Zinn-Justin, J (1977). Critical exponents for the n-vector model in three dimensions from field theory. *Phys Rev Lett*, **39**, 95.

[374] Le Guillou, JC and Zinn-Justin, J (1980). Critical exponents from field theory. *Phys Rev B*, **21**, 3976–3998.

[375] Le Guillou, JC and Zinn-Justin, J (1985). Accurate critical exponents from the ϵ-expansion. *J de Phys Lett*, **46**, 137–141.

[376] Le Guillou, JC and Zinn-Justin, J (1987). Accurate critical exponents for Ising-like systems in non-integer dimensions. *J de Phys*, **48**, 19–24.

[377] Le Guillou, JC and Zinn-Justin, J (1989). Accurate critical exponents from field theory. *J de Phys*, **50**, 1365–1370.

[378] Lee, J and Sohn, J (2005). Some continued fractions in Ramanujan's lost notebook. *Monatshefte Math*, **146**, 37–48.

[379] Leibler, S, Singh, RRP, and Fisher, ME (1987). Thermodynamic behavior of two-dimensional vesicles. *Phys Rev Lett*, **59**, 1989–1992.

[380] Li, B, Madras, N, and Sokal, AD (1995). Critical exponents, hyperscaling, and universal amplitude ratios for two-and three-dimensional self-avoiding walks. *J Stat Phys*, **80**, 661–754.

[381] Lin, KY (1985). Spiral self-avoiding walks on a triangular lattice. *J Phys A: Math Gen*, **18**, L145–L148.

[382] Lin, KY (1991). Exact solution of the convex polygon perimeter and area generating function. *J Phys A: Math Gen*, **24**, 2411–2417.

[383] Lin, KY and Liu, KC (1986). On the number of spiral self-avoiding walks on a triangular lattice. *J Phys A: Math Gen*, **19**, 585–589.

[384] Lin, KY and Tzeng, WJ (1991). Perimeter and area generating functions of the staircase and row-convex polygons on the rectangular lattice. *Int J Mod Phys B*, **5**, 1913–1925.

[385] Lipson, JEG, Gaunt, DS, Wilkinson, MK, and Whittington, SG (1987). Lattice models of branched polymers: combs and brushes. *Macromol*, **20**, 186–190.

[386] Liu, KC and Lin, KY (1985). Spiral self-avoiding walks on a triangular lattice: end-to-end distance. *J Phys A: Math Gen*, **18**, L647–650.

[387] Livne, S and Meirovitch, H (1988). Computer simulation of long polymers adsorbed on a surface I. Corrections to scaling in an ideal chain. *J Chem Phys*, **88**, 4498–4506.

[388] López, JL (1999). Asymptotic expansions of the Whittaker functions for large order parameter. *Methods Appl Anal*, **6**, 249–256.

[389] Lubensky, TC and Isaacson, J (1978). Field theory for the statistics of branched polymers, gelation, and vulcanization. *Phys Rev Lett*, **41**, 829–832.

[390] Lubensky, TC and Isaacson, J (1979). Statistics of lattice animals and dilute branched polymers. *Phys Rev A*, **20**, 2130–2146.

[391] M-P, Delest (1988). Generating functions for column-convex polyominoes. *J Comb Theo, Ser A*, **48**, 12–31.

[392] Ma, SK and Lin, KY (1988). Anisotropic spiral self-avoiding loops. *Int J Mod Phys B*, **2**, 287–299.

[393] MacDonald, D, Hunter, DL, Kelly, K, and Jan, N (1992). Self-avoiding walks in two to five dimensions: exact enumerations and series study. *J Phys A: Math Gen*, **25**, 1429–1440.

[394] Madras, N (1988). End patterns of self-avoiding walks. *J Stat Phys*, **53**, 689–701.

[395] Madras, N (1995). Critical behaviour of self-avoiding walks that cross a square. *J Phys A: Math Gen*, **28**, 1535–1547.

[396] Madras, N (1995). A rigorous bound on the critical exponent for the number of lattice trees, animals, and polygons. *J Stat Phys*, **78**, 681–699.

[397] Madras, N (1999). A pattern theorem for lattice clusters. *Ann Comb*, **3**, 357–384.

[398] Madras, N (2014). A lower bound for the end-to-end distance of self-avoiding walk. *Can Math Bull*, **57**, 113–118.

[399] Madras, N and Slade, G (1993). *The Self-Avoiding Walk*. Birkhaüser.

[400] Madras, N and Sokal, AD (1987). Nonergodicity of local, length-conserving Monte Carlo algorithms for the self-avoiding walk. *J Stat Phys*, **47**, 573–595.

[401] Madras, N and Sokal, AD (1988). The pivot algorithm: a highly efficient Monte Carlo method for the self-avoiding walk. *J Stat Phys*, **50**, 109–186.

[402] Madras, N, Soteros, CE, and Whittington, SG (1988). Statistics of lattice animals. *J Phys A: Math Gen*, **21**, 4617–4635.

[403] Madras, N, Soteros, CE, Whittington, SG, Martin, JL, Sykes, MF, Flesia, S, and Gaunt, DS (1990). The free energy of a collapsing branched polymer. *J Phys A: Math Gen*, **23**, 5327–5350.

[404] Maes, D and Vanderzande, C (1990). Self-avoiding rings at the θ-point. *Phys Rev A*, **41**, 3074–3080.

[405] Marcone, B, Orlandini, E, Stella, AL, and Zonta, F (2004). What is the length of a knot in a polymer? *J Phys A: Math Gen*, **38**, L15–L21.

[406] Martin, JL, Sykes, MF, and Hioe, FT (1967). Probability of initial ring closure for self-avoiding walks on the face-centered cubic and triangular lattices. *J Chem Phys*, **46**, 3478–3481.

[407] Martin, R, Causo, MS, and Whittington, SG (2000). Localization transition for a randomly coloured self-avoiding walk at an interface. *J Phys A: Math Gen*, **33**, 7903–7918.

[408] Matthews, R, Louis, AA, and Yeomans, JM (2011). Confinement of knotted polymers in a slit. *Mol Phys*, **109**, 1289–1295.

[409] Mazo, RM (1963). Free energy of a system with random elements. *J Chem Phys*, **39**, 1224–1225.

[410] McCrea, WH and Whipple, FJW (1940). Random paths in two and three dimensions. In *Proc Royal Soc Edinburgh*, Volume 60, pp. 281–298.

[411] Meirovitch, H (1985). Computer simulation of the free energy of polymer chains with excluded volume and with finite interactions. *Phys Rev A*, **32**, 3709–3715.

[412] Meirovitch, H and Chang, I (1993). Surface critical exponents of self-avoiding walks on a square lattice with an adsorbing linear boundary: a computer simulation study. *Phys Rev E*, **48**, 1960–1969.

[413] Meirovitch, H and Lim, HA (1989). The collapse transition of self-avoiding walks on a square lattice: a computer simulation study. *J Chem Phys*, **91**, 2544–2552.

[414] Meirovitch, H and Lim, HA (1989). θ-Point exponents of polymers in $d = 2$. *Phys Rev Lett*, **62**, 2640–2640.

[415] Meirovitch, H and Lim, HA (1990). Computer simulation study of the θ-point in three dimensions: I. Self-avoiding walks on a simple cubic lattice. *J Chem Phys*, **92**, 5144–5154.

[416] Meirovitch, H and Livne, S (1988). Computer simulation of long polymers adsorbed on a surface II. Critical behavior of a single self-avoiding walk. *J Chem Phys*, **88**, 4507–4515.

[417] Melnichenko, YB, Kiran, E, Heath, KD, Salaniwal, S, Cochran, HD, Stamm, M, Van Hook, WA, and Wignall, GD (2000). Comparison of the behaviour of polymers in supercritical fluids and organic solvents via small-angle neutron scattering. *J Appl Crystallography*, **33**, 682–685.

[418] Metropolis, N, Rosenbluth, AW, Rosenbluth, MN, Teller, AH, and Teller, E (1953). Equation of state calculations by fast computing machines. *J Chem Phys*, **21**, 1087–1092.

[419] Metropolis, N and Ulam, S (1949). The Monte Carlo method. *J Amer Stat Assoc*, **44**, 335–341.

[420] Michels, JPJ and Wiegel, FW (1982). Probability of knots in a polymer ring. *Phys Lett A*, **90**, 381–384.

[421] Michels, JPJ and Wiegel, FW (1986). On the topology of a polymer ring. *Proc Royal Soc London A*, **403**, 269–284.

[422] Millett, KC and Rawdon, EJ (2005). Universal characteristics of polygonal knot probabilities. In *Physical and Numerical Models in Knot Theory, Series on Knots and Everything* (ed. JA Calvo, KC Millett, EJ Rawdon and A Stasiak), Volume 36, pp. 247–274. World Scientific.

[423] Mishna, M (2009). Classifying lattice walks restricted to the quarter plane. *J Comb Theo Ser A*, **116**, 460–477.

[424] Mishna, M and Rechnitzer, A (2009). Two non-holonomic lattice walks in the quarter plane. *Theo Comp Sci*, **410**, 3616–3630.

[425] Moak, DS (1984). The q-analogue of Stirling's formula. *Rocky Mountain J Math*, **14**, 403–413.

[426] Moghaddam, MS (2003). A Monte Carlo study of adsorption of random copolymers on random surfaces. *J Phys A: Math Gen*, **36**, 939–949.

[427] Moghaddam, MS and Whittington, SG (2002). A Monte Carlo study of polymer adsorption: random copolymers and random surfaces. *J Phys A: Math Gen*, **35**, 33–42.

[428] Morita, T (1964). Statistical mechanics of quenched solid solutions with application to magnetically dilute alloys. *J Math Phys*, **5**, 1401–1405.

[429] Motzkin, T (1948). Independence of resultants. *Bull Amer Math Soc*, **54**, 360–365.

[430] Nägele, G, D'Aguanno, B, and Akcasu, AZ (1997). Preface of Proceedings of the Workshop on Colloid Physics held at the University of Konstanz, Germany from 30 November to 2 December 1995. Elsevier.
[431] Narayana, TV (1979). *Lattice Path Combinatorics with Statistical Applications*. University of Toronto Press.
[432] Nemirovsky, AM, Freed, KF, Ishinabe, T, and Douglas, JF (1992). Marriage of exact enumeration and $1/d$-expansion methods: lattice model of dilute polymers. *J Stat Phys*, **67**, 1083–1108.
[433] Nickel, B (1981). Hyperscaling and universality in 3 dimensions. *Physica A*, **106**, 48–58.
[434] Nidras, PP (1996). Grand canonical simulations of the interacting self-avoiding walk model. *J Phys A: Math Gen*, **29**, 7929–7942.
[435] Nidras, PP and Brak, R (1997). New Monte Carlo algorithms for interacting self-avoiding walks. *J Phys A: Math Gen*, **30**, 1457–1469.
[436] Nienhuis, B (1982). Exact critical point and critical exponents of $O(n)$ models in two dimensions. *Phys Rev Lett*, **49**, 1062–1065.
[437] Nienhuis, B (1983). Coulomb gas formulation of two-dimensional phase transitions. In *Phase Transitions and Critical Phenomena* (ed. C Domb and JL Lebowitz), Volume 11, pp. 1–53. Academic Press.
[438] Nienhuis, B (1984). Coulomb gas description of 2-d critical behaviour. *J Stat Phys*, **34**, 731–761.
[439] Nienhuis, B (1990). Critical spin-1 vertex models and $o(n)$ models. *Int J Modn Phys B*, **4**, 929–942.
[440] Nienhuis, B, Riedel, EK, and Schick, M (1980). Magnetic exponents of the two-dimensional q-state Potts model. *J Phys A: Math Gen*, **13**, L189–L192.
[441] O'Brien, GL (1990). Monotonicity of the number of self-avoiding walks. *J Stat Phys*, **59**, 969–979.
[442] Odlyzko, AM and Wilf, HS (1988). The editor's corner: n coins in a fountain. *Amer Math Monthly*, **95**, 840–843.
[443] Olver, FWJ (1974). *Asymptotics and Special Functions*. Academic Press.
[444] Orlandini, E, Janse van Rensburg, EJ, Tesi, MC, and Whittington, SG (1999). Random linking of lattice polygons. *J Phys A: Math Gen*, **27**, 335–346.
[445] Orlandini, E, Janse van Rensburg, EJ, and Whittington, SG (1996). A Monte Carlo algorithm for lattice ribbons. *J Stat Phys*, **82**, 1159–1198.
[446] Orlandini, E, Stella, AL, Einstein, TL, Tesi, MC, Beichl, I, and Sullivan, F (1996). Bending-rigidity-driven transition and crumpling-point scaling of lattice vesicles. *Phys Rev E*, **53**, 5800–5807.
[447] Orlandini, E, Stella, AL, Tesi, MC, and Sullivan, F (1993). Vesicle adsorption on a plane: scaling regimes and crossover phenomena. *Phys Rev E*, **48**, R4203–R4206.
[448] Orlandini, E, Stella, AL, and Vanderzande, C (2004). Loose, flat knots in collapsed polymers. *J Stat Phys*, **115**, 681–700.
[449] Orlandini, E, Stella, AL, and Vanderzande, C (2009). The size of knots in polymers. *Phys Biol*, **6**, 025012.
[450] Orlandini, E and Tesi, MC (1998). Knotted polygons with curvature in $\mathbb{R}^3$. *J Phys A: Math Gen*, **31**, 9441–9454.
[451] Orlandini, E, Tesi, MC, Janse van Rensburg, EJ, and Whittington, SG (1996). Entropic exponents of lattice polygons with specified knot type. *J Phys A: Math Gen*, **29**, L299–L304.
[452] Orlandini, E, Tesi, MC, Janse van Rensburg, EJ, and Whittington, SG (1998). Asymptotics of knotted lattice polygons. *J Phys A: Math Gen*, **31**, 5953–5968.
[453] Orlandini, E, Tesi, MC, and Whittington, SG (1999). A self-avoiding walk model of random copolymer adsorption. *J Phys A: Math Gen*, **32**, 469–477.
[454] Orlandini, E, Tesi, MC, and Whittington, SG (2000). Self-averaging in models of random copolymer collapse. *J Phys A: Math Gen*, **33**, 259–266.
[455] Orlandini, E, Tesi, MC, Whittington, SG, Sumners, DW, and Janse van Rensburg, EJ (1994). The writhe of a self-avoiding walk. *J Phys A: Math Gen*, **27**, L333–L338.

[456] Orlandini, E and Whittington, SG (2007). Statistical topology of closed curves: some applications in polymer physics. *Rev of Modern Phys*, **79**, 611–642.

[457] Osborn, J-A and Prellberg, T (2010). Forcing adsorption of a tethered polymer by pulling. *J Stat Mech: Theo Expr*, **2010**, P09018.

[458] Owczarek, AL, Prellberg, T, Bennett-Wood, D, and Guttmann, AJ (1999). Universal distance ratios for interacting two-dimensional polymers. *J Phys A: Math Gen*, **27**, L919–L925.

[459] Owczarek, AL, Prellberg, T, and Brak, R (1993). New scaling form for the collapsed polymer phase. *Phys Rev Lett*, **70**, 951–953.

[460] Parisi, G and Sourlas, N (1981). Critical behavior of branched polymers and the Lee-Yang edge singularity. *Phys Rev Lett*, **46**, 871–874.

[461] Peard, PJ and Gaunt, DS (1995). $1/d$-Expansions for the free energy of lattice animal models of a self-interacting branched polymer. *J Phys A: Math Gen*, **28**, 6109–6124.

[462] Peierls, R (1936). On Ising's model of ferromagnetism. *Math Proc Camb Phil Soc*, **32**, 477–481.

[463] Perron, O (1954). *Die Lehre von den Kettenbrüchen Bd 1. Elementare Kettenbrüche.* BG Teubner Verlagsgesellschaft.

[464] Peters, HP, Stauffer, D, Hölters, HP, and Loewenich, K (1979). Radius, perimeter, and density profile for percolation clusters and lattice animals. *Z Physik B Cond Mat*, **34**, 399–408.

[465] Pincus, P (1976). Excluded volume effects and stretched polymer chains. *Macromol*, **9**, 386–388.

[466] Pippenger, N (1989). Knots in random walks. *Disc Appl Math*, **25**, 273–278.

[467] Pólya, G (1969). On the number of certain lattice polygons. *J Comb Theo*, **6**, 102–105.

[468] Prellberg, T (1995). Uniform q-series asymptotics for staircase polygons. *J Phys A: Math Gen*, **28**, 1289–1304.

[469] Prellberg, T and Brak, R (1995). Critical exponents from nonlinear functional equations for partially directed cluster models. *J Stat Phys*, **78**, 701–730.

[470] Prellberg, T and Krawczyk, J (2004). Flat histogram version of the pruned and enriched Rosenbluth method. *Phys Rev Lett*, **92**, 120602.

[471] Prellberg, T and Owczarek, AL (1995). Stacking models of vesicles and compact clusters. *J Stat Phys*, **80**, 755–779.

[472] Prentis, JJ (1991). Renormalisation theory of self-avoiding walks which cross a square. *J Phys A: Math Gen*, **24**, 5097–5103.

[473] Prentis, JJ and Sisan, DR (2002). Granular polymer solution. *Phys Rev E*, **65**, 031306.

[474] Pringsheim, A (1898). Ueber die convergenz unendlicher kettenbrüche. *München Bericht*, **28**, 295–324.

[475] Privman, V (1983). Spiral self-avoiding walks. *J Phys A: Math Gen*, **16**, L571–573.

[476] Privman, V (1984). Convergence and extrapolation in finite size scaling renormalisation. *Physica A*, **123**, 428–442.

[477] Privman, V (1986). Study of the θ-point by enumeration of self-avoiding walks on the triangular lattice. *J Phys A: Math Gen*, **19**, 3287–3297.

[478] Privman, V and Švrakić, NM (1988). Difference equations in statistical mechanics I. Cluster statistics models. *J Stat Phys*, **51**, 1091–1110.

[479] Privman, V, Forgacs, G, and Frisch, HL (1988). New solvable model of polymer-chain adsorption at a surface. *Phys Rev B*, **37**, 9897–9900.

[480] Privman, V and Kurtze, DA (1986). Partition function zeros in two-dimensional lattice models of the polymer θ-point. *Macromol*, **19**, 2377–2379.

[481] Privman, V and Rudnick, J (1985). Size of rings in two dimensions. *J Phys A: Math Gen*, **18**, L789–L794.

[482] Rajesh, R and Dhar, D (2005). Convex lattice polygons of fixed area with perimeter-dependent weights. *Phys Rev E*, **71**, 016130.

[483] Rampf, F, Paul, W, and Binder, K (2007). On the first-order collapse transition of a three-dimensional, flexible homopolymer chain model. *Euro Phys Lett*, **70**, 628–634.

[484] Rapaport, DC (1985). On three-dimensional self-avoiding walks. *J Phys A: Math Gen*, **18**, 113–126.
[485] Rawdon, E, Dobay, A, Kern, JC, Millett, KC, Piatek, M, Plunkett, P, and Stasiak, A (2008). Scaling behavior and equilibrium lengths of knotted polymers. *Macromol*, **41**, 4444–4451.
[486] Read, RC (1962). Contributions to the cell growth problem. *Canad J Math*, **14**, 1–20.
[487] Rechnitzer, A and Janse van Rensburg, EJ (2002). Canonical Monte Carlo determination of the connective constant of self-avoiding walks. *J Phys A: Math Gen*, **35**, L605–L612.
[488] Rechnitzer, A and Janse van Rensburg, EJ (2004). Exchange relations, Dyck paths and copolymer adsorption. *Disc Appl Math*, **140**, 49–71.
[489] Rechnitzer, A and Janse van Rensburg, EJ (2008). Generalized atmospheric Rosenbluth methods (GARM). *J Phys A: Math Theo*, **41**, 442002.
[490] Redner, S (1979). Mean end-to-end distance of branched polymers. *J Phys A: Math Gen*, **12**, L239–L244.
[491] Reeve, J and Guttmann, AJ (1981). Renormalisation group calculations of the critical exponents of the n-vector model with a free surface. *J Phys A: Math Gen*, **14**, 3357–3366.
[492] Reidemeister, K (1948). *Knotentheorie*. Chelsea.
[493] Richard, C (2006). Staircase polygons: moments of diagonal lengths and column heights. *J Phys: Conf Ser*, **42**, 239–257.
[494] Robbins, H (1955). A remark on Stirling's formula. *Amer Math Monthly*, **62**, 26–29.
[495] Rolfsen, D (2003). *Knots and Links*. American Mathematical Society.
[496] Rosenbluth, MN and Rosenbluth, AW (1955). Monte Carlo calculation of the average extension of molecular chains. *J Chem Phys*, **23**, 356–359.
[497] Royden, H and Fitzpatrick, P (2010). *Real Analaysis* (4 edn). Prentice-Hall.
[498] Rybenkov, VV, Cozzarelli, NR, and Vologodskii, AV (1993). Probability of DNA knotting and the effective diameter of the DNA double helix. *Proc Nat Acad Sci*, **90**, 5307–5311.
[499] Rybenkov, VV, Vologodskii, AV, and Cozzarelli, NR (1997). The effect of ionic conditions on DNA helical repeat, effective diameter and free energy of supercoiling. *Nucleic Acids Res*, **25**, 1412–1418.
[500] Saleur, H (1987). Conformal invariance for polymers and percolation. *J Phys A: Math Gen*, **20**, 455–470.
[501] Schaefer, HH (1974). *Banach Lattices and Positive Operators*. Springer.
[502] Schafer, L, von Ferber, C, Lehr, U, and Duplantier, B (1992). Renormalization of polymer networks and stars. *Nucl Phys B*, **374**, 473–495.
[503] Scharein, R, Ishihara, K, Arsuaga, J, Diao, Y, Shimokawa, K, and Vazquez, M (2009). Bounds for the minimum step number of knots in the simple cubic lattice. *J Phys A: Math Theo*, **42**, 475006.
[504] Schick, M, Walker, JS, and Wortis, M (1977). Phase diagram of the triangular Ising model: renormalization-group calculation with application to adsorbed monolayers. *Phys Rev B*, **16**, 2205–2219.
[505] Schram, RD, Barkema, GT, and Bisseling, RH (2011). Exact enumeration of self-avoiding walks. *J Stat Mech: Theo Expr*, **2011**, P06019.
[506] Schramm, O (2011). Conformally invariant scaling limits: an overview and a collection of problems. In *Selected Works of Oded Schramm*, pp. 1161–1191. Springer.
[507] Schubert, H (1954). Über eine Numerische Knoteninvariante. *Math Z*, **61**, 245–288.
[508] Seitz, WA and Klein, DJ (1981). Excluded volume effects for branched polymers: Monte Carlo results. *J Chem Phys*, **75**, 5190–5193.
[509] Seno, F and Stella, AL (1988). θ-Point of a linear polymer in 2 dimensions: a renormalization group analysis of Monte Carlo enumerations. *J de Phys*, **49**, 739–748.
[510] Seno, F and Vanderzande, C (1994). Erratum: non-universality in the collapse of two-dimensional branched polymers. *J Phys A: Math Gen*, **27**, 7937–7938.
[511] Seno, F and Vanderzande, C (1994). Non-universality in the collapse of two-dimensional branched polymers. *J Phys A: Math Gen*, **27**, 5813–5830.
[512] Shaw, SY and Wang, JC (1993). Knotting of a DNA chain during ring closure. *Science*, **260**, 533–536.

[513] Slade, G (1995). Bounds on the self-avoiding-walk connective constant. *J Fourier Anal Appl*, **1**, 525–534.

[514] Slater, LJ (1960). *Confluent Hypergeometric Functions*. Cambridge University Press.

[515] Smirnov, S (2006). Towards conformal invariance of 2d lattice models. In *Proceedings of the International Congress of Mathematicians (Madrid, August 22-30, 2006)*, Volume II, pp. 1421–1451. Euro Math Soc Zürich.

[516] Sokal, AD (1994). Static scaling behavior of high-molecular-weight polymers in dilute solution: a reexamination. *EuroPhys Lett*, **27**, 661–666.

[517] Sokal, AD (1996). Monte Carlo methods for the self-avoiding walk. *Nucl Phys B – Proc Suppl*, **47**, 172–179.

[518] Sokal, AD and Thomas, LE (1989). Exponential convergence to equilibrium for a class of random-walk models. *J Stat Phys*, **54**, 797–828.

[519] Soteros, CE (1992). Adsorption of uniform lattice animals with specified topology. *J Phys A: Math Gen*, **25**, 3153–3173.

[520] Soteros, CE (1993). Lattice models of branched polymers with specified topologies. *J Math Chem*, **1**, 91–102.

[521] Soteros, CE (1993). Random knots in uniform branched polymers. *Math Modelling Sci Comp*, **2**, 747–752.

[522] Soteros, CE (1998). Knots in graphs in subsets of $\mathbb{Z}^3$. In *Topology and Geometry in Polymer Science* (ed. SG Whittington, DW Sumners, and T Lodge), pp. 101–133. Springer.

[523] Soteros, CE, Narayanan, KSS, De'Bell, K, and Whittington, SG (1996). Polymers with restricted branching. *Phys Rev E*, **53**, 4745–4753.

[524] Soteros, CE, Sumners, DW, and Whittington, SG (1992). Entanglement complexity of graphs in F^3. *Math Proc Camb Phil Soc*, **111**, 75–91.

[525] Soteros, CE, Sumners, DW, and Whittington, SG (1999). Linking of random p-spheres in zd. *J Knot Theo Ram*, **8**, 49–70.

[526] Soteros, CE and Whittington, SG (1988). Critical exponents for lattice animals with fixed cyclomatic index. *J Phys A: Math Gen*, **21**, 2187–2193.

[527] Soteros, CE and Whittington, SG (1989). Lattice models of branched polymers: effects of geometrical constraints. *J Phys A: Math Gen*, **22**, 5259–5270.

[528] Soteros, CE and Whittington, SG (2004). The statistical mechanics of random copolymers. *J Phys A: Math Gen*, **37**, R279–R325.

[529] Soteros, Chris E and Whittington, Stuart G (1988). Polygons and stars in a slit geometry. *J Phys A: Math Gen*, **21**, L857–L861.

[530] Spitzer, F (2001). *Principles of Random Walk*, Volume 34. Springer Verlag.

[531] Stanley, RP (2011). *Enumerative Combinatorics*, Volume 49. Cambridge University Press.

[532] Stauffer, D (1978). Monte Carlo study of density profile, radius, and perimeter for percolation clusters and lattice animals. *Phys Rev Lett*, **41**, 1333–1336.

[533] Stauffer, D (1979). Scaling theory of percolation clusters. *Phys Rep*, **54**, 1–74.

[534] Stella, AL, Orlandini, E, Beichl, I, Sullivan, F, Tesi, MC, and Einstein, TL (1992). Self-avoiding surfaces, topology, and lattice animals. *Phys Rev Lett*, **69**, 3650–3653.

[535] Stella, AL and Vanderzande, C (1989). Scaling and fractal dimension of Ising clusters at the d= 2 critical point. *Phys Rev Lett*, **62**, 1067–1070.

[536] Stephen, JM (1975). Collapse of a polymer chain. *Phys Lett A*, **53**, 363–364.

[537] Stratychuk, LM and Soteros, CE (1996). Statistics of collapsed lattice animals: rigorous results and Monte Carlo simulations. *J Phys A: Math Gen*, **29**, 7067–7087.

[538] Sumithra, K and Baumgaertner, A (1998). Polymer adsorption on planar random surfaces. *J Chem Phys*, **109**, 1540–1544.

[539] Sumners, DW and Whittington, SG (1988). Knots in self-avoiding walks. *J Phys A: Math Gen*, **21**, 1689–1694.

[540] Sun, ST, Nishio, I, Swislow, G, and Tanaka, T (1980). The coil–globule transition: radius of gyration of polystyrene in cyclohexane. *J Chem Phys*, **73**, 5971–5975.

[541] Swislow, G, Sun, ST, Nishio, I, and Tanaka, T (1980). Coil-globule phase transition in a single polystyrene chain in cyclohexane. *Phys Rev Lett*, **44**, 796–798.

[542] Sykes, MF (1961). Some counting theorems in the theory of the Ising model and the excluded volume problem. *J Math Phys*, **2**, 52–62.

[543] Tasaki, H and Hara, T (1985). Collapse of random surfaces in the connected plaquettes model. *Phys Lett A*, **112**, 115–118.

[544] Temperley, HNV (1956). Combinatorial problems suggested by the statistical mechanics of domains and of rubber-like molecules. *Phys Rev*, **103**, 1–16.

[545] Tesi, MC, Janse van Rensburg, EJ, Orlandini, E, Sumners, DW, and Whittington, SG (1994). Knotting and supercoiling in circular DNA: a model incorporating the effect of added salt. *Phys Rev E*, **49**, 868–872.

[546] Tesi, MC, Janse van Rensburg, EJ, Orlandini, E, and Whittington, SG (1995). Interacting self-avoiding walks and polygons in three dimensions. *J Phys A: Math Gen*, **29**, 2451–2463.

[547] Tesi, MC, Janse van Rensburg, EJ, Orlandini, E, and Whittington, SG (1996). Monte Carlo study of the interacting self-avoiding walk model in three dimensions. *J Stat Phys*, **82**, 155–181.

[548] Tesi, MC, Janse van Rensburg, EJ, Orlandini, E, and Whittington, SG (1997). Torsion of polygons in $\mathbb{Z}^3$. *J Phys A: Math Gen*, **30**, 5179–5194.

[549] Torrie, GM and Valleau, JP (1977). Nonphysical sampling distributions in Monte Carlo free-energy estimation: umbrella sampling. *J Comp Phys*, **23**, 187–199.

[550] Touchette, H (2009). The large deviation approach to statistical mechanics. *Phys Rep*, **478**, 1–69.

[551] Trigueros, S, Arsuaga, J, Vazquez, ME, Sumners, DW, and Roca, J (2001). Novel display of knotted DNA molecules by two-dimensional gel electrophoresis. *Nucleic Acids Res*, **29**, e67–e67.

[552] van den Berg, J and Kesten, H (1985). Inequalities with applications to percolation and reliability. *J Appl Prob*, **22**, 556–569.

[553] van der Hofstad, R, den Hollander, F, and König, W (1997). Central limit theorem for the Edwards' model. *Ann Prob*, **25**, 573–597.

[554] Van Vleck, EB (1901). On the convergence of continued fractions with complex elements. *Trans Amer Math Soc*, **2**, 215–233.

[555] Vanderzande, C (1990). Self-avoiding walks near an excluded line. *J Phys A: Math Gen*, **23**, 563–566.

[556] Vanderzande, C (1993). Vesicles, the tricritical-0-state Potts model, and the collapse of branched polymers. *Phys Rev Lett*, **70**, 3595–3598.

[557] Vanderzande, C (1995). On knots in a model for the adsorption of ring polymers. *J Phys A: Math Gen*, **28**, 3681–3700.

[558] Vanderzande, C (1998). *Lattice Models of Polymers*. Cambridge University Press.

[559] Veal, AR, Yeomans, JM, and Jug, G (1991). The effect of attractive monomer-monomer interactions on adsorption of a polymer chain. *J Phys A: Math Gen*, **24**, 827–849.

[560] Verdier, PH and Stockmayer, WH (1962). Monte Carlo calculations on the dynamics of polymers in dilute solution. *J Chem Phys*, **36**, 227–237.

[561] Viennot, XG (1985). Problemes combinatoires posés par la physique statistique. *Séminaire Bourbaki*, **26**, 225–246.

[562] Viennot, XG (1986). Heaps of pieces, I: basic definitions and combinatorial lemmas. In *Combinatoire Énumérative* (ed. G Labelle and P Leroux), pp. 321–350. Springer.

[563] Viennot, XG (2000). A combinatorial interpretation of the quotient-difference algorithm. In *Formal Power Series and Algebraic Combinatorics* (ed. D Krob, AA Mikhalev and AV Mikalev), pp. 379–390. Springer.

[564] Viennot, XG (2006). Multi-directed animals, connected heaps of dimers and Lorentzian triangulations. *J Phys: Conf Ser*, **42**, 268–280.

[565] Vologodskii, AV, Levene, SD, Klenin, KV, Frank-Kamenetskii, MD, and Cozzarelli, NR (1992). Conformational and thermodynamic properties of supercoiled DNA. *J Mol Bio*, **227**, 1224–1243.

[566] Vologodskii, AV, Lukashin, AV, Frank-Kamenetskii, MD, and Anshelevich, VV (1974). The knot problem in statistical mechanics of polymer chains. *Sov J Expr Theo Phys*, **39**, 1059–1063.

[567] Vrbová, T and Whittington, SG (1996). Adsorption and collapse of self-avoiding walks and polygons in three dimensions. *J Phys A: Math Gen*, **29**, 6253–6264.

[568] Vrbová, T and Whittington, SG (1998). Adsorption and collapse of self-avoiding walks at a defect plane. *J Phys A: Math Gen*, **31**, 7031–7041.

[569] Vrbová, T and Whittington, SG (1998). Adsorption and collapse of self-avoiding walks in three dimensions: a Monte Carlo study. *J Phys A: Math Gen*, **31**, 3989–3998.

[570] Vujic, D (1999). Branched polymers on the two-dimensional square lattice with attractive surfaces. *J Stat Phys*, **95**, 767–774.

[571] Wall, FT and Erpenbeck, JJ (1959). Statistical computation of radii of gyration and mean internal dimensions of polymer molecules. *J Chem Phys*, **30**, 637–641.

[572] Wall, HS (1967). *Analytic Theory of Continued Fractions*. Chelsea.

[573] Wang, JC (2009). A journey in the world of DNA rings and beyond. *Annual Rev Biochem*, **78**, 31–54.

[574] Wang, Y-L, Lee, F, and Kimel, JD (1987). Phase diagrams of the spin-1 Ising Blume-Emery-Griffiths model: Monte Carlo simulations. *Phys Rev B*, **36**, 8945–8947.

[575] Welsh, DJA (1993). Percolation in the random cluster process and q-state Potts model. *J Phys A: Math Gen*, **26**, 2471–2483.

[576] Westwater, J (1980). On Edwards' model for long polymer chains. *Comm Math Phys*, **72**, 131–174.

[577] Westwater, J (1981). On Edwards' model for polymer chains. *Comm Math Phys*, **79**, 53–73.

[578] Westwater, J (1982). On Edwards' model for polymer chains III. Borel summability. *Comm Math Phys*, **84**, 459–470.

[579] Whittington, SG (1970). The excluded volume effect in polymers. *J Phys A: General Phys*, **3**, 28–32.

[580] Whittington, SG (1987). Self-avoiding walks and related systems. In *Interfaces between Mathematics, Chemistry and Computer Science* (ed. RC Lacher), Volume 54 of *1987: International Conference Proceedings (Studies in Physical and Theoretical Chemistry)*, pp. 285–296. Elsevier Science.

[581] Whittington, SG (1993). Statistical mechanics of three-dimensional vesicles. *J Math Chem*, **14**, 103–110.

[582] Whittington, SG (1998). A self-avoiding walk model of copolymer adsorption. *J Phys A: Math Gen*, **31**, 3769–3775.

[583] Whittington, SG, DW, Sumners, and Lodge, T (1998). *Topology and Geometry in Polymer Science*, Volume 103. Springer.

[584] Whittington, SG and Guttmann, AJ (1990). Self-avoiding walks which cross a square. *J Phys A: Math Gen*, **23**, 5601–5609.

[585] Whittington, SG and Janse van Rensburg, EJ (1993). Random knots in ring polymers. *Math Mod and Sci Comp*, **2**, 741–746.

[586] Whittington, SG, Lipson, JEG, Wilkinson, MK, and Gaunt, DS (1986). Lattice models of branched polymers: dimensions of uniform stars. *Macromol*, **19**, 1241–1245.

[587] Whittington, SG and Soteros, CE (1990). Lattice animals: rigorous results and wild guesses. In *Disorder in Physical Systems* (ed. G Grimmett and D Welsh), pp. 321–334. Oxford University Press.

[588] Whittington, SG and Soteros, CE (1991). Polymers in slabs, slits, and pores. *Israel J Chem*, **31**, 127–133.

[589] Whittington, SG and Soteros, CE (1992). Uniform branched polymers in confined geometries. *Macromol Rep*, **29**(S2), 195–199.

[590] Whittington, SG, Valleau, JP, and Torrie, GM (1977). How many figure eights are there? some bounds. *J Phys A: Math Gen*, **10**, L111–L112.

[591] Wiese, KJ and David, F (1995). Self-avoiding tethered membranes at the tricritical point. *Nucl Phys B*, **450**(3), 495–557.

[592] Wilker, JB (1984). Stirling ideas for freshman calculus. *Math Magazine*, **57**, 209–214.
[593] Wilker, JB and Whittington, SG (1979). Extension of a theorem on super-multiplicative functions. *J Phys A: Math Gen*, **12**, L245–L248.
[594] Wilkinson, MK, Gaunt, DS, Lipson, JEG, and Whittington, SG (1986). Lattice models of branched polymers: statistics of uniform stars. *J Phys A: Math Gen*, **19**, 789–796.
[595] Wilson, KG and Fisher, ME (1972). Critical exponents in 3.99 dimensions. *Phys Rev Lett*, **28**, 240–243.
[596] Wright, EM (1968). Stacks. *Quarterly J Math*, **19**, 313–320.
[597] Wright, EM (1971). Stacks (II). *Quarterly J Math*, **22**, 107–116.
[598] Wu, F-Y (1978). Percolation and the Potts model. *J Stat Phys*, **18**, 115–123.
[599] Wu, F-Y (1982). The Potts model. *Rev Mod Phys*, **54**, 235–268.
[600] www.jmol.org (2012). *Jmol, An Open-Source Java Viewer for Chemical Structures in* 3*d*.
[601] Yeomans, JM (1992). *Statistical Mechanics of Phase Transitions*. Oxford University Press.
[602] You, S and Janse van Rensburg, EJ (1998). Critical exponents and universal amplitude ratios in lattice trees. *Phys Rev E*, **58**, 3971–3976.
[603] You, S and Janse van Rensburg, EJ (2000). A lattice tree model of branched copolymer adsorption. *J Phys A: Math Gen*, **33**, 1171–1186.
[604] You, S and Janse van Rensburg, EJ (2001). Adsorbing trees in two dimensions: a Monte Carlo study. *Phys Rev E*, **64**, 046101.
[605] Zeilberger, D (1989). Six etudes in generating functions. *Int J Comp Math*, **29**(2-4), 201–215.

INDEX